T0297917

Measuring Heavy Metal Contaminants in Cannabis and Hemp

Measuring Heavy Metal Contaminants in Cannabis and Hemp

Robert J. Thomas

CRC Press is an imprint of the
Taylor & Francis Group, an **informa** business

First edition published 2020
by CRC Press
6000 Broken Sound Parkway NW, Suite 300, Boca Raton, FL 33487-2742

and by CRC Press
2 Park Square, Milton Park, Abingdon, Oxon, OX14 4RN

CRC Press is an imprint of Taylor & Francis Group, LLC

Library of Congress Cataloging-in-Publication Data
Names: Thomas, Robert J., author.
Title: Measuring heavy metal contaminants in cannabis and hemp / Robert J. Thomas.
Description: First edition. | Boca Raton : Taylor and Francis, 2020. |
Includes bibliographical references and index.
Identifiers: LCCN 2020024670 (print) | LCCN 2020024671 (ebook) |
ISBN 9780367417376 (hardback) | ISBN 9781003004158 (ebook)
Subjects: LCSH: Cannabis—Effect of heavy metals on.
Classification: LCC SB295.C35 T46 2020 (print) | LCC SB295.C35 (ebook) |
DDC 633.7/9—dc23
LC record available at https://lccn.loc.gov/2020024670
LC ebook record available at https://lccn.loc.gov/2020024671

ISBN: 978-0-367-41737-6 (hbk)
ISBN: 978-0-367-55472-9 (pbk)
ISBN: 978-1-003-00415-8 (ebk)

Typeset in Times
by codeMantra

Contents

Foreword

Legalization of cannabis for medical and recreational purposes in several countries and US states has created many exciting new opportunities to fully discover the potential medical benefits of this ancient plant. Separating scientific facts from popular myth and correlating the therapeutic effects of cannabis to its chemical ingredients requires an in-depth understanding of cannabis chemistry, and involves various analytical measurement tools and techniques at molecular and elemental levels.

HISTORICAL PERSPECTIVE OF MEDICAL CANNABIS

Cannabis (*Cannabis sativa* L. from the family of *Cannabaceae*) is a prevalent plant with origins tracing back to the ancient world. Evidence suggests that cannabis and its close relative, hemp, have a long history of medicinal and practical use in the Middle East and Asia dating back to antiquity. Hemp, from the same family as cannabis, but containing less than 0.3% (according to the FDA definition) of the psychoactive ingredient Δ9-tetrahydrocannabinol (THC), was one of the first plants to be used as fiber around 10,000 years ago. Historical records suggest that the medical use of cannabis began as early as 2700 BC in China. Cannabis was also used medicinally and recreationally in other parts of Asia, including Japan and India. Hemp was introduced to Western Asia, Egypt, and Europe between 1000 and 2000 BC and was cultivated in many parts of Europe after 500 AD for industrial purposes. In 1545, the Spanish brought hemp to South America, where it was cultivated and refined into a variety of commercial products such as paper, textiles, and clothing. Around 1600, hemp was introduced to North America, and in the late 1700s, some American medical journals started referring to the potential benefits of hemp to treat various health problems. In the United States, cannabis was listed for the first time in the United States Pharmacopoeia (USP) in 1850 and was consumed as a medicine into the early 20th century.

The passage of the Marihuana Tax Act in 1937 restricted the use and sale of cannabis in the United States, and cannabis was subsequently dropped from the USP in 1942. More legal restrictions and penalties were introduced with the enactment of the Boggs and Narcotic Control Acts of 1951 and 1956, respectively. The passage of the Controlled Substances Act in 1970 further prohibited cannabis under US federal law, listing the plant as a Schedule I Controlled Substance. In Canada, prohibition started when cannabis was added to the Confidential Restricted List in 1923 under the Narcotics Drug Act Amendment Bill. In addition to criminalization, these legislative actions severely limited scientific research by restricting procurement and possession of cannabis for academic purposes and resulted in a public stigma which has continued to this day.

EVOLVING GLOBAL LANDSCAPE

Over the past decade, the legal landscape for medical cannabis has changed dramatically. To date, more than 40 countries have legalized or decriminalized the medical use of cannabis and/or cannabis-derived pharmaceuticals, and more countries are on their way to legalize it.

The legalization of recreational cannabis in some countries (Canada, Uruguay, and some US states, with several other countries pending) also means more government funding and private investment for cannabis research. Public safety concerns regarding the consumption of cannabis and cannabis-derived products (oils, extracts, edibles, topicals, pharmaceuticals, etc.) demand complex and accurate tests to ensure that the level of potency, heavy metals, pesticides, residual solvents, mycotoxins, etc. are appropriately measured and controlled. Such tests require sophisticated instrumentation, validated methods, and clear guidelines and standards.

Meanwhile, more research is starting to show the effectiveness of cannabis-derived drugs in the treatment of epilepsy, post-traumatic stress disorder (PTSD), chemotherapy-induced nausea and vomiting, anorexia, and chronic pain. In fact, the first cannabis-based drug (Epidiolex, from GW Pharmaceuticals) was approved in 2018 by the US FDA for the treatment of severe, childhood-onset epilepsy syndromes in people aged 2 and older. These promising developments open the door to more investment, government funding, research, and perhaps inevitably the legalization of medical cannabis and cannabis-derived pharmaceuticals in other jurisdictions.

PRODUCT DIVERSITY FROM FLOWER TO EXTRACTS, EDIBLES, AND NUTRACEUTICALS

Historically, the most common method of delivery for cannabis has been smoking. When cannabis is smoked, a high temperature flame converts the THCA (tetrahydrocannabinolic acid) present in the plant to Δ^9-THC. Other cannabinoids, such as cannabidiolic acid (CBDA), go through this same process (known as *decarboxylation*) when exposed to high temperature. Note that THCA and CBDA do not interact with the endocannabinoid receptors of the human body as effectively as THC and CBD, and therefore, the use of cannabis plant without decarboxylation (e.g., by ingestion) does not have the same effects as smoking.

More recently, chemical extraction techniques have been developed to isolate the cannabinoids, terpenes, and other chemicals from the plant, creating highly concentrated resins and oils which can be used to produce various edibles (gummy bears, chocolates, etc.), vaporization cartridges, topical creams, and softgel capsules with controlled dosages. The major advantage of these products is that they do not require smoking, which inevitably results in carbon-based contaminants entering the lungs. However, these new types of products introduce several new challenges:

- **Dosage (potency) control**: It is critical that the concentration of cannabinoids such as THC and CBD is precisely controlled in these products to avoid unintended consequences.
- **Degree of decarboxylation**: THCA and CBDA must be converted to THC and CBD to be effective, and therefore, the degree of decarboxylation must be closely monitored during the extraction process.
- **Concentration of heavy metals, pesticide, solvent residues, and other chemical contaminants**: If the original material (usually cannabis flower) contains high levels of heavy metals or other chemical contaminants, these chemicals may become concentrated in the extracts, resulting in dangerous levels of chemical contaminants. In addition, if organic solvents are used for extraction, it becomes imperative to monitor the levels of residual solvents and ensure that they are not at harmful levels. Current guidelines in Canada, for example, suggest the extension of the US and EU Pharmacopeia standards for chemical contaminants in cannabis-based extracts.

CRITICAL CANNABIS TESTS REQUIRED FOR PRODUCT SAFETY AND EFFICACY

Comprehensive analytical testing of cannabis ensures the safety of the products and provides a solid basis for research and product development by quantifying the unique properties of different cultivars. Chemical analysis of cannabis and derived products is particularly challenging due to the complex chemistry of cannabis with more than 500 molecules that could potentially interfere with one another during analysis. The diversity of sample types, especially edibles, where the matrix could have very high levels of sugars, fatty acids, pigments, etc. adds to these challenges. Therefore, developing validated methods from sample preparation to analysis, data processing, and interpretation is a challenging task and requires in-depth understanding of cannabis chemistry.

Cannabinoid and potency profiling: Cannabinoid profiling and potency testing are essential for the development of cannabis products. Small variations in potency and cannabinoid content can dramatically alter both the recreational and therapeutic effects of different cultivars. Various analytical methods based on GC-MS, HPLC, and LC-MS/MS are typically used to provide detailed identification and quantitation of known cannabinoids including THC, THCA, CBD, CBDA, and minor cannabinoids such as CBN and CBG from a wide variety of samples such as fresh and dried cannabis flowers, oils, extracts, edibles, and isolates.

More than 100 cannabinoids have been isolated from the cannabis plant and, for the most part, the therapeutic effects or potential toxicity of these cannabinoids is not yet fully understood. The prevailing hypothesis is that different cannabinoids interact with the endocannabinoid system in human beings simultaneously in what is commonly referred to as the *entourage effect*. This makes extraction, isolation, and quantification of active ingredients from the cannabis plant quite important.

Terpene profiling: As mentioned earlier, terpenoids and flavonoids are the primary organic compounds that give different cultivars of cannabis their varying scents, flavors, and colors. As with the terpenes in wine or beer, these attributes attract consumers to particularly palatable and unique products. In addition, terpenes may also be responsible for enhancing the therapeutic effects of cannabinoids. More importantly, profiling these compounds alongside the cannabinoids provides a chemical fingerprint for each cultivar of cannabis and can be used for provenance purposes. Typically, GC-MS and LC-MS/MS are used for the analysis of known terpenes in cannabis and hemp products.

Pesticide detection: Harmful pesticides can accrue in cannabis samples, both as a residue of legitimate cultivation aids, and due to contamination from products used on nearby crops or on crops sharing transportation and storage space. Different jurisdictions have come up with their own list and levels of pesticides. For example, 96 pesticides are regulated by Health Canada, which could be identified and quantified using LC-MS/MS and GC-MS/MS at low part per billion (ppb) levels.

Mycotoxin detection: Cannabis and hemp plants are susceptible to fungal growth during cultivation, processing, and storage. Some fungi can in turn propagate secondary toxic metabolites, or *mycotoxins*, which are dangerous to immunologically vulnerable people, such as the elderly and the ill. Both HPLC- and LC-MS/MS-based methods are used for the detection toxic mycotoxins such as B1, B2, G1, G2, and Ochratoxin A, which have been directly linked to serious health issues such as liver cancer.

Microbiological analyses: As with all agricultural products, cannabis and hemp crops are subject to a variety of biological maladies. To ensure consumer safety and product purity, a full range of microbial analyses is required, including tests for *Escherichia coli* salmonella, total yeast and mould, and aerobic bacteria.

Residual solvents testing: Cannabis and hemp concentrates are often extracted for use in oils, edibles, and pharmaceutical drugs using organic solvents that can negatively impact product quality, potency, and consumer health. Headspace GC-MS is typically used to analyze trace levels of residual solvent in cannabis-based products.

Heavy metals analysis: Cannabis and hemp plants have been shown to bioaccumulate and concentrate heavy metals absorbed from even lightly contaminated soil, water, air, and fertilizers. Although maximum action limits for heavy metals in cannabis and derived products vary from one jurisdiction to another, almost all of them have issued strict limitations on arsenic, cadmium, lead, and mercury (the "big four"). Because cannabis flower and oil can be vaporized and inhaled, many jurisdictions follow the USP Chapter 232 guidelines for inhalation drug products. Currently, majority of cannabis labs in legalized states and Canada mainly use inductively coupled plasma mass spectroscopy (ICP-MS) and in some cases inductively coupled plasma optical emission spectrometry (ICP-OES) for heavy metal analysis. Because cannabis samples could be in different forms and matrices, developing and validating appropriate sample preparation methods for different sample types are critical. For many sample types such as flower, oil, and gummies, closed vessel microwave digestion is the most suitable method and provides the required precision and accuracy under USP, ICH, and other guidelines. In addition to heavy metal analysis in cannabis, many producers also require full spectrum of ultra-trace elemental analyses for advanced micronutrient profiling and product development.

In my opinion, the timing of Rob's book is perfect. This book provides the valuable insight that the cannabis testing labs and scientists need when analyzing elemental impurities in challenging and diverse cannabis matrices. The lessons that are learned from elemental analysis in pharmaceutical samples, and are shared in this book, will help analytical chemists in cannabis labs choose the right instruments and sample preparation methods, troubleshoot their equipment, and validate their methods. As a former ICP and mass spectrometry designer and scientist, I am also delighted to see the chapters that provide in-depth understanding of ICP-MS and ICP-OES technologies, and critical aspects of operating such advanced instruments. Without understanding the fundamentals, the analysts would be susceptible to make mistakes in choosing the right operating conditions or methods, which could ultimately lead to inaccurate results and jeopardize the safety of consumers and patients (in the case of medical cannabis). On the other hand, considering the detection capabilities of ICP-OES and ICP-MS, properly validated methods on these platforms will play a major role in ensuring the safety of products and advancing the science of cannabis.

Dr. Kaveh Kahen
President and CEO
Sigma Analytical Services Inc.Toronto, Canada
August 2020

Dr. Kaveh Kahen is the President and CEO of Sigma Analytical Services, a GMP-certified market leader in analytical testing services for cannabis and pharmaceutical products. Sigma Analytical Services is headquartered in Toronto, Canada, and is currently undergoing expansion to South America and Western Canada. Prior to establishing Sigma Analytical Services, Dr. Kahen was the Global General Manager of the Mass Spectrometry business unit for PerkinElmer, where he led this business in exponential growth by focusing on food, pharmaceutical, and cannabis testing markets, and offering differentiated and complete solutions. Prior to this, he led the global R&D organization of PerkinElmer by focusing on new technology development. He received his Ph.D. in the field of Analytical Chemistry from the George Washington University, Washington, DC, USA, and thereafter joined the research organization SCIEX, Toronto, Canada, where he designed and developed novel technologies for inductively coupled plasma mass spectrometry (ICP-MS), leading to several US patents. He has published and delivered more than 100 peer-reviewed papers and invited lectures at various international journals and conferences focused on analytical instrumentations and methodologies.

Preface

Measuring Heavy Metal Contaminants in Cannabis and Hemp: A Practical Guide, represents my fifth textbook in 16 years on the topic of using atomic spectroscopic techniques for trace element analysis. Ever since I got downsized (not my choice!) from a leading manufacturer of analytical instrumentation in the early 2000s, my mission has been to educate users on how to get the most out of these techniques. Almost 100 magazine/journal publications and five textbooks later, I'm still trying to do that. Not that I have a great passion for writing (typing with two fingers will do that), but I like to undertake a major project every few years that requires me to focus my energy into working with a blank sheet of paper (actually a pad), carrying out the necessary research, and putting all the components together to address the topic I'm going to write about. It's a process which was daunting with my first book in 2003, but it gradually became easier to manage and even though the writing part requires a great deal of focus and really long days (and a lot of caffeine) of sitting in front of my PC, the end result is one of the most rewarding experiences you can imagine. So with this as background information, allow me take you on a journey over the past 17 years to give you a flavor of what my life as a freelance writer has been like ever since that fateful day, when my career of 25 years at PerkinElmer came to an end.

I had a scary thought last year (2019), as I hit the big 70, that it has been 17 years since I published the first edition of my textbook, *Practical Guide to Inductively Coupled Plasma Mass Spectrometry: A Tutorial for Beginners*. There have since been two further editions of the book, the last one being published in 2014. What was originally a series of tutorials on the basic principles of inductively coupled plasma mass spectrometry (ICP-MS) for *Spectroscopy* magazine in 2001, it quickly grew into a textbook focusing on the practical benefits of the technique. With almost 6,000 copies of the English version sold, and over 3,000 copies of the Chinese (Mandarin) book in print (currently looking for a Spanish publisher to also get it translated into Spanish), I feel honored that the book has gained the reputation of being the reference book of choice for novices and beginners to the technique all over the world. Sales of the book have exceeded my wildest expectations. Of course, it helps when it is "recommended reading" for a Pittsburgh Conference ICP-MS Short Course I teach every year on "How to Select an ICP-MS." It also helps when you get the visibility of your book being displayed at 20 different vendors' booths at the Pittsburgh Conference every year. But there is no question in my mind that the major reason for its success has been that it presents ICP-MS in a way that is very easy to understand for beginners and also shows the practical benefits of the technique for carrying out routine trace element analysis.

APPLICATION LANDSCAPE

During the time the book has been in print, the application landscape has slowly changed. When the technique was first commercialized in 1983, it was mainly the environmental and geological communities that were the first to realize its benefits. Then as its capabilities became better understood, and its performance improved, other application areas such as clinical, toxicological, and semiconductor markets embraced the technique. Today, ICP-MS with all its performance and productivity enhancement tools is the most dominant trace element technique, and besides the traditional application fields mentioned, it is now being utilized in other exciting areas including, nanoparticle research, oil exploration, pharmaceutical manufacturing, and of course cannabis and hemp testing labs. As a result, with a street price tag for a single quadrupole-based ICP-MS in the order of $120, 000, it is being purchased for applications that were previously being carried out by inductively coupled plasma optical emission spectrometry (ICP-OES) and atomic absorption/fluorescence (AA/AV). It is truly an exciting time for users of the ICP-MS technique, but it also means that the other suitable techniques, which are often overlooked, now have lower price tags that make them a lot more attractive for labs with a limited budget, less demanding detection limit requirements, or with lower sample throughput demands.

PHARMACEUTICAL REGULATIONS

To put the cannabis testing marketplace into context, it's important to compare it with the pharmaceutical and dietary supplements industries. There are a lot of similarities with regard to how the pharmaceutical industry was being regulated for heavy metals, which the cannabis industry can learn from.

For over 100 years, pharma manufacturers only monitored for a small group of heavy metals using a semiquantitative sulfide precipitation test. However, recent regulations required the industry to monitor 24 elemental impurities in pharmaceutical raw materials, drug products, and dietary supplements. These guidelines which are described in the new United States Pharmacopeia (USP) Chapters <232>, <2232>, and <233> were implemented in January, 2018. Similar methodology was also approved by the ICH (International Conference on Harmonization of Technical Requirements for Registration of Pharmaceuticals for Human Use), a consortium of global pharmaceutical industries, including the European Pharmacopoeia (PhEu) under the umbrella of the European Medicine Association, the Japanese Pharmacopoeia (JP), and the USP through the Q3D guidelines, which came into effect in December 2017.

As a result of these new regulations, I began the process of writing a new book in 2015 entitled *Measuring Elemental Impurities in Pharmaceuticals: A Practical Guide*, which focused on these new directives and in particular the use of plasma spectrochemistry to carry out these measurements. The book was eventually published in March, 2018 and was quickly embraced by the industry when they were looking for some much-needed guidance for pharma production labs that had never previously used these techniques.

These new directives were driven by the fact that even though the risk factors for heavy metal contamination have changed dramatically, standard methods for their testing and control have changed little for more than 100 years, and as a result, most heavy metals limits have little basis in toxicology. This standard method described in Chapter <231> of the USP's National Formulary (NF) (1) has been in existence since 1908. Unfortunately, the test involving precipitation of the metal sulfides using thioacetamide, an extremely toxic chemical, is also known to be unreliable and prone to errors. As a result, USP has spent over 20 years investigating the use of plasma-based techniques for this analysis, particularly ICP-MS, which has far greater specificity, is more sensitive, and is applicable to a wider range of metals of interest.

Based on these investigations, three brand new chapters were eventually approved. Chapters <232> and <233> specify toxicity limits and analytical procedures for elemental impurities in pharmaceutical products, whereas Chapter <2232> deals with elemental contaminants in neutraceuticals.

THE ROLE OF THE AMERICAN CHEMICAL SOCIETY

In June of 2017, the 11th edition of the American Chemical Society's (ACS) book *Specifications and Procedures for Reagent Chemicals* was officially published, which was the culmination of 7 years work for 25 dedicated analytical chemists and their affiliated organizations who volunteer their time and resources by serving on this ACS committee. This new compendium which is now updated every year with an online edition, reflected new methodology, updated procedures, and more stringent specifications with regard to chemical reagents used for analytical testing purposes.

It was my involvement with the ACS Committee on Reagent Chemicals, which I've served on for the past 20 years that I became involved with the USP. Because the ACS is a valued stakeholder, we had an important role to play in reviewing and commenting on these new chapters, as our book of *Reagent Chemicals* is used by the pharmaceutical community (and other similar organizations where analytical testing is carried out such as ASTM, AOAC, and the Environmental Protection Agency (EPA)). As a result, we have worked very closely with USP to ensure that ACS methods on heavy metals testing align with theirs. In fact, the General Notices and Requirements in the USP-NF book of compendial standards states … *Unless otherwise specified, reagents conforming to the specifications set forth in the current edition of Reagent Chemicals, published by the American Chemical Society (ACS) shall be used.*

I lead the heavy metals task force and with input from other committee members, wrote the new plasma-based spectrochemical methods in the ACS book, which has replaced a similar heavy metals test to USP <Chapter 231>. In addition, for the past 5 years, I have been invited to give talks on the topic at workshops, expositions, and webinars and also taught a PittCon Short Course on the implementation of the new USP chapters on Elemental Impurities. This allowed me to gets lots of feedback from many people in the pharmaceutical manufacturing community. This inside knowledge gave me a unique insight into understanding the heavy metals requirements of the pharmaceutical industry and, as a result, was my major incentive to write a book on the topic.

THE WORLD OF CANNABIS TESTING

So fast forward to early 2019. My pharmaceutical book had been in the public domain for just over 12 months. I was looking for a new writing project, when one of my colleagues told me about the exciting world of cannabis testing. So I did some basic research and found out that it was an industry that was compelling, chaotic, and crazy all at the same time, but in my opinion, a little out of control (the wild west as many people would describe it). However, I thought I could make a significant difference, particularly in the measurement of heavy metals. So I decided to jump in and spent the next 4–6 months interviewing growers, cultivators, processors, manufacturers, testing labs, and state regulators in the United States and Canada to get a solid understanding of the major issues around testing of cannabis and hemp for elemental contaminants. I knew I might have been ruffling a few feathers when a cannabis grower accused me of being an investigative journalist when I asked them if they knew the levels of heavy metals in their cannabis plants.

Based on my research, it was clear the industry was in need of help. It was moving at such an alarming rate that the scientific and analytical testing community was struggling to keep up with it. It is estimated that the demand for medicinal and adult recreational together with pet-related cannabis-based products will exceed $25 billion in the United States by 2025, including an announcement in early 2020 that the Food and Drug Administration (FDA) was considering making cannabidiol (CBD) a legal ingredient in dietary supplements. However, the testing labs that will be required to verify these products are safe for consumption are basically novices without a good grasp of what it takes to generate high-quality data using plasma spectrochemical techniques used to carry out the measurement of elemental contaminants. I immediately saw what I had seen in the pharmaceutical industry 3 years previously … an opportunity to engage with the cannabis industry that needed some guidance.

So I mapped out a plan to learn as much as I could about the cannabis industry with the objective of compiling a similar reference to my pharmaceutical book, covering the fundamental principles and practical benefits of not only ICP-MS but also other suitable techniques such as ICP-OES, AA, and AF (atomic fluorescence) in a reader-friendly format but focusing on the real-world problems faced by novice users in the cannabis industry, including explanations of hardware components, calibration protocols, typical interferences, routine maintenance, and troubleshooting procedures.

There is no question that the surge of interest in cannabis-based products has put an extremely high demand on testing capabilities, particularly for contaminants such as heavy metals, which are naturally taken up through the roots of the plants from the soil, growing medium, and fertilizers, but can also be contaminated by the metallic grinding and processing equipment. So this became my incentive for putting together a very focused reference book on this topic; not only to be a useful training tool for operators of plasma-based spectrochemistry instrumentation in an application area that will be growing very rapidly over the next 2–3 years but also a useful resource for growers and cultivators to give them a better understanding of potential sources of heavy metal contaminants. I also wanted to offer some guidance to cannabis processors to minimize the amount of metallic contaminants from their manufacturing processing in addition to giving some direction to regulators who need to understand that the four heavy metals currently mandated by the vast majority of states have to be expanded to include elemental contaminants that are not currently tested to ensure that all future products are safe for human consumption. So allow me to take you on a quick journey through the book.

A FLAVOR OF WHAT'S IN THE BOOK

In 1996, California became the first state to legalize medical cannabis. Then in 2012, Colorado and Washington became the first states to legalize its recreational use, sparking a trend that quickly spread to the 33 states and Washington, DC that allow its use today. In the early days, there was very little testing of the product. However, it became abundantly clear that with the increasing legalization of both adult recreational and medical cannabis, there was a need for robust and reliable analytical testing to ensure consumer safety. Today, the major analytes of interest currently include cannabinoid potency, terpenes, residual solvents, pesticides, heavy metals, and microorganisms. So with this background information, I wanted to make sure we set the tone for the rest of the book with the first chapter which gives an overview of why the testing of cannabis is so important and what are the most commonly used analytical techniques. For that reason, I asked my colleague Dr. Jack Henion, to write the chapter. Jack highly regarded in this field is the Chief Scientific Officer of Advion Biosciences, Inc., a very prestigious mass spec company and Professor Emeritus of Toxicology at Cornell University.

The next two chapters deal specifically with the importance of measuring and regulating heavy metal contaminants in cannabis and hemp and in particular what the industry can learn from the pharmaceutical community that went through this process over 20 years ago, as they wrestled with changing a 100-year-old culture that carried out the bare minimum of testing for heavy metals in drug products to today's testing protocol which requires the measurement of up to 24 elemental impurities.

Even though this book will be a very useful educational resource for cannabis growers, cultivators, cannabinoid processors, and state regulators, the main focus is to give analytical chemists and technicians in the cannabis and hemp testing communities a better understanding of the atomic spectroscopic analytical techniques that are available for the determination of the heavy metal contaminants. In particular, it gives detailed descriptions of the fundamental principles and practical benefits of ICP-MS, ICP-OES, AA (atomic absorption), and AF (atomic fluorescence) in a reader-friendly format that a novice will find easy to understand. (These chapters have been supplemented with contributions from my colleagues, Dr. Maura Rury from LGC Standards and Debbie Bradshaw from DS Consulting, Inc., both with decades of experience in trace element analysis.) In addition, I have given overviews of other AS techniques such as MP-AES (microwave plasma atomic emission spectrometry), together with sold sampling techniques such as XRF (X-ray fluorescence), LIBS (laser-induced breakdown spectrometry), and LA-LI-TOFMS (laser ablation, laser ionization time-of-flight mass spectrometry) which offer the possibility of measuring the heavy metals directly in the sample without going through an acid digestion.

These chapters will walk a user through the different approaches, including easy-to-read sections on hardware components, calibration and measurement protocols, typical interferences, routine maintenance, and troubleshooting procedures, which should give them the tools to develop robust and bullet-proof methods for cannabis-related samples and also help them to optimize sample preparation and instrumental method development to lower elemental detection limits with the goal of generating higher quality data.

For ICP-MS users in particular, there is also a detailed chapter on sampling accessories either to enhance instrument performance or to maximize productivity. The user should find this chapter very useful, particularly if there is a requirement to determine speciated forms of arsenic and/or mercury at a future date when it might be required, or if they are looking for creative ways to increase sample throughput. Some of these sampling techniques might not be applicable to cannabis materials today, but they are included to give the novice user a broader perspective on the wide applicability of ICP-MS.

I have also included a chapter on how best to assess what elemental contaminants/heavy metals might be present in cannabinoid vaping devices/liquids (a contribution from my colleague, Dr. Steve Pappas from the Tobacco Inorganics Group at the CDC) ... an extremely important topic based on the recently reported lung illnesses and deaths caused by the cutting compound, vitamin E acetate, commonly referred to as EVALI (E-cigarette or vaping product use associated lung injury).

I'm hoping that the cannabis testing community will also find the other chapters useful, including a comparison of the four major atomic spectroscopic techniques for the determination of heavy metals, based on the required maximum action limits for cannabis-related materials. Even though ICP-MS is clearly the most sensitive technique, the use of axial ICP-OES with sampling accessories or electrothermal atomization (ETA) or atomic fluorescence might offer a more cost-effective solution to characterize cannabis samples for heavy metals. And if cost of analysis is a concern, included in this chapter is a comparison of the running costs of these techniques, based on operating costs, including electricity, gases, and instrument consumables. And on a similar theme, I have also written a chapter on offering a set of evaluation guidelines for cannabis testing labs that are thinking of investing in commercial ICP-MS instrumentation and want a better understanding of the selection process. This evaluation guideline is based on my experience of demonstrating the technique and running customer samples for over 15 years.

Other contributed chapters include one on ways of reducing contamination in the analytical testing procedure by Patti Atkins from SPEX CertiPrep and one on the use of reference materials in method validation and verification and the importance of quality assurance programs by Dr. Melissa Phillips and her colleagues from NIST. (FYI, more information can be found about my collaborators in the Acknowledgment section of this book.)

Finally, I have included a glossary of terms for novice users who might not be familiar with these techniques. Something I would have embraced with open arms when I first got into ICP-MS back in 1984. I had come to the technique from an AA and ICP-OES background, but the fact that ICP-MS was measuring ions and not photons introduced me to a myriad of terms and phrases that was completely alien to me. I would have loved something like this when I began learning the fundamentals back then. So I knew when I eventually wrote my first book on ICP-MS in 2004 that an educational glossary was always going to be one of my top priorities. It has since appeared in my two more recent books in 2009 and 2014, and for this book, I have expanded it to include ICP-OES, AA, AF, and the other AS techniques.

FINAL THOUGHTS

I firmly believe that those who are trying to raise the quality bar today will be recognized when the federal government eventually begins to regulate the industry. The recent vitamin E acetate in vaping fluids crisis resulting in serious illness and many deaths is a testament to what happens when illicit products get into the marketplace with little or no oversight.

There is no question that the uncertainty about future regulations in the cannabis industry represents a very challenging opportunity to put together a focused book on this topic of measuring elemental contaminants in cannabis and cannabinoid products. Unfortunately, for a writer who wants to capture the essence of this industry that's a rapidly moving target, it's very difficult to know when to put your stake in the ground. If I wait until there is stability in this industry, I would never finish the book. So the time has come to stop writing and send the manuscript off for publication.

So here is the final product for all to critique. It was written with the cannabis testing community in mind, but I'm convinced it will also be a useful resource for growers, cultivators, processors, testers, regulators, and even consumers who want to be better educated about the potential dangers of heavy metal contaminants in cannabis and hemp.

However, these are very uncertain times as we are currently in the midst of the corona virus pandemic. I'm hoping that by the time you are reading this book (publication planned for September, 2020), that life will be back to normal … whatever normal is after a pandemic? At least, I'm hopeful we will be in a much stronger position to deal with the next one. So, let me finish by wishing you the best of health and good luck with your laboratory endeavors. I'm hoping my book will be placed next to your testing equipment to use as a reference source for all your colleagues … or better still, convince them that they should have their own copy! And as always, thanks for your continued support of my work.

Robert Thomas, CSci, CChem, FRSC

Robert Thomas

April 2, 2020

Acknowledgments

When I began the process of writing this book and planning the book chapters, I realized that I did not have the necessary background experience of working in the cannabis industry. My expertise is in atomic spectroscopic techniques, having worked in the field trace element analysis for over 45 years. So to give this book added credibility to users in the cannabis industry, I decided I needed to collaborate with experts in this field and other related areas. For that reason, I reached out to people with the necessary skillset and background knowledge including experts in the cannabis testing field, academia, federal agencies, analytical instrument companies, and training specialists and asked them to contribute a chapter to this book. I would, therefore, like to acknowledge the contributions of the following collaborators.

Professor **Jack Henion**, PhD, is the author of Chapter 1 "The Importance of Testing Cannabis and the Major Analytical Techniques Used." He is Emeritus Professor of Toxicology at Cornell University where he was a member of the College of Veterinary Medicine commencing in 1976. Dr. Henion was co-founder of Advion BioSciences in 1993 where he served as President and CEO until 2006 when he became CSO of Advion, Inc. Dr. Henion carried out a wide range of research in many application areas involving GC/MS and LC/MS/MS techniques. Professor Henion has received three Doctor Honoris Causa (Honorary Doctorate) degrees in recognition of his international reputation in modern analytical techniques. These were awarded from each of the University of Ghent, Uppsala University, and Albany University. During his tenure at Cornell, Professor Henion conducted research and explored applications in many areas of liquid chromatography/mass spectrometry (LC/MS) employing atmospheric pressure ionization (API) sources.

He has published over 235 peer-reviewed papers in the scientific literature, trained nearly 100 students, post-doctoral scientists, and trainees while receiving 12 patents for inventions developed from his work. He has also received a number of awards which recognize his contributions to analytical chemistry and entrepreneurship. More recently in April 2017, Dr. Henion received the Outstanding Contribution to Anti-Doping Science Award from the Partnership for Clean Competition (PCC) for his development of a novel Book-Type Dried Plasma Spot Card, and in the Fall of 2017, Dr. Henion was the winner of the 2018 Bioanalysis Outstanding Contribution Award (BOSCA).

Patricia (Patti) Atkins has written Chapter 22 "A Practical Guide to Reducing Errors and Contamination"—a critically important topic on knowing contamination sources in the modern trace element laboratory and in particular how to minimize them to ensure the generated data is accurate and representative of what's in the original sample.

Patti is a senior applications scientist at SPEX CertiPrep in New Jersey. She is a graduate of Rutgers University in NJ and was laboratory supervisor for Ciba Specialty Chemicals in the Water Treatment Division. Patti later accepted a position conducting research and managing an air pollution research group within Rutgers University's Civil and Environmental Engineering Department.

In 2008, she joined SPEX CertiPrep as a senior application scientist in their certified reference material's division and spends her time researching industry trends and developing new reference materials. She is a frequent presenter and speaker at numerous conference including NACRW, NEMC, PittCon, and AOAC and published author with her work appearing in various journals and trade publications including spectroscopy, LCGC, and Cannabis Science and Technology where she is a columnist for analytical issues in cannabis testing.

The Chemical Metrology Group of the Chemical Sciences Division at the National Institute of Standards and Technology (NIST) has written Chapter 23 "The Importance of Laboratory Quality Assurance Programs" an extremely important topic of discussion for improving method performance. The following researchers are coauthors of this chapter.

Laura Wood, PhD, is a program coordinator for the food and the dietary supplements Standard Reference Material (SRM) programs at the National Institute of Standards and Technology (NIST). She is also one of the program coordinators for the Health Management Assessment Quality Assurance Program (HAMQAP). While at NIST, she has been involved in the development and certification of numerous SRMs, primarily for nutritional and toxic elements.

Melissa M. Phillips, PhD, has been a research chemist at NIST since 2008. She is involved in the certification efforts for food and dietary supplement SRMs and is a coordinator of the Dietary Supplement Laboratory Quality Assurance Program (DSQAP), the Health Assessment Measurements Quality Assurance Program (HAMQAP), and the Food Reference Materials Program. Her interests include development of new analytical methods for the determination of marker compounds, vitamins, and contaminants in foods and dietary supplements and improving the measurement capabilities of the food and dietary supplement communities using reference materials and quality assurance programs. Melissa obtained a BS in Chemistry, and MS in Forensic Chemistry, and a PhD in Analytical Chemistry from Michigan State University. She is also a Fellow and active member of the AOAC International Official Methods, Editorial Board.

Charles A. Barber (Chuck), PhD, is a research chemist at the NIST. His work involves making nutritional and toxic element measurements for the development and certification of food and dietary supplement reference materials in addition to toxic elements for hemp- and cannabis-related materials. He also serves as a program coordinator for the HAMQAP and Hemp Quality Assurance Program.

Catherine A. Rimmer (Kate), PhD, is a research chemist in the Chemical Sciences Division at the NIST. After earning a dual major undergraduate degree in anthropology and chemistry at the University of Vermont, Kate went to work for a pharmaceutical company. After working for 2 years, she began graduate school at Florida State University and received a PhD in Chemistry. Kate's primary area of focus was related to analytical separations; however, more recently that knowledge has been applied more specifically to the separation and quantitative determination of specific marker compounds in dietary supplements and natural products. The separation methods are then used for the development of NIST SRMs. In addition to reference material work, Kate works with the DSQAP and the HAMQAP. This work is especially rewarding as it involves working closely with the dietary supplement community to determine analytical strengths, weaknesses, and needs.

Walter B. Wilson (Brent), PhD, is a research chemist at the NIST coordinating the development of a Cannabis Quality Assurance Program and Reference Material product line. While at NIST, he has been involved in the development and certification of numerous natural product SRMs such as tobacco and dietary supplements.

The Tobacco Inorganics Group at the Centers for Disease Control and Prevention (CDC) are the authors of Chapter 24 "Measurement of Elemental Constituents of Cannabis Vaping Liquids and Aerosols by ICP-MS"—an important chapter in this book to ensure the safety of consumers of cannabinoid aerosols delivered by vaping devices. The following researchers have authored this chapter.

Richard Steven Pappas (Steve), PhD, team leader of the Tobacco Inorganics Group at CDC earned his BS in Chemistry at Middle Tennessee State University and completed his PhD in Biochemistry at Vanderbilt University. After faculty positions at Middle Tennessee State University and Georgia State University, he was employed at CDC to develop methods for analysis of toxic metals in urine and blood for emergency response and state health department laboratory training. In the second phase of his work at CDC, Steve became responsible for development of methods within the ISO 17025 framework for analysis of toxic metals in tobacco and smoke, during which he became the Tobacco Inorganics Group project lead. More recent responsibilities have included development of methods for analysis of metals in electronic cigarette liquids and aerosols and to characterize particles on tobacco, in smoke, and in electronic cigarette aerosols using ICP-MS, dynamic light scattering, and scanning electron microscopy with energy dispersive X-ray spectroscopy. He oversees method development and validation, ISO 17025 accreditation, interaction with database programmers, and is responsible for publishing and interpreting data in terms of public health risks. Steve has earned three group honor awards for Excellence in Laboratory Research, and an individual Innovation Award for research on characterization of particles in electronic cigarette aerosols.

In addition to authoring application manuscripts, Steve has written Annex 1, Toxic Metals in tobacco and in Cigarette Smoke in World Health Organization (WHO) Technical Report Series 967 on inflammation and sensitization responses in animal and human studies, a Metallomics review on the same topic, and the toxic metals section in *A Report of the Surgeon General: How Tobacco Smoke Causes Disease* (2010).

Nathalie González-Jiménez, PhD, earned her BS in Chemistry at the Interamerican University of Puerto Rico—San German Campus. She participated in various research projects during her undergraduate studies. She is currently pursuing her graduate studies, Masters in Public Health at Mercer University in Atlanta, GA. After graduating with her bachelor's degree, she was employed at the CDC in the Division of Laboratory Sciences, Tobacco Volatiles Branch in the Tobacco Inorganic Group. Nathalie has been part of the Tobacco Inorganic Team for 7.5 years. She has expertise in validation of analytical methods accredited under ISO 17025 requirements and has contributed to the validation of inorganic analytical methods, methods for the analysis of carbonyls and diacetyl. She has received an award for Excellence in Laboratory Research for contributions to the E-Cigarette, Vaping and Lung Injury Response at CDC. Nathalie is coauthor on peer-reviewed manuscripts' inorganic analysis of tobacco and tobacco-related products.

Naudia Gray, PhD, earned her BS in Environmental Science and MS in Environmental Science and Management working under Dr. H. M. "Skip" Kingston at Duquesne University. Naudia has been employed as a research chemist in the Tobacco Inorganics Group at the CDC since 2012. Naudia has extensive experience with tobacco product sample preparation, smoking and vaping machines, microwave digestion, and single quad and triple quad ICP-MS. She has participated in the development and validation of methods within the ISO 17025 framework for analysis of toxic metals in tobacco, smoke, and electronic cigarette liquids and aerosols. She has applied these methods to the analyses of various sample types. Recently, Naudia was involved in leading the efforts for the collection of aerosol from vaping products in the E-cigarette, or Vaping, Product Use Associated

Lung Injury (EVALI) response. Naudia is the first author of a manuscript on the analysis of metals in liquid from electronic cigarettes and is the coauthor of several other manuscripts. She has also has earned two group honor awards for Excellence in Laboratory Research and one group honor award for Excellence in Emergency Response—Domestic.

Mary M. Halstead earned her BS in Chemistry at Le Moyne College. In the initial 10 years of her career, Mary followed USP/NF methodologies in a cGMP environment as it related to pharmaceutical testing, she was employed in the pharmaceutical industry developing, validating, and transferring analytical methods for a variety of pharmaceutical products using a diverse group of analytical techniques (i.e., Ion Selective Electrode titrations, ICP-MS, LC-MS, and GC-MS). The next 6 years in the pharmaceutical industry, she began focusing her analytical expertise in the field of forensic investigation where she was responsible for the analysis of suspected counterfeit samples using a variety of analytical instrumentation (i.e., SEM/EDS, FTIR, XRD, and NIR). Most recently, she has been employed by Battelle Memorial Institute as a government contractor at the CDC in the Division of Laboratory Sciences, Tobacco and Volatile Branch, Tobacco Products Laboratory splitting her time between the management of the ISO 17025 quality system and the Inorganic Analysis Group. Mary is also responsible for the development and testing of tobacco-related products (i.e., tobacco, tobacco smoke, electronic cigarette liquids, and aerosols) by SEM/EDS as well as working with the ICP-MS team on the development of analytical methods. Mary has authored and coauthored peer-reviewed manuscripts on inorganic analysis of tobacco and tobacco-related products utilizing ICP-MS and SEM/EDS.

Maura Rury, PhD, has compiled the very comprehensive Chapter 25 "Fundamental Principles, Method Development, Optimization and Operational Requirements of ICP-Optical Emission" where she has extensive knowledge. She has published widely on atomic spectroscopic applications and has contributed a number of articles on plasma spectrochemistry to my Atomic Perspectives column in *Spectroscopy* magazine.

Maura is a Global Product Manager for the Applied Testing Reference Materials Division of LGC Standards. In this role, Maura is responsible for managing volatile organic and semi-volatile organic reference materials that are produced for food and beverage, environmental, cannabis, and petroleum applications. Her prior roles include portfolio marketing management at PerkinElmer, Inc., and product marketing and technical applications at Thermo Fisher Scientific, supporting atomic absorption and inductively coupled plasma-based instruments for a wide range of applications. This provided her with a significant amount of experience with atomic spectroscopy applications in specific markets including: food and beverage, pharmaceutical, environmental, industrial, and clinical. Maura holds a PhD in analytical chemistry from the University of Massachusetts, Amherst, under the tutelage of Professors Julian Tyson and Ramon Barnes.

Deborah (Debbie) Bradshaw is an analytical chemist who has been working in the field of atomic spectroscopy for over 35 years. She is responsible for writing Chapter 26 "An Overview of Atomic Absorption and Atomic Fluorescence." Working as a chemist in the environmental field, she started using flame atomic absorption in the 1970s and then moved into graphite furnace analysis. Using a Zeeman background corrected instrument, in the early 1980s she developed methods for a number of elements for the analysis of seawater samples. Taking a position with PerkinElmer Corporation in 1989, she migrated into analyzing a variety of sample types using not only AAS but also the plasma techniques, both optical and mass spectrometry. For the past 20 years, Debbie has been working as a consultant in the field of atomic spectroscopy, conducting training classes and giving technical support for AA, ICP-OES, and ICP-MS.

Debbie is currently on the Editorial Advisory Board for *Spectroscopy* magazine. She has been a member of the Society for Applied Spectroscopy (SAS) since 1980 and is now a Fellow of the society. For 14 years, she was the News Column Editor for the *Journal of Applied Spectroscopy*, SAS's monthly publication. From 2002 to 2004, she was Treasurer of SAS and has sat on several of the society committees. Debbie was also the recipient of the SAS Distinguished Service Award in 2008.

In efforts to promote education and support analysts using atomic spectroscopy techniques, she was the FACSS (now SciX) Atomic Spectroscopy Symposia Chair in 2007 and 2008, has organized technical symposia at both FACSS and PittCon, and has been a short course instructor for SAS. Debbie continues to be a short course instructor at the Winter Plasma Conference on Spectrochemical Analysis since 2000.

OTHER CONTRIBUTORS

Besides my chapter collaborators on this book, I want to acknowledge other contributors who have helped educate me about the cannabis and hemp industry and given me the background information and confidence to write a book about a topic that 18 months ago I knew very little about. The support of these people has been invaluable. So I would like to give recognition to the following people/groups (in no particular order):

- Debbie Miran ex Commissioner of the Maryland Medical Cannabis Commission (MMCC) and now a consultant in the field for her expertise and knowledge of the industry ... and for just being available to answer every question I threw at her.
- Lori Dodson, Current Deputy Director of the MMCC for getting me exposure to the local cannabis testing community in MD, DC, PA, DE, and helping me to organize the first workshop on "Measuring Heavy Metals in Cannabis" in October, 2020.
- John Dwan and all his colleagues at Shimadzu Scientific in Columbia, MD for sharing their knowledge of the cannabis industry and generously supporting my MMCC "Heavy Metals in Cannabis" workshop.
- All the sponsors of my MMCC "Heavy Metals in Cannabis" workshop, including Shimadzu, Spex, Spectron, Glass Expansion, Milestone, and MMCC.
- All the speakers at my heavy metals in cannabis workshop, including Andrew Fornadel of Shimadzu, Patti Atkins of Spex, Laura Thompson of Milestone, Ryan Brennan and Justin Masone of Glass Expansion, Lawrence Neufeld of Spectron, Melissa Phillips of NIST, and Steve Pappas of the CDC.
- Kasey Kirby of Atlas District Products for the excellent videoing of my MMC workshop on "Measuring Heavy Metals in Cannabis" and the editing required linking all the talks to the PowerPoint presentations (available on request).
- All the other analytical instrumentation vendors (too many to mention) who have given me access to their cannabis customer base and application material and for displaying my books at various conferences and meetings over the past 20 years.
- The folks at the Spark Networking Group for allowing me to get early access to key decision makers in the cannabis industry.
- Barbara Knott, George Kenney, and Danielle Zarfati at CRC Press/Taylor and Francis for their friendship and supporting the publication of this book and my four other book projects.
- Shannon Hoffman, Daniel Kulakowski, and all the staff at Steep Hill Labs in Columbia, who helped me learn and understand the challenges of testing cannabis.
- Josh Crossney, organizer and CEO of the Cannabis Science Conference, for allowing me to give the very first oral paper on heavy metals at a cannabis conference (special thanks also to COO, Andrea Peraza).
- Jack Rudd, Editor of Analytical Cannabis, for embracing my vision for this book and getting visibility with the global marketplace, by publishing a series of articles in his magazine.

- Meg L'Hereux, Editor of Cannabis Science and Technology, for publishing my very first article on heavy metals.
- Laura Bush, Executive Editor of Spectroscopy and LCGC, for my Atomic Perspectives (AP) columns and supporting my writing projects over the past 10 years.
- Numerous cannabis cultivators and processors, for many interesting and "lively discussions" on the many sources of heavy metals in cannabis, cannabis products, and vaping devices.
- Ashley Cabecinha, Health Canada, Ottawa, for giving me the Canadian perspective, which was invaluable to understand the global industry.
- Heather Klug, State Marijuana Laboratory Sciences Program Manager for Colorado, for helping me understand the many challenges involved with implementing state-based regulations.
- Kaveh Kahen, President and CEO, Sigma Analytical Testing Labs, Toronto, for inviting me to his cannabis testing oral session at an ACS Regional Meeting and for writing the Foreword to my book.
- Chris Hudalla, Founder/Chief Scientific Officer, ProVerde Testing Labs in Milford, MS for having such a unique perspective and understanding of this industry.
- Roy Upton, President of the American Herbal Pharmacopoeia (AHP), for sharing his detailed knowledge of the cannabis and hemp plants.
- All my colleagues on the ACS Reagent Chemicals Committee for giving me a deep knowledge of the pharmaceutical industry, which was critical to exploring the possibility of writing a book on cannabis testing.
- All the folks at AOAC International for allowing me to tap into their knowledge and expertise in writing standards for the food, plant, and herbal supplements testing.

So to all these people who have their fingerprints over this book, I am truly indebted to all of you … thank you so much for your support! hopefully, I haven't missed anyone, but if I have, I apologize … I'll include you in the second edition … stay tuned!

I would also like to give a special thanks to my very talented daughter, Glenna Thomas, who designed the front cover and drew all the sketches for the chapter titles.

Author

Robert (Rob) Thomas is the Principal of Scientific Solutions, a consulting company that serves the training, application, marketing, and writing needs of the trace element user community. He has worked in the field of atomic and mass spectroscopy for more than 45 years, including 24 years for a manufacturer of atomic spectroscopic instrumentation. He has served on the ACS Committee on Analytical Reagents (CAR) for the past 20 years as Leader of the plasma spectrochemistry, heavy metals task force, where he has worked very closely with the United States Pharmacopeia (USP) to align ACS heavy metal testing procedures with pharmaceutical guidelines. Rob has written almost 100 technical publications, including a 15-part tutorial series on ICP-MS. He is also the Editor and frequent Contributor of the Atomic Perspectives column in *Spectroscopy* magazine. In addition, Rob has authored three textbooks on ICP-MS and in 2018 completed his fourth publication, entitled *Measuring Elemental Impurities in Pharmaceuticals: A Practical Guide*. He has spent the past 2 years researching and writing this new book, *Measuring Heavy Metal Contaminants in Cannabis and Hemp*. Rob has an advanced degree in analytical chemistry from the University of Wales, United Kingdom, and is also a Fellow of the Royal Society of Chemistry (FRSC) and a Chartered Chemist (CChem).

1 The Importance of Testing Cannabis

An Overview of the Analytical Techniques Used

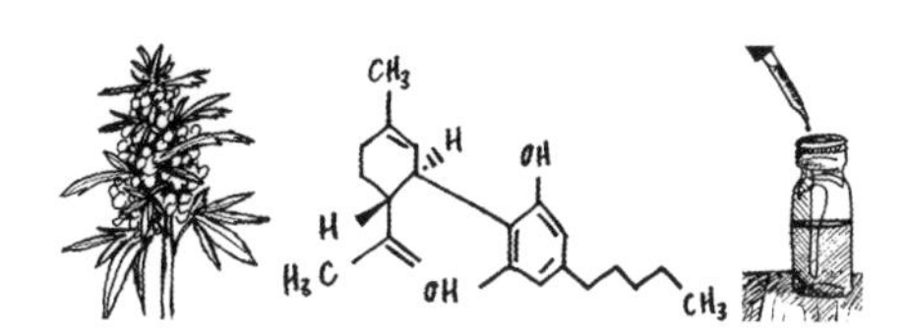

The cannabis industry has shown a great deal of promise and business opportunities! Instead of growing corn at a few hundred dollars per acre, there are reports that growers can make well over $1,000/acre by growing hemp. With marijuana flowers selling for $60/gram (personal observations of author from visiting a cannabis dispensary in Anchorage, AK in 2018) which can bring $27,240/lb, it is no wonder there is a huge financial interest in the cannabis industry. Although it is very exciting to have an entirely new "agricultural crop" that can provide this level of economic opportunity, it is sobering to know there is so little federal or state regulatory guidance on the products derived from a cannabis harvest. Cannabis products range from edibles to medicine and can now be produced and sold by anyone interested in doing so without any government guidance or regulatory oversight.

The focus of this chapter compiled by Jack Henion, CSO of Advion Inc., explains why there is a need for analytical testing of cannabis and its multitude of cannabinoid products with specific reference to the analytical horsepower of mass spectrometry (MS) as a detector. The various "tests" can range from traditional simple color tests to the use of more sophisticated analytical instrumentation carried out in modern, accredited laboratories by well-trained technologists. This chapter will highlight why there are a variety of related analytical chemical "tests" which are important for the determination of different chemical constituents in cannabis products. This is intended to be an important overview of the current analytical techniques and instrumentation employed today and in the future for the safety of cannabis-derived products.

It has been less than 100 years since more than 500 chemical compounds have been identified in cannabis and where some of the chemicals in cannabis (many of which are unique to the *Cannabaceae* family) may be responsible for the pharmacological effects many of us are awareof.[1] Particularly noteworthy advances have been the discovery that cannabis is the source of a family of over 100 compounds known as phyto cannabinoids, and that one of these compounds is delta-9-tetrahydrocannabinol (THC). This is the main psychoactive constituent of cannabis which for some people gives the term "cannabis" a bad name. Despite a long history of human and even animal use of cannabis, its use is illegal to this day in many states and countries.

The effectiveness of cannabis for treating various medical conditions has been reported in certain publications, but systematic, carefully conducted clinical trials are limited due to the stigma associated with the recent history of cannabis in today's world.[2] There are increasing numbers of anecdotal testimonials to the effectiveness of "medicinal marijuana" to treat rare seizure disorders in young children and relieve symptoms of cancer therapy drugs, as well as relieving symptoms of anorexia and immune deficiency syndrome (HIV/AIDS).[3] Of course, cannabis has a popular following by those interested in the psychoactive effects of delta-9-THC which reportedly makes one feel relaxed, happy, and even "giddy". In contrast to the considerable negative stigma associated with cannabis, the Food and Drug Administration (FDA) recently approved the drug, Epidiolex, which is obtained by an efficient extraction of cannabidiol (CBD) from *Cannabis sativa*.[4] Some believe this recent development will be a huge boost for the acceptance of the non-psychoactive nature and benefits of CBD.

HOW IS CANNABIS BEING USED?

CBD has attracted considerable interest from those experiencing insomnia, anxiety, and pain in addition to a growing list of other ailments.[5] Even beauty products have been reported such as mascara, lip balm, and eye cream that contain certain non-psychoactive cannabis components. One of the seemingly most dramatic recent reports is that the Coca-Cola Company is reportedly in discussions with the Canadian cannabis industry to infuse CBD into future Coca-Cola products.[6] This has been preceded by *Cannavines*, a California wine company which now sells a non-alcoholic red wine which has been infused with CBD.[7] Not to be outdone, *Sovereign Vines* in New York State has produced a novel wine which has been infused with certain "special tastes and flavors" from hemp.[8] From these reports, we can likely expect a wide diversity of new products containing cannabis and hemp[9] constituents going forward.

Another area of increasing interest and importance is veterinary use of hemp-based oils and products for cherished family pets.[10] The increasing popularity of cannabis products for human use has prompted shared interests of pet owners and veterinarians who are curious to know how marijuana products may be used to treat predominant pet ailments, including anxiety, joint pain, and other ailments.[11] Not surprising is the rapidly increasing variety of new CBD "products" becoming available without any regulatory scrutiny as to the safety or efficacy of the products. A recent report described a study of 13 commercially available CBD oils intended for veterinary use had inaccurate information on the labels regarding the contents in the bottle.[10] A recent rigorous veterinary clinical study at Cornell University described substantial improvement with the oral administration of a hemp-based CBD oil efficacy of a CBD-based oil in dogs with osteoarthritis (OA).[12] It was reported that routine nonsteroidal anti-inflammatory drug (NSAID) treatments, though efficacious, may not provide adequate relief of pain due to OA and might have potential side effects that preclude its use, particularly in geriatric patients with certain comorbidities, such as kidney or gastrointestinal pathologies. In contrast, the Cornell study concluded the pharmacokinetic and clinical study results suggested that oral administration of 2 mg/kg of CBD twice daily can help increase comfort and activity in dogs with OA. In contrast to this systematic clinical study, accidental ingestion of discarded cannabis products by dogs is increasing with frequent toxic effects where veterinarians can struggle with an accurate diagnosis since they may not know the origin of the patient's problem. It has been suggested that dogs have a higher number of cannabinoid receptors in the brain compared to humans and thus may be more susceptible to the toxic effects than are humans.[13] Given the tendency for dogs to ingest many things they "encounter" as they sniff their territory, it is increasingly important that cannabis products are not carelessly discarded. With the increasing use of medical cannabis for recreational use, the interest in using it for pets is certainly increasing. As a result, veterinarians must also become aware of and familiar with the various aspects, effects, and treatments for companion pets that may have been exposed to cannabis products.

CANNABIS OR MARIJUANA?

The Latin word for marijuana is cannabis which was classified as a Schedule I substance in 1970 in the United States under the Controlled Substances Act of that year. The stigma associated with the controlled substances notation has thwarted research on the potential medicinal value while producing the opportunity for abuse by those interested in its psychoactive properties. The term marijuana is a common name for the cannabis plant, *C. sativa* L. which is a widespread species in nature.[14] Although the word marijuana is often used, it is important to understand its meaning generally refers to the cultivar (species) of the cannabis plant that contains high levels of delta-9-THC. It should be noted that the term cannabis is the correct word for describing the plant and its products. The term hemp refers to the use of cannabis as a fiber which is very low in the psychoactive compound, delta-9-THC (<0.3%) and has been used for many different products for thousands of years. The pharmacological effects of cannabis have been exploited for thousands of years for recreational, medicinal, or religious purposes.[15]

The current range of cannabis products extends from recreational use to medicines, edibles (foods), beverages, pet products, and beyond. The quality standards employed to assure product integrity and safety in the cannabis industry are lacking compared to those currently used by the pharmaceutical, food, and beverage industries. This chapter will overview the important analytical techniques available for measuring chemicals present from the growth and harvest of the cannabis plant to the production of a wide variety of its products. Currently the topics of interest for safety in cannabis testing where analytical chemistry can play an important role include what are currently referred to as potency, residual solvents, pesticides, terpenes, heavy metals, and mycotoxins from mold. Since each state in the United States as well as several countries has its own regulations, the analytical opportunities and challenges vary depending upon which jurisdiction a laboratory is supporting. One of the takeaways from this report will be that among the many available analytical techniques, a combination of modern chromatography coupled with MS as a detector is the preferred combination of analytical techniques for each of the above "tests."

THE ROLE OF ANALYTICAL CHEMISTRY

For those analytical chemists who are familiar with modern analytical technologies, it might be assumed that the techniques of gas chromatography (GC)/MS as well as liquid chromatography (LC)/MS would be the preferred analytical techniques for the analysis of cannabis and cannabis-derived products for organic chemicals contained within. As we will see later (*vide infra*), this is the case, but for some analysts, procedures and techniques from earlier times are still endorsed. As an example, the Duquenois–Levine test is a simple chemical color reaction test initially developed in the 1930s as a screening test for marijuana by the French biochemist, Pierre Duquenois and was adopted in the 1950s by the United Nations as the preferred test for cannabis.[16] To administer a forensic test, a police officer simply breaks a seal on a tiny micropipette of chemicals and inserts a particle of the suspected substance to be tested; if the chemicals turn purple, this supposedly indicates the possibility of marijuana. But the color variations can be subtle, and readings can vary by examiner.

During its early days, the test was reported to be specific to cannabis. After several modifications, its use continued until the 1960s. However, following this period, various studies showed that the test was not specific to cannabis such that in 1973 the Supreme Court of Wisconsin ruled the Duquenois–Levine test insufficient evidence for demonstrating that a substance was cannabis specifically noting that the tests used are not exclusive or specific for marijuana (https://en.wikipedia.org/wiki/Cannabis_drug_testing).

It is often desired that a "test" be very easy and inexpensive to carry out. Color tests such as the above-described Duquenois–Levine test were very popular early in the 20th century because they satisfied these criteria. Unfortunately, most of these easy tests have been abandoned for the same reasons described above; they lack specificity. The desire continues, however, for easy tests, for example, for the determination of cannabis potency. The potency term generally refers to the percentage of THC and or CBD in the plant material but in addition at least three other cannabinoids routinely monitored including tetrahydrocannabinolic acid (THCA), cannabidiolic acid (CBDA), and cannabinol (CBN) are of interest. It is asking a lot of a "simple test" to selectively detect with good sensitivity and accuracy each of these chemical entities in a complex sample matrix as well as to quantify them at the fractional percent level. Recall that for hemp to be legal in the United States currently, its THC content must be less than 0.3% by weight. If we fast-forward to the present a possible solution to having a relatively easy analytical test that is better than a color test, we have the mid-infrared spectroscopy technique.[17] A recent report by the referenced author compares the accuracy (without reference to selectivity) of chromatography versus spectroscopy.[18] The conclusion from this publication is that chromatography, whether it is GC or LC, is more accurate than, for example, infrared spectroscopy. However, the latter is faster (2 min vs. 20 min/test), easier because a lay person can potentially perform the test, cheaper and potentially field portable. It is worth noting that currently, although spectroscopic instrumentation is commercially available, it has not yet been widely adopted for determining the potency of cannabis. In the view of this author, it is uncertain whether the subtle differences in the infrared absorption bands as a function of concentration differences are sufficiently different as a function of THC and other cannabinoid concentrations to be reliably quantified by this technique.

SAMPLING PROTOCOL FOR CANNABIS IN THE FIELD

Before a sample can be prepared for analysis as will be described below, the sample must be collected and submitted to the laboratory. If the initial sample collection and preparation are flawed then the final answers will be biased. Collecting a representative sample of blood or urine from a patient for chemical analysis is fairly easy to do. Typically, a sample of urine is collected or blood via venous puncture, the sample is stored in the cold to mitigate chemical instability whereupon sample preparation followed by chemical analysis is performed. But how does one get a representative sample of cannabis grown in a 100-acre field or a greenhouse where hundreds of plants are growing? A key point here is the representative sample! Representative samples are selected to accurately reflect the larger group and should represent the characteristics of the group as a whole. Ideally, representative samples are homogeneous or similar in nature, but when that is not possible, the best attempts must be made to achieve samples that represent the majority of the characteristics of the larger grouping.[19] It is well known that the chemical composition of cannabis plants can differ significantly within a field as well as within a given plant itself. For example, the cannabinoids of interest are often present at much higher levels in the flower of the plant compared to the stem or seeds. Cannabis is a very complicated plant. So, what portion of the plant does one collect for chemical analysis? Different amounts of compounds can occur in different locations within the plant. In some cases, it has been reported that higher THC concentrations are found in buds located high on the plant as opposed to buds located lower in the plant.[20] Adding to the challenge is the fact that THC content of hemp generally peaks as the plant ripens, so the timing of when sampling occurs is important to accurately measure THC concentration and monitor compliance with the US Department of Agriculture (USDA) hemp production program.[21] One of the most common complaints of growers of cannabis is the long delays from the time of collecting a sample to the availability of the laboratory results (author's personal correspondence). Often this can be from 15 to 30 days depending upon the state jurisdiction and the availability of laboratory facilities.[21] If a hemp grower wishes to harvest when the crop's CBD content is at a maximum, how does he or she know when that optimal time is for maximum profits? Finally, there is the question of how many plant samples and what portion of the plant should be collected per unit area or agricultural plot of cannabis. The Massachusetts sampling protocol suggests that a sample is more likely to accurately represent the production batch if the material is homogenous (i.e., well mixed).[22] Mixing or other homogenization steps help to homogenize the product before sample collection. The latter jurisdiction also suggests that because production batches may vary in scale (i.e., volume or weight), varying numbers or sizes of samples may be required to promote representativeness. In section 6.0 of the Massachusetts guidance, a nine-step detailed sample collection protocol is provided.[22] However, there is little guidance to the person collecting the samples as to how many plant samples per unit area, how much and what portion of the plant to collect, or how best to maintain a chain of custody, etc. Suffice it to say those people currently collecting cannabis samples are provided some form of local guidance from their particular jurisdiction and or should follow the recommendations of the recent USDA guidance until further updates are published.[21] In the end, sampling protocols are not yet well defined and each state tends to use their own interpretation of best practices.

SAMPLE PREPARATION OF CANNABIS PLANT MATERIALS AND CANNABIS-DERIVED PRODUCTS

Before we move on to overview some of the modern analytical techniques used for the analysis of cannabis and cannabis-derived samples, it is worth highlighting the importance of sample preparation prior to chemical analyses. As noted above, the breadth of sample types relevant to modern cannabis analyses is very broad. It is well known by those experienced in the chemical analysis of plant materials, biological samples, and commercial products such as edibles that an optimized sample preparation method is very important for accurate and precise chemical measurements. Thus, there is a very big difference in the interfering sample matrix components within a cannabis plant flower compared to the matrix of a chocolate brownie. Thus, optimized sample preparation for each of the different sample types must be employed to obtain accurate quantitative determination of the targeted cannabinoids, pesticides, etc. that are of interest. Although there are currently many different methods used by participating laboratories, the one by Giese et al. represents perhaps one of the preferred approaches for the preparation of cannabis-derived samples.[23] Despite this suggested sample preparation procedure in practice, there are currently no standard sample preparation procedures. Each laboratory tends to develop its own method or tweak a referenced procedure to suit its needs. Consequently, there can be considerable variability with interlaboratory results even on the same samples. There remains much work to be done going forward to develop more standardized sample preparation as well as analysis procedures.

The above introductory discussion on the chemical analyses applicable to cannabis brings us to the modern day where we will describe the current methods and techniques being used. Given that this book focusses on the analytical horsepower of inductively coupled plasma (ICP)-MS and other atomic spectroscopic techniques for the ultratrace determination of heavy metals, we will begin with why this important modern analytical technique is very important to the field of cannabis analyses.

HEAVY METALS AND ICP-MS

Cannabis plants, both marijuana and hemp, are hyper-accumulators of heavy metals from the soil in which they grow. Phytoremediation refers to procedures which employ plants to scavenge heavy metals from the environment.[24] Figure 1.1 shows a cartoon of the concept of a cannabis plant preferentially absorbing heavy metals from the soil in which it is growing.

For agricultural crops, this procedure is sometimes used prior to seeding the farmland for food products. This technique depends upon the ability of certain plants to sequester heavy metals from the soil so that when food-producing crops are planted there will ideally be no heavy metals in that crop. As it turns out, cannabis plants are remarkably capable of removing heavy metals from the soil without affecting their own heartiness.[25] This of course concentrates these heavy metals in the cannabis plant materials where they can later reside in products derived from the cannabis plant. While some metals are beneficial and essential for life, others are highly toxic and have negative

FIGURE 1.1 Cartoon depiction of cannabis plant roots sequestering heavy metals from the soil.

effects on good health. Whether the dry cannabis plant's leaves are smoked or medicinal products are produced, the user may be exposed to unhealthy levels of potentially toxic heavy metals. As a result, it has become standard practice to analyze cannabis plant materials as well as products derived from them for an increasing number of these heavy metals. Among those heavy metals, most commonly determined are lead, cadmium, mercury, and arsenic although depending upon which states in the United States are involved this list may be larger. Several studies suggest that cannabis is an active accumulator of additional metals including magnesium, copper, nickel, chromium, vanadium, manganese, and cobalt. These metals are often found if the plant is grown near a mining, smelting, sewage sludge, or automobile emissions.[25]

Cannabis testing laboratories are adding sample preparation and analysis procedures to their services which can quantify a growing list of heavy metals found in plant and their derived products. It is worth noting that in contrast to the notes on sample preparation for cannabinoids, terpenes, and pesticides, sample preparation for heavy metal analysis by ICP-MS is very different. The latter typically involves microwave digestion at elevated temperatures in rather concentrated nitric acid followed by aqueous dilution of the digested sample and continuous introduction via a pump into the ICP-MS torch.[26] (Refer to Chapter 19 "Sampling and Sample Preparation Techniques" for more detailed information.) Cannabis plant materials can be tested for heavy metals in many ways, including various forms of atomic spectrometry, including atomic absorption (AA), atomic fluorescence (AF) ICP-optical emission spectroscopy (OES), and ICP-MS. For AA, the method used would have to be the more sensitive graphite furnace atomic absorption (GFAA) since the flame AA method would not be sensitive enough for most elements. Also, mercury must be measured by a method called cold vapor atomic absorption spectroscopy (CVAAS) due to the lower sensitivity by AA. (Refer to Chapters 25–28 for more details.)

In general, flame techniques can measure elements at low parts per million, and GFAA goes down to low parts per billion. In addition, AF combined with hydride generation achieves limits somewhere in between. Also, the AA method usually measures one element at a time, although multielement lamps can be used. ICP-OES and ICP-MS are techniques capable of measuring multiple elements simultaneously. Nonetheless, using ICP-OES to test cannabis for heavy metals often requires a way to enhance its sensitivity, such as introducing the sample with an ultrasonic nebulizer (USN) or an electrothermal vaporizer (ETV). Reports have shown that USN can increase sensitivity up to a factor of 10 in many cases. Experienced analytical chemists generally report that ICP-MS offers the best sensitivity and is the method of choice in many modern laboratories. For some important organometallic compounds, it can be useful to employ LC/ICP-MS techniques which can allow detection and quantitation of the individual organometallic compounds.[27] The FDA and United States Pharmacopeia have standardized methods for heavy metals analysis, which are very useful resources in the fledgling cannabis testing industry, where regulation is slow to catch up.[28]

ORGANIC COMPOUNDS IN CANNABIS

There are hundreds or perhaps thousands of organic compounds that can be found to be present in cannabis plants. In the view of this author, it is important to understand the important role MS can play as a sensitive and selective analytical technique for the selective and sensitive determination of these compounds. It is tempting to expect that it would be the technique of choice in this new industry based upon its abundant acceptance in the pharmaceutical, clinical, food, and environmental industries. Although LC/MS and GC/MS techniques are used in some cannabis applications (*vide infra*), other less definitive techniques such as LC/photo diode detector (PDA), GC/flame ionization detector (FID), and near-infrared (NIR) spectroscopy[17] are currently popular approaches for some practitioners.[29] The focus of this chapter is to contrast and compare the relative merits of these techniques applied to the needs of the cannabis industry and society as a whole. The sample types range from botanical plant materials and their oil products to edible foods including wines and beverages as mentioned above. The chemical constituents range from endogenous organic compounds including cannabinoids, terpenes, and flavonoids to exogenous compounds which include mycotoxins from mold growth to a wide range of pesticides and herbicides used illegally in the cultivation of the plants. In addition, the determination of a substantial range of heavy metals as described above must be measured as well as an impressive array of bacteria, molds, etc. To serve the growing need for chemical analysis of the many cannabis products, numerous contract service laboratories are springing up across the United States and elsewhere. Since each state has its own regulations on this industry, there is considerable inconsistency interstate on how things are done. The author's purpose is to present recommendations as to why mass spectrometric techniques should be an important component for the chemical analysis of cannabis and its many products. What follows is specific discussion on possible preferred analytical techniques for each of the chemical classes currently being measured.

PESTICIDES

Pesticides are not approved for the cultivation of cannabis in any of the US states where cannabis is grown. Since cannabis is legalized for medical and recreational use on a state-by-state basis, safety regulations relating to cannabis products are becoming increasingly important. Pesticides rank high on the list of safety concerns in cannabis plants and the products derived from them. The level of pesticides and herbicides of course should be very low since these could be present and/or concentrated in subsequent products as a result of the methods employed to produce the products. Cannabis samples from shops and dispensaries in various states that have legalized the sale of recreational marijuana invariably reveal products contaminated with insecticides, fungicides, rodenticides, and other compounds used to eliminate or prevent insect or bacterial infestations.[30] Even though much cannabis is grown indoors within high-technology greenhouses with careful control of light, humidity, and temperature, a variety of insect and bacterial pests can adversely affect the growth and quality of the plant before harvest. As a result, growers are tempted to use pesticides to control these problems even though it is unlawful in each state to do so.

The US Environmental Protection Agency (EPA) has long been faced with setting and enforcing legal safe limits on toxic chemicals. Cannabis product safety is particularly relevant since medicinal cannabis often is used for young and immunocompromised patients as the intended consumers. Pesticide residues have evolved into a significant problem for the cannabis industry. A number of studies have reported high levels of pesticides on cannabis samples taken from medical markets in Colorado and Washington.[30] Recalls of cannabis products in both states as well as Canada (which recently legalized cannabis use nationwide) highlight the need for regulatory control. All this begs for routine chemical analysis of cannabis plants and products derived from them. An example of the toxic effects possible from pesticides comes from myclobutanil, a persistent fungicide used by some growers. When cannabis plant material containing myclobutanil[31] is smoked via a cigarette, it releases the very toxic gas, hydrogen cyanide. It is therefore important to provide reliable chemical analysis of plant material before it is consumed to ensure the absence of such pesticides. Each state has its own list of those pesticides which must be monitored in cannabis samples. California, for example, currently has a list approaching 100 pesticides which must be monitored with "safe levels" being at low part per billion lower limits of quantitation (LLOQ).[32]

The chemical complexity of cannabis plant materials and products produced from them coupled with the regulations for low ppb levels can benefit from the analytical capabilities of modern separation science coupled with MS. Although GC/FID analyses have been employed in the past for measuring some pesticides, more recently GC/MS, LC/MS, or LC/MS/MS techniques are preferred. Figure 1.2 shows the chemical structures for 15 representative pesticides which were artificially sprayed onto cannabis plant flowers in an effort to approximate the agricultural field application of pesticides to cannabis plants.

These same 15 pesticides are also listed in Table 1.1 along with their selected ion monitoring (SIM) LC/MS chromatographic retention times observed in this work.

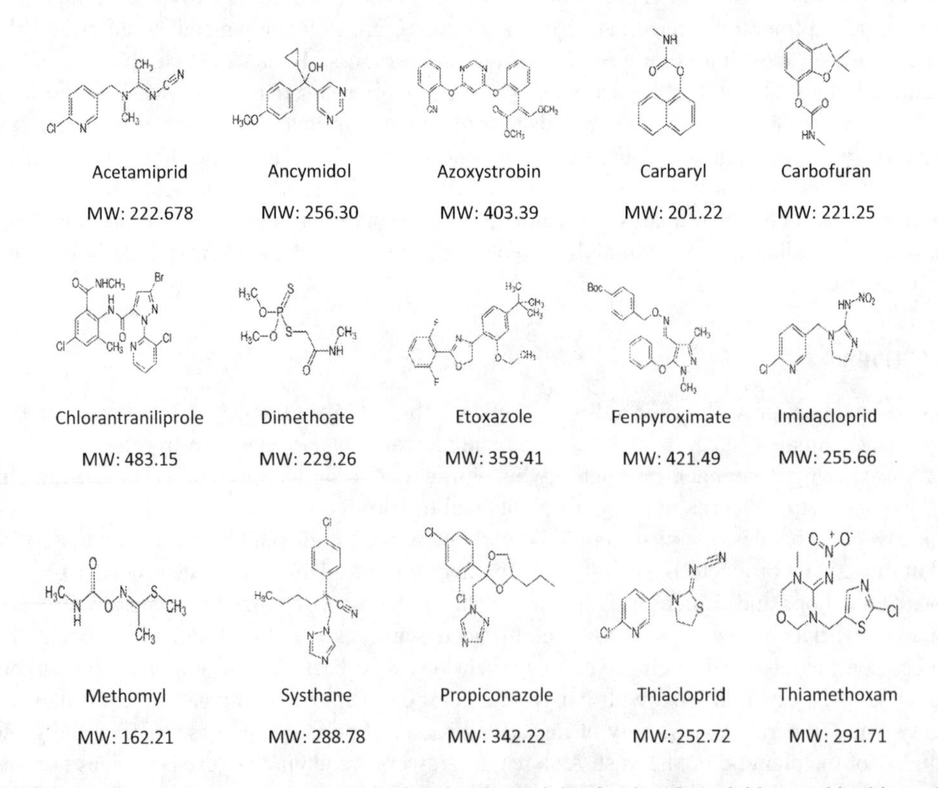

FIGURE 1.2 Chemical structures, names, and molecular weights for the 15 pesticides used in this study.

TABLE 1.1
Names of 15 Pesticides and Their Retention Times Reported in This Work

Pesticides	RT (min)
Methomyl	7.87
Thiamethoxam	8.26
Imidacloprid	8.83
Dimethoate	9.01
Acetamiprid	9.17
Thiacloprid	9.60
Carbofuran	10.90
Ancymidol	11.24
Carbaryl	11.59
Chlorantraniliprole	13.31
Azoxystrobin	13.70
Systhane	14.41
Propiconazole	15.59
Etoxazole	17.29
Fenpyroximate	17.64

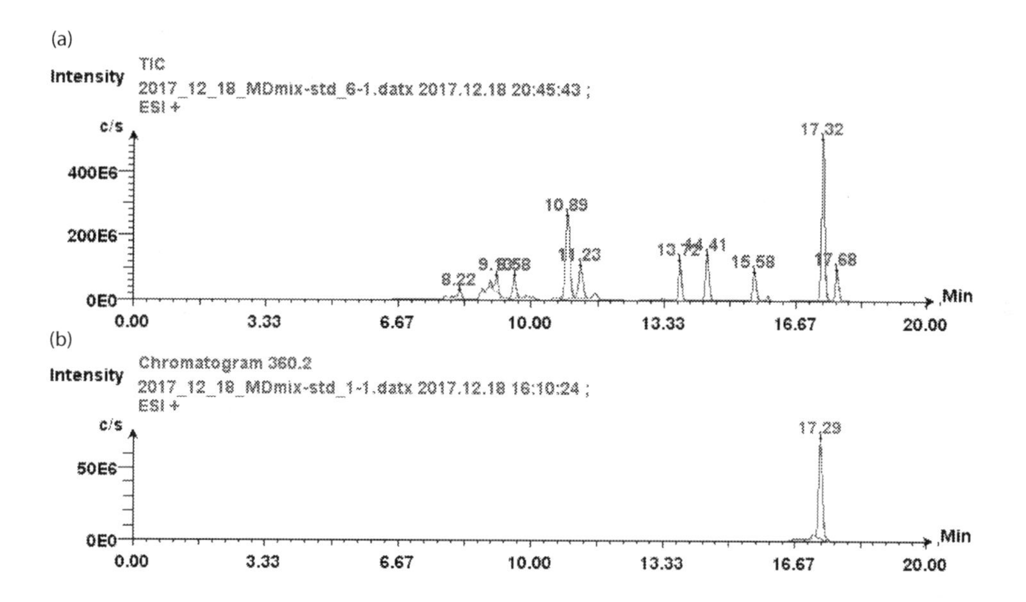

FIGURE 1.3 (a) Full-scan LC/MS analysis of a hemp plant extract where the hemp flower had been previously sprayed with a mixture of 15 pesticides (see Table 1.1). See text for experimental details. (b) Positive ion ESI from the SIM LC/MS analysis of a hemp plant flower which contained Etoxazole at the 1 ppb level.

Figure 1.3 is an example of the SIM LC/MS analysis of a plant extract containing the 15 pesticides shown in Figure 1.2 where positive ion electrospray ionization (ESI) SIM LC/MS was employed to analyze the sample for the determination of the pesticides in cannabis plant matrix extracts at low ppb levels.

In this example, a flower portion of a cannabis plant which had been artificially sprayed with the 15 pesticides was followed by sample preparation using the QUECHERS procedure.[33] The extract was fortified with two stable isotope-incorporated internal standards (d3 thiamethoxam and d3 carbofuran, structures not shown here) followed by SIM LC/MS analysis. Figure 1.3a is the full-scan LC/MS total ion chromatogram (TIC) analysis of a cannabis flower which had been sprayed with a solution containing 100 μg/mL of the 15 pesticides. The single quadrupole mass spectrometer (expression compact mass spectrometer, Advion, Inc.) used in this example can be rapidly switched between full-scan and SIM LC/MS. Thus, the TIC shown in Figure 1.3a shows each of the 15 pesticides present at a fortified higher level such that each could be observed via full-scan LC/MS acquisition. The lower chromatogram (Figure 1.3b) was obtained from different sample of cannabis QUECHERS extract where the pesticide, Etoxazole, was detected at the 1 ppb level. These results suggest it is possible to quantify low ppb levels of some pesticides in cannabis plant extracts although for particularly challenging applications the preferred analytical strategy may be selected reaction monitoring (SRM) LC/MS which provides the added analytical capabilities of tandem mass spectrometry (MS/MS). The latter is helpful for so-called mixture analysis such as when an endogenous matrix component or some other interference chemical co-elutes with a targeted pesticide.[34]

It is interesting to note some of the cannabis flowers we analyzed as negative control samples; for example, cannabis plant samples believed to be free of pesticides, already contained detectable levels of some pesticides. This is not surprising in the wake of a report by an accomplished laboratory which reported that 83% of samples they analyzed in 2017 contained one or more pesticides.[39] With increased testing of cannabis-derived samples for pesticides, numerous detections of low-to-moderate levels of pesticide contamination (10s of ppb or less) from a number of growers claiming to use "organic" or "clean green" growing methods prompted the investigation of possible sources. The most prevalent pesticide detected in these cases was myclobutanil, most commonly used as a treatment for mold infestations.[35] In many cases where pesticide levels are in the mid-to-low ppb range, this SIM LC/MS approach may be appropriate. For those samples containing very low ppb levels of unknown pesticides, it may be preferred to employ LC/MS/MS techniques. A recent report compared the range of pesticides amenable to GC/MS versus those amenable to LC/MS/MS which concluded that GC/MS only covered a few of more than 500 pesticides studied due to the thermal instability or in volatility[36] of many pesticides. Electrospray LC/MS/MS techniques covered most of those pesticides studied, although some analyses may still benefit from modern GC/MS techniques to compliment the LC/MS/MS approach. In general, most state laboratories, including New York State, suggest the use of LC/MS/MS techniques for the quantitative determination of pesticides in cannabis, hemp, and the products derived from them.[37] This technology coupled with recommended sample preparation provides broad coverage with high sensitivity and selectivity for the quantitative determination of over 100 pesticides.

POTENCY: CANNABINOIDS

One of the frequent measurements made for cannabis and hemp plant materials is "potency." The potency term generally refers to the percentage of THC and/or CBD in the plant material but in addition at least three other cannabinoids routinely monitored including THCA, CBDA and CBN. Figure 1.4 shows the structures of these and several related cannabinoid compounds.

There are remarkable similarities between them, some being isobaric yet with dramatic differences in their pharmacological properties. More recently, a total of 11 different cannabinoids have been measured in cannabis plant materials,[38] and the list grows as new developments occur.

With consideration to the commercial "enthusiasm" for new cannabis products, it is important to recognize that the various cannabis strains or "cultivars" contain a wide variety of other chemical constituents which are present in different ratios relative to each other as a function of the cultivar.[39] Thus, care must be taken with assuming the properties between different strains or cultivars are the same since the ratios of many constituents differ between similar cultivars so chemical analysis must be performed to know the composition which can assure the expected medicinal effects.

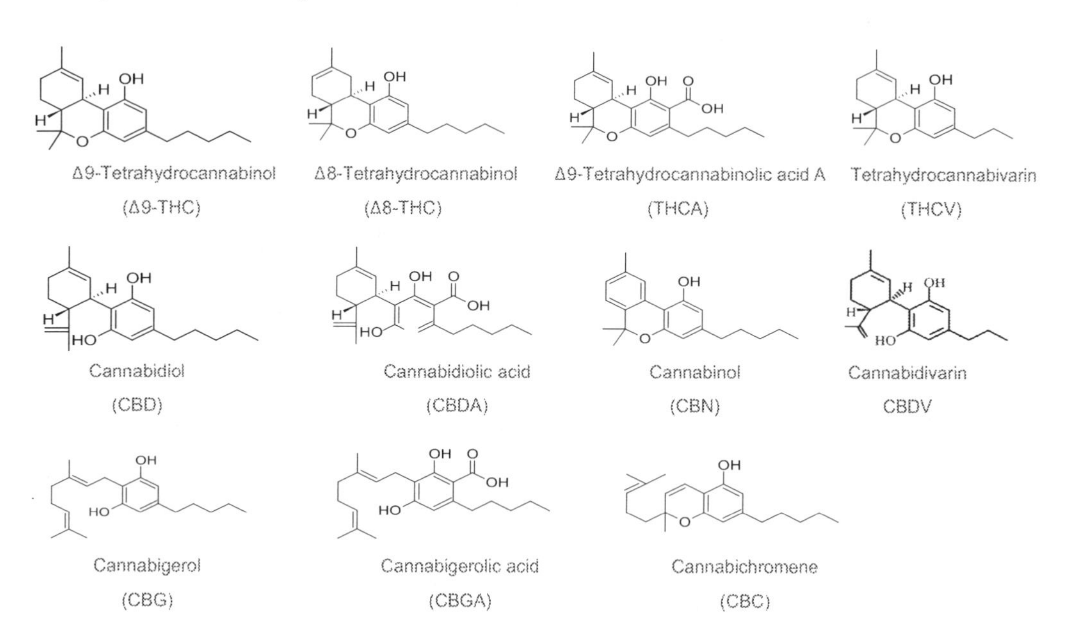

FIGURE 1.4 Chemical structures for 11 common cannabinoids found in cannabis plant materials.

There is growing evidence that the ratio of CBD relative to other cannabinoids as well as endogenous chemicals such as terpenes and flavonoids exhibits a desirable "entourage" effect[40] for medicinal benefits. Thus, it is the synergy between these other chemicals with the cannabinoids which may be responsible for the desirable medicinal outcome. There are at least 554 identified compounds in *C. sativa* L. plants which include 113 identified phyto cannabinoids and 120 terpenes.[41] Although it is believed that many of the cannabis plant constituents are important for a variety of reasons, the first question is the relative quantities of the major cannabinoids present. In contrast to modern pharmaceutical LC/MS/MS bioanalytical trace analysis techniques, some of the cannabinoids are present in the plant at relatively high percent levels. It is common to dilute a cannabis or hemp plant extract by 100- to 1000-fold or more to reduce the concentration of targeted cannabinoids to levels appropriate for analytical quantitative determination. Analytical procedures need to be amended to maximize the information content obtained.

Thus, an important question for a cannabis grower is the "potency" of the plant which will be a matter of the intended product market of the plant. A key component of accurate and precise analyses is specificity or "selectivity" of the detector used. In terms of analytical method, validation selectivity is defined as a method's ability to measure and differentiate targeted analytes in the presence of other components that may be expected to be present.[38] The current popular analytical technique used in support of the cannabis industry for determining potency is high-performance liquid chromatography (HPLC)/PDA or HPLC using a photodiode array detector. This technology is accepted by the cannabis analytical community because it is relatively easy and inexpensive to employ. Although some may suggest that a PDA detector is relatively selective, there is growing evidence that it may not reveal potential co-eluting or potentially interfering endogenous plant chemicals. An experienced plant botanist knows that a crude extract of a plant substance will contain a large number of chemical constituents. The HPLC/PDA analysis of this extract will potentially include co-eluting chemical constituent(s) under the HPLC/PDA chromatographic peak(s). If this should occur, the area under that peak will be larger than it would be if only the targeted compound was contained in that peak. This increased peak area will produce a reported quantity of the cannabinoid that is higher than actually present and thus produce an incorrect result.

Figure 1.5a and B shows a comparison analysis by HPLC/UV for two different samples.

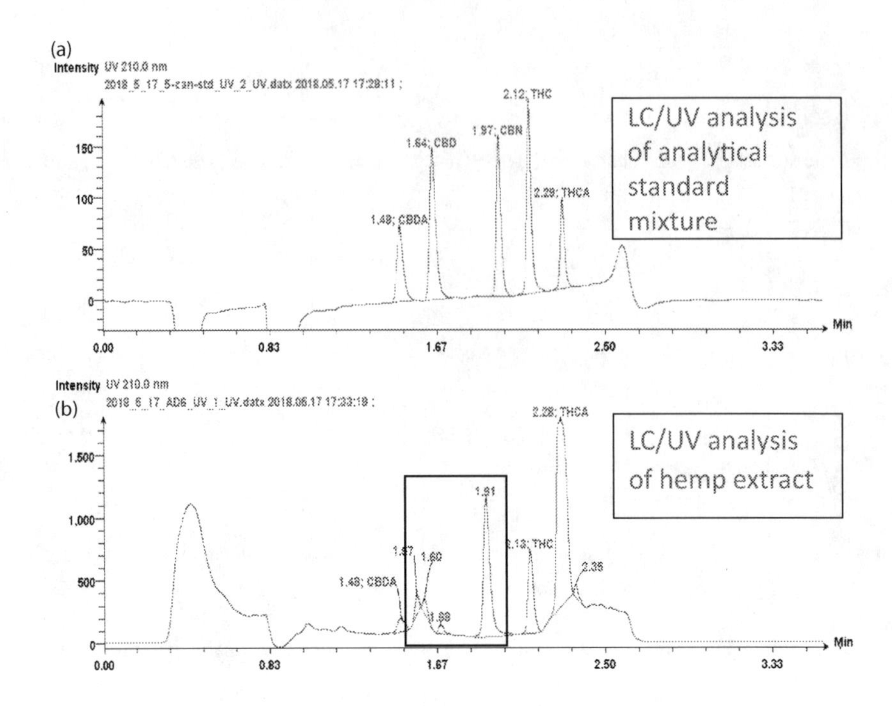

FIGURE 1.5 LC/UV analysis of a synthetic mixture of five cannabinoids (a) and LC/UV analysis of a methanol extract of a hemp flower (b).

Figure 1.5a is a synthetic mixture containing five cannabinoids analyzed by LC/UV versus an LC/UV chromatogram from the analysis of a hemp sample extract (Figure 1.5b). Each LC peak is clearly resolved from the others in Figure 1.5a with no evidence for other interfering chemical constituents because the sample is a mixture of only authentic standards. In contrast, Figure 1.5b shows an overlap of several possible endogenous chemical interferences in the cannabis extract centered at the 1.57-min retention time region. This presents two possible problems. One is the confident identification of these other LC peaks and the other is the challenge of accurately determining the area under each of these unresolved peaks for the purpose of quantitative analysis. In the absence of adequate selectivity from the UV detector, it is uncertain what these other peaks are which are not present in the standard mixture chromatogram observed in Figure 1.5a. If a PDA detector was used rather than a fixed wavelength detector, it might be possible to determine what the unresolved unknown peaks are but even that could be a challenge. This situation poses somewhat of a dilemma for the accurate quantitative determination for each of the LC peaks observed in Figure 1.5b.

As an alternative to LC/UV as an analytical technique for potency analysis, one might employ SIM LC/MS which offers additional selectivity and sensitivity. It may be suggested that high sensitivity is often not required for potency analysis, although added selectivity can often be very worthwhile. As shown in Figure 1.6a, the same LC/UV chromatogram is shown as in Figure 1.5b followed by the full-scan TIC from a single quadrupole LC/MS system shown in Figure 1.6b.

Essentially, the same chromatogram is observed in Figure 1.6a and b. However, with the added capabilities of LC/MS software, an extracted ion chromatogram (XIC) can be obtained for selected ions as shown in Figure 1.6c–f. The astute observer could suspect a low concentration of CBD may be buried under the PDA signals in the 1.64- to 1.74-min retention time region (Figure 1.5a and b). In the absence of additional selectivity, this uncertainty could go unanswered. An inexperienced technologist could mistake the interference from the retention time peak at 1.68 or 1.75 min as CBD using this LC/UV data. It is also worthy of note that in Figure 1.6e and f the XICs for *m/z* 287.3 and *m/z* 313.2 essentially co-elute. However, since we are using an LC/MS technique these two

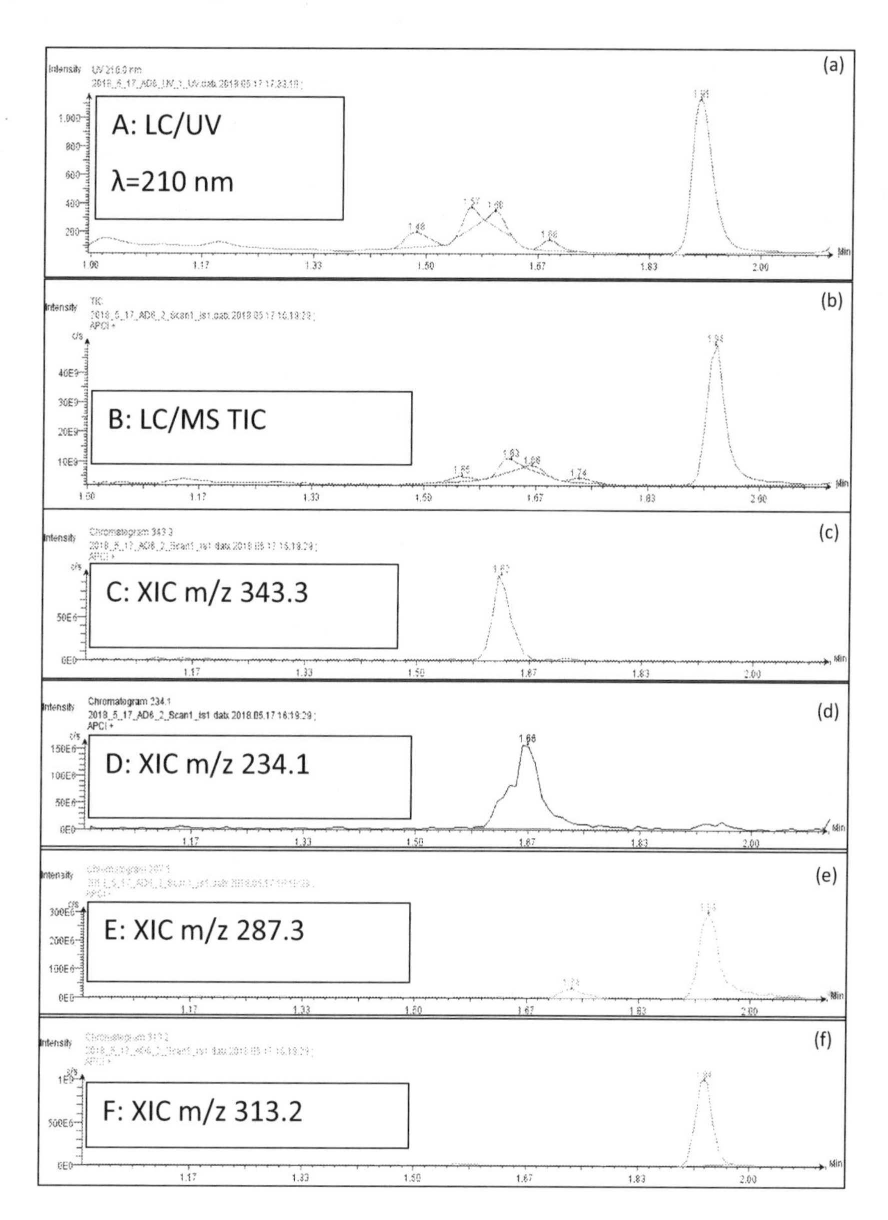

FIGURE 1.6 LC/UV (a) and SIM LC/MS (b) chromatograms of a hemp plant extract, (c–f) are representative XICs for ions observed in the LC/MS analysis of this hemp extract sample.

different, yet co-eluting chemical entities can be observed even though they do in fact co-elute. It is unlikely this could be accomplished using LC/UV or LC/PDA analytical techniques.

When SIM LC/MS analysis is performed on the same sample, the chromatograms can be much more definitive for the purpose of obtaining the areas under the various LC peaks often revealing much improved selectivity and the absence of interfering chromatographic components. This is due to the increased selectivity of SIM LC/MS techniques where only the protonated molecules of the targeted cannabinoid(s) are monitored. Even higher selectivity coupled with mixture analysis capability could be obtained if LC SRM MS/MS techniques were employed, although the added cost and complexity of this more sophisticated analytical technique is more than really needed for potency analyses. The use of SIM LC/MS techniques is a little more expensive with higher technologist understanding than LC/UV but can reduce the potential for unknown interferences due to co-elution which may go unnoticed when HPLC/PDA techniques are used. It is worth suggesting that the use of ultra-high-performance liquid chromatography (UHPLC) 1.7-μm columns could provide higher separation efficiency for potency analyses, but the limited selectivity of the PDA detector would still pose the potential for not differentiating between targeted and untargeted compounds. In addition, some of these chemical interferences are not yet fully characterized and lack commercial availability of certified reference standards. In contrast to HPLC/PDA, if either LC/MS or LC/MS/MS analysis of the same sample was performed, the considerably higher selectivity of either of these techniques could facilitate qualitative identification of unknown compounds as well as mitigate the potential for unexpected chemical interferences for the quantitative determination of the targeted compound.[42]

The astute analytical chemist might suggest that prior HPLC/PDA analysis of a negative control plant sample could show where possible co-eluting plant chemical constituents may be observed such as those described in Figure 1.5a. The challenge here is the lack of an available negative control plant tissue since any cannabis or hemp sample will contain many of the targeted compounds of interest. In fact, most current methods used for potency testing by HPLC/PDA use a calibration curve with standards and quality control (QC) samples prepared in water or aqueous alcohol solvents which lack any plant matrix components. This approach to generating a calibration curve for the purpose of quantitative analysis is contrary to modern accepted regulated bioanalytical techniques for the pharmaceutical or environmental industries and exposes the data to a variety of potential errors.[43,44]

TERPENES

Trichomes are an important part of the plant kingdom for the biosynthesis of botanical chemicals and are an important part of cannabinoids.[15] These tiny botanical factories produce a wide range of terpenes which contain only carbon and hydrogen (these are the predominant terpenes in cannabis) as well as oxygenated neutral compounds including terpene alcohols and ketones in addition to some aldehydes and esters.[15] Representative examples of these compounds are shown in Figure 1.7 where it can perhaps be understood why they have a relatively high vapor pressure and thus give off a pleasant odor from the plant due to their low molecular weight and unsaturated nature.

The familiar smell of a fresh cannabis or hemp plant is due in part to these terpene-related compounds.[15] Terpenoids are not unique to cannabis, but various types of cannabis plants produce unique terpene profiles which in combination with the unique ratios of cannabinoids may determine some of the preferred medicinal effects.[5] It is the determination of these chemical profiles that requires chemical analysis which has generally been done by capillary GC/FID or more recently GC/MS for this complex mixture of over 200 different terpene compounds. The very high separation efficiency of capillary GC columns coupled with the generally universal ionization efficiency of EI MS would appear to be a preferred analytical technique for the determination of these compounds. In contrast, however, is a recent report employing LC/MS/MS techniques with atmospheric pressure chemical ionization (APCI).[45] This approach may not have the optimal chromatographic separation efficiency but when combined with the mixture analysis capabilities of tandem MS may produce the results needed while also being amenable to some of the other important cannabis compounds.

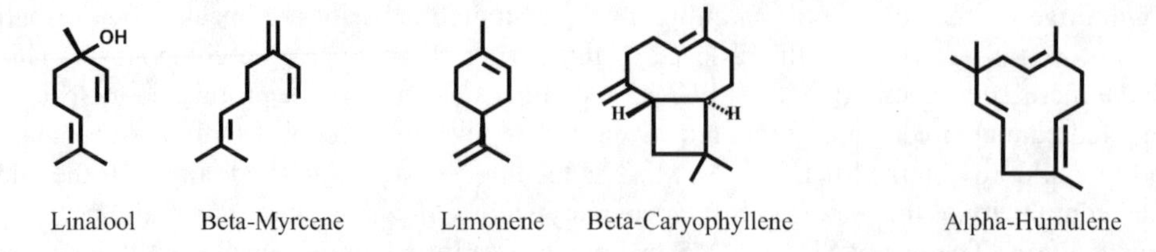

FIGURE 1.7 Representative terpene structures.

Since terpenes are not cannabinoids, some feel these are not important constituents of a cannabis plant or product. However, there is increasing evidence that other chemicals in the plant work synergistically with the cannabinoids that produce therapeutic responses ranging from lowering of inflammation and regulation of blood glucose to protection of neurons and gastric cells.[5] This is the so-called "entourage" effect[40] where other chemicals may enhance or moderate the effects of the THC, CBD, etc. For example, one report suggests that CBD may be an "entourage" compound in cannabis that modulates the effects of THC.[40] Therefore, the quantitative determination of terpenes and sesquiterpenes is often desired and perhaps best done by either GC/MS or LC/MS/MS instead of some of the earlier reported techniques.[15]

MICROBIOLOGY

Cannabis is often grown in greenhouses with carefully controlled growing conditions which are warm with relatively high humidity; these are excellent conditions for the growth of a wide variety of bacteria and fungi. As a result, the final cannabis product may be contaminated with these organisms, which can adversely affect the safety of the eventual products produced from these plants. Accordingly, most states now require testing laboratories to analyze samples for microbial growth and the by-products of this growth. Microbiological contamination of cannabis plants and products is typically detected using culture growth or quantitative polymerase chain reaction (qPCR) techniques. Currently, MS is not generally used for identifying bacteria and fungi in complex samples, although matrix-assisted laser desorption (MALDI) MS is potentially well suited for the qualitative identification of these organisms in pure culture or simple matrices. Further studies could perhaps demonstrate the potential for MALDI identification of bacteria and fungi in cannabis samples, as is currently used in the clinical laboratory environment.[46]

MYCOTOXINS

Mycotoxins are toxic secondary metabolites of mold which result from long-term storage of organic plant material. They are a concern in cannabis plants because of the suspected carcinogens causing acute and chronic toxicity.[5] Aflatoxins are a subset of mycotoxins produced by *Aspergillus flavus* and *Aspergillus parasiticus*. Aflatoxin B1 is considered the most toxic, but the presence of B2, G1, and G2 must also be considered as they result from decaying vegetation and soils when warm, moist conditions exist. Figure 1.8a shows the structures for aflatoxins B1, B21, G1, G2, and OTA (ochratoxin A) which are often targeted in the chemical analysis of cannabis and its associated products.

These compounds can be determined via a variety of analytical techniques including LC/UV, LC/MS, and LC/MS/MS. Although capillary GC/MS may be used, it is generally preferred to do LC separations coupled with MS for the quantitative determination of these compounds. Figure 1.8b shows the SIM LC/MS analysis of a synthetic mixture of the five aflatoxins referenced above and summarized in Table 1.2 along with their chemical formulae, protonated molecules, and retention times observed. There are regulations now that set limits on the allowable levels of these toxins in foods. Both recreational and medicinal cannabis must now be screened for mycotoxins derived from microbial contamination. This can be done by a number of analytical techniques, but due to the chemical complexity of the sample extracts and the polar and/or labile nature of some mycotoxins they are good candidates for LC/MS analysis.[43]

RESIDUAL SOLVENTS

Due to a variety of extraction processes used for isolating cannabinoids from the cannabis plant, residual solvents often remain in the sample. Some of these solvents include hexane, ethanol, butane, propane, and in some cases chlorinated solvents.[5] Many states require testing for these residual solvents which is often done by headspace GC/FID. The sample preparation for these samples is

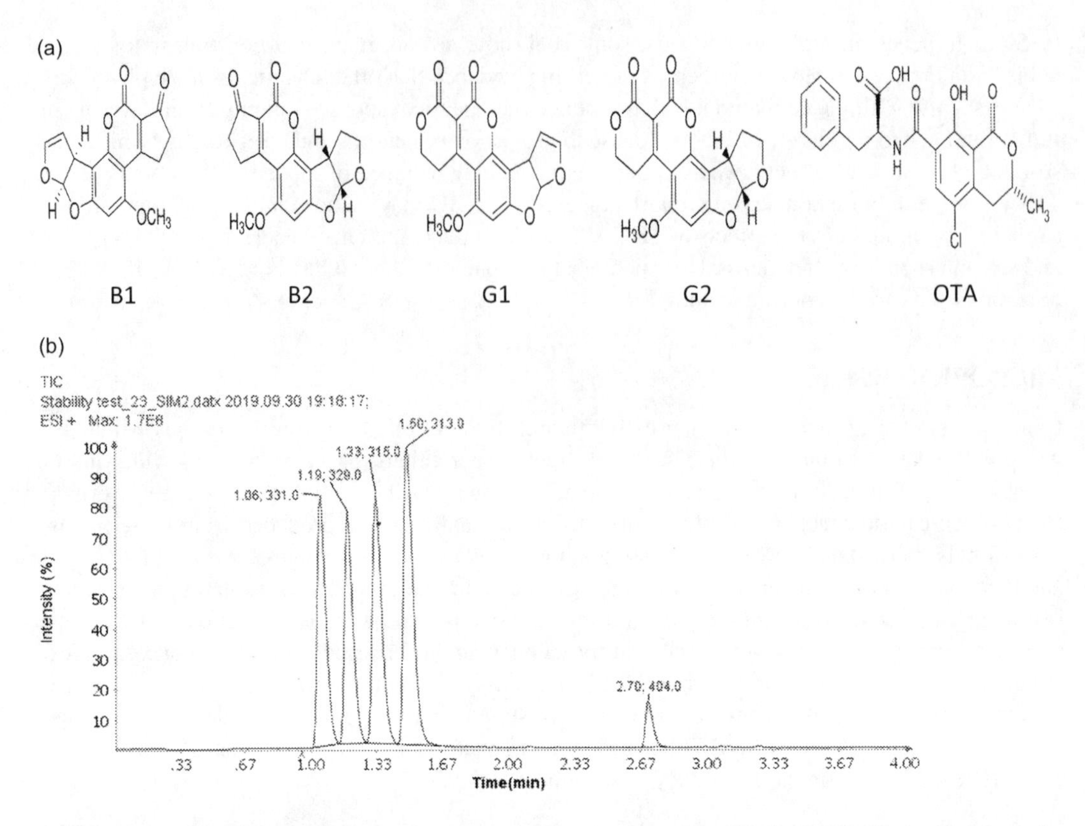

FIGURE 1.8 Chemical structures for five targeted aflatoxins (a) and the SIM LC/MS TIC obtained from the LC/MS analysis of a synthetic mixture of these five aflatoxins (b).

TABLE 1.2
Names, Chemical Formulae, Protonated Molecules, and SIM LC/MS Retention Times Observed for the Five Aflatoxins Studied in This Work

Name	Formula	$[M + H]^+$	RT (min)
Aflatoxin B1	$C_{17}H_{12}O_6$	313.07	1.5
Aflatoxin B2	$C_{17}H_{14}O_6$	315.08	1.33
Aflatoxin G1	$C_{17}H_{12}O_7$	329.07	1.19
Aflatoxin G2	$C_{17}H_{14}O_7$	331.08	1.06
Orchratoxin A	$C_{20}H_{18}ClNO_6$	404.1	2.70

relatively simple where a small amount of the extract is placed into a vial and heated to mimic the natural evaporation process. The amount of solvent that is evaporated from the sample and into the air represents the "headspace." An aliquot of this can be analyzed by GC/FID. Although an alternative technique could be EI GC/MS analysis of such headspace samples, the reduced equipment costs of GC/FID may likely prevail for the foreseeable future. As is customary the retention time of the residual solvents is the means of "identifying" these constituents, it would appear reasonable to instead employ GC/MS instead since the identification of each GC peak could be best confirmed by its mass spectrum irrespective of its retention time. However, many laboratories lean towards GC/FID primarily due to its reduced cost. It is the view of the author that the difference in cost is worth it given the positive selectivity of the MS detection in the event of any potential GC retention time deviation.

DETERMINATION OF ORIGIN OF GROWTH OF CANNABIS USING IRMS

Federal and local agencies are often interested in a drug's origin and in determining large-scale trafficking routes for potential legal/enforcement solutions. *C. sativa* L. can reveal its photosynthesis activities by recording the isotope ratio of carbon.[48] Environmental factors such as climate, water availability, temperature, and light intensity have an impact on the assimilation of carbon 13 in the plant and its contents. Plant tissues are depleted with respect to the atmospheric source because photosynthesis and enzymatic fixation are discriminating against the heavier isotope. The photosynthetic pathway for cannabis is known depending upon the growth conditions. The ratio of CO_2 concentration in the plant to the concentration in the air is controlled as part of the photosynthetic activity. Isotope ratio mass spectrometry (IRMS) has been used to accurately determine the carbon 13 content of cannabis plants and can be used as a forensic tool to establish the geographic origin of the plant's growth.[49] Although this analytical technology is not commonly employed in routine analytical services for the cannabis industry, it has certainly proven itself in those instances where it is important to know the cannabis plant's geographic origin if needed.[49]

BIOLOGICAL SAMPLE ANALYSIS: FORENSIC TOXICOLOGY

Currently, most of the state cannabis testing laboratories are not performing analyses of biological samples (blood, urine, etc.) for the quantitative determination of cannabinoids or their metabolites. These applications could be used for clinical toxicology purposes to determine if cannabis contributed to observed toxicity or for forensic toxicology purposes to determine unlawful use of performance-altering substances. The topic of driving under the influence of drugs (DUID) raises increasing concerns by authorities such as the National Safety Council's Alcohol, Drugs and Impairment Division (NSC-ADID),[46] the National Highway Transportation Safety Administration (NHTSA)[47] as well as the general public. Recent studies on driving impairment due to cannabis use and/or combined with alcohol consumption showed increased risks of driver-induced automobile crashes.[50–52]

Blood THC and metabolite concentrations are greatly hampered by the delay in time required to collect the samples after a police stop or automobile accident. Historically, sample analysis has been performed by GC/MS which required chemical derivatization and relatively long run times in a central laboratory. Since urine analysis for delta-9-THC and its metabolite, carboxy-THC, has not generally shown an association between cannabis and automobile crash risk, blood or plasma analyses as well as oral fluid analyses[53–55] have been studied for DUID applications. Several challenges exist with both approaches. The non-psychoactive cannabinoid, THC-COOH can be detected for several days depending upon the dose and mode of administration while psychoactive THC can last for many days from chronic smokers in the blood, but there is little correlation between THC in blood and driving impairment.[55] With current delays due to the lack of facile on-site blood collection, the THC level in a smoker's blood will be very low several hours after the incident making it difficult to relate "under the influence" with the biological fluid levels at the time of analysis.

In an effort to provide a potential solution to these issues, we and others have considered the potential for microsampling with dried blood spots (DBSs)[56,57] and dried plasma spots (DPSs).[58] In other studies, the potential for oral fluids[54] as well as breath analysis[59] continues to be evaluated as a means for facile sample collection coupled with efficient transport to a forensic laboratory. An example of a facile finger prick collection of blood onto a DPS card is shown in Figure 1.9a where a single drop of blood from regular cannabis users was collected on-site at a Cannabis Conference.

The book-type DPS card[58] filters the red blood cells from a drop of blood to produce a separate small sample of red blood cells along with a separate DPS from the same sample. The latter may be seen in Figure 1.9b where the "book-type" DPS card is "opened" revealing the red blood spot on the upper half of the card along with the pale yellow DPS which resides on the lower surface of the cellulose portion of the DPS card. This approach provides a fast, on-site sampling technique

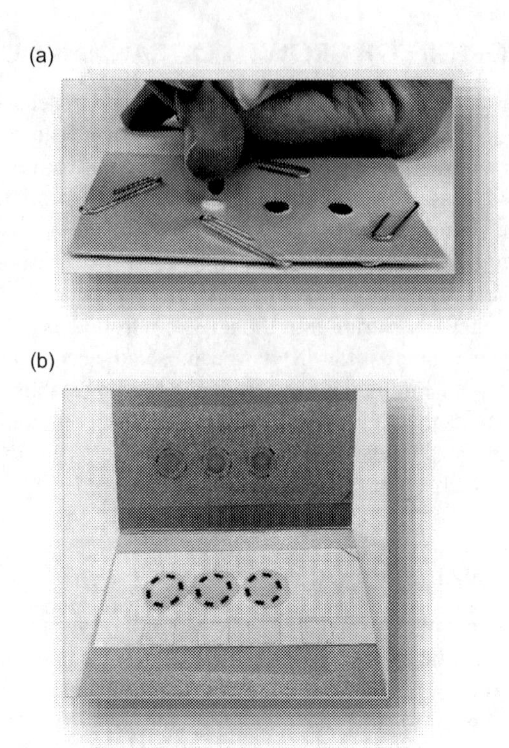

FIGURE 1.9 Photo of a prototype book-type DPS card for collecting finger prick blood (a) and a view of the book-type device opened showing both the dried red blood cells and the DPS (b).

that minimizes analysis delays and stabilizes THC and its metabolites because the samples are dry. After a 20 min drying period with the two "spots" separated from each other by opening the device 180° via the hinge (see Figure 1.9b), the device can be sent via courier to a laboratory where the DPS samples may be analyzed. Given the chemical complexity of a DPS, it is usually preferred to employ SRM LC/MS techniques to obtain a quantitative determination of the THC level in such a sample.

Figure 1.10a shows a photo of the on-site collection of finger prick blood onto the book-type DPS card.

Figure 1.10b shows the SRM LC/MS ion current chromatogram for delta-9-THC via the SRM transition from *m/z* 315.10 to *m/z* 193.05 for THC determined from the finger prick of a person who had recently smoked a marijuana cigarette. The lower magenta trace shown in Figure 1.10b shows the SRM LC/MS chromatogram for the corresponding deuterated internal standard for THC and its SRM transition from *m/z* 318.20 to *m/z* 195.95. The calibration curve for THC in this study is shown in Figure 1.11 which provided a quantitative analysis result for 455.6 ng/mL of THC in the finger prick of plasma.

These results suggest the potential for on-site facile collection of blood onto a DPS card which can then be shipped without cooling to a central laboratory for chemical analysis. It would appear possible that future adaptations for sample collection of biological fluids will evolve with the preferred chemical bioanalysis technique being LC/MS/MS.

THE DETERMINATION OF THC IN THE BREATH OF MOTORISTS

To further extend the ease of collecting human biological samples for the chemical determination of THC as it may affect driving impairment, we have recently explored the potential for breath sample analyses. Using a second-generation breath sampling device (BreathExplor, Munkplast AB,

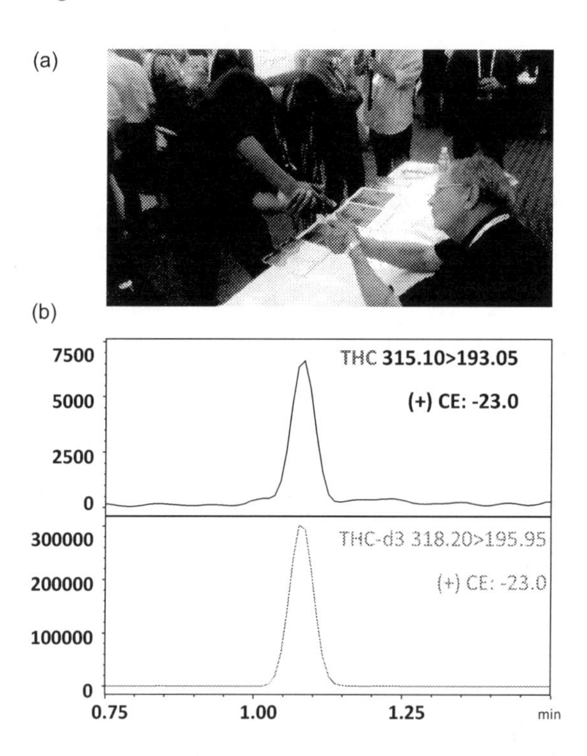

FIGURE 1.10 Photo of on-site finger prick blood sample collection at a major cannabis conference (a) and the SRM LC/MS ion current chromatograms for the THC transition from *m/z* 315.10 to *m/z* 193.05 and the stable isotope internal standard, THC-d3 *m/z* 318.20 to *m/z* 195.95 (b).

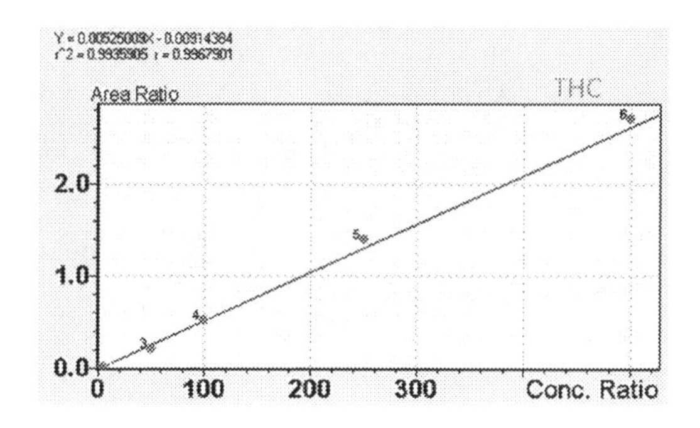

FIGURE 1.11 Calibration curve for the SRM LC/MS determination of incurred THC in a DPS collected at a cannabis conference.

Sweden), we have performed single quadrupole SIM LC/MS analysis of methanol extracts of the collector(s) within the BreathExplor device (Figure 1.12a).

Following the collection of, for example, 12 breath exhales from non-smokers as well as known volunteer marijuana smokers from 20 min post smoking up to 6 h post-smoking, we have demonstrated these samples may be screened for presumptive positives by SIM LC/MS. The sample preparation is relatively simple in that the inert polymer collectors (Figure 1.12b) upon which the breath sample is deposited may be extracted with a 2 mL of methanol containing the d3-THC stable isotope internal standard followed by concentration to dryness via a vacuum evaporator with reconstitution of the residue in 40 μL of 80% methanol/water. Following injection of 25 μL of this sample under UHPLC SIM MS acquisition conditions, THC may be detected and tentatively quantified up to 6 h post-smoking of a marijuana cigarette.

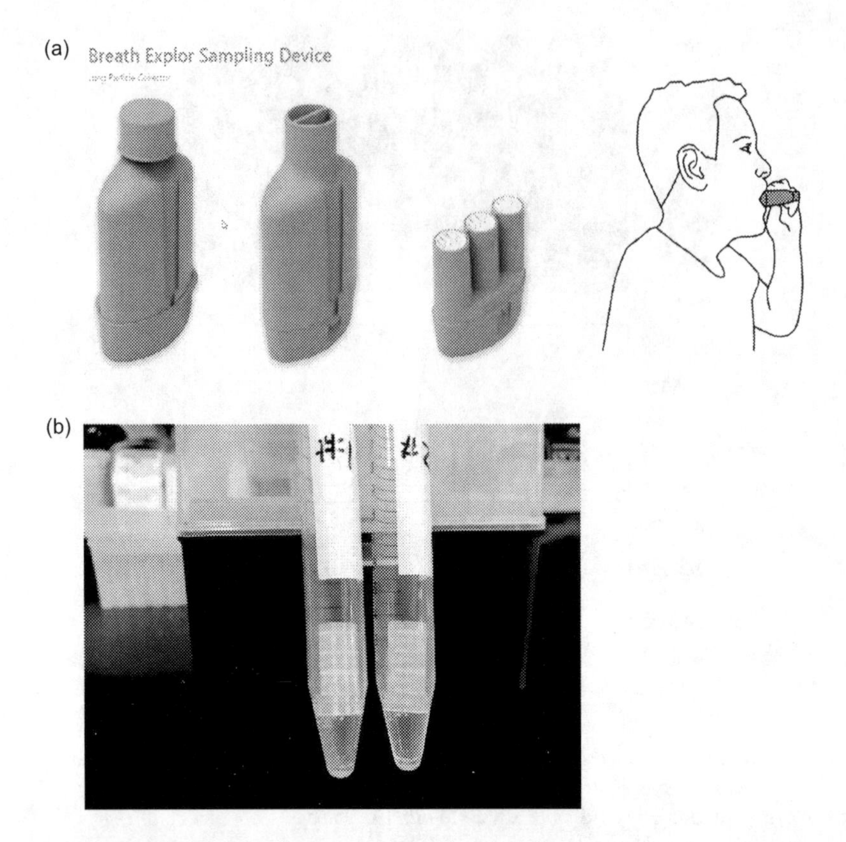

FIGURE 1.12 Depiction of breath collection device for collection of human breath samples followed by SIM LC/MS determination of THC in breath samples. (a) The breath collection device in its closed, opened, and disassembled format. The latter view of the device is that for removal of the three collectors coupled with a sketch of how a person collects 12 breath exhales in the device (a). Shown in Figure 1.12b are two of the breath collectors in a tube ready for extraction with methanol as described in the text.

Figure 1.13a shows the calibration curve for the SIM UHPLC/MS quantitative determination of THC in breath samples prepared as described above.

The blue points in the calibration curve shown in Figure 1.13a are from calibration standards of THC fortified into negative control breath samples collected with the BreathExplor device (Figure 1.12a). The calibration range shown is from 10 pg/breath sample up to 10,000 pg/breath sample. Since the literature had suggested that THC in the breath of marijuana smokers does not exceed 1,000 pg/breath sample, we initially generated a calibration curve to accommodate an upper limit of quantitation (ULOQ) of 1,000 pg/breath sample. However, some of the breath samples analyzed by this described method exceeded this level significantly, so we generated another calibration curve with an ULOQ of 10,000 pg/sample as shown in Figure 1.13a. In addition to the blue calibration standards shown in the calibration curve of Figure 1.13a, we also included four QC samples prepared as replicates in a similar way as the standards albeit from a different source of certified reference material (SRM) THC. The quantitative SIM UHPLC/MS analysis included two sets of calibration standards with one set analyzed with increasing concentration at the beginning of the sample set with the second set of calibration standards analyzed at the end of the sample set also in the order of increasing concentration. The QC samples were placed randomly within the set of samples. Figure 1.13b shows the SIM UHPLC/MS signal-to-noise ratio (*S*/*N*) for the lowest standard at 10 pg of THC while Figure 1.13c shows the *S*/*N* for the high standard at 10,000 pg THC injected. From Figure 1.13a, red points for the two unknowns are observed on the calibration curve.

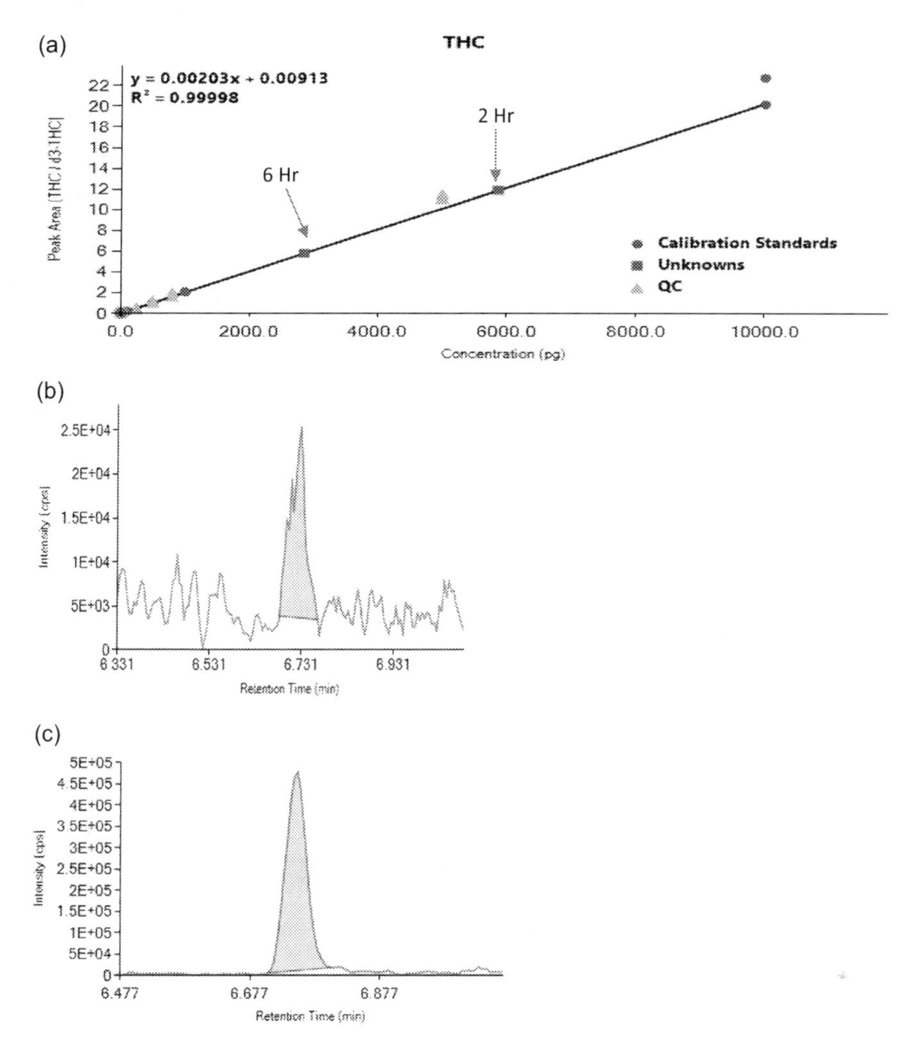

FIGURE 1.13 (a) The SIM LC/MS calibration curve for THC in breath samples from 10 pg to 10,000 pg/sample. (b and c) The SIM LC/MS chromatograms for the 10 pg and 10,000 pg/sample, respectively.

The lower point is from the 6-h smoker of marijuana who had a level of 2,860 pg/breath sample while the higher red point on the calibration curve is from the 2-h smoker with a level of 5,862 pg/breath sample.

Although these preliminary studies appear promising for a preliminary means of determining the level of THC in breath samples employing the merits of SIM UHPLC/MS analysis, more work is needed. In particular, since the BreathExplor device contains three separate collectors such that there is an A, B, and C sample, it should be possible to employ the above-described analytical method with a comparison of results from a central laboratory using the trusted SRM LC/MS technology in a central laboratory on the B-sample. If these experiments can show successful comparison quantitative results with both experimental methods, it could lend credence to the feasibility of using the less expensive yet relatively good selectivity and sensitivity of the SIM UHPLC/MS method. If one is to dream even further, it may be possible in the future to employ the described SIM UHPLC/MS method to roadside testing for DRUD for THC and possibly other drugs. A compelling feature of this approach is the ease of collecting the breath sample which certainly a police officer could administer and collect in real time.

FINAL THOUGHTS

Movement towards acceptance of cannabis in its many potential uses has been glacially slow, but it is beginning to appear that a tidal wave of the industry's growth is occurring as recently evidenced by Canada legalizing cannabis nationwide. There is an urgent need for accurate and precise chemical analysis of samples ranging from cannabis plant materials to final products including medicinal products, foods, and biological fluids. Although MS techniques are currently used to some extent in the cannabis industry, this report suggests it should be used more widely in the future to provide the superior selectivity and sensitivity demanded by the chemical complexity and diversity of the plant and the many products likely to be produced. This will be necessary for the production and sale of safe products and to earn customer confidence in those products. An important component of this is the use of modern analytical chemistry techniques by accredited laboratories with procedures which include standard operating procedures (SOPs) and rigorously validated methods. This will require well-trained analytical chemists who practice sound scientific procedures common to laboratories following good laboratory practices (GLP) and anti-doping laboratories that abide by well recognized accredited laboratory practices.[60] Good examples to follow are those recommended by GLP as well as approved methods and guidance provided by well-established organizations such as the ISO/IEC 17025,[61] AOAC (Association of Official Agricultural Chemists), A2LA (American Association of Laboratory Accreditation),[60] and individual state health departments.[37] When well-qualified analytical service laboratories are in place for monitoring the chemical safety and integrity of cannabis materials and products, the industry may thrive beyond our current level of appreciation.

ACKNOWLEDGMENTS

The author wishes to thank Prof. Olof Beck of the Karolinska Institute and Mr. Peter Stambeck of BreathExplor for helpful guidance in collecting and analyzing the breath samples described in this work. The author also very much appreciates the help from his colleagues, Drs. Daniel Eikel, Changtong Hao, and Robert Forties of Advion, Inc., for very helpful support during the SIM LC/MS quantitative analysis of samples described in this work.

FURTHER READING

1. Iversen, L. (ed.) *Handbook of Cannabis. (Handbooks in Psychopharmacology).* Oxford, UK: Oxford University Press, 2014: 781.
2. Andre, C.M., J.F. Hausman, and G. Guerriero, *Cannabis sativa: The plant of the thousand and one molecules. Front Plant Sci.*, 2016. **7**: p. 19.
3. Olaizola, O.A., et al., Evolution of the cannabinoid and terpene content during the growth of *Cannabis sativa* plants from different chemotypes. *J. Nat. Prod.*, 2016. 79: pp. 324–331.
4. FDA, FDA approves first drug comprised of an active ingredient derived from marijuana to treat rare, severe forms of epilepsy. 2018; Available from: https://www.fda.gov/NewsEvents/Newsroom/PressAnnouncements/ucm611046.htm.
5. Louis, B.W.-S. Clinical rationale for CBD use on mood, depression, anxiety, brain function, and optimal aging. *Cannabis*: *A Clinician's Guide*, ed. B.W.-S. Louis. 2018: Boca Rotan, FL: CRC Press, 288.
6. Skerritt, J. and C. Giammona, Coca-Cola is eyeing the cannabis market, in *Bloomberg*. 2018.
7. Cannavines. 2018; Available from: https://cannavines.com/.
8. Castetter, K. Sovereign Vines. 2018; Available from: https://www.linkedin.com/in/kaelan-castetter-642bb5b8/.
9. Giroud, C., Analysis of cannabinoids in hemp plants. *Chemia*, 2002. 56: pp. 80–83.
10. Nie, B., J. Henion, and J. Wakshlag, Analysis of veterinary hemp-based oils for product integrity by LC/MS. *Cannabis Sci. Tech.*, 2019. 2(3): pp. 36–45.
11. Carrazzo, A., Medical marijuana research remains top priority for veterinarians. *Am. Vet.*, 2018.
12. Gamble, L.J., et al., Pharmacokinetics, safety, and clinical efficacy of cannabidiol treatment in osteoarthritic dogs. *Front. Vet. Sci.*, 2018. 5(165): pp. 1–9.

13. Gyles, C., Marijuana for pets? *CVJ*, 2016. 57: pp. 1215–1217.
14. Hazekamp, A., et al., Preparative isolation of cannabinoids from *Cannabis sativa* by centrifugal partition chromatography. *J. Liq. Chromatogr. Relat. Technol.*, 2009. **27**(15): pp. 2421–2439.
15. Pertwee, R.G., *Handbook of Cannabis*. Oxford, UK: Oxford University Press, 2014: 1–781.
16. Kelly, J.F., K. Addanki, and O. Barasra, The non-specificity of the Duquenois-Levine field test for marijuana. *Open Forensic Sci. J.*, 2012. 5: pp. 4–8.
17. Smith, B., Quantitation of cannabinoids in dried ground hemp by mid-infrared spectroscopy. *Cannabis Sci. Technol.*, 2019. 2(6): pp. 5–6.
18. Smith, B., LC vs spectroscopy for potency. *Cannabis Sci. Tech.*, 2019. 2(6): pp. 10–14.
19. Atkins, P., Right from the grow: A look at sampling and sample preparation methods of solid cannabis for analysis. *Cannabis Sci. Tech.*, 2019. 2(2): pp. 26–34.
20. Sexton, M. and J. Ziskind, *Sampling Cannabis for Analytical Purposes* Michelle Sexton, ND, Steep HIll Laboratory, 2013.
21. USDA, <USDA Sampling Guidelines for Hemp.pdf>.
22. Massachusetts, C.o., Protocol for sampling and analysis of finished medical marijuana products and marijuana-infused products for massachusetts registered medical marijuana dispensaries. Cannabis Control Commission, 2017: pp. 1–21.
23. Giese, M.W., et al., Development and validation of a reliable and robust method for the analysis of cannabinoids and terpenes in cannabis. *J. AOAC Int.*, 2015. 98(6): pp. 1503–1522.
24. Irshid, M., et al., Phytoaccumulation of heavy metals in natural plants thriving on wastewater effluent at Hattar Industrial Estate. *Pakistan Intem. J. Phydoremed.*, 2015. 17: pp. 154–158.
25. Gauvin, D.V., J. Yoder, and R. Trapp, Marijuana toxicity: Heavy metal exposure through state-sponsored access to la Fee Verte. *Pharmaceut. Reg. Affairs: Open Access*, 2018. **7**(1).
26. Enamorado-Báez, S.M., J.M. Abril, and J.M. Gómez-Guzmán, Determination of 25 trace element concentrations in biological reference materials by ICP-MS following different microwave-assisted acid digestion methods based on scaling masses of digested samples. *ISRN Anal. Chem.*, 2013. 2013: pp. 1–14.
27. Wang, T., Liquid chromatography–inductively coupled plasma mass spectrometry (LC–ICP–MS). *J. Liq. Chromatogr Related Technol.*, 2007. 30(5–7): pp. 807–831.
28. Filipiak-Szok, A., M. Kurzawa, and E. Szlyk, Determination of toxic metals by ICP-MS in Asiatic and European medicinal plants and dietary supplements. *J. Trace Elem. Med. Biol.*, 2015. 30: pp. 48–58.
29. Steimling, J. and T. Kahler, Liquid chromatography's complementary role to gas chromatography in cannabis testing. *LCGC Chromacademy*, 2018. 36(6): pp. 36–40.
30. Schaneman, B., Cost of new mandatory marijuana pesticide testing tough to absorb for Colorado's growers, in *Marijuana Business Daily*. 2018: US.
31. Dalmia, A., et al., LC–MS/MS with ESI and APCI sources for meeting california cannabis pesticide and mycotoxin residue regulatory requirements. *Cannabis Sci. Technol.*, 2018. 1(3): p. 38–50.
32. Regulations, C.C.O., Bureau of cannabis control proposed text of regulations, in *16*, D.B.O.C. CONTROL, Editor. 2017: California. pp. 1–136.
33. Morris, B.D. and R.B. Schriner, Development of an automated column solid-phase extraction cleanup of QuEChERS extracts, using a zirconia-based sorbent, for pesticide residue analyses by LC-MS/MS. *J. Agric. Food Chem.*, 2015. 63(21): pp. 5107–5109.
34. Zweigenbaum, J., et al., High-throughput bioanalytical LC/MS/MS determination of benzodiazepines in human urine: 1000 samples per 12 hours. *Anal. Chem.*, 1999. 71: pp. 2294–2300.
35. Torres, A., et al., Study of pesticides in clones: research summary. 2016, Steep Hill.
36. Alder, L., et al., Residue analysis of 500 high priority pesticides. *Mass Spectrom. Rev.*, 2006. 25: pp. 838–865.
37. (ELAP), E.L.A.P. Requirements for Laboratory Certification/Certification Manual, Determination of the Plant Growth Regulator Indole-3-butyric Acid and Pesticides in Medical Marijuana using LC-MS/MS NYS DOH MML-306. 2018; Available from: https://www.wadsworth.org/regulatory/elap/requirements-for-laboratory-certification-certification.
38. Rigdon, A., et al., Method validation for cannabis analytical labs: approaches to addressing unique industry challenges, in *Cannabis Science Conference*. 2017: Portland, OR.
39. Hazekamp, A. and J.T. Fischedick, Cannabis - from cultivar to chemovar. *Drug Test Anal.*, 2011. 4: pp. 660–667.
40. Russo, E.B., Taming THC: potential cannabis synergy and phytocannabinid-terpeoid entourage effects. *British J. Pharm.*, 2011. 163: pp. 1344–1364.
41. Thomas, B.F. and M.A. ElSohly, The botany of *Cannabis sativa* L. *Anal. Chem. Cannabis*, 2016. pp. 1–26.

42. Meng, Q., et al., A reliable and validated LC-MS/MS method for the simultaneous quantification of 4 cannabinoids in 40 consumer products. *PLoS One*, 2018. 13(5): pp. 1–16.
43. *White Paper* May 2018 FDA *BA Guidance.pdf.* http://www.fda.gov/Drugs/GuidanceCompliance RegulatoryInformation/Guidances/default.htm
44. Jones, B.R., et al., Surrogate matrix and surrogate analyte approaches for definitive quantitation of endogenous biomolecules. *Bioanalysis*, 2012. 4(19): pp. 2343–2356.
45. Di Lorenzo, R. and P. Winkler, Comprehensive cannabis analysis: Pesticides, aflatoxins, terpenes, and high linear dynamic range potency. 2018; Available from: https://sciex.com/applications/food-and-beverage-testing/cannabis-testing.
46. Faron, M., B.W. Buchan, and N.A. Ledeboer, Matrix-assisted desorption ionization time of flight mass spectrometry for the use with positive blood cultures: methodology, performance, and optimization. *J. Clin. Microbiol.*, 2017. 55(12): pp. 3328–3338.
47. Dzuman, Z., et al., A rugged high-throughput analytical approach for the determination and quantification of multiple mycotoxins in complex feed matrices. *Talanta*, 2014. 121: p. 263.
48. Muccio, Z., et al., Comparison of bulk and compound-specific d13C isotope ratio analyses for the discrimination between cannabis samples. *J. Forensic Sci.*, 2012. 57(3): pp. 757–764.
49. Shibuya, E.K., et al., Sourcing Brazilian marijuana by applying IRMS analysis to seized samples. *Forensic Sci. Int.*, 2006. 160: pp. 35–43.
50. Logan, B.K., et al., Recommendations for toxicological investigation of drug-impaired driving and motor vehicle fatalities: 2017 update. *J. Anal. Tox.*, 2017. 12: pp. 1–6.
51. NHTSA (ed). 2018. *National Highway Transportation Safety Administration.* Drug-Impaired Driving, ed. U.S.D.o. Transportation, Washington, DC.
52. Concheiro, M., et al., Simultaneous quantification of Δ9-tetrahydrocannabinol, 11- nor-9-carboxy-tetrahydrocannabinol, cannabidiol and cannabinol in oral fluid by microflow-liquid chromatography-high resolution mass spectrometry. *J. Chromatogr. A*, 2013. 1257: pp. 123–130.
53. Heustis, M.A., A. Barnes, and M.L. Smith, Estimating the time of last cannabis use from plasma delta9-tetrahydrocannabinol and 11-nor-9-carboxy-delta9-tetrahydrocannabinol concentrations. *Clin. Chem.*, 2005. 51(12): pp. 2289–2295.
54. Concheiro, M., et al., Simultaneous quantification of Delta(9)-tetrahydrocannabinol, 11-nor-9-carboxy-tetrahydrocannabinol, cannabidiol and cannabinol in oral fluid by microflow-liquid chromatography-high resolution mass spectrometry. *J Chromatogr. A*, 2013. 1297: pp. 123–130.
55. Willie, S.M.R., et al., Conventional and alternative matrices for driving under the influence of cannabis: recent progress and remaining challenges. *Bioanalysis*, 2010. 2(4): pp. 791–806.
56. Verplaetse, R. and J. Henion, Quantitative determination of opioids in whole blood using fully automated dried blood spot desorption coupled to on-line SPE-LC-MS/MS. *Drug Test Anal.*, 2016. 8(1): pp. 30–38.
57. Velghe, S., L. Delahaye, and C.P. Stove, Is the hematocrit still an issue in quantitative dried blood spot analysis? *J. Pharm. Biomed. Anal.*, 2018. 163: pp. 188–196.
58. Ryona, I. and J. Henion, A book-type dried plasma spot card for automated flow-through elution coupled with online spe-lc-ms/ms bioanalysis of opioids and stimulants in blood. *Anal. Chem.*, 2016. 88: pp. 11229–11237.
59. Ullah, S., S. Sandqvist, and O. Beck, Measurement of lung phosphatidylcholines in exhaled breath particles by a convenient collection procedure. *Anal. Chem.*, 2015. 87(22): pp. 11553–11560.
60. American Association for Laboratory Accreditation, AL2A, Editor. 2018.
61. ISO/IEC 17025. *General Requirements for the Competence of Testing and Calibration Laboratories.* Geneva, Switzerland: ISO, 2017.

2 Importance of Measuring Elemental Contaminants in Cannabis and Hemp

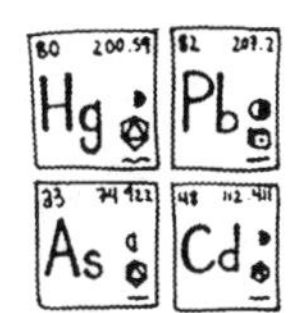

Cannabis and hemp are known to be hyper-accumulators of contaminants in the soil. That is why they have been used to clean up toxic waste sites where other kinds of remediation attempts have failed. In the aftermath of the Chernobyl nuclear meltdown in the Ukraine in 1986, industrial hemp was planted to clean up the radioactive isotopes that had leaked into the soil and groundwaters.[1] Of course Chernobyl is an extreme example of heavy metal and radionuclide contamination, but as a result of normal human (anthropogenic) activities over the past few decades including mining, smelting, electroplating, gasoline exhaust, energy production, use of fertilizers, pesticides, waste treatment plants, and lead-based paint and plumbing materials, heavy metal pollution has become one of the most serious environmental problems today. As a result, considerations about where cannabis and hemp are grown are going to be critically important because it could have serious implications on the level of heavy metals that are absorbed by the plant and its resulting safety for human consumption. Historically, much of the cannabis in the United States has been grown indoors in greenhouses under controlled growing environments, so the absorption of heavy metals into the plant has been kept in check reasonably well. However in 2020, it will be legal to grow hemp for CBD production anywhere in the United States. It could, therefore, become a lot more challenging to keep the levels low, because most of the hemp plants will be grown outdoors on farms where the soil might be an additional source of contamination. This chapter will look at all potential sources of elemental contaminants in cannabis and hemp, not just from the growing medium standpoint but also from the cannabinoid manufacturing processes and delivery systems used together with those contaminants derived from the analytical testing procedure.

Phytoremediation using certain plants is emerging as a cost-effective technology to concentrate and remove elements, compounds, and pollutants from the environment.[2] Within this field of phytoremediation, the use of cannabis and hemp plants to concentrate metals from the soil into the harvestable parts of roots and above-ground shoots (phytoextraction) has great potential as a viable alternative to traditional contaminated land remediation methods.[3] However, the natural inclination of these plants to absorb heavy metals from the soil and growing environment could potentially limit its commercial use for the production of medicinal cannabinoid-based compounds. A number of studies have now been carried out on cannabis and hemp that provide convincing evidence that they are active accumulators of heavy metals such as lead, cadmium, arsenic, mercury, magnesium, copper, chromium, nickel, manganese, and cobalt as a result of human activities.[4,5]

An added complication is that cannabis and hemp plants can absorb heavy metals not only from the soil but also from contaminants in fertilizers, nutrients, pesticides, and the growing medium as well as from other environmental pathways.[6,7] Additionally, the process of cutting, grinding, and preparing the cannabis/hemp flowers for extraction can often pick up elemental contaminants from the manufacturing equipment. It's also worth pointing out that the equipment used to deliver these products to consumers such as inhalers, vaporizers, and transdermal patches can mean the user is exposed to additional sources of elemental contaminants, apart from what's in the cannabinoid compound itself.

Finally, the extraction/evaporation/concentration process has the potential to extract heavy metals, depending on the solvents and the super/sub-critical extraction temperatures/pressures used which could possibly end up in the finished cannabinoid products. It is therefore critically important that an optimized extraction process is carried out in order to minimize the carryover of heavy metals.[8]

MAIN FACTORS FOR METAL UPTAKE FROM SOIL

The health and growth of all plants rely on the absorption of essential nutrients and minerals being available in the dissolved, ionic form in the soil. To maintain enough water to survive and thrive, the plant's primary means of facilitating the movement of water is through transpiration, which is a highly efficient means of drawing a concentrated solution of minerals and nutrients out of the soil. Transpiration works by the evaporative loss of water from the shoots, which is controlled by the opening and closing of specialized pores (known as stomata) embedded in the surface of the leaves that initiates gas exchange. When the stomata are open, the pressure potential of the plant becomes very negative, creating a vacuum effect that draws water and nutrients into the plant, moving it from the roots to the shoots. Unfortunately, this is also the mechanism that the plant draws in heavy metal contaminants in addition to the nutrients. Bode wrote an excellent white paper on the underlying mechanisms of heavy metal uptake by cannabis plants.[9]

The rhizosphere is the region of soil in the vicinity of plant roots in which the chemistry and microbiology is influenced by their growth, respiration, and nutrient exchange. Unfortunately, under certain conditions, the plant's root system will preferentially absorb heavy metals over other minerals, which cannot be explained exclusively by passive ion uptake. The hyper-accumulating properties of cannabis and hemp are not fully understood but are partially dependent upon other factors, including soil pH, availability of other metal/mineral ions in solution, the N/P/K (nitrogen/phosphorus/potassium) ratio, and the ability of the plant's natural metalloproteins which act as chelating compounds to bind with the heavy metals to reduce the rate of absorption and overcome their toxic effects. It should also be noted that the plants' natural polyamine compounds will also strengthen the defense response of plants and enhance their defense against diverse environmental stressors including toxicity and oxidative stress.[10]

Based on evidence in the public domain, there are approximately 15–20 heavy metals found in natural ecosystems (soil, water, air) and industrial anthropogenic activities that could be potential sources of contaminants accumulated by the plant, including Pb, As, Hg, Cd, Ni, V, Co, Cu, Se, B, Tl, Ba, Sb, Ag, Au, Zn, Sn, Mn, Mo, W, Fe, and U. In addition, many of these elements exist as different species (e.g., metalloids), based on their oxidation state, organic/inorganic/ionic form,[10] or more recently as engineered nanoparticles that could find their way into effluent and waste water streams.[11] Their levels of toxicity would need to be investigated further, but there is a compelling case to be made that many of these elements, metalloids, and speciated forms could be the future basis of a federally regulated panel of elemental contaminants in cannabis and hemp.

Let's take a more detailed look at the most common of the natural and anthropogenic sources of heavy metals and their toxicity impact on human health. It's not an exhaustive list, but at least it offers guidance on the most likely metals of concern. The bulk of this information has been sourced from these five references,[4,9–12] and supported by information from the CDC's Agency for Toxic Substances and Disease Registry (ATSDR) (https://www.atsdr.cdc.gov/).

LEAD

Lead has no known biological or physiological purpose in the human body but is avidly absorbed into the system by ingestion, inhalation, or skin absorption. Children are particularly susceptible, because of their playing and eating habits. Lead is absorbed more easily if there is a calcium/iron deficiency or if the child has a high fat, inadequate mineral and/or low protein diet. When absorbed, lead is distributed within the body in three main areas—bones, blood, and soft tissue. About 90% is contained distributed in the bones, while the majority of the rest gets absorbed into the bloodstream where it gets taken up by porphyrin molecules in the red blood cells and readily crosses the blood–brain barrier where it impacts the growth of brain neurons. The long-term exposure of children to lead will result in neurodegenerative diseases, such as attention deficit disorders, resulting in lower IQs and significant learning deficiencies. It is therefore clear that the repercussions and health risks are potentially enormous, if consumers of cannabis products are exposed to abnormally high levels of lead.

Automobile emission residues particularly from leaded gasoline before it was banned in 1996 are one of major reason for extremely high levels of lead in soil, particularly around major highways. In addition, decades of using lead-based paint before it was banned in 1978, particularly in older buildings, has contributed to high levels of lead in household dust and particulates. The use of lead water pipes has also contributed to higher lead levels in the drinking water supplies if it is not chemically treated correctly. The people of Flint Michigan, who were drinking water contaminated with over 10,000 ppb of lead (Maximum Contaminant Level (MCL) in drinking water is 15 ppb), are a testament to this. Heavy metals, including lead, have also been deposited by sludge water run-off, excavated earth materials dumped as mine depth increased, and by the presence of US military bases and nuclear test facilities that deposited radioactive elements such as cobalt, uranium, strontium, and cesium, as well as heavy metals from weapon testing, such as lead, chromium, and nickel. There is no question that these types of activities in the United States are the leading cause of environmental heavy metal pollution in soil and water. It should also be emphasized that Pb-based solder is commonly used to connect the battery terminals inside an electronic vaping device (commercial solder is mainly made from Pb and Sn).

Fortunately, lead has a low solubility in soils and is typically not readily bioavailable to plants. It tends to form complexes with phosphate and sulfate fertilizers, which are commonly found in the roots since they are essential to plant nutrition. It appears that soil composition is the most important factor for the uptake of lead and that nutrient depleted soils are more likely to result in greater lead uptake in plants. In addition, high levels of lead have been observed on the surface of the roots, which is a defense mechanism to protect them from soil abrasion.

MERCURY

Metallic mercury has wide industrial uses, including the production of chlorine gas, caustic soda, batteries, thermometers, and industrial electrical switches. Inorganic mercury (including metallic and inorganic forms) compounds enter the environment from mining ore deposits, metal smelters, burning coal, and industrial wastes from manufacturing plants that use mercury. It also enters the water or soil from natural deposits, erosion of certain rocks, and volcanic activity. Once in the environment, it finds its way into soils and potable water systems where it is ingested by small organisms and bacteria that convert the mercury into the more toxic form, methyl mercury (CH_3Hg). If it enters ponds, lakes, or rivers, these small organisms are then a source of food for the crustaceans that live on the bottom, which are consumed by the small fish which are in turn eaten by the large fish.

This process which is known as bio-magnification means that the larger, older predator fish, which humans tend to eat, contain the highest levels of mercury in their tissue.

Mercury affects virtually every system in the human body and often occurs with no distinctive symptoms. It impacts the central nervous system, kidneys, reproductive organs as well as causing chromosomal damage and severe skin reactions, and at extremely higher levels, can cause coma, convulsions, and even death. The long-term effects of even low levels of methyl mercury on the developing brain of young children are significant. It will cross over the blood–brain barrier as well as the placental, impairing the growth of brain neurons in the fetus resulting in cognitive damage, lower intelligence, and reduced brain development. The Minamata disaster in Japan in the late 1950s, where the Chisso Chemical Corporation dumped hundreds of tons of mercury into the Bay of Minamata for over 30 years, is a devastating testament to what happens when mercury pollution is allowed to occur with no oversight.

The US Geological Survey (USGS) has reported that the total mercury production in California between 1850 and 1981 was more than 200 million pounds, which was used primarily for gold recovery, leading to its relatively high contribution to the total eco-burden of soil-based heavy metals. In addition, US industries, including coal-fired power plants, petrochemicals, cement, and metal refineries, emit almost 100 tons of mercury into the atmosphere every year, which eventually finds its way into the soil, aquatic systems, and vegetation. It's also important to mention the three western states in the United States that lead the country in the agricultural growth, processing, and sales of cannabis for recreational and therapeutic use are California, Oregon, and Washington… states where the mining industry has been its most expansive.

CADMIUM

Cadmium is a serious industrial pollutant. It is emitted into the environment by ferrous and nonferrous metal mining and refining, production of phosphate fertilizers, combustion of fossil fuels in power plants, and industrial incineration and disposal from N-Cd battery production. Cadmium, as the oxide, chloride, and/or sulfate from these high temperature processes, will exist in air as particles or vapors and be transported long distances in the atmosphere, where it will deposit onto soils and water surfaces. Cadmium and its compounds may travel through soil, but its mobility depends on several factors such as pH and amount of organic matter, which will vary depending on the local environment. Generally, cadmium binds strongly to organic matter where it will be immobile in soil and be taken up by plants eventually, entering the food supply. Cadmium exists as the hydrated ion or as ionic complexes with other inorganic or organic substances in soil.

Studies have shown that inhalation of cadmium appears to be the most serious, particularly chronic occupational exposure to cadmium fumes and dusts with increased risk of chronic obstructive lung disease and emphysema. There have also been studies examining the role of cadmium in the development of chronic obstructive pulmonary disease (COPD) in smokers. The most recent studies showed that current and former smokers had higher body burdens of cadmium than nonsmokers and that within smokers; the body burden of cadmium was related to lung damage related to smoking. This study concluded that cadmium might be important in the development of tobacco-related lung disease. Chronic cadmium inhalation is also suspected to be a possible cause of lung cancer. Other respiratory effects of chronic exposure to cadmium include rhinitis, affecting the ability to smell, as well as the development of bronchitis.

Cadmium is known to be one of the most toxic of the soil-based heavy metals. Besides the roots, the highest concentrations can be found in the cannabis leaves, whereas the lowest are typically observed in seeds. Cannabis has been found to be highly cadmium-tolerant and very useful in bioaccumulation studies. The risk to public health is the plant's ability to accumulate metals and unlike most plants used in bioremediation, it is not destroyed but used for medicinal and recreational purposes. Cadmium concentrations of up to 70 mg/kg in soil have been reported to have no negative effect on germination of cannabis, and exposures of up to 100 ppm in solution have no significant effects on the growth of the cannabis plant. The highest concentration of cadmium tolerance has been shown in the roots of cannabis at a maximum 830 mg/kg without adversely affecting the growth of the plant. Most plants generally do not accumulate elements beyond near-term metabolic needs, which are typically in the range of 10–15 ppm of most trace elements. However, hyper-accumulators like cannabis are exceptions to the rule and have been reported to accumulate toxic metals up to a few hundred parts per million (ppm).

ARSENIC

Arsenic is a unique element with distinct physical characteristics and toxicity whose importance in public health is well recognized. The toxicity of arsenic varies across its different forms. While the carcinogenicity of arsenic has been confirmed, the mechanisms behind the diseases occurring after acute or chronic exposure to arsenic are not well understood. Inorganic arsenic has been confirmed as a human carcinogen that can induce skin, lung, and bladder cancer, whereas organic forms of arsenic (e.g., arseno sugars found in marine algae) are relatively innocuous. There are also reports of its significant association to liver, prostate, and bladder cancer. Recent studies have also suggested a relationship with diabetes, neurological effects, cardiac disorders, and reproductive organs, but further studies are required to confirm these associations. The majority of research to date has examined cancer incidence after a long exposure to high concentrations of arsenic. However, numerous studies have reported various health effects caused by chronic exposure to low concentrations of arsenic.

Potable water contaminated with arsenic is very common in the United States. This is particularly the case where gold mining is carried out. Gold in the mines is often bound with arseno-pyrite (iron arsenic sulfide rock), which is pulverized and cleaned to isolate the precious metal. In addition, copper smelting is known to produce significant amounts of arsenic waste, because arsenic is known to be present in copper sulfide ores. Cannabis and hemp plants naturally absorb arsenic and store it in their tissues. In addition, the speciated form of arsenic directly determines its bioavailability. Trivalent arsenic (As^{3+}) is much less mobile in the soil, and, therefore, less bioavailable than pentavalent arsenic (As^{5+}) which has greater soil mobility. Soil pH also impacts the chemical form of arsenic in the soil and thus its bioavailability. Acidic soil tends to increase plant uptake of arsenic by enhancing its ability to complex with transition elements such iron and manganese.

The fact that some soils could be contaminated with arsenic from an industrial source, such as metal refining, was highlighted by the scare a few years ago when high levels of arsenic were found in apple juice which was made from a source of apples grown in arsenic-contaminated soil. Fortunately, it was the relatively innocuous organic species (arseno sugars) and not the highly toxic inorganic form. For this reason, it is always a good idea to test wells or groundwaters for elemental contaminants if they are being used to irrigate and/or grow cannabis or hemp plants. It's also worth pointing out that recently (Nov, 2019) a number of dispensaries in Michigan were closed down because high levels of arsenic were found in some of their cannabis flowers.

COBALT

Cobalt and its compounds have been recently listed as carcinogenic by the American Association for Cancer Research. As a constituent element of marijuana grown in contaminated soils of northern California, Oregon, and Washington, cobalt represents one of the hidden toxic dangers because it is not on the mandatory list of metals to be tested in any of these states. In fact it has been reported that many of California's medical cannabis patients are susceptible to the toxic effects of cobalt due to their compromised immune systems particularly those with hepatic illnesses.

Cobalt is a naturally occurring element found in rocks, soil, water, plants, and animals. Cobalt is used to produce alloys used in the manufacture of aircraft engines, magnets, grinding and cutting tools, and artificial hip and knee joints. Cobalt compounds are also used to color glass, ceramics, and paints and used as a drier for porcelain enamel and paints. Radioactive cobalt is used for commercial and medical purposes. ^{60}Co is used for sterilizing medical equipment and consumer products, radiation therapy for treating cancer patients, manufacturing plastics, and irradiating food, while ^{57}Co is used in medical and scientific research.

NICKEL

Nickel and chromium are elements found naturally together in rocks and present in the soil in a range from trace amounts to relatively high concentrations. Particularly high Ni and Cr levels are found in serpentine rocks (metallosilicates) and soils, which are often intense green in color. Cannabis and hemp that are grown in these types of soils are likely to accumulate quite high levels of nickel and chromium. In recent decades, the large release of Cr and Ni by industrial activities, mainly in the manufacture of stainless steel, as well as the use of sewage sludge in agricultural soils have caused a significant increase in the levels of these two metals in the soil and other ecological matrices. This has led to increasing environmental concern because, while relatively low concentrations of Ni and Cr are essential for plants and other living organisms including humans, both elements are toxic for all living organisms if present in excessive concentrations.

The most common harmful health effect of nickel in humans is an allergic reaction. Approximately 10%–20% of the population is sensitive to nickel. A person can become sensitive to nickel when jewelry or other items containing nickel are in direct contact and prolonged contact with the skin. In fact wearing nickel jewelry in ears or other pierced body parts may also sensitize a person to nickel. Once a person is sensitized to nickel, further contact with the metal may produce a reaction. The most common reaction is a skin rash at the site of contact. In some sensitized people, dermatitis may develop in an area of the skin that is away from the site of contact. For example, hand eczema is fairly common among people sensitized to nickel. Some workers exposed to nickel by inhalation can become sensitized and have persistent asthma attacks.

CHROMIUM

Major factors governing the toxicity of chromium compounds are oxidation state and solubility. Hexavalent (Cr^{6+}) compounds, which are powerful oxidizing agents and thus tend to be irritating and corrosive, appear to be much more toxic systemically than trivalent (Cr^{3+}) compounds, given similar amounts and solubilities. Hexavalent chromium was the focus of the 2000 movie Erin Brockovich, which highlighted the contamination around a power plant owned by Pacific Gas and Electric (PG&E) in Southern California. PG&E had used hexavalent chromium in a cooling tower system to fight corrosion. The waste water was then discharged into unlined ponds at the site where some percolated into the groundwater, polluting a two square mile area around the power plant.

Cr^{6+} can be absorbed by the lung and gastrointestinal tract and even to a certain extent by intact skin. Although mechanisms of biological interaction are uncertain, the toxicity may be related to the ease with which Cr^{6+} can pass through cell membranes and its subsequent intracellular reduction to reactive intermediates. When inhaled, chromium compounds are respiratory tract irritants and can cause pulmonary sensitization. Chronic inhalation of Cr^{6+} compounds increases the risk of lung, nasal, and sinus cancer. In addition, severe dermatitis and painless skin ulcers can result from contact with Cr^{6+} compounds.

VANADIUM

Vanadium is a naturally occurring element, which is found in nature as a gray–white metallic oxide and ore and is common in rocks and soils where chromium and nickel are prevalent. It is also found in over 60 different minerals, which are widely distributed in the earth's crust. Vanadium metal is used in producing rust-resistant, high-speed tool steels and is an important additive to stainless steels. Vanadium in the form of pentoxide is used in ceramics, as a catalyst in the refining of crude oil, and in the production of superconductive magnets. Small amounts of vanadium are released into the environment from volcanic emissions, but most of the emissions are associated with industrial sources, particularly oil refineries and power plants, which burn fuel oil and coal.

Breathing air containing vanadium particles can result in coughing which can last a number of days after exposure. Damage to the lungs, throat, and nose has been observed in rats and mice exposed to vanadium pentoxide. In addition, exposure to vanadium will cause nausea, mild diarrhea, and stomach cramps, which have been reported in people taking sodium vanadate salts for the experimental treatment of diabetes.

A number of effects have been found in rats and mice ingesting several vanadium compounds, including decrease in the number of red blood cells, elevated blood pressure, mild neurological damage, and developmental growth. In addition, the International Agency for Research on Cancer (IARC) has determined that vanadium is carcinogenic to humans.

MANGANESE

Manganese is a neurotoxic metal which can induce a dysfunction in the human motor system, impacting a person's movement. It has been suggested that the neurotoxicity of inhaled manganese is related to an uptake of the metal directly into the brain via the olfactory pathways. In this way, manganese can circumvent the vascular system and the impermeable blood–brain barrier and gain direct access to the central nervous system.

Manganese makes up about 0.1% (1,000 ppm) of the earth's crust, making it the 12th most abundant element. Some soil may contain up 10,000 ppm of manganese with an average content of approximately 400 ppm. Manganese occurs principally as pyrolusite (manganese oxide), braunite (manganese silicate), psilomelane (manganese oxide containing barium and potassium), and to a lesser extent as rhodochrosite (manganese carbonate).

OTHER ELEMENTS OF CONCERN

Let's take a look at some additional elements of concern that could potentially have an impact on human health, particularly from the metallic components, such as stainless steel, nickel alloy, brass, or solder found inside cannabinoid electronic vaping devices.

IRON (Fe) is an essential micronutrient, but not toxic by ingestion, unless it is inhaled. So if stainless steel components are found inside electronic vaping device, there is a strong possibility that iron aerosol particulates could be delivered to the lungs as a consequence of corrosion (as well as Ni and Cr). In fact, Fe particles, when dissolved, can potentially produce reactive oxygen species when inhaled.

COPPER (Cu) is also a micronutrient but not intended to be inhaled. Copper particles are not very toxic by ingestion but quite inflammatory when dissolved and enter the human lungs. Similar to Fe, Cu is also possibly involved in producing a reactive oxygen species when inhaled. Copper is a component of brass, which is quite common in electronic vaping/inhaling devices.

ZINC (Zn) is not very toxic but worth investigating because of high concentrations in aerosols from electronic vaping devices as a consequence of brass component corrosion. Inhalation of undissolved particles is a potential cause of inflammation for any element, not so much the toxicity of zinc itself.

TIN (Sn) is a component of solder, usually combined with Pb, but not always. High levels of Sn (with no Pb) have been found in some vaping liquids, which infers that an Sn-based solder has been use to solder the battery terminals. Because inorganic tin compounds usually enter and leave your body rapidly after ingestion, they do not usually cause harmful effects. The widespread use of tin cans to store foodstuffs means that the element is relatively innocuous if ingested. However, breathing or inhaling Sn fumes and dusts over the long term can cause pneumoconiosis, a disease that causes inflammation of the lungs.

It's also worth noting that other metallic forms could potentially be a source of contamination in cannabis and hemp, other than dissolved ions in the growing medium or nutrients. Two of the most common forms are described below.

ELEMENTAL SPECIES/METALLOIDS

Many elements found in natural ecosystems are present in different metalloid or speciated forms, some of which are toxic, while others are relatively harmless. For example, because of the many potential competing reactive species in soil including natural biological compounds, inorganic/organic pollutants, and residential/industrial wastewaters and effluents, major and minor elemental constituents can be altered to another species through a reduction/oxidation process resulting in a different organic, inorganic, or oxidation/valence state of the element being taking up by the root system of the cannabis plant. Some of these include species such as hexavalent chromium (Cr^{6+}), which is extremely toxic, while trivalent chromium (Cr^{3+}) is relatively innocuous. Inorganic forms (pentavalent and trivalent) of arsenic (As^{5+}, As^{3+}) are known to be toxic, while organic forms of arsenic (arseno sugars) are quite harmless. Elemental mercury discharged from industrial activity (power plants, metal refining, cement production) is converted to methyl mercury (CH_3Hg) by marine animals and plants and is far more toxic than the elemental or inorganic forms. The pharmaceutical and nutraceutical industries have been acutely aware of this, particularly for elements like arsenic and mercury, by having limits for both the inorganic and organic forms of the elements if the maximum permitted daily exposure (PDE) limit for the total element is exceeded.

METALLIC NANOPARTICLES

It is well recognized that engineered or manufactured nanomaterials (ENMs/MNMs) can exhibit unique physical, chemical, and biological characteristics, which are not exhibited by the bulk materials with the same composition. This is particularly true of metal-containing nanomaterials which are used for a variety of consumer, healthcare, and industrial applications. Although many of them are incorporated into the product, significant numbers are used in applications where they are either intentionally released or somehow find their way into the environment including: fabrics containing nano-silver bactericides, cosmetics containing nano-titanium and zinc sunscreen pigments, and diesel fuel containing nano-cerium to improve engine combustion. Unfortunately, many of these ENMs find their way into soils through effluent discharges and wastewater streams and could have a serious effect on the ecosystem they are released into.

Much of the early studies in this field have focused on understanding how nanoparticles enter and impact watershed and hydrologic systems by assessing the size, distribution, surface characteristics, shape, and chemical composition of the particles under investigation. Different manufactured nanoparticles (MNPs) will have different properties and will therefore behave very differently when they enter the environment. Understanding all these metrics for hydrologic systems is very important, but now researchers are going a step further and beginning to investigate the impact of nanoparticles on soil fertility and their uptake by crops grown for human consumption. Some of the many questions they are trying to answer are whether the nanoparticles remain as the metal or metallic oxide or whether are they converted into a biomolecular form as they get absorbed through the root system into the vegetable, leaf, or fruit, which eventually gets consumed by the human population. As it's well recognized that cannabis and hemp are natural hyper-accumulators of metals in the soil, the industry needs to have a solid understanding of not only how the ionic form of these metals get absorbed by the root system of the plant but also how nanoparticles affect the growth of the plant. There is current ongoing research being conducted by the Environmental Protection Agency (EPA) and Food and Drug Administration (FDA) on nanoparticle characterization, so it is only a matter of time before they become standard regulated methodologies for environmental and food-based assays… when that happens, the pharmaceutical industry and eventually the cannabis industry will not be far behind.

STATE-BASED HEAVY METAL LIMITS

It's well recognized that state-based elemental toxicological guidelines to regulate the cannabis industry are being taken very loosely from a combination of methods and limits derived by the pharmaceutical, dietary supplements, food, and environmental and cosmetics industries. Even though the process of manufacturing cannabis products might be similar in some cases to drugs and herbal medicines, the consumers of cannabis and hemp products (especially food and beverage-based products) are using them very differently and in very different quantities, particularly compared to pharmaceuticals, which typically have a maximum daily dosage. The bottom line is that heavy metal toxicological data generated for pharmaceuticals over a number of decades cannot simply be transferred to cannabis, hemp, and their multitude of products.

Medical cannabis is legal in 34 states, while 12 states including Washington, DC, allow its use for adult recreational consumption. However, there are many inconsistencies with heavy metal limits in different states where cannabis is legal. The vast majority of states define four heavy metals (Pb, As, Cd, and Hg), while Missouri (MO) and Michigan (MI) add Cr to the list (MI also includes Cu and Ni for vapes). Maryland (MD) adds Cr, Se, Ag, and Ba, while New York (NY) adds Cr, Zn, Cu, Ni, Sb, and Ba. Some base their limits directly in the cannabis, while others are based on human consumption per day. Others take into consideration the body weight of the consumer, while some states do not even have heavy metal limits. Certain states only require heavy metals in the cannabis plant/flower, while some give different metals and limits for the delivery method (oral, inhalation, or transdermal).

Clearly, from a federal perspective, it doesn't make any sense to have such inconsistent and confusing regulations. So let's take a more detailed look at the action limits of a few selected states, so we can get a clearer picture of how confusing and inconsistent they are. Table 2.1 shows cannabis heavy metal action limits for seven states (CA, CO, CT, MA, MD, NY, and OR) compared with the American Herbal Pharmacopoeia (AHP) guidelines, United States Pharmacopeia (USP) Chapter <232> PDEs for elemental impurities in drug compounds, and USP Chapter <2232> for contaminants in dietary supplements (refer to Chapter 3 for more information about the USP chapters).

It's worth providing more detailed information about this table to better understand the variations in the data across state lines and how they compare with limits for pharmaceuticals and dietary/herbal supplements.

- **USP Chapter <232>** PDE levels are based on maximum limits of the drug compound per day. For a suggested maximum daily dosage of 10 g, these PDE limits should be divided by ten. However, even though 15 elements are listed here, the actual Chapter <232> list includes 24 elements to reflect all the potential sources of elemental impurities.[14] Parental (intravenous) PDEs have not been included, because there is no equivalent delivery method for cannabis products. Note: Chapter <232> limits are the same as the ones defined in ICH Q3D Guidelines implemented by the European Medicines Agency (EMA).[15] For more detailed information on these directives, please refer to Chapter 3 " What the Cannabis Industry Can Learn from Pharmaceutical Regulations."
- **USP Chapter <2232>** PDE levels specifically give maximum limits per day for dietary supplements only. The arsenic PDE (15 μg/day) may be measured using a non-speciation procedure under the assumption that all arsenic contained in the supplement are in the inorganic form. Where the limit is exceeded using a non-speciation procedure, compliance with the limit for inorganic arsenic shall be demonstrated on the basis of a speciation procedure. Methyl mercury determination is not necessary when the content for total mercury (15 μg/day) is less than the limit for methyl mercury (2 μg/day).[16] Note: High-performance

TABLE 2.1
Heavy Metal Limits for Cannabis in Selected States (N/S = Not Specified)[13]

	USP Chapter 232 PDEs (µg/day)	USP Chapter 2232 (µg/day)	AHP (µg/ Daily Dose)	CA (µg/g)	CO (ppm)	OR	MD (ppm)	NY (µg/dose)	CT (µg/Kg of Body Weight/day	MA (mg/Kg)
Element	Oral/Inhalation	Oral	Oral	Inhalation/All Other Products	Oral/Inhalation/ Topical	No Limits	All products	Oral/ Inhalation	All products	Oral/All products
Lead	5.0/5.0	10	6	0.5/0.5	1.0/0.5/10	N/S	1	0.5/0.5	0.29	1,000/500
Arsenic	15/2.0	15 (Inorganic)	10 (Inorganic)	0.2/1.5	1.5/0.2/3.0	N/S	0.4	1.5/1.5	0.14	1,500/200
Cadmium	5.0/2.0	5	4.1	0.2/0.5	0.5/0.2/3	N/S	0.4	0.5/0.2	0.09	500/200
Mercury	30/1.0	15 (2) Total (methyl)	2.0 (methyl)	0.1/3.0	1.5/1.0/1.0	N/S	0.2	3.0/0.3	0.29	1,500/100
Nickel	200/5.0	N/S	N/S	N/S	N/S	N/S	N/S	20/2.0	N/S	N/S
Silver	150/7.0	N/S	N/S	N/S	N/S	N/S	1.4	N/S	N/S	N/S
Selenium	150/130	N/S	N/S	N/S	N/S	N/S	26	N/S	N/S	N/S
Chromium	11,000/3.0	N/S	N/S	N/S	N/S	N/S	0.6	1,100/110	N/S	N/S
Barium	1,400/300	N/S	N/S	N/S	N/S	N/S	60	N/S	N/S	N/S
Copper	3,000/30	N/S	N/S	N/S	N/S	N/S	N/S	300/30	N/S	N/S
Antimony	1,200/20	N/S	N/S	N/S	N/S	N/S	N/S	120/2	N/S	N/S
Zinc	N/S	N/S	N/S	N/S	N/S	N/S	N/S	4,000/4000	N/S	N/S

liquid chromatography (HPLC) coupled with inductively coupled plasma mass spectrometry (ICP-MS) is the traditional way of carrying out elemental species studies today.

- **American Herbal Pharmacopoeia (AHP):** Standards of Identity, Analysis, and Quality Control (2014),[17] which are derived from the American Herbal Products Association (AHPA) *Heavy Metals Analysis and Limits of Herbal Dietary Supplements* (*2009*) and provides manufacturers of herbal products with general recommendations for maximum heavy metal levels in herbal products, based on the daily product intake amount. (These limits were originally sourced from California Proposition 65 Safe Drinking Water and Toxic Enforcement Act of 1986. For an updated version of these limits, refer to Guidance on California Proposition 65 and Herbal Products (AHPA, 2017).)[18]
- **California** limits are defined in microgram per gram (µg/g) in oral and inhalation cannabis products and are based on a maximum amount of 10g of intended daily consumption.
- **Colorado** limits are defined in parts per million (ppm) and also include limits for transdermal (via the skin) products. The limits for inhaled products are based on USP Chapter <232> PDE levels for drug compounds, but the limits for oral (including rectal and vaginal delivery) products are based on USP Chapter <2232> for dietary supplements. However, the topical limits are based on FDA's testing of heavy metals in cosmetics. They further provided a list of products to be tested including medical marijuana-infused products, water-based, food-based, heat/pressure-based, and solvent-based medical marijuana concentrates. Note these limits are based on acceptable limits per one gram of intended use.
- **Oregon** does not have a list of limits for heavy metal contaminants (as of April, 2020).
- **Maryland** limits for all cannabis products are defined in ppm, based on 5g of intended consumption per day. They are based on Chapter <232> PDEs for inhaled products, irrespective of how the products are consumed. Note, they also include four additional elements, silver, selenium, barium, and chromium, but do not explain why these contaminants have been added. So it's important to understand that besides chromium, the other elements do not have any significant toxicological impact at these levels. Barium is environmentally ubiquitous. Wherever you find calcium, you will find barium, which is not a carcinogen, and the toxicity relative to concentrations in plants is very low. Selenium is a micronutrient, and at levels taken up by plants, the potential for toxicity is very low. Silver is not a carcinogen and is found at extremely low levels in plants, and its toxicity relative to these concentrations in plants is very low.
- **New York** acceptable limits are defined in µg per dose (no mention of what a dose is) in oral and inhalation products. Note: The oral limits are based on Chapter <232> PDEs for oral drug compounds based on a daily consumption of 10g per day. However, the inhalation limits are based mostly on Chapter <232> PDEs for parenteral (intravenous) drugs, with the exception of antimony, which is based on the inhalation PDE (there is no clear explanation of why they did this). It's also important to note that five additional elements are included in the NY list (Sb, Cr, Cu, Ni, and Zn), with no explanation of why they have been added. In fact, Zn is in this list, which is not even included in USP Chapter <232>.
- **Connecticut** limits are a little unusual because they are based on µg of contaminant per kg of user body weight per day, so it's very difficult to get a good understanding what this means with regard to allowable levels in the product without knowing the weight of the consume.
- **Massachusetts** limits are defined in µg/kg (and not µg/g or ppm as with the other state limits) for oral and all other products, assuming that no more than 10g of product has been consumed per day. The oral limits are derived from USP Chapter <2232> PDEs for dietary supplements and includes plant material, cannabis resin, and concentrates intended for ingestion only, while limits for all other products are derived from USP Chapter <232> inhalation drug PDEs and includes plant material, cannabis resin, and concentrates

intended for inhalation and dermal applications. Interestingly, it also includes a statement that all products intended for oral consumption must also be labeled to indicate that this product has been evaluated for impurities based on oral consumption and "should NOT be inhaled." It's also worth pointing out that Massachusetts is also the only state that defines the limit for arsenic as the inorganic form and mercury as total mercury. Clearly, this has been taken from the USP Chapter <2232> for dietary supplements (see previous explanation). However, it mentions nothing about how to carry out elemental speciation of either arsenic or mercury.

- **Canadian** limits are not listed, because Health Canada does not publish definitive maximum action limits for heavy metals in cannabis. In fact they state in *Cannabis Regulations Product Safety Quality Requirements* that "cannabis products cannot contain any substances that may be injurious to health and that any contaminants must not exceed established limits taking into account how the product is consumed." Established limits refer to those set by the USP or European Pharmacopeia. In other words, they are not restricted to requesting 4 heavy metals limits required by most US states but can ask for any number of the 24 elements defined in USP 232/ICH Q3D directives, based on the potential impact to health of the method of consumption/delivery. However, currently they are requiring a minimum of the "big four" heavy metals at inhalation levels, irrespective of the mode of delivery to the consumer.

Note: On April 14, 2020, the United States Pharmacopeia (USP) published an article on the quality attributes of cannabis in the *Journal of Natural Products*,* which included setting elemental contaminant limits for cannabis products. Even though the article appeared in the public domain, after I submitted my manuscript for publication, I felt it was important to at least reference it here. The following paragraph has been taken from that article:

> **Elemental Contaminants**. Cannabis has been identified as a hyper-accumulator for heavy metals. These elemental impurities may be introduced from soils, water, and other inputs, and exposure of consumers to cannabis products containing such contaminants is an important quality and safety consideration. Toxicologically based limits for elemental contaminants are described in the USP general chapter <232> Elemental Impurities–Limits, while analytical methodologies are discussed in the USP general chapter <233> Elemental Impurities–Procedures. Considering the potential inhalation use of cannabis inflorescence, the panel had suggested adoption of acceptance criteria from the USP general chapter <232> for inhalation products:
>
> - Arsenic: NMT 0.2 μg/g
> - Cadmium: NMT 0.2 μg/g
> - Lead: NMT 0.5 μg/g
> - Mercury: NMT 0.1 μg/g
>
> When contamination with other elemental impurities may be possible (e.g., due to past or nearby industrial activities), in addition to the above specifications, USP general chapter <232> also requires that "when additional elemental impurities are known to be present, have been added, or have the potential for introduction, assurance with the specified levels is required".

POTENTIAL OF "REAL-WORLD" SOURCES OF ELEMENTAL CONTAMINANTS IN CANNABIS

This list in Table 2.1 represents only seven of the 34 states that allow the use of medical cannabis. However, it is legal in another 27 states, which all have their own variations of these regulations.

* N. D, Sama, et al. Cannabis Inflorescence for Medical Purposes: USP Considerations for Quality Attributes, *ACS Journal of Natural Products*. https://pubs.acs.org/action/showCitFormats?doi=10.1021/acs.jnatprod.9b01200&ref=pdf.

It is therefore highly likely that federal regulators would pose serious questions as to how individual states are regulating heavy metals in cannabis products. In addition, it's clearly insufficient to be only testing for the big four heavy elements, as it has been pointed out previously that there is enough evidence in the public domain that many other elemental contaminants are potentially just as serious, especially if there is ignorance and denial as to where they come from and how they can be reduced or even eliminated. This is particularly the case with the cannabis plant that can absorb elemental contaminants through its roots system from the growing medium, soil, water, nutrients, and fertilizers, etc., and through absorption from atmospheric environmental pollution via its foliage.[9] In addition, the biological and organic components in the soil have the potential to change the species or the oxidation state of the elemental contaminant, and under the right environmental conditions, the possibility of various types of nanoparticle could end up in the soil. So with all the diverse and varied conditions used for growing cannabis, it is very difficult to eliminate all the potential sources of elemental contaminants and heavy metal pollutants in order to reduce their impact on the plant They might not all have a negative impact on the health of the plant during cultivation, but the chances that they will end up in the flowers and the final manufactured products are very high. It's therefore worth listing some of the many "real-world" sources of elemental contamination, both from a cannabis/hemp plant cultivation perspective and the cannabinoid manufacturing process and how they can possibly be reduced or even eliminated. This not an exhaustive list, but at least it represents a good baseline to begin investigating the problem.

OUTDOOR GROWING SOURCES

- Many heavy metals are found in soil, and growing media particularly for plants that are cultivated outdoors. Characterize the soil to make sure that the elemental contaminants are at acceptable levels. Explore the use of natural chelating agent such as humic acid or synthetic ones like biochar to bind with the harmful metallic contaminants to minimize their uptake by the plant's root system.[19]
- Although there are no elemental species of As and Hg defined in the current state-based limits for heavy metals, the pharmaceutical industry has shown that inorganic and organic forms of these elements should be monitored if the maximum limits for the total amount are exceeded. In addition, depending on where the cannabis/hemp plants are grown, will also dictate whether other speciated forms should be monitored.
- Abandoned gold and silver mines, particularly in the western regions of the United States, used mercury for extraction purposes.
- Any metal smelting plant will experience heavy metal contamination in surrounding areas. Copper ores in particular contain high levels of arsenic.
- EPA Superfund sites, such as those involved in the manufacture of weapons, could have high levels of elemental contaminants in the soil.
- Coal-fired power plants where fly ash waste from the combustion process has been dumped.
- Decades of using leaded gasoline will have contaminated the soil close to major highways and roads.
- Industries that are known to emit elemental mercury into the atmosphere... these industries include coal-fired power plants, metal refineries/smelters, petrochemical plants, and cement works. It is well documented in the Clean Air Act that approximately 100 tons of mercury are emitted by US industries annually.
- Wood preservation chemicals contain high levels of copper, arsenic, and chromium.
- Low-grade fertilizers/nutrients made from phosphate rocks contain significant amounts of metallic impurities.
- Nickel has been promoted as a bud-enhancer and silicon as a way of increasing shoot size, which means that higher levels of Ni and Si will invariably end up in the cannabis product. These elements are typically not required by the vast majority of state regulators.

In fact currently, only NY has a requirement to test for nickel in cannabis products, while no states test for silicon, so both these elements would escape the scrutiny of most state regulators.

- Inorganic pesticides contain, arsenic, copper, lead, and mercury.
- Although there are no elemental species of As and Hg defined in the current state-based limits for heavy metals, the pharmaceutical industry has shown that inorganic and organic forms of these elements should be monitored if the maximum limits for the total amount are exceeded. In addition, depending on where the cannabis/hemp plants are grown will also dictate whether other speciated forms should be monitored.
- At some point in the future, nanoparticle characterization may be needed particularly when there are regulated methods for environmental and food-based assays

INDOOR GROWING SOURCES

Although a greenhouse growing environment is far more controlled than cultivating plants outdoors, there are still many opportunities for picking up elemental contaminants. Nutrients, fertilizers, and potable water are three of the potential sources of metal contamination. As a result, indoor plants cultivated in a soil-based growing medium or hydroponically grown are highly dependent on the nutrients, minerals, and water used. For that reason, high-quality fertilizers and a source of clean water are essential for healthy plants. Here are three areas of concern when growing plants indoors.

- Last year, the EPA estimated that 30 million people in the United States live in areas where drinking water violated safety standards (https://time.com/longform/clean-water-access-united-states/). The EPA defines a list of primary and secondary elemental MCLs (maximum contaminants levels) in drinking water and overseas all local water authorities/municipalities in the United States to ensure compliance. It's important to know these levels in your region to make sure they are well below the limits for metal impurities (they will be posted online by your local water authority). Also keep in mind that these are levels for samples taken at the water treatment plant and NOT the levels measured at your home or growing site. EPA mandates that a municipality only has to measure water quality at its customer's sites every 2 years and has to take remedial action only if 10% of those samples are above the MCL. Look what happened with Pb-contaminated drinking water in Flint, MI, because the source was changed from lake water (Lake Huron) to the local Flint River, without understanding the chemistry and its corrosion properties. As a result, the water dissolved the inside of the old lead pipes and ended up in the drinking water supply.
- Decades of using lead, cadmium, and arsenic-based pigments in residential and commercial paint has meant that many of the older homes and buildings still contain these types of paints, which have often been painted over but still produce dust/particles that can potentially be problematic.
- Some drywall/plasterboard/partitions used in the construction industry are made from flue gas desulfurization (FGD) waste products, sometimes called "clinker," which is produced by scrubbing particulate emissions from coal-fired power plants, which are notoriously high in heavy metals.

MANUFACTURING/PROCESSING SOURCES

The pharmaceutical industry has been forced to spend the past 20+ years investigating all potential sources of elemental impurities in drug compounds. With the approval of USP Chapter <232>, <233>, and ICH Q3D guidelines, they were mandated to fully understand elemental pathways of the entire manufacturing process, including the raw materials, excipients, active ingredients, organic synthesis method, water quality, manufacturing equipment, mixing vessels, containers, and packaging. As a result, to comply with these directives, they had to show convincing evidence (data and/or risk assessment studies) to the regulatory agency that up to 24 elemental impurities of toxicological concern are below certain maximum PDE limits for three different drug delivery methods (oral, parenteral, inhalation). It was challenging, painful, and sometimes confusing, but they eventually accepted that they needed to generate these data to show the drugs were safe to use.

Unfortunately, the cannabis and hemp industries are moving so quickly that no one is taking the time to investigate potential sources of heavy metals throughout the entire manufacturing process of the multitude of cannabis products on the market today. It is well accepted that cannabis and hemp will absorb heavy metals from the growing medium, soil, nutrients, and fertilizers but is less understood what heavy metals (and how much) are carried over to the pure cannabinoids from the extraction, evaporation, concentration, or distillation processes. It makes sense that there is some degree of concentration from the plant to the extract, but how does the extraction solvent have an impact?

If a sub/supercritical extraction process is used, does the temperature or pressure of the extraction play a role? And if there is a distillation step, does the distillation method and/or temperature have an impact? We know that grinding the cannabis flowers with stainless steel grinding equipment can contaminate the sample, with Fe, Cr, and Ni, but do any metallic components leach out from the extraction/distillation equipment used? Clearly these are all questions that require an answer at some point and will need to be clarified before the regulatory agency starts to take an interest.

Perhaps the GW Pharmaceuticals patent for Epidiolex (prescription CBD drug for seizures in young children) might give us a clue? Their extraction method uses CO_2 with subcritical conditions for the fluid extraction at a temperature of 5°C–15°C and a pressure of 50–70 bar. To explain their logic, the following text is lifted directly from their patent[8]

> … the density of sub-critical CO_2 is low, and remains low even as pressure is increased until the critical point of the system is reached. Thus, whilst the solvating power of sub-critical CO_2 is reduced, a high degree of selectivity can be achieved, as only the most soluble components are efficiently dissolved by the CO_2; in this case the cannabinoid fraction. The result is the production of a relatively simple extract containing, as well as the cannabinoids, only a limited number of non-target compounds (inc. heavy metals), many of which can be removed relatively easily by a simple clean-up step. In contrast, at higher temperatures there is a significant increase in the density of the CO_2 as it now exists in a supercritical fluid state. This has the effect of greatly increasing the solvating power of the solvent, which whilst generally advantageous in that more cannabinoids are solubilized, thereby giving high yields, in fact proves disadvantageous because the decreased selectivity of the more powerful solvent results in increased solubility of a range of non-target compounds which makes the resulting extract more difficult to purify.

They claim that by using this extraction method, the total amount of heavy metals in the extract is kept low. However, it's important to emphasize that they grow the plants indoors using pesticide-free compost in a controlled environment using potable, high-quality water and fertilizers, without the use of any synthetic pesticides or herbicides. So even under these strict indoor growing conditions, where do the heavy metals come from? Imagine how difficult it would be to control all these variables with plants gown outdoors!

Anyway, below are some areas worth investigating in order to reduce the number and levels of heavy metals from the cannabis manufacturing process.

- Try to minimize the use of metal-based manufacturing/storage equipment. In particular, if stainless steel cutting blades and grinding equipment are used to chop, cut, and grind the cannabis flowers into fine particles, chromium, nickel, and iron will invariably find their way into the processed cannabis flowers and eventually into the extracted oils and concentrates.
- Use clean solvents, and chemicals (low in heavy metal specs) used in the extraction, distillation, concentration, and infusion of cannabinoids from the plants.

- Use an optimized solvent and extraction process to minimize the amount of heavy metal ending up in the extracted products. (Refer to the extraction process patent for Epidiolex, the only CBD-based prescription drug regulated by the FDA in the United States which has to show compliance with USP Chapter <232>.)
- Leaching of heavy metals from delivery devices, such as vaping sticks, inhalation devices, and infused transdermal patches. In particular, ensure that internal parts of vaping devices such as battery connections, atomizer, tank, and mouthpiece are not made from metallic components that could corrode at elevated vaping temperatures and deliver elemental contaminants such as Pb, Cr, Ni, Co, and Fe to the consumer.
- Raw materials used in cannabinoid tablets, gel caps, creams, oils, edibles, and drinks formulations that could possibly be contaminated with heavy metals. An example of this could be fillers and mineral-based raw materials which are added to dilute or bind the tablet formulations or diluent/cutting oils used in vaping liquids.
- Any products or ingredients that come from Asia can potentially be a source of contamination (think melamine in infant formula/pet food, lead paint on toys, and lead in fake silver jewelry). In particular, it has been shown that inexpensive vaping cartridges that are sourced in China contain metallic components that corrode when vaped.

THE SMOKING/INHALING OF CANNABIS

It's worth pointing out that historically, most consumers of recreational cannabis use it by the inhalation/smoking route. Smoke chemistry has been predominantly investigated in tobacco products, but many studies over the past 10 years have highlighted the qualitatively similar carcinogenic chemicals contained within both tobacco and cannabis smoke.[20] In a recent study, the International Organization for Standardization (ISO) and Health Canada analyzed tobacco and cannabis cigarettes. The heavy metals contained in both smoked products included: mercury, cadmium, lead, chromium, nickel, arsenic, manganese, and selenium.[21] Quantitatively, there were lower heavy metal concentrations in cannabis smoke condensates, due mainly to the fact that the cannabis supply was grown hydroponically. In addition, the soil-less growth medium of the cannabis plants required water and water-soluble hydroponic vegetable fertilizers which contain nitrogen in the form of nitrates. So with no soil-based heavy metals to be extracted during the growth cycle of the cannabis, it was the liquid fertilizers used in the hydroponic systems that contributed mostly to the heavy metal levels.

It should also be emphasized that in any hydroponic growing process, the elemental impurities in the water supply should be below the EPA MCLs, otherwise the plant will pick up heavy metals from the water. This could be real concern with old buildings that perhaps have been using lead pipes or copper/iron pipes connected with lead-based solder. There is a great deal of information in the public domain about the uptake of heavy metals into tobacco and the resulting content in tobacco products, such as nicotine and electronic nicotine delivery (END) devices.[20,22]

However, the more common way of inhaling cannabis products today is via vaping sticks or pens. These are devices where the cannabis extract or oil is heated up to about 300°F and the aerosol is vaporized into the consumer's mouth, very much like an inhaler for an asthma sufferer. The problem with this mode of delivery is that the components inside these vaping devices are typically metal, including the liquid tank, coil, mouth piece, and battery terminals, which are usually made from materials such as stainless steel (Fe, Cr, Ni, Co), brass (Cu, Zn), chromel (Cr, Ni), Inconel (Ni, Cr, and Fe), nichrome (Ni, Cr), and soldered battery connectors (Pb, Sb, Sn). This means that at these kinds of temperatures, dissolved metals or even fine metallic particles will almost certainly be delivered to the consumer's air pathways and lungs via the mouth. This is of grave concern, because unlike oral delivery through the mouth and gastrointestinal digestion system, the lungs and respiratory system were designed to allow us to breathe. They bring oxygen from the air into our bodies and send carbon dioxide out. Air enters the respiratory system through the nose or the mouth. If it goes in through the nostrils, there are tiny hairs called cilia that protect the nasal passageways and other parts of the respiratory tract, filtering out dust and other particles that enter the nose. There is no such filtering system in the mouth, so if any metal particulates are inhaled, there is no mechanism to stop them entering through the respiratory system into the lungs, where they can do serious damage, particularly if vaping is carried out on a regular basis over extended periods of time. The recent nationwide outbreak of lung injuries and deaths by consumers using e-cigarettes, and/or cannabis vaping product that contains vitamin E acetate is a tragic testament to this fact. It's also worth pointing out that most state-based regulations only specify Pb, As, Cd, and Hg and, as a result, would not find out if the other elements (Fe, Cu, Zn, Cr, Ni, Sb, and Sn) were present in the vaping liquid because there is no requirement to test for them. Note: Chapter 24 gives more detailed information about how to test for elemental contaminants in vaping liquids and aerosols.

TESTING PROCEDURES

As a result of the high probability of some heavy metals being present in hemp and cannabis products, the testing for elemental contaminants is absolutely critical. The most suitable and widely used technique is considered to be ICP-MS, which is a very sophisticated multielement analytical technique that can easily measure down to ppt (parts per trillion) detection levels. However, it does not preclude the use of other techniques such as ICP-OES (optical emission spectrometry), flame atomic absorption (FAA), electrothermal atomization atomic absorption (ETAA), or even atomic fluorescence (AF) as long as the detection capability is low enough for the required maximum limits defined by the state. However, it's also worth pointing out that ICP-MS, with its superior sensitivity over these other techniques is probably the best suited, because it is multielement, and even though today there are just four heavy metals required by most states, there is a strong possibility that when the FDA comes knocking on the door of the cannabis industry, that list will be expanded to at least ten elemental contaminants ... and maybe more if the pharmaceutical industry is any indicator.

However, one of the disadvantages of ICP-MS is that it requires an analytical chemist with a reasonably high level of knowledge and expertise to develop and run methods and to fully understand the nuances of ultratrace elemental analysis, including lab cleanliness, sources of contamination, sample preparation, digestion techniques, instrumental method development, interference corrections, calibration routines, use of reference materials, and validation procedures. In other words, in the hands of an inexperienced user it could easily generate erroneous results.

For that reason, the expertise of the testing lab and the people running the instrumentation is of prime consideration and in particular to have an intimate knowledge of working in the ultratrace environment, to be aware of all the potential sources of elemental contaminants outlined and to use robust validation procedures to ensure high integrity data.[24] This is exactly what the pharmaceutical industry did when they abandoned the old USP <Chapter 231> sulfide colorimetric test and wrote a brand new method (Chapter <233>) to measure all 24 elemental impurities which covered the entire analytical procedure including selecting and validating the most suitable atomic spectroscopic technique, optimum sample digestion procedure to make sure the samples are completely digested and robust validation protocols using standardized spike recovery testing procedures, together with certified reference materials if available. The backbone of this new USP Chapter <233> is USP Chapter <730>,[23] which describes the use of plasma spectrochemistry for pharmaceutical-type samples and USP Chapter *1225 for the Validation of Compendial Procedures.*[24]

LABORATORY TESTING PROTOCOLS

Before we venture into what's required from a cannabis testing lab to be proficient in using highly sophisticated analytical equipment for measuring ultratrace levels of heavy metals in cannabis, let's take a closer look at the inherent weakness in the cannabis industry today ... the lack of consistency in quality control. Consumers are asked to trust product ingredients, dosing suggestions, and claims based on what the producers tell them. An informed consumer might ask for a Certificate of Analysis (CoA) from the producer to show third-party lab test results. However, the problem with this is that there is very little required standardization across testing facilities. Many of the labs that have emerged to fill the need for specialized cannabis testing are not regulated themselves, meaning there is no set standardized protocols for equipment, operating procedures, certifications, or qualifications of lab personnel.

Cannabis testing labs currently don't come under the umbrella of the FDA, so they have to be guided by state regulators who often don't have the necessary inspection expertise. Without standardization across all facilities, test results can be wildly inconsistent. What is the value of a CoA from a third-party lab that does not meet any kind of standards themselves? However, as they begin to gain more experience, state regulators are putting in place basic accreditation measures for cannabis labs to be licensed. One such approach is to implement the ISO laboratory competence certification accreditation system, which in conjunction with the International Electrotechnical Commission (IEC), is responsible for ISO/IEC 17025:2017, a standard for calibration and testing laboratories across the globe, ensuring technical competence and ability to produce precise and accurate test and calibration data. In addition, ISO/IEC 17043:2010 specifies general requirements for the competence of providers to develop and operate proficiency testing schemes using well-established interlaboratory comparison studies.

However, cannabis testing requires specialized sample preparations and methods of analysis that can be extremely challenging due to its complex composition and various concentration levels of different compounds in the plant. ISO/IEC accreditation standards can provide confidence to cannabis consumers that testing is being performed correctly and to a universally accepted standard.

To qualify for the accreditation, laboratories must conform to the ISO standards in all areas, including analytical procedures, calibration of instruments and equipment, as well as properly staffed and trained technicians who have met specific academic credentials. Overall, these required qualifications can be costly and time-consuming, which may deter many of the start-up cannabis labs to invest in the necessary infrastructure to ensure they meet these high standards. So to satisfy the demands of the industry, specialized organization like Emerald Scientific (https://pt.emeraldscientific.com/) are developing interlaboratory comparison and proficiency test (ILC/PT) programs specifically for cannabis and hemp testing facilities, establishing much needed standardized protocols for testing. Through the participation of labs around the world, organizations like Emerald Scientific and the universally recognized Emerald Test™ are helping to raise the industry standard for cannabis and hemp testing.

So let's now take a closer look at some of the most important factors for a testing lab to consider when measuring elemental contaminants in cannabis and hemp, which should help them become more proficient at working in the ultratrace environment.

- It is important to adopt a robust quality assurance program (QAP) based on recognized quality management systems such as ISO and GLP which utilize official mandated methods defined by federal agencies such as FDA, EPA, National Institute of Standards and Technology (NIST), and US Department of Agriculture (USDA) or consensus methods put out by organizations such as American Society for Testing Materials (ASTM), Association of Official Agricultural Chemists (AOAC), USP, and AHPA. This QAP aspect is absolutely critical, because without currently having a choice of certified reference materials of cannabis and cannabis extracts to validate the accuracy of the method, the analytical procedures must involve recognized spike recovery procedures where all standards, blanks, and samples are spiked with known concentrations based on the regulated maximum limits for the cannabis product before the sample preparation step to ensure that no false positive or negative results are generated. This is outlined in the section on J-values in USP Chapter <233> which should be a critical component of all heavy metal testing of cannabis products.[25] Note: The FDA EPA, USDA, NIST, USP, ASTM, AOAC, and AHPA have all made public announcements about their positions on regulating cannabis and hemp and are in various stages of publishing standardized methods and reference materials to cover the measurement of elemental contaminants.[26–33,37]

- Understand the error and uncertainty involved with your analytical methodology including volumetric glassware, analytical balances, calibration standards, reference materials, sample preparation procedures, and precision/repeatability of the instrumental technique used.[34]
- Understand the practical real-world detection capability of the analytical method and in particular the potential sources of contamination that can impact the limit of quantitation (LOQ), some of which are outlined below:
  - Cleanliness of sample preparation test area and equipment
  - Laboratory dust/dirt of unknown origin (e.g., old Pb-based paint)
  - Cleanliness of sample digestion procedure
  - Quality of the deionized water
  - Purity of analytical reagents, acids, and solvents
  - Impurities in laboratory glassware and plastic vessels/containers
  - Purity of the plastic tubing used in delivering the sample to the instrument
  - Contaminants from the analyst, including clothing, cosmetics, lotions, perfumes, shampoo, jewelry, and smoke
  - (Note: Refer to Chapter 22 for more detailed information.)
- If you are tasked with characterizing cannabis vaping pens, get familiar with methodology for measuring toxic metals in vaping liquids, as well as determining toxic metals in the aerosol being delivered to the consumer, because the requirements are very different (see Chapter 24 " Measurement of Elemental Constituents of Cannabis Vaping Liquids and Aerosols by ICP-MS." It's relatively straightforward to aspirate the liquid into the spectrochemical measuring technique, using a conventional sample introduction system and quantitate the elemental contaminants, as long as suitable calibration standards are used. However, how do you aspirate an aerosol from the vaping device directly into the instrument and be able to quantitate each puff or spray unless you have specialized sampling tools and/or smoking machines. This becomes very challenging unless you have had experience in carrying out this type of analysis. However, Halstead and Gray and coworkers have recently published two very pertinent papers, which measure a suite of toxic metals in both liquids and aerosols in nicotine-based electronic cigarettes.[35,36]
- Finally, a word of advice for regulators who are tasked with inspecting testing labs that are carrying out heavy metals analysis. You should become familiar with analytical procedures described in USP Chapter <233> and, in particular, the strict validation protocols described in the chapter. These protocols, which are based on strict spike recoveries, have been developed and tested by the USP over the past 20 years. They are the very essence of the FDA inspection process of pharmaceutical labs to ensure the data generated is of the highest quality. If a lab is not carrying out these procedures thoroughly and correctly, there is a very good chance the sample data will be flawed. However, providing data for regulatory inspection and product release documentation in any industry is only part of the challenge facing the modern analytical laboratory. For example in the pharmaceutical industry, external inspections of facilities and procedures mean that a great of infrastructure is required to support the actual analytical operation including (but not limited to) people, equipment specifications, GLP capability, lab facilities, SOPs, validation, and data management. The pharmaceutical analytical laboratory has taken this aspect very seriously, because they are a part of a highly regulated industry. It's unlikely that cannabis state regulators will have the necessary background and experience to carry out this kind of detailed inspection of cannabis testing labs. However, it is a given that when the FDA eventually comes knocking, they will know exactly what they are looking for and will leave

no stone unturned, having had 4–5 years' experience of regulating the pharmaceutical industry. My previous publication, *Measuring Elemental Impurities in Pharmaceuticals*, covers this in great detail,[42] but an excellent resource for getting ready for an FDA inspection can also be found in these references.[37,38]

A WORD ABOUT HEMP REGULATIONS

The challenges of regulating the cannabis industry will be compounded when CBD products derived from hemp begin to hit the marketplace in 2020. Up to now much of the cannabis grown in the United States is cultivated in greenhouses where the growing environment is controlled. However, when the growing of hemp for research purposes was legalized in the Agricultural Act of 2014 (also known as the 2014 Farm Bill), it's clear the majority of it was being grown outdoors with much less control over growing conditions and a higher risk of heavy metal contamination from the soil. It also doesn't help that hemp is being presented as a wonder crop, especially to struggling tobacco farmers. As a result, most of the information in this chapter has focused on cannabis, mainly because it has been legalized by many US states for medicinal purposes for a number of years, even before the enactment of the 2014 Farm Bill.

However, it's also worth emphasizing the current status of regulations for hemp in the United States. Because of the revamped Hemp Farming Act of 2018 (also known as the 2018 Farm Bill), the federal government has made it legal to grow hemp (defined as *Cannabis* containing less than 0.3% THC) by removing it from the controlled substance list. As a result, growers do not need a permit from the Drug Enforcement Administration (DEA). This adds a new twist to the state-based regulations for cannabis, because the bill directed the USDA to issue regulations and guidance to implement a program for the commercial production of hemp which could either be used for a variety of industrial processes (including fuels, textiles, fibers, and plastics), as a health food supplement (e.g., hemp seeds) or as a source of CBD products from the hemp flowers (oil, supplements edibles, etc.).[39]

USDA has already begun the process to gather information for rulemaking. Once complete, this information will be used to formulate regulations that will include specific details for both federally regulated hemp production and a process for how individual states should submit their plans to the USDA. In the short term, it is likely that the USDA will look to the states' departments of agriculture who will be assisted by specialized hemp analytical testing laboratories which must be registered with the DEA to handle controlled substances. Currently, potency is the most important requirement because hemp is not allowed to have >0.3% THC, but it is likely they will be guided by state cannabis commissions for maximum levels of contaminants such as pesticides and heavy metals. And as previously mentioned, it could potentially be more complicated to regulate hemp grown for CBD products, because the vast majority of hemp plants will be grown outdoors where there is less control over the growing environment.

Regulations for states who submit plans will include procedures and information collections regarding: land to be used for planting; testing; effective disposal of plants and products; compliance with law enforcement; annual inspections; submission of information to USDA; and certification that resources and personnel are available to carry out these procedures. States do not need to submit plans for approval until regulations are in place. However, should a state submit a plan, USDA will hold that submission until regulations have been implemented.

The Farm Bill allowed states and institutions of higher education (universities, research organizations) to continue operating under authorities of the hemp research program for the 2019 planting season. Of course, the 2018 Farm Bill also extended the 2014 Farm Bill authority for 1 year. States thus may continue operating pilot programs until October 31, 2020. This means states may also apply existing hemp laws and regulations during the 2020 growing season.

The USDA also established a plan to monitor and regulate the production of hemp in those states that do not have an approved plan, as well as issuing regulations in the fall of 2019 to accommodate the 2020 planting season (the US Domestic Hemp Production Program comment period ended on January 29, 2020, to allow stakeholders time to provide feedback, but USDA has indicated it will give itself 2 years until November 1, 2021, to issue final regulations). As soon as the regulations have been approved, they will be published in the Federal Register, when it will be legal to grow hemp anywhere in the United States for the production of CBD-based products. So it will be interesting to see how the Department of Agriculture regulates the industry at the federal level, particularly with regard to heavy metals when cannabis is currently regulated by the individual states.[40]

It's also worth pointing out that the Farm Bill also explicitly preserved the authority of the US FDA to regulate hemp products under the Federal Food, Drug, and Cosmetic Act, which states that products containing cannabinoid-derived compounds are subject to the same authorities and requirements as FDA-regulated drug products containing any other substance.

FINAL THOUGHTS

Our environment has been severely polluted by heavy metals, which has compromised the ability of our natural ecosystems to sustain and foster life. Heavy metals are known to be naturally occurring compounds, but anthropogenic activities introduce them in extremely large quantities into our agricultural growing and cultivation systems. Nowhere is this more evident than in the delicate balance of growing cannabis and hemp for commercial, medicinal, and recreational uses, not only in the United States but also in other parts of the world that are supplying us with cannabis and hemp. This is exemplified by a recent editorial which highlighted the gross soil contamination produced by metal refineries in many parts China,[41] which is extremely troubling as Yunnan Province in Southern China is now producing CBD products for the US market.[42]

Unfortunately, the demand for cannabinoid-based products is moving so fast that the scientific community is not keeping up with it, whether it's the testing of the products to make sure they are safe for human consumption or the medical research required to understanding the biochemistry that is fundamental to treating a particular disease or ailment. The industry is both exciting and chaotic at the same time, but because of its unparalleled growth, there appears to be very little incentive to bring in sensible regulations. There clearly needs to be a more comprehensive suite of elemental contaminants tested and to set the maximum limits on toxicological data based on the manner and the quantity that cannabis products are consumed, particularly as CBD products processed from hemp gown outdoors will start to appear in the marketplace.

Unfortunately, there are more growing variables and uncertainty with outdoor hemp farms compared to indoor cannabis cultivation where the conditions are far more controlled. This makes it even important that a wider suite of elemental contaminants are included in any future regulations. In fact, standard organizations such as ASTM have fifteen elements in its ICP-MS heavy metals method for cannabis and hemp, while the AOAC has specified four but with the option of adding another eight.

I'm firmly convinced that key educators who are trying to raise the bar now will be rewarded when the FDA eventually starts to get involved. Clearly, the cannabis industry needs to have a much better understanding of the many sources of heavy metal contamination before they can hope to satisfy federal regulators, as the 20+ years saga of regulating elemental impurities in the pharmaceutical industry has shown us.[43] However, in the meantime, I'm committed to not only offering guidance to state regulators on contamination issues in cannabis- and hemp-based products but also helping independent testing laboratories improve the quality of their analytical results. The main objective of this book is not only to focus on the broader issues of testing but also to use my 45 years of experience in trace element analytical procedures to give testing labs a better understanding of the fundamental principles of the techniques they are using, which will hopefully help them to generate data of the highest quality and ultimately ensure that cannabis and hemp and its many products are safe for the consumer.

FURTHER READING

1. P. Soudek, et al. From Laboratory Experiments to Large Scale Application – An Example of the Phytoremediation of Radionuclides, Advanced Science and Technology for Biological Decontamination of Sites Affected by Chemical and Radiological Nuclear Agents, pp. 139–158, 2007.
2. M. Lasat. *The Use of Plants for the Removal of Toxic Metals from Contaminated Soil*, AAAS Report, Washington, DC, 2004.
3. R. Ahmad, et al. Phytoremediation potential of hemp (*Cannabis sativa* L.): Identification and characterization of heavy metals responsive genes, *Clean* 44(2): 195–201, 2016.
4. D.V. Gauvin, et al. A budding cannabis cottage-industry has set the stage for an impending public health crisis, *Pharmaceut. Reg. Affairs* 7(1): 199, 2018.
5. D.V. Gauvin, et al. Marijuana toxicity: Heavy metal exposure through state-sponsored access to "la Fee Verte", *Pharmaceut. Reg. Affairs* 7: 1, 2018.
6. V. Masindi and K.L. Muedi. Environmental contamination by heavy metals, *IntechOpen* doi: 10.5772/intechopen.76082.
7. W. Chen, et al. Arsenic, cadmium, and lead in California cropland soils: Role of phosphate and micronutrient fertilizers, *J. Environ. Qual.* 37(2): 689–695, 2008.
8. B. Whittle, et al. Wheatley Extraction of pharmaceutically active components from plant materials, US Patent Number, US7344736B2, Inventors, GW Pharmaceuticals, 2008.
9. G. Bode. Back to the root—The role of botany and plant physiology in cannabis testing, part I: Understanding mechanisms of heavy metal uptake in plants, *Cannabis Sci. Technol.* 3(2): 26–29, 2020.
10. B.J. Alloway, ed. *Heavy Metals in Soils: Trace Metals and Metalloids in Soils and their Bioavailability*, Department of Soil Science, University of Reading, UK, Springer Science, Dordrecht, 2013, ISBN: 978-94-00-4469-1.
11. C. Stephan and R. Thomas. Single-particle ICP-MS: A key analytical technique for characterizing nanoparticles, *Spectrosc. Mag.* 32(3): 12–25, 2017.
12. M.A. Khan, et al. Effect of soil contamination on some heavy metals content of *Cannabis sativa*, *J. Chem. Soc. Pak.*, 30(6): 805–809, 2008.
13. Marijuana Policy by State: Medical Marijuana Policy Project, https://www.mpp.org/states/.
14. USP. United States Pharmacopeia General Chapter <232> Elemental Impurities–Limits: First Supplement to USP 40-NF 35, 2017, https://www.usp.org/chemical-medicines/elemental-impurities-updates.
15. ICH. ICH Q3D Step 5 Guidelines, ICH Website: http://www.ich.org/products/guidelines/quality/article/quality-guidelines.html (Q3D).
16. USP. United States Pharmacopeia General Chapter <2232>: Elemental Contaminants in Dietary Supplements, PF 38(3), May–June 2012, https://www.uspnf.com/notices/general-chapter-elemental-contaminants-dietary-supplements.
17. AHP. American Herbal Pharmacopoeia (AHP): Standards of Identity, Analysis and Quality Control, 2014, https://herbal-ahp.org/online-ordering-cannabis-inflorescence-qc-monograph/.
18. AHP. Guidance on California Proposition 65 and Herbal Products, American Herbal Products Association (AHP), http://www.ahpa.org/News/Alerts/TabId/100/ArtMID/1052/ArticleID/861/Guidance-on-California-Proposition-65-for-herbal-products.aspx.
19. X. Zhang, et al. Using biochar for remediation of soils contaminated with heavy metals and organic pollutants, *Environ. Sci. Pollut. Res.* doi: 10.1007/s11356-013-1659-0, 2013.
20. R. Pappas, et al. Toxic metal concentrations in mainstream smoke from cigarettes available in the USA, *J. Anal. Toxicol.* 38: 204–211, 2014.
21. J. Pariati. Analysis of heavy metals in cigarette tobacco, *J. Med. Discovery* 15(1): 636–644, 2017.
22. P. Olmedo, et al. Metal concentrations in E-cigarette liquid and aerosol samples: The contribution of metallic coils, *Environ. Health Perspect.* 26(2), 2018.
23. USP. United States Pharmacopeia General Chapter <730> Use of Plasma Spectrochemistry for Pharmaceutical Analysis.
24. USP. United States Pharmacopeia General Chapter <1225> for Validation of Compendial Procedures.
25. USP. United States Pharmacopeia General Chapter <233> Elemental Impurities – Procedures: Second Supplement to USP 38-NF 33, 2015, https://www.usp.org/chemical-medicines/elemental-impurities-updates.
26. FDA. Regulation of cannabis and cannabis-derived products, including cannabidiol (CBD), https://www.fda.gov/news-events/public-health-focus/fda-regulation-cannabis-and-cannabis-derived-products-including-cannabidiol-cbd.

27. EPA. Seeks public comment on pesticide applications for hemp, https://www.epa.gov/newsreleases/epa-seeks-public-comment-pesticide-applications-hemp.
28. NIST. Tools for cannabis laboratory quality assurance, https://www.nist.gov/programs-projects/nist-tools-cannabis-laboratory-quality-assurance.
29. USP. Scientific Data and Information About Products Containing Cannabis or Cannabis-Derived Compounds; Public Hearing; Request for Comments [Docket No. FDA-2019-N-1482] 84 Fed. Reg. 12969, April 3, 2019, https://www.usp.org/sites/default/files/usp/document/about/public-policy/2019-07-05-comment-from-usp-cannabis.pdf.
30. AHPA. Cannabis Committee, http://www.ahpa.org/AboutUs/Committees/CannabisCommittee.aspx.
31. AHPA. Comments of the AHPA on FDA-related scientific data and information about products containing cannabis or cannabis derived products, http://www.ahpa.org/Portals/0/PDFs/Advocacy/19_0716_AHPA_Comments_FDA_Cannabis.pdf.
32. ASTM. Committee D37 on Cannabis, https://www.astm.org/COMMITTEE/D37.htm.
33. AOAC. Cannabis Analytical Science Program, https://www.aoac.org/scientific-solutions/casp/.
34. P. Atkins. Holding data to a higher standard, Part II: when every peak counts—a practical guide to reducing contamination and eliminating error in the analytical laboratory. *Cannabis Sci. Technol.* 1(4): 40–49, 2018.
35. N. Gray, et al. Analysis of toxic metals in liquid from electronic cigarettes, *Int. J. Environ. Res. Public Health* 16(22): 4450, 2019.
36. M. Halstead, et al. Analysis of toxic metals in electronic cigarette aerosols using a novel trap design, *J. Anal. Toxicol.* 001: 1–7, 2019.
37. FDA. Inspection readiness: A guide to preparing subject matter experts to face the FDA, J. Larsen, The Food and Drug Newsletter, 2013, https://www.fdanews.com/ext/resources/files/The_Food_And_Drug_Letter/2013/Inspection-Readiness-ExecSeries.pdf.
38. FDA. Q3D Elemental Impurities Guidance for Industry, U. S. Department of Health and Human Services, FDA, Drug Evaluation and Research (CDER), Center for Biologics Evaluation and Research (CBER), March 2020, http://www.fda.gov/regulatory-information/search-fda-guidance-documents/q3dr1-elemental-impurities.
39. USDA. U.S. Domestic Hemp Production Program, USDA Website, https://www.ams.usda.gov/rules-regulations/hemp.
40. USDA. Releases Long-Awaited Industrial Hemp Regulations, https://www.fb.org/market-intel/usda-releases-long-awaited-industrial-hemp-regulations.
41. G. Shih. China: Toxic trails from metal production harms health of poor communities amid soaring global demand for gadgets, Washington Post, January 5, 2020, https://www.business-humanrights.org/en/china-toxic-trails-from-metal-production-harms-health-of-poor-communities-amid-soaring-global-demand-for-gadgets.
42. S. Meyers. China Cashes In on the Cannabis Boom, New York Times editorial, May 4, 2019, https://www.nytimes.com/2019/05/04/world/asia/china-cannabis-cbd.html.
43. R. Thomas. Regulating Heavy Metal Contaminants in Cannabis: What Can be Learned from the Pharmaceutical Industry? Parts 2-5 Analytical Cannabis, May 14, 2020, https://www.analyticalcannabis.com/articles/regulating-heavy-metal-contaminants-in-cannabis-what-can-be-learned-from-the-pharmaceutical-312401.

3 What the Cannabis Industry Can Learn from Pharmaceutical Regulations

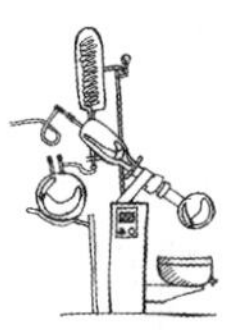

This chapter will take a closer look at how the pharmaceutical industry changed its 100-year-old semiquantitative colorimetric test for a small group of heavy metals, to finally arrive at a list of 24 elemental impurities using plasma spectrochemical techniques. The cannabis and hemp industries can learn a great deal from this process to understand not only the many potential sources of heavy metal contamination during its growing and cultivation but also how the final products can be contaminated by the manufacturing process, the extraction procedure, and the cannabinoid delivery system. In addition, as a result of inconsistencies in heavy metal action limits across state lines in the United States, regulators would be well advised to look to the pharmaceutical community for much-needed guidance on how best to prepare the cannabis industry for federal oversight.

The lack of federal oversight with regard to medicinal cannabis and hemp products in the United States has meant that it has been left to the individual states to regulate its use. Medical marijuana is legal in 34 states, while 12 states allow its use for adult recreational consumption. The sale of these products is strictly regulated by the tetrahydrocannabinol (THC) and cannabidiol (CBD) content, depending on their use. However, it's also critical to monitor levels of contaminants such heavy metals, as the cannabis plant is known to be a hyper-accumulator of heavy metals in the soil, the growing medium, manufacturing equipment, and other external sources. This will be compounded when CBD products derived from hemp which is grown outdoors with less control of growing conditions begin to hit the marketplace in 2020. Unfortunately, there are many inconsistencies with heavy metal limits in different states where medical cannabis is legal. As mentioned in Chapter 2, some states define four heavy metals while others specify up to nine. Some are based on limits directly in the cannabis, while others are based on consumption per day. Others take into consideration the body weight of the consumer; while some states do not even have heavy metal limits (refer to Chapter 2 for more detailed information about state regulations).

So clearly there is a need for more harmony across state lines, in order that consumers know they are using products which are safe to use. The toxicity effects of heavy metals have been well documented in the public domain, because they have such a serious impact on human health, particularly for young children and adults with compromised immune systems. For that reason, it is critically important to measure elemental contaminants (heavy metals) in cannabis and hemp and in particular to ensure that they are adequately regulated so products are safe for human (and pet) consumption. So let's take a historical look at how the pharmaceutical industry navigated these challenges.

ELEMENTAL IMPURITIES IN PHARMACEUTICALS: A HISTORICAL PERSPECTIVE

The identification and quantification of elemental impurities in pharmaceuticals has been a challenge for many years. In the late 1800s and early 20th century, lead (Pb) was the major issue since it was used in the production of solder and pipes and without controls and environmental regulations that are in place today, metals from manufacturing and processing equipment posed significant problems, together with excipients and fillers that came from minerals such as talcum, gypsum, and calcite which were known to contain many heavy metal contaminants.

The United States Pharmacopeia (USP) first addressed this issue in the early 1900s with the publication of general chapter <231>—Heavy Metals.[1] This test involves the high-temperature acid digestion of the pharmaceutical material followed by sulfide precipitation of the metals of interest. The specification is based on comparison of the color of the resulting precipitate to the color of a lead sulfide precipitate prepared at the level of interest (typically 10 parts per million). There are many limitations with this test, which generates hydrogen sulfide, through the use of thioacetamide (an organic sulfide compound), which in itself is highly toxic. The test was initially designed to detect common heavy metals including lead, mercury, arsenic, and cadmium but was not intended for metals that were found in manufacturing equipment such as nickel, chromium, vanadium, and iron and platinum group metals used for catalytic hydrogenation.

In addition, Chapter <231> is a screening test and not element specific as each precipitate has a different color, which is usually different than the color of the lead sulfide standard, as exemplified in Figure 3.1. Another limitation of the test is that it is based on a visual color comparison, so its result was often determined by the experience of the person who was assessing the color change. And the final concern was that it was not indicating the identity of the metals which are known to be of toxicological concern.

THE PROCESS FOR CHANGE

Given all these issues, the test remained as an official test in the USP until January, 2018, approximately 110 years after its introduction; the primary reason being that there was no industry-wide, globally accepted test to replace it. For all its faults, the methodology of general Chapter <231> was well understood, routinely used by the industry, and was harmonized with the European and Japanese Pharmacopoeias. The winds of change began with the publication in 1995 of Stimuli to the Revision Process article in the USP Pharmacopeial Forum.[2] In this article, it was well understood that almost half the metals of interest were lost in the ashing process, and that there was essentially zero recovery for mercury, one of the more toxic and common elements of interest. In 2000, Wang made similar observations in an article in the *Journal of Pharmaceutical and Biomedical Analysis*, where he noted "Although still widely accepted and used in the pharmaceutical industry, these methods based on the intensity of color of sulfide precipitation are non-specific, labor intensive, and more often than not, yield low recoveries or no recoveries at all."[3] Then in 2004, Lewen and coworkers directly compared the recoveries of 14 different elements using the USP <231> Method with inductively coupled plasma mass spectrometry (ICP-MS). Consistent with the other two articles, this study showed a number of recoveries around 50%, with recoveries of 5% or less for Se, Sn, Sb, and Ru, with, as anticipated, no recovery at all for Hg, as exemplified in Figure 3.2.[4]

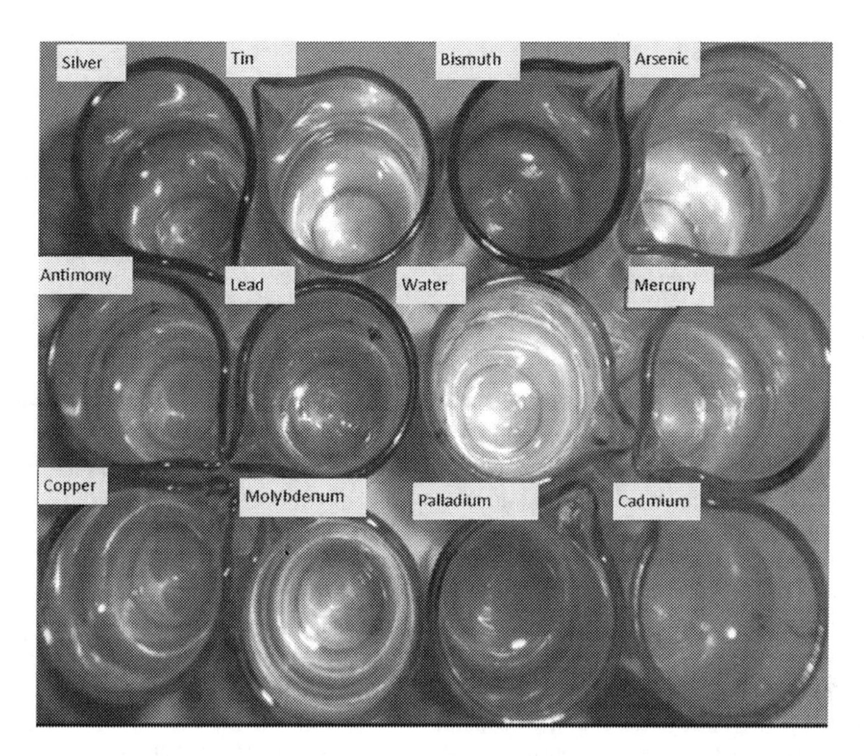

FIGURE 3.1 A group of common elements that produce insoluble sulfides, exemplifying how difficult it is for an operator to compare a sample containing many elements with slightly different colored precipitates against a lead reference standard. (Used with permission from PerkinElmer Inc., All Rights Reserved.)

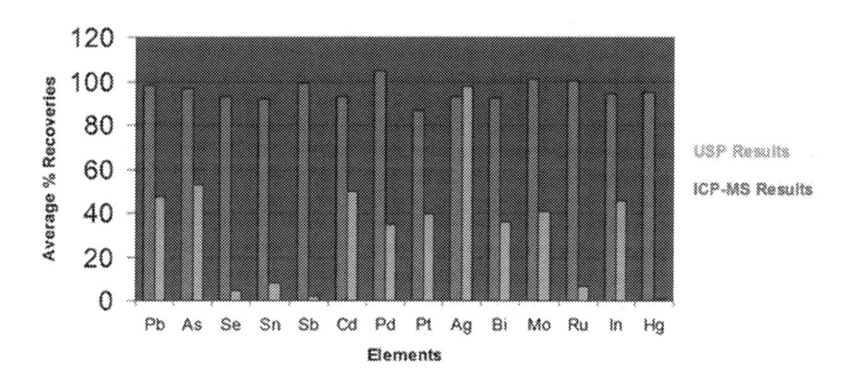

FIGURE 3.2 Comparisons between ICP-MS and the method described in USP Chapter <231>.[4]

USP IMPLEMENTATION PROCESS

This work set the stage for the establishment at USP of a series of committees, beginning in 2005 to update the concept of heavy metals testing. The process built on an Institute of Medicine meeting commissioned by USP in 2008 and on a document issued by the European Medicines Agency (EMA) which developed a guideline on the control of residual catalysts in pharmaceuticals with the goal of establishing limits based on toxicological safety assessments of common catalytic elements. This work began in 1998 and resulted in a guideline that was officially implemented in 2008.[5]

Building on this, USP published two draft chapters in 2008. These were general Chapter <232>—Elemental Impurities—Limits,[6] and Chapter <233>—Elemental Impurities—Procedures.[7] This effort led to the desire by the pharmaceutical industry to have a chapter that was harmonized across the pharmacopeias in the ICH (The International Council for Harmonization of Technical Requirements for Pharmaceuticals for Human Use) regions of the world (United States, Europe, and Japan). The result was the ICH Q3D expert working group, and the final output was the ICH Q3D guideline on elemental impurities, with the step 4 document published on the ICH website in 2014.[8] The guideline was implemented by the FDA for new drug products in July, 2017, and for all existing drug products in January, 2018, concurrent with the removal of USP general Chapter <231>. (Note: The final version of ICH Q3D (R1) was adopted by the EMA's Committee for Human Medicinal Products (CHMP) on March 28, 2019.)

ICH Q3D focused strictly on the establishment of a list of elements of toxicological concern, and their limits in oral, parenteral, and inhalational dosage forms. An extensive body of information is also available to help practitioners understand the scope of the guideline and how to think about issues such as the calculations of concentrations in the drug product and extension to other dosage form.[8] The entire process is risk based, and testing is only one of many ways by which compliance with the standard can be demonstrated. The full list of Permitted Daily Exposure (PDE) elemental impurity limits defined in USP Chapter <232> and ICH Q3D (R1) guidelines are shown in Table 3.1. Note: The USP Group responsible for dietary and herbal supplements generated its own limits for elemental contaminants which were described in USP Chapter <2232>.[9]

Note: The new methodology no longer referred to heavy metals but as elemental impurities or contaminants. The reason being that heavy metal is a loose term that typically indicates "environmentally harmful" but is scientifically meaningless as defined by IUPAC (International Union of Pure and Applied Chemistry), the world authority on chemical nomenclature,[10] which stated … *You may know what you mean when you talk about a heavy metal but someone else may mean something entirely different.* However, in order to use terminology understood by the cannabis industry, we will continue to use the term heavy metals which refers to elemental impurities or elemental contaminants, for the purpose of clarity.

It should also be emphasized that PDE limits are defined as the maximum acceptable intake of elemental impurity in pharmaceutical products per day. So if the suggested daily dosage of a drug is 10 g, these PDE levels must be divided by 10 to determine the allowable elemental concentration in the drug compound per day. It's also important to emphasize that classification of each elemental impurity shown in Table 3.1 is based on the toxicological impact of each element, with Class 1 and 2A elements considered the most toxic, followed by Class 2B and finally Class 3 being the least toxic. However, it is very important to mention that Chapter <232> clearly states that it's critical to monitor Class 3 element impurities in drugs that are delivered via inhalation, because the impact of these heavy metals on the lungs and repertory system is more serious.

It's also important to point out that neither USP nor the ICH specifically addresses PDE levels for transdermal (via the skin) applications in their guidelines, mainly because elemental impurities are omnipresent throughout life and the skin, as the outer barrier of the body is in direct contact with these metals on a regular basis particularly from metal-containing particles that might be in the surrounding environment. For that reason, the dermal uptake of elemental impurities in the human body via the skin from topically applied drug product is expected to be low or even negligible due to the excellent barrier properties of the skin itself.

Ultimately, all available evidence and data to date support the view that dermal exposure to most elemental impurities is unlikely to represent a substantive toxicological concern. Currently, the establishment of limits for elemental impurities in cutaneous and transdermal drug products is under examination and is expected to be announced in mid-2020. In the meantime, oral PDEs given in ICH guidelines may be considered a suitable point of reference as the intestine may be regarded as a comparable absorption mechanism to the skin.

TABLE 3.1
Permitted Daily Exposure (PDE) Limits as Defined in USP Chapter <232> and ICH Q3D Step 4 Guidelines[6,8]

Element	Class	Oral PDE (μg/day)	Parenteral PDE (μg/day)	Inhalational PDE (μg/day)
Cd	1	5	2	3[a]
Pb	1	5	5	5
As	1	15	15	2
Hg	1	30	3	1
Co	2A	50	5	3
V	2A	100	10	1
Ni	2A	200	20	5
Tl	2B	8	8	8
Au	2B	100	100	1
Pd	2B	100	10	1
Ir	2B	100	10	1
Os	2B	100	10	1
Rh	2B	100	10	1
Ru	2B	100	10	1
Se	2B	150	80	130
Ag	2B	150	10	7
Pt	2B	100	10	1
Li	3	550	250	25
Sb	3	1,200	90	20
Ba	3	1,400	700	300
Mo	3	3,000	1,500	10
Cu	3	3,000	300	30
Sn	3	6,000	600	60
Cr	3	11,000	1,100	3

[a] The inhalation PDE for Cd was initially set at 2 μg/day in January, 2018, but was reassessed in March, 2019 to 3 μg/day, because of an error in the original calculation.[11]

Testing procedures of drug products and excipients were not discussed in detail by the ICH Q3D working group. Rather, it was left to the individual pharmacopeias to work out their own procedures, with the goal of long-term harmonization of the final results. However, the USP wrote general Chapter <233>, which describes two procedures, ICP-OES and ICP-MS, along with system suitability and validation requirements for these tests. Let's take a closer look at the validation protocols, which are described in Chapter <233>.

VALIDATION PROCEDURES

Validation procedures are an integral part of ensuring high-quality data using the chosen analytical method when carrying out the measurement of elemental impurities in drug compounds. Because certified reference materials are typically not available for common pharmaceutical materials, a spike recovery testing protocol must be adopted where all samples are spiked with known analyte concentrations at the level of the PDE, based on the sample weight and sample dilution before the sample digestion step is carried out. This concentration value is commonly known as the J-value, which is defined as the concentration (w/w) of the element(s) of interest at the PDE target limit, appropriately diluted to the working range of the instrument. So let's take Pb as an example. The PDE limit for Pb in an oral medication defined in Chapter <232> is 5 μg/day.

Based on a suggested dosage of 10 g of the drug product/day, that's equivalent to 0.5 µg/g Pb. If 1.0 g of sample is digested/dissolved and made up to 500 mL, that's a 500-fold dilution, which is equivalent to 1.0 µg/L. So the J-value for Pb in this example is equal to1.0 µg/L. The method then suggests using a calibration made up of two standards: Standard 1= 1.5 J and Standard 2 = 0.5 J. So for Pb, that's equivalent to 1.5 µg/L for Standard 1 and 0.5 µg/L for Standard 2.

The suitability of the analytical methodology is first determined by measuring the calibration drift and comparing results for Standard 1 before and after the analysis of all the sample solutions under test. This calibration drift should be <20% for each target element. However, once the suitability of the technique has been determined, further validation protocols described in detail in <Chapter 233> must be carried out to show compliance to the regulatory agency if required.

Meeting the validation protocol is critical when selecting the best technique for this application, because all aspects of the analytical procedures, including the instrumental technique and sample dissolution process must be validated and shown to be acceptable, in accordance with the validation protocol described in Chapter <233>. The requirements for the validation of a procedure for elemental impurities are then determined by following a set of quality protocols, which cover a variety of performance tests, including:

- Detectability
- Precision
- Specificity
- Accuracy
- Ruggedness
- Limit of quantification
- Linear range.

Meeting these performance requirements defined in these tests must be demonstrated experimentally using an appropriate system suitability procedure and reference materials. The suitability of the method must be determined by conducting studies with the material under test supplemented/spiked with known concentrations of each target element of interest at the appropriate acceptance limit concentration. It should also be emphasized that the materials under test must be spiked before any sample preparation steps are performed. This spike recovery procedure is a very important part of validating the method and the generated data has to be shown to the regulatory agency as proof that the analysis has been carried out correctly.

Even though the entire process to update the analytical procedures has taken over 20 years, the implementation of the ICH Q3D guideline and the two USP general chapters and the elimination of USP <231> are a major step forward in the analysis of elemental impurities in pharmaceuticals. As a result, there is now broad agreement across the pharmaceutical industry and across many countries on the elements of toxicological concern and the levels at which they should be controlled. Today, the PDE limits defined in Table 3.1 are the basis by which the FDA regulates all drugs manufactured in the United States.

TOXICITY CLASSIFICATION

It's also important to emphasize that all the elemental impurities defined in USP Chapter <232> and ICH Q3D (R1) guidelines are separated into three main categories based on their toxicity and likelihood of occurrence in the drug product.

- **Class 1:** The elements arsenic (As), cadmium (Cd), mercury (Hg), and lead (Pb) are human toxicants that have limited or no use in the manufacture of pharmaceuticals. They should be evaluated in all risk assessments.

- **Class 2:** Route-dependent human toxicants. Further divided in sub-classes 2A and 2B based on their relative likelihood of occurrence in the drug product.
  - **Class 2A:** Elements have relatively high probability of occurrence in the drug product and should be evaluated in all risk assessments
  - **Class 2B:** Elements have a reduced probability of occurrence in the drug product related to their low abundance and low potential to be co-isolated with other materials. As a result, they can be excluded from the risk assessment unless they are intentionally added during the manufacture of drug substances, excipients, or other components of the drug product
- **Class 3:** Relatively low toxicities by the oral route of administration but should definitely be considered in the risk assessment for inhalation and parenteral routes.

HOW ARE PDEs CALCULATED?

So how are these PDEs calculated because state medical cannabis commissioners and regulators across the United States have taken selected data from these documents and applied them to cannabis and hemp, without fully explaining how they have been derived?

PDE limits are based on decades of historical toxicological data together with factors based on well-established animal studies and human exposure to carcinogenic and hazardous substances.[8] They are calculated according to the procedures for setting exposure limits in pharmaceuticals and the method adopted by International Program for Chemical Safety (IPCS) for Assessing Human Health Risk of Chemicals (AHHRC). These methods are similar to those used by the United States Environmental Protection Agency (US EPA) Integrated Risk Information System, the United States Food and Drug Administration (US FDA) and others. A brief overview of the methodology used is outlined below to give a better understanding of the origin of the PDE values. First of all, it's important to understand the term **Minimal Risk Level (MRL)**, which is an estimate of the daily human exposure to a hazardous substance that is likely to be without appreciable risk as defined by the Agency for Toxic Substances and Disease Registry (ATSDR).

When an MRL is used to set the PDE, no additional modifying factors need to be used, as they are incorporated into the derivation of the MRL. For carcinogenic elements unit, risk factors are used to set the PDE using a 1:100,000 risk level; these are described in the individual monographs for the specific elemental impurity in ICH Q3D (R1) guidelines. Some PDEs for inhalation were derived using occupational exposure limits, applying modifying factors, and considering any specific effects to the respiratory system.

The PDE is then derived from the No-Observed Adverse Effect Level (NOAEL) or the Lowest-Observed Adverse Effect Level (LOAEL) in the most relevant animal studies (including rat, mouse, monkey, dog, rabbit, guinea pig, and human) as follows[8]:

$$\text{PDE} = \frac{\text{NOAEL (or LOAEL)} \times \text{mass adjustment}}{[\text{F1} \times \text{F2} \times \text{F3} \times \text{F4} \times \text{F5}](\text{modifying factors})}$$

where

NOAEL (No-Observed Adverse Effect Level) is defined by IUPAC as the greatest concentration or amount of a substance, found by experiment or observation, which causes no detectable adverse alteration of morphology, functional capacity, growth, development, or life span of the target organism under defined conditions of exposure.

LOAEL (Lowest-Observed Adverse Effect Level) is defined by IUPAC as the lowest concentration or amount of a substance (dose), found by experiment or observation, that causes an adverse effect on morphology, functional capacity, growth, development, or life span of a target organism distinguishable from normal (control) organisms of the same species and strain under defined conditions of exposure.

Mass adjustment is based on an arbitrary adult human body mass for either sex of 50 kg. This relatively low mass provides an additional safety factor against the standard masses of 60 kg or 70 kg that are often used in this type of calculation. It is recognized that some patients weigh less than 50 kg; these patients are considered to be accommodated by the built-in safety factors used to determine a PDE and that lifetime studies were often used.

Modifying factor is an individual factor determined by professional judgment of a toxicologist and applied to bioassay data to relate that data to human safety.

NOAEL data is preferably used, but if no NOAEL is available, the LOAEL may be used. Modifying factors proposed here, for relating the data to humans, are the same kind of "uncertainty factors" used in Environmental Health Criteria, and "modifying factors" or "safety factors" as defined in the Pharmacopeial Forum (Dec, 1989).[12]

The modifying factors used (on a scale of 1–10) are as follows:

- F1 = Factor to account for study in humans
- F2 = Factor to account for differences between individual humans
- F3 = Factor to account for duration of the study
- F4 = Factor to account for severity of toxicity
- F5 = Factor to account for whether NOAEL or LOAEL is used

So let's take a look at cadmium as an example. A summary of the most PDE limits for cadmium is shown in Table 3.2

For clarity purposes, we'll just examine the oral and inhalation PDEs because cannabis and hemp are not typically delivered to the patient or consumer intravenously (parenteral).

CADMIUM PDE—ORAL EXPOSURE

A meaningful assessment for oral exposure to cadmium and cadmium salts is renal toxicity.[13] Skeletal and renal effects are observed at similar exposure levels and are a sensitive marker of cadmium exposure (ATSDR).[14] A number of oral exposure studies of cadmium in rats and mice showed no evidence of carcinogenicity. Therefore, the renal toxicity assessment was used to establish the oral PDE for cadmium, following the recommendations of ATSDR; an MRL of 0.1 μg/kg for chronic exposure is used to set the oral PDE. This is consistent with the WHO drinking water limit of 0.003 mg/L/day (WHO, 2011).[15]

$$\text{PDE} = 0.1\ \mu\text{g/kg/day} \times 50\ \text{kg} = 5.0\ \mu\text{g/day}$$

No modifying factors were applied because they are incorporated into the derivation of the MRL.

TABLE 3.2
PDE Limits for Cadmium

Element	Class	Oral PDE (μg/day)	Parenteral PDE (μg/day)	Inhalational PDE (μg/day)
Cd	1	5.0	1.7	3.4

CADMIUM PDE—INHALATION EXPOSURE

The United States Department of Labor Occupational Safety and Health Administration (OSHA) has developed an inhalation permitted exposure level (PEL) of 5 $\mu g/m^3$ for cadmium. Taking into account the modifying factors, the inhalation PDE is calculated as follows:

$$\text{For continuous dosing} = \frac{5\ \mu g/m^3 \times 8\ \text{hour/day} \times 5\ \text{day/week}}{24\ \text{hour/day} \times 7\ \text{day/week}} = \frac{1.19\ \mu g/m^3}{1{,}000\ L/m^3} = 0.00119\ \mu g/L$$

$$\text{Daily dose} = \frac{0.00119\ \mu g/L \times 28{,}800\ L}{50\ kg} = 0.685\ \mu g/kg$$

$$\text{PDE} = 0.685\ \mu g/kg \times 50\ kg/1 \times 10 \times 1 \times 1 \times 1 = 3.43\ \mu g/day$$

A modifying factor for F4 of 1 was chosen based on the potential for toxicity to be mitigated by the possible species specificity of tumorigenesis (production or formation of a tumor), including uncertainty of human occupational tumorigenesis, ambient exposure levels not expected to be a health hazard, and workplace exposure levels expected to be safe. A larger factor F4 was not considered necessary as the PDE is based on the PEL.

POTENTIAL SOURCES OF ELEMENTAL IMPURITIES IN DRUG COMPOUNDS

ICH Q3D (R1) recommends that manufacturers conduct a product risk assessment by first identifying known and potential sources of elemental impurities. Manufacturers should consider all potential sources of elemental impurities, such as elements intentionally added, elements potentially present in the materials used to prepare the drug product, and elements potentially introduced from manufacturing equipment or container closure systems. Manufacturers should then evaluate each elemental impurity likely to be present in the drug product by determining the observed or predicted level of the impurity and comparing it with the established PDE. If the risk assessment fails to show that an elemental impurity level is consistently less than the PDE*, additional controls should be established to ensure that the elemental impurity level does not exceed the PDE in the drug product. These additional controls could be included as in-process controls or in the specifications of the drug product or components.

*Note: The control threshold for risk assessment is generally accepted as 30% of the PDE which is explained in the Q3D Elemental Impurities Guidance for Industry published by the FDA as shown below:[16]

> If the total elemental impurity level from all sources in the drug product is expected to be consistently less than 30 percent of the PDE, then additional controls are not required, provided that the applicant has appropriately assessed the data and demonstrated adequate controls on elemental impurities.

In other words, ICH Q3D (R1) provides a clear structure for companies to follow in designing their risk assessment process, which is summarized as a fishbone diagram shown in Figure 3.3.

With these five likely routes for the introduction of elemental impurities it is clear that if the inputs are known to be "clean" with respect to the relevant elemental impurity limits, then the drug product will be acceptable without testing for elemental impurities. The ICH has provided guidance on this as follows:

> The applicant's risk assessment can be facilitated with information about the potential elemental impurities provided by suppliers of drug substances, excipients, container closure systems, and manufacturing equipment. The data that support this risk assessment can come from a number of sources that include, but are not limited to:

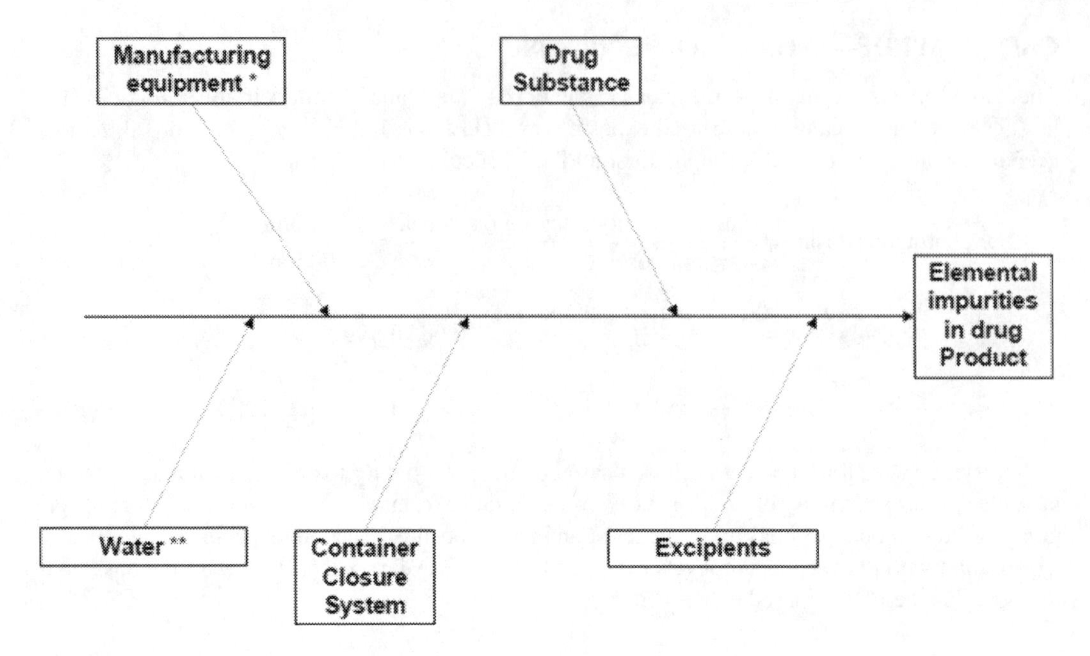

FIGURE 3.3 Risk assessment fishbone diagram.

- Prior knowledge
- Published literature
- Data generated from similar processes
- Supplier information or data including certificates of analysis (CoA)
- Testing of the components of the drug product
- Testing of the drug product

The bottom line is that testing is not the mandated requirement but data is absolutely critical. This means that as a pharma company becomes more familiar with the elemental impurity risk assessment process, they will be able to leverage information gained in the development of previous products enabling them to streamline the process. Similarly, as information is published in the scientific literature, the reliance on analytical testing will be reduced.

If we consider the five sources of elemental impurities, it is clear that some are more likely to be problematic than others. It is therefore critical component of understanding the source of elemental impurities in pharmaceutical materials, because it has driven manufacturers to have detailed knowledge of all the potential sources of contamination including raw materials, active pharmaceutical ingredients, and every step of the manufacturing process and packaging materials. The cannabis industry should pay careful attention to this approach to risk assessment, because they currently do not have a solid understanding of how and what heavy metal contaminants end up in their final products. So clearly at some point in time, PDE limits have to eventually be determined and optimized for cannabis and cannabis products to adopt any meaningful regulations.

This is confirmed by the fact that Epidiolex (manufactured by G.W. Pharmaceuticals), the only legal CBD-based prescription drug, is regulated by the FDA in the United States and has to show compliance according to USP Chapter <232> or ICH Q3D (R1) directives. This means that the manufacturer must either show that the 24 elemental impurities are below maximum PDE levels for oral drugs or show compelling evidence that they have adopted a satisfactory risk assessment strategy to ensure compliance. Federal regulations, of course take precedence over the four heavy metals (Pb, Cd, As, Hg), required by most state regulators. This is a strong indicator that when cannabinoid-based medications eventually come under the oversight of the federal government, they will be regulated according to USP Chapter <232> or ICH Q3D (R1) guidelines. Refer to the formal response from the Division of Drug Information, Center for Drug Evaluation and Research at the FDA below for clarity about how Epidiolex is being regulated for elemental impurities.

> This information is regarding control of elemental impurities in Epidiolex (cannabidiol) oral solution and compliance with USP Chapter <232> Elemental Impurities. During the review of a new drug application, FDA evaluates the adequacy of the applicant's proposed control strategy for each product to ensure consistency with the International Conference on Harmonization of Technical Requirements for Registration of Pharmaceuticals for Human Use (ICH) guidance Q3D Elemental Impurities guidance[16] and compliance with the requirements of USP <232> prior to approval of the application. Even though this manufacturer complied with these regulations, we are unable to provide information regarding control strategies for elemental impurities in specific products, as the information is considered a trade secret by the manufacturer.

FINAL THOUGHTS

The cannabis industry should take a close look at industries that have gone through the process of adopting meaningful methods and regulations for elemental contaminant, including food and beverages, dietary/herbal supplements, cosmetics, and drug products. However, there is no question that the pharmaceutical industry represents the most logical direction to follow, because they went through the same challenges back in the late 1990s to update its 100-year-old sulfide precipitation test for an unknown suite of heavy metals to the current list of 24 elemental impurities today. Even though the source of elemental contaminants is very different with cannabis products, the pharmaceutical industry has developed a well-established investigative path to follow to fully understand how they end up in the final products. They first determine the likelihood of the elemental impurity being present in the raw materials, active ingredient, or manufacturing/packaging process. If the chances are high, they then determine the toxicity impact of that element and categorize it accordingly. By using well-established risk assessment procedures, they then decide which elements to focus their investigations on. In addition, there are recommended daily doses for drug products, which determine their efficacy based on whether they are taken orally, via inhalation (vaping), or via the skin (transdermal). The cannabis industry would be well advised to follow this direction as currently there is very little understanding of where the elemental contaminants are coming from and what heavy metals they should be testing for. This problem is also compounded by the fact that we don't know how often a person consumes cannabis products and in what quantities. There is clearly a great deal more we need to understand before we can adopt sensible and meaningful regulations.

This chapter has shown that the pharmaceutical community can offer much guidance and direction to understanding what additional elemental contaminants might be of concern to the cannabis industry. By changing its 100-year-old semiquantitative test for a small group of heavy metals, the drug industry finally arrived at a list of 24 elemental impurities, through a complete understanding of the entire production cycle from raw materials to the manufacturing process and packaging of the final drug products as described in my previous book.[17] In addition, the pharmaceutical industry has also shown that testing can be kept to a minimum, by adopting a risk-based approach by having a complete understanding of all the potential sources of contamination from the cultivation, extraction, and manufacturing processes.

FURTHER READING

1. USP. United States Pharmacopeia General Chapter <231> Heavy Metals Test in USP National Formulary (NF), https://www.usp.org/chemical-medicines/elemental-impurities-updates.
2. K. Blake. Harmonization of the USP, EP, and JP heavy metals testing procedures. *Pharmacopeial Forum* 21(6), 1995, 1632–1637.
3. T. Wang. *Journal of Pharmaceutical and Biomedical Analysis* 23, 2000, 867–890.
4. N. Lewen, et al. *Journal of Pharmaceutical and Biomedical Analysis* 35, 2004, 739–752.

5. European Medicines Agency. Committee for Medicinal Products for Human Use, 2008, Guideline on the specification limits for residues of metal catalysts or metal reagents. Doc. Ref. EMEA/CHMP/SWP/4446/20007.
6. USP. United States Pharmacopeia General Chapter <232> Elemental Impurities – Limits: First Supplement to USP 40-NF 35, 2017, https://www.usp.org/chemical-medicines/elemental-impurities-updates.
7. USP. United States Pharmacopeia General Chapter <233> Elemental Impurities – Procedures: Second Supplement to USP 38-NF 33, 2015, https://www.usp.org/chemical-medicines/elemental-impurities-updates.
8. ICH. International Conference on Harmonization Website Q3D. http://www.ich.org/products/guidelines/quality/article/quality-guidelines.html (Q3D).
9. USP. United States Pharmacopeia General Chapter <2232>: Elemental Contaminants in Dietary Supplements, PF 38(3), May–June 2012, https://www.uspnf.com/notices/general-chapter-elemental-contaminants-dietary-supplements.
10. J.H. Duffus. Heavy metals—A meaningless term? *Pure Applied Chemistry* 74, 2002, 793–807. Available on the IUPAC website at: http://www.iupac.org/publications/pac/2002/7405/7405x0793.html.
11. Q3D(R1). Elemental Impurities: Revision for Cadmium, https://www.fda.gov/media/114618/download.
12. Pharmacopeial Forum, USP Pharmacopeial Forum. https://www.uspnf.com/pharmacopeial-forum.
13. J.P. Buchet, et al. Renal effects of cadmium body burden of the general population, *Lancet* 337(8756), 1991, 1554.
14. Agency for Toxic Substances and Disease Registry (ATSDR). ATSDR is a Govt agency. https://www.atsdr.cdc.gov/.
15. World Health Organization. WHO – Self Explanatory. https://www.who.int/.
16. Q3D Elemental Impurities Guidance for Industry, U. S. Department of Health and Human Services, FDA, Drug Evaluation and Research (CDER), Center for Biologics Evaluation and Research (CBER), FDA Guidance for Industry. March 2020, http://www.fda.gov/regulatory-information/search-fda-guidance-documents/q3dr1-elemental-impuritiesJournal.
17. R.J. Thomas. *Measuring Elemental Impurities in Pharmaceuticals: A Practical Guide*, CRC Press, Boca Raton, FL, 2018, ISBN: 9781138197961.

4 An Overview of ICP Mass Spectrometry

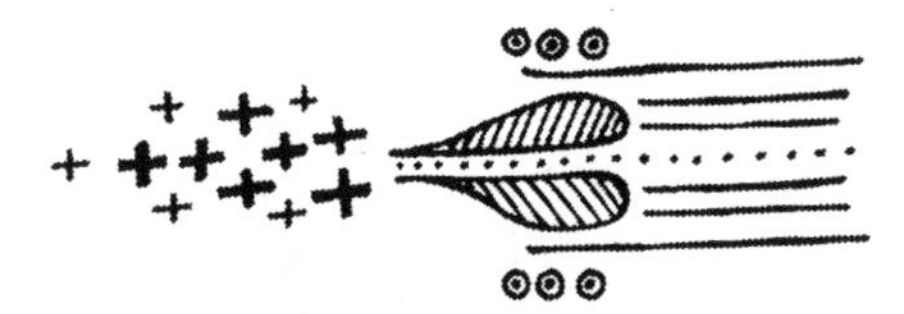

Inductively coupled plasma mass spectrometry (ICP-MS) not only offers extremely low detection limits in the sub-parts per trillion (ppt) range but also enables quantitation at the high parts per million (ppm) level. This unique capability makes the technique very attractive compared to other trace metal techniques such as graphite furnace atomic absorption (AA), which is limited to determinations at the trace level, or flame atomic absorption spectrometry (AAS) and inductively coupled plasma optical emission spectrometry (ICP-OES), which are traditionally used for the detection of higher concentrations. In Chapter 4, we will present an overview of ICP-MS and explain how its characteristic low detection limits are achieved.

ICP-MS is undoubtedly the fastest growing trace element technique available today. Since its commercialization in 1983, over 25,000 systems have been installed worldwide for many varied and diverse applications. The most common ones, which represent approximately 90% of the ICP-MS analysis being carried out today, include environmental, geological, semiconductor, biomedical, nuclear, academia, and pharmaceutical market segments. However, over the past few years, with new heavy metal regulations being implemented, the analysis of cannabis and cannabis products is rapidly catching them up. There is no question that the major reason for its unparalleled growth is its ability to carry out rapid multielement determinations at the ultratrace level. Even though it can broadly determine the same suite of elements as other atomic spectroscopic techniques, such as flame atomic absorption (FAA), electrothermal atomization (ETA), atomic fluorescence (AF), and ICP-OES, ICP-MS has clear advantages in its multielement characteristics, speed of analysis, detection limits, and isotopic capability. Figure 4.1 shows approximate detection limits of all the elements that can be detected by ICP-MS, together with their isotopic abundance.

PRINCIPLES OF OPERATION

There are a number of different ICP-MS designs available today that share many similar components, such as nebulizer, spray chamber, plasma torch, interface cones, vacuum chamber, ion optics, mass analyzer, and detector. However, the engineering design and implementation of these components can vary significantly from one instrument to another. Instrument hardware is described in greater detail in the subsequent chapters on the basic principles of the technique. So let us begin here by giving an overview of the principles of operation of ICP-MS. Figure 4.2 shows the basic components that make up an ICP-MS system. The sample, which usually must be in a liquid form, is pumped at 1 mL/min, usually with a peristaltic pump into a nebulizer, where it is converted into a fine aerosol with argon gas at about 1 L/min. The fine droplets of the aerosol, which represent only 1%–2% of the sample, are separated from larger droplets by means of a spray chamber. The fine aerosol then emerges from the exit tube of the spray chamber and is transported into the plasma torch via a sample injector.

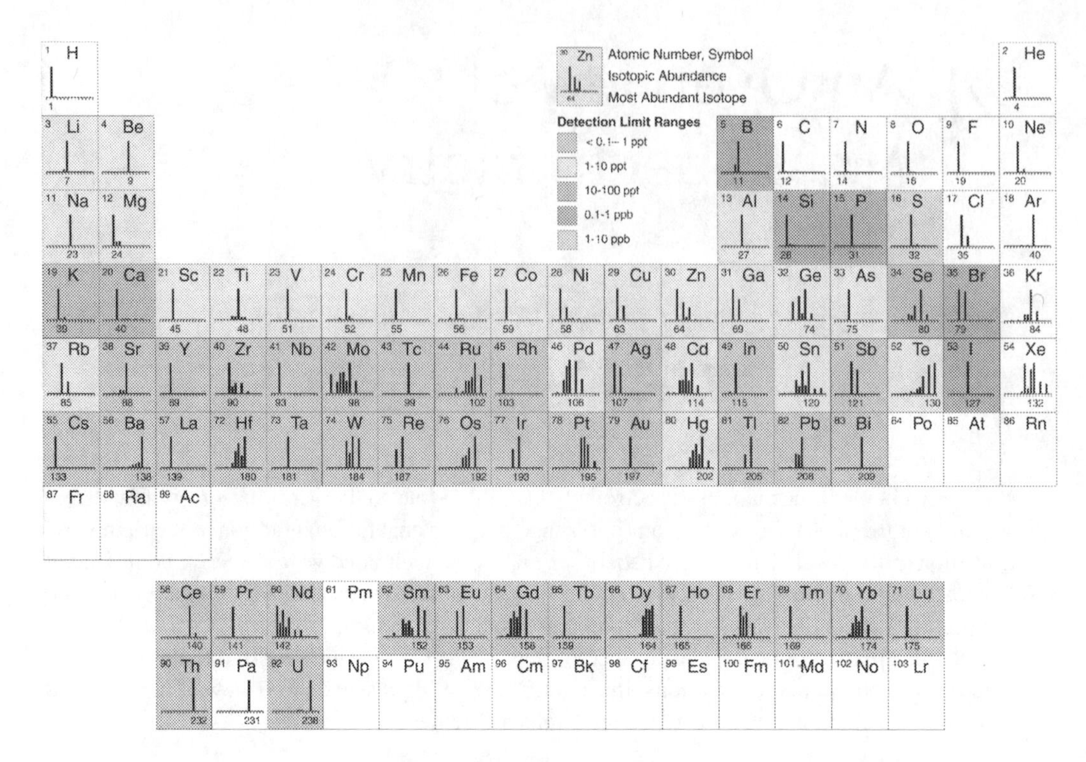

FIGURE 4.1 Approximate detection capability of ICP-MS, together with elemental isotropic abundances. (Reproduced with permission from Perkin Elmer Inc., All Rights Reserved.)

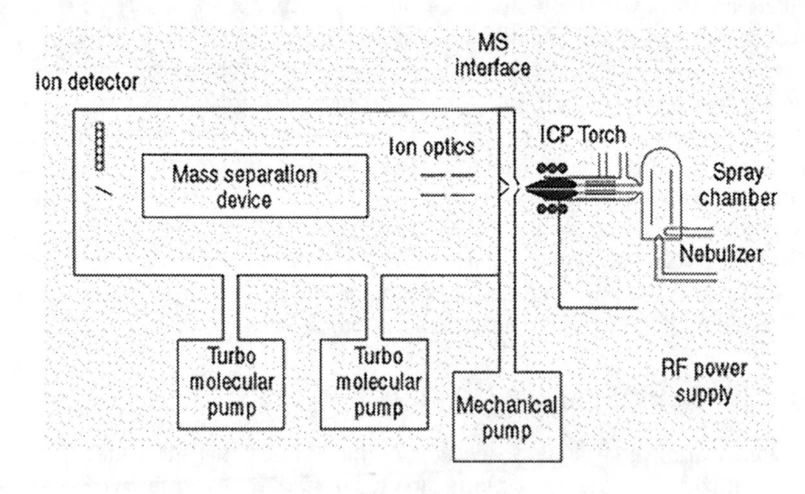

FIGURE 4.2 Basic instrumental components of an ICP mass spectrometer.

It is important to differentiate between the roles of the plasma torch in ICP-MS compared to ICP-OES. The plasma is formed in exactly the same way, by the interaction of an intense magnetic field (produced by radio frequency (RF) passing through a copper coil) on a tangential flow of gas (normally argon), at about 12–15 L/min flowing through a concentric quartz tube (torch). This has the effect of ionizing the gas, which when seeded with a source of electrons from a high-voltage spark, forms a very-high-temperature plasma discharge (~10,000 K) at the open end of the tube. However, this is where the similarity ends. In ICP-OES, the plasma, which is normally vertical (but can be horizontal with dual-view designs), is used to generate photons of light by the excitation of electrons of a ground-state atom to a higher energy level. When the electrons "fall" back to ground state, wavelength-specific photons are emitted that are characteristic of the element of interest. In ICP-MS, the plasma torch, which is positioned horizontally, is used to generate positively charged ions and not photons. In fact, every attempt is made to stop the photons from reaching the detector because they have the potential to increase signal noise. It is the production and detection of large quantities of these ions that gives ICP-MS its characteristic low-ppt detection capability—about three to four orders of magnitude lower than ICP-OES.

Once the ions are produced in the plasma, they are directed into the mass spectrometer via the interface region, which is maintained at a vacuum of 1–2 torr with a mechanical roughing pump. This interface region consists of two or three metallic cones (depending on the design), called the sampler and a skimmer cone, each with a small orifice (0.6–1.2 mm) to allow the ions to pass through to the ion optics, where they are guided into the mass separation device.

The interface region is one of the most critical areas of an ICP mass spectrometer, because the ions must be transported efficiently and with electrical integrity from the plasma, which is at atmospheric pressure (760 torr), to the mass spectrometer analyzer region, which is at approximately 10^{-6} torr. Unfortunately, there is the likelihood of capacitive coupling between the RF coil and the plasma, producing a potential difference of a few hundred volts. If this is not eliminated, an electrical discharge (called a secondary discharge or pinch effect) between the plasma and the sampler cone would occur. This discharge would increase the formation of interfering species and also dramatically affect the kinetic energy of the ions entering the mass spectrometer, making optimization of the ion optics very erratic and unpredictable. For this reason, it is absolutely critical that the secondary charge be eliminated by grounding the RF coil. There have been a number of different approaches used over the years to achieve this, including a grounding strap between the coil and the interface, balancing the oscillator inside the RF generator circuitry, a grounded shield or plate between the coil and the plasma torch, or the use of a double interlaced coil where RF fields go in opposing directions. They all work differently but basically achieve a similar result, which is to reduce or eliminate the secondary discharge.

Once the ions have been successfully extracted from the interface region, they are directed into the main vacuum chamber by a series of electrostatic lens, called ion optics. The operating vacuum in this region is maintained at about 10^{-3} torr with a turbomolecular pump. There are many different designs of the ion optic region, but they serve the same function, which is to electrostatically focus the ion beam towards the mass separation device, while stopping photons, particulates, and neutral species from reaching the detector.

The ion beam containing all the analytes and matrix ions exits the ion optics and now passes into the heart of the mass spectrometer—the mass separation device, which is kept at an operating vacuum of approximately 10^{-6} torr with a second turbomolecular pump. There are many different mass separation devices, all with their strengths and weaknesses. Three of the most common types are discussed in this book—quadrupole, magnetic sector, and time-of-flight technology—but they basically serve the same purpose, which is to allow analyte ions of a particular mass-to-charge ratio through the detector and to filter out (reject) all the non-analyte, interfering, and matrix ions. Depending on the design of the mass spectrometer, this is either a scanning process where the ions arrive at the detector in a sequential manner or a simultaneous process where the ions are either sampled or detected at the same time. Most quadrupole instruments nowadays are also sold with collision/reaction cells or interfaces. This technology offers a novel way of minimizing polyatomic spectral interferences by bleeding a gas into the cell or interface and using ion–molecule collision and reaction mechanisms to reduce the impact of the ionic interference.

The final process is to convert the ions into an electrical signal with an ion detector. The most common design used today is called a discrete dynode detector, which contains a series of metal dynodes along the length of the detector. In this design, when the ions emerge from the mass filter, they impinge on the first dynode and are converted into electrons. As the electrons are attracted to the next dynode, electron multiplication takes place, which results in a very high stream of electrons emerging from the final dynode. This electronic signal is then processed by the data-handling system in the conventional way and converted into analyte concentration using ICP-MS calibration standards. Most detection systems can handle up to ten orders of dynamic range, which means they can be used to analyze samples from low/sub-ppt levels up to a few hundred ppm, depending on the analyte mass.

It is important to emphasize that because of the enormous interest in the technique, most ICP-MS instrument companies have very active R&D programs in place, in order to get an edge in a very competitive marketplace. This is obviously very good for the consumer, because not only does it drive down instrument prices but also the performance, applicability, usability, and flexibility of the technique are being improved at a dramatic rate. Although this is extremely beneficial for the ICP-MS user community, it can pose a problem for a writer of textbooks who is attempting to present a snapshot of instrument hardware and software components at a particular moment in time. Hopefully, I have struck the right balance in not only presenting the fundamental principles of ICP-MS but also making the reader aware of what the technique is capable of achieving and where new developments might be taking it, particularly with regard to analyzing cannabis and cannabis-related products.

5 Principles of Ion Formation

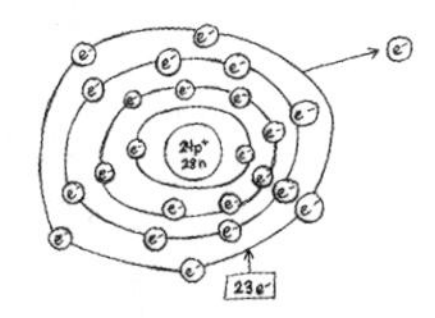

Chapter 5 gives a brief overview of the fundamental principles of ion formation in inductively coupled plasma mass spectrometry (ICP-MS)—the use of a high-temperature argon plasma to generate positively charged ions. The highly energized argon ions that make up the plasma discharge are used to first produce analyte ground-state atoms from the dried sample aerosol and then to interact with the atoms to remove one or more electrons and generate positively charged ions, which are then steered into the mass spectrometer for detection and measurement.

In ICP-MS, the sample, which is usually in liquid form, is delivered into the sample introduction system, comprising of a spray chamber and nebulizer. It emerges as an aerosol, where it eventually finds its way, via a sample injector, into the base of the plasma. As it travels through the different heating zones of the plasma torch, it is dried, vaporized, atomized, and ionized. During this time, the sample is transformed from a liquid aerosol to solid particles and then into a gas. When it finally arrives at the analytical zone of the plasma, at approximately 6,000–7,000 K, it exists as ground-state atoms and ions, representing the elemental composition of the sample. The excitation of the outer electrons of a ground-state atom to produce wavelength-specific photons of light is the fundamental basis of atomic emission. However, there is also enough energy in the plasma to remove one or more electrons from its orbitals to generate a free ion. The energy available in an argon plasma is ~15.8 eV, which is high enough to ionize most of the elements in the periodic table (the majority have first ionization potentials on the order of 4–12 eV). It is the generation, transportation, and detection of significant numbers of positively charged ions that gives ICP-MS its characteristic ultratrace detection capabilities. It is also important to mention that although ICP-MS is predominantly used for the detection of positive ions, negative ions are also produced in the plasma. However, because the extraction and transportation of negative ions is different from that of positive ions, most commercial instruments are not designed to measure them. The process of the generation of positively charged ions in the plasma is conceptually shown in greater detail in Figure 5.1.

ION FORMATION

The actual process of conversion of a neutral ground-state atom to a positively charged ion is shown in Figures 5.2 and 5.3. Figure 5.2 shows a very simplistic view of the chromium atom Cr^0, at mass 52 consisting of a nucleus with 24 protons (p^+) and 28 neutrons (n), surrounded by 24 orbiting electrons (e^-). It must be emphasized that this is not meant to be an accurate representation of the electron's shells and subshells, but just a conceptual explanation for the purpose of clarity. From this, we can conclude that the atomic number of chromium is 24 (number of protons) and its atomic mass is 52 (number of protons + neutrons).

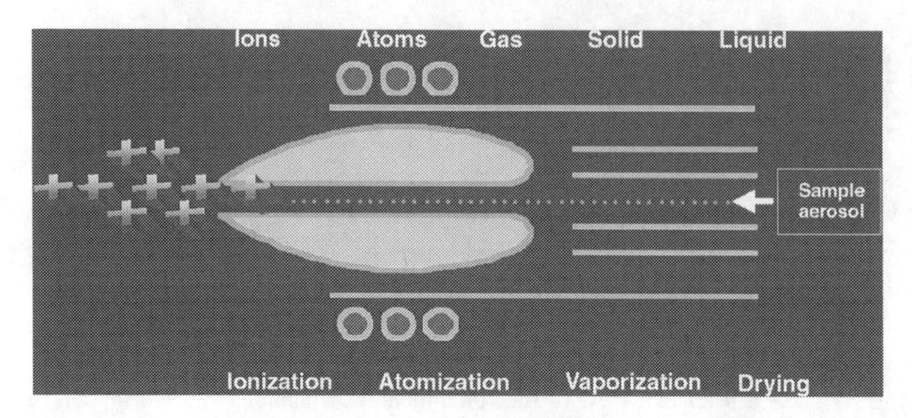

FIGURE 5.1 Generation of positively charged ions in the plasma.

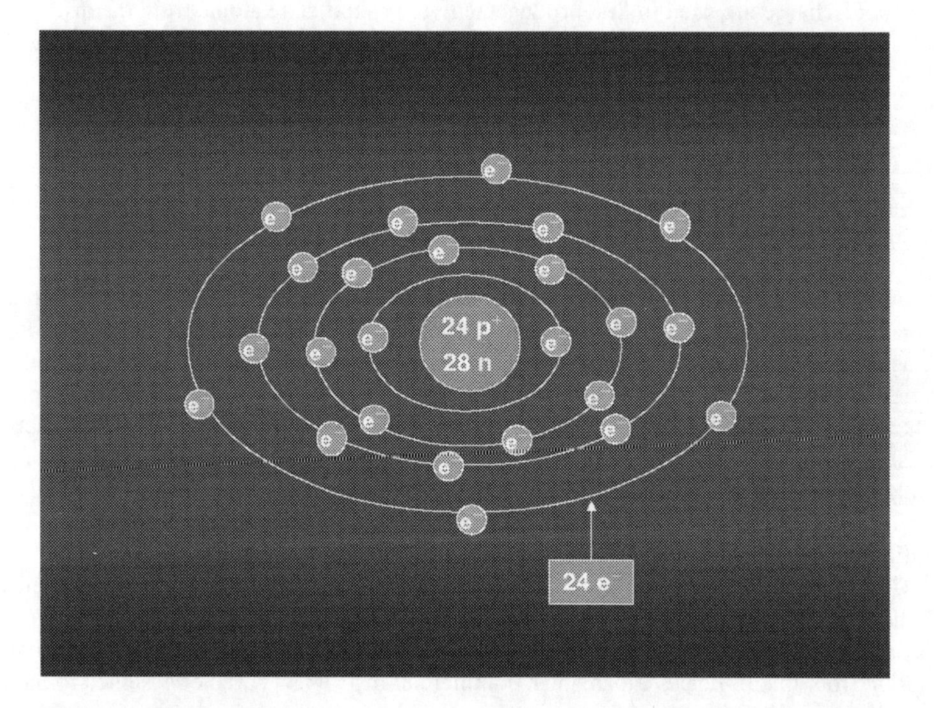

FIGURE 5.2 Simplified schematic of a chromium ground-state atom (Cr^0).

If energy is then applied to the chromium ground-state atom in the form of heat from a plasma discharge, one or more orbiting electrons will be stripped off the outer shell. This will result in only 23 (or less) electrons left orbiting the nucleus. Because the atom has lost a negative charge (e^-) but still has 24 protons (p^+) in the nucleus, it is converted into an ion with a net positive charge. It still has an atomic mass of 52 and an atomic number of 24 but is now a positively charged ion and not a neutral ground-state atom. This process is shown in Figure 5.3.

NATURAL ISOTOPES

This is a very basic look at the process, because most elements have more than one form (isotope). In fact, chromium has four naturally occurring isotopes, which means that the chromium atom exists in four different forms, all with the same atomic number of 24 (number of protons) but with

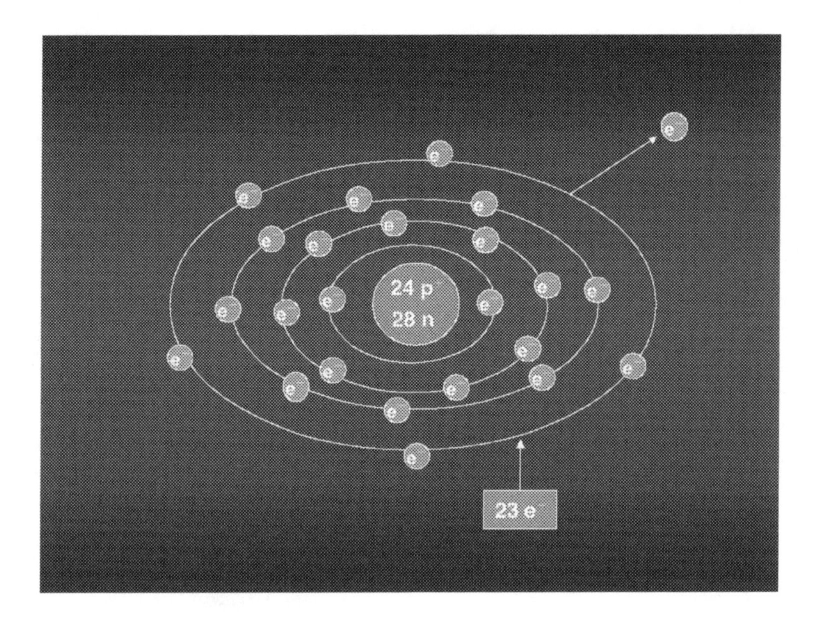

FIGURE 5.3 Conversion of a chromium ground-state atom (Cr^0) to an ion (Cr^+).

different atomic masses (numbers of neutrons). So besides the $^{52}Cr^+$ ion, chromium can also generate the $^{50}Cr^+$, $^{53}Cr^+$, and $^{54}Cr^+$ positively charged ions.

To make this a little easier to understand, let us take a closer look at an element such as copper, which only has two different isotopes—one with an atomic mass of 63 (^{63}Cu) and another with an atomic mass of 65 (^{65}Cu). They both have the same number of protons and electrons but differ in the number of neutrons in the nucleus. The natural abundances of ^{63}Cu and ^{65}Cu are 69.1% and 30.9%, respectively, which gives copper a nominal atomic mass of 63.55—the value you see for copper in atomic weight reference tables. Details of the atomic structure of the two copper isotopes are shown in Table 5.1.

When a sample containing naturally occurring copper is introduced into the plasma, two different ions of copper, $^{63}Cu^+$ and $^{65}Cu^+$, are produced that generate two different masses—one at mass 63 and another at mass 65. This can be seen in Figure 5.4, which is an actual ICP-MS spectral scan of a sample containing copper, showing a peak for the $^{63}Cu^+$ ion on the left, which is 69.17% abundant, and a peak for $^{65}Cu^+$ at 30.83% abundance on the right.

TABLE 5.1
Breakdown of the Atomic Structure of Copper Isotopes

	^{63}Cu	^{65}Cu
Number of protons (p^+)	29	29
Number electrons (e^-)	29	29
Number of neutrons (n)	34	36
Atomic mass (p^+ + n)	63	65
Atomic number (p^+)	29	29
Natural abundance	69.17%	30.83%
Nominal atomic weight	63.55[a]	

[a] The nominal atomic weight of copper is calculated using the formula 0.6917n (^{63}Cu) + 0.3083n (^{65}Cu) + p^+ and referenced to the atomic weight of carbon.

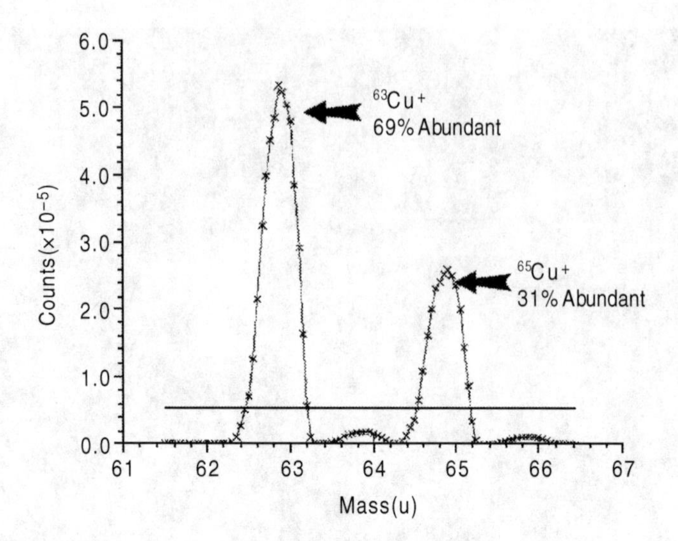

FIGURE 5.4 Mass spectra of the two copper isotopes—$^{63}Cu^+$ and $^{65}Cu^+$.

You can also see small peaks for two Zn isotopes at mass 64 ($^{64}Zn^+$) and mass 66 ($^{66}Zn^+$). (Zn has a total of five isotopes at masses 64, 66, 67, 68, and 70.) In fact, most elements have at least two or three isotopes, and many elements including zinc and lead have four or more isotopes. Figure 5.5 is a chart showing the relative abundance of the naturally occurring isotopes of all the elements.

Relative Abundance of the Natural Isotopes

Isotope	%	%	%
1	H 99.985		
2	H 0.015		
3		He 0.000137	
4		He 99.999863	
5			
6			Li 7.5
7			Li 92.5
8			
9	Be 100		
10		B 19.9	
11		B 80.1	
12			C 98.90
13			C 1.10
14	N 99.643		
15	N 0.366		
16		O 99.762	
17		O 0.038	
18		O 0.200	
19			F 100
20	Ne 90.48		
21	Ne 0.27		
22	Ne 9.25		
23		Na 100	
24			Mg 78.99
25			Mg 10.00
26			Mg 11.01
27	Al 100		
28		Si 92.23	
29		Si 4.67	
30		Si 3.10	
31			P 100
32	S 95.02		
33	S 0.75		
34	S 4.21		
35		Cl 75.77	
36	S 0.02		Ar 0.337
37		Cl 24.23	
38			Ar 0.063
39	K 93.2581		
40	K 0.0117	Ca 96.941	Ar 99.600
41	K 6.7302		
42		Ca 0.647	
43		Ca 0.135	
44		Ca 2.086	
45			Sc 100
46	Ti 8.0	Ca 0.004	
47	Ti 7.3		
48	Ti 73.8	Ca 0.187	
49	Ti 5.5		
50	Ti 5.4	V 0.250	Cr 4.345
51		V 99.750	
52			Cr 83.789
53			Cr 9.501
54	Fe 5.8		Cr 2.365
55		Mn 100	
56	Fe 91.72		
57	Fe 2.2		
58	Fe 0.28		Ni 68.077
59		Co 100	
60			Ni 26.223

Isotope	%	%	%
61			Ni 1.140
62			Ni 3.634
63	Cu 69.17		
64		Zn 48.6	Ni 0.926
65	Cu 30.83		
66		Zn 27.9	
67		Zn 4.1	
68		Zn 18.8	
69			Ga 60.108
70	Ge 21.23	Zn 0.6	
71			Ga 39.892
72	Ge 27.66		
73	Ge 7.73		
74	Ge 35.94	Se 0.89	
75			As 100
76	Ge 7.44	Se 9.36	
77		Se 7.63	
78	Kr 0.35	Se 23.78	
79			Br 50.69
80	Kr 2.25	Se 49.61	
81			Br 49.31
82	Kr 11.6	Se 8.73	
83	Kr 11.5		
84	Kr 57.0	Sr 0.56	
85			Rb 72.165
86	Kr 17.3	Sr 9.86	
87		Sr 7.00	Rb 27.835
88		Sr 82.58	
89			Y 100
90	Zr 51.45		
91	Zr 11.22		
92	Zr 17.15	Mo 14.84	
93			Nb 100
94	Zr 17.38	Mo 9.25	
95		Mo 15.92	
96	Zr 2.80	Mo 16.68	Ru 5.52
97		Mo 9.55	
98		Mo 24.13	Ru 1.88
99			Ru 12.7
100		Mo 9.63	Ru 12.6
101			Ru 17.0
102	Pd 1.02		Ru 31.6
103		Rh 100	
104	Pd 11.14		Ru 18.7
105	Pd 22.33		
106	Pd 27.33	Cd 1.25	
107			Ag 51.839
108	Pd 26.46	Cd 0.89	
109			Ag 48.161
110	Pd 11.72	Cd 12.49	
111		Cd 12.80	
112	Sn 0.97	Cd 24.13	
113		Cd 12.22	In 4.3
114	Sn 0.65	Cd 28.73	
115	Sn 0.34		In 95.7
116	Sn 14.53	Cd 7.49	
117	Sn 7.68		
118	Sn 24.23		
119	Sn 8.59		
120	Sn 32.59	Te 0.096	

Isotope	%	%	%
121			Sb 57.36
122	Sn 4.63	Te 2.603	
123		Te 0.908	Sb 42.64
124	Sn 5.79	Te 4.816	Xe 0.10
125		Te 7.139	
126		Te 18.95	Xe 0.09
127	I 100		
128		Te 31.69	Xe 1.91
129			Xe 26.4
130	Ba 0.106	Te 33.80	Xe 4.1
131			Xe 21.2
132	Ba 0.101		Xe 26.9
133		Cs 100	
134	Ba 2.417		Xe 10.4
135	Ba 6.592		
136	Ba 7.854	Ce 0.19	Xe 8.9
137	Ba 11.23		
138	Ba 71.70	Ce 0.25	La 0.0902
139			La 99.9098
140		Ce 88.48	
141			Pr 100
142	Nd 27.13	Ce 11.08	
143	Nd 12.18		
144	Nd 23.80	Sm 3.1	
145	Nd 8.30		
146	Nd 17.19		
147		Sm 15.0	
148	Nd 5.76	Sm 11.3	
149		Sm 13.8	
150	Nd 5.64	Sm 7.4	
151			Eu 47.8
152	Gd 0.20	Sm 26.7	
153			Eu 52.2
154	Gd 2.18	Sm 22.7	
155	Gd 14.80		
156	Gd 20.47	Dy 0.06	
157	Gd 15.65		
158	Gd 24.84	Dy 0.10	
159			Tb 100
160	Gd 21.86	Dy 2.34	
161		Dy 18.9	
162	Er 0.14	Dy 25.5	
163		Dy 24.9	
164	Er 1.61	Dy 28.2	
165			Ho 100
166	Er 33.6		
167	Er 22.95		
168	Er 26.8	Yb 0.13	
169			Tm 100
170	Er 14.9	Yb 3.05	
171		Yb 14.3	
172		Yb 21.9	
173		Yb 16.12	
174		Yb 31.8	Hf 0.162
175	Lu 97.41		
176	Lu 2.59	Yb 12.7	Hf 5.206
177			Hf 18.606
178			Hf 27.297
179			Hf 13.629
180	Ta 0.012	W 0.13	Hf 35.100

Isotope	%	%	%
181	Ta 99.988		
182		W 26.3	
183		W 14.3	
184	Os 0.02	W 30.67	
185			Re 37.40
186	Os 1.58	W 28.6	
187	Os 1.6		Re 62.60
188	Os 13.3		
189	Os 16.1		
190	Os 26.4		Pt 0.01
191		Ir 37.3	
192	Os 41.0		Pt 0.79
193		Ir 62.7	
194			Pt 32.9
195			Pt 33.8
196	Hg 0.15		Pt 25.3
197		Au 100	
198	Hg 9.97		Pt 7.2
199	Hg 16.87		
200	Hg 23.10		
201	Hg 13.18		
202	Hg 29.86		
203			Tl 29.524
204	Hg 6.87	Pb 1.4	
205			Tl 70.476
206		Pb 24.1	
207		Pb 22.1	
208		Pb 52.4	
209	Bi 100		
210			
211			
212			
213			
214			
215			
216			
217			
218			
219			
220			
221			
222			
223			
224			
225			
226			
227			
228			
229			
230			
231	Pa 100		
232	Th 100		
233			
234	U 0.0055		
235	U 0.7200		
236			
237			
238	U 99.2745		

"Isotopic Compositions of the Elements 1989" Pure Appl. Chem., Vol. 63, No. 7, pp. 991-1002, 1991. © 1991 IUPAC

FIGURE 5.5 Relative abundance of the naturally occurring isotopes of the elements. (From UIPAC Isotopic Composition of the Elements, *Pure and Applied Chemistry* 75(6), 683–799, 2003.)

6 ICP-MS Sample Introduction

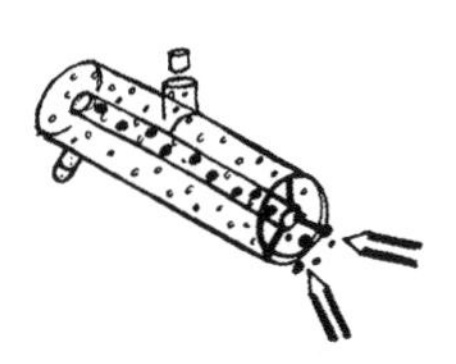

This chapter examines one of the most critical areas of the inductively coupled plasma mass spectrometry (ICP-MS) instrument—the sample introduction system. It discusses the basic principles of converting a liquid into a fine-droplet aerosol suitable for ionization in the plasma and presents an overview of the different types of commercially available nebulizers and spray chambers. Although this chapter briefly touches upon some of the newer sampling components introduced in the past few years, such as microflow nebulizers and aerosol dilution systems, the new advancements in desolvating nebulizers, chilled spray chambers, online chemistry approaches, autodilution/auto calibration, intelligent autosamplers, and productivity enhance systems are specifically described in a Chapter 20.

The majority of current ICP-MS applications involve the analysis of liquid samples. Even though the technique has been adapted over the years to handle solids and slurries, it was developed in the early 1980s primarily to analyze solutions. There are many different ways of introducing a liquid into an ICP mass spectrometer, but they all basically achieve the same result, which is to generate a fine aerosol of the sample so that it can be efficiently ionized in the plasma discharge. The sample introduction area has been called the Achilles' heel of ICP-MS, because it is considered the weakest component of the instrument. Only about 2% of the sample finds its way into the plasma, depending on the matrix and method of introducing the sample.[1] Although there has recently been significant innovation in this area, particularly in instrument-specific components custom built by third-party vendors, the fundamental design of a traditional ICP-MS sample introduction system has not dramatically changed since the technique was first introduced in 1983.

Before I discuss the mechanics of aerosol generation in greater detail, let us look at the basic components of a sample introduction system. Figure 6.1 shows the location of the sample introduction area relative to the rest of the ICP mass spectrometer, whereas Figure 6.2 represents a more detailed view showing the individual components.

The traditional way of introducing a liquid sample into an analytical plasma can be considered as two separate events: aerosol generation using a nebulizer and droplet selection using a spray chamber.[2]

AEROSOL GENERATION

As mentioned previously, the main function of the sample introduction system is to generate a fine aerosol of the sample. It achieves this with a nebulizer and a spray chamber. The sample is normally pumped at a maximum of 1 mL/min (for low pressure nebulizers) or between 100 and 600 µL/min (for more efficient higher pressure nebulizers) via a peristaltic or syringe pump into the nebulizer. A peristaltic pump is a small pump with mini rollers mounted between circular disks that are turned by a motor with adjustable speed. The pump speed is adjusted depending on the pump tubing diameter to deliver the desired liquid flow rate. The constant motion and pressure of the rollers on the pump tubing feeds the sample through to the nebulizer.

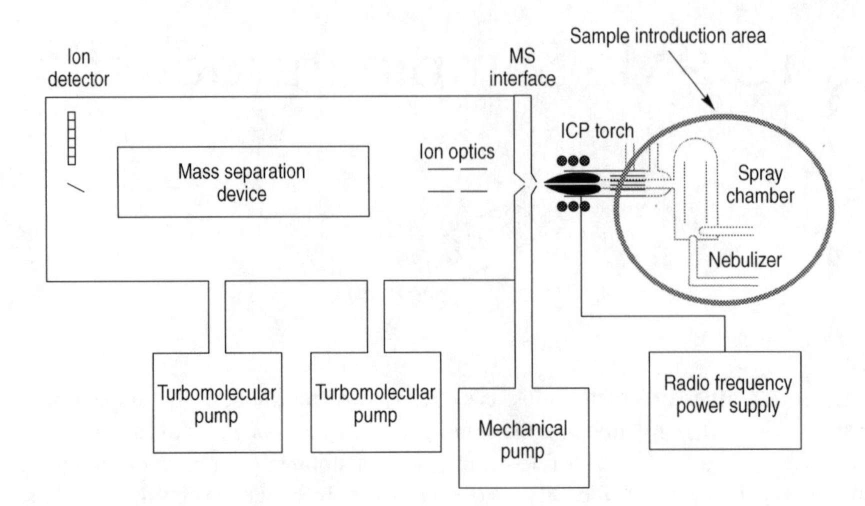

FIGURE 6.1 Location of the ICP-MS sample introduction area.

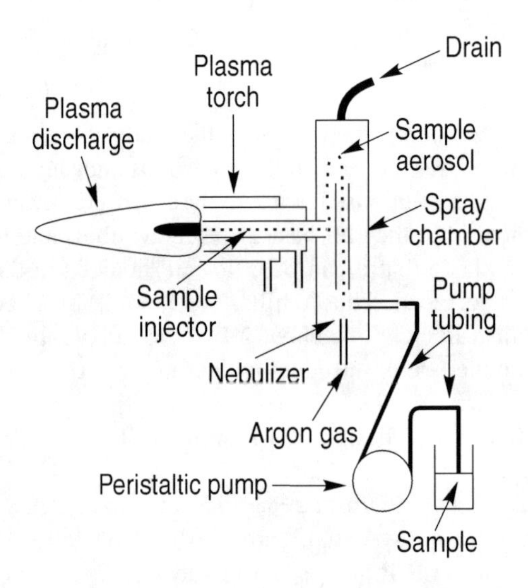

FIGURE 6.2 More detailed view of the ICP-MS sample introduction area.

A syringe pump delivers the sample via a pneumatic piston, which imparts smoother flow in the absence of rollers. Both syringe and peristaltic pumps can also be equipped with a switching valve to switch the liquid flow path so that sample uptake path rinse may take place during previous sample analysis, followed by rapid sequential sample loading in the sample loop. This type of system is typically two to three times faster than standard autosampler rinse sequences, which means that not only is sample throughput increased but also faster rinse-out is achieved and therefore less carryover to the next sample. The benefits of using a syringe-type pump for autodilution and the online addition of internal standards are well recognized in ICP-MS and will be discussed in greater detail in Chapter 20 "Performance and Productivity Enhancement Tools." However, in its basic configuration, a syringe pump with switching valve significantly improves precision by eliminating the pulsations of a peristaltic pump, allowing shorter measurement times to achieve the same performance levels. The stability of a continuous ICP-MS signal with a syringe pump compared to a peristaltic pump is shown in Figure 6.3.

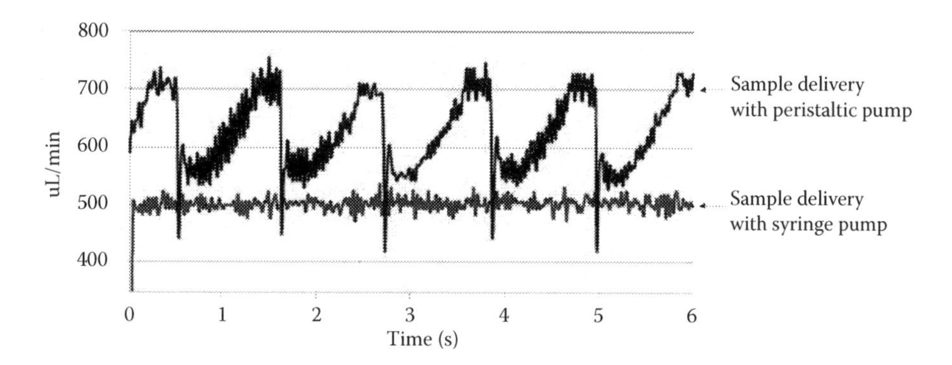

FIGURE 6.3 The stability of a continuous ICP-MS signal with a syringe pump compared to a peristaltic pump.

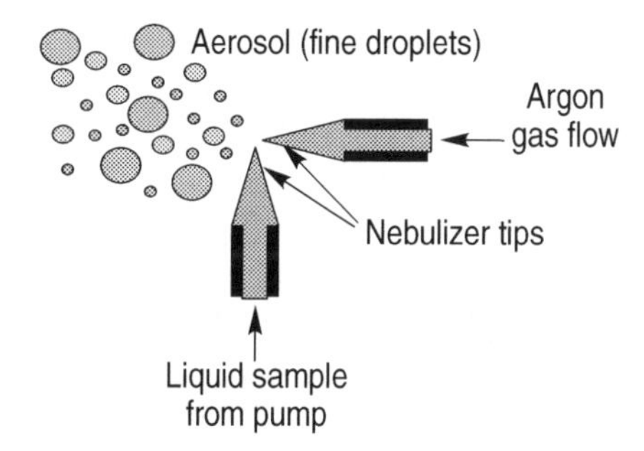

FIGURE 6.4 Conceptual representation of aerosol generation using a cross-flow nebulizer.

Once the sample enters the nebulizer, the liquid is then broken up into a fine aerosol by the pneumatic action of a flow of gas (~1 L/min) "smashing" the liquid into tiny droplets, very similar to the spray mechanism in a can of deodorant. It should be noted that although pumping the sample is the most common approach to introducing the sample, some pneumatic designs such as concentric nebulizers do not require a pump, because they rely on the natural "venturi effect" of the positive pressure of the nebulizer gas to suck the sample through the tubing. This sample uptake without the use of a pump is called "self-aspiration." Solution nebulization is conceptually represented in Figure 6.4, which shows aerosol generation using a cross-flow-designed nebulizer.

DROPLET SELECTION

Because the plasma discharge is not very efficient at dissociating large droplets, the function of the spray chamber is primarily to allow only the small droplets to enter the plasma. Its secondary purpose is to smooth out pulses that occur during the nebulization process, mainly from the peristaltic pump. Spray chambers are discussed in greater detail later in this chapter, but the most common type is the double-pass design, where the aerosol from the nebulizer is directed into a central tube running the entire length of the chamber. The droplets then travel the length of this tube, where the large droplets (>10 µm dia.) which have greater momentum than small droplets will hit the wall of the spray chamber and exit through the drain tube at the end of the spray chamber. The fine droplets (<10 µm dia.) then pass between the outer wall and the central tube, where they eventually emerge from the spray chamber and are transported into the sample injector of the plasma torch.[3] Although there are many different designs available, the spray chamber's main function is to allow only the smallest droplets into the plasma for dissociation, atomization, and, finally, ionization of the sample's elemental components. A simplified schematic of this process using a double-pass-designed spray chamber is shown in Figure 6.5.

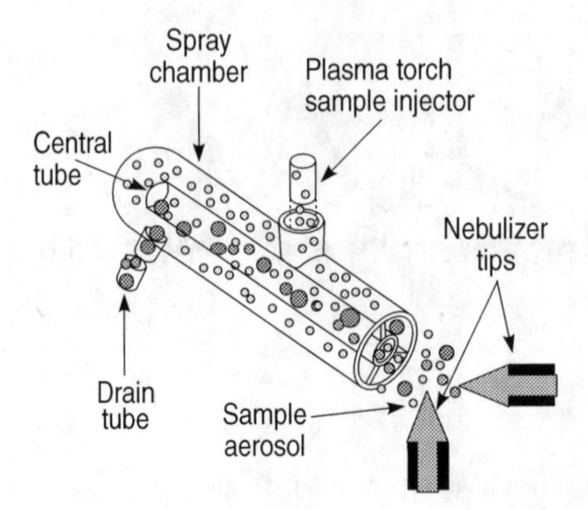

FIGURE 6.5 Simplified representation of the separation of large and fine droplets in a double-pass spray chamber.

Let us now look at the most common nebulizers and spray chamber designs used in ICP-MS. We cannot cover every conceivable design, because over the past few years there has been a huge demand for application-specific solutions, which has generated a number of third-party manufacturers that sell sample introduction components directly to ICP-MS users.

NEBULIZERS

By far, the most common design used for ICP-MS is the pneumatic nebulizer, which uses mechanical forces of a gas flow (normally argon at a pressure of 20–30 psi) to generate the sample aerosol. This type of nebulizer is generally used for higher dissolved solids but unfortunately is very inefficient. Some of the most popular designs of pneumatic nebulizers include the concentric, microconcentric, microflow, and cross-flow. They are usually made from glass or quartz, but other nebulizer materials, such as various kinds of polymers and plastics, are becoming more popular, particularly for highly corrosive samples and specialized applications.

It should be emphasized at this point that nebulizers designed for use with ICP-OES are far from ideal for use with ICP-MS. This is the result of a limitation in the quantity of total dissolved solids (TDS) that can be put into the ICP-MS interface area. Because the orifice sizes of the sampler and skimmer cones used in ICP-MS are so small (~0.6–1.2 mm), the matrix components must generally be kept below 0.2%, although higher concentrations of some matrices can be tolerated (refer to Chapter 5).[4] This means that general-purpose ICP-OES nebulizers that are designed to aspirate 1%–2% dissolved solids or high-solid nebulizers such as the Babbington, V-groove, or cone-spray, which are designed to handle up to 20% dissolved solids, are not ideally suited to analyzing solutions by ICP-MS.

Some researchers have attempted to analyze slurries by ICP-MS using this approach. However, this is not recommended for high-throughput, routine work because of the potential of blocking the interface cones, but as long as the particle size of the slurry is kept below 10 μm in diameter, some success has been achieved using these types of nebulizers.[5] Additionally, there are researchers who are attempting to characterize engineered nanoparticles by using a technique called single-particle (SP) ICP-MS. This is in its early stages, but it is a very exciting development, which uses a novel way to separate out individual nanoparticles in suspension, typically in environmental samples and then detect them by optimizing the measurement electronics of the ICP-MS detection system.[6] This technique will be described in greater detail in Chapter 20.

The most common of the pneumatic nebulizers used in commercial ICP mass spectrometers are the concentric and cross-flow design types. The concentric design is the most widely used nebulizer, whereas the cross-flow is generally more tolerant to samples containing higher solids and particulate matter. However, the cross-flow design is not widely used or provided with instruments anymore unless specifically requested, because, recent advances in the concentric design have allowed for the aspiration of many different types of samples, including those high solids.

CONCENTRIC DESIGN

In traditional concentric nebulization, a solution is introduced through a fine-bore capillary tube, where it comes into contact with a rapidly moving flow of argon gas at a pressure of approximately 30–50 psi. The high-speed gas and the lower-pressure sample combine to create a venturi effect, which results in the sample being sucked through to the end of the capillary, where it is broken up into a fine-droplet aerosol. Most concentric nebulizers being used today are manufactured from borosilicate glass or quartz. However, polymer-based materials are now being used for applications that require corrosion resistance. Typical sample flow rates for a standard concentric nebulizer are on the order of 0.1–1 mL/min, although lower flows can be used to accommodate more volatile sample matrices, such as organic solvents. A schematic of a glass concentric nebulizer with the different parts labeled is shown in Figure 6.6, and the aerosol generated by the nebulization process is shown in Figure 6.7.

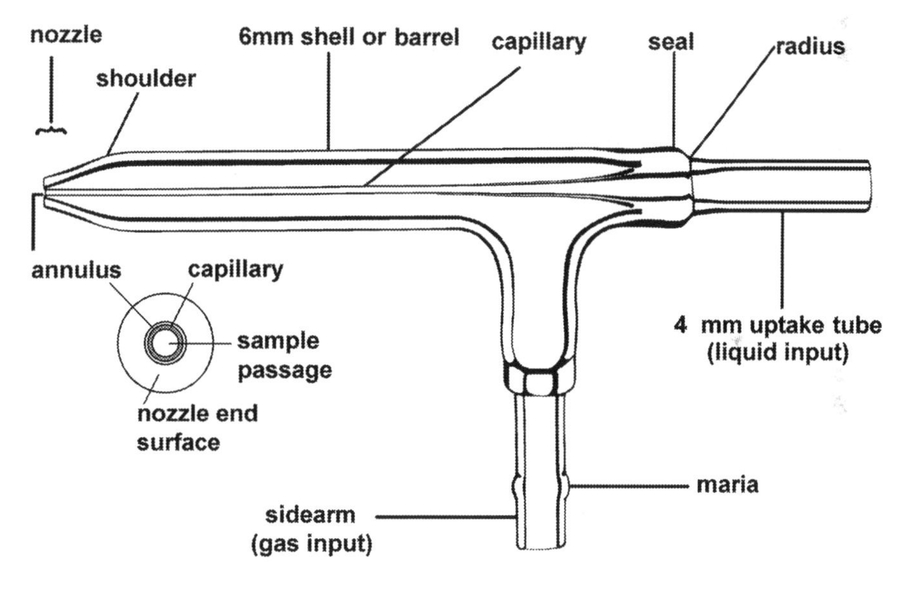

FIGURE 6.6 Schematic of a glass concentric nebulizer. (Courtesy of Meinhard Glass Products, a part of Elemental Scientific Inc.)

FIGURE 6.7 Aerosol generated by a concentric nebulizer. (Courtesy of Meinhard Glass Products, a part of Elemental Scientific Inc.)

The standard concentric pneumatic nebulizer will give excellent sensitivity and stability, particularly with clean solutions. However, the narrow capillary can be plagued by blockage problems, especially if heavier-matrix samples are being aspirated. For that reason, manufacturers of concentric nebulizers offer modifications to the basic design utilizing different size capillary tubing and recessed tips to allow aspiration of samples with higher dissolved solids and particulate matter. There are even specially designed concentric nebulizers with a smaller-bore input capillary to significantly reduce the dead volume for better coupling of a high-performance liquid chromatography (HPLC) system to the ICP-MS when carrying out trace element speciation studies.

CROSS-FLOW DESIGN

For the routine analysis of samples that contain a heavier matrix, or maybe small amounts of undissolved matter, the cross-flow design is the more rugged design. With this nebulizer, the argon gas is directed at right angles to the tip of a capillary tube, in contrast to the concentric design, where the gas flow is parallel to the capillary. The solution is either drawn up through the capillary tube via the pressure created by the high-speed gas flow, or, as is most common with cross-flow nebulizers, fed through the tube with a peristaltic pump. In either case, contact between the high-speed gas and the liquid stream causes the liquid to break up into an aerosol. Cross-flow nebulizers are generally not as efficient as concentric nebulizers at creating the very small droplets needed for ionization in the plasma. However, the larger-diameter liquid capillary and longer distance between liquid and gas injectors reduce the potential for clogging problems. Some analysts feel that the penalty to be paid in analytical sensitivity and precision with cross-flow nebulizers compared to the concentric design is compensated by the fact that they are better suited for high-throughput, routine applications. In addition, they are typically manufactured from plastic materials, which make them far more rugged than a glass concentric nebulizer. A cross section of a cross-flow nebulizer is shown in Figure 6.8.

MICROFLOW DESIGN

More recently, microflow or high-efficiency nebulizers have been designed for ICP-MS to operate at much lower sample flows. Whereas conventional nebulizers have a sample uptake rate of about 1 mL/min, microflow/high-efficiency nebulizers typically run at less than 0.1 mL/min. They are based on the concentric principle but usually operate at higher gas pressure to accommodate the lower sample flow rates. The extremely low uptake rate makes them ideal for applications where sample volume is limited or where the sample or analyte is prone to sample introduction memory effects. The additional benefit of this design is that it produces an aerosol with smaller droplets, and as a result, it is generally more efficient than a conventional concentric nebulizer.

These nebulizers and their components are typically constructed from polymer materials, such as polytetrafluoroethylene (PTFE), perfluoroalkoxy (PFA), or polyvinylfluoride (PVF), although some

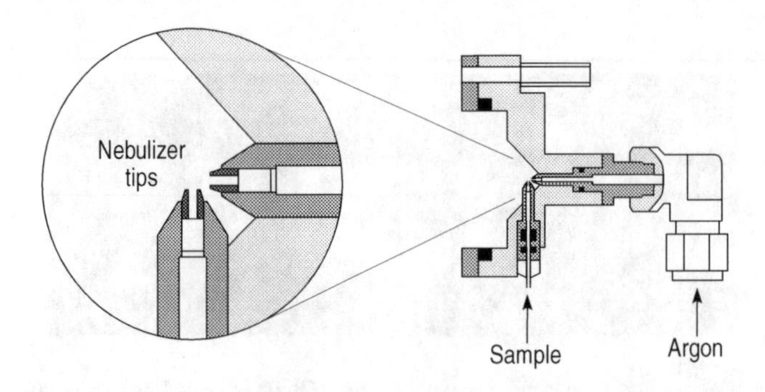

FIGURE 6.8 Schematic of a cross-flow nebulizer. (Copyright 2013, all rights reserved, Perkin Elmer Inc.)

designs are available in borosilicate glass or quartz. The excellent corrosion resistance of the polymer nebulizers means they have naturally low blank levels. This characteristic, together with their ability to handle small sample volumes found in applications such as vapor phase decomposition (VPD), makes them an ideal choice for semiconductor laboratories that are carrying out ultratrace element analysis.[6,7] A microflow concentric nebulizer made from PFA is shown in Figure 6.9, and a typical spray pattern of the nebulization process is shown in Figure 6.10.

The disadvantage of microconcentric nebulizers is that they use an extremely fine capillary, which makes them not very tolerant to high concentrations of dissolved solids or suspended particles. Their high efficiency also means that most of the sample make it into the plasma, but the low liquid flow decreases plasma loading and tends to provide greater freedom from matrix suppression problems. For these reasons, they have been found to be most applicable for the analysis of aqueous-type samples or samples containing low levels of dissolved solids. However, the development of higher

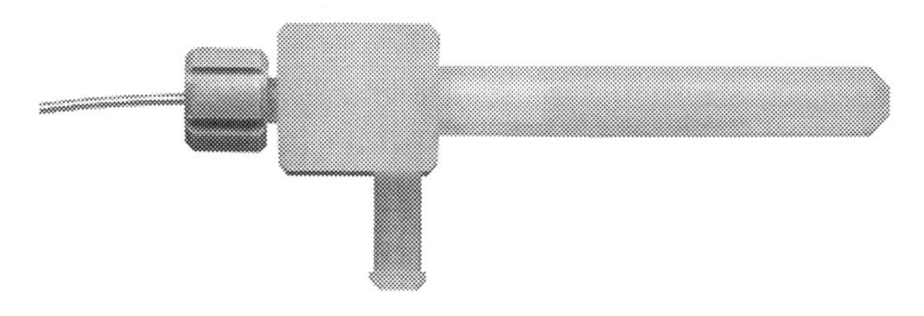

FIGURE 6.9 The OpalMist™ microflow concentric nebulizer made from PFA. (Courtesy of Glass Expansion Inc.)

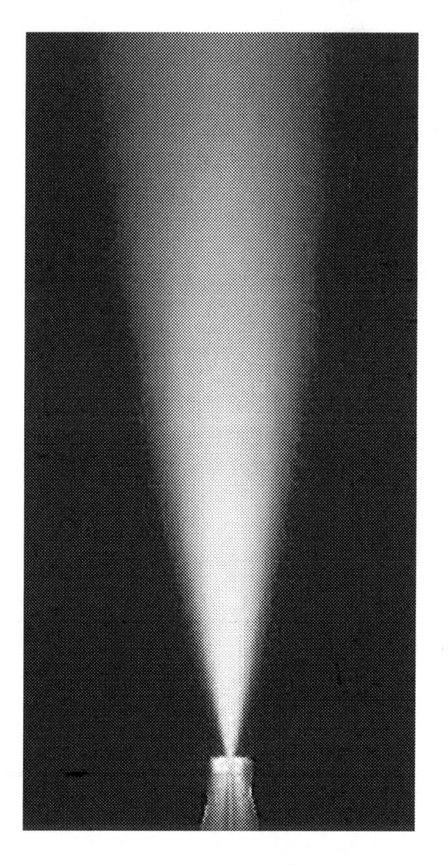

FIGURE 6.10 Spray pattern of a PFA microflow concentric nebulizer. (Courtesy of Elemental Scientific Inc.)

pressure (60–70 psi) concentric nebulizers with wider capillary internal diameters has resulted in higher tolerance to dissolved solids with higher nebulization efficiency as well.

One of the application areas that high-efficiency nebulizers are well suited is in the handling of extremely small volumes being eluted from an HPLC or flow injection analyzer (FIA) system into an ICP-MS for doing speciation or microsampling work. The analysis of discrete sample volumes encountered in these types of applications allows for detection limits equivalent to a standard concentric nebulizer, while consuming 10–20 times less sample.

SPRAY CHAMBERS

Let us now turn our attention to spray chambers. There are basically two designs that are used in today's commercial ICP-MS instrumentation: double-pass and cyclonic spray chambers. The double-pass is by far the most common, with the cyclonic type rapidly gaining in popularity. As mentioned earlier, the function of the spray chamber is to reject the larger aerosol droplets and also to smooth out nebulization pulses produced by the peristaltic pump, if it is used. In addition, some ICP-MS spray chambers are externally cooled for thermal stability of the sample and to reduce the amount of solvent going into the plasma. This can have a number of beneficial effects, depending on the application, but the main advantages are to reduce oxide species, minimize signal drift, and reduce the solvent loading on the plasma, particularly when aspirating volatile organic solvents.

DOUBLE-PASS SPRAY CHAMBER

By far, the most common design of the double-pass spray chamber is the Scott design, which selects the small droplets by directing the aerosol into a central tube. The larger droplets impact the spray chamber wall and exit the spray chamber via a drain tube. The liquid in the central tube is kept at positive pressure (usually by way of a loop), which forces the small droplets back between the outer wall and the central tube, and emerges from the spray chamber into the sample injector of the plasma torch. Double-pass spray chambers come in a variety of shapes, sizes, and materials and are generally considered the most rugged design for routine use. Figure 6.11 shows a Scott double-pass spray chamber made of a polysulfide-type material, coupled to a cross-flow nebulizer.

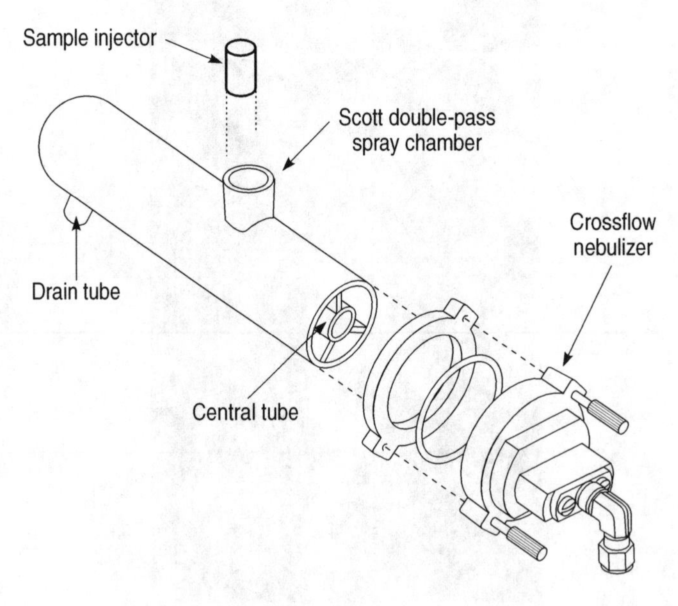

FIGURE 6.11 A Scott double-pass spray chamber with cross-flow nebulizer. (Copyright 2013, all rights reserved, Perkin Elmer Inc.).

CYCLONIC SPRAY CHAMBER

The cyclonic spray chamber operates by centrifugal force. Droplets are discriminated according to their size by means of a vortex produced by the tangential flow of the sample aerosol and argon gas inside the chamber. Smaller droplets are carried with the gas stream into the ICP-MS, whereas the larger droplets impinge on the walls and fall out through the drain. It is generally accepted that a cyclonic spray chamber has a higher sampling efficiency, which for clean samples translates into higher sensitivity and lower detection limits. However, the droplet size distribution appears to be different from a double-pass design and for certain types of samples can give slightly inferior precision. Beres and coworkers published a very useful study describing the capabilities of a cyclonic spray chamber.[8] Figure 6.12 shows a cyclonic spray chamber connected to a concentric nebulizer.

The cyclonic spray chamber is growing in popularity, particularly as its potential is getting realized in more and more application areas. Just as there is a wide selection of nebulizers available for different applications, there is also a wide choice of customized cyclonic spray chambers, manufactured from glass, quartz, and different polymer materials. Depending on the application being carried out, modifications to the cyclonic design are available for low sample flows, high dissolved solids, fast sample washout, corrosion resistance, and organic solvents. Figure 6.13 shows one of the many variations of cyclonic spray chamber, called the jacketed Cinnabar™, which is a water-cooled borosilicate glass spray chamber optimized for aspirating small sample volumes with a microflow concentric nebulizer.

It is worth emphasizing that cooling the spray chamber is generally beneficial in ICP-MS, because it reduces the solvent loading on the plasma. This has three major benefits. First, because very little plasma energy is wasted vaporizing the solvent, more is available to excite and ionize the analytes. Second, if there is less water being delivered to the plasma, there is less chance of forming oxide and hydroxide species, which can potentially interfere with other analytes. Finally, if the spray chamber is kept at a constant temperature, it leads to better long-term signal stability, especially if there are environmental temperature changes over the time period of the analysis. For these reasons, some manufacturers supply cooled spray chambers as standard, whereas others offer the capability as an option. There is also a wide variety of cooled and chilled spray chambers available from third-party vendors.

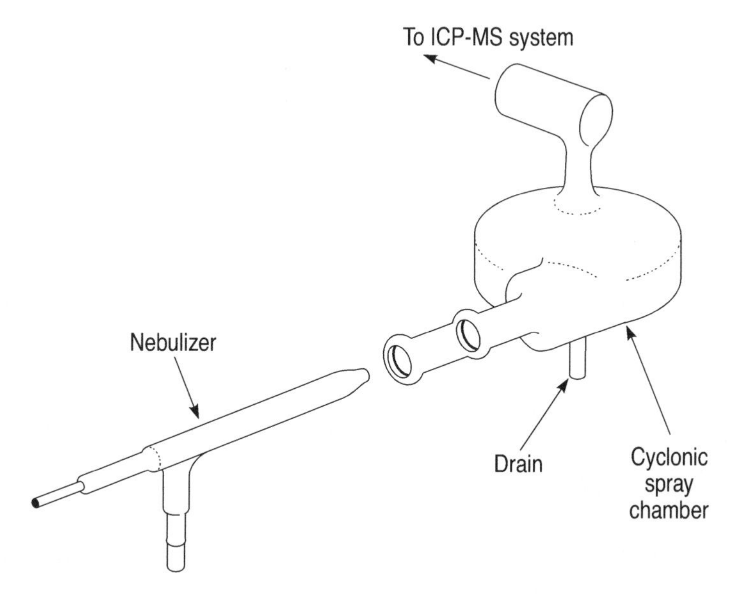

FIGURE 6.12 A cyclonic spray chamber (shown with a concentric nebulizer). (From S. A. Beres, P. H. Bruckner, and E. R. Denoyer, *Atomic Spectroscopy*, 15(2), 96–99, 1994.)

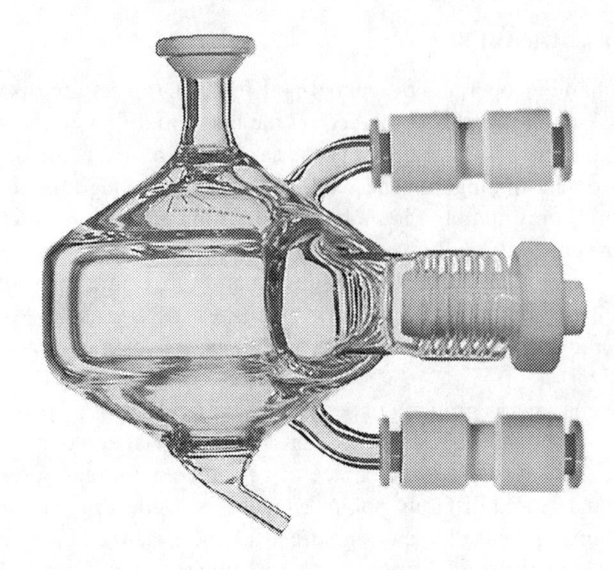

FIGURE 6.13 The low-flow Cinnabar™, a water-cooled cyclonic spray chamber for use with a microflow concentric nebulizer. (Courtesy of Glass Expansion Inc.)

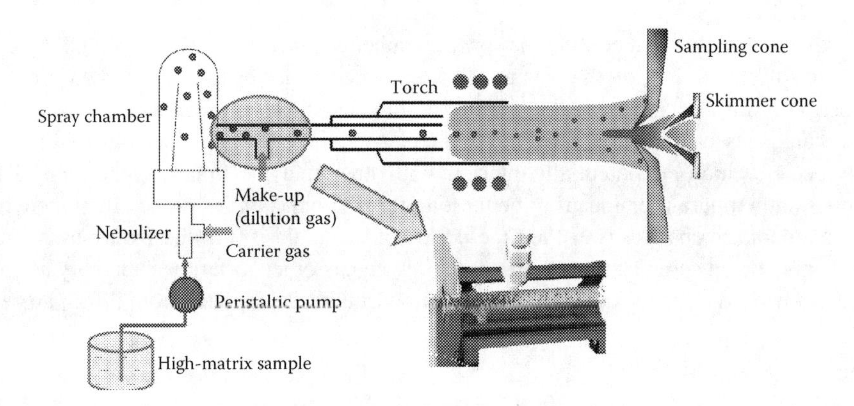

FIGURE 6.14 Principle of aerosol dilution. (From "Today's Agiient–New Atomic Spectroscopy Solutions for Environmental Laboratories" online webinar.)

AEROSOL DILUTION

To address the limitation of the TDS capability of ICP-MS, some vendors offer an aerosol dilution system, which introduces a flow of argon gas between the nebulizer and the torch to carry our aerosol dilution of the sample. This has the effect of reducing the sample's solvent loading on the plasma, so it can tolerate much higher TDS levels than the <0.2%, which is typical for most ICP-MS instrumentation. However, it's important to emphasize that the dilution is done after the nebulizer, so care must be taken in selecting the optimum nebulizer if the sample contains high levels of dissolved solids. The principles of aerosol dilution are shown in Figure 6.14. Some of the benefits of this novel type of dilution include:

- Enables the direct analysis of samples containing medium to high percentage levels of dissolve solids, assuming the nebulizer can handle them
- Significantly improves plasma robustness compared to conventional sample introduction methods.

- Because less solvent/matrix is entering the plasma, it reduces oxide interferences to very low levels, providing better accuracy and more stable sampling conditions
- Eliminates the need for conventional liquid dilution of high matrix samples prior to analysis, which has several disadvantages, including increased risk of sample contamination, dilution errors, and sample prep time.

FINAL THOUGHTS

There are many other nonstandard sample introduction devices such as laser ablation, ultrasonic nebulizers, desolvation devices, direct injection nebulizers, flow injection systems, enhanced productivity systems, autodilution, and online chemistry techniques, which are not described in this chapter. However, because they are becoming more and more important, particularly as ICP-MS users are demanding higher performance, more productivity, and greater flexibility, they are covered in greater detail in Chapter 20.

FURTHER READING

1. R. A. Browner and A. W. Boorn, *Analytical Chemistry*, **56**, 786–798A, 1984.
2. B. L. Sharp, *Analytical Atomic Spectrometry,* **3**, 613, 1980.
3. L. C. Bates and J. W. Olesik, *Journal of Analytical Atomic Spectrometry*, **5**(3), 239, 1990.
4. R. S. Houk, *Analytical Chemistry*, **56**, 97A, 1986.
5. J. G. Williams, A. L. Gray, P. Norman, and L. Ebdon, *Journal of Analytical Atomic Spectrometry*, **2**, 469–472, 1987.
6. J. Ranville, K. Neubauer, and R. Thomas, *Spectroscopy*, **27**(8), 20–27, 2012.
7. E. Debrah, S. A. Beres, T. J. Gluodennis, R. J. Thomas, and E. R. Denoyer, *Atomic Spectroscopy*, **16**(7), 197–202, 1995.
8. R. A. Aleksejczyk and D. Gibilisco, *Micro*, 5, 39–42, 1997.
9. S. A. Beres, P. H. Bruckner, and E. R. Denoyer, *Atomic Spectroscopy*, **15**(2), 96–99, 1994.

7 Plasma Source

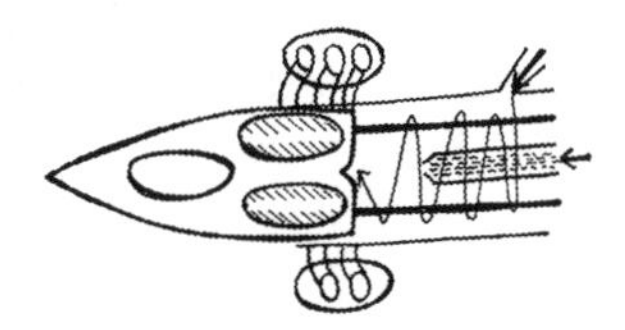

This chapter takes a look at the region of the ICP-MS where the ions are generated—the plasma discharge. It gives a brief historical perspective of some of the common analytical plasmas used over the years and discusses the components used to create the inductively coupled plasma (ICP). It also explains the fundamental principles of formation of a plasma discharge and how it is used to convert the sample aerosol into a stream of positively charged ions of low kinetic energy required by the ion-focusing system and the mass spectrometer.

ICPs are by far the most common type of plasma sources used in today's commercial ICP optical emission (ICP-OES) and ICP mass spectrometric (ICP-MS) instrumentation. However, it was not always that way. In the early days, when researchers were attempting to find the ideal plasma source to use for spectrometric studies, it was not clear which approach would prove to be the most successful. In addition to ICPs, some of the other novel plasma sources developed were direct current plasmas (DCPs) and microwave-induced plasmas (MIPs). Before I go on to describe the ICP, let us first take a closer look at these other two excitation sources.

A DCP is formed when a gas (usually argon) is introduced into a high current flowing between two or three electrodes. Ionization of the gas produces a Y-shaped plasma. Unfortunately, early DCP instrumentation was prone to interference effects and also had some usability and reliability problems. For these reasons, the technique never became widely accepted by the analytical community.[1] However, its one major benefit was that it could aspirate high dissolved or suspended solids because there was no restrictive sample injector for the solid material to block. This feature alone made it very attractive for some laboratories, and once the initial limitations of DCPs were better understood, the technique became more accepted. Limitations in the DCP approach led to the development of electrodeless plasma, of which the MIP was the simplest form. MIP technology has mainly been used as an ion source for mass spectrometry (MS)[2] and also as emission-based detectors for gas chromatography. It is only recently that the technology has advanced and been viewed as a possible alternative to the ICP for elemental analysis.

An MIP basically consists of a quartz tube surrounded by a microwave waveguide or cavity. Microwaves produced from a magnetron fill the cavity and cause the electrons in the plasma support gas to oscillate. The oscillating electrons collide with other atoms in the flowing gas to create and maintain a high-temperature plasma. As in the ICPs, a high voltage spark is needed to create the initial electrons to create the plasma, which achieves temperature of approximately 5,000 K.

The limiting factor to their use was that with the low power and high frequency of the MIP, it was very difficult to maintain the stability of the plasma when aspirating liquid samples containing high levels of dissolved solids. Various attempts had been made over the years to couple desolvation techniques to the MIP but only managed to achieve limited success. However, an MIP-atomic emission spectrometry (AES) system using nitrogen gas has been developed which appears to have overcome many of the limitations of the earlier designs and is achieving detection limit better than flame atomic absorption and only slightly inferior to ICP-OES for many elements.[3]

Because of the limitations of the DCP and MIP approaches, ICPs became the dominant area of research for both optical emission and mass spectrometric studies. As early as 1964, Greenfield and coworkers reported that an atmospheric pressure ICP coupled with OES could be used for elemental analysis.[4] Although crude by today's standards, it showed the enormous possibilities of the ICP as an excitation source and most definitely opened the door in the early 1980s to the even more exciting potential of using the ICP to generate ions.[5]

THE PLASMA TORCH

Before we take a look at the fundamental principles behind the creation of an ICP used in ICP-MS, let us take a look at the basic components used to generate the source—a plasma torch, radio frequency (RF) coil, and power supply. Figure 7.1 shows their proximity compared to the rest of the instrument, and Figure 7.2 is a more detailed view of the plasma torch and RF coil relative to the MS interface.

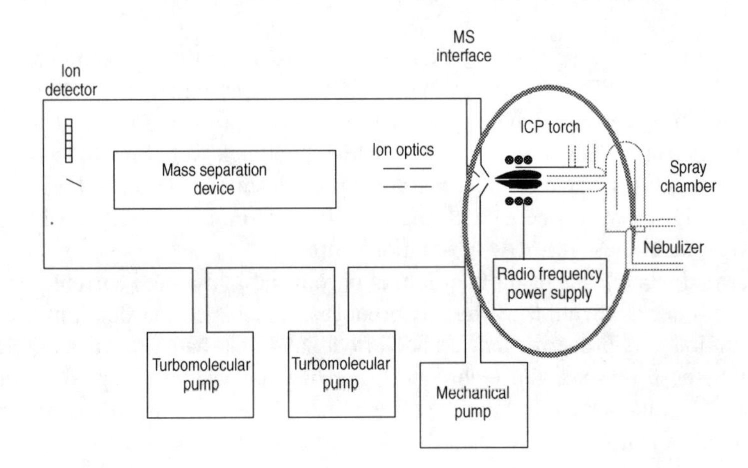

FIGURE 7.1 ICP-MS system showing location of the plasma torch and RF power supply.

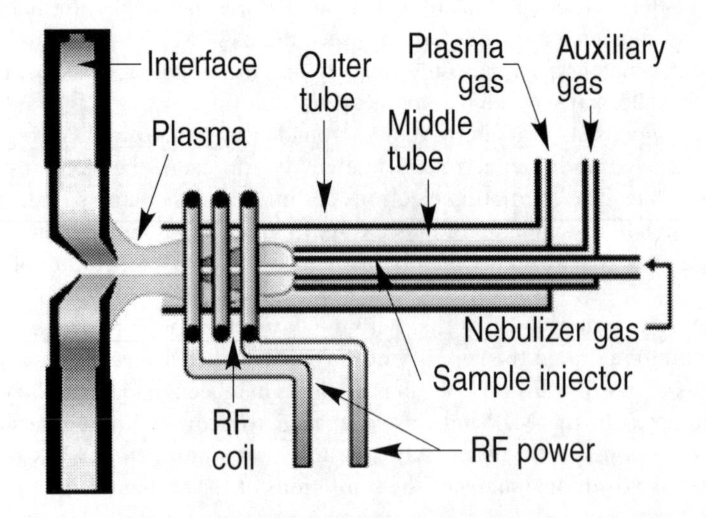

FIGURE 7.2 Detailed view of plasma torch and RF coil relative to the ICP-MS interface.

The plasma torch consists of three concentric tubes, which are normally made from quartz. In Figure 7.2, these are shown as the outer tube, middle tube, and sample injector. The torch can either be one piece, in which all three tubes are connected, or it can employ a demountable design in which the tubes and the sample injector are separate. The gas (usually argon) that is used to form the plasma (plasma gas) is passed between the outer and middle tubes at a flow rate of ~12–17 L/min. A second gas flow (auxiliary gas) passes between the middle tube and the sample injector at ~1 L/min and is used to change the position of the base of the plasma relative to the tube and the injector. A third gas flow (nebulizer gas), also at ~1 L/min, brings the sample, in the form of a fine-droplet aerosol, from the sample introduction system and physically punches a channel through the center of the plasma. The sample injector is often made from other materials besides quartz, such as alumina, platinum, and sapphire—if highly corrosive materials need to be analyzed. It is worth mentioning that although argon is the most suitable gas to use for all three flows, there are analytical benefits in using other gas mixtures, especially in the nebulizer flow.[6] The plasma torch is mounted horizontally and positioned centrally in the RF coil, approximately 10–20 mm from the interface. This can be seen in Figure 7.3, which shows a photograph of a plasma torch mounted in an instrument.

Ceramic components are also available for most ICP-MS torches. The outer, inner, and sample injector tubes are normally made of quartz, but for some applications, it is beneficial to consider using an alternative material like ceramic. And if a demountable torch is being used, any or all of the tubes can be replaced. Some of the applications that might benefit from a ceramic torch include:

- When silicon is one of the analytes, because quartz outer tubes often produce high Si background signals
- For the analysis of fusion mixtures or samples with high levels of dissolved solids which might cause devitrification of quartz tubes
- For the analysis of organic-based samples where quartz outer tubes often suffer from short lifetime.

A fully demountable ceramic torch is shown in Figure 7.4.

FIGURE 7.3 Photograph of a plasma torch mounted in an instrument. (Copyright 2013, all rights reserved, PerkinElmer Inc.)

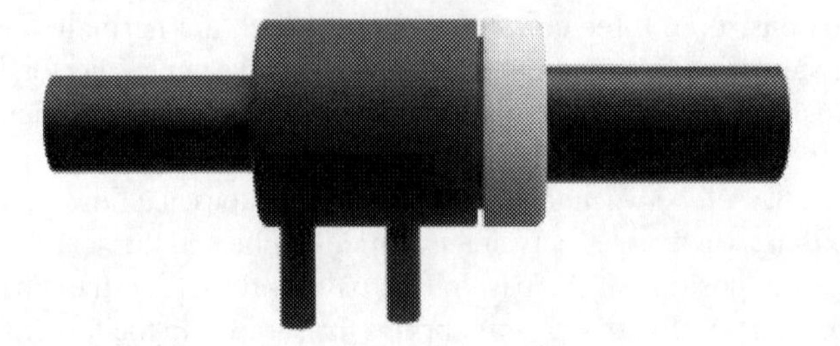

FIGURE 7.4 A fully demountable ceramic torch. (Courtesy of Glass Expansion Inc.)

It's also worth emphasizing that the coil used in an ICP-MS plasma is slightly different from the one used in ICP-OES; the reason being that in a plasma discharge, there is a potential difference of a few hundred volts produced by capacitive coupling between the RF coil and the plasma. In an ICP mass spectrometer, this would result in a secondary discharge between the plasma and the interface cone, which can negatively affect the performance of the instrument. To compensate for this, the coil must be grounded to keep the interface region as close to zero potential as possible. The full implications of this are discussed in greater detail in Chapter 8.

FORMATION OF AN ICP DISCHARGE

Let us now discuss the mechanism of formation of the plasma discharge in greater detail. First, a tangential (spiral) flow of argon gas is directed between the outer and middle tube of a quartz torch. A load coil (usually copper) surrounds the top end of the torch and is connected to an RF generator. When RF power (typically 750–1,500 W, depending on the sample) is applied to the load coil, an alternating current oscillates within the coil at a rate corresponding to the frequency of the generator. In most ICP generators, this frequency is either 27 or 40 MHz (commonly known as megahertz or million cycles/second). This RF oscillation of the current in the coil causes an intense electromagnetic field to be created in the area at the top of the torch. With argon gas flowing through the torch, a high-voltage spark is applied to the gas, causing some electrons to be stripped from their argon atoms. These electrons, which are caught up and accelerated in the magnetic field, then collide with other argon atoms, stripping off still more electrons. This collision-induced ionization of the argon continues in a chain reaction, breaking down the gas into argon atoms, argon ions, and electrons forming what is known as ICP discharge. The ICP discharge is then sustained within the torch and load coil as RF energy is continually transferred to it through the inductive coupling process. The amount of energy required to generate argon ions in this process is on the order of 15.8 eV (first ionization potential), which is enough energy to ionize the majority of the elements in the periodic table. The sample aerosol is then introduced into the plasma through a third tube called the sample injector. The entire process is conceptually shown in Figure 7.5.[7]

THE FUNCTION OF THE RF GENERATOR

Although the principles of an RF power supply have not changed since the work of Greenfield, the components have become significantly smaller. Some of the early generators that used nitrogen or air required 5–10 kW of power to sustain the plasma discharge—and literally took up half the room. Most of today's generators use solid-state electronic components, which means that vacuum power amplifier tubes are no longer required. This makes modern instruments significantly smaller and, because vacuum tubes were notoriously unreliable and unstable, far more suitable for routine operation.

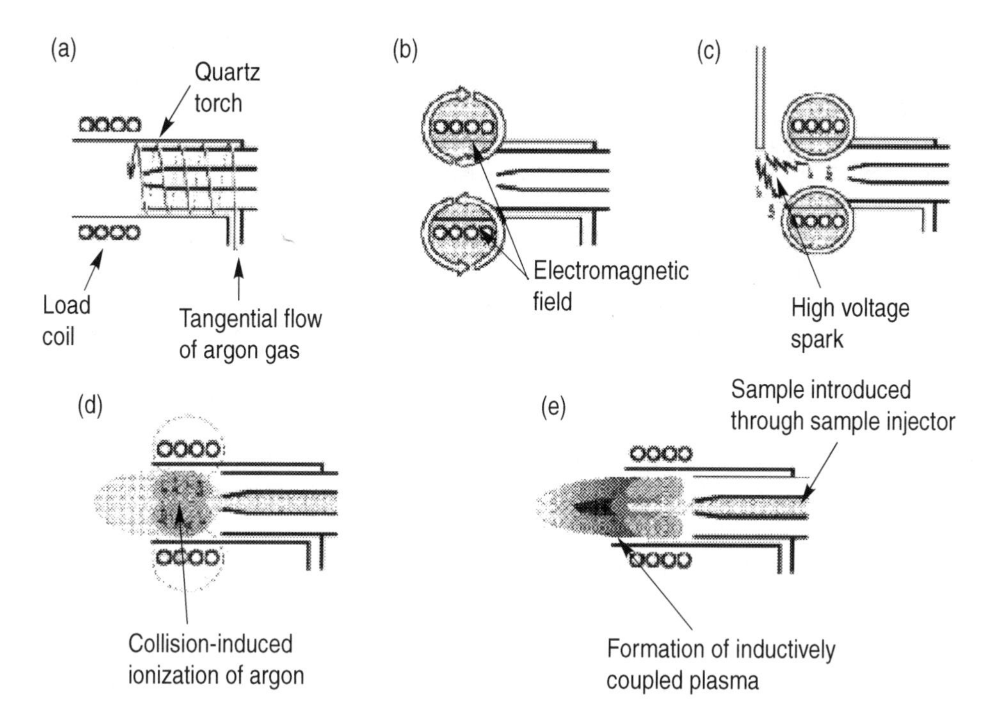

FIGURE 7.5 Schematic of an ICP torch and load coil showing how the ICP is formed. (a) A tangential flow of argon gas is passed between the outer and middle tube of the quartz torch. (b) RF power is applied to the load coil, producing an intense electromagnetic field. (c) A high-voltage spark produces free electrons. (d) Free electrons are accelerated by the RF field, causing collisions and ionization of the argon gas. (e) The ICP is formed at the open end of the quartz torch. The sample is introduced into the plasma via the sample injector. (From C. B. Boss and K. J. Fredeen, *Concepts, Instrumentation and Techniques in Inductively Coupled Plasma Optical Emission Spectrometry*, 2nd edition, PerkinElmer Corporation, Norwalk, 1997.)

As mentioned previously, two frequencies have typically been used for ICP RF generators—27 and 40 MHz. These frequencies have been set aside specifically for RF applications of this kind, so that they will not interfere with other communication-based frequencies. There has been much debate over the years as to which frequency gives the best performance.[8,9] I think it is fair to say that although there have been a number of studies, no frequency appears to give a significant analytical advantage over the other. In fact, of all the commercially available ICP-MS systems, there seems to be roughly an equal number of 27 and 40 generators. More recently, a 34 MHz generator has been developed.

The more important consideration is the coupling efficiency of the RF generator to the coil. The majority of modern solid-state RF generators are on the order of 70%–75% efficient, which means that 70%–75% of the delivered power actually makes it into the plasma. This was not always the case, and some of the older vacuum tube-designed generators were notoriously inefficient, with some of them experiencing over a 50% power loss. Another important criterion to consider is the way the matching network compensates for changes in impedance (a material's resistance to the flow of an electric current) produced by the sample's matrix components or differences in solvent volatility, or both. In earlier-designed crystal-controlled generators, this was usually done with servo-driven capacitors. They worked very well with most sample types but, because they were mechanical devices, struggled to compensate for very rapid impedance changes produced by some samples. As a result, it was fairly easy to extinguish the plasma, particularly when aspirating volatile organic solvents.

These problems were partially overcome by the use of free-running RF generators, in which the matching network was based on electronic tuning of small changes in frequency brought about by the sample solvent or matrix components or both. The major benefit of this approach was that compensation for impedance changes was virtually instantaneous, because there were no moving parts. This allowed for the successful analysis of many sample types, which would most probably have extinguished the plasma of a crystal-controlled generator. However, because of improvements in electronic components over the years, the more recent crystal-controlled generators appear to be as responsive as free-running designs.

However, it should be mentioned that the recent development of a novel 34 MHz free-running-designed RF generator using solid-state electronics has enhanced the capability of ICP-MS to analyze some real-world samples, particularly when using cool plasma conditions (see Chapter 17 "Review of ICP-MS Interferences"). This new design, which is based on an air-cooled plasma load coil, allows the matching network electronics to rapidly respond to changes in the plasma impedance produced by different sampling conditions and sample matrices, while still maintaining low plasma potential at the interface region.[10]

IONIZATION OF THE SAMPLE

To better understand what happens to the sample on its journey through the plasma source, it is important to understand the different heating zones within the discharge. Figure 7.6 shows a cross-sectional representation of the discharge along with the approximate temperatures for different regions of the plasma.

As mentioned previously, the sample aerosol enters the injector via the spray chamber. When it exits the sample injector, it is moving at such a velocity that it physically punches a hole through the center of the plasma discharge. It then goes through a number of physical changes, starting at the preheating zone and continuing through the radiation zone, before it eventually becomes a positively charged ion in the analytical zone. To explain this in a very simplified way, let us assume that the element exists as a trace metal salt in solution. The first step that takes place is desolvation of the droplet. With the water molecules stripped away, it then becomes a tiny solid particle. As the sample moves further into the plasma, the solid particle changes first into gaseous form and then into a ground-state atom. The final process of conversion of an atom to an ion is achieved mainly by collisions of energetic argon electrons (and to a lesser extent by argon ions) with the ground-state atom.[11] The ion then emerges from the plasma and is directed into the interface of the mass spectrometer (for details on the mechanisms of ion generation, refer to Chapter 4). This process of conversion of droplets into ions is represented in Figure 7.7.

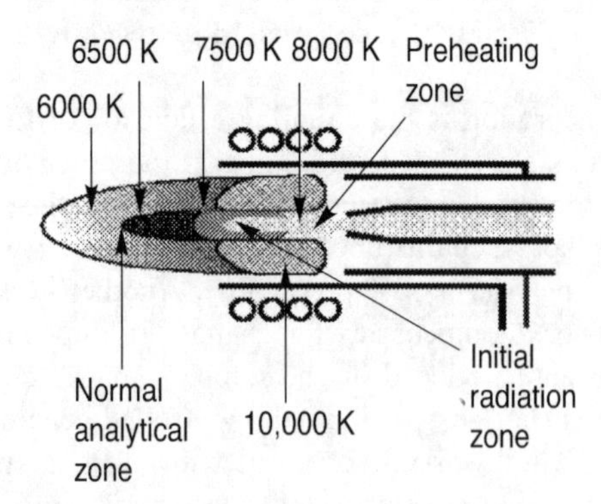

FIGURE 7.6 Different temperature zones in the plasma. (From C. B. Boss and K. J. Fredeen, *Concepts, Instrumentation and Techniques in Inductively Coupled Plasma Optical Emission Spectrometry*, 2nd edition, Perkin Elmer Corporation, Norwalk, 1997.)

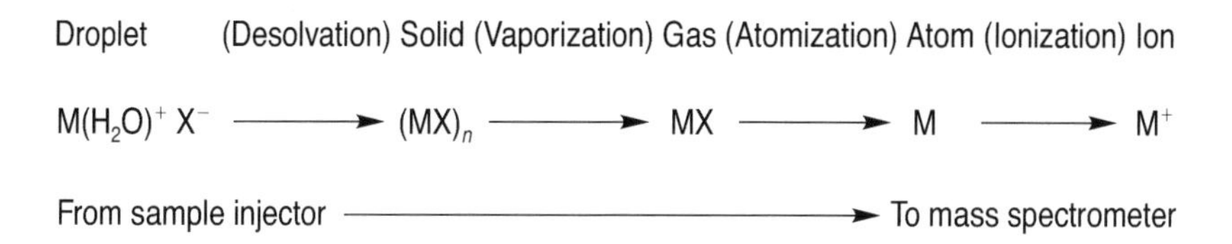

FIGURE 7.7 Mechanism of conversion of a droplet to a positive ion in the ICP.

FURTHER READING

1. A. L. Gray, *Analyst*, **100**, 289–299, 1975.
2. D. J. Douglas and J. B. French, *Analytical Chemistry*, **53**, 37–41, 1981
3. R. J. Thomas, Emerging Technology Trends in Atomic Spectroscopy Are Solving Real-World Application Problems, *Spectroscopy*, **29** (3), 42–51, 2014
4. S. Greenfield, I. L. Jones, and C. T. Berry, *Analyst*, **89**, 713–720, 1964.
5. R. S. Houk, V. A. Fassel, and H. J. Svec, *Dynamic Mass Spectrometry*, **6**, 234–238, 1981.
6. J. W. Lam and J. W. McLaren, *Journal of Analytical Atomic Spectrometry*, **5**, 419–424, 1990.
7. C. B. Boss and K. J. Fredeen, *Concepts, Instrumentation and Techniques in Inductively Coupled Plasma Optical Emission Spectrometry*, 2nd edition, Perkin Elmer Corporation, Norwalk, 1997.
8. K. E. Jarvis, P. Mason, T. Platzner, and J. G. Williams, *Journal of Analytical Atomic Spectrometry*, **13**, 689–696, 1998.
9. G. H. Vickers, D. A. Wilson, and G. M. Hieftje, *Journal of Analytical Atomic Spectrometry*, **4**, 749–754, 1989.
10. K. Neubaeur, *Spectroscopy*, 373–440, 2017.
11. T. Hasegawa and H. Haraguchi, *ICPs in Analytical Atomic Spectrometry*, A. Montasser, D. W. Golightly (eds), 2nd edition, VCH, New York, 1992.

8 Interface Region

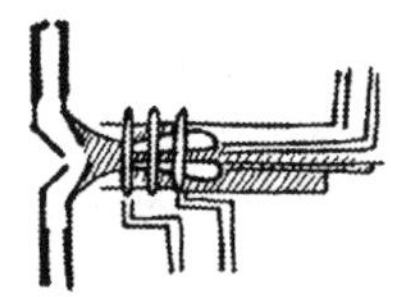

This chapter takes a look at the inductively coupled plasma mass spectrometer (ICP-MS) interface region, which is probably the most critical area of the entire ICP-MS system. It gave the early pioneers of the technique the most problems to overcome. Although we take all the benefits of ICP-MS for granted, the process of taking a liquid sample, generating an aerosol that is suitable for ionization in the plasma and then sampling a representative number of analyte ions, transporting them through the interface, focusing them via the ion optics into the MS, and finally ending up with detection and conversion to an electronic signal is not a trivial task. Each part of the journey has its own unique problems to overcome, but probably the most challenging is the extraction of the ions from the plasma into the MS.

The role of the interface region, which is shown in Figure 8.1, is to transport the ions efficiently, consistently, and with electrical integrity from the plasma, which is at atmospheric pressure (760 torr), to the MS analyzer region at approximately 10^{-6} torr.

This is first achieved by directing the ions into the interface region. The interface consists of two or three metallic cones (depending on the design) with very small orifices, which are maintained at a vacuum of ~1–2 torr with a mechanical roughing pump. After the ions are generated in the plasma, they pass into the first cone, known as the sampler cone, which has an orifice of 0.8–1.2 mm i.d.

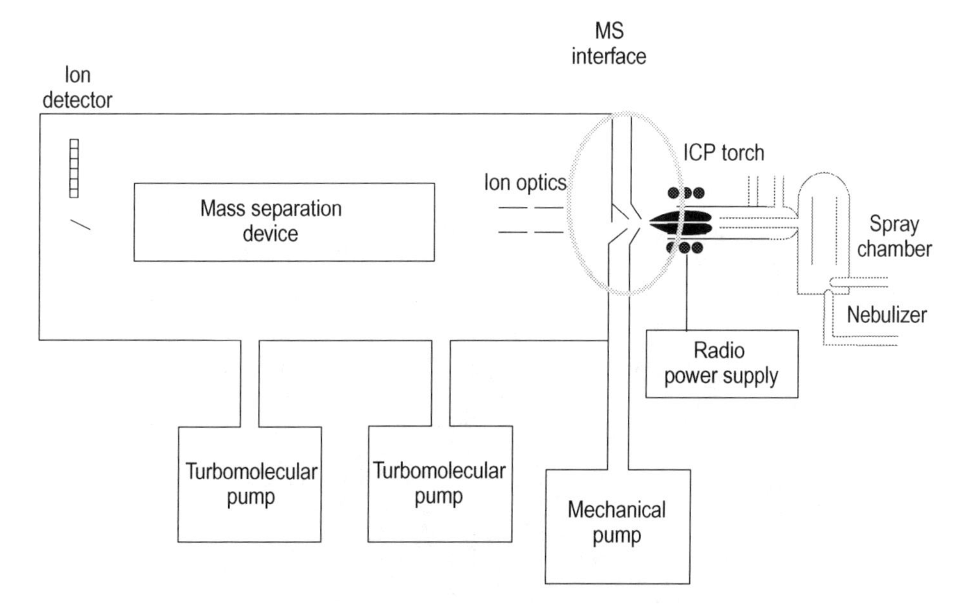

FIGURE 8.1 Schematic of an ICP-MS, showing proximity of the interface region.

From there, they travel a short distance to the skimmer cone, which is generally smaller and more pointed than the sampler cone. The skimmer also has a much smaller orifice (typically 0.4–0.8 mm i.d.) than the sampler cone. In some designs, there is a third cone called the hyper-skimmer cone, which is used to reduce the vacuum in smaller steps and provide less dispersion of the ion beam. Whether the system uses two cones or incorporates a triple cone interface, they are usually made of nickel but can be made of other materials such as platinum, which is far more tolerant to corrosive liquids. To reduce the effects of high-temperature plasma on the cones, the interface housing is water-cooled and made from a material that dissipates heat easily, such as copper or aluminum. The ions then emerge from the skimmer cone, where they are directed through the ion optics and, finally, guided into the mass separation device. Figure 8.2 shows the interface region in greater detail, and Figure 8.3 shows a close-up of a platinum sampler cone on the left and a platinum skimmer cone on the right.

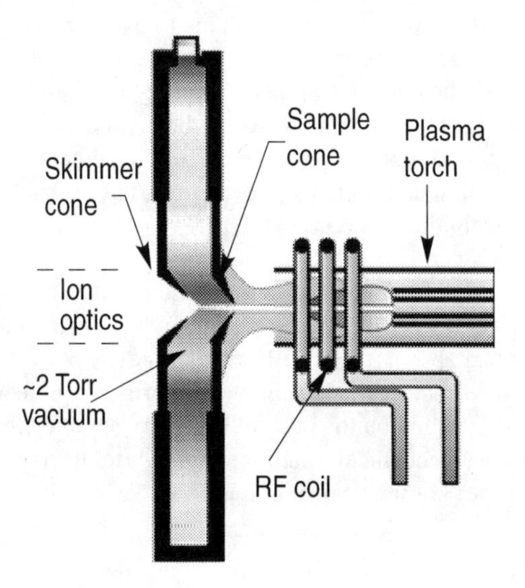

FIGURE 8.2 Detailed view of the interface region.

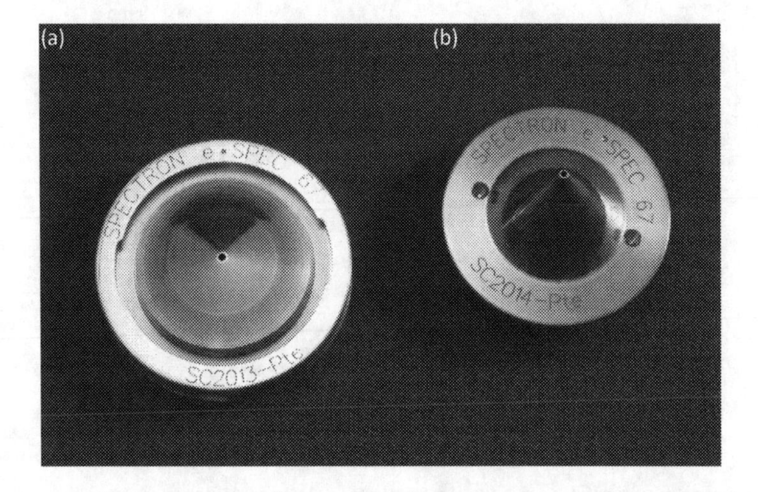

FIGURE 8.3 Close-up of a platinum sampler cone (a) and a platinum skimmer cone (b). (Courtesy of Spectron Inc.)

It should be noted that for most sample matrices, it is desirable to keep the total dissolved solids (TDS) below 0.2%, because of the possibility of deposition of the matrix components around the sampler cone orifice. This is not such a serious problem with short-term use, but it can lead to long-term signal instability if the instrument is being run for extended periods of time. The TDS levels can be higher (0.5%–1%) when analyzing a matrix that forms a volatile oxide such as sodium chloride because, once deposited on the cones, the volatile sodium oxide tends to revaporize without forming a significant layer that could potentially affect the flow through the cone orifice. By careful optimization of the plasma radio frequency (RF) power, sampling depth, and extraction lens voltage, it is therefore possible to run a 1:1 dilution of seawater (1.5% NaCl) for extended periods with no significant cone blockage.

More recently, a novel technique for the handling of high-matrix samples has been developed that introduces a make-up gas between the spray chamber and the torch. This technique, which has been termed "aerosol dilution," does not require the introduction of more water to the system and protects the MS components from the high levels of matrix components in the sample. This technique enables the ICP-MS system to directly aspirate 2%–3% total dissolved solids, with the added benefit of reducing oxide levels, because the sample is not being diluted with water in the traditional way. This technique was described in greater detail in Chapter 6 "ICP-MS Sample Introduction."

CAPACITIVE COUPLING

The coupling of the plasma to the MS proved to be very problematic during the early development of ICP-MS because of an undesirable electrostatic (capacitive) coupling between the voltage on the load coil and the plasma discharge, producing a potential difference of 100–200 V. Although this potential is a physical characteristic of all ICP discharges, it was more serious in an ICP-MS, because the capacitive coupling created an electrical discharge between the plasma and the sampler cone. This discharge, commonly called the "pinch effect" or secondary discharge, shows itself as arcing in the region where the plasma is in contact with the sampler cone.[1] This is shown in a simplified manner in Figure 8.4.

If not addressed, this arcing can cause all kinds of problems, including an increase in doubly charged interfering species, a wide kinetic energy spread of sampled ions, formation of ions generated from the sampler cone, and decreased orifice lifetime. These were all problems reported by many of the early researchers into the technique.[2,3] In fact, because the arcing increased with sampler cone orifice size, the source of the secondary discharge was originally thought to be the result of an electro-gas-dynamic effect, which produced an increase in electron density at the orifice.[4] After many experiments, it was eventually realized that the secondary discharge was a result of

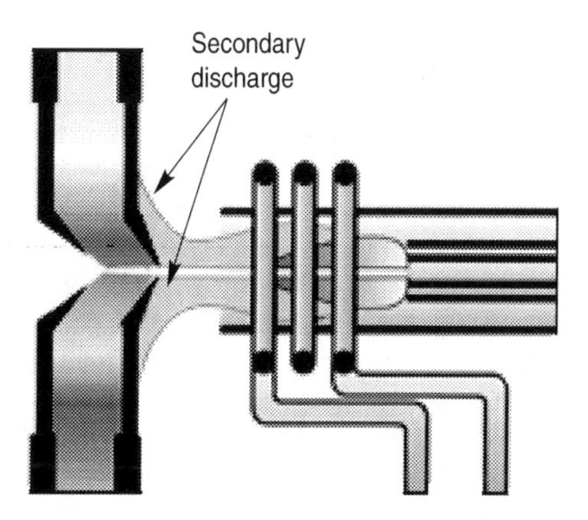

FIGURE 8.4 Interface showing area affected by a secondary discharge.

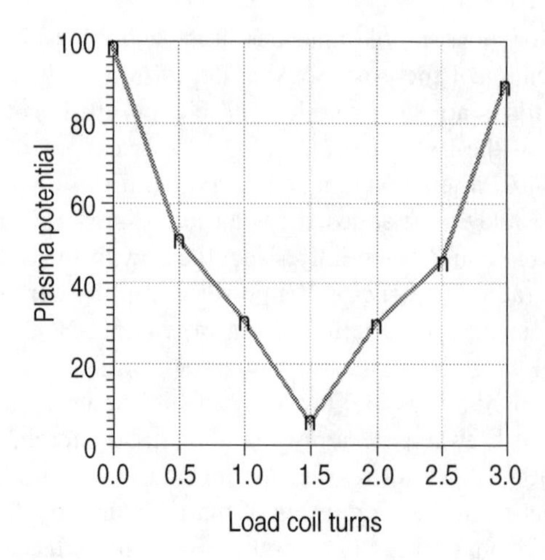

FIGURE 8.5 Reduction in plasma potential as the load coil is grounded at different positions (turns) along its length. (From A. L. Gray, R. S. Houk, and J. G. Williams, *Journal of Analytical Atomic Spectrometry*, **2**, 13–20, 1987.)

electrostatic coupling of the load coil to the plasma. The problem was first eliminated by grounding the induction coil in the center, which had the effect of reducing the RF potential to a few volts. This can be seen in Figure 8.5, which is taken from one of the early papers and shows the reduction in plasma potential as the coil is grounded at different positions (turns) along its length.[5]

This work has since been supported by other researchers who carried out Langmuir probe measurements, the results indicating that plasma potential was lowest with a center-tapped coil as opposed to the grounding being elsewhere on the coil.[6,7] In today's instrumentation, the "grounding" is implemented in a number of different ways, depending on the design of the interface. Some of the most popular designs include balancing the oscillator inside the circuitry of the RF generator,[8] positioning a grounded shield or plate between the coil and the plasma torch,[9] and using two interlaced coils where the RF fields go in opposite directions.[10] They all work differently, but many experts believe that the center-tapped coil and the interlaced coil achieve the lowest plasma potential compared to the other designs. However, they all appear to work equally well when it comes to using cool plasma conditions requiring higher RF power and lower nebulizer gas flow. Further details about cool and cold plasma technology can be found in Chapter 17 "Review of ICP-MS Interferences."

ION KINETIC ENERGY

The impact of a secondary discharge cannot be overemphasized with respect to its effect on the kinetic energy of the ions being sampled. It is well documented that the energy spread of the ions entering the MS must be as low as possible to ensure they can all be focused efficiently and with full electrical integrity by the ion optics and the mass separation device. When the ions emerge from the argon plasma, they will all have different kinetic energies, depending on their mass-to-charge ratio. Their velocities should all be similar, because they are controlled by rapid expansion of the bulk plasma, which will be neutral as long as it is maintained at zero potential. As the ion beam passes through the sampler cone into the skimmer cone or cones, expansion will take place, but its composition and integrity will be maintained, assuming the plasma is neutral. This can be seen in Figure 8.6.

Electrodynamic forces do not play a role as the ions enter the sampler or the skimmer, because the distance over which the ions exert an influence on one another (known as the Debye length) is small (typically 10^{-3} to 10^{-4} mm) compared to the diameter of the orifice (0.5–1.0 mm),[5] as shown in Figure 8.7.

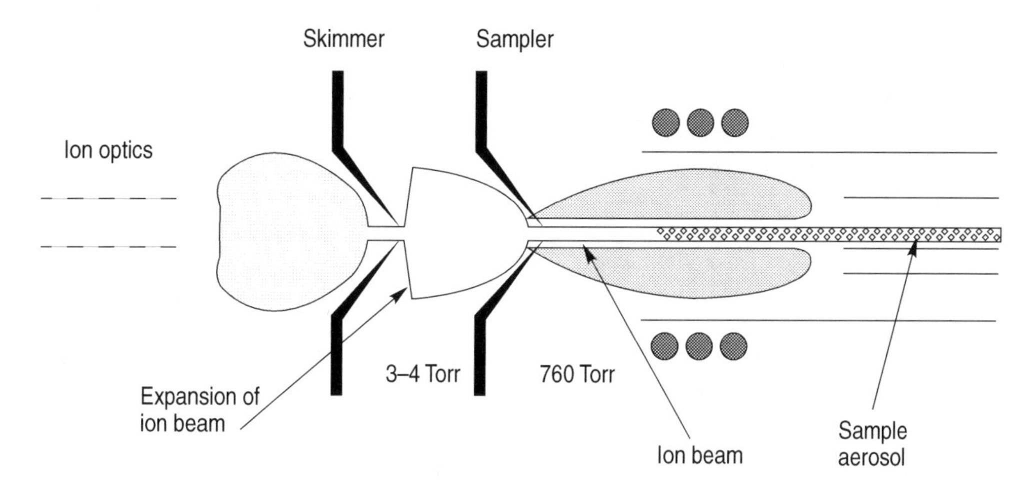

FIGURE 8.6 The composition of the ion beam is maintained as it passes through the interface, a neutral plasma being assumed.

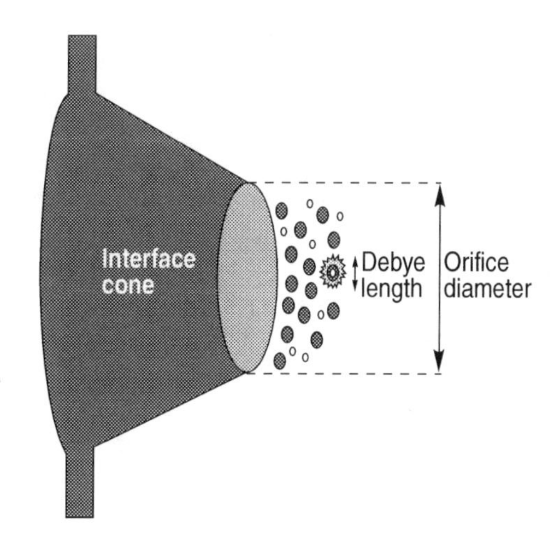

FIGURE 8.7 Electrodynamic forces do not affect the composition of the ion beam entering the sampler or the skimmer cone.

It is therefore clear that maintaining a neutral plasma is of paramount importance to guarantee the electrical integrity of the ion beam as it passes through the interface region. If a secondary discharge is present, the electrical characteristics of the plasma change, which will affect the kinetic energy of the ions differently, depending on their mass-to-charge ratio. If the plasma is at zero potential, the ion energy spread is on the order of <5 eV. However, if a secondary discharge is present, it results in a much wider spread of ion energies entering the MS (typically 20–40 eV), which makes ion focusing far more complicated.[5]

BENEFITS OF A WELL-DESIGNED INTERFACE

The benefits of a well-designed interface are not readily obvious if simple aqueous samples are being analyzed using only one set of operating conditions. However, it becomes more apparent when many different sample types are being handled, requiring different operating parameters. The design of the interface is really put to the test when plasma conditions need to be changed, when the sample matrix changes, or when ICP-MS is being used to analyze solid materials. Analytical scenarios such as these have the potential to induce a secondary discharge, change the kinetic energy of the ions entering the MS, and affect the tuning of the ion optics. It is therefore critical that the interface grounding mechanism be able to handle these types of real-world analytical situations, including:

- **Using cool plasma conditions:** Although not utilized so much (but still useful for some applications) since the development of collision/reaction cells and interfaces, all instruments today have the ability to use cool plasma conditions. By using a combination of one or more of the following: reducing RF power to 500–700 W, reducing plasma coolant flow to ~10 L/min, and increasing nebulizer gas flow to 1.0–1.3 L/min, the plasma temperature is lowered, which reduces argon-based polyatomic interferences such as $^{40}Ar^{16}O^+$, $^{40}Ar^+$, and $^{38}ArH^+$ in the determination of difficult elements such as $^{56}Fe^+$, $^{40}Ca^+$, and $^{39}K^+$. Such dramatic deviations from normal operating conditions (RF power: 1,000 W, plasma coolant flow: 15 L/min, neb flow: 0.8 L/min) will affect the electrical characteristics of the plasma. (Note: For some applications, it can be beneficial to combine cool/cold plasma conditions with collision/reaction cell technology to reduce some polyatomic spectral interferences.)
- **Running organic solvents:** Analyzing oil or organic-based samples requires a chilled spray chamber or a membrane desolvation system to reduce the solvent loading on the plasma. In addition, higher RF power (~1300–1500 W) and lower nebulizer gas flow (~0.4–0.8 L/min) are required to dissociate the organic components in the sample. A reduction in the amount of solvent entering the plasma combined with higher power and lower nebulizer gas flow translates into a hotter plasma and a change in its ionization mechanism.
- **Optimizing conditions for low oxides:** The formation of oxide species can be problematic in some sample types. For example, in geochemical applications, it is quite common to sacrifice sensitivity by lowering the nebulizer gas flow and increasing the RF power to reduce the formation of rare earth oxides, which can spectrally interfere with the determination of other analytes. Unfortunately, these conditions will change the electrical characteristics of the plasma, which can induce a secondary discharge.
- **Using sampling accessories:** Sampling accessories such as membrane desolvators and laser ablation systems are being used more routinely to improve performance/productivity and enhance the flexibility of ICP-MS. Some of these sampling devices such as laser ablation or membrane desolvation generate a "dry" sample aerosol, which requires completely different operating conditions compared to conventional "wet" plasma. An aerosol that contains no solvent can have a dramatic effect on the ionization conditions in the plasma.

FINAL THOUGHTS

Even though most modern ICP-MS interfaces have been designed to minimize the effects of the secondary discharge, it should not be taken for granted that they can all handle changes in operating conditions and matrix components with the same ease. The most noticeable problems that have been reported include spectral peaks of the cone material appearing in the blank, erosion/discoloration of the sampling cones, widely different optimum plasma conditions (neb flow/RF power) for different masses, and frequent retuning of the ion optics.[11,12] Chapter 30 on how best to evaluate ICP-MS instrumentation goes into this subject in greater detail, but there is no question that the plasma discharge, interface region, and ion optics have to be designed in concert to ensure the instrument can handle a wide range of operating conditions and sample types. In addition, the interface cones have to be cleaned on a regular basis to ensure the secondary discharge is kept to a minimum. Neufeld wrote an excellent article on how to clean interface cones that may have been impacted by long-term aspiration of matrix components or discolored by acid erosion.[13]

FURTHER READING

1. A. L. Gray and A. R. Date, *Analyst*, **108**, 1033–1041, 1983.
2. R. S. Houk, V. A. Fassel, and H. J. Svec, *Dynamic Mass Spectrometry*, **6**, 234–238, 1981.
3. A. R. Date and A. L. Gray, *Analyst*, **106**, 1255–1259, 1981.
4. A. L. Gray and A. R. Date, *Dynamic Mass Spectrometry*, **6**, 252–256, 1981.
5. D. J. Douglas and J. B. French, *Spectrochimica Acta*, **41B** (3), 197–201, 1986.
6. A. L. Gray, R. S. Houk, and J. G. Williams, *Journal of Analytical Atomic Spectrometry*, **2**, 13–20, 1987.
7. R. S. Houk, J. K. Schoer, and J. S. Crain, *Journal of Analytical Atomic Spectrometry*, **2**, 283–286, 1987.
8. S. D. Tanner, *Journal of Analytical Atomic Spectrometry*, **10**, 905–911, 1995.
9. K. Sakata and K. Kawabata, *Spectrochimica Acta*, **49B**, 1027, 1994.
10. S. Georgitus and M. Plantz, *Winter Conference on Plasma Spectrochemistry*, **FP4**, Fort Lauderdale, 1996.
11. D. J. Douglas, *Canadian Journal of Spectroscopy*, **34**, 2–8, 1989.
12. J. E. Fulford and D. J. Douglas, *Applied Spectroscopy*, **40**, 7–12, 1986.
13. L. Neufeld, *Spectroscopy*, **34** (7), 12–17, 2019.

9 Ion-Focusing System

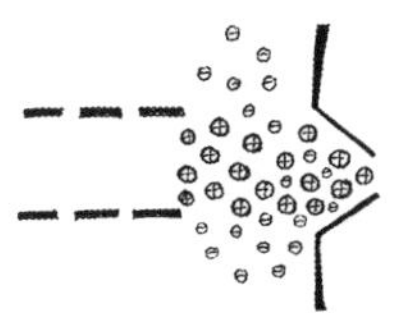

This chapter takes a detailed look at the inductively coupled plasma mass spectrometry (ICP-MS) ion-focusing system—a crucial area of the ICP mass spectrometer—where the ion beam is focused before it enters the mass analyzer. Sometimes known as the ion optics, it comprises one or more ion lens components, which electrostatically steer the analyte ions in an axial (straight) or orthogonal (right-angled) direction from the interface region into the mass separation device. The strength of a well-designed ion-focusing system is its ability to produce a flat signal response over the entire mass range, low background levels, good detection limits, and stable signals in real-world sample matrices.

Although the detection capability of ICP-MS is generally recognized as being superior to any of the other atomic spectroscopic techniques, it is probably most susceptible to the sample's matrix components. The inherent problem lies in the fact that ICP-MS is relatively inefficient—out of a million ions generated in the plasma, only a few ions actually reach the detector. One of the main contributing factors to the low efficiency is the higher concentration of matrix elements compared to the analyte, which has the effect of defocusing the ions and altering the transmission characteristics of the ion beam. This is sometimes referred to as a space charge effect and can be particularly severe when the matrix ions are of a heavier mass than the analyte ions.[1] The role of the ion-focusing system is therefore to transport the maximum number of analyte ions from the interface region to the mass separation device, while rejecting as many of the matrix components and non-analyte-based species as possible. Let us now discuss this process in greater detail.

ROLE OF THE ION OPTICS

The ion optics, shown in Figure 9.1, are positioned between the skimmer cone (or cones) and the mass separation device. They typically consist of one or more electrostatically controlled lens components, maintained at a vacuum of approximately 10^{-3} torr with a turbomolecular pump.

They are not traditional optics that we associate with ICP emission or atomic absorption but are made up of a series of metallic plates, barrels, or ion mirrors, which have a voltage placed on them. A recent commercial design implements a quadrupole deflector, which turns the ion beam at right angles into the mass spectrometer. Whatever the design, the function of the ion optic system is to take ions from the hostile environment of the plasma at atmospheric pressure via the interface cones and steer them into the mass analyzer, which is under high vacuum. The nonionic species such as particulates, neutral species, and photons are prevented from reaching the detector by using some kind of physical barrier, positioning the mass analyzer off axis relative to the ion beam, or electrostatically bending the ions by 90° into the mass analyzer.

As mentioned in previous chapters, the plasma discharge and interface region have to be designed in concert with the ion optics. It is absolutely critical that the composition and electrical

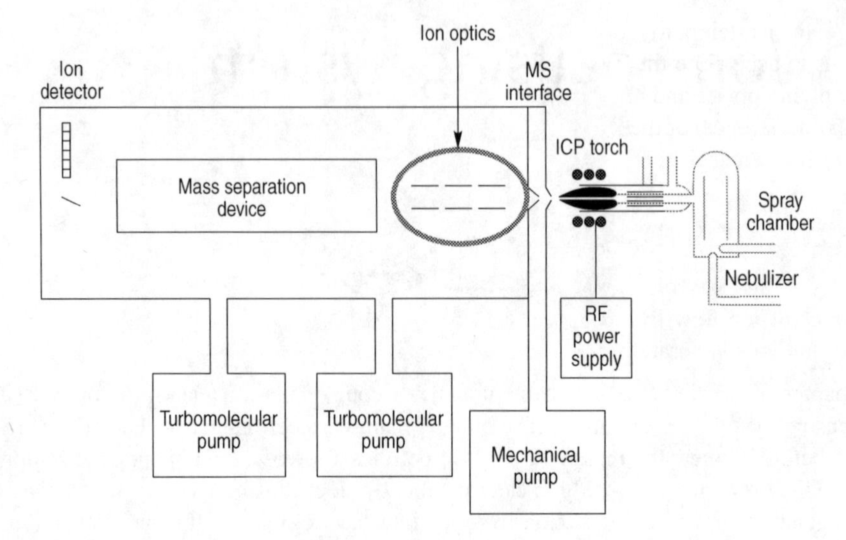

FIGURE 9.1 Position of ion optics relative to the plasma torch and interface region.

integrity of the ion beam be maintained as it enters the ion optics. For this reason, it is essential that the plasma be at zero potential to ensure that the magnitude and spread of ion energies are as low as possible.[2]

A secondary, but also very important, role of the ion optic system is to stop particulates, neutral species, and photons from getting through to the mass analyzer and the detector. These species cause signal instability and contribute to background levels, which ultimately affect the performance of the system. For example, if photons or neutral species reach the detector, they will elevate the noise of the background and therefore degrade detection capability. In addition, if particulates from the matrix penetrate further into the mass spectrometer region, they have the potential to deposit on lens components, and in extreme cases, get into the mass analyzer. In the short term, this will cause signal instability and, in the long term, increase the frequency of cleaning and routine maintenance.

There are basically four different approaches of reducing the chances of these undesirable species entering the mass spectrometer. The first method is to place a grounded metal stop (disk) behind the skimmer cone. This stop allows the ion beam to move around it and physically block the particulates, photons, and neutral species from traveling "downstream."[3] Although implemented very successfully in earlier instruments, this design has not been utilized for a number of years, mainly because it can have a negative impact on instrument sensitivity, by reducing the number of ions reaching the detector. The second approach is to set the mass analyzer off axis to the ion lens system (in some systems this is called a chicane design). The positively charged ions are then steered with the lens components into the mass analyzer, while the photons, neutral, and nonionic species are ejected out of the ion beam.[4] The third development is to deflect the ion beam 90° with anion mirror.[5] This allows the photons, neutrals, and solid particles to pass through, whereas the ions are deflected at right angles into an off-axis mass analyzer that incorporates curved fringe rod technology.[6] The fourth and most recent approach is to deflect the ion beam emerging from the plasma by 90°. This has the effect of changing direction and focusing the ion beam into the mass spectrometer, while allowing the neutral species, photons, and particulate matter to go straight through and be ejected.

It is also worth mentioning that some lens systems incorporate an extraction lens after the skimmer cone to electrostatically "pull" the ions from the interface region. This has the benefit of improving the transmission and detection limits of the low-mass elements (which tend to be pushed out of the ion beam by the heavier elements), resulting in a more uniform response across the full mass range. In an attempt to reduce these space charge effects, some older designs have utilized lens components to accelerate the ions downstream. Unfortunately, this can have the effect of degrading the resolving power and abundance sensitivity (ability to differentiate an analyte peak from the wing of an interference) of the instrument, because of the much higher kinetic energy of the accelerated ions as they enter the mass analyzer.[7]

DYNAMICS OF ION FLOW

To fully understand the role of the ion optics in ICP-MS, it is important to get an appreciation of the dynamics of ion flow from the plasma through the interface region into the mass spectrometer. When the ions generated in the plasma emerge from the skimmer cone (or cones), there is a rapid expansion of the ion beam as the pressure is reduced from 760 torr (atmospheric pressure) to approximately 10^{-3} to 10^{-4} torr in the lens chamber with a turbomolecular pump. The composition of the ion beam immediately behind the cone is the same as in front of the cone because the expansion at this stage is controlled by normal gas dynamics and not by electrodynamics. One of the main reasons for this is that in the ion-sampling process, the Debye length (the distance over which ions exert influence on one another) is small compared to the orifice diameter of the sampler or skimmer cone. Consequently, there is little electrical interaction between the ion beam and the cone and relatively little interaction between the individual ions in the beam. In this way, the compositional integrity of the ion beam is maintained throughout the interface region.[8] With the rapid drop in pressure in the lens chamber, electrons diffuse out of the ion beam. Because of the small size of the electrons relative to the positively charged ions, the electrons diffuse further from the beam than the ions, resulting in an ion beam with a net positive charge. This is represented schematically in Figure 9.2.

The generation of a positively charged ion beam is the first stage in the charge separation process. Unfortunately, the net positive charge of the ion beam means that there is now a natural tendency for the ions to repel each other. If nothing is done to compensate for this, ions of higher mass-to-charge

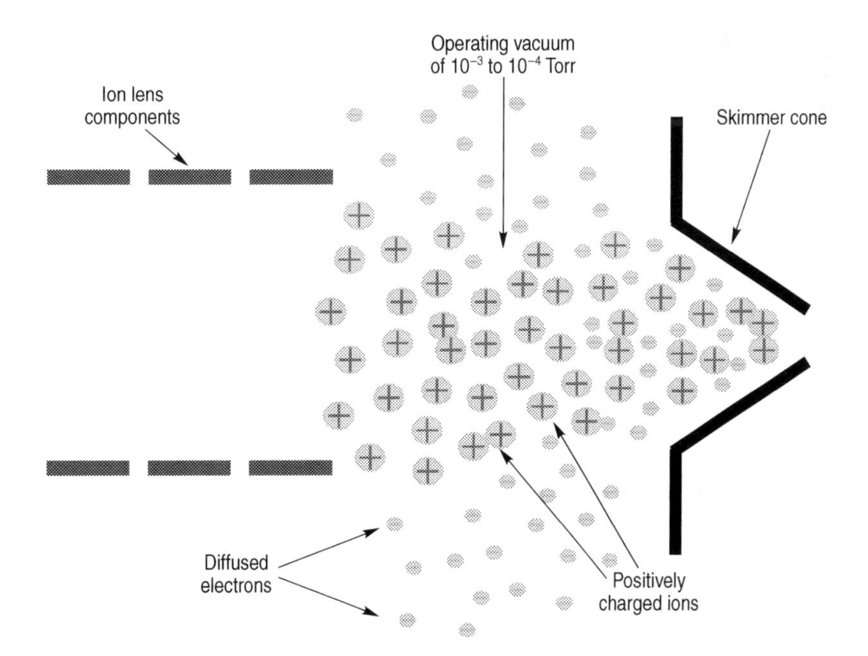

FIGURE 9.2 Extreme pressure drop in the ion optic chamber produces diffusion of electrons, resulting in a positively charged ion beam.

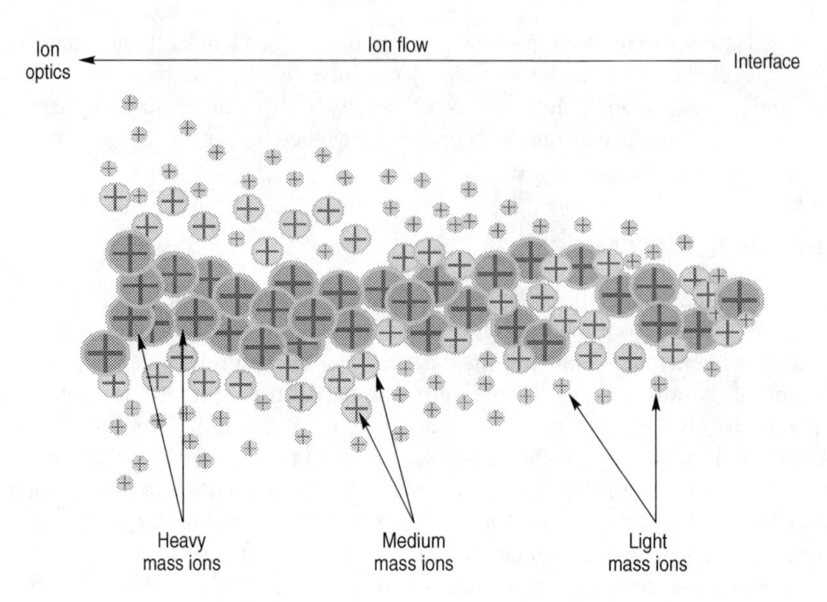

FIGURE 9.3 The degree of ion repulsion will depend on the kinetic energy of the ions—those with high kinetic energy (heavy masses) will be transmitted in preference to those with medium (medium masses) or low kinetic energy (light masses).

ratio will dominate the center of the ion beam and force the lighter ions to the outside. The degree of loss will depend on the kinetic energy of the ions—those with high kinetic energy (high-mass elements) will be transmitted in preference to ions with medium (mid-mass elements) or low kinetic energy (low-mass elements). This is shown in Figure 9.3.

The second stage of charge separation, therefore, consists of electrostatically steering the ions of interest back into the center of the ion beam using the ion lens system. It should be emphasized that this is only possible if the interface is kept at zero potential, which ensures a neutral gas dynamic flow through the interface, maintaining the compositional integrity of the ion beam. It also guarantees that the average ion energy and energy spread of each ion entering the lens systems are at levels optimum for mass separation. If the interface region is not grounded correctly, stray capacitance will generate a discharge between the plasma and sampler cone and increase the kinetic energy of the ion beam, making it very difficult to optimize the ion lens voltages (refer to Chapter 7 for details).

COMMERCIAL ION OPTIC DESIGNS

Over the years, there have been many different ion optic designs. Although they have their own individual characteristics, they perform the same basic function of allowing the maximum number of analyte ions through to the mass analyzer, while at the same time rejecting the undesirable matrix- and solvent-based ions. The oldest and most mature design of ion optics in use today consists of several lens components, all of which have a specific role to play in the transmission of the analyte ions with a minimum of mass discrimination. With these multicomponent lens systems, the voltage can be optimized on every lens of the ion optics to achieve the desired ion specificity. This type of lens configuration has been used in commercial instrumentation for almost 30 years and has proved to be very durable. One of its main benefits is that it produces a uniform response across the mass range with very low background levels, particularly when combined with an off-axis mass analyzer.[9] A schematic of a commercially available multicomponent lens systems is shown in Figure 9.4.

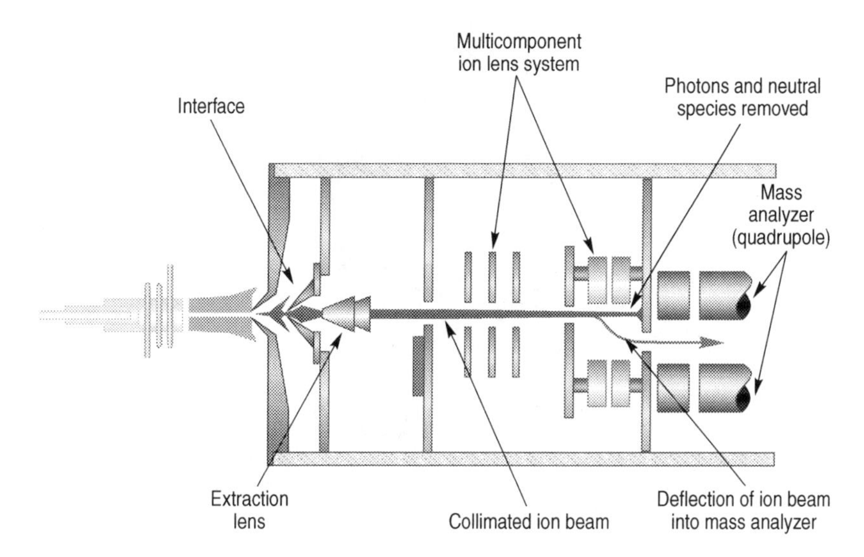

FIGURE 9.4 Schematic of a multicomponent lens system. (From Y. Kishi, *Agilent Technologies Application Journal*, August 1997.)

It should be emphasized that because of the interactive nature of parameters that affect the signal response, the more complex the lens system, the more the variables that have to be optimized. For this reason, if many different sample types are being analyzed, extensive lens optimization procedures have to be carried out for each matrix or group of elements. This is not such a major problem, because most of the lens voltages are computer controlled and methods can be stored for every new sample scenario. However, it could be a factor if the instrument is being used for the routine analysis of many diverse sample types, all requiring different lens settings.

Another well-established approach was the use of a cylinder lens, combined with a grounded stop—positioned just inside the skimmer cone. With this design, the voltage is dynamically ramped "on the fly," in concert with the mass scan of the analyzer. The benefit is that the optimum lens voltage is placed on every mass in a multielement run to allow the maximum number of analyte ions through, while keeping the matrix ions down to an absolute minimum.[10] This design is typically used in conjunction with a grounded stop to act as a physical barrier to reduce the chances that particulates, neutral species, and photons will reach the mass analyzer and detector. Although this design does not generate such a uniform mass response across the full range as an off-axis multilens system with an extraction lens, it appears to offer better long-term stability with real-world samples. It works well for many sample types but is most effective when low-mass elements are being determined in the presence of high-mass matrix elements.

A more recent variation of the single lens approach uses a right-angled cylinder lens. This is a very simple design that utilizes a single, fixed-voltage ion lens to eliminate particulates and photons from reaching the detector. This is done by deflecting the positive ion beam 90°, thus allowing the neutral species to go straight through and be pumped out of the mass spectrometer. The benefit of this design is simplicity, reduced maintenance, easy to clean, and very low background levels. This lens system is unique to one instrument design and is used in conjunction with a low mass cut-off collision/reaction cell. A photograph of this ion optic lens system is shown in Figure 9.5.

Another design in ion-focusing optics utilizes a parabolic electrostatic field created with an ion mirror to reflect and refocus the ion beam at 90° to the ion source.[5] This ion mirror incorporates a hollow structure, which allows photons, neutrals, and solid particles to pass through it, while allowing ions to be reflected at right angles into the mass analyzer. The major benefit of this design is the very efficient way the ions are refocused, offering the capability of extremely high sensitivity across

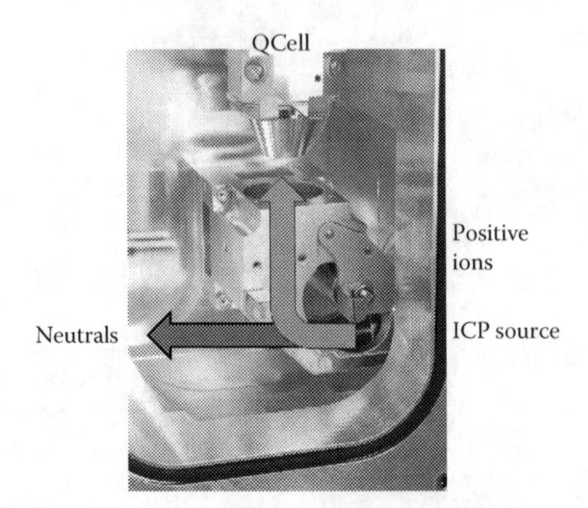

FIGURE 9.5 Right-angled cylinder lens showing how it deflects the positive ion beam 90° into the collision/reaction cell, while allowing the neutral species to go straight through.

the mass range, with very little sacrifice in oxide performance. In addition, there is very little contamination of the ion optics, because a vacuum pump sits behind the ion mirror to immediately remove these particles before they have a chance to penetrate further into the mass spectrometer. Removing these undesirable species and photons before they reach the detector, in addition to incorporating curved fringe rods prior to an off-axis mass analyzer, means that background levels are very low. Figure 9.6 shows a schematic of a quadrupole-based ICP-MS that utilizes a 90° ion optic design.[6,11]

The most novel commercial development in ion optic design utilizes a quadrupole ion deflector (QID), which utilizes a miniaturized quadrupole. This novel filtering technology bends the ion beam 90°, focusing ions of a specified mass into the mass separation device, while discarding all neutral species, photons, and particulates into the turbo pump. The major benefit of removing these nonionic species means they won't be deposited on component surfaces in the mass analyzer region, which significantly minimizes drift and ensures good signal stability, even when running the most challenging sample matrices. This QID approach to ion filtering is shown in Figure 9.7.

It is also worth emphasizing that a number of ICP-MS systems offer what is called a high-sensitivity option. All these work slightly differently but share similar components. By using a combination of slightly different cone geometry, higher vacuum at the interface, one or more extraction lenses, and slightly modified ion optic design, they offer up to ten times the sensitivity of a traditional interface. However, in some systems, this increased sensitivity sometimes comes with slightly worse stability

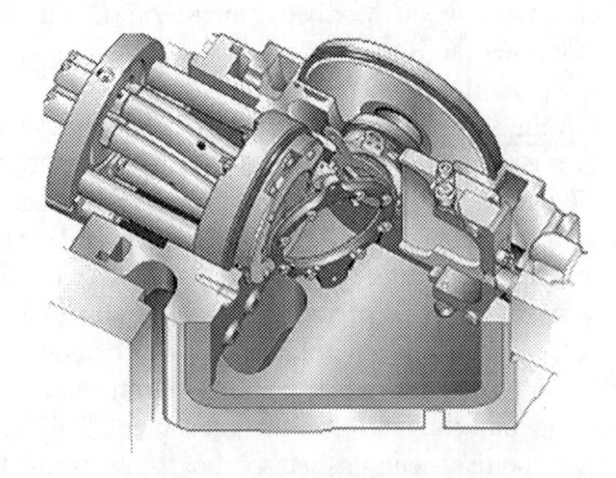

FIGURE 9.6 A 90° ion optic design used with curved fringe rods and an off-axis quadrupole mass analyzer. (Courtesy of Analytik Jena.)

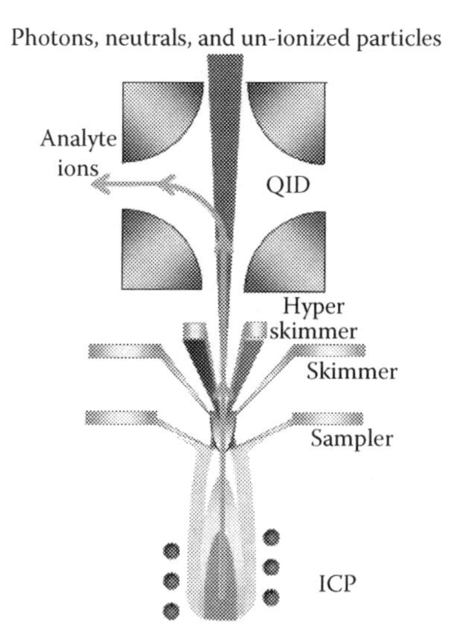

FIGURE 9.7 A novel commercial development in ion optic design utilizes a QID. (Copyright 2013, all rights reserved, PerkinElmer Inc.)

and an increase in background levels, particularly for samples with a heavy matrix. To get around this, these kinds of samples typically need to be diluted before analysis—which has somewhat limited their applicability to real-world samples with high dissolved solids.[12] However, they have found a use in non-liquid-based applications in which high sensitivity is crucial—for example, in the analysis of small spots on the surface of a geological specimen using laser ablation ICP-MS. For this application, the instrument must offer high sensitivity, because a single laser pulse is often used to ablate very small amounts of the sample, which is then swept into the ICP-MS for analysis.

The importance of the ion-focusing system cannot be overemphasized, because it has a direct bearing on the number of ions that find their way to the mass analyzer. In addition to affecting background levels and instrument response across the entire mass range, it has a huge impact on both long- and short-term signal stability, especially in real-world samples. However, there are many different ways of achieving this. It is almost irrelevant whether the design of the ion optics is based on a multicomponent lens system, an RF-multipole guide, or a right-angled deflection approach using an ion mirror or a quadrupole. The most important consideration when evaluating any ion lens system is not the actual design but its ability to perform well with real sample matrices.

FURTHER READING

1. J. A. Olivares and R. S. Houk, *Analytical Chemistry*, **58**, 20, 1986.
2. D. J. Douglas and J. B. French, *Spectrochimica Acta*, **41B** (3), 197–201, 1986.
3. S. D. Tanner, L. M. Cousins, and D. J. Douglas, *Applied Spectroscopy*, **48**, 1367–1372, 1994.
4. D. Potter, *American Lab*, 34–38, July, 1994.
5. I. Kalinitchenko, Ion Optical System for a Mass Spectrometer, Patent Number 750860, 1999.
6. S. Elliott, M. Plantz, and L. Kalinitchenko, Oral Paper 1360-8, *Pittsburgh Conference*, Orlando, FL, 2003.
7. P. Turner, Paper at *2nd International Conference on Plasma Source Mass Spec*, Durham, UK, 1990.
8. S. D. Tanner, D. J. Douglas, and J. B. French, *Applied Spectroscopy*, **48**, 1373–1378, 1994.
9. Y. Kishi, *Agilent Technologies Application Journal*, 5(2), 10–14, August, 1997.
10. E. R. Denoyer, D. Jacques, E. Debrah, and S. D. Tanner, *Atomic Spectroscopy*, **16** (1), 1–6, 1995.
11. I. Kalinitchenko, Mass Spectrometer Including a Quadrupole Mass Analyzer Arrangement, Patent Applied for—WO 01/91159 A1.
12. B. C. Gibson, Paper at *Surrey International Conference on ICP-MS*, London, UK, 1994.

10 Mass Analyzers
Quadrupole Technology

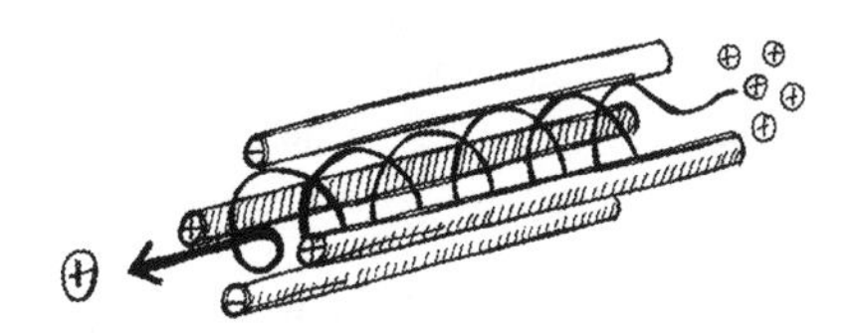

Chapters 10–13 deal with the heart of the inductively coupled plasma mass spectrometry (ICP-MS) system—the mass separation device. Sometimes called the mass analyzer, it is the region of the ICP mass spectrometer that separates the ions according to their mass-to-charge ratio. This selection process is achieved in a number of different ways, depending on the mass separation device, but they all have one common goal, which is to separate the ions of interest from all other non-analyte, matrix, solvent, and argon-based ions. Quadrupole mass filters are described in this chapter, followed by magnetic sector systems in Chapter 11, time-of-flight mass spectrometers in Chapter 12, and finally collision/reaction cell/interface technology in Chapter 13.

Although ICP-MS was commercialized in 1983, the first 10 years of its development utilized a traditional quadrupole mass analyzer to separate the ions of interest. These worked exceptionally well for most applications but proved to have limitations when determining difficult elements or dealing with more complex sample matrices. This led to the development of alternative mass separation devices that allowed ICP-MS to be used for applications that were previously beyond the capabilities of quadrupole-based technology. Before we discuss these different mass spectrometers in greater detail, let us take a look at the proximity of the mass analyzer in relation to the ion optics and detector. Figure 10.1 shows this in greater detail.

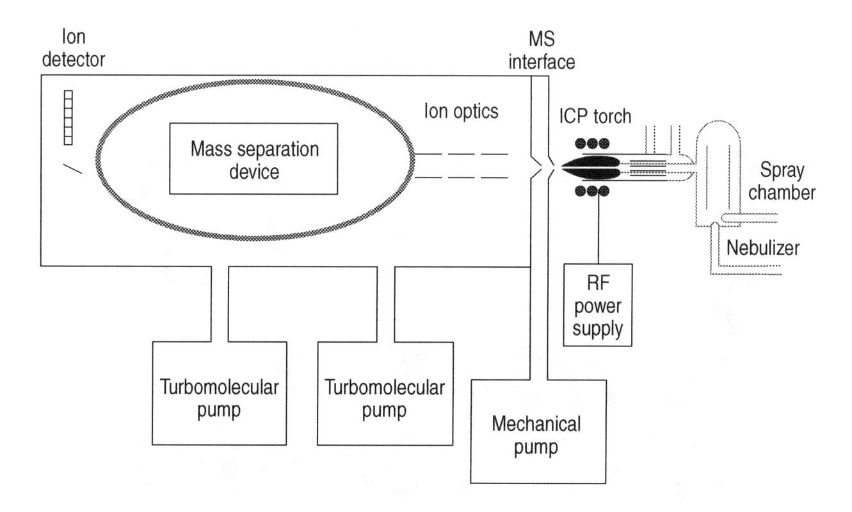

FIGURE 10.1 The mass separation device is positioned between the ion optics and the detector.

As can be seen, the mass analyzer is positioned between the ion optics and detector, and it is maintained at a vacuum of approximately 10^{-6} torr with an additional turbomolecular pump to the one that is used for the lens chamber. Assuming the ions are emerging from the ion optics at the optimum kinetic energy, they are ready to be separated according to their mass-to-charge ratio (m/z) by the mass analyzer. There are basically three different kinds of commercially available mass analyzers: quadrupole mass filters, double-focusing magnetic sectors, and time-of-flight mass spectrometers. It should be noted that although collision/reaction cell/interface technology has been given its own chapter in this book, it is not utilized on its own to carry out mass separation. It is typically used in conjunction with a quadrupole (or quadrupoles) to reduce the impact of polyatomic spectral interferences before the ions are passed into the mass analyzer for separation and detection. They are very powerful enhancements to the technique, but they are not considered primary separation devices on their own.

They all have their own strengths and weaknesses, which will be discussed in greater detail over the next four chapters. Let us first begin with the most common type of mass separation device used in ICP-MS—the quadrupole mass filter.

QUADRUPOLE TECHNOLOGY

Developed in the early 1980s for ICP-MS, quadrupole-based systems represent approximately 90% of all ICP mass spectrometers used today. This design was the first to be commercialized, and as a result, today's quadrupole ICP-MS technology is considered a very mature, routine trace element technique. A quadrupole usually consists of four cylindrical or hyperbolic metallic rods of the same length and diameter, although in one design of collision/reaction cell a quadrupole with flat sides is used (this "flatapole" design is described in greater detail in Chapter 13 "Mass Analyzers: Collision/Reaction Cell and Interface Technology"). Quadrupole rods are typically made of stainless steel or molybdenum and sometimes coated with a ceramic coating for corrosion resistance. When used for ICP-MS, they are typically 15–25 cm in length, about 1 cm in diameter, and operate at a frequency of 2–3 MHz. Figure 10.2 shows a photograph of a quadrupole system mounted in its housing.

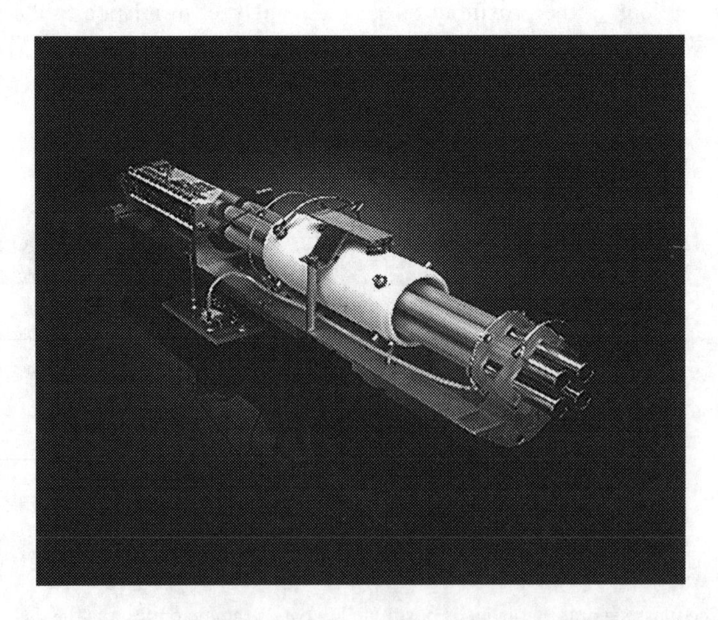

FIGURE 10.2 Photograph of a quadrupole system mounted in its housing. (Copyright 2013, all rights reserved, PerkinElmer Inc.)

BASIC PRINCIPLES OF OPERATION

A quadrupole operates by placing both a direct current (DC) field and a time-dependent alternating current (AC) of radio frequency on opposite pairs of the four rods. By selecting the optimum AC/DC ratio on each pair of rods, ions of a selected mass are allowed to pass through the rods to the detector, whereas the others are unstable and ejected from the quadrupole. Figure 10.3 shows this in greater detail.

In this simplified example, the analyte ion (black) and four other ions (gray) have arrived at the entrance to the four rods of the quadrupole. When a particular AC/DC potential is applied to the rods, the positive or negative bias on the rods will electrostatically steer the analyte ion of interest down the middle of the four rods to the end, where it will emerge and be converted to an electrical pulse by the detector. The other ions of different *m/z* values will be unstable, pass through the spaces between the rods, and be ejected from the quadrupole. This scanning process is then repeated for another analyte with a completely different mass-to-charge ratio until all the analytes in a multielement analysis have been measured. The process for the detection of one particular mass in a multielement run is represented in Figure 10.4.

It shows a $^{63}Cu^{+}$ ion emerging from the quadrupole and being converted to an electrical pulse by the detector. As the AC/DC voltage of the quadrupole—corresponding to $^{63}Cu^{+}$—is repeatedly scanned, the ions are stored and counted by a multichannel analyzer as electrical pulses. This multichannel data acquisition system typically has 20 channels per mass. As the electrical pulses are counted in each channel, a profile of the mass is built up over the 20 channels, corresponding to the spectral peak of $^{63}Cu^{+}$. In a multielement run, repeated scans are made over the entire suite of analyte masses as opposed to just one mass represented in this example.

Quadrupole scan rates are typically on the order of 2,500–5,000 amu per second (depending on the commercial design) and can cover the entire mass range of 0–300 amu in about one tenth of a second. However, real-world analytical speed and sample throughput are much slower than this, and in practice, 25 elements can be determined in duplicate with good precision in 1–2 min, depending on the analytical requirements.

QUADRUPOLE PERFORMANCE CRITERIA

There are two very important performance specifications of a mass analyzer that govern its ability to separate an analyte peak from a spectral interference. The first is the resolving power (R), which in traditional mass spectrometry is represented by the equation $R = m/\Delta m$, where m is the

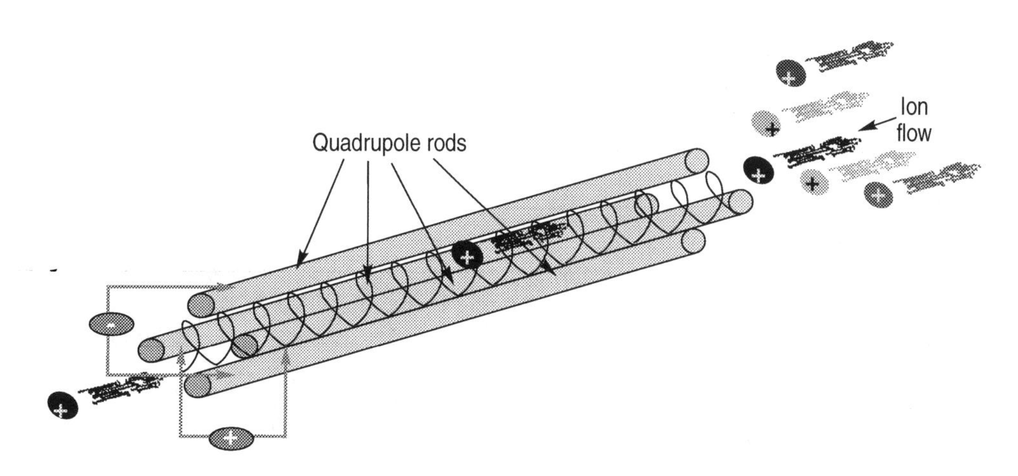

FIGURE 10.3 Schematic representation showing principles of mass separation using a quadrupole mass filter.

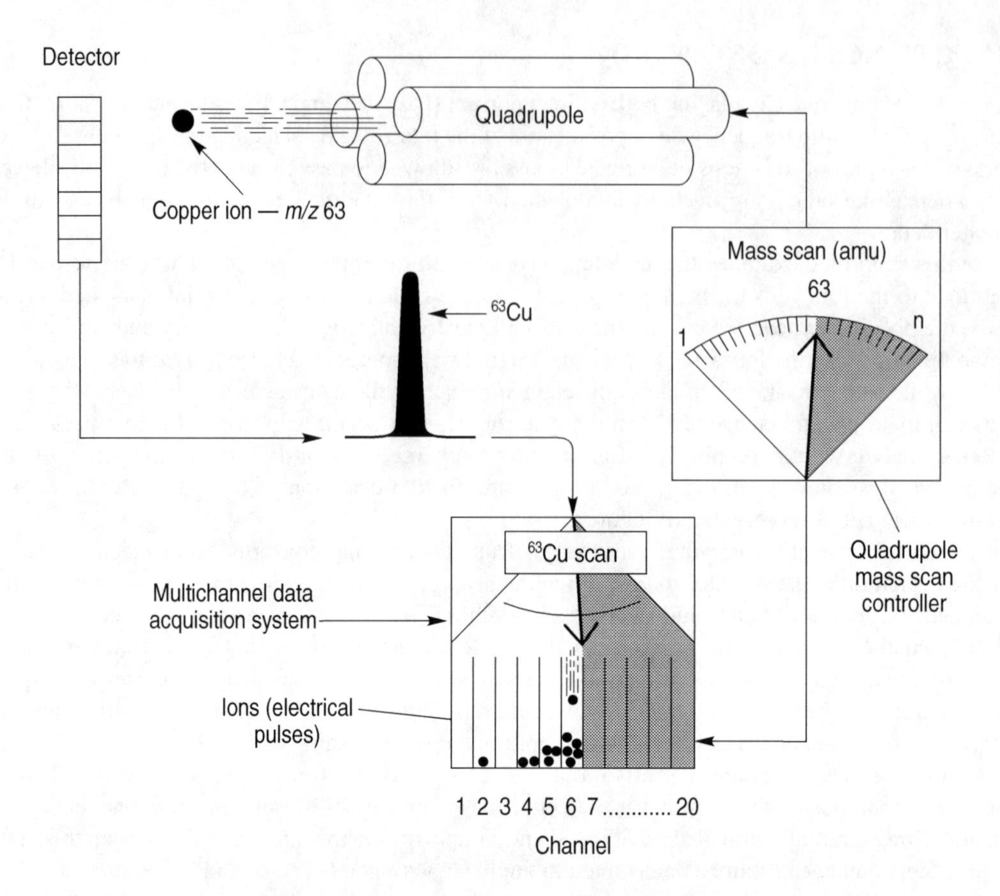

FIGURE 10.4 Profiles of different masses are built up using a multichannel data acquisition system. (Copyright 2013 all rights reserved, PerkinElmer Inc.)

nominal mass at which the peak occurs and Δm is the mass difference between two resolved peaks.[1] However, for quadrupole technology, the term *resolution* is more commonly used and is normally defined as the width of a peak at 10% of its height. The second specification is abundance sensitivity, which is the signal contribution of the tail of an adjacent peak at one mass lower and one mass higher than the analyte peak.[2] Even though they are somewhat related and both define the quality of a quadrupole, the abundance sensitivity is probably the most critical. If a quadrupole has good resolution but poor abundance sensitivity, it will often prohibit the measurement of an ultratrace analyte peak next to a major interfering mass.

Resolution

Let us now discuss this area in greater detail. The ability to separate different masses with a quadrupole is determined by a combination of factors, including shape, diameter, and length of the rods; frequency of quadrupole power supply; operating vacuum; applied RF/DC voltages; and the motion and kinetic energy of the ions entering and exiting the quadrupole. All these factors will have a direct impact on the stability of the ions as they travel down the middle of the rods and, therefore, the quadrupole's ability to separate ions with differing *m/z* values. This is represented in Figure 10.5, which shows a simplified version of the Mathieu mass stability plot of two separate masses (A and B) entering the quadrupole at the same time.[3]

Any of the RF/DC conditions shown under the peak on the left will only allow mass A to pass through the quadrupole, whereas any combination of RF/DC voltages under the peak on

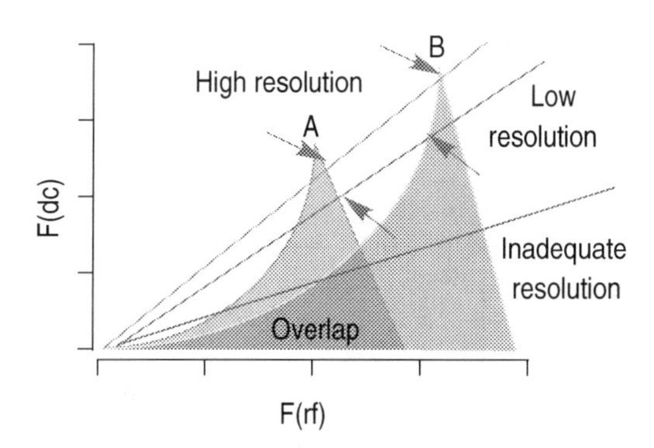

FIGURE 10.5 Simplified Mathieu stability diagram of a quadrupole mass filter, showing separation of two different masses A and B. (From P. H. Dawson, Ed., *Quadrupole Mass Spectrometry and Its Applications*, Elsevier, Amsterdam, 1976; reissued by AIP Press, Woodbury, NY, 1995.)

the right will only allow mass B to pass through the quadrupole. If the slope of the RF/DC scan rate is steep, represented by the top line (high resolution), the spectral peaks will be narrow and masses A and B will be well separated. However, if the slope of the scan is shallow, represented by the middle line (low resolution), the spectral peaks will be wide and masses A and B will not be well separated. On the other hand, if the slope of the scan is too shallow, represented by the lower line (inadequate resolution), the peaks will overlap each other and both masses A and B will pass through the quadrupole without being separated. Theoretically, the resolution of a quadrupole mass filter can be varied between 0.3 and 3.0 amu but is normally kept at 0.7–1.0 amu for most applications. However, improved resolution is always accompanied by a sacrifice in sensitivity, as seen in Figure 10.6, which shows a comparison of the same mass at resolutions of 3.0, 1.0, and 0.3 amu.

It can be seen that the peak height at 3.0 amu is much larger than that at 0.3 amu, but as expected, it is also much wider. This would prohibit using a resolution of 3.0 amu with spectrally complex samples. Conversely, the peak width at 0.3 amu is very narrow, but the sensitivity is low. For this reason, a compromise between peak width and sensitivity normally has to be reached, depending on the application. This can clearly be seen in Figure 10.7, which shows a spectral overlay of two copper isotopes—$^{63}Cu^+$ and $^{65}Cu^+$—at resolution settings of 0.70 and 0.50 amu. In practice, the quadrupole is normally operated at a resolution of 0.7–1.0 amu for the majority of applications.

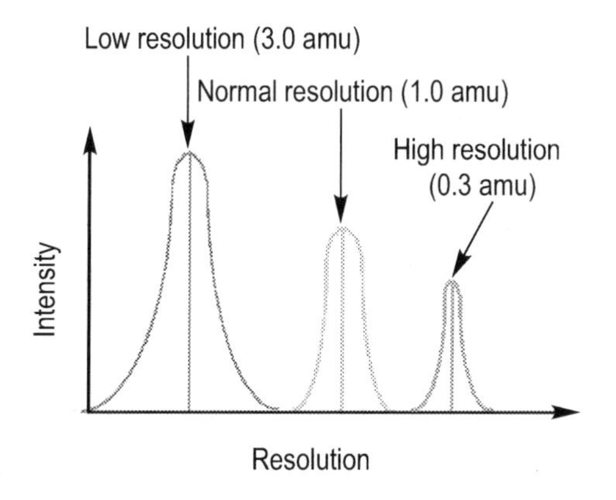

FIGURE 10.6 Sensitivity comparison of a quadrupole operated at 3.0, 1.0, and 0.3 amu resolutions.

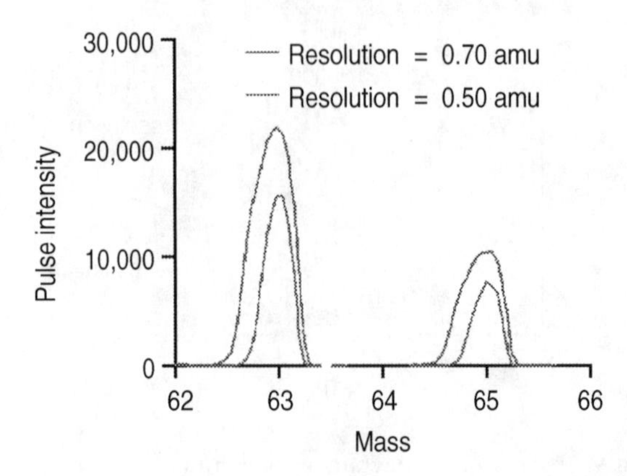

FIGURE 10.7 Sensitivity comparison of two copper isotopes—$^{63}Cu^+$ and $^{65}Cu^+$—at resolution settings of 0.70 and 0.50 amu.

It is worth mentioning that most quadrupoles are operated in the first stability region, where resolving power is typically on the order of 500–700. If the quadrupole is operated in the second or third stability regions, resolving powers of 4,000[4] and 9,000,[5] respectively, have been achieved. However, improving resolution using this approach has resulted in a significant loss of signal. Although there are ways of improving sensitivity, other problems have been encountered. As a result, to date there are no commercial quadrupole instruments available that use higher stability regions.

Some instruments can vary the peak width "on the fly," which means that the resolution can be changed between 3.0 and 0.3 amu for every analyte in a multielement run. Although this appears to offer some benefits, in reality they are few and far between, and for the vast majority of applications, it is adequate to use the same resolution setting for every analyte. Even though quadrupoles can be operated at a higher resolution (in the first stability region), up to now the slight improvement has not been shown to be of practical benefit for most routine applications.

Abundance Sensitivity

It can be seen in Figure 10.7 that the tail of the spectral peaks drops off more rapidly at the high-mass end of the peak compared to the low-mass end. The overall peak shape, particularly its low-mass and high-mass tail, is determined by the abundance sensitivity of the quadrupole, which is impacted by a combination of factors, including the design of the rods, frequency of the power supply, and operating vacuum.[6] Even though they are all important, probably the biggest impact on abundance sensitivity is the motion and kinetic energy of the ions as they enter and exit the quadrupole. If the Mathieu stability plot in Figure10.5 is examined, it can be seen that the stability boundaries of each mass are less defined (not so sharp) on the low-mass side compared to the high-mass side.[3] As a result, the characteristic of ion motion at the low-mass boundary is different from that at the high-mass boundary and is therefore reflected in poorer abundance sensitivity at the low-mass side compared to the high-mass side. The velocity, and therefore the kinetic energy, of the ions entering the quadrupole will affect the ion motion and, as a result, will have a direct impact on the abundance sensitivity. For that reason, factors that affect the kinetic energy of the ions, such as high plasma potential and the use of lenses to accelerate the ion beam, could have a negative effect on the instrument's abundance sensitivity.[7]

These are the fundamental reasons why the peak shape is not symmetrical with a quadrupole and explains why there is always a pronounced shoulder at the low-mass side of the peak compared to the high-mass side—as represented in Figure 10.8, which shows the theoretical peak shape of a nominal mass *M*.

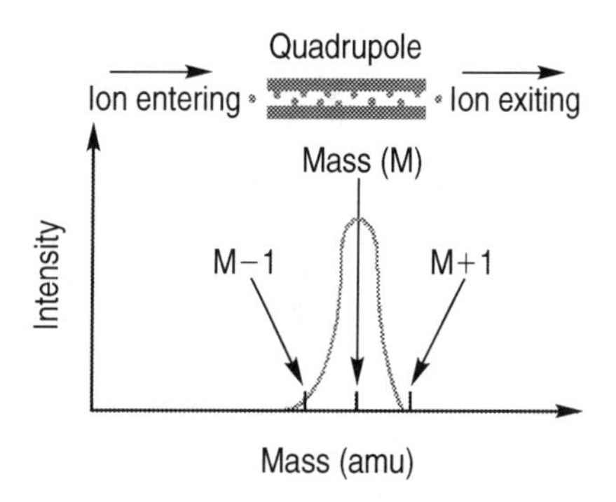

FIGURE 10.8 Ions entering the quadrupole are slowed down by the filtering process and produce peaks with a pronounced tail or shoulder at the low-mass end ($M - 1$) compared to the high-mass end ($M + 1$).

It can be seen that the shape of the peak at one mass lower ($M - 1$) is slightly different from the other side of the peak at one mass higher ($M + 1$) than the mass M. For this reason, the abundance sensitivity specification for all quadrupoles is always worse on the low-mass side than the high-mass side and is typically 1×10^{-6} at $M - 1$ and 1×10^{-7} at $M + 1$. In other words, an interfering peak of 1 million counts per second (mcps) at $M - 1$ would produce a background of 1 cps at M, whereas it would take an interference of 10 million cps at $M + 1$ to produce a background of 1 cps at M.

Benefit of Good Abundance Sensitivity

An example of the importance of abundance sensitivity is shown in Figure 10.9. Figure 10.9a is a spectral scan of 50 ppm of the doubly charged europium ion, $^{151}Eu^{++}$, at 75.5 amu (a doubly charged ion is one with two positive charges, as opposed to a normal singly charged positive ion, and exhibits an *m/z* peak at half its mass). It can be seen that the intensity of the peak is so great that its tail

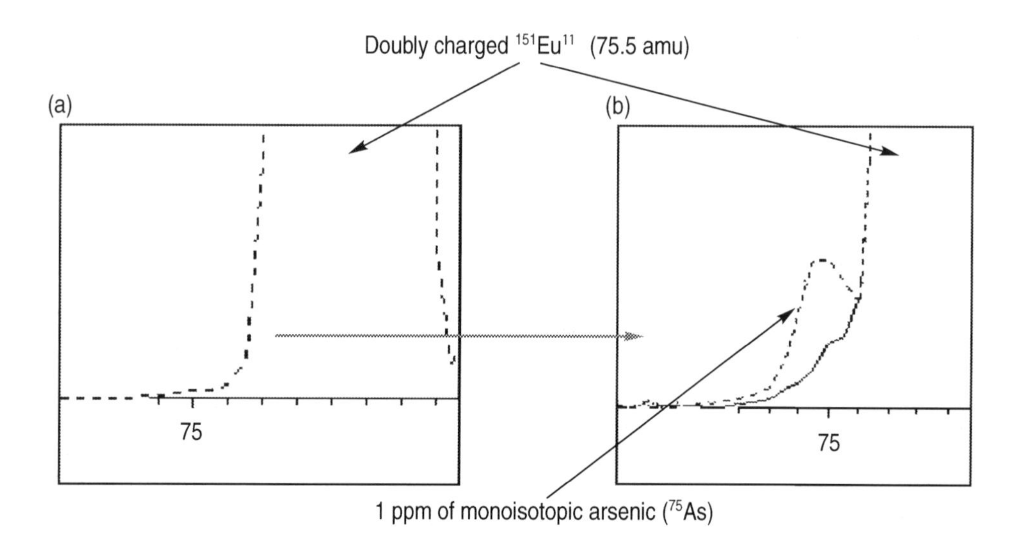

FIGURE 10.9 A low abundance sensitivity specification is critical to minimize spectral interferences, as shown by (a), which represents a spectral scan of 50 ppb of $^{151}Eu^{++}$ at 75.5 amu, and (b), which shows how the tail of the $^{151}Eu^{++}$ elevates the spectral background of 1 ppb of As at mass 75. (Copyright 2013, all rights reserved, PerkinElmer Inc.)

overlaps the adjacent mass at 75 amu, which is the only available mass for the determination of arsenic. This is highlighted in Figure 10.9b, which shows an expanded view of the tail of the $^{151}Eu^{++}$ together with a scan of 1 ppb of As at mass 75. It can be seen very clearly that the $^{75}As^{+}$ signal lies on the sloping tail of the $^{151}Eu^{++}$ peak. Measurement on a sloping background similar to this would result in a significant degradation in the arsenic detection limit, particularly as the element is monoisotopic and no alternative mass is available. In this particular example, a slightly higher resolution setting was also used (0.5 amu instead of 0.7 amu) to enhance the separation of the arsenic peak from the europium peak but nevertheless still emphasizes the importance of good abundance sensitivity in ICP-MS.

There are many different designs of quadrupole used in ICP-MS, all made from different materials with varied dimensions, shape, and physical characteristics. In addition, they are all maintained at a slightly different vacuum chamber pressure and operate at different frequencies. Theoretically, these hyperbolic rods should generate a better hyperbolic (elliptical) field than cylindrical rods, resulting in higher transmission of ions at higher resolution. It also tells us that a higher operating frequency means a higher rate of oscillation—and therefore separation—of the ions as they travel down the quadrupole. Finally, it is very well accepted that a higher vacuum produces fewer collisions between gas molecules and ions, resulting in a narrower spread in kinetic energy of the ions and, therefore, a reduction in the tail at the low-mass side of a peak. Given all these theoretical differences, in reality the practical capabilities of most modern quadrupoles used in ICP-MS are very similar. However, there are some subtle differences in each instrument's measurement protocol and the software's approach to peak quantitation. This will be discussed in greater detail in Chapter 15 "Peak Measurement Protocol."

FURTHER READING

1. F. Adams, R. Gijbels, and R. Van Grieken, *Inorganic Mass Spectrometry*, John Wiley & Sons, New York, 1988.
2. E. Montasser, Ed., *Inductively Coupled Plasma Mass Spectrometry*, Wiley-VCH, Berlin, 1998.
3. P. H. Dawson, Ed., *Quadrupole Mass Spectrometry and Its Applications*, Elsevier, Amsterdam, 1976; reissued by AIP Press, Woodbury, NY, 1995.
4. Z. Du, T. N. Olney, and D. J. Douglas, *Journal of American Society of Mass Spectrometry*, **8**, 1230–1236, 1997.
5. P. H. Dawson and Y. Binqi, *International Journal of Mass Spectrometry*, **56**, 25–32, 1984.
6. D. Potter, Agilent Technologies Application Note: 228–349, January, 1996.
7. E. R. Denoyer, D. Jacques, E. Debrah, and S. D. Tanner, *Atomic Spectroscopy*, **16** (1), 1–6, 1995.

11 Mass Analyzers *Double-Focusing Magnetic Sector Technology*

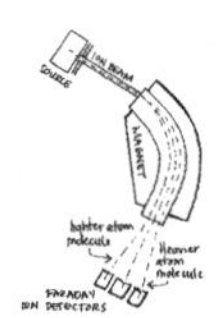

Although quadrupole mass analyzers represent approximately 80%–90% of all inductively coupled plasma mass spectrometry (ICP-MS) systems installed worldwide, limitations in their resolving power have led to the development of high-resolution spectrometers based on the double-focusing magnetic sector design. In this chapter, we take a detailed look at this very powerful mass separation device, which has found its niche in solving challenging application problems that require excellent detection capability, exceptional resolving power, and very high precision.

As discussed in Chapter 10, a quadrupole-based ICP-MS system typically offers a resolution of 0.7–1.0 amu. This is quite adequate for most routine applications but has proved to be inadequate for many elements that are prone to argon-, solvent-, and/or sample-based spectral interferences. These limitations in quadrupoles drove researchers in the direction of traditional high-resolution magnetic sector technology to improve quantitation by resolving the analyte mass away from the spectral interference.[1] These ICP-MS instruments that were first commercialized in the late 1980s offered resolving power of up to 10,000, compared to that of a quadrupole, which for most applications is in the order of ~500. This dramatic improvement in resolving power allowed difficult elements such as Fe, K, As, V, and Cr to be determined with relative ease, even in complex sample matrices.

MAGNETIC SECTOR MASS SPECTROSCOPY: A HISTORICAL PERSPECTIVE

Mass spectrometers, using separation based on velocity focusing[2,3] and magnetic deflection,[4,5] were first developed over 80 years ago, primarily to investigate isotopic abundances and calculate atomic weights. Even though these designs were combined into one instrument in the 1930s to improve both sensitivity and resolving power,[6,7] they were still considered rather bulky and expensive to build. For that reason, in the late 1930s and 1940s, magnetic field technology, and in particular the small radius sector design of Nier,[8] became the preferred method of mass separation. Because Nier was a physicist, most of the early work carried out with this design was used for isotope studies in the disciplines of earth and planetary sciences. However, it was the oil industry that accelerated the commercialization of MS, because of its demand for the fast and reliable analysis of complex hydrocarbons in oil refineries.

Once scanning magnetic sector technology became the most accepted approach for high-resolution mass separation in the 1940s, the challenges that lay ahead for mass spectroscopists were in the design of the ionization source—especially as the technique was being used more and more for the analysis of solids. The gas discharge ion source that was developed for gases and high-vapor-pressure liquids proved to be inadequate for most solid materials. For this reason, one of the first successful methods of ionizing solids was carried out using the hot anode method,[9] where the previously dissolved material was deposited onto a strip of platinum foil and evaporated by passing an electric current through it. Unfortunately, although there were variations of this approach that all worked reasonably well, the main drawback of a thermal evaporation technique was selective ionization. In other words, because of the different volatilities of the elements, it could not be guaranteed that the ion beam properly represented the compositional integrity of the sample.

It was finally the work carried out by Dempster in 1946,[10] using a vacuum spark discharge and a high-frequency, high-voltage spark, that led researchers to believe that it could be applied to sample electrodes and used as a general-purpose source for the analysis of solids. The breakthrough came in 1954 with the development of the first modern spark source mass spectrometer (SSMS) based on the Mattauch–Herzog mass spectrometer design which separated the ions in the same flat plane, so they could be detected by a linear detector such as a photographic plate.[11] Using this design, Hannay and Ahearn showed that it was possible to determine sub-ppm impurity levels directly in a solid material.[12]

Over the years, as a result of a demand for more stable ionization sources, lower detection capability, and higher precision, researchers were led in the direction of other techniques such as secondary ion mass spectrometry (SIMS),[13] ion microprobe mass spectrometry (IMMS),[14] and laser ionization mass spectrometry (LIMS).[15] Although they are considered somewhat complementary to SSMS, they all had their own strengths and weaknesses, depending on the analytical objectives for the solid material being analyzed. However, it should be emphasized that these techniques were predominantly used for microanalysis because only a very small area of the sample is vaporized. This meant that it could only provide meaningful analytical data of the bulk material if the sample was sufficiently homogeneous. For that reason, other ionization sources that sampled a much larger area, such as the glow discharge, became a lot more practical for the bulk analysis of solids by MS.[16]

USE OF MAGNETIC SECTOR TECHNOLOGY FOR ICP-MS

Even though magnetic sector technology was the most common mass separation device for the analysis of inorganic compounds using traditional ion sources, it lost out to quadrupole technology when ICP-MS was first developed in the early 1980s. However, it was not until the mid- to late-1980s that the analytical community realized that quadrupole ICP-MS suffered from serious limitations in its ability to resolve troublesome polyatomic spectral interferences. In addition, because quadrupoles were scanning devices, the technique was not suitable for applications that required high precision, such as isotope ratio measurements. As a result, analytical chemists began to look at double-focusing magnetic sector technology to alleviate these kinds of limitations.

Initially, this technology was found to be unsuitable as a separation device for an ICP because of the high voltage required to accelerate the ions into the mass analyzer. This high potential at the interface region dramatically changed the energy of the ions entering the mass spectrometer and therefore made it very difficult to steer the ions through the ion optics and still maintain a narrow spread of ion kinetic energies. For this reason, basic changes had to be made to the ion acceleration mechanism for magnetic sector technology to be successfully used as a separation device for ICP-MS. This was a significant challenge when magnetic sector systems were first developed in the late 1980s.

However, by the early 1990s, this problem was solved by moving the high-voltage components away from the plasma and positioning the interface closer to the mass spectrometer. Modern

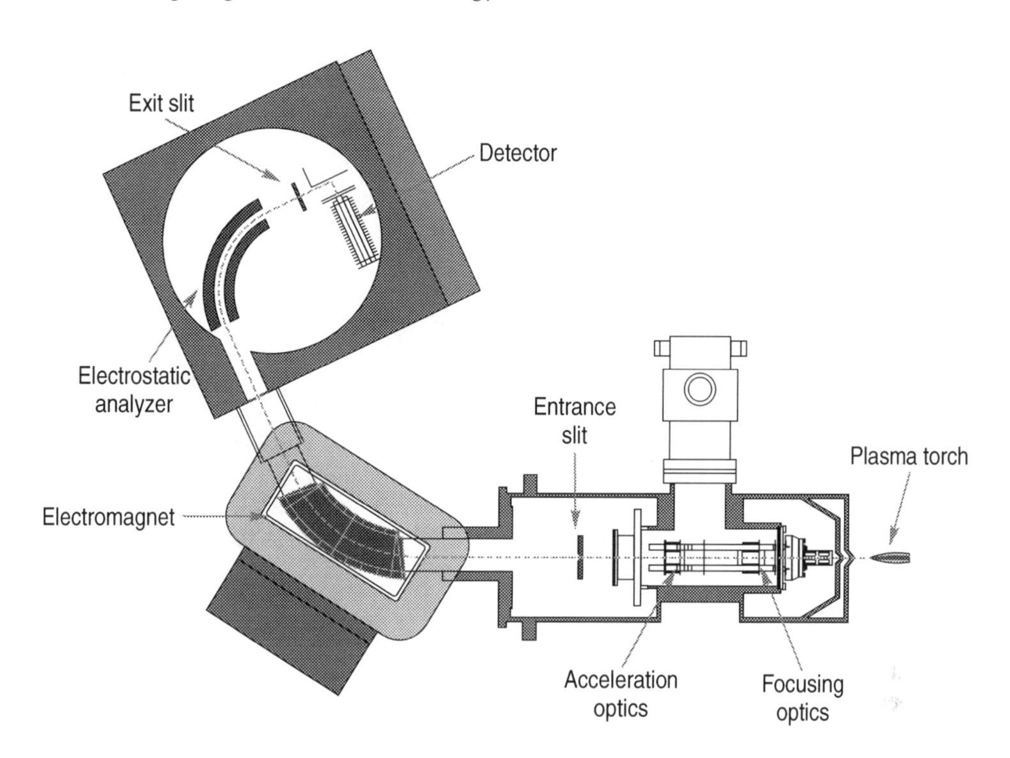

FIGURE 11.1 Schematic of a reverse Nier–Johnson double-focusing magnetic sector mass spectrometer. (From U. Geismann and U. Greb, *Fresnius' Journal of Analytical Chemistry*, **350**, 186–193, 1994.)

instrumentation has typically been based on two different approaches, the "standard" and "reverse" Nier–Johnson geometry. Both these designs, which use the same basic principles, consist of two analyzers: a traditional electromagnet and an electrostatic analyzer (ESA). In the standard (sometimes called forward) design, the ESA is positioned before the magnet, and in the reverse design, it is positioned after the magnet. A schematic of the reverse Nier–Johnson spectrometer is shown in Figure 11.1.

PRINCIPLES OF OPERATION OF MAGNETIC SECTOR TECHNOLOGY

The original concept of magnetic sector technology was to scan over a large mass range by varying the magnetic field over time with a fixed acceleration voltage. During a small window in time, which was dependent on the resolution chosen, ions of a particular mass-to-charge ratio are swept past the exit slit to produce the characteristic flat-top peaks. As the resolution of a magnetic sector instrument is independent of mass, ion signals, particularly at low mass, are far apart. Unfortunately, this results in a relatively long time being spent scanning and settling the magnet. This was not such a major problem for qualitative analysis or mass spectral fingerprinting of unknown compounds but proved to be impractical for rapid trace element analysis, where you had to scan to individual masses, slow down, settle the magnet, stop, take measurements, and then scan to the next mass.

However, by using the double-focusing approach, the ions are sampled from the plasma in a conventional manner and then accelerated in the ion optic region to a few kilovolts (kV) before they enter the mass analyzer. The magnetic field, which is dispersive with respect to ion energy and mass, then focuses all the ions with diverging angles of motion from the entrance slit. The ESA, which is only dispersive with respect to ion energy, then focuses all the ions onto the exit slit, where the detector is positioned. If the energy dispersion of the magnet and ESA is equal in magnitude but opposite in direction, they will focus both ion angles (first focusing) and ion energies (second or double focusing) when combined together. Changing the electric field in the opposite direction during the cycle time of the magnet (in terms of the mass passing the exit slit) has the effect of "freezing" the mass for detection. Then, as soon as a certain magnetic field strength is passed, the electric field is set to its original value and the next mass is "frozen." The voltage is varied on a per-mass basis, allowing the operator to scan only the mass peaks of interest rather than the full mass range.[17,18]

It should be pointed out that although this approach represents an enormous time savings over traditional magnet scanning technology, it is still slower than quadrupole-based instruments. The inherent problem lies in the fact that a quadrupole can be electronically scanned faster than a magnet. Typical speeds for a full mass scan (0–250 amu) of a magnet are on the order of 200 ms compared to approximately 50 ms for a quadrupole. In addition, it takes much longer for a magnet to slow down, settle, and stop to take measurements—typically, 20 ms compared to <1 ms for a quadrupole. So, even though in practice the electric scan dramatically reduces the overall analysis time, modern double-focusing magnetic sector ICP-MS systems are still slower than state-of-the-art quadrupole instruments, which make them less than ideal for fast, high-throughput multielement applications or the characterization of rapid transient peaks.

Resolving Power

As mentioned previously, most commercial magnetic sector ICP-MS systems offer up to 10,000 resolving power (10% valley definition), which is high enough to resolve the majority of spectral interferences. It is worth emphasizing that resolving power is represented by the equation $R = m/\Delta m$, where m is the nominal mass at which the peak occurs and Δm is the mass difference between two resolved peaks.[19] In a quadrupole, the resolution is selected by changing the ratio of the radio frequency (RF) and DC voltages on the quadrupole rods.

However, because a double-focusing magnetic sector instrument involves focusing ion angles and ion energies, mass resolution is achieved by using two mechanical slits—one at the entrance to the mass spectrometer and another at the exit, prior to the detector. Varying resolution is achieved by scanning the magnetic field under different entrance and exit slit width conditions. Similar to optical systems, low resolution is achieved by using wide slits, whereas high resolution is achieved with narrow slits. Varying the width of both the entrance and exit slits effectively changes the operating resolution.

However, it should be emphasized that, similar to optical spectrometry, as the resolution is increased, the transmission decreases. So even though extremely high resolution is available, detection limits will be compromised under these conditions. This can be seen in Figure 11.2, which shows a plot of resolution against ion transmission.

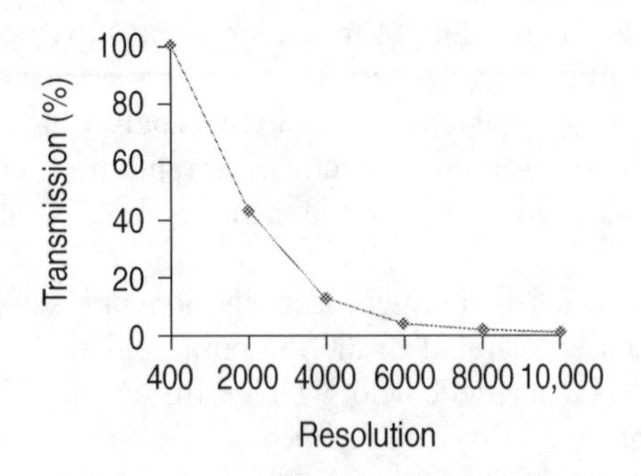

FIGURE 11.2 Ion transmission with a magnetic sector instrument decreases as the resolution increases.

TABLE 11.1
Resolution Required to Separate Some Common Polyatomic Interferences from a Selected Group of Isotopes

Isotope	Matrix	Interference	Resolution	Transmission (%)
$^{39}K^+$	H_2O	$^{38}ArH^+$	5,570	6
$^{40}Ca^+$	H_2O	$^{40}Ar^+$	199,800	0
$^{44}Ca^+$	HNO_3	$^{14}N^{14}N^{16}O^+$	970	80
$^{56}Fe^+$	H_2O	$^{40}Ar^{16}O^+$	2,504	18
$^{31}P^+$	H_2O	$^{15}N^{16}O^+$	1,460	53
$^{34}S^+$	H_2O	$^{16}O^{18}O^+$	1,300	65
$^{75}As^+$	HCl	$^{40}Ar^{35}Cl^+$	7,725	2
$^{51}V^+$	HCl	$^{35}Cl^{16}O^+$	2,572	18
$^{64}Zn^+$	H_2SO_4	$^{32}S^{16}O^{16}O^+$	1,950	42
$^{24}Mg^+$	Organics	$^{12}C^{12}C^+$	1,600	50
$^{52}Cr^+$	Organics	$^{40}Ar^{12}C^+$	2,370	20
$^{55}Mn^+$	HNO_3	$^{40}Ar^{15}N^+$	2,300	20

It can be seen that a resolving power of 400 produces 100% transmission, but at a resolving power of 10,000, only ~2% is achievable. This dramatic loss in sensitivity could be an issue if low detection limits are required in spectrally complex samples that require the highest possible resolution. However, spectral demands of this nature are not very common. Table 11.1 shows the resolution required to resolve fairly common polyatomic interferences from a selected group of elemental isotopes, together with the achievable ion transmission.

Figure 11.3 is a comparison between a quadrupole and a magnetic sector instrument of one of the most common polyatomic interference—$^{40}Ar^{16}O^+$ on $^{56}Fe^+$, which requires a resolution of 2,504 to separate the peaks. Figure 11.3a shows a spectral scan of $^{56}Fe^+$ using a quadrupole instrument. What it does not show is the massive polyatomic interference $^{40}Ar^{16}O^+$ (produced by oxygen ions from the water combining with argon ions from the plasma) completely overlapping the $^{56}Fe^+$. It shows very clearly that these two masses are not resolvable with a quadrupole. If that same spectral scan is carried out on a magnetic sector-type instrument, the result is the scan shown in Figure 11.3b.[20] It should be pointed out that in order to see the spectral scan on the same scale, it is necessary to examine a much smaller range. For this reason, a 0.100amu window was taken, as indicated by the dotted lines.

OTHER BENEFITS OF MAGNETIC SECTOR INSTRUMENTS

Besides high-resolving power, another attractive feature of magnetic sector instruments is their very high sensitivity combined with extremely low background levels. High ion transmission in low-resolution mode translates into sensitivity specifications of up to 1 billion counts per second (1 Bcps) per ppm (parts per million), whereas background levels resulting from extremely low dark current noise are typically 0.1–0.2 counts per second (cps). This compares to typical sensitivity levels of 100 million cps (Mcps) and background levels of 2–5 cps for a quadrupole technology, although the newer instruments are now capable of generating in excess of 200 Mcps and up to 500 Mcps when optimized for high sensitivity, with background (BG) levels in the order of 1 cps. However, because background levels are usually significantly lower on a magnetic sector system, detection limits, especially for high-mass elements such as uranium where high resolution is generally not required, are typically five to ten times better than a quadrupole-based instrument.

Besides good detection capability, another of the recognized benefits of the magnetic sector approach is its ability to quantitate with excellent precision. Measurement of the characteristically

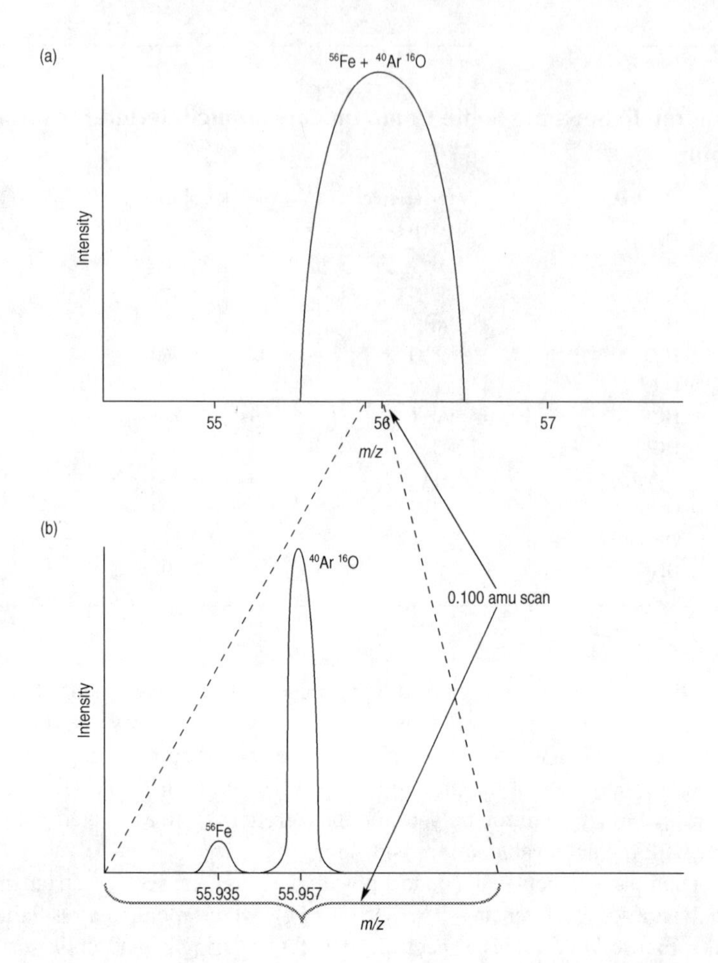

FIGURE 11.3 Comparison of resolution between (a) a quadrupole and (b) a magnetic sector instrument for the polyatomic interference of $^{40}Ar^{16}O^+$ on $^{56}Fe^+$. (From U. Greb and L. Rottman, *Labor Praxis*, August 1994.)

flat-topped spectral peaks translates directly into high-precision data. As a result, in the low-resolution mode relative standard deviation (RSD) values of 0.01%–0.05% are fairly common, which makes it an ideal approach for carrying out high-precision isotope ratio work.[21] Although precision is usually degraded as resolution is increased, modern instrumentation with high-speed electronics and low mass bias are still capable of precision values of <0.1% RSD in medium- or high-resolution mode.[22]

The demand for ultra-high-precision data, particularly in the field of geochemistry, has led to the development of instruments dedicated to isotope ratio analysis. These are based on the double-focusing magnetic sector design, but instead of using just one detector, these instruments use multiple detectors. Often referred to as multicollector systems, they offer the capability of detecting and measuring multiple ion signals at exactly the same time. As a result of this simultaneous measurement approach, they are recognized as producing extremely low isotope ratio precision.[23]

SIMULTANEOUS MEASUREMENT APPROACH USING ONE DETECTOR

The scanning limitations of a single detector magnetic sector instrument have been addressed by the development of spectrometers based on Mattauch–Herzog geometry and simultaneous measurement of the ions using linear plane array detectors.[24–27] The early designs were limited in their applicability to the entire mass spectrum and the use of an ICP excitation source, but they eventually led to the development of a commercially available instrument in 2010.[28] This particular instrument is made up of an entrance slit, ESA, energy slit, and a permanent magnet. In this design, the magnetic field focuses all of the ions onto a flat linear focal plane, without having to adjust the voltage of the mass analyzer or the strength of the magnetic field. In this way, no scanning is involved, and as a result, all the ions generated can be collected on a solid-state ion detector for the simultaneous detection of all the elemental masses in the sample, similar to using a CCD (charge-coupled detector) or CID (charge-injection detector) for optical emission work. The principles of this design are exemplified in Figure 11.4. The ion detector has over 4,800 channels, which is enough to record all 210 isotopes of the 75 elements that can be determined by ICP-MS, using an average of 20 channels per isotope. Every channel is a combination of two detector arrays with different signal amplifiers. In this way, it is possible to cover six orders of dynamic range plus an additional three orders by optimizing the measurement times. More details will be given about this type of detector in Chapter 14 "Ion Detectors."

The major benefit of this approach is that no scanning of the magnet is required. For this reason, it is ideal for any application that benefits from a rapid simultaneous measurement of the analyte masses. Some of these applications include:

- Analysis that requires high precision, such as isotope ratio/dilution studies or the real-time measurement of internal standards
- Multielement analysis of a fast transient event such as laser ablation or chromatographic separation techniques coupled with ICP-MS
- Applications where rapid speed of analysis is important, such as in high-throughput environmental contract labs
- Because the complete mass spectrum is collected every time, any mass can be interrogated after the analysis to check for any unexpected interferences.

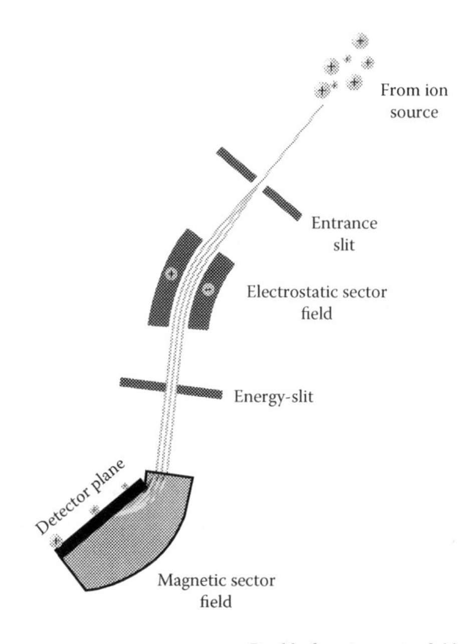

FIGURE 11.4 The principles of the Mattauch–Herzog double-focusing magnet sector mass spectrometer with a simultaneous detector plane.

However, it should be emphasized that this type of magnetic sector mass spectrometer does not offer the high-resolving power of a traditional Nier–Johnson scanning system, which is typically in the order of 10,000. The resolution of this design of Mattauch–Herzog mass spectrometer is only slightly better than a quadrupole system, which limits its use for the analysis of spectrally complex sample matrices.

FINAL THOUGHTS

There is no question that magnetic sector ICP-MS instruments are no longer novel analytical techniques. They have proved themselves to be a valuable addition to the trace element tool box, particularly for challenging applications that require good detection capability, exceptional resolving power, and very high precision. Even though they are not trying to compete with quadrupole instruments when it comes to rapid, high-sample-throughput applications, the measurement speed of modern systems, particularly the newer Mattauch–Herzog simultaneous detection approach, has opened up the technique to brand new application areas, which were traditionally beyond the scope of earlier designs of this technology.

FURTHER READING

1. N. Bradshaw, E. F. H. Hall, and N. E. Sanderson, *Journal of Analytical Atomic Spectrometry*, **4**, 801–803, 1989.
2. F. W. Aston, *Philosophical Magazine*, **38**, 707, 1919.
3. J. L. Costa, *Annals of Physics*, **4**, 425–428, 1925.
4. A. J. Dempster, *Physical Review*, **11**, 316–320, 1918.
5. W. F. G. Swann, *Journal of the Franklin Institute*, **210**, 751–756, 1930.
6. A. J. Dempster, *Proceedings of the American Philosophical Society*, **75**, 755–760, 1935.
7. K. T. Bainbridge and E. B. Jordan, *Physical Review*, **50**, 282–288, 1936.
8. A. O. Nier, *Review of Scientific Instruments*, **11**, 252–256, 1940.
9. G. P. Thomson, *Philosophical Magazine*, **42**, 857–860, 1921.
10. A. J. Dempster, MDDC 370, U. S. Department of Commerce, Washington, DC, 1946.
11. J. Mattauch and R. Herzog, *Zeitschrift für Physik*, **89**, 786–791, 1934.
12. N. B. Hannay and A. J. Ahearn, *Analytical Chemistry*, **26**, 1056–1062, 1954.
13. R. E. Honig, *Journal of Applied Physics*, **29**, 549–553, 1958.
14. R. Castaing and G. Slodzian, *Journal of Microscopy*, **1**, 394–401, 1962.
15. R. E. Honig and J. R. Wolston, *Applied Physics Letters*, **2**, 138–143, 1963.
16. J. W. Coburn, *Review of Scientific Instruments*, **41**, 1219–1223, 1970.
17. R. Hutton, A. Walsh, D. Milton, and J. Cantle, *ChemSA*, **17**, 213–215, 1991.
18. U. Geismann and U. Greb, *Fresnius' Journal of Analytical Chemistry*, **350**, 186–193, 1994.
19. F. Adams, R. Gijbels, and R. Van Grieken, *Inorganic Mass Spectrometry*, John Wiley and Sons, New York, 1988.
20. U. Greb and L. Rottman, *Labor Praxis*, **23**(2), 3–8, August 1994.
21. F. Vanhaecke, L. Moens, R. Dams, and R. Taylor, *Analytical Chemistry*, **68**, 565–571, 1996.
22. M. Hamester, D. Wiederin, J. Willis, W. Keri, and C. B. Douthitt, *Fresnius' Journal of Analytical Chemistry*, **364**, 495–497, 1999.
23. J. Walder and P. A. Freeman, *Journal of Analytical Atomic Spectrometry*, **7**, 571–575, 1992.
24. J. H. Barnes, R. P. Sperline, M. B. Denton, C. J. Barinaga, D. W. Koppenaal, E.T. Young, G. M. Hieftje, *Analytical Chemistry*, **74** (20), 5327–5332, 2002.
25. J. H. Barnes, G. D. Schilling, R. P. Sperline, M. B. Denton, E.T. Young, C. J. Barinaga, D. W. Koppenaal, G. M. Hieftje, *Analytical Chemistry*, **76** (9), 2531–2536, 2004.
26. G. D. Schilling, F. J. Andrade, J. H. Barnes, R. P. Sperline, M. B. Denton, C. J. Barinaga, D. W. Koppenaal, G. M. Hieftje, *Analytical Chemistry*, **78** (13), 4319–4325, 2006.
27. G. D. Schilling, S. J. Ray, A. A. Rubinshtein, J. A. Felton, R. P. Sperline, M. B. Denton, C. J. Barinaga, D. W. Koppenaal, G. M. Hieftje, *Analytical Chemistry*, **81** (13), 5467–5473, 2009.
28. A New Era in Mass Spectrometry: Spectro Analytical Instruments, http://www.spectro.com/pages/e/p010402tab_overview.htm

12 Mass Analyzers
Time-of-Flight Technology

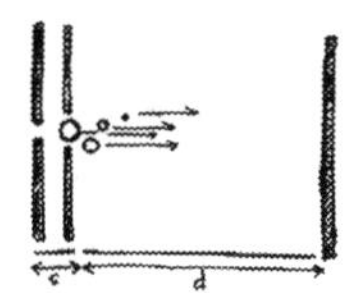

Let us turn our attention to time-of-flight (TOF) technology in this chapter. Although the first TOF mass spectrometer was first described in the literature in the late 1940s,[1] it has taken over 50 years to adapt it for use with a commercial inductively coupled plasma (ICP) mass spectrometer. The interest in TOF ICP-MS (mass spectrometry) instrumentation has come about because of its ability to sample all ions generated in the plasma at exactly the same time, which is ideally suited for multielement determinations of rapid transient signals, high-precision isotope ratio analysis, fast data acquisition for high-throughput workloads, and the rapid semiquantitative fingerprinting of unknown samples.

BASIC PRINCIPLES OF TOF TECHNOLOGY

The simultaneous nature of sampling ions in TOF offers distinct advantages over traditional scanning (sequential) quadrupole technology for ICP-MS applications in which large amounts of data need to be captured in a short span of time. To understand the benefits of this mass separation device, let us first take a look at its fundamental principles. All TOF mass spectrometers are based on the same principle: the kinetic energy (*KE*) of an ion is directly proportional to its mass (*m*) and velocity (*V*). This can be represented by the following equation:

$$KE = \frac{1}{2} mV^2$$

Therefore, if a population of ions—all with different masses—is given the same *KE* by an accelerating voltage (*U*), the velocities of the ions will all be different, depending on their masses. This principle is then used to separate ions of different mass-to-charge ratios (*m*/*e*) in the time (*t*) domain over a fixed flight path distance (*D*) represented by the following equation:

$$m/e = \frac{2Ut^2}{D^2}$$

This is schematically shown in Figure 12.1, which shows three ions of different mass-to-charge ratios being accelerated into a "flight tube" and arriving at the detector at different times. It can be seen that, depending on their velocities, the lightest ion arrives first, followed by the medium mass ion, and finally the heaviest one. Using flight tubes of 1 m length, even the heaviest ions typically take less than 50 μs to reach the detector. This translates into approximately 20,000 mass spectra per second—three orders of magnitude faster than the sequential scanning mode of a quadrupole system.

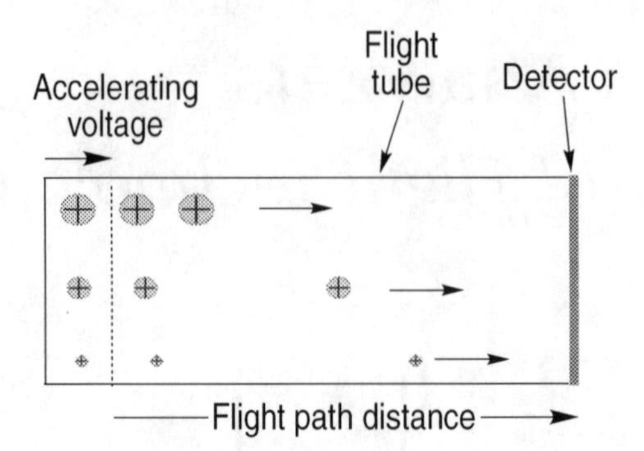

FIGURE 12.1 Principles of ion detection using TOF technology, showing separation of three different masses in the time domain.

COMMERCIAL DESIGNS

Even though this process sounds fairly straightforward, it is not a trivial task to sample the ions in a simultaneous manner from a continuous source of ions being generated in the plasma discharge. There are basically two different sampling approaches used—the orthogonal design,[2] in which the flight tube is positioned at right angles to the sampled ion beam, and the axial design,[3] in which the flight tube is along the same axis as the ion beam.

In both designs, all ions that contribute to the mass spectrum are sampled through the interface cones, but instead of being focused into the mass filter in the conventional way, packets (groups) of ions are electrostatically injected into the flight tube at exactly the same time. With the orthogonal approach, an accelerating potential is applied at right angles to the continuous ion beam from the plasma source. The ion beam is then "chopped" by using a pulsed voltage supply coupled to the orthogonal accelerator to provide repetitive voltage "slices" at a frequency of a few kilohertz. The "sliced" packets of ions, which are typically tall and thin in cross section (in the vertical plane), are then allowed to "drift" into the flight tube, where the ions are temporally resolved according to their differing velocities. This is shown schematically in Figure 12.2.

In the axial approach, an accelerating potential is applied axially (in the same axis) to the incoming ion beam as it enters the extraction region. Because the ions are in the same plane as the detector, the beam has to be modulated using an electrode grid to repel the "gated" packet of ions into the flight tube. This kind of modulation generates an ion packet that is long and thin in cross section (in the horizontal plane). The different masses are then resolved in the time domain in a similar manner to the orthogonal design. An on-axis TOF system is schematically shown in Figure 12.3.

Figures 12.2 and 12.3 offer a rather simplified explanation of the TOF principles of operation. In practice, there are many complex ion-focusing components in a TOF mass analyzer that ensure that the maximum number of analyte ions reach the detector, and also that undesired photons, neutral species, and interferences are ejected from the ion beam. Some of these components are seen in Figure 12.4, which shows a more detailed view of a commercial orthogonal TOF ICP-MS system.

It can be seen in this design that an orthogonal accelerator is used to inject packets of ions at right angles from the ion beam emerging from the MS interface. They are then directed towards an ion blanker, where unwanted ions are rejected from the flight path by deflection plates. The packets of ions are then directed into an ion reflectron where they do a U-turn and are deflected back 180°, where they are detected by a channel electron multiplier or discrete dynode detector. The reflectron, which is a type of ion mirror, functions as an energy compensation device so that different ions of the same mass arrive at the detector at the same time.[4]

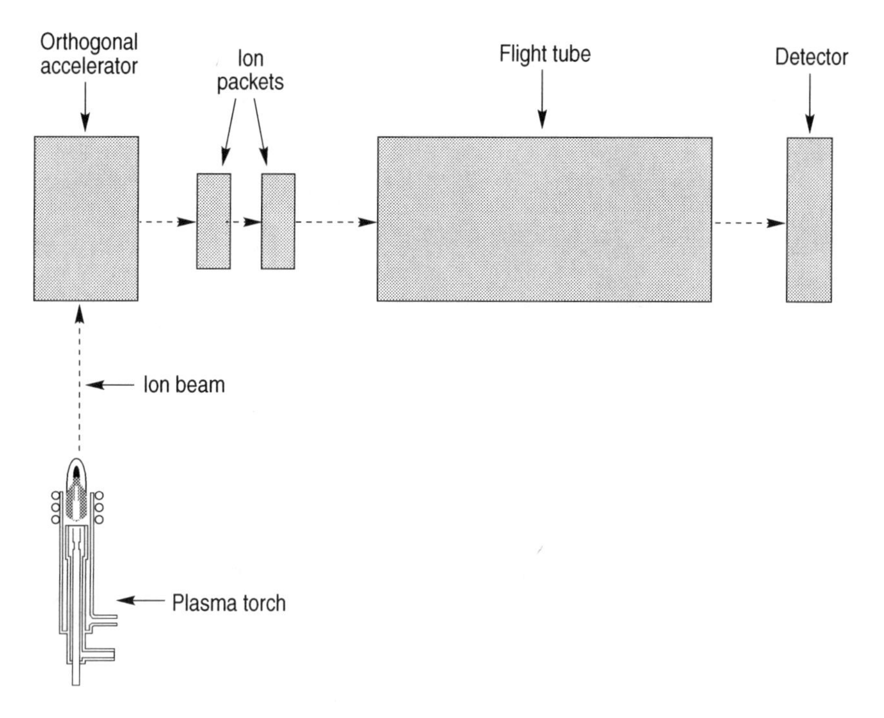

FIGURE 12.2 A schematic of an orthogonal acceleration TOF analyzer. (From Technical Note: 001-0877-00, GBC Scientific, February 1998.)

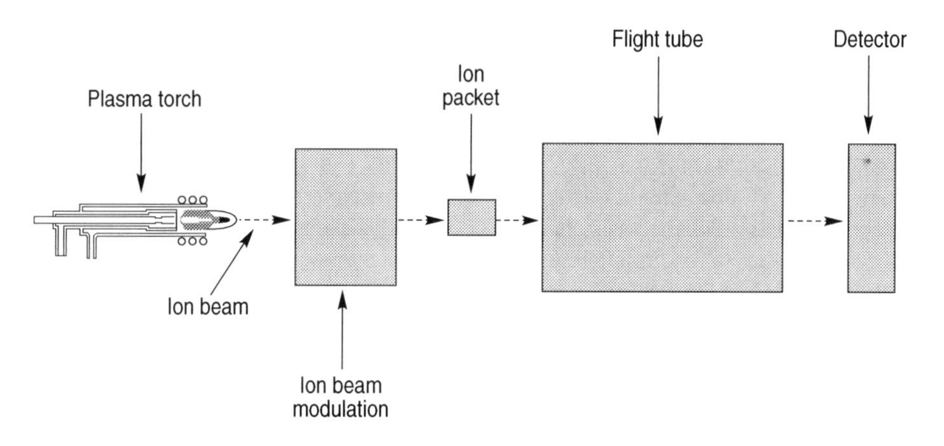

FIGURE 12.3 A schematic of an axial acceleration TOF analyzer. (From Technical Note: 001-0877-00, GBC Scientific, February 1998.)

DIFFERENCES BETWEEN ORTHOGONAL AND ON-AXIS TOF

Although there are real benefits of using TOF over quadrupole technology for some ICP-MS applications, there are also subtle differences in the capabilities of each type of TOF design. Some of these differences include:

- **Sensitivity:** The axial approach tends to produce higher ion transmission because the steering components are in the same plane as the ion generation system (plasma) and the detector. This means that the direction and magnitude of greatest energy dispersion is along

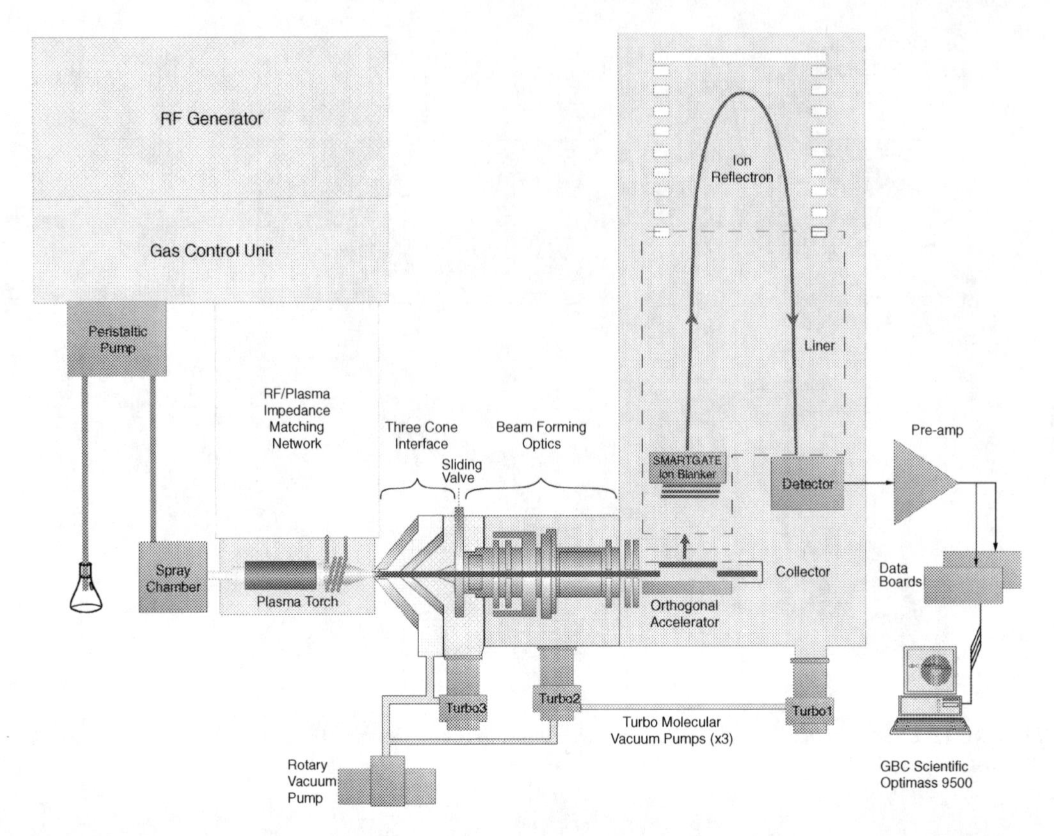

FIGURE 12.4 A more detailed view of an orthogonal (right-angled) TOF ICP-MS system, showing some of the ion-steering components. (Courtesy of GBC Scientific Equipment Pty Ltd.)

the axis of the flight tube. In addition, when ions are extracted orthogonally, the energy dispersion can produce angular divergence of the ion beam, resulting in poor transmission efficiency. However, on the basis of current evidence, the sensitivity of both TOF designs is generally an order of magnitude lower than the latest commercial quadrupole instruments.

- **Background levels:** The axial design tends to generate higher background levels because neutral species and photons stand a better chance of reaching the detector. This results in background levels on the order of 20–50 cps—approximately one order of magnitude higher than the orthogonal design. However, because the ion beam in the axial design has a smaller cross section, a smaller detector can be used, which generally has better noise characteristics. In comparison, most commercial quadrupole instruments offer background levels of 1–10 cps, depending on the design.
- **Duty cycle:** This is usually defined as the fraction (percentage) of extracted ions that actually make it into the mass analyzer. Unfortunately, with a TOF ICP mass spectrometer, which has to use "pulsed" ion packets from a continuous source of ions generated in the plasma, this process is relatively inefficient. It should also be emphasized that even though the ions are sampled at the same time, detection is not simultaneous because of different masses arriving at the detector at different times. The difference between the sampling mechanisms of the orthogonal and axial TOF designs translates into subtle differences in their duty cycles.
    - With the orthogonal design, duty cycle is defined by the width of the extracted ion packets, which are typically tall and thin in cross section, as shown in Figure 12.2. In comparison, the duty cycle of an axial design is defined by the length of the extracted

ion packet, which is typically wide and thin in cross section, as shown in Figure 12.3. The duty cycle can be improved by changing the cross-sectional area of the ion packet, but, depending on the design, it is generally at the expense of resolution. However, this is not a major issue, because TOF instruments are generally not used for high-resolution ICP-MS applications. In practice, the duty cycles for both orthogonal and axial designs are on the order of 15%–20%.

- **Resolution:** The resolution of the orthogonal approach is slightly better because of its two-stage extraction/acceleration mechanism. Because a pulse of voltage pushes the ions from the extraction area into the acceleration region, the major energy dispersion lies along the axis of ion generation. For this reason, the energy spread is relatively small in the direction of extraction compared to spread with the axial approach, resulting in better resolution. However, the resolving power of commercial TOF ICP-MS systems is typically on the order of 500–2,000, depending on the mass region, which makes them inadequate to resolve many of the problematic polyatomic species encountered in ICP-MS. In comparison, commercial high-resolution systems based on the double-focusing magnetic sector design offer resolving power up to 10,000, whereas commercial quadrupoles can typically achieve 400–500.
- **Mass bias:** This is also known as mass discrimination and is the degree to which ion transport efficiency varies with mass. All instruments show some degree of mass bias, which is usually compensated for by measuring the difference between the theoretical and observed ratio of two different isotopes of the same element. In TOF, the velocity (energy) of the initial ion beam will affect the instrument's mass bias characteristics. In theory, it should be less with the axial design because the extracted ion packets do not have any velocity in a direction perpendicular to the axis of the flight tube, which could potentially impact their transport efficiency.

BENEFITS OF TOF TECHNOLOGY FOR ICP-MS

It should be emphasized that these performance differences between the two designs are subtle and should not detract from the overall benefits of the TOF approach for ICP-MS. As mentioned earlier, a scanning device such as a quadrupole can only detect one mass at a time, which means that there is always a compromise between number of elements, detection limits, precision, and the overall measurement time. However, with the TOF approach, the ions are sampled at the same moment in time, which means that multielement data can be collected with no significant deterioration in quality. The ability of a TOF system to capture a full mass spectrum, approximately three orders of magnitude faster than a quadrupole, translates into three major benefits—multielement determinations in a fast transient peak, improved precision, especially for isotope ratioing techniques, and rapid data acquisition for carrying out qualitative or semiquantitative scans.[4] Let us look at these in greater detail.

RAPID TRANSIENT PEAK ANALYSIS

Probably, the most exciting potential for TOF ICP-MS is in the multielement analysis of a rapid transient signal generated by sampling accessories such as laser ablation (LA),[5] and flow injection systems.[6] Even though a scanning quadrupole can be used for this type of analysis, it struggles to produce high-quality multielement data when the transient peak lasts only a few seconds. The simultaneous nature of TOF instrumentation makes it ideally suited for this type of analysis, because the entire mass range can be collected in less than 50 μs. In particular, when used with an ETV system, the high-acquisition speed of TOF can help to reduce matrix-based spectral overlaps by resolving them from the analyte masses in the temperature domain. There is no question that TOF technology is ideally suited (probably more than any other design of ICP mass spectrometer) for the analysis of transient peaks.

Improved Precision

To better understand how TOF technology can help improve precision in ICP-MS, it is important to know the major sources of instability. The most common source of noise in ICP-MS is the flicker noise associated with the sample introduction process (peristaltic pump pulsations, nebulization mechanisms, plasma fluctuations, etc.) and the shot noise derived from photons, electrons, and ions hitting the detector. Shot noise is based on counting statistics and is directly proportional to the square root of the signal. It, therefore, follows that as the signal intensity increases, the shot noise has less of an impact on the precision (% RSD) of the signal. This means that at high ion counts, the most dominant source of imprecision in ICP-MS is derived from the flicker noise generated in the sample introduction area.

One of the most effective ways to reduce instability produced by flicker noise is to use a technique called internal standardization, where the analyte signal is compared and ratioed to the signal of an internal standard element (usually of similar mass or ionization characteristics) that is spiked into the sample. Even though a quadrupole-based system can do an adequate job of compensating for these signal fluctuations, it is ultimately limited by its inability to measure the internal standard at precisely the same time as the analyte isotope. So, in order to compensate for sample introduction and plasma-based noise and achieve high precision, the analyte and internal standard isotopes need to be sampled and measured simultaneously. For this reason, the design of a TOF mass analyzer is perfect for simultaneous internal standardization required for high-precision work. It therefore follows that TOF is also well suited for high-precision isotope ratio analysis, where its simultaneous nature of measurement is capable of achieving precision values close to the theoretical limits of counting statistics. Also, unlike a scanning-quadrupole-based system, it can measure ratios for as many isotopes or isotopic pairs as needed—all with excellent precision.[7]

Rapid Data Acquisition

As with a scanning ICP-OES system, the speed of a quadrupole ICP mass spectrometer is limited by its scanning rate. To determine ten elements in duplicate with good precision and detection limits, an integration time of 3 s per mass is normally required. When overhead scanning and settling times are added for each mass and replicate, this translates into approximately 2 min per sample. With a TOF system, the same analysis would take significantly less time, because all the data are captured simultaneously. In fact, detection limit levels in a TOF instrument are typically achieved within a 10–30 s integration time, which translates into a fivefold to tenfold improvement in data acquisition time over a quadrupole instrument. The added benefit of a TOF instrument is that the speed of analysis is not impacted by the number of analytes being determined: it would not matter if the method contained 10 or 70 elements—the measurement time would be virtually the same. However, there is one point that must be stressed. A large portion of the overall analysis time is taken up for flushing an old sample out of and pumping a new sample into the sample introduction system. This can be as much as 2 min per sample for real-world matrices. So, when this is taken into account, the difference between the sample throughput of a quadrupole and a TOF ICP mass spectrometer is not so evident.

Another benefit of the fast acquisition time is that qualitative or semiquantitative analysis is relatively seamless compared to scanning quadrupole technology, because every multielement scan contains data for every mass. This also makes spectral identification much easier by comparing the spectral fingerprint of unknown samples against a known reference standard. This is particularly useful for forensic work, where the evidence is often an extremely small sample.

Commercial TOF ICP-MS instruments have been available since the late 1990s, mainly being used for the multielement characterization of fast transient signals and high-precision isotope ratio studies. However, two recent novel developments of TOF technology are widening its applicability and extending its real-world analytical capabilities.

HIGH-SPEED MULTIELEMENTAL IMAGING USING LASER ABLATION COUPLED WITH TOF ICP-MS

A more recent example of the benefits of TOF technology for the characterizing rapid transients is in the field of high-speed multielement imaging using LA. Since its first appearance in the mid-1980s, LA-ICP-MS has established itself as a routine method for the quantitation of trace elements in solid samples. In recent times, LA-ICP-MS has further become an important tool for elemental imaging applied in geological, biological, and medical research studies. To date, the most common imaging approach is to run the laser in continuous-scan mode which involves firing the laser continuously with a specific repetition rate (usually 1–10 Hz) while slowly moving the stage with the mounted sample underneath the laser beam. The image is then constructed by ablating parallel lines on the sample surface. The ablated aerosol is washed out of the air-tight ablation chamber (cell) to the ICP with a continuous flow of inert gas.

The washout times for conventional ablation cells are on the order of 0.5–30 s. After ionization in the ICP, the ablation signal is measured in a time-resolved manner, most commonly using a quadrupole or a sector field mass spectrometer. These instruments operate sequentially, resulting in the measurement of only one isotope at a time. For the analysis of multiple isotopes/elements, sequential peak hopping must be performed which, depending on the number of elements of interest, can become very restrictive and time-consuming. The conventional approach to LA imaging is associated with drawbacks on spatial resolution and speed of analysis which is usually limited to 1–2 pixels per second. The recent advent of fast-washout ablation cells and high-speed TOF ICP-MS allowed some of these limitations to be overcome by enabling spot-resolved imaging.

In order to capitalize on the benefits of fast-washout ablation cells, a mass spectrometer capable of extremely fast and simultaneous multielement data acquisition is required, and in particular, its ability to handle rapid transient signals from these cells. The TOF ICP-MS is well suited for the shot-resolved imaging approach due to its simultaneous and fast multielement analysis, because the entire mass spectrum is measured simultaneously with each ion package extraction. Different isotopes from this package are then separated based on their TOF traversing the region from extraction to detector, which is directly related to their mass-to-charge ratio. Thus, in contrast to conventional sequential mass spectrometers, it is not necessary to pre-define or to limit a range of isotopes for analysis. Moreover, the TOF ICP-MS can record complete mass spectra at a rate of 30 μs, which amounts to a maximum of ~33,000 spectra per second. This allows short transient signals from single laser pulses to be temporally resolved with sufficient sampling density (e.g., up to ~1,000 spectra per 30 ms pulse). In practice, however, multiple spectra are integrated to improve counting statistics and to simplify data processing. Due to this simultaneous measurement of complete mass spectra, TOF ICP-MS can have a higher total ion utilization than sequential mass spectrometers, and as a result, a greater proportion of the ablated sample is actually used for analysis. This work has been presented in a recent publication by Bussweiler and coworkers.[8]

LASER ABLATION LASER IONIZATION TIME-OF-FLIGHT MASS SPECTROMETRY

Laser ablation, laser ionization coupled with TOF mass spectrometry (LALI-TOFMS) is a brand new development that offers virtually the entire periodic table of the elements at mass spec detection capability. It has broad applicability for any type of lab that is looking to avoid sample digestion procedures to carry out multielement analysis directly on solid samples.

LALI-TOFMS utilizes dual-laser technology to first extract and ionize material in two discrete steps—first ablation of the sample and then ionization of the analyte elements. The first step uses a focused laser beam to ablate material from a solid sample surface using a Nd:YAG laser with an adjustable wavelength. A short delay (<1 μs) allows dispersion of plasma-generated ions before a second Nd:YAG laser is triggered for ionization.

By ionizing elements across a wide range of ionization energies, this technique serves as a highly efficient ion source and replaces the Ar plasma of ICP-MS instruments. Once ionized, an ion funnel collects and focuses the ions into a quadrupole ion deflector (QID) and turns the ion beam 90° through an einzel lens stack into a reflectron TOF mass spectrometer which completes the mass separation and detection. More detailed information about this very exciting technology can be found in Chapter 27 "Other Traditional and Emerging Atomic Spectroscopy Techniques."

FINAL THOUGHTS

There is no question that TOF ICP-MS, with its rapid, simultaneous mode of measurement, excels at multielement applications that generate fast transient signals, such as LA or speciation studies, using chromatographic techniques. In addition, it offers excellent precision, for isotope ratioing techniques, and also has the potential for rapid data acquisition when carrying out multielement semiquantitative analysis. And when coupled with novel excitation/ionization sources, it has the potential to be utilized for the determination of trace elements directly on solid samples. So although TOF mass separation devices are a fairly recent commercial development compared to quadrupole technology, they definitely should be considered as an option if the application demands it.

FURTHER READING

1. A. E. Cameron and D. F. Eggers, *The Review of Scientific Instruments*, **19** (9), 605–608, 1948.
2. D. P. Myers, G. Li, P. Yang, and G. M. Hieftje, *Journal of American Society of Mass Spectrometry*, **5**, 1008–1016, 1994.
3. D. P. Myers, *12th Asilomar Conference on Mass Spectrometry*, Pacific Grove, CA, September 20–24, 1996.
4. GBC Scientific 9600 TOF-ICP-MS Product Brochure: http://www.gbcsci.com/wp-content/uploads/2019/10/01-1070-00-OptiMass-Brochure-September-2019-Website.pdf
5. P. Mahoney, G. Li, and G. M. Hieftje, *Journal of American Society of Mass Spectrometry*, **11**, 401–406, 1996.
6. R. E. Sturgeon, J. W. H. Lam, and A. Saint, *Journal of Analytical Atomic Spectrometry*, **15**, 607–616, 2000.
7. F. Vanhaecke, L. Moens, R. Dams, L. Allen, and S. Georgitis, *Analytical Chemistry*, **71**, 3297–3303, 1999.
8. Y. Bussweiler, O. Borovinskaya, M. Tanner, *Spectroscopy Magazine*, **32** (5), 14–22, 2017.

13 Mass Analyzers
Collision/Reaction Cell and Interface Technology

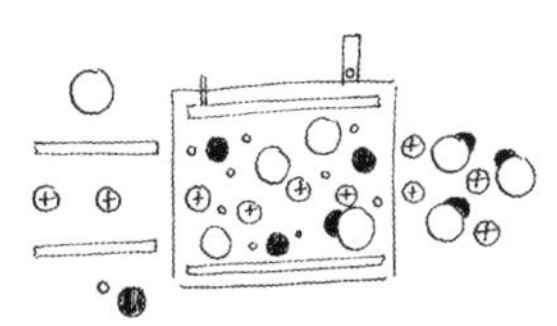

The detection capability for some elements using traditional quadrupole mass analyzer technology is severely compromised because of the formation of polyatomic spectral interferences generated by a combination of argon, solvent, and matrix-derived ions. Although there are ways to minimize these interferences, including correction equations, cool plasma technology, and matrix separation, they cannot be completely eliminated. However, a novel approach using collision/reaction cell (CRC) and interface technology has been developed that significantly reduces the formation of many of these harmful species before they enter the mass analyzer. Chapter 13 takes a detailed look at this very powerful technique and the exciting potential it has to offer and how it might be beneficial for the analysis of cannabis and hemp.

There are a small number of elements that are recognized as having poor detection limits by inductively coupled plasma mass spectrometry (ICP-MS). These are predominantly elements that suffer from major spectral interferences generated by ions derived from the plasma gas, matrix components, or the solvent/acid used in sample preparation. Examples of these interferences include the following:

- $^{40}Ar^{16}O^+$ in the determination of $^{56}Fe^+$
- $^{38}ArH^+$ in the determination of $^{39}K^+$
- $^{40}Ar^+$ in the determination of $^{40}Ca^+$
- $^{40}Ar^{40}Ar^+$ in the determination of $^{80}Se^+$
- $^{40}Ar^{35}Cl^+$ in the determination of $^{75}As^+$
- $^{40}Ar^{12}C^+$ in the determination of $^{52}Cr^+$
- $^{35}Cl^{16}O^+$ in the determination of $^{51}V^+$

The cold/cool plasma approach, which uses a lower temperature to reduce the formation of the argon-based interferences, has been a very effective way to get around some of these problems.[1] However, this approach can sometimes be difficult to optimize, is only suitable for a few of the interferences, is susceptible to more severe matrix effects, and can be time-consuming to change back and forth between normal and cool plasma conditions. These limitations and the desire to improve performance have led to the commercialization of CRCs and collision/reaction interfaces (CRIs). (Note: CRIs have more recently been referred to as the integrated Collision Reaction Cell (iCRC) design.) Designs for CRC and CRI were based on the early work of Rowan and Houk, who used Xe and CH_4 in the late 1980s to reduce the formation of ArO^+ and Ar_2^+ species in the determination of Fe and Se with a modified tandem mass spectrometer.[2] This research was investigated further by Koppenaal and coworkers in 1994, who carried out studies using an ion trap for the determination of Fe, V, As, and Se in a 2% hydrochloric acid matrix.[3] However, it was not until 1996 that studies describing the coupling of a CRC with a traditional quadrupole ICP mass spectrometer were published. Eiden and coworkers experimented using hydrogen as a collision gas,[4] whereas Turner and coworkers based their investigations on using helium gas.[5] These studies and the work of other groups at the time[6,7] proved to be the basis for modern collision and reaction cells and interfaces that are commercially available today.

Let's take a look at the fundamental principles of CRCs and interfaces.

BASIC PRINCIPLES OF COLLISION/REACTION CELLS

With all CRCs, ions enter the interface in the normal manner and then are directed into a CRC positioned prior to the analyzer quadrupole. A collision/reaction gas (e.g., helium, hydrogen, ammonia, or oxygen, depending on the design) is then bled via an inlet aperture into the cell containing a multipole (a quadrupole, hexapole, or octapole), usually operated in the radio frequency (RF)-only mode. The RF-only field does not separate the masses as a traditional quadrupole mass analyzer does, but instead, it has the effect of focusing the ions, which then collide and react with molecules of the collision/reaction gas. By a number of different ion–molecule collision and reaction mechanisms, polyatomic interfering ions such as $^{40}Ar^+$, $^{40}Ar^{16}O^+$, and $^{38}ArH^+$ will either be converted to harmless noninterfering species or the analyte will be converted to another ion that is not interfered with. This process is exemplified by the equation below, which shows the use of hydrogen as a reaction gas to reduce the $^{40}Ar^+$ interference in the determination of $^{40}Ca^+$.

$$H_2 + {}^{40}Ar^+ = Ar + H_{2^+}$$

$$H_2 + {}^{40}Ca^+ = {}^{40}Ca^+ + H_2 \left(\text{no reaction}\right)$$

It can be seen that the hydrogen molecule interacts with the argon interference to form atomic argon and the harmless H_2^+ ion. However, there is no interaction between the hydrogen and the calcium. As a result, the $^{40}Ca^+$ ions, free of the argon interference, emerge from the CRC through the exit aperture where they are directed towards the quadrupole analyzer for normal mass separation. Other gases are better suited to reduce the $^{40}Ar^+$ interference, but this process at least demonstrates the principles of the reaction mechanisms in a CRC. The layout of a typical CRC within the instrument is shown in Figure 13.1.

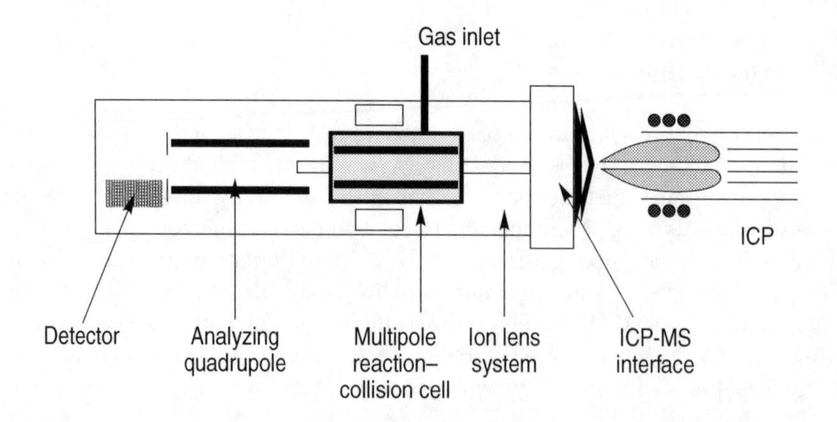

FIGURE 13.1 Layout of a typical CRC instrument.

The equation above is an example of an ion–molecule reaction using the process of charge transfer. By the transfer of a positive charge from the argon ion to the hydrogen molecule, an innocuous neutral Ar atom is formed, which is invisible to the mass analyzer. There are many other reaction and collisional mechanisms that can take place in the cell, depending on the nature of the analyte ion, the interfering species, the reaction/collision gas, and the type of multipole used. Other possible mechanisms that can occur in the cell, in addition to charge transfer, include the following:

- **Proton transfer:** The interfering polyatomic species gives up a proton, which is then transferred to the reaction gas molecule to form a neutral atom.
- **Hydrogen atom transfer:** A hydrogen atom is transferred to the interfering ion, which is converted to an ion at one mass higher.
- **Molecular association reactions:** An interfering ion associates with a neutral species (atom or molecule) to form a molecular ion.
- **Collisional fragmentation:** The polyatomic ion is broken apart or fragmented by the process of multiple collisions with the gaseous atoms.
- **Collisional retardation:** The gas atoms/molecules undergo multiple collisions with the polyatomic interfering ion in order to retard or lower its kinetic energy. Because the interfering ion has a larger cross-sectional area than the analyte ion, it undergoes more collisions and, as a result, can be separated or discriminated from the analyte ion based on kinetic energy differences.
- **Collisional focusing:** Analyte ions lose energy as they collide with the gaseous molecules and, depending on the molecular weight of the gas, will either enhance ion transmission as the ions migrate towards the central axis of the cell or decrease sensitivity if ion scattering takes place.

The CRI, which will be discussed later in this chapter, uses a slightly different principle to remove the interfering ions. It does not use a pressurized cell before the mass analyzer but, instead, injects a reaction/collision gas directly into the aperture of interface skimmer cone. The injection of the collision/reaction into this region of the ion beam produces collisions between the argon gas and the injected gas molecules, and as a result, argon-based polyatomic interferences are destroyed or removed before they are extracted into the ion optics.

DIFFERENT COLLISION/REACTION CELL APPROACHES

All these possible interactions between ions and molecules indicate that many complex secondary reactions and collisions can take place, which generate undesirable interfering species. If these species are not eliminated or rejected, they could potentially lead to additional spectral interferences. There are basically two different approaches used to reduce the formation of polyatomic interferences and discriminate the products of these unwanted side reactions from the analyte ion. They are:

- **Collisional mechanisms** using non- or low-reactive gases and kinetic energy discrimination (KED)[8]
- **Reaction mechanisms** using highly reactive gases and discrimination by selective band-pass mass filtering.[9]

The major differences between the two approaches are how the gaseous molecules interact with the interfering species and what type of multipole is used in the cell. These dictate whether it is an ion–molecule collision or reaction mechanism taking place. Let us take a closer look at each process because there are distinct differences in the way the interference is rejected and separated from the analyte ion.

Collisional Mechanisms Using Nonreactive Gases and Kinetic Energy Discrimination

The collisional mechanisms approach was adapted from collision-induced dissociation (CID) technology, which was first used in the early to mid-1990s in the study of organic molecules using tandem mass spectrometry. The basic principle relies on using a nonreactive gas in a hexapole collision cell to stimulate ion–molecule collisions. The more collision-induced daughter species that are generated, the better the chance of identifying the structure of the parent molecule.[10,11] However, this very desirable CID characteristic for identifying and quantifying biomolecules was a disadvantage in inorganic mass spectrometry, where uncontrolled secondary reactions are generally something to be avoided. The limitation restricted the use of hexapole-based collision cells in ICP-MS to inert gases such as helium or low-reactivity gases such as hydrogen because of the potential to form undesirable reaction by-products, which could spectrally interfere with other analytes. Unfortunately, higher-order multipoles have little control over these secondary reactions because their stability boundaries are very diffuse and not well defined like a quadrupole. As a result, they do not provide adequate mass separation capabilities to suppress the unwanted secondary reactions. Thus, the need is to rely mainly on collisional mechanisms and a process called *kinetic energy discrimination* (KED) to distinguish the interfering ions and by-product species from the analyte ions. So, what is KED?

KED relies on the principle of separating ions depending on their different ion energies. As ions enter the interface region, they all have differing kinetic energies based on the ionization process in the plasma and their mass-to-charge ratio. When the analyte and plasma/matrix-based interfering ions (sometimes referred to as *precursor ions*) enter the pressurized cell, they will undergo multiple collisions with the collision gas. Because the collisional cross-sectional area of the precursor ions and other collision-induced by-product ions are usually larger than the analyte ion, they will undergo more collisions. This has the effect of lowering the kinetic energy of these interfering species compared to the analyte ion. If the collision cell rod offset potential is set slightly more negative than the mass filter potential, the polyatomic ions with lower kinetic energy are rejected or discriminated by the potential energy barrier at the cell exit. On the other hand, the analyte ions, which have a higher kinetic energy, are transmitted to the mass analyzer. Figure 13.2 shows the principles of KED.

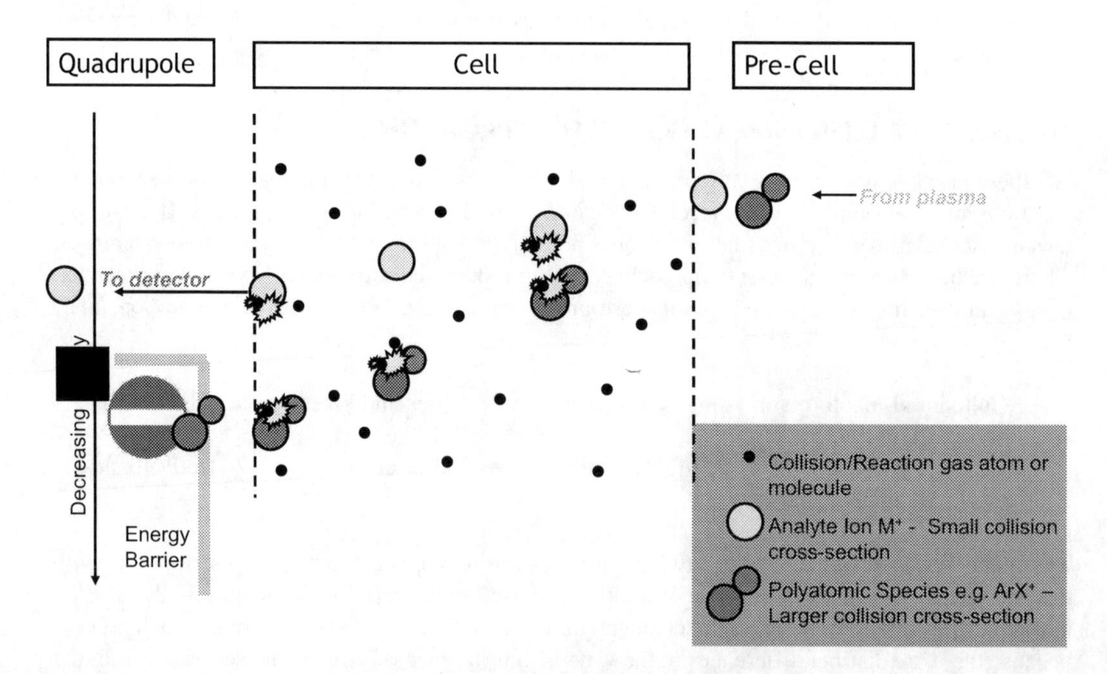

FIGURE 13.2 Principles of KED. (Courtesy of Thermo Scientific.)

In addition to the kinetic energy generated by the ionization process, the spread of ion energies will also be dictated by the efficiency of the RF-grounding mechanism (refer to Chapter 8 “Interface Region”). Therefore, for KED to work properly, the ion energy spread of ions generated in the plasma must be as narrow as possible to ensure that there is very little overlap between the analyte and the polyatomic interfering ion as they enter the mass analyzer. This means that it is absolutely critical that the RF-grounding mechanism is working efficiently to guarantee a low potential at the interface. If this is not the case, and there is a secondary discharge between the plasma and the interface, it will increase the ion energy spread of ions entering the collision cell and make it extremely difficult to separate the polyatomic interfering ion from the analyte of interest based on their kinetic energy difference. The relevance of having a narrow spread of ion energies is shown in Figure 13.3.

It can be seen that the ion energy spread of the analyte ion (gray peak) and a polyatomic ion (black peak) is very similar as they enter the collision cell. This allows the collision process and KED system to easily separate the ions as they exit the cell. If the ion energy spread is larger, there would be more of an overlap as the ions enter the mass analyzer and, therefore, compromise the detection limit for that analyte.

KED using helium as the collision gas works very well when the interfering polyatomic ion is physically larger than the analyte ion. This is exemplified in Figure 13.4, which shows helium flow optimization plots for six elements in 1:10 diluted seawater.[12]

It can be seen that the signal intensity for the analytes—Cr, V, Co, Ni, Cu, and As—are all at a maximum, whereas their respective matrix, argon, and solvent-based polyatomic interferences are all at a minimum at a similar helium flow rate of 5–6 mL/min. All the analytes show very good sensitivity because the KED process has allowed the analytes to be efficiently separated from their respective polyatomic interfering ions. The additional benefit of using helium is that it is inert and even if it is not being used in an interference reduction mode, it can have a beneficial effect on the other elements in a multielement run by increasing sensitivity via the process of collisional focusing. This makes the use of helium and KED very useful for both quantitative and semiquantitative multielement analysis using one set of tuning conditions.

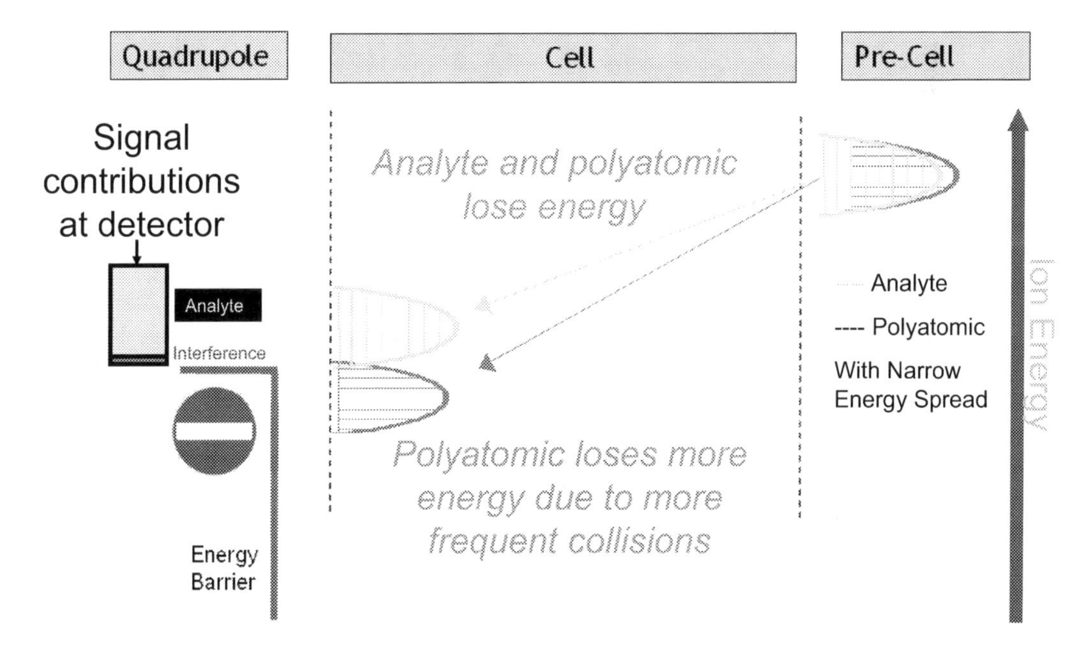

FIGURE 13.3 For optimum separation of analyte ion from the interfering species, it is important to have a narrow spread of ion energies entering the collision cell. (Courtesy of Thermo Scientific.)

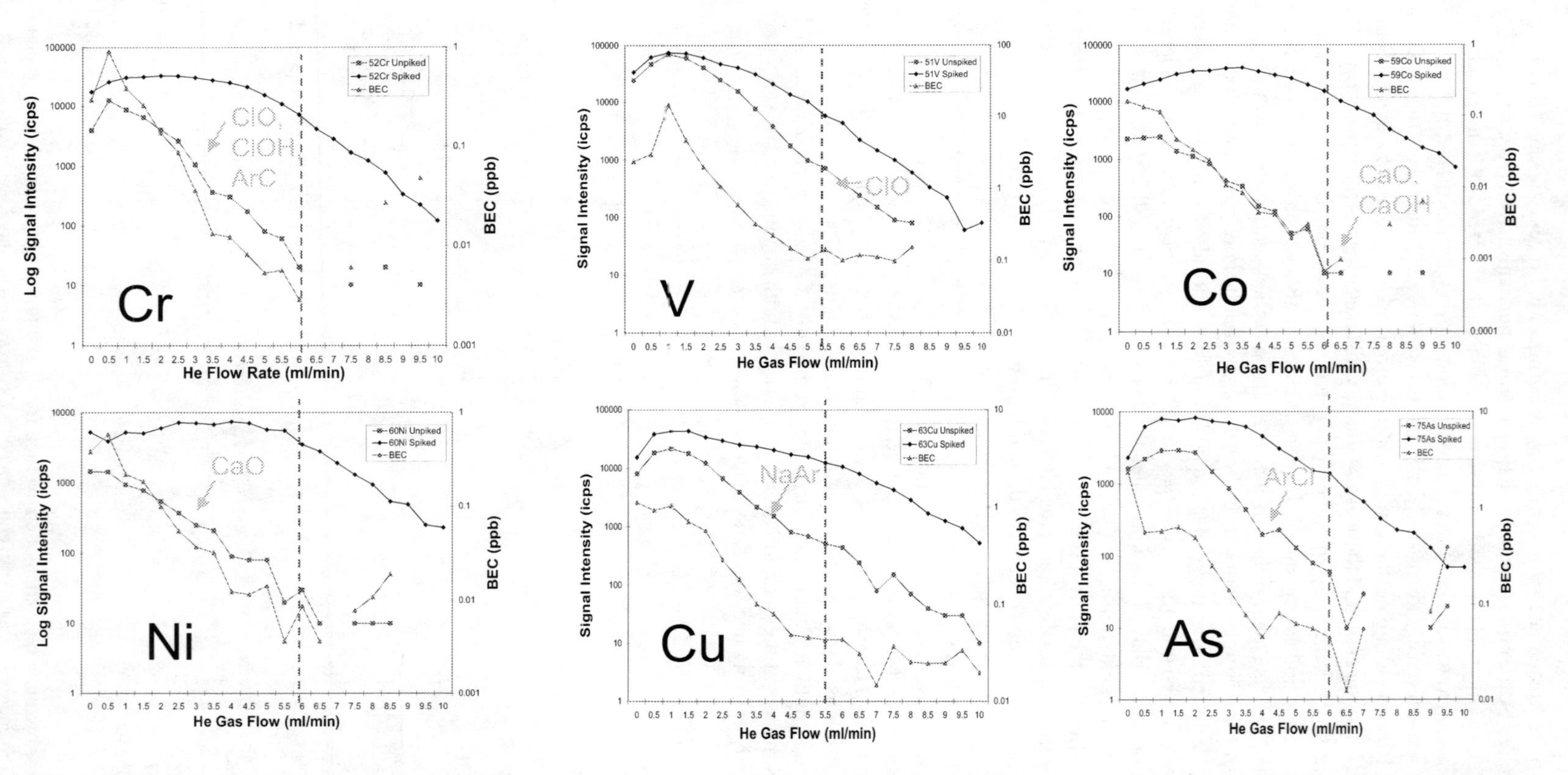

FIGURE 13.4 Helium cell gas flow optimization plots for Cr, V, Co, Ni, Cu, and As in 1:10 diluted seawater, showing that all polyatomic interfering ions are reduced to an acceptable level under one set of cell gas flow conditions. (Courtesy of Thermo Scientific.)

For the KED process to work efficiently, there must be a distinct difference between the kinetic energy of the analyte compared to the interfering ion. In most cases where the polyatomic ion is large, this is not a problem. However, in many cases where the interfering and analyte ions have a similar physical size (cross section), the process requires an extremely large number of collisions, which will have an impact not only on the attenuation of the interfering ion but also on the analyte ion. This means that for some situations, especially if the requirement is for ultratrace detection capability, the collisional process with KED is not enough to reduce the interference to acceptable levels. That is why most collision cells also have basic reaction mechanism capabilities, which allows low-reactivity gases such as hydrogen and, in some cases, small amounts of highly reactive gases such as ammonia or oxygen mixed with helium to be used.[13] But it should be pointed out that even though this initiates a basic reaction mechanism, rejection of the reaction by-product ions is still handled through the process of KED, which in some applications might not offer the most efficient way of reducing the interfering species. Using the collision cell as a basic reaction cell with low-reactivity gases is described later in this chapter.

One further point to keep in mind is that a KED-based cell relies on interactions of the interfering ion with an inert or low-reactivity gas, such that it can be separated from the analyte based on their differences in kinetic energy. If the gas contains impurities such as organic compounds or water vapor, the impurity could be the dominant reaction or collision pathway as opposed to the predicted collision/reaction with bulk gas. In addition, other unexpected ion–molecule reactions can readily occur if there are chemical impurities in the gas. This could also pose a secondary problem because of the formation of unexpected cluster ions such as metal oxide and hydroxide species, which have the potential to interfere with other analyte ions. Fortunately, many of these new ions formed in the cell as a result of reactions with the impurities have low energy and are adequately handled by KED. However, depending on the level of the impurity, some of the ions formed have higher energies and are therefore too high to be attenuated by the KED process, which could negatively impact the performance of the interference reduction process. For this reason, it is strongly advised that the highest-purity collision/reaction gases be used. If this is not an option, it is recommended that a gas-purifier (getter) system be placed in the gas line to cleanse the collision/reaction gas of impurities such as H_2O, O_2, CO_2, CO, or hydrocarbons. If you want to learn more about this process, Yamada and coworkers published a very interesting paper describing the effects of cell–gas impurities and KED in an octapole-based collision cell.[14]

Let us now go on to discuss the other major way of interference rejection in a CRC using highly reactive gases and mass (bandpass) filtering discrimination.

Reaction Mechanisms with Highly Reactive Gases and Discrimination by Selective Bandpass Mass Filtering

Another way of rejecting polyatomic interfering ions and the products of secondary reactions/collisions is to discriminate them by mass. As mentioned previously, higher-order multipoles cannot be used for efficient mass discrimination because the stability boundaries are diffuse and sequential secondary reactions cannot be easily intercepted. The only way this can be done is to utilize a quadrupole (instead of a hexapole or octapole) inside the reaction/collision cell and use it as a selective bandpass (mass) filter. There are a number of commercial designs using this approach, so let's take a look at them in greater detail in order to better understand how they work and how they differ.

Dynamic Reaction Cell

The first commercial instrument to use this approach was called *dynamic reaction cell* (DRC) *technology*.[15] Similar in appearance to the hexapole and octapole CRCs, the DRC is a pressurized multipole positioned prior to the analyzer quadrupole. However, this is where the similarity ends. In DRC technology, a quadrupole is used instead of a hexapole or octapole. A highly reactive gas such as ammonia, oxygen, or methane is bled into the cell, which is a catalyst for ion molecule chemistry to take place. By a number of different reaction mechanisms, the gaseous molecules react with the interfering ions to convert them into either an innocuous species different from the analyte mass or a harmless neutral species. The analyte mass then emerges from the DRC free of its interference and is steered into the analyzer quadrupole for conventional mass separation.

The advantage of using a quadrupole in the reaction cell is that the stability regions are much better defined than higher-order multipoles, so it is relatively straightforward to operate the quadrupole inside the reaction cell as a mass or bandpass filter and not just as an ion-focusing guide. Therefore, by careful optimization of the quadrupole electrical fields, unwanted reactions between the gas and the sample matrix or solvent, which could potentially lead to new interferences, are prevented. It means that every time an analyte and interfering ions enter the DRC, the bandpass of the quadrupole can be optimized for that specific problem and then changed on the fly for the next one. This is shown schematically in Figure 13.5, where an analyte ion $^{56}Fe^+$ and an isobaric interference $^{40}Ar^{16}O^+$ enter the DRC. As can be seen, the reaction gas NH_3 picks up a positive charge from the $^{40}Ar^{16}O^+$ ion to form atomic oxygen, argon, and a positive NH_3 ion (this is known as a "charge transfer reaction"). There is no reaction between the $^{56}Fe^+$ and the NH_3, as predicted by thermodynamic reaction kinetics. The quadrupole's electrical field is then set to allow the transmission of the analyte ion $^{56}Fe^+$ to the analyzer quadrupole, free of the problematic isobaric interference, $^{40}Ar^{16}O^+$. In addition, the NH_3^+ is prevented from reacting further to produce a new interfering ion.

The practical benefit of using highly reactive gases is that they increase the number of ion–molecule reactions taking place inside the cell, which results in a faster, more efficient removal of the interfering species. Of course, they will also generate more side reactions which, if not prevented, will lead to new polyatomic ions being formed and could possibly interfere with other analyte masses. However, the quadrupole reaction cell is well characterized by well-defined stability boundaries. So, by careful selection of bandpass parameters, ions outside the mass/charge (*m/z*) stability boundaries are efficiently and rapidly ejected from the cell. It means that additional reaction chemistries, which could potentially lead to new interferences, are successfully interrupted. In addition, the bandpass of the reaction cell quadrupole can be swept in concert with the bandpass of the quadrupole mass analyzer. This allows a dynamic bandpass to be defined for the reaction cell

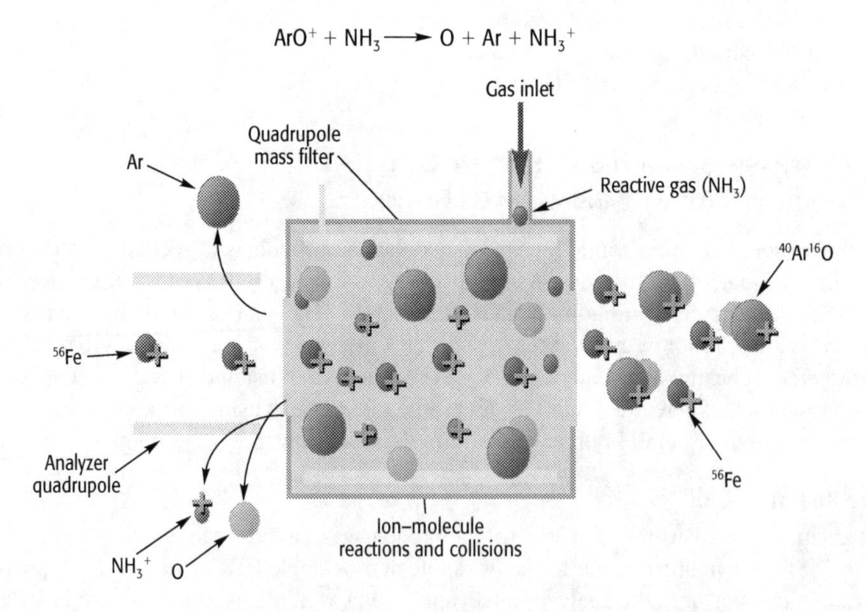

FIGURE 13.5 Elimination of the $^{40}Ar^{16}O^+$ interference with a DRC. (Copyright 2013, all rights reserved, PerkinElmer Inc.)

so that the analyte ion can be efficiently transferred to the analyzer quadrupole. The overall benefit is that within the reaction cell, the most efficient thermodynamic reaction chemistries can be used to minimize the formation of plasma- and matrix-based polyatomic interferences, in addition to simultaneously suppressing the formation of further reaction by-product ions.

The process described can be exemplified by the elimination of $^{40}Ar^+$ by NH_3 gas in the determination of $^{40}Ca^+$. The reaction between NH_3 gas and the $^{40}Ar^+$ interference, which is predominantly charge transfer/exchange, occurs because the ionization potential of NH_3 (10.2 eV) is low compared to that of Ar (15.8 eV). This makes the reaction extremely exothermic and fast. However, as the ionization potential of Ca (6.1 eV) is significantly less than that of NH_3, the reaction, which is endothermic, is not allowed to proceed.[15] This can be seen in greater detail in Figure 13.6.

Of course, other secondary reactions are probably taking place, which you would suspect with such a reactive gas as ammonia, but by careful selection of the cell quadrupole electrical fields, the optimum bandpass only allows the analyte ion to be transported to the analyzer quadrupole, free of the interfering species. This highly efficient reaction mechanism and selection process translates into a dramatic reduction of the spectral background at mass 40, which is shown graphically in Figure 13.7. It can be seen that at the optimum NH_3 flow, a reduction in the $^{40}Ar^+$ background signal of about eight orders of magnitude is achieved, resulting in a detection limit of approximately 0.1 ppt for $^{40}Ca^+$.

One final thing to point out is that when highly reactive gases are used, the purity of the gas is not so critical because the impurity is almost insignificant in determining the ion–molecule reaction mechanism. On the other hand, with collision and low-reactivity gases that contain impurities, such as carbon dioxide, hydrocarbons, or water vapor, the impurity could be the dominant reaction pathway as opposed to the predicted collision/reaction with the bulk gas. In addition, the formation of unexpected by-product ions or other interfering species, which have the potential to interfere with other analyte ions in a KED-based collision cell, is not such a serious problem with the DRC system because of its ability to intercept and stop these side reactions using the bandpass mass-filtering discrimination process.

These observations were, in fact, made by Hattendorf and Günter, who attempted to quantify the differences between KED and bandpass tuning with regard to suppression of interferences

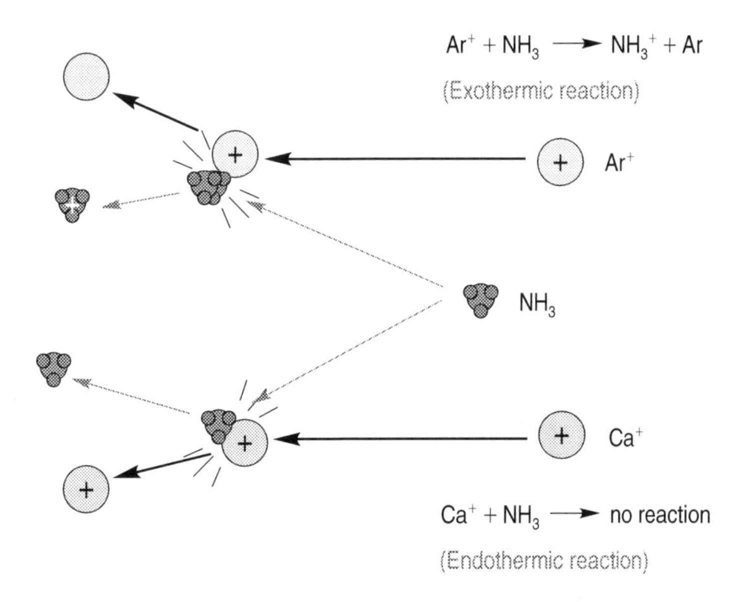

FIGURE 13.6 The reaction between NH_3 and Ar^+ is exothermic and fast, whereas there is no reaction between NH_3 and Ca^+ in the DRC. (Copyright 2013, all rights reserved, PerkinElmer Inc.)

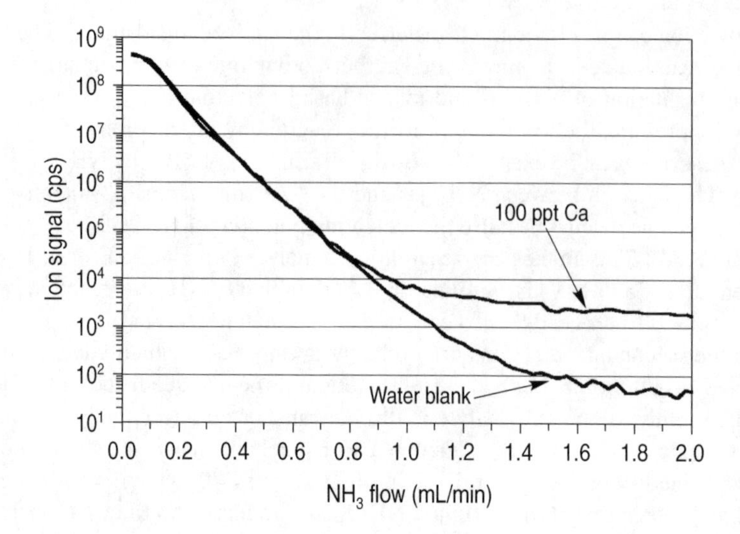

FIGURE 13.7 A reduction of eight orders of magnitude in the $^{40}Ar^+$ background signal is achievable with the DRC, resulting in a 0.1 ppt detection limit for $^{40}Ca^+$. (Copyright 2013, all rights reserved, PerkinElmer Inc.)

generated in a CRC for a group of mainly monoisotopic element (Sc, Y, La, Th) oxides.[16] They observed that when the collision/reaction gas contains impurities such as water vapor or ammonia, a broad range of additional interferences are produced in the cell. Depending on the relative mass of the precursor ions (ions that are formed in the plasma) compared to the by-product ions formed in the cell, there will be significant differences in the way these interferences are suppressed. They concluded that unless the mass (or energy) differences between the precursor and by-product ions are large, there will be significant overlap of kinetic energy distribution, making it very difficult to separate them, which limits the effectiveness of KED to suppress the cell-generated ions. On the other hand, bandpass tuning can tolerate a much smaller difference in mass between the precursor and by-product interfering ions because of its ability to set the optimum mass/charge cut-off at the point where these interfering ions are rejected. In addition, they found that the bandpass tuning method can use a heavier or denser collision gas if desired, without suffering a loss of sensitivity due to scattering observed with the KED method. The overall conclusion of their study was that "under optimized conditions, the bandpass tuning approach provides superior analytical performance because it retains a significantly higher elemental sensitivity and provides more efficient suppression of cell-generated oxide ions, when compared to kinetic energy discrimination."

Low Mass Cut-Off Collision/Reaction Cell

A variation on bandpass filtering uses slightly different control of the filtering process. By operating the cell in the RF-only mode, the quadrupole's stability boundaries can be tuned to cut off low masses where the majority of the interferences occur.[17] This stops many of the problematic argon-, matrix- and solvent-based ions from entering the CRC, therefore reducing the likelihood of creating new precursor ions in the cell which has the potential to negatively impact the determination of the analytes. The basic principles of this technique are shown in Figure 13.8, which shows a typical Mathieu stability plot for a quadrupole. It can be seen with a fixed DC electrical field (a) of zero (thick horizontal line on baseline shown with rf-only arrow) and an RF electrical field scan (q) of 0–0.9 (large triangle); all masses above the thick horizontal line and inside the triangle are stable and will pass through the quadrupole rods. While all masses that are outside the triangle will be unstable and be ejected out of the electrical field. So, for example, a q-value of 0.47 might be equivalent to 24 amu, which would mean ^{24}Mg is stable and allowed to be transmitted, while a q-value of 0.95 will be equivalent to mass 12, which would mean ^{12}C is unstable and be rejected. And if ^{12}C

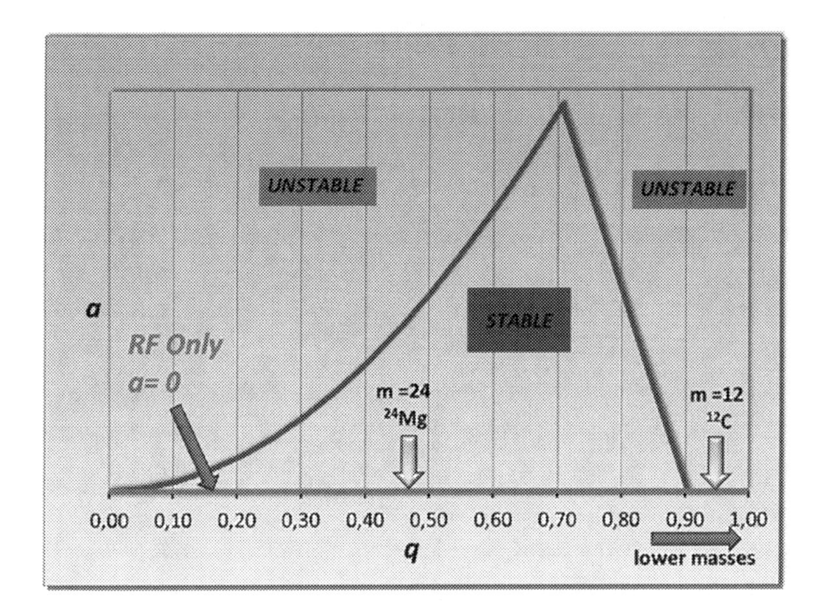

FIGURE 13.8 Basic principles of a low mass cut-off CRC.

is not transmitted, the resulting $^{12}C^{12}C$ polyatomic ion at mass 24 would not be present to interfere with ^{24}Mg.

The benefit of this approach is seen in Figure 13.9, which shows a list of analytes in the far left-hand column and the suggested cut-off mass in the next column. By using this cut-off mass, the potential polyatomic interferents in the third column won't be created because the majority of the precursor ions shown in the final column will be unstable and won't be present to react and combine with the other matrix or solvent-based ions in the cell. For some elements, this can translate into a fourfold to fivefold improvement in detection capability over a traditional collision cell, particularly elements such as $^{31}P^+$ and $^{34}S^+$, which are prone to interferences from the $^{16}O^+$ ion.

The major difference of this technology is that it doesn't use a traditional quadrupole with spherical or elliptical-shaped rods. It uses a flatapole, which has beveled or straight edges as opposed to round edges. The benefit is that the transmission efficiency tends to be similar to higher-order multipoles, thus allowing more ions through. The difference between the fields of a flatapole and quadrupole is shown in Figure 13.10.

"Triple Quadrupole" Collision/Reaction Cell

It should be noted that historically single multipole-based CRCs dominated the commercial landscape, but in the late 2000s, multiple multipole technology was developed which placed an additional quadrupole prior to the CRC multipole and the analyzer quadrupole. This first quadrupole acts as a simple mass filter to allow only the analyte masses to enter the cell, while rejecting all other masses. With all non-analyte, plasma, and sample matrix ions excluded from the cell, sensitivity and interference removal efficiency is significantly improved compared to traditional CRC technology coupled with a single quadrupole mass analyzer.

This very exciting collision/reaction technology is known as a "triple quadrupole" CRC[18]—a name derived from the liquid chromatography–mass spectrometry–mass spectrometry (LC–MS–MS) technique, where three quadrupoles are used to separate, detect, and confirm the presence of organic molecules such as proteins and peptides in biological samples. However, the term is not technically correct in this configuration used for ICP-MS. Even though the first quadrupole (Q1) is a mass filter and the second quadrupole (Q2) is the analyzer quadrupole, the middle multipole device is actually an octapole CRC. This means that the cell cannot be used as a conventional bandpass

Analyte	Cutoff mass	Potential interferent	Precursors
^{45}Sc	29	$^{13}C^{16}O_2$, $^{12}C^{16}O_2H$, ^{44}CaH, $^{32}S^{12}CH$, $^{32}S^{13}C$, $^{33}S^{12}C$	H, C, O, S, Ca
^{47}Ti	32	$^{31}P^{16}O$, ^{46}CaH, $^{35}Cl^{12}C$, $^{32}S^{14}NH$, $^{33}S^{14}N$	H, C, N, O, P, S, Cl, Ca
^{49}Ti	33	$^{31}P^{18}O$, ^{48}CaH, $^{35}Cl^{14}N$, $^{37}Cl^{12}C$, $^{32}S^{16}OH$, $^{33}S^{16}O$	H, C, N, O, P, S, Cl, Ca
^{50}Ti	34	$^{34}S^{16}O$, $^{32}S^{18}O$, $^{35}Cl^{14}NH$, $^{37}Cl^{12}CH$	H, C, N, O, S, Cl
^{51}V	35	$^{35}Cl^{16}O$, $^{37}Cl^{14}N$, $^{34}S^{16}OH$	H, O, N, S, Cl
^{52}Cr	36	$^{36}Ar^{16}O$, $^{40}Ar^{12}C$, $^{35}Cl^{16}OH$, $^{37}Cl^{14}NH$, $^{34}S^{18}O$	H, C, O, N, S, Cl, Ar
^{55}Mn	39	$^{37}Cl^{18}O$, $^{23}Na^{32}S$, $^{23}Na^{31}PH$,	H, O, Na, P, S, Cl, Ar
^{56}Fe	39	$^{40}Ar^{16}O$, $^{40}Ca^{16}O$	O, Ar, Ca
^{57}Fe	40	$^{40}Ar^{16}OH$, $^{40}Ca^{16}OH$	H, O, Ar, Ca
^{58}Ni	41	$^{40}Ar^{18}O$, $^{40}Ca^{18}O$, $^{23}Na^{35}Cl$	O, Na, Cl, Ar, Ca
^{59}Co	42	$^{40}Ar^{18}OH$, $^{43}Ca^{16}O$, $^{23}Na^{35}ClH$	H, O, Na, Cl, Ar, Ca
^{60}Ni	43	$^{40}Ca^{16}O$, $^{23}Na^{32}Cl$	O, Na, Cl, Ca
^{61}Ni	44	$^{44}Ca^{16}OH$, $^{38}Ar^{23}Na$, $^{23}Na^{37}ClH$	H, O, Na, Cl, Ca
^{63}Cu	45	$^{44}Ar^{23}Na$, $^{12}C^{15}O^{35}Cl$, $^{12}C^{14}N^{37}Cl$, $^{31}P^{32}S$, $^{31}P^{16}O_2$	C, N, O, Na, P, S, Cl
^{64}Zn	46	$^{32}S^{16}O_2$, $^{32}S_2$, $^{36}Ar^{12}C^{16}O$, $^{38}Ar^{12}C^{14}N$, $^{48}Ca^{16}O$	C, N, O, S, Ar, Ca
^{65}Cu	47	$^{32}S^{16}O_2H$, $^{32}S_2H$, $^{14}N^{15}O^{35}Cl$, $^{48}Ca^{16}OH$	H, N, O, S, Cl, Ca
^{66}Zn	47	$^{34}S^{16}O$, $^{32}S^{34}S$, ^{33}S,^{48}C,^{18}O	O, C, S
^{67}Zn	47	$^{32}S^{34}SH$, $^{33}S_2H$, $^{48}Ca^{18}OH$, $^{14}N^{16}O^{37}Cl$,$^{35}Cl^{16}O_2$	H, N, O, S, Cl, Ca
^{68}Zn	47	$^{32}S^{18}O_2$, $^{34}S_2$	O, S
^{69}Ga	47	$^{32}S^{18}O_2H$, $^{34}S_2H$, $^{37}Cl^{16}O_2$	H, O, S, Cl
^{70}Zn	47	$^{34}S^{18}O_2$, $^{35}Cl_2$	O, S, Cl
^{75}As	47	$^{40}Ar^{34}SH$, $^{40}Ar^{35}Cl$, $^{40}Ca^{35}Cl$, $^{37}Cl_2H$	H, S, Cl, Ca, Ar
^{77}Se	47	$^{40}Ar^{37}Cl$, $^{40}Ca^{37}Cl$	Cl, Ca, Ar
^{78}Se	47	$^{40}Ar^{38}Ar$	Ar
^{80}Se	47	$^{40}Ar_2$, $^{40}Ca_2$, $^{40}Ar^{40}Ca$, $^{32}S_2$$^{16}O$, $^{32}S^{16}O_3$	O, S, Ar, Ca

FIGURE 13.9 A list of elements with potential interferences that would benefit from the low mass cut-off approach. (Note: Elements in grey are above the low-mass cut-off and could still contribute to a polyatomic interference.)

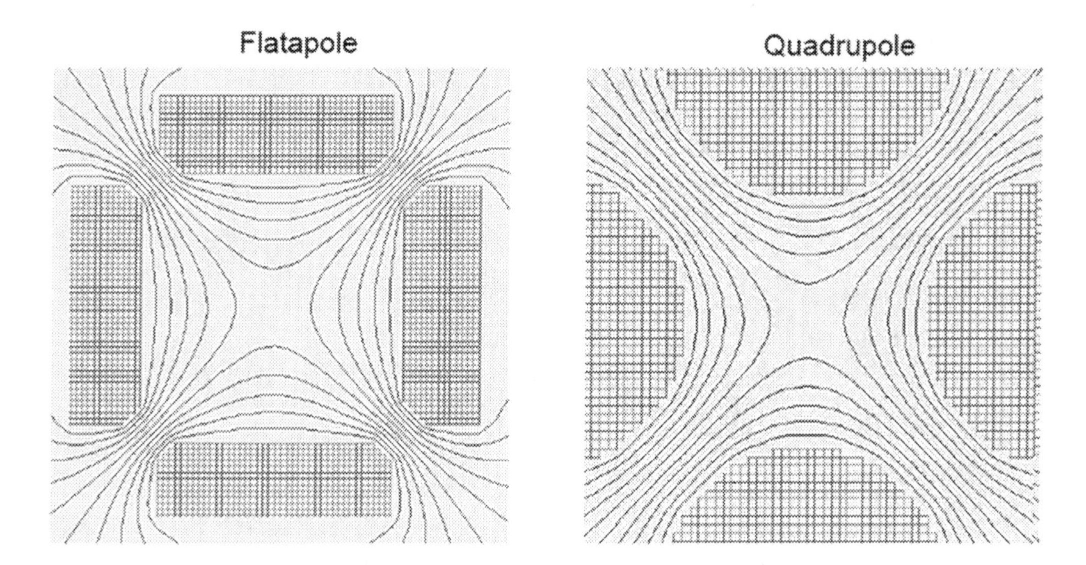

FIGURE 13.10 The difference between the electric fields of a flatapole compared to a quadrupole.

filter, like a single quadrupole-based DRC or low-mass cut-off device. Reactive gases can be used to initiate ion–molecule chemistry in the octopole cell, but it is then used just to pass all the product ions formed into the analyzer quadrupole (Q2), which is used to separate and select the mass or masses of interest. The principles of this technology are shown in Figure 13.11.

The capability and flexibility of the "triple quad" CRC has enormous potential. There are a number of different ways it can be utilized depending on the severity of the analytical problem. The two most common modes of analysis are:

- M/S mode
- MS/MS mode.

M/S Mode

In its most basic configuration, the instrument can be utilized in the single M/S mode, where Q1 acts as a simple ion guide allowing all ions through to the CRC, similar to a traditional single-quad ICP-MS system that uses an octopole-based collision cell. It can also be used in single M/S mode where Q1 acts as a bandpass filter, allowing a "window" of masses through, above/below the Q2 mass range selected by the user. In this mode, it functions in a similar way to a single-quad with a "scanning" bandpass filter-type CRC, except masses outside the bandpass window are rejected before they can enter the cell.

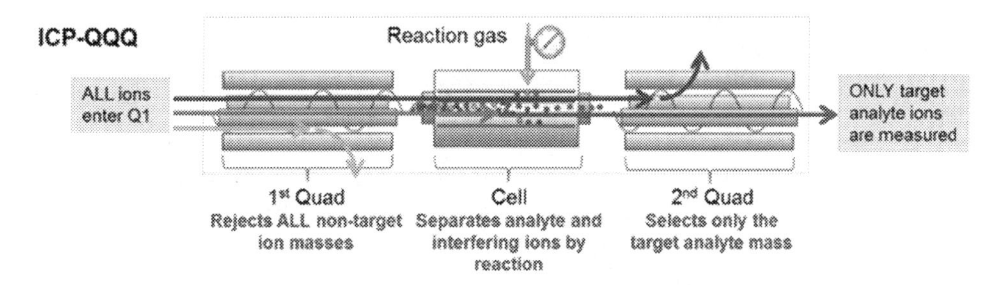

FIGURE 13.11 The fundamental principles of "triple quadrupole" technology used in ICP-MS.

MS/MS Mode

The instrument can also be used in the MS/MS mode, where the first quadrupole is operated with a 1 amu fixed bandpass window, allowing only the target ions to enter the CRC. This process can be implemented in two different ways:

- On-mass mode
- Mass shift mode.

On-Mass MS/MS Mode

In this configuration, Q1 and Q2 are both set to the target mass. Q1 allows only the precursor ion mass to enter the cell (analyte and on-mass polyatomic interfering ions). The octopole CRC then separates the analyte ion from the interferences using the reaction chemistry of a reactive gas, while Q2 measures the analyte ion at the target mass after the on-mass interferences have been removed by reactions in the cell.

An example of this is in the removal of sulfur-based interferences using ammonia (NH_3) gas in the determination of vanadium in the presence of a sulfuric acid matrix. The major isotope of vanadium is $^{51}V^+$. However, in the presence of high concentrations of H_2SO_4, the interfering ions, $^{33}S^{18}O^+$ and $^{34}S^{16}OH^+$, will overlap the $^{51}V^+$ ion. By using NH_3 as the reaction gas, which reacts very quickly with S-based polyatomic interferences, but is virtually unreactive with the vanadium ion, the $^{33}S^{18}O^+$ and $^{34}S^{16}OH^+$ ions are removed, thus allowing the $^{51}V^+$ ion to be detected free of any interferences. So in this example, Q1 and Q2 would be set at mass 51 to take advantage of the reactive properties of ammonia to effectively remove the SO^+/SOH^+ interferences. No new analyte- or matrix-based NH_3 cluster ions can be created at mass 51, as no other ions are able to enter the cell. A detection limit in the order of 13 ppt is achievable for the determination of vanadium in percentage levels of sulfuric acid as demonstrated by the calibration seen in Figure 13.12. This detection capability is similar to what is achievable in an aqueous solution.

Mass Shift MS/MS Mode

In this configuration, Q1 and Q2 are set to different masses. Similar to the "on-mass mode," Q1 is set to the precursor ion mass (analyte and on-mass polyatomic interfering ions), controlling the ions that enter octapole CRC. However, in this mode, Q2 is then set to mass of a target reaction product ion containing the original analyte. Mass-shift mode is typically used when the analyte ion is reactive, while the interfering ions are unreactive with a particular CRC gas. The basic principles of this approach are shown in Figure 13.13.

An example of this is in the determination of arsenic in the presence of transition metal oxides. Because As is monoisotopic, it can only be determined at 75 amu, which is overlapped by the $^{59}Co^{16}O$ polyatomic interference. However, by reacting ^{75}As with oxygen gas in the cell, the $^{75}As^{16}O$ species at 91 amu can be used for quantitation. This is achieved by setting a bandpass to only allow ions of mass 75 amu to pass through Q1. All ions at 75 amu will then enter the octapole CRC where they will interact with the oxygen cell gas. The ^{75}As will react with O_2 to form $^{75}As^{16}O$ species at 91 amu, while the $^{59}Co^{16}O$ at mass 75 amu will be unreactive. Q2 is then set at 91 amu to only allow the $^{75}As^{16}O$ ion through to be detected and quantified.

For more advanced applications, this technology offers other modes of interference reduction, including:

- **Precursor ion scan**, where Q2 is set to the target ion mass, while Q1 scans over a user-defined mass range to select the precursor ions that enter the cell and react with the collision/reaction gas. An example of this is the monitoring of the by-product $^{14}NH_4$ ion in the determination of Hg using NH_3 as a reaction gas for a multielement method. In this example, NH_3 is the optimum reaction gas for the majority of the other elements, but by measuring the $^{14}NH_4$ ion, Hg can also be determined in the same suite.

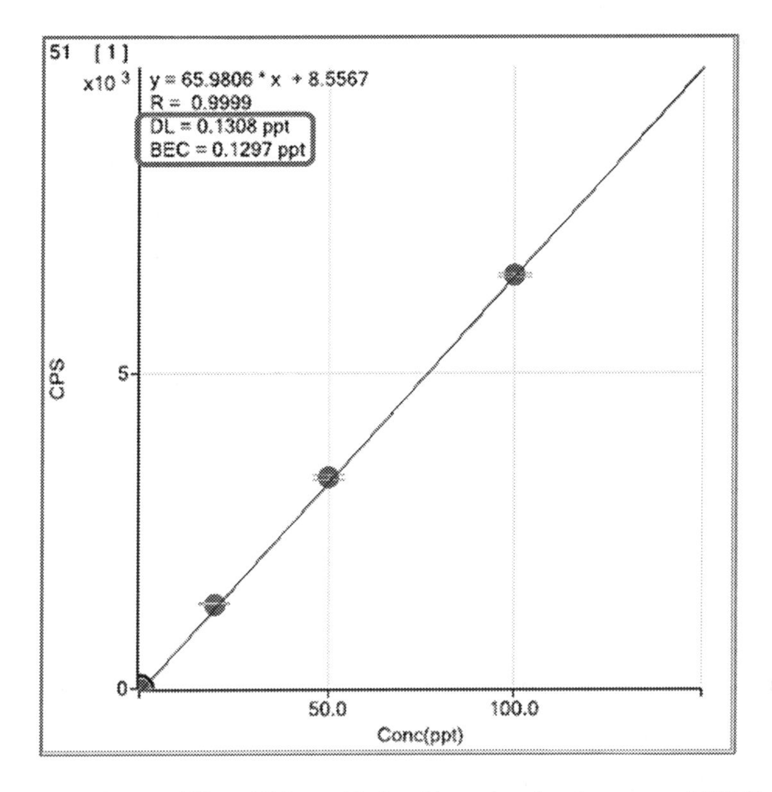

FIGURE 13.12 Detection capability of V in a sulfuric acid matrix using the on-mass MS/MS mode is in the order of 13 ppt as shown in this calibration plot.

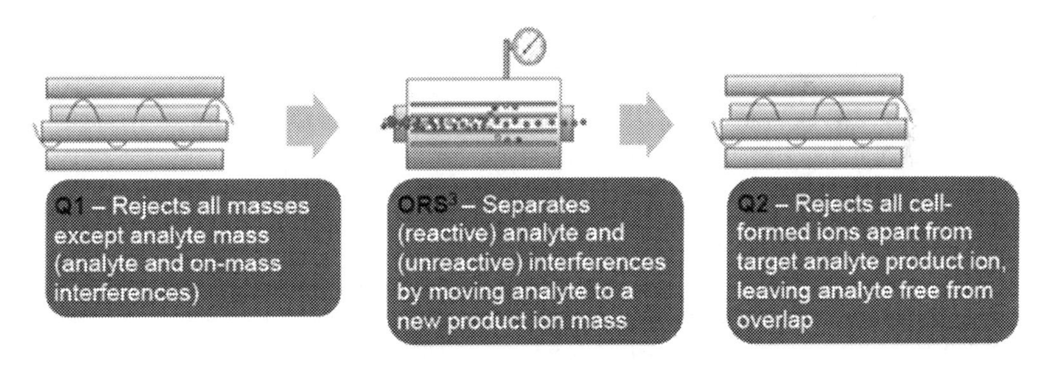

FIGURE 13.13 The basic principles of the mass-shift MS/MS mode.

- **Product ion scan**, where Q1 is set to allow only the target precursor ion mass to enter the cell, while Q2 scans to measure all the product ions formed in the cell, including controlled cluster ion analysis. An example of this is the use of NH_3 gas to create cluster ions of an analyte like titanium. By allowing only ^{48}Ti through Q1, only titanium cluster complex ions are formed in the cell and not other potentially interfering transition metal cluster ions as with a traditional CRC.
- **Neutral gain scan**, where Q1 and Q2 are scanned together, with a user-defined mass difference. This mode allows monitoring of the product ions from a particular transition for all ions in the Q1 scan range. An example of this is in the determination of titanium using oxygen as the collision/reaction gas. By only scanning Q1 at the masses for titanium

(^{46}Ti, ^{47}Ti, ^{48}Ti, ^{49}Ti, ^{50}Ti) and Q2 at 16 amu higher ($^{46}Ti^{16}O$, $^{47}Ti^{16}O$, $^{48}Ti^{16}O$, $^{49}Ti^{16}O$, $^{50}Ti^{16}O$), the isotopic abundances of the titanium oxide ions can be unambiguously measured, without the presence of other transition metals in the mass spectrum.

There is no question that the potential of this "triple quad" approach to reducing interferences using a CRC analysis is a truly very exciting addition. However, it is probably more suited for research-type applications, or in an academic environment, where non-routine investigations are being carried out. In my opinion, it will be competing with the double-focusing, magnetic sector technology as a problem-solving tool and for the analysis of more complex sample matrices. The traditional, single-quadrupole ICP-MS instrumentation will still represent the vast majority of instruments sold into the marketplace, for carrying out high-throughput, routine applications. Although, for companies that are purchasing a second instrument, a triple quad probably represents a good investment.

Let us now go on to discuss a slightly different approach to reduce interferences using a CRC—the CRI.

THE COLLISION/REACTION INTERFACE

Unlike collision and reaction cells, the CRI does not use a pressurized multipole-based cell before the mass analyzer. Instead, it injects a reaction/collision gas (typically He, or H_2) at relatively high flow rates (100–150 mL/min) into the plasma through the aperture of the interface skimmer cone, where the plasma density is high.[19] This increases the rate of interactions between the introduced gas and interfering ions giving improved attenuation of interfering ions. In addition, the reaction/collision gas is supplied directly to the plasma, which means that the plasma electrons are still available to assist in attenuating the interfering ions through electron–ion recombination. The presence of plasma electrons also significantly reduces the generation of secondary by-product ions produced from the interference attenuation process. The overall result is that most argon-based polyatomic interferences are destroyed or removed before they are extracted into the ion optics. Figure 13.14 shows the basic principles of the CRI.

The limitations of the earlier designs restricted their use for real-world samples because there appeared to be no way to effectively focus the ions, and therefore, there was very little control over the collision process. So, even though the addition of a collision/reaction gas helped reduce plasma-based spectral interferences, it did virtually nothing for matrix-induced spectral interferences. In addition, there appeared to be no way to carry out KED in the interface region, and as a result, it made it very difficult to take advantage of collisional mechanisms using an inert gas such as helium.

However, in the most recent commercial design, all the reactions/collisions processes are actually taking place inside the tip of the skimmer and not between the sampler and skimmer cone as with earlier designs. Because of this subtle difference, simple, loosely bonded polyatomic species can receive sufficient energy through collisional (vibrational and rotational) excitation mechanisms to bring about the dissociation of the interference, whereas the analyte ions simply lose energy as they collide with the gas molecules. And where the collisional impact is not suitable for interference reduction, as in the removal of the argon dimer ($^{40}Ar_2^+$) in the determination of $^{80}Se^+$, or the elimination of the $^{40}Ar^{12}C^+$ interference in the determination of $^{52}Cr^+$, a low-reactivity gas such as hydrogen can be used to initiate an ion–molecule reaction.

This can be seen in Figure 13.15, which shows the reduction of the $^{40}Ar_2^+$ and $^{40}Ar^{12}C^+$ interfering ions using hydrogen as the collision/reaction gas, in the determination of ^{80}Se and ^{52}Cr, respectively. The sensitivities of three internal standard elements, ^{45}Sc, ^{89}Y, and ^{115}In, were monitored at the same time as monitoring the two interfering ions. It should be noted that there is no Se or Cr in this solution, so the signals at mass 80 and 52 amu are contributions from the $^{40}Ar^{40}Ar$ and $^{40}Ar^{12}C$ polyatomic ions, respectively. It can be seen very clearly that there is sharper decrease of the interferent signals as compared to those of the internal standards, showing evidence of the removal of the $ArAr^+$ and ArC^+ polyatomic interferences with increasing H_2 flow rate. The optimization plot shows

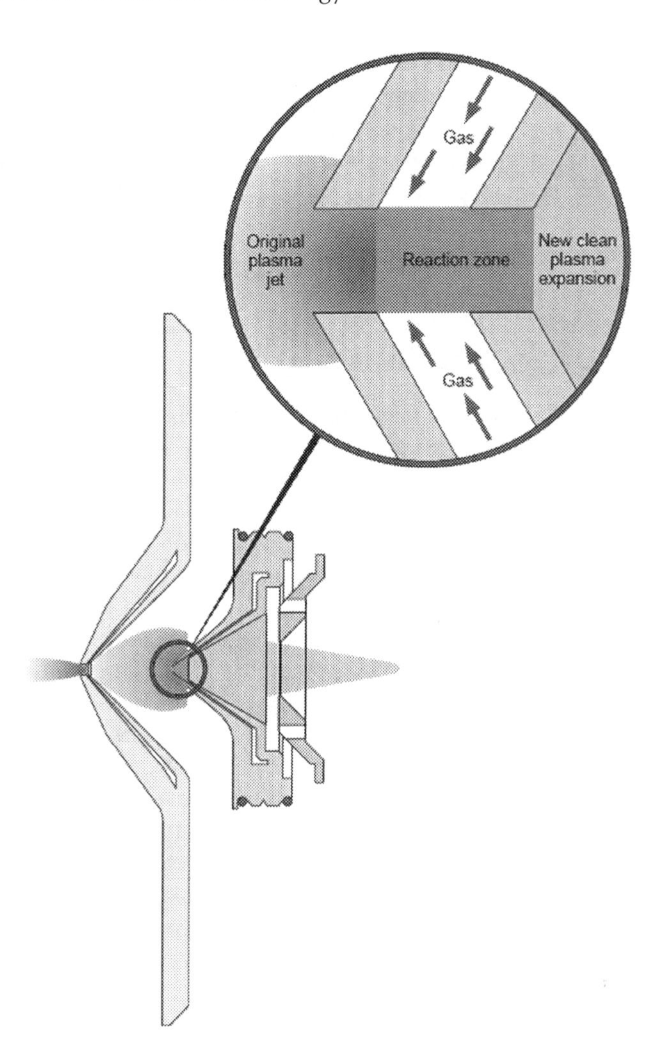

FIGURE 13.14 Principles of the CRI. (Courtesy of Analytic Jena.)

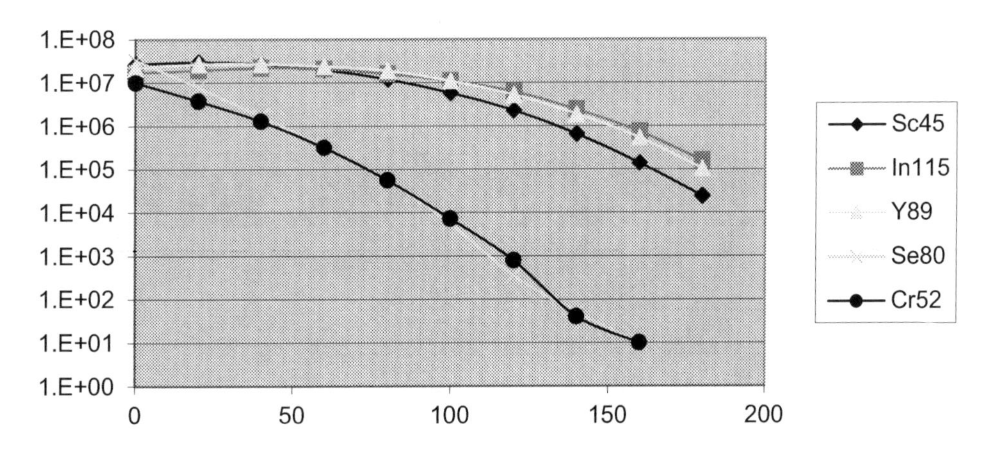

FIGURE 13.15 Optimization of hydrogen gas flow rate, showing three Internal Standard (Sc, In, Y) signals while monitoring the interferences $^{40}Ar^{40}Ar$ (^{80}Se) and $^{40}Ar^{12}C$ (^{52}Cr). (Courtesy of Analytic Jena.)

that at a flow rate of 140 mL/min, signals of the interfering ions decrease by six orders of magnitude while that of the Sc, Y, and In are only reduced by two orders of magnitude.[20]

The CRI looks to be a very interesting concept, which appears to offer a relatively straightforward, non-cell-based solution to minimizing plasma- and matrix-based spectral interferences in ICP-MS. Each year more and more challenging applications appear in the public domain showing the capabilities of CRI systems. If there is an interest in this approach, it is worth checking out the vendor application notes, which show the capabilities of the CRI in many different, real-world matrices.[21]

USING REACTION MECHANISMS IN A COLLISION CELL

After almost 20 years of solving real-world application problems, the practical capabilities of both hexapole and octapole collision cells using KED are fairly well understood. It is clear that the majority of applications are being driven by the demand for routine multielement analysis of well-characterized matrices, where a rapid sample turnaround is required. This technique has also been promoted by the vendors as a fast, semiquantitative tool for unknown samples. The fact that it requires very little method development, just one collision gas, and one set of tuning conditions, makes it very attractive for these kinds of applications.[22]

However, it is well recognized that this approach will not work well for many of the more complex interfering species, especially if the analyte is at ultratrace levels. For example, the collision mode using helium is not the best choice for quantifying selenium using its two major isotopes ($^{80}Se^+$, $^{78}Se^+$) because it requires a large number of collisions to separate the argon dimers ($^{40}Ar_2^+$ and $^{40}Ar^{38}Ar^+$) from the analyte ions using KED. As a result of the inefficient interference reduction process, the signal intensity for the analyte ion is also suppressed. For this reason, it is well accepted that the best approach is to use a reaction gas in order to initiate some kind of charge-transfer mechanism. This can be seen in Figure 13.16, which shows the calibration for $^{78}Se^+$ using hydrogen as the reaction gas to minimize the impact of the argon dimer, $^{40}Ar^{38}Ar^+$. On the left is the 0, 1.0, 2.5, 5.0, and 10.0 ppb calibration using helium and on the right is the same calibration using hydrogen as the reaction gas. It can be seen very clearly that the signal intensity for the calibration standards using hydrogen gas is approximately five times higher than the calibration standards using helium,

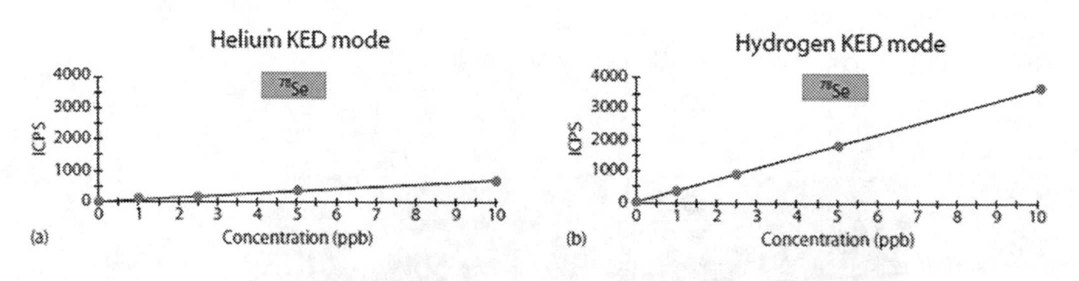

Mode	He	H_2
Sensitivity (icps/ppm)	72,000	374,000
BEC (ppt0)	22	7
Est. IDL (ppt)	15	1
RSD % at 1 ppb level	8	1

(c)

FIGURE 13.16 Comparison of $^{78}Se^+$ calibration plots and detection limits using the collision mode with helium (left) and reaction mode with hydrogen (right). (Courtesy of Thermo Scientific.)

producing a selenium detection limit of 15 times lower (1 ppt compared to 15 ppt); this is also seen in the table following the calibration graphs.[23]

Likewise, helium has very little effect on reducing the $^{40}Ar^+$ interference in the determination of $^{40}Ca^+$. So, when using a collision cell with helium, the quantification of Ca must be carried out using the $^{44}Ca^+$ isotope, which is about 50 times less sensitive than $^{40}Ca^+$. For this reason, in order to achieve the lowest detection limits for calcium, a low-reactivity gas such as pure hydrogen is the better option.[22] By initiating an ion–molecule reaction, it allows the most sensitive calcium isotope at 40 amu to be used for quantitation. In fact, even though the use of hydrogen significantly improves the detection limit for calcium, the best interference reduction is achieved using a mixture of ammonia and helium.

Another example of the benefits of using more reactive gases, such as an ammonia and helium mixture over pure helium or hydrogen gas, is in the determination of vanadium in a high-concentration chloride matrix. The collision mode using helium works reasonably well on the reduction of the $^{35}Cl^{16}O^+$ interference at 51 amu. However, when 1% NH_3 in helium is used, the interference is dramatically reduced by the process of charge/electron transfer. This allows the most abundant isotope, $^{51}V^+$, to be used for the quantitation of vanadium in matrices such as seawater or hydrochloric acid. Vanadium detection capability in a chloride matrix is improved by a factor of 50–100 times using the reaction chemistry of NH_3 in helium compared to pure helium in the collision mode.[23]

In addition to ammonia–helium mixtures, oxygen is sometimes the best reaction gas to use because it offers the possibility of either moving the analyte ion to a region of the mass spectrum where the interfering ion does not pose a problem or moving the interfering species away from the analyte ion by forming an oxygen-derived polyatomic ion 16 amu higher. An example of changing the mass of the interfering ion is in the determination of $^{114}Cd^+$ in the presence of high concentrations of molybdenum. In the plasma, the molybdenum forms a very stable oxide species $^{98}Mo^{16}O^+$ at 114 amu, which interferes with the major isotope of cadmium, also at mass 114. By using pure oxygen as the reaction gas, the $^{98}Mo^{16}O^+$ interference is converted to the $^{98}Mo^{16}O^{16}O^+$ complex at 16 amu higher than the analyte ion, allowing the $^{114}Cd^+$ isotope to be used for quantitation.[23]

The benefits of using reaction mechanisms for the determination of calcium, vanadium, and cadmium are seen in Figure 13.17, which shows (a) 0–300 ppt calibration plot for $^{40}Ca^+$ using hydrogen gas, (b) 0–500 ppt calibration plot for $^{51}V^+$ in 2% hydrochloric acid using 1% NH_3 in helium, and (c) 0–10 ppb calibration plot for $^{114}Cd^+$ in a molybdenum matrix using pure oxygen. Background equivalent concentration (BEC) values of 4, 2, and 3 ppt were achieved for calcium, vanadium, and cadmium, respectively.[23]

However, it is important to point out that even though a higher-order multipole collision cell with KED can use low-reactivity gases, it usually requires significantly more interactions than a reaction cell that uses a highly reactive gas.[23] Take for example, the reduction of the $^{40}Ar^+$ interferences in the determination of $^{40}Ca^+$. In a collision cell, even though the kinetic energy of the $^{40}Ar^+$ ion will be reduced by reactive collisions with molecules of hydrogen gas, the $^{40}Ca^+$ will also lose kinetic energy because it, too, will collide with the reaction gas. Now, even though these interactions are basically nonreactive with respect to $^{40}Ca^+$, it will experience the same number of collisions because it has a similar cross-sectional area as the $^{40}Ar^+$ ion. So, in order to achieve many orders of interference reduction with a low-reactivity gas such as hydrogen, a high number of collisions are required. This means that in addition to the interference being suppressed, the analyte will also be affected to a similar extent. As a result, the energy distribution of both the interfering species and the analyte ion at the cell exit will be very close, if not overlapping, resulting in compromised interference reduction capabilities compared to a reaction cell that uses highly reactive gases and discrimination by mass filtering. This compromise translates into a detection limit for $^{40}Ca^+$ with a KED-based collision cell using hydrogen being approximately 50–100 times worse than a DRC using pure ammonia.[24]

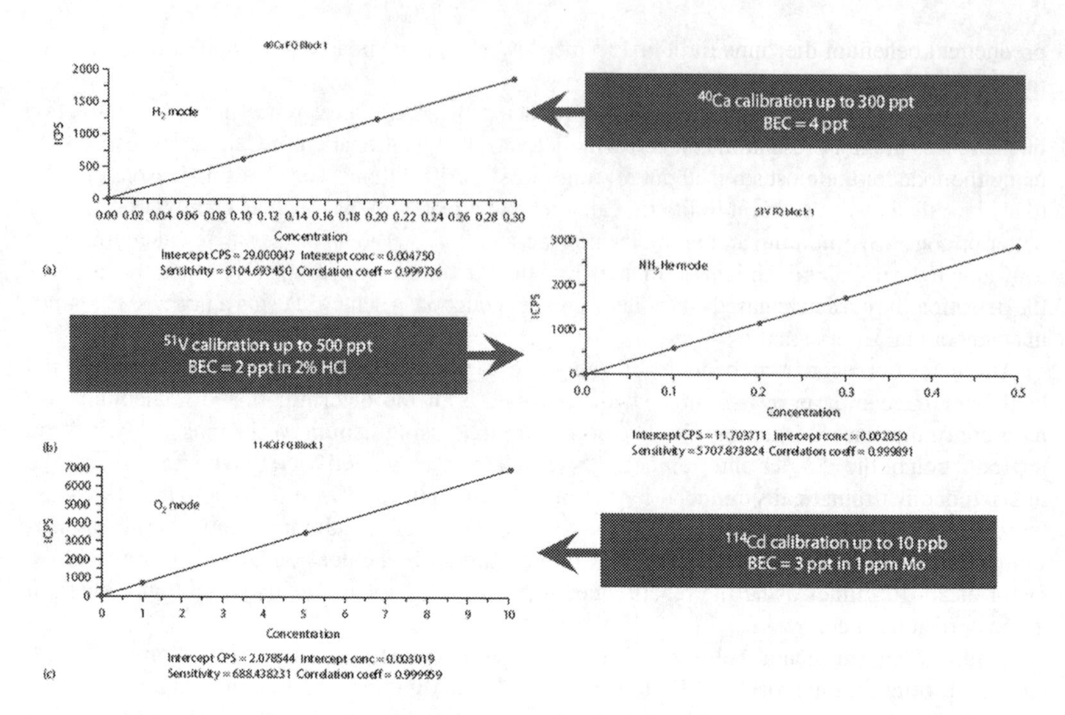

FIGURE 13.17 Calibration plots for (a) $^{40}Ca^+$ using hydrogen, (b) $^{51}V^+$ in 2% HCl using 1% NH_3 in helium, and (c) $^{114}Cd^+$ in high concentrations of molybdenum using oxygen. (Courtesy of Thermo Scientific.)

THE "UNIVERSAL" CELL

A recent development is the commercialization of an instrument that can be used in both the collision mode and reaction mode. More commonly known as the Universal Cell, this innovative design can be utilized as both a simple collision cell using KED and a dynamic reaction using a pure reaction gas.[25] When operating in the collision cell mode, the universal cell works on the principle that the interfering ion is physically larger than the analyte ion. If both ions are allowed to pass through the cell containing the inert gas molecules, the interfering ion will collide more frequently with the inert gas atoms than the analyte ion due to its larger size. This results in a greater loss of kinetic energy by the interferent compared to the analyte ion. An energy barrier is then placed at the exit of the cell, so that the higher-energy analyte ions are allowed to pass through it, while the lower-energy interferences are not. As mentioned earlier in this chapter, this process is commonly referred to as KED.

When the universal cell is operating in the reaction mode, it takes advantage of whether the interaction of the analyte and interfering species with the reaction gas is either exothermic (fast) or endothermic (slow). The reactivity will be based on the relative ionization potential of the analyte and interfering species compared to the reaction gas. Typically interfering ions tend to react exothermally with a reactive gas like ammonia, while analyte ions react endothermally. This means if both interferent ions and analyte ions enter the cell, the interferent ions will be reacting with the reactive gas and be converted to a new species with a different mass, while the analyte ions will be unaffected and pass through the reaction cell into the filtering quadrupole, free of the interference.

But it should be emphasized that while a reactive gas efficiently removes interferences, it is also capable of creating new interferences if not properly controlled. For that reason, an optimized reaction cell (coupled with a single mass analyzer) requires the use of a scanning quadrupole to prevent these new interferences from forming through the creation of a unit-resolution bandpass filter, to allow only a single mass to pass through.

The real benefit of the "universal" CRC approach is that it can be used in both the collision cell and the reaction cell modes. This means the operator has the flexibility of operating the system in three different modes all in the same multielement method—the standard mode for elements where interferences are not present, in collision mode for removal of minor interference, and in dynamic reaction mode for the most severe polyatomic spectral interferences.

DETECTION LIMIT COMPARISON

In general, highly reactive gases are recognized as being more efficient at reducing the interference and generating better signal-to-noise ratio than either an inert gas such as helium, or low-reactivity gases such as hydrogen or mixtures of ammonia and helium. However, it's very difficult to make a detection limit comparison with this technology because detection capability tends to be application specific. In other words, depending on the interference reduction capability of the CRC/interface device, it might offer better detection capability than another instrument in one particular sample matrix but offer inferior detection limits in another. In addition, vendors' instrument detection limits are typically carried out in simple aqueous solutions, which is unlikely to show a significant difference in the performance of the different CRC/interface technology. The only way the interference reduction capability of a particular device can be truly evaluated is in a real sample containing matrix and solvent components. For that reason, refer to the cited references, which are a selection of technical/data sheets and application literature showing performance characteristics of the different CRC/interface approaches.[17–24,26]

FINAL THOUGHTS

There is no question that CRCs and interfaces have given a new lease of life to quadrupole mass analyzers used in ICP-MS. They have enhanced its performance and flexibility and most definitely opened up the technique to more demanding applications that were previously beyond its capabilities. This is most definitely the case with new triple-quad system, which will be competing with the magnetic sector, high-resolution systems for the more difficult, research-type applications. It should also be noted that when I wrote my ICP-MS textbook in 2014, there was only one vendor of triple-quad ICP-MS instrumentation. Today, there is a second vendor,[25] and very recently a third one entered the market place.[27]

However, it must be emphasized that when assessing CRC technology, it is critical that you fully understand the capabilities of the different approaches, especially how they match up to your application objectives. The KED-based collision cell using an inert gas such as helium is probably better suited to doing multielement analysis in a routine environment. However, you have to be aware that its detection capability is compromised for some elements depending on the type of samples being analyzed. By using ion–molecule reactions as opposed to collisions, detection limits for many of these elements can be improved quite significantly. Of course, if two or even three different gases have to be used, the convenience of using one gas goes away.

On the other hand, using highly reactive gases with discrimination by mass filtering appears to offer the best performance and the most flexibility of all the different commercial approaches. By careful matching of the reaction gas with the analyte ion and polyatomic interference, extremely low detection limits can be achieved by ICP-MS, even for many of the notoriously difficult elements. It should be emphasized that selection of the optimum reaction gas and selection of the best quadrupole bandpass parameters can sometimes translate into quite lengthy method development, especially if there is very little application data available. However, most vendors do a very good job of generating application studies for some of the more routine applications. If you analyze out-of-the-ordinary or complex samples, you might initially need to spend the time to develop an analytical method that is both robust and routine. If you are uncertain, which is the best approach for your application, you should consider investing in a system that offers both collision cell and reaction cell capabilities in the same instrument. The simple collision cell approach using helium could be used for your routine samples, while the more powerful and DRC could be utilized for the more difficult sample matrices. Even if you don't need both, it will at least give you peace of mind that you have the flexibility to tackle the most demanding applications if needed. And finally, if you are investing in a second instrument, a triple-quad system might be the best option.

So when evaluating this technique, particularly for its capabilities with cannabis-related samples, pay attention not only to what the technique can do for your application problem but also to what it cannot do, which is equally important. In other words, make sure you evaluate its capabilities on the basis of all your present and future analytical requirements. This is particularly important, because it might be suitable for reducing interferences on Pb, Cd, As, and Hg, but when you are required to expand the elemental suite, based on future regulations, make sure it is capable of mitigating potential interferences on other heavy metals. For that reason, when assessing vendor-generated data, make sure the performance is achievable in your laboratory today as well as on all future elemental and sample workload demands.

FURTHER READING

1. K. Sakata and K. Kawabata, *Spectrochimica Acta*, **49B**, 1027–1032, 1994.
2. J. T. Rowan and R. S. Houk, *Applied Spectroscopy*, **43**, 976–980, 1989.
3. D. W. Koppenaal, C. J. Barinaga, and M. R. Smith, *Journal of Applied Analytical Chemistry*, **9**, 1053–1058, 1994.
4. G. C. Eiden, C. J. Barinaga, and D. W. Koppenaal, *Journal of Applied Analytical Chemistry*, **11**, 317–322, 1996.
5. P. Turner, T. Merren, J. Speakman, and C. Haines, Plasma Source Mass Spectrometry: Developments and Applications, Royal Society of Chemistry, Washington, DC, ISBN 0-85404-727-1, 28–34, 1996.
6. D. J. Douglas and J. B. French, *Journal of American Society of Mass Spectrometry*, **3**, 398, 1992.
7. B. A. Thomson, D. J. Douglas, J. J. Corr, J. W. Hager, and C. A. Joliffe, *Analytical Chemistry*, **67**, 1696–1704, 1995.
8. X Series ICP-MS: Enhanced Collision Cell Technology CCT, Thermo Scientific Product Specifications, July 2004, http://www.thermo.com/eThermo/CMA/PDFs/Articles/articlesFile_24138.pdf.
9. E. R. Denoyer, S. D. Tanner, and U. Voellkopf, *Spectroscopy*, **14**, 2, 1999.
10. H. H. Willard, L. L. Merritt, J. A. Dean, and F. A. Settle, Spectroscopic and Spectrometric Techniques (Chapters 5–13), Instrumental Methods of Analysis, Wadsworth Publishing Co., Belmont, CA, 465–507, 1988.
11. E. De Hoffman, J. Charette, and V. Stroobant, *Mass Spectrometry, Principles and Applications*, John Wiley and Sons, Paris, 1996.
12. Analysis of Ultra-trace Levels of Elements in Seawaters Using 3rd-Generation Collision Cell Technology, Thermo Scientific Product Application Note: 40718, April 2007, http://www.thermo.com/eThermo/CMA/PDFs/Articles/articlesFile_26161.pdf.
13. J. Takahashi, Determination of Impurities in Semiconductor Grade Hydrochloric Acid Using the Agilent 7500 cs ICP-MS, Agilent Technologies Application Note 5989-4348EN, January 2006.
14. N. Yamada, J. Takahashi, and K. Sakata, *Journal of Analytical Atomic Spectrometry*, **17**, 1213–1222, 2002.
15. S. D. Tanner and V. I. Baranov, *Atomic Spectroscopy*, **20** (2), 45–52, 1999.
16. B. Hattendorf and D. Günter, *Journal of Analytical Atomic Spectrometry*, **19**, 600–606, 2004.
17. PlasmaQuant MS ICP-MS, Analytik Jena, https://www.analytik-jena.us/products/elemental-analysis/icp-ms/plasmaquant-ms/.
18. R. Chemnitzer, Strategies for Achieving the Lowest Possible Detection Limits in ICP-MS, Spectroscopy Magazine, Analytik Jena, Oct 01, 2019, http://www.spectroscopyonline.com/column-atomic-perspectives.
19. Product Overview of the Agilent Technologies 8900 Triple Quadrupole ICP-MS, https://www.agilent.com/en/products/icp-ms/icp-ms-systems/8900-triple-quadrupole-icp-ms.
20. Agilent Technologies 7900 ICP-MS Product Overview, https://www.agilent.com/en/products/icp-ms/icp-ms-systems/7900-icp-ms.
21. S. D. Tanner, V. I. Baranov, and D. R. Bandura, *Spectrochimica Acta*, **57B** (9), 1361–1452, 2002.

22. K. Kawabata, Y. Kishi, and R. Thomas, *Spectroscopy*, **18** (1), 16–31, 2003.
23. Overview of the NexION 300 ICP-MS Using the "Universal" Collision/Reaction Cell, PerkinElmer Inc., http://www.perkinelmer.com/Catalog/Family/ID/NexION.
24. Product Specifications of the Thermo Scientific iCAP Q, A Dramatically Different ICP-MS, https://static.thermoscientific.com/images/D20721~.pdf.
25. The 30-Minute Guide to ICP-MS, ICP-Mass Spectrometry Technical Note, PerkinElmer Inc., https://www.perkinelmer.com/CMSResources/Images/44-74849tch_icpmsthirtyminuteguide.pdf
26. Thermo Fisher Scientific: iCAP TQ Tripe Quad ICP-MS Landing Page, http://www.thermofisher.com/order/catalog/product/731546.
27. NexION 5000 Multi Quadrupole ICP-MS Product Technical Note https://www.perkinelmer.com/Product/nexion-5000-icp-ms-n8160010

14 Ion Detectors

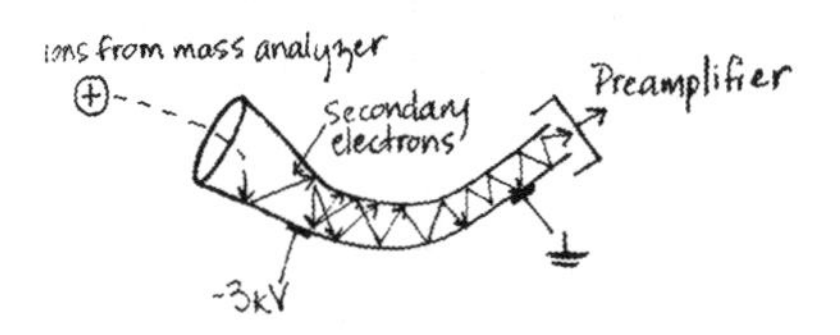

This chapter looks at the detection system—an important area of the mass spectrometer that detects and quantifies the number of ions emerging from the mass analyzer. The detector converts the ions into electrical pulses, which are then counted using its integrated measurement circuitry. The magnitude of the electrical pulses corresponds to the number of analyte ions present in the sample, which is then used for trace element quantitation by comparing the ion signal with known calibration or reference standards. In this section, we will take a look at conventional dynode detection, which monitors discrete ions emerging from the mass separation device in a sequential manner, in addition to describing the new breed of array detectors, which can monitor the entire mass spectrum simultaneously.

Since inductively coupled plasma mass spectrometry (ICP-MS) was first introduced in the early 1980s, a number of different ion detection designs have been utilized, the most popular being electron multipliers for low ion count rates and Faraday collectors for high count rates. Today, the majority of ICP-MS systems that are used for ultratrace analysis use detectors that are based on the active film or discrete dynode electron multiplier. They are very sophisticated pieces of equipment and are very efficient at converting ion currents emerging from the mass analyzer into electrical signals. The location of the detector in relation to the mass analyzer is shown in Figure 14.1.

Before I go on to describe discrete dynode detectors in greater detail, it is worth looking at two of the earlier designs—the channel electron multiplier (Channeltron®)[1] and the Faraday cup—to get a basic understanding of how the ICP-MS ion detection process works.

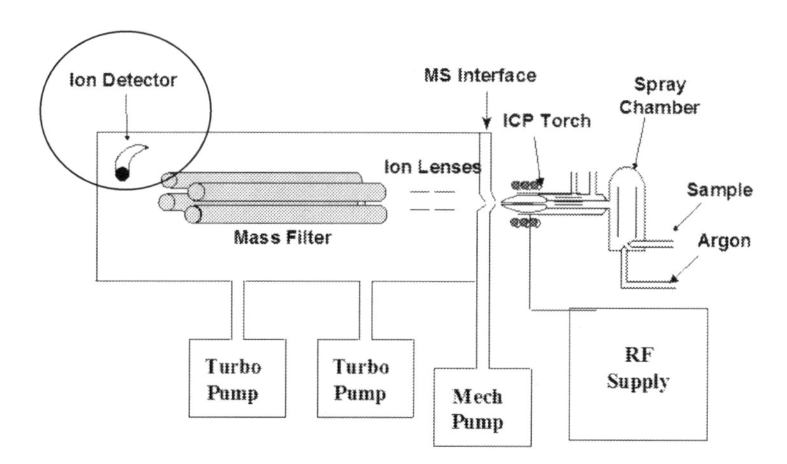

FIGURE 14.1 The location of the detector in relation to the mass analyzer.

CHANNEL ELECTRON MULTIPLIER

The operating principles of the channel electron multiplier are similar to a photomultiplier tube used in ICP-OES. However, instead of using individual dynodes to convert photons to electrons, the Channeltron is an open glass cone (coated with a semiconductor-type material) to generate electrons from ions that impinge on its surface. For the detection of positive ions, the front of the cone is biased at a negative potential while the far end near the collector is kept at ground. When the ion emerges from the quadrupole mass analyzer, it is attracted to the high negative potential of the cone. When the ion hits this surface, one or more secondary electrons are formed. The potential gradient inside the tube varies based on position, so the secondary electrons move further down the tube. As these electrons strike new areas of the coating, more secondary electrons are emitted. This process is repeated many times. The result is a discrete pulse, which contains many millions of electrons generated from an ion that first hits the cone of the detector.[1] This process is shown simplistically in Figure 14.2.

This pulse is then sensed and detected by a very fast preamplifier. The output pulse from the preamplifier then goes to a digital discriminator and counting circuitry that only counts pulses above a certain threshold value. This threshold level needs to be high enough to discriminate against pulses caused by spurious emission inside the tube, from any stray photons from the plasma itself, or photons generated from fast-moving ions striking the quadrupole rods.

It is worth pointing out that the rate at which ions hit the detector is sometimes too high for the measurement circuitry to handle in an efficient manner. This is caused by ions arriving at the detector during the output pulse of the preceding ion and not being detected by the counting system. This "dead time," as it is known, is a fundamental limitation of the multiplier detector and is typically 30–50 ns, depending on the detection system. Compensation in the measurement circuitry has to be made for this dead time to count the maximum number of ions hitting the detector.

FARADAY CUP

For some applications, where ultratrace detection limits are not required, the ion beam from the mass analyzer is directed into a simple metal electrode or Faraday cup. With this approach, there is no control over the applied voltage (gain), so they can only be used for high ion currents. Their lower working range is on the order of 10^4 cps, which means that if they are to be used as the only detector, the sensitivity of the ICP mass spectrometer will be severely compromised. For this reason, they are normally used in conjunction with a Channeltron or discrete dynode detector to extend the dynamic range of the instrument. An additional problem with the Faraday cup is that because of the time constant used in the DC amplification process to measure the ion current, they are limited to

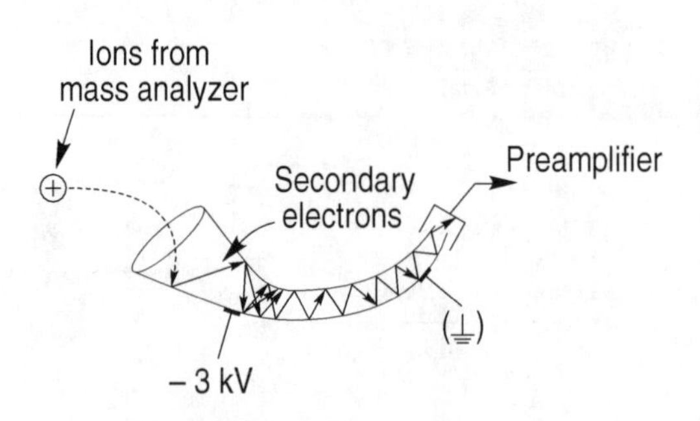

FIGURE 14.2 Basic principles of a channel electron multiplier. (From Channeltron®, *Electron Multiplier Handbook for Mass Spectrometry Applications*, Galileo Electro-Optic Corp., Sturbridge, MA, 1991. Channeltron is a registered trademark of Galileo Corp.)

relatively low scan rates. This limitation makes them unsuitable for the fast scan rates required for traditional pulse counting used in ICP-MS and also limits their ability to handle fast transient peaks.

The Faraday cup was never sensitive enough for quadrupole ICP-MS technology, because it was not suitable for very low ion count rates. An attempt was made in the early 1990s to develop an ICP-MS system using a Faraday cup detector for the environmental market, but its sensitivity was compromised and, as a result, was considered more suitable for applications requiring ICP-OES trace-level detection capability. However, Faraday cup technology is still utilized in some magnetic sector instruments, particularly where high ion signals are encountered in the determination of high-precision isotope ratios, using a multicollector detection system.

DISCRETE DYNODE ELECTRON MULTIPLIER

These detectors, which are often called *active film multipliers*, work in a similar way to the Channeltron but utilize discrete dynodes to carry out the electron multiplication.[2] Figure 14.3 illustrates the principles of operation of this device.

The detector is positioned off-axis to minimize the background noise from stray radiation and neutral species coming from the ion source. When an ion emerges from the quadrupole, it sweeps through a curved path before it strikes the first dynode. On striking the first dynode, it liberates secondary electrons. The electron-optic design of the dynode produces acceleration of these secondary electrons to the next dynode where they generate more electrons. This process is repeated at each dynode, generating a pulse of electrons that are finally captured by the multiplier collector or anode. Because of the materials used in the discrete dynode detector and the difference in the way electrons are generated, it is typically 50%–100% more sensitive than Channeltron technology and with much lower dead time (typically 5–10 ns).

Although most discrete dynode detectors are very similar in the way they work, there are subtle differences in the way the measurement circuitry handles low and high ion count rates. When ICP-MS was first commercialized, it could only handle up to five orders of dynamic range. However, when attempts were made to extend the dynamic range, certain problems were encountered. Before we discuss how modern detectors deal with this issue, let us first look at how it was addressed in earlier instrumentation.

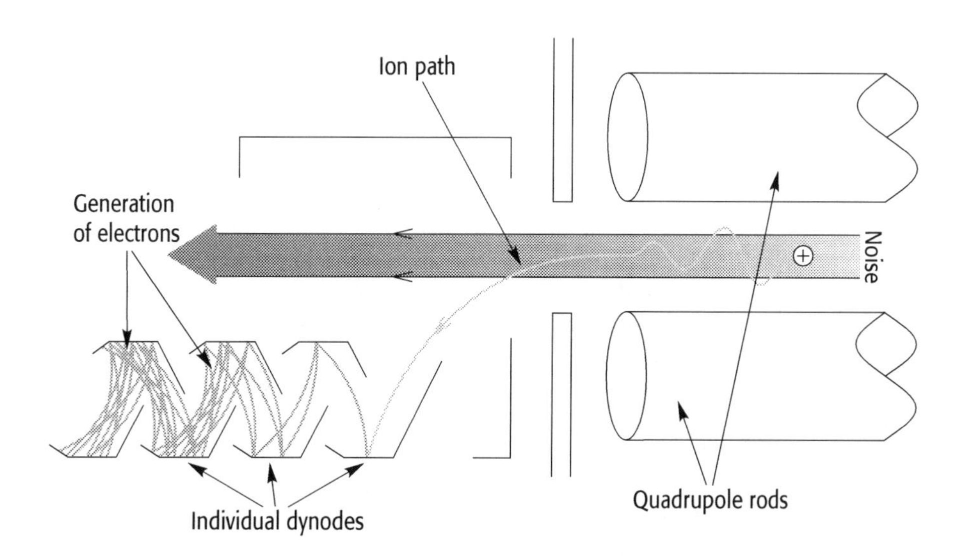

FIGURE 14.3 Schematic of a discrete dynode electron multiplier. (From K. Hunter, *Atomic Spectroscopy*, **15** (1), 17–20, 1994.)

EXTENDING THE DYNAMIC RANGE

Traditionally, ICP-MS using the pulse-counting measurement is capable of about five orders of linear dynamic range. This means that ICP-MS calibration curves, generally speaking, are linear from detection limit up to a few hundred parts per billion. However, there are a number of ways of extending the dynamic range of ICP-MS another four to five orders of magnitude and working from sub-ppt levels up to hundreds of parts per million. Here is a brief overview of some of different approaches that have been used.

Filtering the Ion Beam

One of the very first approaches to extending the dynamic range in ICP-MS was to filter the ion beam. This was achieved by putting a non-optimum voltage on one of the ion lens components or the quadrupole itself, to limit the number of ions reaching the detector. This voltage offset, which was set on an individual mass basis, acted as an energy filter to electronically screen the ion beam and reduce the subsequent ion signal to within a range covered by pulse counting ion detection. The main disadvantage with this approach was that the operator had to have prior knowledge of the sample to know what voltage to apply to the high concentration masses.

Using Two Detectors

Another technique that was used on some of the early ICP-MS instrumentation was to utilize two detectors, such as a channel electron multiplier and a Faraday cup, to extend the dynamic range. With this technique, two scans would be made. In the first scan, it would measure the high-concentration masses using the Faraday cup; in the second scan, it would skip over the high-concentration masses and carry out pulse counting of the low-concentration masses with a channel electron multiplier. This worked reasonably well but struggled with applications that required rapid switching between the two detectors, because the ion beam had to be physically deflected to select the optimum detector. Not only did this degrade the measurement duty cycle, but detector switching and stabilization times of several seconds also precluded fast transient signal detection.

Using Two Scans with One Detector

A more recent approach is to use just one detector to extend the dynamic range. This has typically been done by using the detector both in pulse and analog mode, so high and low concentrations can be determined in the same sample. There are basically three approaches to using this type of detection system: two of them involve carrying out two scans of the sample, whereas the third only requires one scan.

The first approach uses an electron multiplier operated in both digital and analog modes.[3] Digital counting provides the highest sensitivity, whereas operation in the analog mode (achieved by reducing the high voltage applied to the detector) is used to reduce the sensitivity of the detector, thus extending the concentration range for which ion signals can be measured. The system is implemented by scanning the spectrometer twice for each sample. The first scan, in which the detector is operated in the analog mode, provides signals for elements present at high concentrations. A second scan in which the detector voltage is switched to digital pulse-counting mode provides high-sensitivity detection for elements present at low levels. A major advantage of this technology is that the user does not need to know in advance whether to use analog or digital detection, because the system automatically scans all elements in both modes. However, one of the drawbacks is that two independent mass scans are required to gather data across an extended signal range. This not only results in degraded measurement efficiency and slower analyses, but it also means that the system is not ideally suited for fast transient signal analysis, because mode switching is generally too slow.

An alternative way of extending the dynamic range is similar to the first approach, except that the first scan is used as an investigative tool to examine the sample spectrum before analysis.[4] This first prescan establishes the mass positions at which the analog and pulse modes will be used for subsequently collecting the spectral signal. The second analytical scan is then used for data collection, switching the detector back and forth rapidly between pulse and analog modes at each analytical mass.

Even though these approaches worked very well, their main disadvantage was that two separate scans are required to measure high and low levels. With conventional nebulization, this is not such a major problem except that it can impact sample throughput. However, it does become a concern when it comes to working with transient peaks found in laser sampling (LS), flow injection analysis (FIA), or electrothermal vaporization (ETV) ICP-MS. Because these transient peaks often only last a few seconds, all the available time must be spent measuring the masses of interest to get the best detection limits. When two scans have to be made, time is wasted collecting data, which is not contributing to the analytical signal.

Using One Scan with One Detector

The limitation of having to scan the sample twice led to the development of an improved design using a dual-stage discrete dynode detector.[5] This technology utilizes measurement circuitry that allows both high and low concentrations to be determined in one scan. This is achieved by measuring the ion signal as an analog signal at the midpoint dynode. When more than a threshold number of ions are detected, the signal is processed through the analog circuitry. When fewer than the threshold number of ions is detected, the signal cascades through the rest of the dynodes and is measured as a pulse signal in the conventional way. This process, which is shown in Figure 14.4, is completely automatic and means that both the analog and the pulse signals are collected simultaneously in one scan.[6]

The pulse-counting mode is typically linear from zero to about 10^6 cps, whereas the analog circuitry is suitable from 10^4 to 10^{10} cps. To normalize both ranges, a cross-calibration is carried out to cover concentration levels, which produces a pulse and an analog signal. This is possible because the analog and pulse outputs can be defined in identical terms of incoming pulse counts per second,

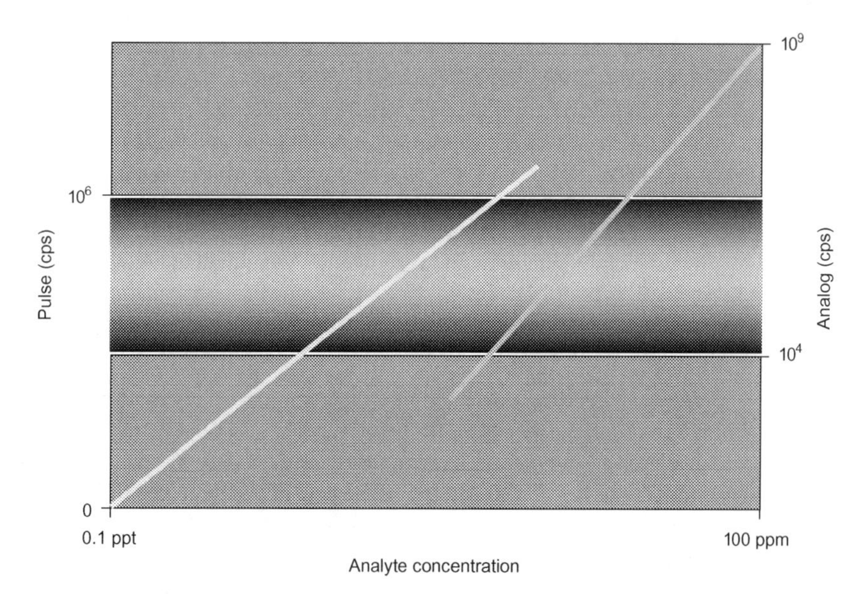

FIGURE 14.4 Dual-stage discrete dynode detector measurement circuitry. (From E. R. Denoyer, R. J. Thomas, and L. Cousins, *Spectroscopy*, **12** (2), 56–61, 1997. Covered by U.S. Patent Number 5,463,219.)

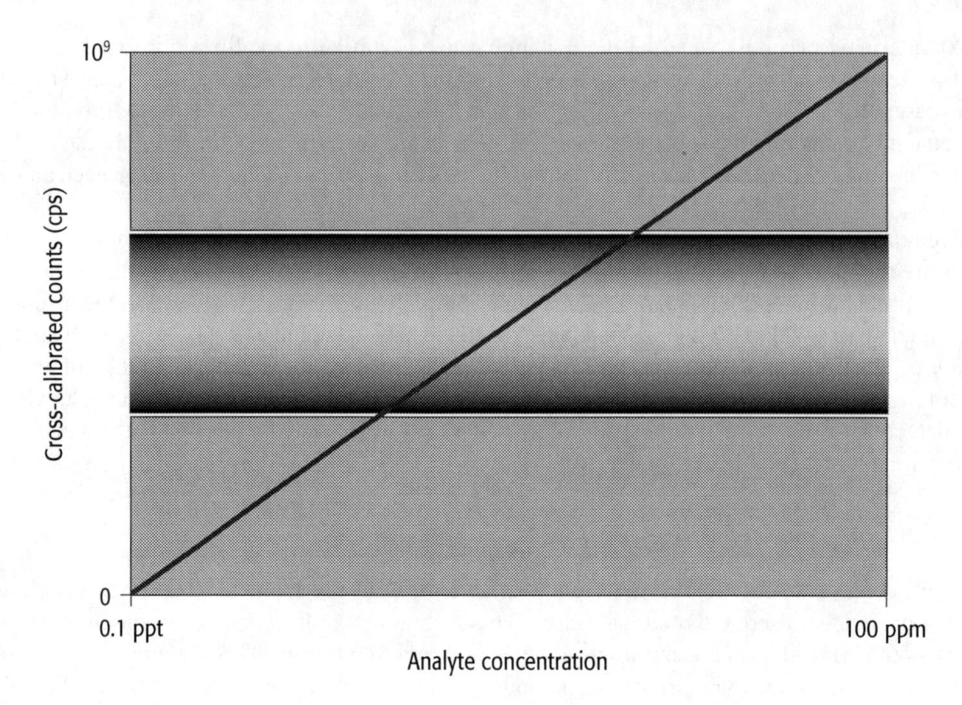

FIGURE 14.5 The pulse counting mode covers up to 10^6 cps, while the analog circuitry is suitable from 10^4 to 10^9 cps, with a dual-mode discrete dynode detector. (From E. R. Denoyer, R. J. Thomas, and L. Cousins, *Spectroscopy*, **12** (2), 56–61, 1997. Covered by U.S. Patent Number 5,463,219.)

based on knowing the voltage at the first analog stage, the output current, and a conversion factor defined by the detection circuitry electronics. By carrying out a cross-calibration across the mass range, a dual mode detector of this type is capable of achieving approximately nine to ten orders of dynamic range in one simultaneous scan. This can be seen in Figures 14.5 and 14.6. Figure 14.5 shows that the pulse counting calibration curve (left-hand line) is linear up to 10^6 cps, whereas the analog calibration curve (right-hand line) is linear from 10^4 to 10^9 cps. Figure 14.6 shows that after cross-calibration, the two curves are normalized, which means the detector is suitable for concentration levels between 0.1 ppt and 100 ppm and above—typically nine to ten orders of magnitude for most elements.[5]

There are subtle variations of this type of detection system, but its major benefit is that it requires only one scan to determine both high and low concentrations. It, therefore, not only offers the potential to improve sample throughput but also means that the maximum data can be collected on a transient signal that only lasts a few seconds. This is described in greater detail in Chapter 15, where we discuss different measurement protocols and peak integration routines.

EXTENDING THE DYNAMIC RANGE USING PULSE-ONLY MODE

Another recent development in extending the dynamic range is to use the pulse-only signal. This is achieved by monitoring the ion flux at one of the first few dynodes of the detector (before extensive electron multiplication has taken place) and then attenuating the signal up to 10,000:1 by applying a control voltage. Electron pulses passed by the attenuation section are then amplified to yield pulse heights that are typical in normal pulse-counting applications.

There are basically three ways of implementing this technology based on the types of samples being analyzed. It can be run in conventional pulse-only mode for normal low-level work. It can also be run using an operator-selected attenuation factor if the higher level elements being determined are known and similar in concentration. If the samples are complete unknowns and have not been

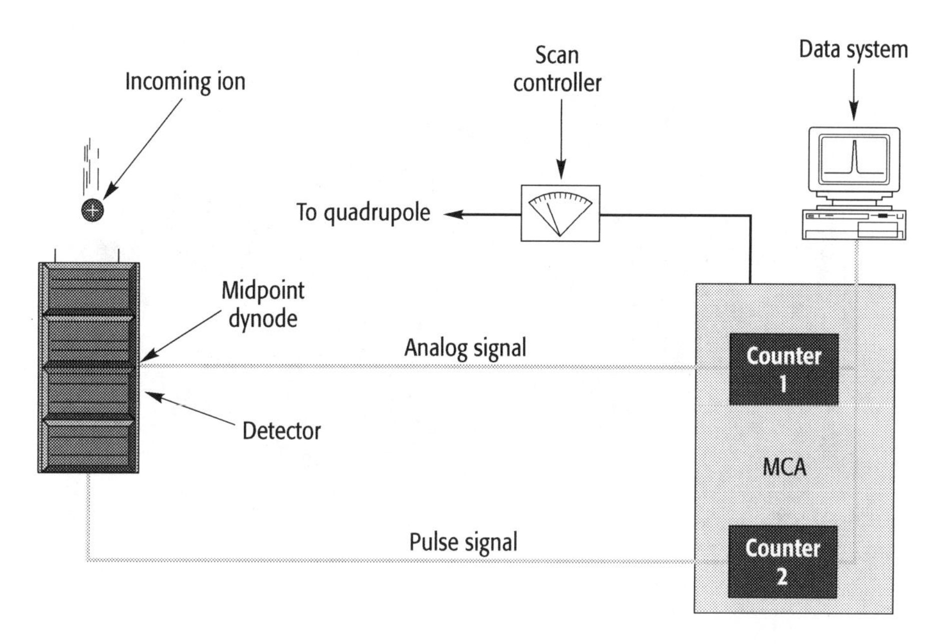

FIGURE 14.6 Using cross-calibration of the pulse and analog modes, quantitation from sub-ppt to high-ppm levels is possible. (From E. R. Denoyer, R. J. Thomas, and L. Cousins, *Spectroscopy*, **12** (2), 56–61, 1997. Covered by U.S. Patent Number 5,463,219.)

well characterized beforehand, a dynamic attenuation mode of operation is available. In this mode, an additional pre-measurement time is built into the quadrupole settling time to determine the optimum detector attenuation for the selected dwell times used.

This novel, pulse-only approach to extending the dynamic range looks to be a very interesting development, which does not have the limitation of having to calibrate where pulse and analog signals cross over. However, it does require a pre-analysis attenuation calibration to be carried out on a fairly frequent basis to determine the extent of signal attenuation required. The frequency of this calibration will vary depending on sample workload but is expected to be on the order of once every 4 weeks.

However, it should be strongly emphasized that irrespective of which extended range technology is used, if low and high concentrations of the same analyte are expected in a suite of samples, it is unrealistic to think you can accurately quantitate down at the low end and at the top end of the linear range with the same calibration graph. If you want to achieve accurate and precise data at or near the limit of quantitation, you must run a set of appropriate calibration standards to cover your low-level samples. In addition, if you are expecting high and low concentrations in the same suite of samples, you have to be absolutely sure that a high concentration sample has been thoroughly washed out from the spray chamber/nebulizer system, before a low-level sample in introduced. For this reason, caution must be taken when setting up the method with an autosampler, because if the read delay/integration times are not optimized for a suite of samples, erroneous results can be generated, which might necessitate a rerun under the manual supervision of the instrument operator.

SIMULTANEOUS ARRAY DETECTORS

Discrete dynode detectors are designed to handle a sequential stream of separated ions emerging from the mass spectrometer. Similar to a photomultiplier tube, that converts photons from an optical emission signal into a pulse of electrons, these detectors cannot capture the entire mass spectrum at the same time. However, a new breed of ion detectors have recently been developed that are based

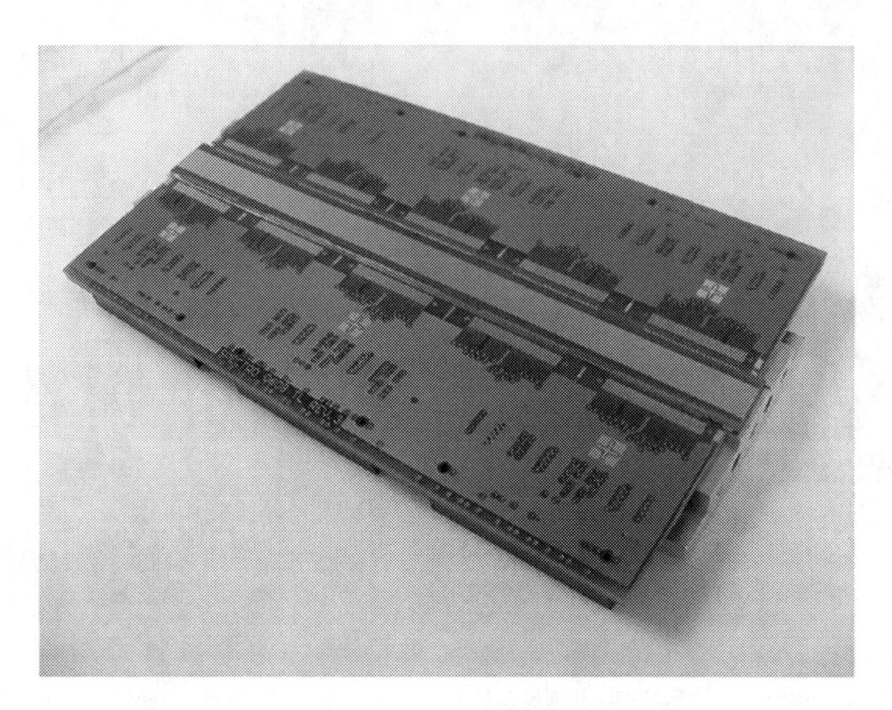

FIGURE 14.7 A DCD used for simultaneous measurement of ions separated by a Mattauch–Herzog double-focusing magnetic sector mass spectrometer.

on solid-state, direct charge arrays, similar to collision-induced dissociation/charge coupled device (CID/CCD) technology used in ICP optical emission. By projecting all the separated ions from a mass separation device onto a two-dimensional array, these detectors can view the entire mass spectrum simultaneously.[7] Designed specifically for the Mattauch–Herzog double-focusing magnetic sector technology described in Chapter 8, this CMOS (complementary metal oxide semiconductor) ion-sensitive device is a 12 cm long array that covers the entire mass range simultaneously in 4,800 separate channels.[8] The basic principle of the detector is similar to that of a Faraday cup: when a charged ion arrives at a detector array, it is discharged by receiving an electron, which generates a signal. The detector is referred to as a direct charge detector (DCD) because every ion arriving at the detector contributes to the signal. Furthermore, with 4,800 detector arrays covering the mass range (from 5 to 240 amu), every mass unit is covered on average by 20 separate channels, resulting in a true mass spectrum rather than a single point for each atomic mass unit. A photograph of this DCD is shown in Figure 14.7.

To cover linearity of up to eight orders of magnitude required in ICP-MS measurements, each detector channel incorporates separate high- and low-gain detector elements, allowing it to independently handle a wide range of signal levels in the basic integration cycle. The dynamic range can be further extended by optimizing the readout process. In the basic integration cycle, each channel is monitored by the electronics every 20 ms. If the signal integrated in this time interval nears the threshold of the channel, the integrated signal of that channel is automatically logged and the channel is reset and its measurement cycle repeated. This is repeated until the end of the defined measurement time, when all the collected data is integrated to produce the final signal. This means that the detector is always working within its linear response range and that longer integration times can be used without fear of detector saturation. For the benefits of simultaneous detection in ICP mass spectrometry, refer to Chapter 10 "Mass Analyzers" technology and Chapter 15 "Peak Measurement Protocol."

FURTHER READING

1. Channeltron®, Electron Multiplier Handbook for Mass Spectrometry Applications, Galileo Electro-Optic Corp., Sturbridge, MA, 1991. Channeltron is a registered trademark of Galileo Corp.
2. K. Hunter, *Atomic Spectroscopy*, **15** (1), 17–20, 1994.
3. R. C. Hutton, A. N. Eaton, and R. M. Gosland, *Applied Spectroscopy*, **44** (2), 238–242, 1990.
4. Y. Kishi, *Agilent Technologies Application Journal*, **3**(2), 6–8, August 1997.
5. E. R. Denoyer, R. J. Thomas, and L. Cousins, *Spectroscopy*, **12** (2), 56–61, 1997. Covered by U.S. Patent Number 5,463,219.
6. J. Gray, R. Stresau, and K. Hunter, Ion Counting Beyond 10 GHz, *Poster Presentation Number 890-6P*, Pittsburgh Conference and Exposition, Orlando, FL, 2003.
7. G. D. Schilling, S. J. Ray, A. A. Rubinshtein, J. A. Felton, R. P. Sperline, M. B. Denton, C. J. Barinaga, D. W. Koppenaal, and G. M. Hieftje, *Analytical Chemistry*, **81** (13), 5467–5473, 2009.
8. A New Era in Mass Spectrometry: Spectro Analytical Instruments: http://www.spectro.com/pages/e/p010402tab_overview.htm

15 Peak Measurement Protocol

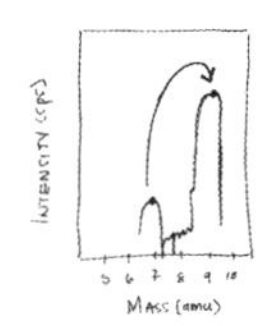

With its multielement capability, superb detection limits, wide dynamic range, and high sample throughput, inductively coupled plasma spectrometry (ICP-MS) is proving to be a compelling technique for more and more diverse application areas. However, it is very unlikely that two different application areas have identical analytical requirements. For example, cannabis, environmental, and clinical contract/testing laboratories, although wanting reasonably low detection limits, are not really pushing the technique to its extreme detection capability. Their main requirement usually is high sample throughput, because the number of samples these laboratories can analyze in a day directly impacts their revenue. On the other hand, a semiconductor fabrication plant or a supplier of high-purity chemicals to the electronics industry is interested in the lowest detection limits the technique can offer, because of the contamination problems associated with manufacturing high-performance electronic devices. Chapter 15 looks at the many different measurement protocols associated with identifying and quantifying the analyte peak in ICP-MS and how they impact sample throughput and the quality of the data generated.

To meet such diverse application needs, modern ICP-MS instrumentation has to be very flexible if it is to keep up with the increasing demands of its users. Nowhere is this more important than in the area of peak integration and measurement protocol. The way the analytical signal is managed in ICP-MS has a direct impact on its multielement characteristics, isotopic capability, detection limits, dynamic range, and sample throughput—the five major strengths that attracted the trace element community to the technique over 35 years ago. To understand signal management in greater detail and its implications on data quality, we will discuss how measurement protocol is optimized based on the application's analytical requirements and its impact on both continuous signals generated by traditional nebulization devices and transient signals produced by alternative sample introduction techniques such as automated sample delivery systems, chromatographic separation studies, and particle size measurements.

MEASUREMENT VARIABLES

There are many variables that affect the quality of the analytical signal in ICP-MS. The analytical requirements of the application will often dictate this, but there is no question that instrumental detection and measurement parameters can have a significant impact on the quality of data in ICP-MS. Some of the variables that can potentially impact the quality of the data, particularly when carrying out multielement analysis, are as follows:

- A continuous or transient signal
- The temporal length of the sampling event
- Volume of sample available

- Number of samples being analyzed
- Number of replicates per sample
- Number of elements being determined
- Detection limits required
- Precision/accuracy expected
- Dynamic range needed
- Integration time used
- Peak quantitation routines.

Before we go on to discuss these in greater detail and how these parameters affect the data, it is important to remind ourselves how a scanning device like a quadrupole mass analyzer works. Although we will focus on quadrupole technology, the fundamental principles of measurement protocol will be very similar for all types of mass spectrometers that use a sequential approach for multielement peak quantitation.

MEASUREMENT PROTOCOL

The principles of scanning with a quadrupole mass analyzer are shown in Figure 15.1. In this simplified example, the analyte ion in front (black) and four other ions have arrived at the entrance to the four rods of the quadrupole. When a particular RF/DC voltage is applied to each pair of rods, the positive or negative bias on the rods will electrostatically steer the analyte ion of interest down the middle of the four rods to the end, where it will emerge and be converted to an electrical pulse by the detector. The other ions of different mass-to-charge ratio will pass through the spaces between the rods and be ejected from the quadrupole. This scanning process is then repeated for another analyte at a completely different mass-to-charge ratio until all the analytes in a multielement analysis have been measured.

The process for the detection of one particular mass in a multielement run is represented in Figure 15.2. It shows a $^{63}Cu^{+}$ ion emerging from the quadrupole and being converted to an electrical pulse by the detector. As the optimum RF-to-DC ratio is applied for $^{63}Cu^{+}$ and repeatedly scanned, the ions as electrical pulses are stored and counted by a multichannel analyzer. This multichannel data acquisition system typically has 20 channels per mass and as the electrical pulses are counted in each channel, a profile of the mass is built up over the 20 channels, corresponding to the spectral

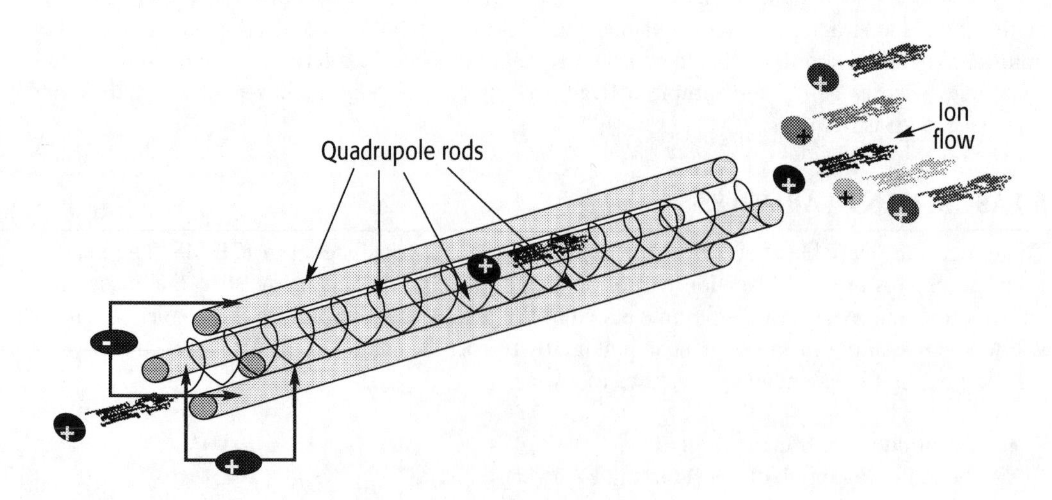

FIGURE 15.1 Principles of mass selection with a quadrupole mass filter.

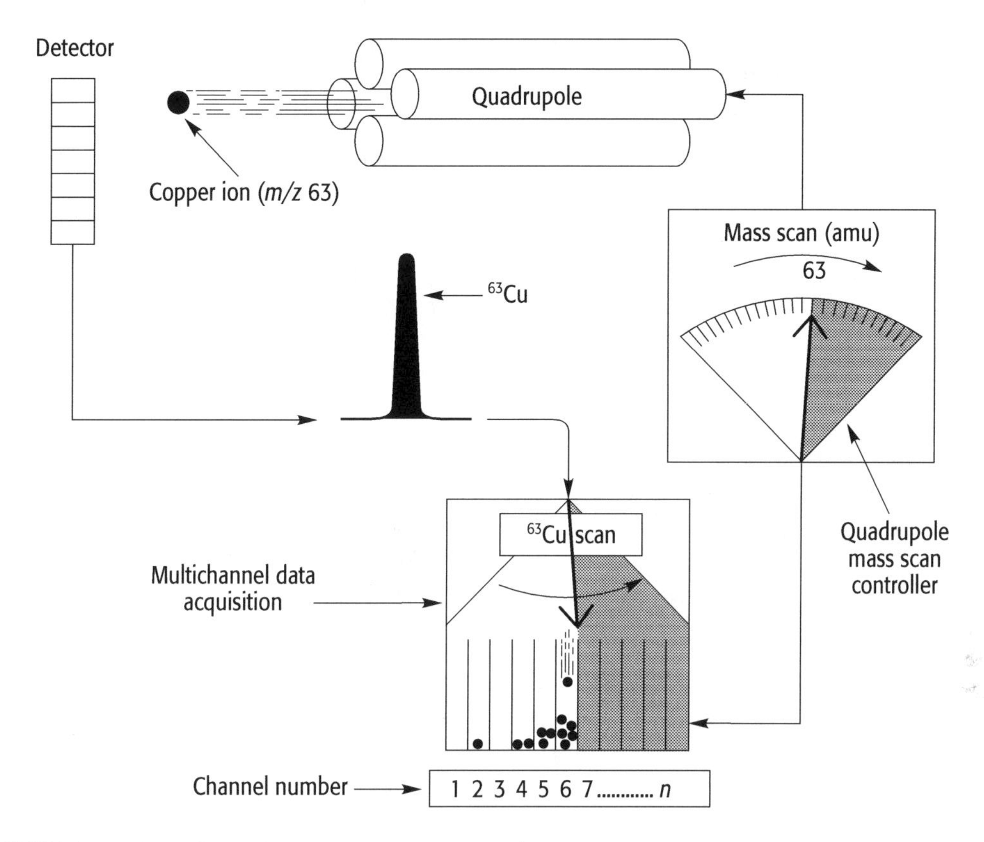

FIGURE 15.2 Detection and measurement protocol using a quadrupole mass analyzer. (From Integrated MCA Technology in the ELAN ICP-Mass Spectrometer, Application Note TSMS-25, PerkinElmer Instruments, 1993.)

peak of $^{63}Cu^{+}$. In a multielement run, repeated scans are made over the entire suite of analyte masses, as opposed to just one mass represented in this example.

The principles of multielement peak acquisition are shown in Figure 15.3. In this example, signal pulses for two masses are continually collected as the quadrupole is swept across the mass spectrum, shown by sweeps 1–3. After a fixed number of sweeps (determined by the user), the total number of signal pulses in each channel is obtained, resulting in the final spectral peak.[1]

When it comes to quantifying an isotopic signal in ICP-MS, there are basically two approaches to consider. One is the multichannel ramp scanning approach, which uses a continuous smooth ramp of 1–*n* channels (where *n* is typically 20) per mass across the peak profile. This is shown in Figure 15.4.

Also, there is the peak hopping approach, in which the quadrupole power supply is driven to a discrete position on the peak (normally the maximum point), allowed to settle, and a measurement is taken for a fixed amount of time. This is represented in Figure 15.5.

The multipoint scanning approach is best for accumulating spectral and peak shape information when doing mass scans. It is normally used for doing mass calibration and resolution checks and as a classical qualitative method development tool to find out what elements are present in the sample and to assess their spectral implications on the masses of interest. Full peak profiling is not normally used for doing rapid quantitative analysis, because valuable analytical time is wasted taking data on the wings and valleys of the peak, where the signal-to-noise ratio is poorest.

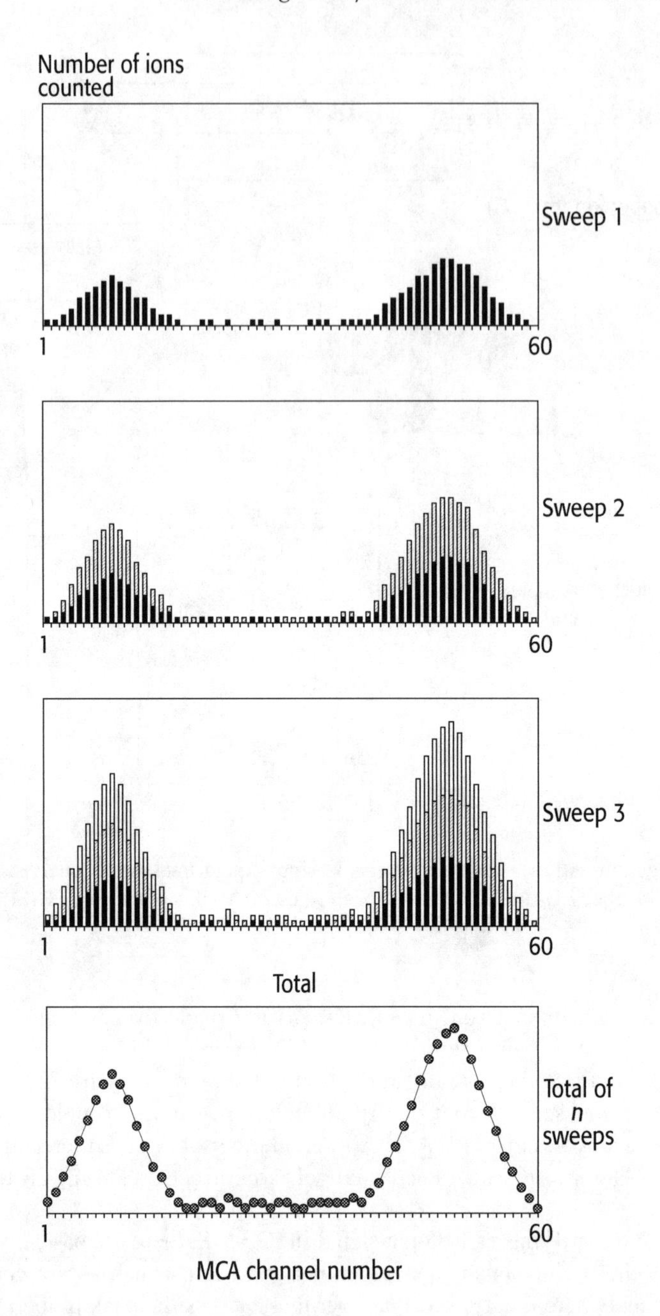

FIGURE 15.3 A profile of the peak is built up by continually sweeping the quadrupole across the mass spectrum. (From Integrated MCA Technology in the ELAN ICP-Mass Spectrometer, Application Note TSMS-25, PerkinElmer Instruments, 1993.)

When the best possible detection limits are required, the peak-hopping approach is best. It is important to understand that to get the full benefit of peak hopping, the best detection limits are achieved when single point–peak hopping at the peak maximum is chosen. However, to carry out single point–peak hopping, it is essential that the mass stability be good enough to reproducibly go to the same mass point every time. If good mass stability can be guaranteed (usually by thermostating the quadrupole power supply), measuring the signal at the peak maximum will always give the

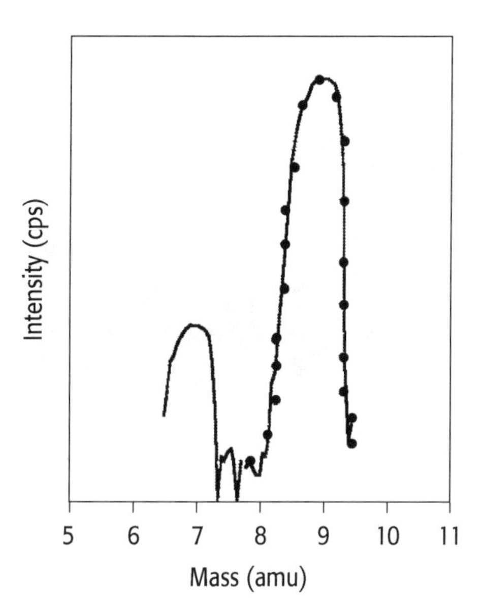

FIGURE 15.4 Multichannel ramp scanning approach using 20 channels per amu.

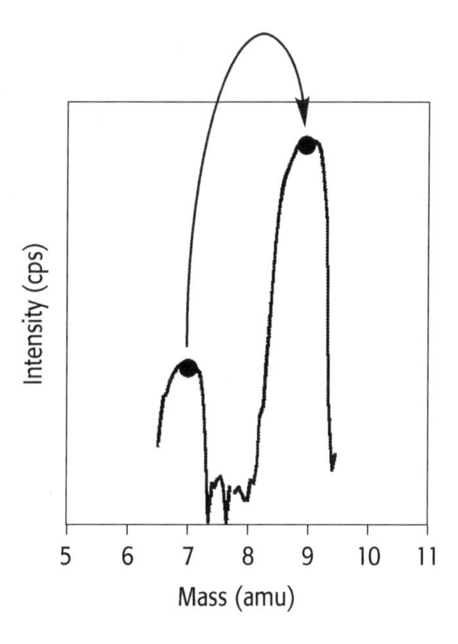

FIGURE 15.5 Peak-hopping approach.

best detection limits for a given integration time. It is well documented that there is no benefit in spreading the chosen integration time over more than one measurement point per mass. If time is a major consideration in the analysis, then using multiple points is wasting valuable time on the wings and valleys of the peak, which contribute less to the analytical signal and more to the background noise. This is shown in Figure 15.6, which demonstrates the degradation in signal-to-background noise of 10 ppb Rh with an increase in the number of points per peak, spread over the same total integration time. Detection limit improvement for a selected group of elements using 1 point/peak compared to 20 points/peak is shown in Figure 15.7.

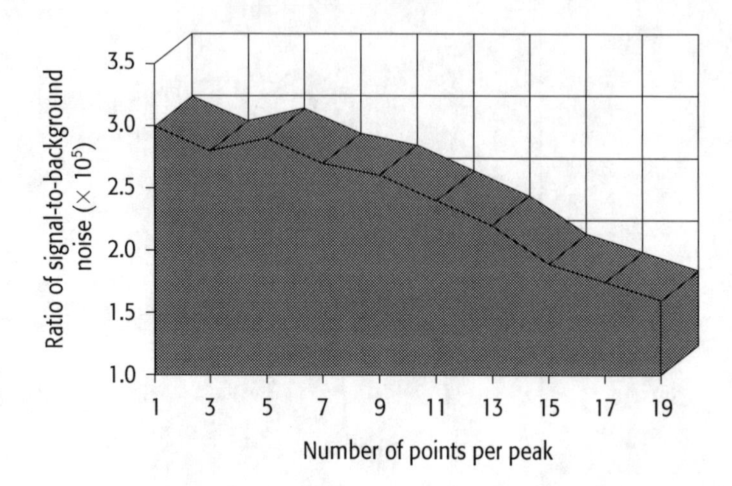

FIGURE 15.6 Signal-to-background noise degrades when more than one point, spread over the same integration time, is used for peak quantitation.

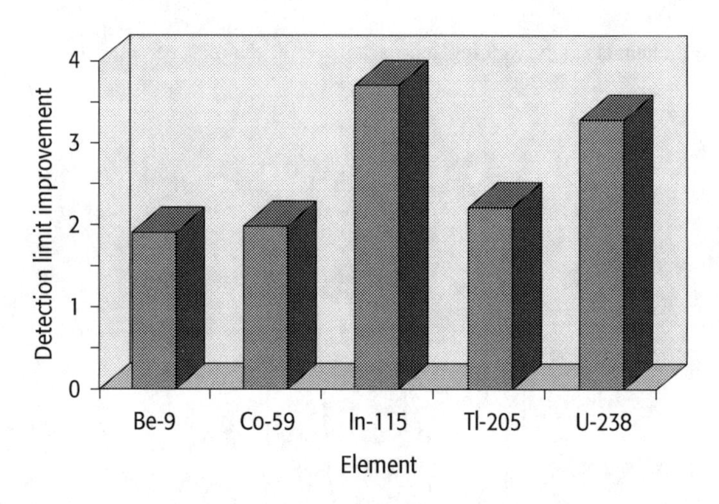

FIGURE 15.7 Detection limit improvement using one point per peak compared to 20 points per peak over the mass range. (From E. R. Denoyer, *Atomic Spectroscopy*, **13** (3), 93–98, 1992.)

OPTIMIZATION OF MEASUREMENT PROTOCOL

Now that the fundamentals of the quadrupole measuring electronics have been described, let us go into more detail on the impact of optimizing the measurement protocol based on the requirements of the application. When multielement analysis is being carried out by ICP-MS, there are a number of decisions that need to be made. First, we need to know if we are dealing with a continuous signal from a nebulizer or a transient signal from a sampling accessory such as the laser ablation system of a chromatographic separation device. If it is a transient event, how long will the signal last? Another question that needs to be addressed is how many elements are going to be determined? With a continuous signal, this is not such a major problem but could be an issue if we are dealing with a transient signal that lasts only a few seconds. We also need to be aware of the level of detection capability required. This is a major consideration with a short laser pulse of a few seconds duration and even more critical when characterizing nanoparticles which only last for a few milliseconds. But it is also an issue with a continuous signal produced by a concentric nebulizer, where we might have to accept a compromise of detection limit based on the speed of analysis requirements or amount of sample available. What analytical precision is expected? If isotope ratio/dilution work is being done, how many ions do we have to count to guarantee good precision? Does increasing the integration time of the measurement help the precision? Finally, is there a time constraint on the analysis? A high-throughput laboratory might not be able to afford to use the optimum sampling time to get the ultimate in detection limit. In other words, what compromises need to be made between detection limit, precision, and sample throughput? It is clear that before the measurement protocol can be optimized, the major analytical requirements of the application need to be defined. Let us look at this in greater detail.

MULTIELEMENT DATA QUALITY OBJECTIVES

Because multielement detection capability is probably the major reason why most laboratories invest in ICP-MS, it is important to understand the impact of measurement criteria on detection limits. We know that in a multielement analysis, the quadrupole's RF-to-DC ratio is "driven" or scanned to mass regions, which represent the elements of interest. The electronics are allowed to settle and then "sit" or dwell on the peak and take measurements for a fixed period of time. This is usually performed a number of times until the total integration time is fulfilled. For example, if a dwell time of 50 ms is selected for all masses and the total integration time is 1 s, then the quadrupole will carry out 20 complete sweeps per mass, per replicate. It will then repeat the same routine for as many replicates that have been built into the method. This is shown in a simplified manner in Figure 15.8, which displays the scanning protocol of a multielement scan of three different masses.

In this example, the quadrupole is scanned to mass *A*. The electronics are allowed to settle (settling time), left to dwell for a fixed period of time at one or multiple points on the peak (dwell time), and intensity measurements taken (based on the dwell time). The quadrupole is then scanned to masses *B* and *C*, and the measurement protocol is repeated. The complete multielement measurement cycle (sweep) is repeated as many times as needed to make up the total integration per peak. It should be emphasized that this is a generalization of the measurement routine—management of peak integration by the software will vary slightly based on different instrumentation.

It is clear from this that during a multielement analysis, there is a significant amount of time spent scanning and settling the quadrupole, which does not contribute to the quality of the analytical

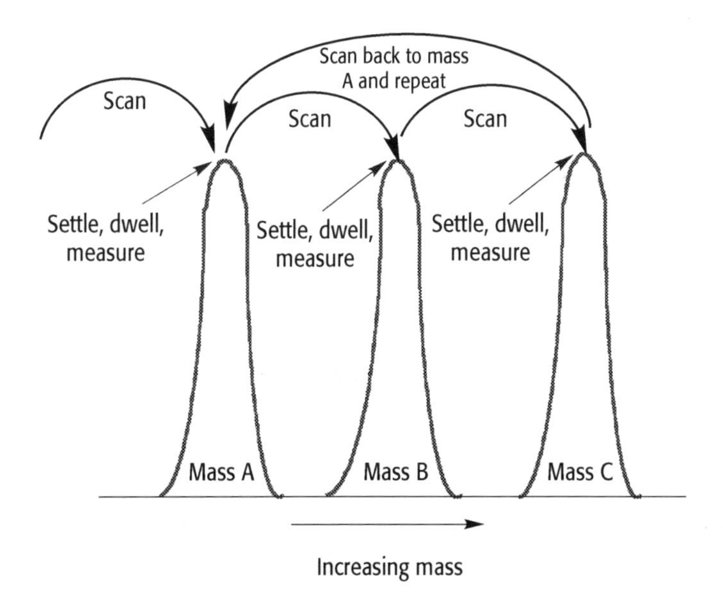

FIGURE 15.8 Multielement scanning and peak measurement protocol used in a quadrupole.

signal. Therefore, if the measurement routine is not optimized carefully, it can have a negative impact on data quality. The dwell time can usually be selected on an individual mass basis, but the scanning and settling times are normally fixed because they are a function of the quadrupole and detector electronics. For this reason, it is essential that the dwell time, which ultimately affects detection limit and precision, dominate the total measurement time, compared to the scanning and settling times. It follows, therefore, that the measurement duty cycle (percentage of actual measuring time compared to total integration time) is maximized when the quadrupole and detector electronics settling times are kept to an absolute minimum. This can be seen in Figure 15.9, which shows a plot of percent measurement duty cycle against dwell time for four different quadrupole settling times—0.2, 1.0, 3.0, and 5.0 ms for one replicate of a multielement scan of five masses, using one point per peak. In this example, the total integration time for each mass was 1 s, with the number of sweeps varying depending on the dwell time used. For this exercise, the percentage of duty cycle is defined by the following equation:

$$\frac{\text{Dwell time} \times \text{\# sweeps} \times \text{\# elements} \times \text{\# replicates}}{\left\{\left(\text{Dwell time} \times \text{\# sweeps} \times \text{\# elements} \times \text{\# replicates}\right) + \left(\text{Scanning/settling time} \times \text{\# sweeps} \times \text{\# elements} \times \text{\# reps}\right)\right\}} \times 100$$

To achieve the highest duty cycle, the non-analytical time must be kept to an absolute minimum. This leads to more time being spent counting ions and less time scanning, and settling, which do not contribute to the quality of the analytical signal. This becomes critically important when a rapid transient peak is being quantified, because the available measuring time is that much shorter.[2] It is therefore a good rule of thumb, when setting up your measurement protocol in ICP-MS, to avoid using multiple points per peak and long settling times, because it ultimately degrades the quality of the data for a given integration time.

It can also be seen in Figure 15.9 that shorter dwell times translate into a lower duty cycle. For this reason, for normal quantitative analysis work, it is probably desirable to carry out multiple sweeps with longer dwell times (typically, 50 ms) to get the best detection limits. So, if an integration time of 1 s is used for each element, this would translate into 20 sweeps of 50 ms dwell time per mass. Although 1 s is long enough to achieve reasonably good detection limits, longer integration times generally have to be used to reach the lowest possible detection limits. This is shown in Figure 15.10, which shows detection limit improvement as a function of integration time for $^{238}U^{+}$.

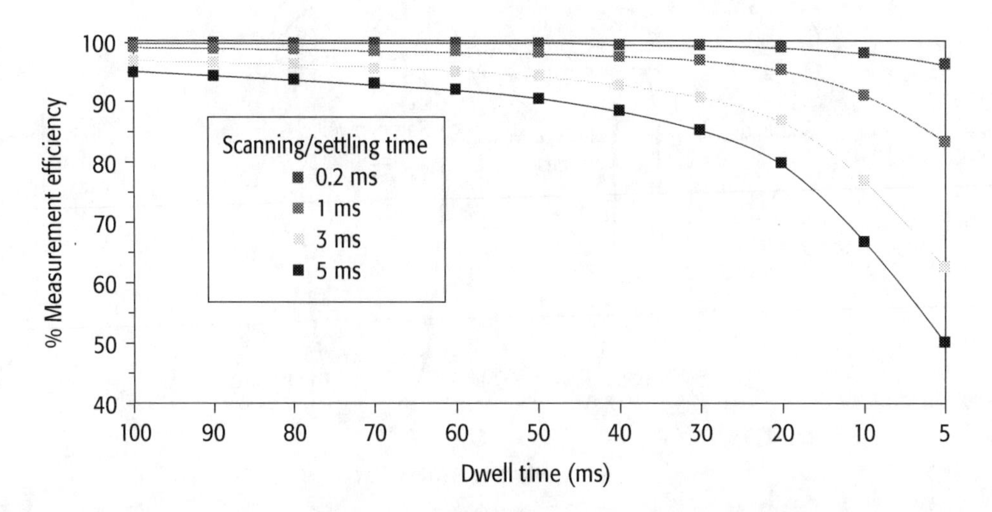

FIGURE 15.9 Measurement duty cycle as a function of dwell time with varying scanning/settling times.

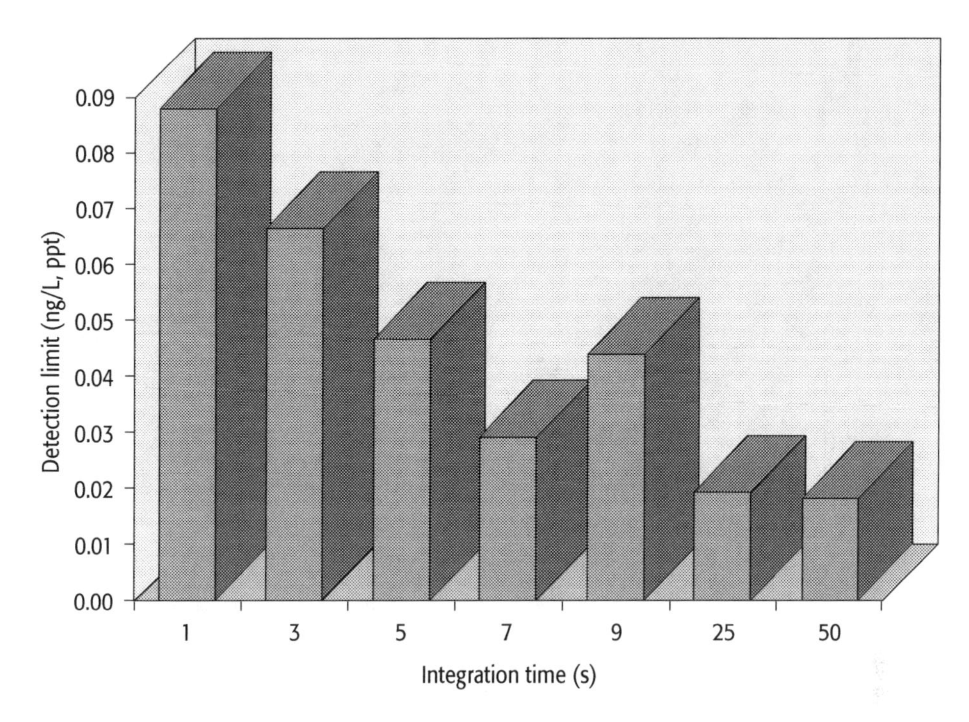

FIGURE 15.10 Plot of detection limit against integration time for $^{238}U^{+}$. (Copyright 2013, all rights reserved, PerkinElmer Inc.)

As would be expected, there is a fairly predictable improvement in the detection limit as the integration time is increased because more ions are being counted without an increase in the background noise. However, this only holds true up to the point where the pulse-counting detection system becomes saturated and no more ions can be counted. In the case of $^{238}U^{+}$, it can be seen that this happens at around 25 s, because there is no obvious improvement in D/L at a higher integration time. So from this data, we can say that there appears to be no real benefit in using longer than a 7 s integration time. When deciding the length of the integration time in ICP-MS, you have to weigh up the detection limit improvement against the time taken to achieve that improvement. Is it worth spending 25 s measuring each mass to get 0.02 ppt detection limit, if 0.03 ppt can be achieved using a 7 s integration time? Alternatively, is it worth measuring for 7 s when 1 s will only degrade the performance by a factor of 3? It really depends on your data quality objectives.

For some applications like isotope dilution/ratio studies, high precision is also a very important data quality objective.[3] However, to understand what is realistically achievable, we have to be aware of the practical limitations of measuring a signal and counting ions in ICP-MS. Counting statistics tells us that the standard deviation of the ion signal is proportional to the square root of the signal. It follows, therefore, that the relative standard deviation (RSD) or precision should improve with an increase in the number (N) of ions counted as shown by the following equation:

$$\%\mathrm{RSD} = \sqrt{\frac{N}{N}} \times 100$$

In practice, this holds up very well as can be seen in Figure 15.11. In this plot of standard deviation as a function of signal intensity for $^{208}Pb^{+}$, the black dots represent the theoretical relationship as predicted by counting statistics. It can be seen that the measured standard deviation (black bars) follows theory very well up to about 100,000 cps. At that point, additional sources of noise (e.g., sample introduction pulsations/plasma fluctuations) dominate the signal, which lead to poorer standard deviation values.

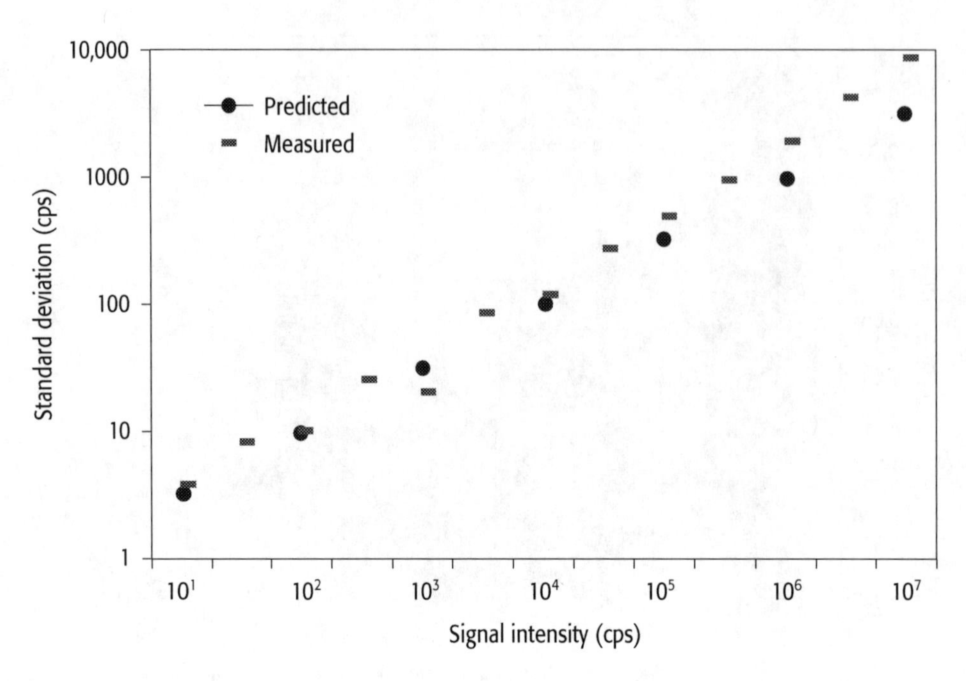

FIGURE 15.11 Comparison of measured standard deviation of a $^{208}Pb^+$ signal against that predicted by counting statistics. (From E. R. Denoyer, *Atomic Spectroscopy*, **13** (3), 93–98, 1992.)

So, based on counting statistics, it is logical to assume that the more ions are counted, the better the precision will be. To put this in perspective, it means that at least 1 million ions need to be counted to achieve an RSD of 0.1%. In practice, of course, these kinds of precision values are very difficult to achieve with a scanning quadrupole system because of the additional sources of noise. If this information is combined with our knowledge of how the quadrupole is scanned, we begin to understand what is required to get the best precision. This is confirmed by the spectral scan in Figure 15.12, which shows the predicted precision at all 20 channels of a 5 ppb $^{208}Pb^+$ peak.[4]

This tells us that the best precision is obtained at the channels where the signal is highest, which as we can see are the ones at or near the center of the peak. For this reason, if good precision is a fundamental requirement of your data quality objectives, it is best to use single-point peak hopping with integration times on the order of 5–10 s. On the other hand, if high-precision isotope ratio or isotope dilution work is being done, where analysts would like to achieve precision values approaching counting statistics, then much longer measuring times are required. That is why integration times on the order of 5–10 min are commonly used for determining isotope ratios involving environmental pollutants[5] or clinical metabolism studies.[6] For this type of analysis, when two or more isotopes are being measured and ratioed to each other, it follows that the more simultaneous the measurement, the better the precision becomes. Therefore, the ability to make the measurement as simultaneous as possible is considered more desirable than any other aspect of the measurement. This is supported by the fact that the best isotope ratio precision data is achieved with multicollector, magnetic sector ICP-MS technology that carries out many isotopic measurements at the same time using multiple detectors.[7] Also, time-of-flight (TOF) technology, which simultaneously samples all the analyte ions in a slice of the ion beam, offers excellent precision, particularly when internal standardization measurement is also carried out in a simultaneous manner.[8] So, the best way to approximate simultaneous measurement with a rapid scanning device such as a quadrupole is to use shorter dwell times (but not too short that insufficient ions are counted) and keep the scanning/settling times to an absolute minimum, which results in more sweeps for a given measurement time. This can be seen in Table 15.1, which shows the precision of Pb isotope ratios at different

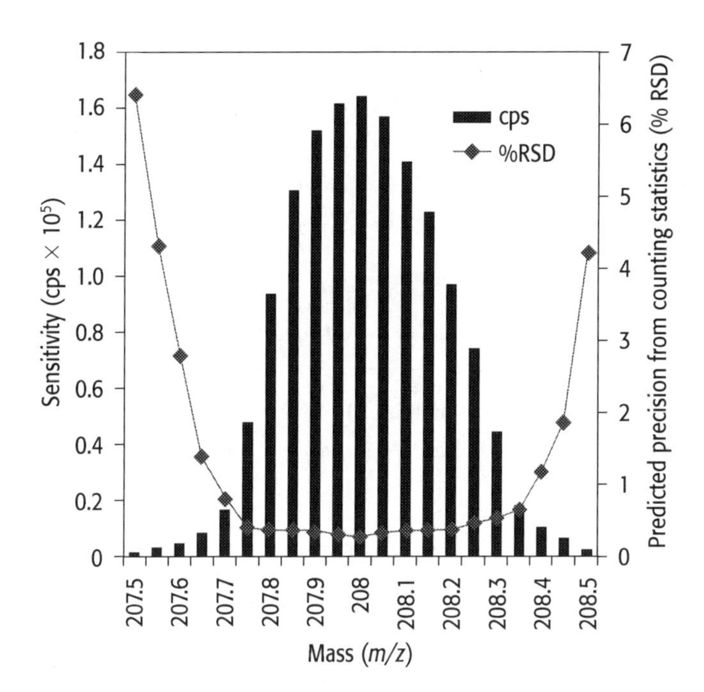

FIGURE 15.12 Comparison of % RSD with signal intensity across the mass profile of a $^{208}Pb^+$ peak. (From E. R. Denoyer, *Atomic Spectroscopy*, **13** (3), 93–98, 1992.)

TABLE 15.1
Precision of Pb Isotope Ratio Measurement as a Function of Dwell Time Using a Total Integration Time of 5.5 s

Dwell Time (ms)	% RSD $^{207}Pb^+/^{206}Pb^+$	% RSD $^{208}Pb^+/^{206}Pb^+$
2	0.40	0.36
5	0.38	0.36
10	0.23	0.22
25	0.24	0.25
50	0.38	0.33
100	0.41	0.38

Source: L. Halicz, Y. Erel, and A. Veron, *Atomic Spectroscopy*, **17** (5), 186–189, 1996.

dwell times carried out by researchers at the Geological Survey of Israel.[9] The data are based on nine replicates of an NIST SRM-981 (75 ppb Pb) solution, using 5.5 s integration time per isotope.

From these data, the researchers concluded that a dwell time of 10 or 25 ms offered the best isotope ratio precision measurement (quadrupole settling time was fixed at 0.2 ms). They also found that they could achieve slightly better precision by using a 17.5 s integration time (700 sweeps at 25 ms dwell time) but felt the marginal improvement in precision for nine replicates was not worth spending the approximately three-and-a-half times longer analysis time. This can be seen in Table 15.2.

This work shows the benefit of being able to optimize the dwell time, settling time, and the number of sweeps to get the best isotope ratio precision data. It also helps to be working with relatively healthy ion signals for the three Pb isotopes, ^{206}Pb, ^{207}Pb, and ^{208}Pb (24.1%, 22.1%, and 52.4% abundance, respectively). If the isotopic signals were dramatically different as in the measurement of two

TABLE 15.2
Impact of Integration Time on the Overall Analysis Time for Pb Isotope Ratios

Dwell Time (ms)	No. of Sweeps	Integration Time (s)/mass	%RSD $^{207}Pb^+/^{206}Pb^+$	% RSD $^{207}Pb^+/^{206}Pb^+$	Time for 9 reps (min/s)
25	220	5.5 s	0.24	0.25	2 min 29 s
25	500	12.5 s	0.21	0.19	6 min 12 s
25	700	17.5 s	0.20	0.17	8 min 29 s

Source: L. Halicz, Y. Erel, and A. Veron, *Atomic Spectroscopy*, **17** (5), 186–189, 1996.

of the uranium isotopes, ^{235}U to ^{238}U, which are 0.72% and 99.2745% abundant, respectively, then the ability to optimize the measurement protocol for individual isotopes becomes of even greater importance to guarantee good precision data.[10]

DATA QUALITY OBJECTIVES FOR SINGLE-PARTICLE ICP-MS STUDIES

The ability to optimize the measurement protocol is even more critical when using ICP-MS to characterize nanomaterials using the single-particle mode of analysis. Engineered nanomaterials, as they are known, refer to the process of producing materials that contain particles <100 nm in size. They often possess different properties compared to bulk materials of the same composition, making them of great interest to a broad spectrum of industrial and commercial applications. Unfortunately many of these nanomaterials, once they get into the environment, have proved to be harmful to humans. So in order to better understand the impact of nanoparticles on human health, several key properties need to be assessed, such as concentration, composition, particle size, and shape. Recent studies have shown that ICP-MS is proving to be a critical tool to characterize nanoparticles using the single particle technique.[11] However, these nanoparticles only exist for a few milliseconds, so the ability to measure them with good accuracy and precision is dependent on optimizing the dwell time, settling time, and the speed of the data acquisition electronics. If the measurement protocol is not handled correctly, it can significantly impact the quality of data collected.

FINAL THOUGHTS

It is clear that the analytical demands put on ICP-MS are probably higher than any other trace element technique, because it is continually being asked to solve a wide variety of application problems at increasingly lower levels. However, by optimizing the measurement protocol to fit the analytical requirement, ICP-MS has shown that it has the unique capability to carry out rapid trace element analysis, with superb detection limits and good precision on both continuous and transient signals, and still meet the most stringent data quality objectives.

FURTHER READING

1. Integrated MCA Technology in the ELAN ICP-Mass Spectrometer, Application Note TSMS-25, PerkinElmer Instruments, 1993.
2. E. R. Denoyer and Q. H. Lu, *Atomic Spectroscopy*, **14** (6), 162–169, 1993.
3. T. Catterick, H. Handley, and S. Merson, *Atomic Spectroscopy*, **16** (10), 229–234, 1995.
4. E. R. Denoyer, *Atomic Spectroscopy*, **13** (3), 93–98, 1992.
5. T. A. Hinners, E. M. Heithmar, T. M. Spittler, and J. M. Henshaw, *Analytical Chemistry*, **59**, 2658–2662, 1987.
6. M. Janghorbani, B. T. G. Ting, and N. E. Lynch, *Microchemica Acta*, **3**, 315–328, 1989.
7. J. Walder and P. A., Freeman, *Journal of Analytical Atomic Spectrometry*, **7**, 571–576, 1992.

8. F. Vanhaecke, L. Moens, R. Dams, L. Allen, and S. Georgitis, *Analytical Chemistry*, **71**, 3297–3303, 1999.
9. L. Halicz, Y. Erel, and A. Veron, *Atomic Spectroscopy*, **17** (5), 186–189, 1996.
10. D. R. Bandura and S. D. Tanner, *Atomic Spectroscopy*, **20** (2), 69–72, 1999.
11. D.M. Mitrano, E.K. Leshner, A. Bednar, J. Monserud, C. P. Higgins, and J.F. Ranville, *Environmental Toxicology and Chemistry*, **31** (1), 115–141, 2014.

16 Methods of Quantitation

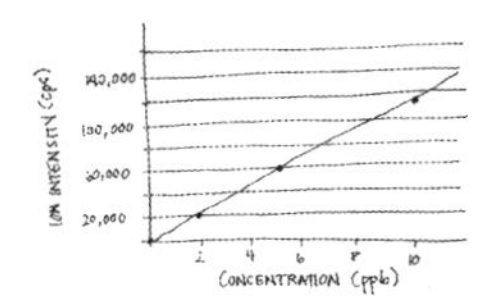

There are many different ways to carry out trace element analysis by inductively coupled plasma mass spectrometry (ICP-MS), depending on your data quality objectives. Such is the flexibility of the technique that it allows detection from sub-ppt up to high-ppm levels using a wide variety of calibration methods from full quantitative and semiquantitative analyses to one of the very powerful isotope ratioing techniques. This chapter looks at the most common quantitation methods available in ICP-MS.

This ability of ICP-MS to carry out isotopic measurements allows the technique to carry out quantitation methods that are not available to any other trace element technique. They include the following:

- Quantitative analysis
- Semiquantitative routines
- Isotope dilution
- Isotope ratio
- Internal standardization.

Each of these techniques offers varying degrees of accuracy and precision; so, it is important to understand their strengths and weaknesses to know which one will best meet the data quality objectives of the analysis. In this chapter, we will focus on using methods of quantitation for carrying out the analysis of liquids using continuous nebulization. However, even though the principles of calibration are similar, we covered the issues of quantitation of transient peaks in Chapter 15 "Peak Measurement Protocol" and will deal with specific sampling accessories in Chapter 20 "Performance and Productivity Enhancement Techniques."

Anyway, let's first take a look at each of the methods of quantitation in greater detail.

QUANTITATIVE ANALYSIS

As in other more trace element techniques such as atomic absorption (AA) and inductively coupled plasma optical emission spectrometry (ICP-OES), quantitative analysis in ICP-MS is the fundamental tool used to determine analyte concentrations in unknown samples. In this mode of operation, the instrument is calibrated by measuring the intensity for all elements of interest in a number of known calibration standards that represent a range of concentrations likely to be encountered in your unknown samples. When the full range of calibration standards and blank have been run, the software creates a calibration curve of the measured intensity versus concentration for each element in the standard solutions. Once calibration data are acquired, the unknown samples are analyzed by plotting the intensity of the elements of interest against the respective calibration curves. The software then calculates the concentrations for the analytes in the unknown samples.

This type of calibration is often called external standardization and is usually used when there is very little difference between the matrix components in the standards and the samples. However, when it is difficult to closely match the matrix of the standards with the samples, external standardization can produce erroneous results, because matrix-induced interferences will change analyte sensitivity based on the amount of matrix present in the standards and samples. When this occurs, better accuracy is achieved by using the method of standard addition or a similar approach called addition calibration. Let us look at these three variations of quantitative analysis to see how they differ.

External Standardization

As explained earlier, this involves measuring a blank solution followed by a set of standard solutions to create a calibration curve over the anticipated concentration range. Typically, a blank and three standards containing different analyte concentrations are run. Increasing the number of points on the calibration curve by increasing the number of standards may improve accuracy in circumstances where the calibration range is very broad. However, it is seldom necessary to run a calibration with more than five standards. After the standards have been measured, the unknown samples are analyzed and their analyte intensities read against the calibration curve. Over extended analysis times, it is common practice to update the calibration curve, either by recalibrating the instrument with a full set of standards or by running one midpoint standard. The following protocol summarizes a typical calibration using external standardization:

1. Blank >
2. Std. 1 >
3. Std. 2 >
4. Std. 3 >
5. Sample 1 >
6. Sample 2 >
7. Sample…*n*
8. Recalibrate
9. Sample $n + 1$, etc.

This can be seen more clearly in Figure 16.1, which shows a typical calibration curve using a blank and three standards of 2, 5, and 10 ppb. This calibration curve shows a simple *linear regression*, but usually other modes of calibration are also available like *weighted linear* to emphasize measurements at the low concentration region of the curve and *linear through zero*, where the linear regression is

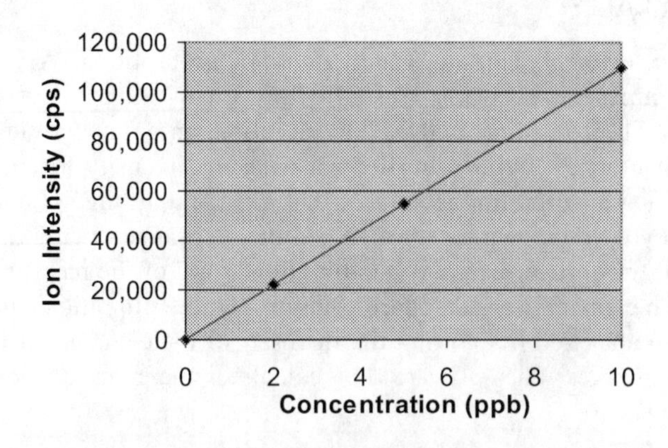

FIGURE 16.1 A simple linear regression calibration curve.

forced through zero. Whatever approach is used, it is critically important to select your range of standards, based on the expected concentration levels in your samples, if you want to ensure to optimum accuracy.

It should be emphasized that this graph represents a single-element calibration. However, because ICP-MS is usually used for multielement analysis, multielement standards are typically used to generate calibration data. For that reason, it is absolutely essential to use multielement standards that have been manufactured specifically for ICP-MS. Single-element AA standards are not suitable, because they usually have only been certified for the analyte element and not for any others. The purity of the standard cannot be guaranteed for any other element and as a result cannot be used to make up multielement standards for use with ICP-MS. For the same reason, ICP-OES multielement standards are not advisable either, because they are only certified for a group of elements and could contain other elements at higher levels, which will affect the ICP-MS multielement calibration.

Standard Additions

This mode of calibration provides an effective way to minimize sample-specific matrix effects by spiking samples with known concentrations of analytes.[1,2] In standard addition calibration, the intensity of a blank solution is first measured. Next, the sample solution is "spiked" with known concentrations of each element to be determined. The instrument measures the response for the spiked samples and creates a calibration curve for each element for which a spike has been added. The calibration curve is a plot of the blank subtracted intensity of each spiked element against its concentration value. After creating the calibration curve, the unspiked sample solutions are then analyzed and compared to the calibration curve. Based on the slope of the calibration curve and where it intercepts the *x*-axis, the instrument software determines the unspiked concentration of the analytes in the unknown samples. This can be seen in Figure 16.2, which shows a calibration of the sample intensity and the sample spiked with 2 and 5 ppb of the analyte. The concentration of sample is where the calibration line intercepts the negative side of the *x*-axis.

The following protocol summarizes a typical calibration using the method of standard additions:

1. Blank >
2. Spiked Sample 1 (Spike Conc. 1) >
3. Spiked Sample 1 (Spike Conc. 2) >
4. Unspiked Sample 1 >
5. Blank >
6. Spiked Sample 2 (Spike Conc. 1) >

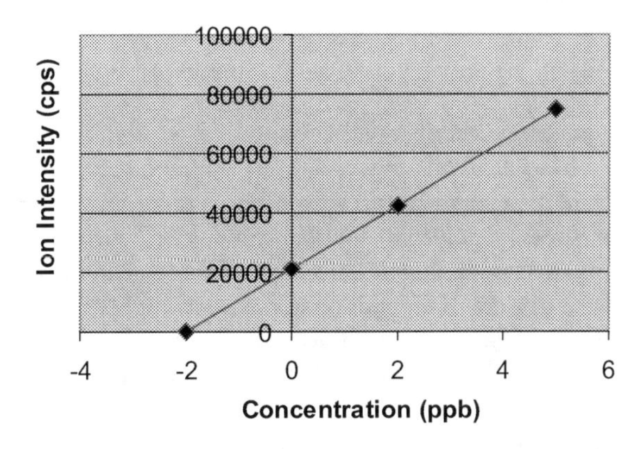

FIGURE 16.2 A typical "method of additions" calibration curve.

7. Spiked Sample 2 (Spike Conc. 2) >
8. Unspiked Sample 2 >
9. Blank >
10. Etc.

Addition Calibration

Unfortunately, with the method of standard additions, each and every sample has to be spiked with all the analytes of interest, which becomes extremely labor intensive when many samples have to be analyzed. For this reason, a variation of standard additions called *addition calibration* is more widely used in ICP-MS. However, this method can only be used when all the samples have a similar matrix. It uses the same principle as standard additions, but only the first (or representative) sample is spiked with known concentrations of analytes and then analyzes the rest of the sample batch against the calibration, assuming all samples have a similar matrix to the first one. The following protocol summarizes a typical calibration using the method of addition calibration:

1. Blank >
2. Spiked Sample 1 (Spike Conc. 1) >
3. Spiked Sample 1 (Spike Conc. 2) >
4. Unspiked Sample 1 >
5. Unspiked Sample 2 >
6. Unspiked Sample 3 >
7. Etc.

SEMIQUANTITATIVE ANALYSIS

If your data quality objectives for accuracy and precision are less stringent, ICP-MS offers a very rapid semiquantitative mode of analysis. This technique enables you to automatically determine the concentrations of up to 75 elements in an unknown sample, without the need for calibration standards.[3,4] This is an approach that could be extremely useful for initially screening samples, before quantitative analysis is carried out.

There are slight variations in the way different instruments approach semiquantitative analysis, but the general principle is to measure the entire mass spectrum, without specifying individual elements or masses. It relies on the principle that each element's natural isotopic abundance is fixed. By measuring the intensity of all their isotopes, correcting for common spectral interferences, including molecular, polyatomic, and isobaric species and applying heuristic, knowledge-driven routines in combination with numerical calculations, a positive or negative confirmation can be made for each element present in the sample. Then, by comparing the corrected intensities against a stored isotopic response table, a good semiquantitative approximation of the sample components can be made.

Semiquant, as it is often called, is an excellent approach to rapidly characterizing unknown samples. Once the sample has been characterized, you can choose to either update the response table with your own standard solutions to improve analytical accuracy or switch to the quantitative analysis mode to focus on specific elements and determine their concentrations with even greater accuracy and precision. Whereas a semiquantitative determination can be performed without using a series of standards, the use of a small number of standards is highly recommended for improved accuracy across the full mass range. Unlike traditional quantitative analysis, in which you analyze standards for all the elements you want to determine, semiquant calibration is achieved using just a few elements distributed across the mass range. This calibration process, shown more clearly in Figure 16.3, is used to update the reference response curve data that correlates measured ion

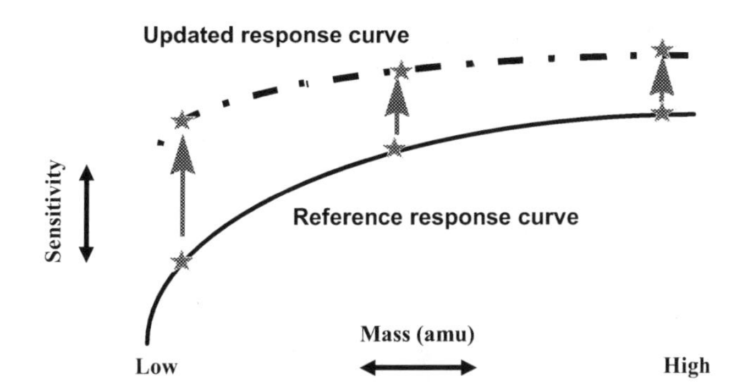

FIGURE 16.3 In semiquantitative analysis, a small group of elements are used to update the reference response curve to improve the accuracy as the sample matrix changes.

intensities to the concentrations of elements in a solution. During calibration, this response data is adjusted to account for changes in the instrument's sensitivity due to variations in the sample matrix.

This process is often called semiquantitative analysis using external calibration and, like traditional quantitative analysis using external standardization, works extremely well for samples that have a similar matrix. However, if you are analyzing samples containing widely different concentrations of matrix components, external calibration does not work very well because of matrix-induced suppression effects on the analyte signal. If this is the case, semiquant using a variation of standard addition calibration should be used. Similar to standard addition calibration used in quantitative analysis, this procedure involves adding known quantities of specific elements to every unknown sample before measurement. The major difference with semiquant is that the elements you add must not already be present in significant quantities in the unknown samples, because they are being used to update the stored reference response curve. As with external calibration, the semiquant software then adjusts the stored response data for all remaining analytes relative to the calibration elements. This procedure works very well but tends to be very labor intensive because the calibration standards have to be added to every unknown sample.

The other factor to be wary of with semiquantitative analysis is the spectral complexity of unknown samples. If you have a spectrally rich sample and are not making any compensation for spectral overlaps close to the analyte peaks, it could possibly give you a false-positive for that element. Therefore, you have to be very cautious when reporting semiquantitative results on completely unknown samples. They should be characterized first, especially with respect to the types of spectral interferences generated by the plasma gas, the matrix, and the solvents/acids/chemicals used for sample preparation. Collision/reaction cells/interfaces can help in the reduction of some of these interferences, but extreme care should be taken, as these devices are known to have no effect on some polyatomic interferences and, in some cases, can increase the spectral complexity by generating other interfering complexes.

ISOTOPE DILUTION

Although quantitative and semiquantitative analysis methods are suitable for the majority of applications, there are other calibration methods available, depending on your analytical requirements. For example, if your application requires even greater accuracy and precision, the "isotope dilution" technique may offer some benefits. Isotope dilution is an absolute means of quantitation based on altering the natural abundance of two isotopes of an element by adding a known amount of one of the isotopes and is considered one of the most accurate and precise approaches to elemental analysis.[5–8]

For this reason, a prerequisite of isotope dilution is that the element must have at least two stable isotopes. The principle works by spiking a known weight of an enriched stable isotope into your sample solution. By knowing the natural abundance of the two isotopes being measured, the abundance of the spiked enriched isotopes, the weight of the spike, and the weight of the sample, the original trace element concentration can be determined by using the following equation:

$$C = \frac{\left[\text{Aspike} - \left(R \times \text{Bspike}\right)\right] \times \text{Wspike}}{\left[R \times \left(\text{Bsample} - \text{Asample}\right)\right] \times \text{Wsample}}$$

where

C = Concentration of trace element
Aspike = Percentage of higher abundance isotope in spiked enriched isotope
Bspike = Percentage of lower abundance isotope in spiked enriched isotope
Wspike = Weight of spiked enriched isotope
R = Ratio of the percentage of higher abundance isotope to lower abundance isotope in the spiked sample
Bsample = Percentage of higher natural abundance isotope in sample
Asample = Percentage of lower natural abundance isotope in sample
Wsample = Weight of sample.

This might sound complicated, but in practice, it is relatively straightforward. This is illustrated in Figure 16.4, which shows an isotope dilution method for the determination of copper in a 250 mg sample of orchard leaves, using the two copper isotopes ^{63}Cu and ^{65}Cu.

In the bar graph on top it can be seen that the natural abundance of the two isotopes are 69.09% and 30. 91%, respectively, for ^{63}Cu and ^{65}Cu. In the middle graph, it shows that 4 μg of an enriched isotope of 100% ^{65}Cu (and 0% ^{63}Cu) is spiked into the sample, which now produces a spiked sample containing 71.4% of ^{65}Cu and 28.6% of ^{63}Cu, as seen in the bottom plot.[9] If we plug these data into the preceding equation, we get:

$$C = \frac{\left[100 - \left(71.4/28.6 \times 0\right)\right] \times 4\ \mu\text{g}}{\left[\left(71.4/28.6 \times 69.09\right) - 30.91\right] \times 0.25\ \text{g}}$$

$$C = 400/35.45 = 11.3\ \mu\text{g/g}$$

The major benefit of the isotope dilution technique is that it provides measurements that are extremely accurate because you are measuring the concentration of the isotopes in the same solution as your unknown sample and not in a separate external calibration solution. In addition, because it is a ratioing technique, loss of solution during the sample preparation stage has no influence on the accuracy of the result. The technique is also extremely precise, because using a simultaneous detection system such as a magnetic sector multicollector or a simultaneous ion sampling device such as a time-of-flight (TOF) ICP-MS, the results are based on measuring the two isotope solution at the same instant in time, which compensates for imprecision of the signal due to sources of sample introduction-related noise, such as plasma instability, peristaltic pump pulsations, and nebulization fluctuations. Even when using a scanning mass analyzer such as a quadrupole, the measurement protocol can be optimized to scan very rapidly between the two isotopes and achieve very good precision. However, isotope dilution has some limitations, which makes it suitable only for certain applications. These limitations include the following:

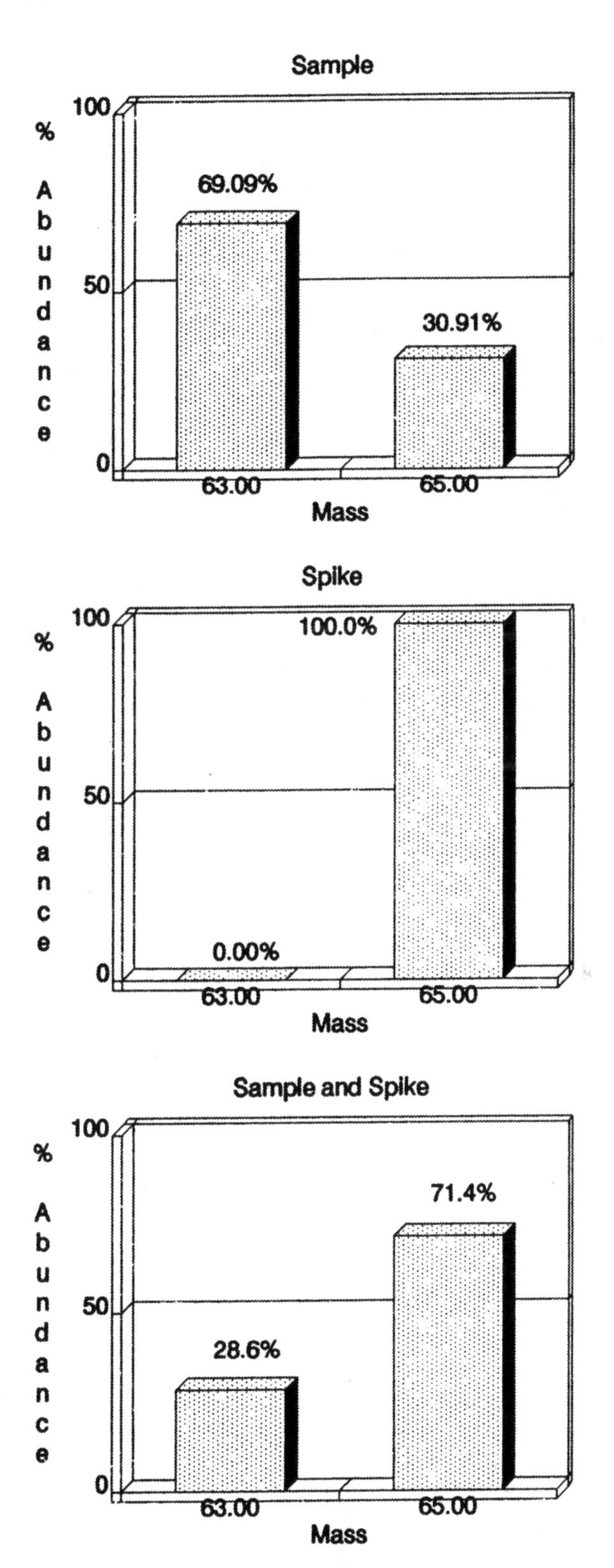

FIGURE 16.4 Quantitation of trace levels of copper in a sample of standard reference material (SRM) orchard leaves using isotope dilution methodology. (From *Multi-elemental Isotope Dilution Using the Elan* ICP-MS *Elemental Analyzer*, ICP-MS *Technical Summary TSMS-1*, PerkinElmer Instruments, 1985.)

- The element you are determining must have more than one isotope, because calculations are based on the ratio of one isotope to another isotope of the same element; this makes it unsuitable for approximately 15 elements that can be determined by ICP-MS.
- It requires certified enriched isotopic standards, which can be very expensive, especially those that are significantly different from the normal isotopic abundance of the element.
- It compensates for interferences due to signal enhancement or suppression but does not compensate for spectral interferences. For this reason, an external blank solution must always be run.

ISOTOPE RATIOS

The ability of ICP-MS to determine individual isotopes also makes it suitable for another isotopic measurement technique called "isotope ratio" analysis. The ratio of two or more isotopes in a sample can be used to generate very useful information, including an indication of the age of a geological formation, a better understanding of animal metabolism, and to help identify sources of environmental contamination.[10–14] Similar to isotope dilution, isotope ratio analysis uses the principle of measuring the exact ratio of two isotopes of an element in the sample. With this approach, the isotope of interest is typically compared to a reference isotope of the same element. For example, you might want to compare the concentration of ^{204}Pb to that of ^{206}Pb. Alternatively, the requirement might be to compare one isotope to all remaining reference isotopes of an element, for example, the ratio of ^{204}Pb to ^{206}Pb, ^{207}Pb, and ^{208}Pb. The ratio is then expressed in the following manner:

$$\text{Isotope ratio} = \frac{\text{Intensity of isotope of interest}}{\text{Intensity of reference isotope}}$$

As this ratio can be calculated from within a single sample measurement, classic external calibration is not normally required. However, if there is a large difference between the concentrations of the two isotopes, it is recommended to run a standard of known isotopic composition. This is done to verify that the higher concentration isotope is not suppressing the signal of the lower concentration isotope and biasing the results. This effect, called mass discrimination, is less of a problem if the isotopes are relatively close in concentration, for example, ^{107}Ag to ^{109}Ag, which are 51.839% and 48.161% abundant, respectively. However, it can be an issue if there is a significant difference in their concentration values, for example, ^{235}U to ^{238}U, which are 0.72% and 99.275% abundant, respectively. Mass discrimination effects can be reduced by running an external reference standard of known isotopic concentration, comparing the isotope ratio with the theoretical value, and then mathematically compensating for the difference. The principles of isotope ratio analysis and how to achieve optimum precision values are explained in greater detail in Chapter 15 "Peak Measurement Protocol."

INTERNAL STANDARDIZATION

Another method of standardization commonly employed in ICP-MS is called *internal standardization*. It is not considered an absolute calibration technique but instead is used to correct for changes in analyte sensitivity caused by variations in the concentration and type of matrix components found in the sample. An internal standard is a non-analyte isotope that is added to the blank solution, standards, and samples before analysis. It should also be noted that the chosen internal standard element must not be present in the sample matrix. It is typical to add three or four internal standard elements to the samples to cover the analyte elements of interest. The software adjusts the analyte concentration in the unknown samples by comparing the intensity values of the internal standard intensities in the unknown sample to those in the calibration standards.

The implementation of internal standardization varies according to the analytical technique that is being used. For quantitative analysis, the internal standard elements are selected on the basis of the similarity of their ionization characteristics to the analyte elements. Each internal standard is bracketed with a group of analytes. The software then assumes that the intensities of all elements within a group are affected in a similar manner by the matrix. Changes in the ratios of the internal standard intensities are then used to correct the analyte concentrations in the unknown samples.

For semiquantitative analysis that uses a stored response table, the purpose of the internal standard is similar but a little different in implementation to quantitative analysis. A semiquant internal standard is used to continuously compensate for instrument drift or matrix-induced suppression over a defined mass range. If a single internal standard is used, all the masses selected for the determination are updated by the same amount based on the intensity of the internal standard. If more than one internal standard is used, which is recommended for measurements over a wide mass range, the software interpolates the intensity values based on the distance in mass between the analyte and the nearest internal standard element.

It is worth emphasizing that if you do not want to compare your intensity values to a calibration graph, most instruments allow you to report raw data. This enables you to analyze your data using external data-processing routines, to selectively apply a minimum set of ICP-MS data processing methods or to just view the raw data file before reprocessing it. The availability of raw data is primarily intended for use in non-routine applications such as chromatography separation techniques and laser sampling devices that produce a time-resolved transient peak or by users whose sample set requires data processing using algorithms other than those supplied by the instrument software.

FURTHER READING

1. D. Beauchemin, J. W. McLaren, A. P. Mykytiuk, and S. S. Berman, *Analytical Chemistry*, **59**, 778–783, 1987.
2. E. Pruszkowski, K. Neubauer, and R. Thomas, *Atomic Spectroscopy*, **19** (4), 111–115, 1998.
3. M. Broadhead, R. Broadhead, and J. W. Hager, *Atomic Spectroscopy*, **11** (6), 205–209, 1990.
4. E. Denoyer, *Journal of Analytical Atomic Spectrometry*, **7**, 1187–1193, 1992.
5. J. W. McLaren, D. Beauchemin, and S. S. Berman, *Analytical Chemistry*, **59**, 610–615, 1987.
6. H. Longerich, *Atomic Spectroscopy*, **10** (4), 112–115, 1989.
7. A. Stroh, *Atomic Spectroscopy*, **14** (5), 141–143, 1993.
8. T. Catterick, H. Handley, and S. Merson, *Atomic Spectroscopy*, **16** (10), 229–234, 1995.
9. *Multi-elemental Isotope Dilution Using the Elan ICP-MS Elemental Analyzer, ICP-MS Technical Summary TSMS-1*, PerkinElmer Instruments, 1985.
10. B. T. G. Ting and M. Janghorbani, *Analytical Chemistry*, **58**, 1334–1343, 1986.
11. M. Janghorbani, B. T. G. Ting, and N. E. Lynch, *Microchemica Acta,* **3**, 315–328, 1989.
12. T. A. Hinners, E. M. Heithmar, T. M. Spittler, and J. M. Henshaw, *Analytical Chemistry*, **59**, 2658–2662, 1987.
13. L. Halicz, Y. Erel, and A. Veron, *Atomic Spectroscopy*, **17** (5), 186–189, 1996.
14. M. Chaudhary-Webb, D. C. Paschal, W. C. Elliott, H. P. Hopkins, A. M. Ghazi, B. C. Ting, and I. Romieu, *Atomic Spectroscopy*, **19** (5), 156–166, 1998.

17 Review of ICP-MS Interferences

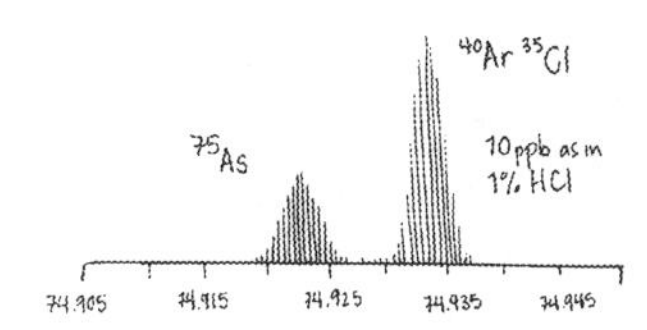

Now that we have covered the fundamental principles of inductively coupled plasma mass spectrometry (ICP-MS) and its measurement and calibration routines, this chapter focuses on the most common interferences found in ICP-MS and the methods that are used to compensate for them. Although interferences are reasonably well understood in ICP-MS, it can often be difficult and time-consuming to compensate for them, particularly in complex sample matrices. Prior knowledge of the interferences associated with a particular set of samples will often dictate the sample preparation steps and the instrumental methodology used to analyze them.

Interferences in ICP-MS are generally classified into three major groups: spectral, matrix, and physical. Each of them has the potential to be problematic in its own right, but modern instrumentation and good software combined with optimized analytical methodologies have minimized their negative impact on trace element determinations by ICP-MS. Let us look at these interferences in greater detail and describe the different approaches used to compensate for them.

SPECTRAL INTERFERENCES

Spectral overlaps are probably the most serious types of interferences seen in ICP-MS. The most common are known as a polyatomic or molecular spectral interference and are produced by the combination of two or more atomic ions. They are caused by a variety of factors but are usually associated with the plasma/nebulizer gas used, matrix components in the solvent/sample, other elements in the sample, or entrained oxygen/nitrogen from the surrounding air. For example, in the argon plasma, spectral overlaps caused by argon ions and combinations of argon ions with other species are very common. The most abundant isotope of argon is at mass 40, which dramatically interferes with the most abundant isotope of calcium at mass 40, whereas the combination of argon and oxygen in an aqueous sample generates the $^{40}Ar^{16}O^+$ interference, which has a significant impact on the major isotope of Fe at mass 56. The complexity of these kinds of spectral problems can be seen in Figure 17.1, which shows a mass spectrum of deionized water from mass 40 to mass 90.

In addition, argon can form polyatomic interferences with elements found in the acids used to dissolve the sample. For example, in a hydrochloric acid medium, $^{40}Ar^+$ combines with the most abundant chlorine isotope at 35 amu to form $^{40}Ar^{35}Cl^+$, which interferes with the only isotope of arsenic at mass 75, whereas in an organic solvent matrix, argon and carbon combine to form $^{40}Ar^{12}C^+$, which interferes with $^{52}Cr^+$, the most abundant isotope of chromium. Sometimes, matrix or solvent ions combine to form spectral interferences of their own. A good example is in a sample that contains sulfuric acid. The dominant sulfur isotope, $^{32}S^+$, combines with two oxygen ions to form an $^{32}S^{16}O^{16}O^+$ molecular ion, which interferes with the major isotope of Zn at mass 64. In the analysis

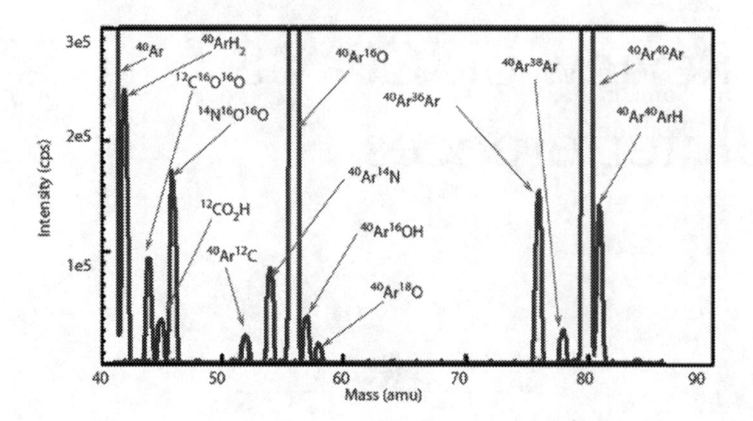

FIGURE 17.1 ICP mass spectrum of deionized water from mass 40 to mass 90.

TABLE 17.1
Some Common Plasma/Matrix/Solvent-Related Polyatomic Spectral Interferences Seen in ICP-MS

Element/Isotope	Matrix/Solvent	Interference
$^{39}K^{+}$	H_2O	$^{38}ArH^{+}$
$^{40}Ca^{+}$	H_2O	$^{40}Ar^{+}$
$^{56}Fe^{+}$	H_2O	$^{40}Ar^{16}O^{+}$
$^{80}Se^{+}$	H_2O	$^{40}Ar^{40}Ar^{+}$
$^{51}V^{+}$	HCl	$^{35}Cl^{16}O^{+}$
$^{75}As^{+}$	HCl	$^{40}Ar^{35}Cl^{+}$
$^{28}Si^{+}$	HNO_3	$^{14}N^{14}N^{+}$
$^{44}Ca^{+}$	HNO_3	$^{14}N^{14}N^{16}O^{+}$
$^{55}Mn^{+}$	HNO_3	$^{40}Ar^{15}N^{+}$
$^{48}Ti^{+}$	H_2SO_4	$^{32}S^{16}O^{+}$
$^{52}C^{+}r$	H_2SO_4	$^{34}S^{18}O^{+}$
$^{64}Zn^{+}$	H_2SO_4	$^{32}S^{16}O^{16}O^{+}$
$^{63}Cu^{+}$	H_3PO_4	$^{31}P^{16}O^{16}O^{+}$
$^{24}Mg^{+}$	Organics	$^{12}C^{12}C^{+}$
$^{52}Cr^{+}$	Organics	$^{40}Ar^{12}C^{+}$
$^{65}Cu^{+}$	Minerals	$^{48}Ca^{16}OH^{+}$
$^{64}Zn^{+}$	Minerals	$^{48}Ca^{16}O^{+}$
$^{63}Cu^{+}$	Seawater	$^{40}Ar^{23}Na^{+}$

of samples containing high concentrations of sodium, such as seawater, the most abundant isotope of Cu at mass 63 cannot be used because of interference from the $^{40}Ar^{23}Na^{+}$ molecular ion. There are many more examples of these kinds of polyatomic and molecular interferences, which have been comprehensively reviewed.[1] Table 17.1 represents some of the most common matrix–solvent spectral interferences seen in ICP-MS.

OXIDES, HYDROXIDES, HYDRIDES, AND DOUBLY CHARGED SPECIES

Another type of spectral interference is produced by elements in the sample combining with H^{+}, $^{16}O^{+}$, or $^{16}OH^{+}$ (either from water or air) to form molecular hydrides (+H^{+}), oxides (+$^{16}O^{+}$), and hydroxides (+$^{16}OH^{+}$), which occur at 1, 16, and 17 mass units, respectively, higher than the element's mass.[2]

TABLE 17.2
Some Elements That Readily Form Oxides, Hydroxides, Hydrides, and Doubly Charged Species in the Plasma, Together with the Analytes Affected by the Interference

Oxide, Hydroxide, Hydride, Doubly Charged Species	Analyte Affected by Interference
$^{40}Ca^{16}O^+$	$^{56}Fe^+$
$^{48}Ti^{16}O^+$	$^{64}Zn^+$
$^{98}Mo^{16}O^+$	$^{114}Cd^+$
$^{138}Ba^{16}O^+$	$^{154}Sm^+$, $^{154}Gd^+$
$^{139}La^{16}O^+$	$^{155}Gd^+$
$^{140}Ce^{16}O^+$	$^{156}Gd^+$, $^{156}Dy^+$
$^{40}Ca^{16}OH^+$	$^{57}Fe^+$
$^{31}P^{18}O^{16}OH^+$	$^{66}Zn^+$
$^{79}BrH^+$	$^{80}Se^+$
$^{31}P^{16}O^2H^+$	$^{64}Zn^+$
$^{138}Ba^{2+}$	$^{69}Ga^+$
$^{139}La^{2+}$	$^{69}Ga^+$
$^{140}Ce^{2+}$	$^{70}Ge^+$, $^{70}Zn^+$

These interferences are typically produced in the cooler zones of the plasma, immediately before the interface region. They are usually more serious when rare earth or refractory-type elements are present in the sample, because many of them readily form molecular species (particularly oxides), which create spectral overlap problems on other elements in the same group. If the oxide species is mainly derived from entrained air around the plasma, it can be reduced by either using an elongated outer tube to the torch or a metal shield between the plasma and the radio frequency (RF) coil.

Associated with oxide-based spectral overlaps are doubly charged spectral interferences. These are species that are formed when an ion is generated with a double positive charge as opposed to a normal single charge and produces an isotopic peak at half its mass. Similar to the formation of oxides, the level of doubly charged species is related to the ionization conditions in the plasma and can usually be minimized by careful optimization of the nebulizer gas flow, RF power, and sampling position within the plasma. It can also be impacted by the severity of the secondary discharge present at the interface,[3] which was described in greater detail in Chapter 8 "Interface Region." Table 17.2 shows a selected group of elements that readily form oxides, hydroxides, hydrides, and doubly charged species, together with the analytes affected by them.

ISOBARIC INTERFERENCES

The final classification of spectral interferences is called isobaric overlaps, produced mainly by different isotopes of other elements in the sample creating spectral interferences at the same mass as the analyte. For example, vanadium has two isotopes at 50 and 51 amu. However, mass 50 is the only practical isotope to use in the presence of a chloride matrix because of the large contribution from the $^{16}O^{35}Cl^+$ interference at mass 51. Unfortunately, mass 50 amu, which is only 0.25% abundant, also coincides with isotopes of titanium and chromium, which are 5.4% and 4.3% abundant, respectively. This makes the determination of vanadium in the presence of titanium and chromium very difficult unless mathematical corrections are made. Figure 17.2 shows all the possible naturally occurring isobaric spectral overlaps in ICP-MS.[4]

Relative Abundance of the Natural Isotopes

Isotope	%	%	%	Isotope	%	%	%	Isotope	%	%	%	Isotope	%	%	%
1	H 99.985			61			Ni 1.140	121			Sb 57.36	181	Ta 99.988		
2	H 0.015			62			Ni 3.634	122	Sn 4.63	Te 2.603		182		W 26.3	
3		He 0.000137		63	Cu 69.17			123		Te 0.908	Sb 42.64	183		W 14.3	
4		He 99.999863		64		Zn 48.6	Ni 0.926	124	Sn 5.79	Te 4.816	Xe 0.10	184	Os 0.02	W 30.67	
5				65	Cu 30.83			125		Te 7.139		185			Re 37.40
6			Li 7.5	66		Zn 27.9		126		Te 18.95	Xe 0.09	186	Os 1.58	W 28.6	
7			Li 92.5	67		Zn 4.1		127	I 100			187	Os 1.6		Re 62.60
8				68		Zn 18.8		128		Te 31.69	Xe 1.91	188	Os 13.3		
9	Be 100			69			Ga 60.108	129			Xe 26.4	189	Os 16.1		
10		B 19.9		70	Ge 21.23	Zn 0.6		130	Ba 0.106	Te 33.80	Xe 4.1	190	Os 26.4		Pt 0.01
11		B 80.1		71			Ga 39.892	131			Xe 21.2	191		Ir 37.3	
12			C 98.90	72	Ge 27.66			132	Ba 0.101		Xe 26.9	192	Os 41.0		Pt 0.79
13			C 1.10	73	Ge 7.73			133		Cs 100		193		Ir 62.7	
14	N 99.643			74	Ge 35.94	Se 0.89		134	Ba 2.417		Xe 10.4	194			Pt 32.9
15	N 0.366			75			As 100	135	Ba 6.592			195			Pt 33.8
16		O 99.762		76	Ge 7.44	Se 9.36		136	Ba 7.854	Ce 0.19	Xe 8.9	196	Hg 0.15		Pt 25.3
17		O 0.038		77		Se 7.63		137	Ba 11.23			197		Au 100	
18		O 0.200		78	Kr 0.35	Se 23.78		138	Ba 71.70	Ce 0.25	La 0.0902	198	Hg 9.97		Pt 7.2
19			F 100	79			Br 50.69	139			La 99.9098	199	Hg 16.87		
20	Ne 90.48			80	Kr 2.25	Se 49.61		140		Ce 88.48		200	Hg 23.10		
21	Ne 0.27			81			Br 49.31	141			Pr 100	201	Hg 13.18		
22	Ne 9.25			82	Kr 11.6	Se 8.73		142	Nd 27.13	Ce 11.08		202	Hg 29.86		
23		Na 100		83	Kr 11.5			143	Nd 12.18			203			Tl 29.524
24			Mg 78.99	84	Kr 57.0	Sr 0.56		144	Nd 23.80	Sm 3.1		204	Hg 6.87	Pb 1.4	
25			Mg 10.00	85			Rb 72.165	145	Nd 8.30			205			Tl 70.476
26			Mg 11.01	86	Kr 17.3	Sr 9.86		146	Nd 17.19			206		Pb 24.1	
27	Al 100			87		Sr 7.00	Rb 27.835	147		Sm 15.0		207		Pb 22.1	
28		Si 92.23		88		Sr 82.58		148	Nd 5.76	Sm 11.3		208		Pb 52.4	
29		Si 4.67		89			Y 100	149		Sm 13.8		209	Bi 100		
30		Si 3.10		90	Zr 51.45			150	Nd 5.64	Sm 7.4		210			
31			P 100	91	Zr 11.22			151			Eu 47.8	211			
32	S 95.02			92	Zr 17.15	Mo 14.84		152	Gd 0.20	Sm 26.7		212			
33	S 0.75			93			Nb 100	153			Eu 52.2	213			
34	S 4.21			94	Zr 17.38	Mo 9.25		154	Gd 2.18	Sm 22.7		214			
35		Cl 75.77		95		Mo 15.92		155	Gd 14.80			215			
36	S 0.02		Ar 0.337	96	Zr 2.80	Mo 16.68	Ru 5.52	156	Gd 20.47	Dy 0.06		216			
37		Cl 24.23		97		Mo 9.55		157	Gd 15.65			217			
38			Ar 0.063	98		Mo 24.13	Ru 1.88	158	Gd 24.84	Dy 0.10		218			
39	K 93.2581			99			Ru 12.7	159			Tb 100	219			
40	K 0.0117	Ca 96.941	Ar 99.600	100		Mo 9.63	Ru 12.6	160	Gd 21.86	Dy 2.34		220			
41	K 6.7302			101			Ru 17.0	161		Dy 18.9		221			
42		Ca 0.647		102	Pd 1.02		Ru 31.6	162	Er 0.14	Dy 25.5		222			
43		Ca 0.135		103		Rh 100		163		Dy 24.9		223			
44		Ca 2.086		104	Pd 11.14		Ru 18.7	164	Er 1.61	Dy 28.2		224			
45			Sc 100	105	Pd 22.33			165			Ho 100	225			
46	Ti 8.0	Ca 0.004		106	Pd 27.33	Cd 1.25		166	Er 33.6			226			
47	Ti 7.3			107			Ag 51.839	167	Er 22.95			227			
48	Ti 73.8	Ca 0.187		108	Pd 26.46	Cd 0.89		168	Er 26.8	Yb 0.13		228			
49	Ti 5.5			109			Ag 48.161	169			Tm 100	229			
50	Ti 5.4	V 0.250	Cr 4.345	110	Pd 11.72	Cd 12.49		170	Er 14.9	Yb 3.05		230			
51		V 99.750		111		Cd 12.80		171		Yb 14.3		231	Pa 100		
52			Cr 83.789	112	Sn 0.97	Cd 24.13		172		Yb 21.9		232	Th 100		
53			Cr 9.501	113		Cd 12.22	In 4.3	173		Yb 16.12		233			
54	Fe 5.8		Cr 2.365	114	Sn 0.65	Cd 28.73		174		Yb 31.8	Hf 0.162	234	U 0.0055		
55		Mn 100		115	Sn 0.34		In 95.7	175	Lu 97.41			235	U 0.7200		
56	Fe 91.72			116	Sn 14.53	Cd 7.49		176	Lu 2.59	Yb 12.7	Hf 5.206	236			
57	Fe 2.2			117	Sn 7.68			177			Hf 18.606	237			
58	Fe 0.28		Ni 68.077	118	Sn 24.23			178			Hf 27.297	238	U 99.2745		
59		Co 100		119	Sn 8.59			179			Hf 13.629				
60			Ni 26.223	120	Sn 32.59	Te 0.096		180	Ta 0.012	W 0.13	Hf 35.100				

"Isotopic Compositions of the Elements 1989" Pure Appl. Chem., Vol. 63, No. 7, pp. 991-1002, 1991. © 1991 IUPAC

FIGURE 17.2 Relative isotopic abundances of the naturally occurring elements, showing all the potential isobaric interferences. (From UIPAC Isotopic Composition of the Elements, *Pure and Applied Chemistry* **75**(6), 683–799, 2003.)

WAYS TO COMPENSATE FOR SPECTRAL INTERFERENCES

Let us now look at the different approaches used to compensate for spectral interferences. One of the very first ways used to get around severe matrix-derived spectral interferences was to remove the matrix somehow. In the early days, this involved precipitating the matrix with a complexing agent and then filtering off the precipitate. However, more recently, this has been carried out by automated matrix removal/analyte preconcentration techniques using chromatography-type equipment. In fact, this is the preferred method for carrying out trace metal determinations in seawater, because of the matrix and spectral problems associated with such high concentrations of sodium and chloride ions.[5]

MATHEMATICAL CORRECTION EQUATIONS

Another method that has been successfully used to compensate for isobaric interferences and some less severe polyatomic overlaps (when no alternative isotopes are available for quantitation) is to use mathematical interference correction equations. Similar to interelement corrections (IECs) in inductively coupled plasma optical emission spectrometry (ICP-OES), this method works on the principle of measuring the intensity of the interfering isotope or interfering species at another mass, which is ideally free of any interferences. A correction is then applied by knowing the ratio of the intensity of the interfering species at the analyte mass to its intensity at the alternate mass. Let us look at a "real-world" example to exemplify this type of correction. The most sensitive isotope for cadmium is at mass 114. However, there is also a minor isotope of tin at mass 114. This means that if there is any tin in the sample, quantitation using $^{114}Cd^+$ can only be carried out if a correction is made for $^{114}Sn^+$. Fortunately, Sn has a total of ten isotopes, which means that probably at least one of them is going to be free of a spectral interference. Therefore, by measuring the intensity of Sn at one of its most abundant isotopes (typically, $^{118}Sn^+$) and ratioing it to $^{114}Sn^+$, a correction is made in the method software in the following manner:

$$\text{Total counts at mass } 114 = {}^{114}\text{Cd}^+ + {}^{114}\text{Sn}^+$$

Therefore,

$${}^{114}\text{Cd}^+ = \text{Total counts at mass } 114 - {}^{114}\text{Sn}^+$$

To find out the contribution from $^{114}Sn^+$, it is measured at the interference-free isotope of $^{118}Sn^+$ and a correction of the ratio of $^{114}Sn^+/^{118}Sn^+$ is applied, which means $^{114}Cd^+$ = Counts at mass 114 – ($^{114}Sn^+/^{118}Sn^+$) × ($^{118}Sn^+$).

Now, the ratio ($^{114}Sn^+/^{118}Sn^+$) is the ratio of the natural abundances of these two isotopes (065%/24.23%) and is always constant.

Therefore,

$${}^{114}\text{Cd}^+ = \text{mass } 114 - (0.65\%/24.23\%) \times \left({}^{118}\text{Sn}^+\right)$$

or

$${}^{114}\text{Cd}^+ = \text{mass } 114 - (0.0268) \times \left({}^{118}\text{Sn}^+\right)$$

An interference correction for $^{114}Cd^+$ would then be entered in the software as

$$-(0.0268) \times \left({}^{118}\text{Sn}^+\right)$$

This is a relatively simple example but explains the basic principles of the process. In practice, especially in spectrally complex samples, corrections often have to be made to the isotope being used for the correction in addition to the analyte mass, which makes the mathematical equation far more complex.

This approach can also be used for some less severe polyatomic-type spectral interferences. For example, in the determination of V at mass 51 in diluted brine (typically 1,000 ppm NaCl), there is a substantial spectral interference from $^{35}Cl^{16}O^+$ at mass 51. By measuring the intensity of the $^{37}Cl^{16}O^+$ at mass 53, which is free of any interference, a correction can be applied in a similar way to the previous example.

COOL/COLD PLASMA TECHNOLOGY

If the intensity of the interference is large, and analyte intensity is extremely low, mathematical equations are not ideally suited as a correction method. For that reason, alternative approaches have to be considered to compensate for the interference. One such approach, which has helped to reduce some of the severe polyatomic overlaps, is to use cold/cool plasma conditions. This technology, which was reported in the literature in the late 1980s, uses a low-temperature plasma to minimize the formation of certain argon-based polyatomic species.[6] Under normal plasma conditions (typically 1,000–1,400 W RF power, 15–20 L/min coolant flow, and 0.8–1.0 L/min of nebulizer gas flow), argon ions combine with matrix and solvent components to generate problematic spectral interferences such as $^{38}ArH^+$, $^{40}Ar^+$, and $^{40}Ar^{16}O^+$, which impact the detection limits of a small number of elements including K, Ca, and Fe. By using cool plasma conditions (e.g., 500–800 W RF power, 10 L/min coolant flow, and 1.5–1.8 L/min nebulizer gas flow), the ionization conditions in the plasma are changed so that many of these interferences are dramatically reduced. The result is that detection limits for this group of elements are significantly enhanced.[7] An example of this improvement is shown in Figure 17.3. It shows a spectral scan of 100 ppt of $^{56}Fe^+$ (its most sensitive isotope) using cool plasma conditions. It can be clearly seen that there is virtually no contribution from $^{40}Ar^{16}O^+$, as indicated by the extremely low background for deionized water, resulting in single figure ppt detection limits for iron. Under normal plasma conditions, the $^{40}Ar^{16}O^+$ intensity is so large that it would completely overlap the $^{56}Fe^+$ peak.[8]

Unfortunately, even though the use of cool plasma conditions is recognized as being a very useful tool for the determination of a small group of elements, its limitations are well documented.[9] A summary of the limitations of cool plasma technology includes the following:

- As a result of less energy being available in a cool plasma, elements that form a strong bond with one of the matrix/solvent ions cannot be easily decomposed, and as a result, their detection limits are compromised.
- Elements with high-ionization potentials cannot be ionized, because there is much less energy compared to a normal, high-temperature plasma.
- Elemental sensitivity is severely affected by the sample matrix; so, cool plasma often requires the use of standard additions or matrix matching to achieve satisfactory results.
- When carrying out multielement analysis, normal plasma conditions must also be used. This necessitates the need for stabilization times on the order of 3 min to change from a normal to a cool plasma, which degrades productivity and results in higher sample consumption.

For this reason, it is not ideally suited for the analysis of complex samples, but it does offer real detection limit improvement for elements with low ionization potential, such as sodium and lithium, which benefit from the ionization conditions of the cooler plasma. However, recent advances in solid-state, free-running RF generators appear to have enhanced the capability of cold plasma technology. Researchers have reported that by using the combination of highly reactive gases in a collision/reaction cell with cold plasma conditions, efficient reduction of some polyatomic interferences was achieved. This new approach allowed ultratrace levels of a suite of elements to be accurately measured in high-purity hydrochloric acid using both cold and normal plasma operating conditions.[10]

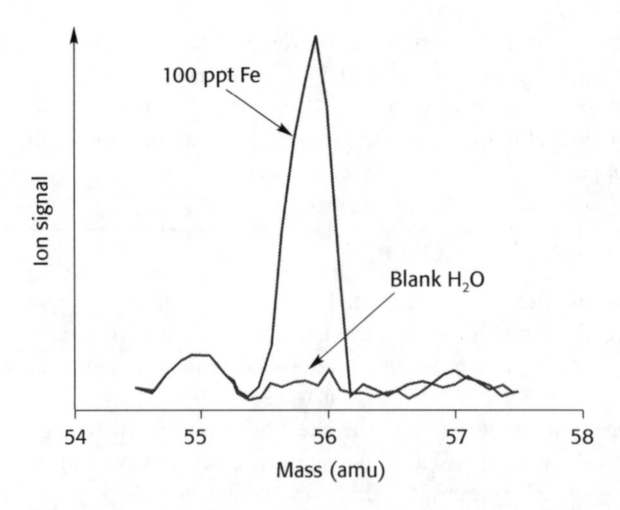

FIGURE 17.3 Spectral scan of 100 ppt ^{56}Fe and deionized water using cool plasma conditions. (From S. D. Tanner, M. Paul, S. A. Beres, and E. R. Denoyer, *Atomic Spectroscopy*, **16**(1), 16, 1995.)

COLLISION/REACTION CELLS

The earlier limitations of cool plasma technology have led to the development of collision/reaction cells and interfaces, which utilize ion–molecule collisions and reactions to cleanse the ion beam of harmful polyatomic and molecular interferences before they enter the mass analyzer. Quadrupole mass analyzers fitted with these devices are showing enormous potential to eliminate many spectral interferences, allowing the use of the most sensitive elemental isotopes that were previously unavailable for quantitation.

A full description and review of collision/reaction cell and interface technology and how they handle spectral interferences were given in Chapter 13 "Mass Analyzers: Collision/Reaction Cell and Interface Technology." However, it's worth noting that one of the unique features of using reaction chemistry in a collision/reaction cell is that it can be used in the mass shift mode, which is particularly beneficial if a spectral interference is encountered with an analyte which is monoisotopic (only one mass is available for quantitation). An example of this is that when high levels of chloride are present in the sample (whether it's from the matrix itself or from hydrochloric acid used in the sample digestion procedure), it can be problematic to quantitate arsenic and vanadium at low levels, because of polyatomic spectral interferences $^{40}Ar^{75}Cl$ and $^{35}Cl^{16}O$, which interfere with the major isotopes of ^{75}As and ^{51}V, respectively. A collision cell with kinetic energy discrimination (KED) can reduce these interferences to a certain level, but if the requirement is for ultratrace determinations of these analytes, reaction chemistry is the better option. Pure ammonia works extremely well to reduce the $^{35}Cl^{16}O$ interference when determining trace levels of vanadium, while oxygen gas is best suited for the determination of low levels of arsenic because only one mass (^{75}As) is available for quantitation. When using oxygen to determine arsenic, reaction chemistry works by implementing the mass shift mode. In this approach, the arsenic reacts rapidly with oxygen to form $^{75}As^{16}O$ at mass 91, 16 amu away from the $^{40}Ar^{35}Cl$ interference at mass 75. Because $^{40}Ar^{35}Cl$ does not react with oxygen, $^{75}As^{16}O$ is measured free of interferences at 91 amu. Figure 17.4 exemplifies this showing the conversion of ^{75}As to $^{75}As^{16}O$ as a function of oxygen flow. As the O_2 flow increases, the signal for ^{75}As decreases (lower plot), while the signal for $^{75}As^{16}O$ increases (upper plot), demonstrating complete conversion.[10]

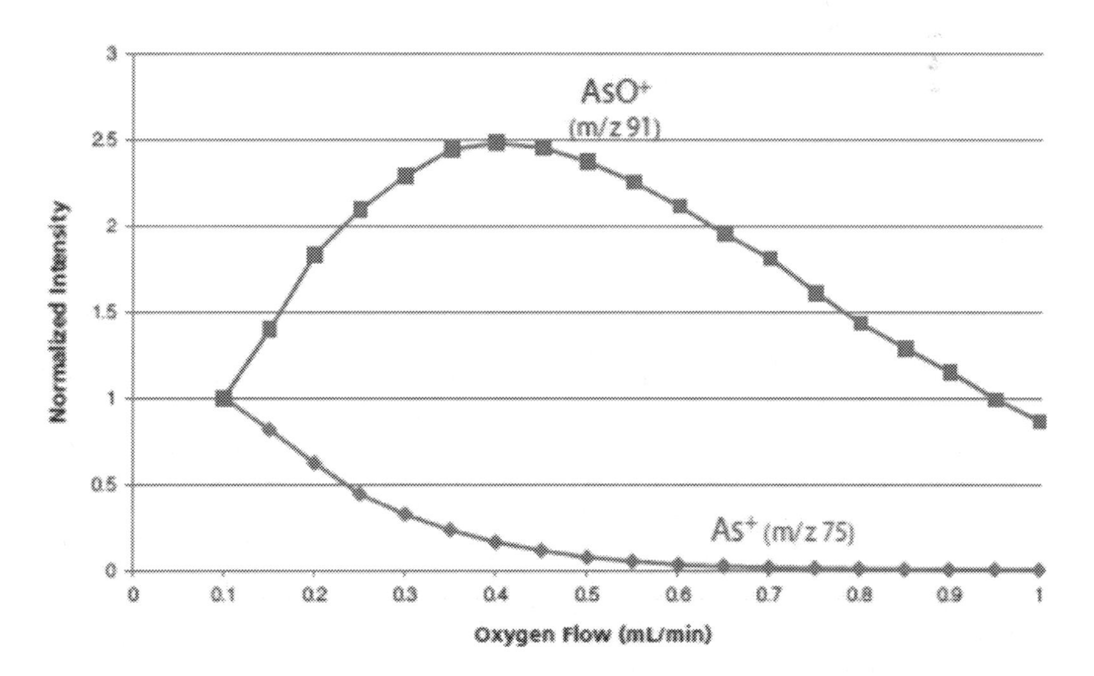

FIGURE 17.4 Conversion of ^{75}As to $^{75}As\ ^{16}O$ as a function of oxygen flow.

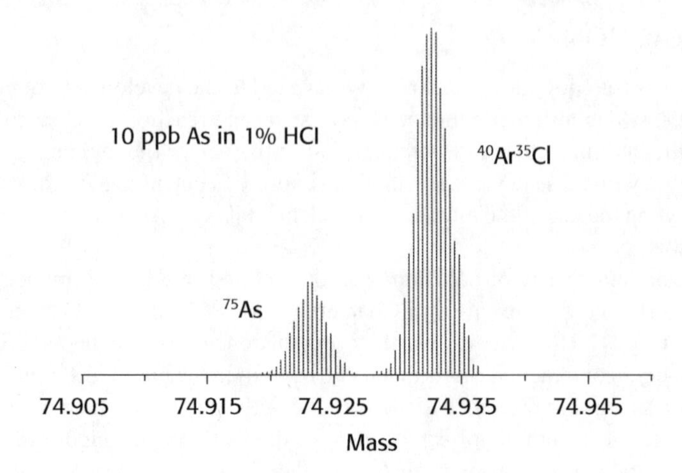

FIGURE 17.5 Separation of $^{75}As^+$ from $^{40}Ar^{35}Cl^+$ using high-resolving power (5,000) of a double-focusing magnetic sector instrument. (From W. Tittes, N. Jakubowski, and D. Stuewer, Poster Presentation at *Winter Conference on Plasma Spectrochemistry*, San Diego, 1994.)

HIGH-RESOLUTION MASS ANALYZERS

The best and probably most efficient way to remove spectral overlaps is to resolve them using a high-resolution mass spectrometer.[11] Over the past 10 years, this approach, particularly double-focusing magnetic sector mass analyzers, has proved to be invaluable for separating many of the problematic polyatomic and molecular interferences seen in ICP-MS, without the need to use cool plasma conditions or collision/reaction cells. This can be seen in Figure 17.5, which shows a spectral peak for 10 ppb of $^{75}As^+$ resolved from the $^{40}Ar^{35}Cl^+$ interference in a 1% hydrochloric acid matrix, using a resolution setting of 5,000.[12]

Although their resolving capability is far more powerful than quadrupole-based instruments, there is a sacrifice in sensitivity if extremely high resolution is used, which can often translate into a degradation in detection capability for some elements, compared to other spectral interference correction approaches. A full review of magnetic sector technology for ICP-MS is given in Chapter 11 "Mass Analyzers: Double-Focusing Magnetic Sector Technology."

MATRIX INTERFERENCES

Let us now look at the other class of interference in ICP-MS—suppression of the signal by the matrix itself. There are basically three types of matrix-induced interferences. The first and simplest to overcome is often called a *sample transport effect* and is a physical suppression of the analyte signal, brought on by the level of dissolved solids or acid concentration in the sample. It is caused by the sample's impact on droplet formation in the nebulizer or droplet size selection in the spray chamber. In the case of organic matrices, it is usually caused by variations in the pumping rate of solvents with different viscosities. The second type of matrix suppression is caused when the sample affects the ionization conditions of the plasma discharge. This results in the signal being suppressed by varying amounts, depending on the concentration of the matrix components. This type of interference is exemplified when different concentrations of acids are aspirated into cool plasma. The ionization conditions in the plasma are so fragile that higher concentrations of acid result in severe suppression of the analyte signal. This can be seen very clearly in Figure 17.6, which shows sensitivity for a selected group of elements in varying concentrations of nitric acid in a cool plasma.[9]

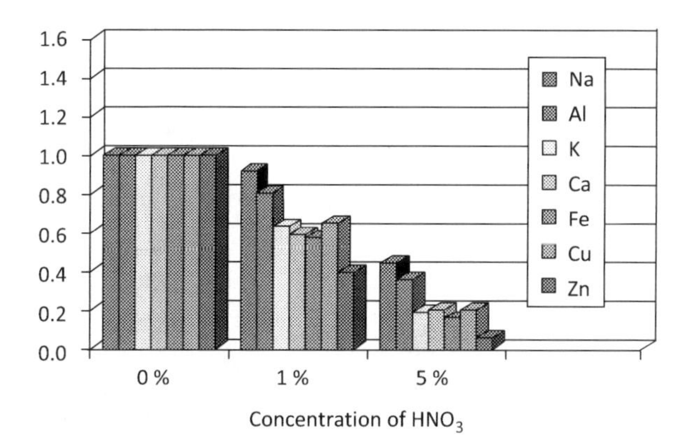

FIGURE 17.6 Matrix suppression caused by increasing concentrations of HNO_3, using cool plasma conditions (RF power: 800 W, nebulizer gas: 1.5 L/min). (From J. M. Collard, K. Kawabata, Y. Kishi, and R. Thomas, *Micro*, 54–56 2002.)

COMPENSATION USING INTERNAL STANDARDIZATION

The classic way to compensate for a physical interference is to use internal standardization (IS). With this method of correction, a small group of elements (usually at the ppb level) are spiked into the samples, calibration standards, and blank to correct for any variations in the response of the elements caused by the matrix. As the intensities of the internal standards change, the element responses are updated every time a sample is analyzed. The following criteria are typically used for selecting an internal standard:

- It is not present in the sample.
- The sample matrix or analyte elements do not spectrally interfere with it.
- It does not spectrally interfere with the analyte masses.
- It should not be an element that is considered an environmental contaminant.
- It is usually grouped with analyte elements of a similar mass range. For example, a low-mass internal standard is grouped with the low-mass analyte elements, and so on, up the mass range.
- It should be of a similar ionization potential to the group of analyte elements, so it behaves in a similar manner in the plasma.

Some of the most common elements/masses reported to be good candidates for internal standards include ^{9}Be, ^{45}Sc, ^{59}Co, ^{74}Ge, ^{89}Y, ^{103}Rh, ^{115}In, ^{169}Tm, ^{175}Lu, ^{187}Re, and ^{232}Th. An internal standard is also used to compensate for long-term signal drift as a result of matrix components slowly blocking the sampler and skimmer cone orifices. Even though total dissolved solids are usually kept below 0.2% in ICP-MS, this can still produce instability of the analyte signal over time with some sample matrices. It should also be emphasized that the difference in intensities of the internal standard elements across the mass range will indicate the flatness of the mass response curve. The flatter the mass response curve (i.e., less mass discrimination), the easier it is to compensate for matrix-based suppression effects using IS.

SPACE CHARGE-INDUCED MATRIX INTERFERENCES

Many of the early researchers reported that the magnitude of signal suppression in ICP-MS increased with decreasing atomic mass of the analyte ion.[12] More recently, it has been suggested that the major cause of this kind of suppression is the result of poor transmission of ions through the ion optics due

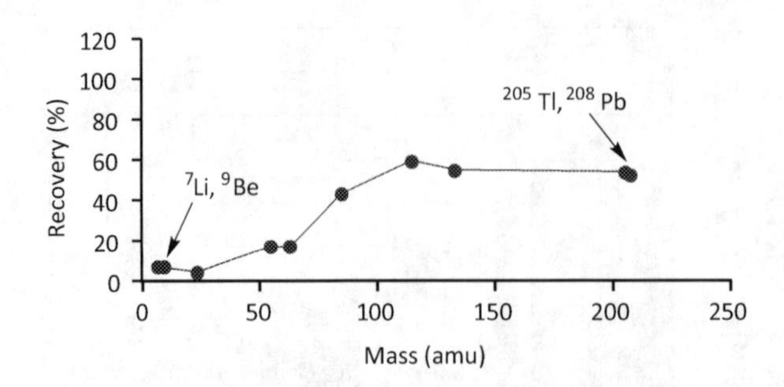

FIGURE 17.7 Space charge matrix suppression caused by 1,000 ppm uranium is significantly higher on low-mass elements such as Li and Be than it is with the high-mass elements such as Tl and Pb.[15]

to matrix-induced space charge effects.[13] This has the effect of defocusing the ion beam, which leads to poor sensitivity and detection limits, especially when trace levels of low-mass elements are being determined in the presence of large concentrations of high-mass matrices. Unless any compensation is made, the high-mass matrix element will dominate the ion beam, pushing the lighter elements out of the way.[14] This can be seen in Figure 17.7, which shows the classic space charge effects of a uranium (major isotope $^{238}U^{+}$) matrix on the determination of $^{7}Li^{+}$, $^{9}Be^{+}$, $^{24}Mg^{+}$, $^{55}Mn^{+}$, $^{85}Rb^{+}$, $^{115}In^{+}$, $^{133}Cs^{+}$, $^{205}Tl^{+}$, and $^{208}Pb^{+}$. It can clearly be seen that the suppression of the low-mass elements such as Li and Be is significantly higher than with the high-mass elements such as Tl and Pb in the presence of 1,000 ppm uranium.[15]

There are a number of ways to compensate for space charge matrix suppression in ICP-MS. IS has been used but unfortunately does not address the fundamental cause of the problem. The most common approach used to alleviate or at least reduce space charge effects is to apply voltages to individual lens components of the ion optics. This is achieved in a number of different ways, but irrespective of the design of the ion-focusing system, its main function is to reduce matrix-based suppression effects by steering as many of the analyte ions through to the mass analyzer while rejecting the maximum number of matrix ions.[15] For more details on space charge effects and different designs of ion optics, refer to Chapter 9 “Ion Focusing System.”

FURTHER READING

1. M. A. Vaughan and G. Horlick, *Applied Spectroscopy*, **41**(4), 523–530, 1987.
2. S. N. Tan and G. Horlick, *Applied Spectroscopy*, **40**(4), 445–4453, 1986.
3. D. J. Douglas and J. B. French, *Spectrochimica Acta*, **41B** (3), 197–200, 1986.
4. UIPAC Isotopic Composition of the Elements, *Pure and Applied Chemistry* **75**(6), 683–799, 2003.
5. S. N. Willie, Y. Iida, and J. W. McLaren, *Atomic Spectroscopy,* **19**(3), 67–73, 1998.
6. S. J. Jiang, R. S. Houk, and M. A. Stevens, *Analytical Chemistry,* **60**, 1217–1226, 1988.
7. K. Sakata and K. Kawabata, *Spectrochimica Acta*, **49B**, 1027–1033, 1994.
8. S. D. Tanner, M. Paul, S. A. Beres, and E. R. Denoyer, *Atomic Spectroscopy*, **16**(1), 16–22, 1995.
9. J. M. Collard, K. Kawabata, Y. Kishi, and R. Thomas, *Micro*, **20**, 39–46, 2002.
10. K. Neubauer, *Spectroscopy*, **32**(10), 22–30, 2017.
11. R. Hutton, A. Walsh, D. Milton, and J. Cantle, *ChemSA*, **17**, 213–215, 1992.
12. W. Tittes, N. Jakubowski, and D. Stuewer, Poster Presentation at *Winter Conference on Plasma Spectrochemistry*, San Diego, 1994.
13. J. A. Olivares and R. S Houk, *Analytical Chemistry*, **58**, 20–23, 1986.
14. S. D. Tanner, D. J. Douglas, and J. B. French, *Applied Spectroscopy*, **48**, 1373–1281, 1994.
15. S. D. Tanner, *Journal of Analytical Atomic Spectrometry*, **10**, 905–910, 1995.

18 Routine Maintenance

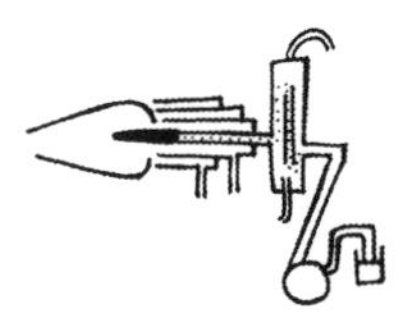

The components of an inductively coupled plasma (ICP) mass spectrometer are generally more complex than other atomic spectroscopic techniques, and as a result, more time is required to carry out routine maintenance to ensure that the instrument is performing to the best of its ability. Some tasks involve a simple visual inspection of a part, whereas others involve cleaning or changing components on a regular basis. However, routine maintenance is such a critical part of owning an inductively coupled plasma mass spectrometry (ICP-MS) system that it can impact both the performance and the lifetime of the instrument, particularly with cannabis-related samples. This chapter covers this topic in greater detail.

The fundamental principle of ICP-MS, which gives the technique its unequalled isotopic selectivity and sensitivity, also unfortunately contributes to some of its weaknesses. The fact that the sample "flows into" the spectrometer and is not "passed by it" at right angles, such as flame atomic absorption (AA) and radial inductively coupled plasma optical emission spectroscopy (ICP-OES), means that the potential for thermal problems, corrosion, chemical attack, blockage, matrix deposits, and drift is much higher than with the other atomic spectrometry (AS) techniques. However, being fully aware of this fact and carrying out regular inspection of instrumental components can reduce and sometimes eliminate many of these potential problem areas. There is no question that a laboratory which initiates a routine maintenance schedule stands a much better chance of having an instrument ready and available for analysis whenever it is needed, compared to a laboratory that basically ignores these issues and assumes the instrument will look after itself.

Let us now look at the areas of the instrument that a user needs to pay attention to. I will not go into great detail but just give a brief overview of what is important, so you can compare it with maintenance procedures of trace element techniques you are more familiar with. These areas should be very similar with all commercial ICP-MS systems, but depending on the design of the instrument and the types of samples being analyzed, the regularity of changing or cleaning components might be slightly different (particularly if the instrument is being used for laser ablation work). The main areas that require inspection and maintenance on a routine or semi-routine basis include the following:

- Sample introduction system
- Plasma torch
- Interface region
- Ion optics
- Roughing pumps
- Air/water filters.

Other areas of the instrument require less attention, but nevertheless the user should also be aware of maintenance procedures required to maximize their lifetime. They will be discussed at the end of this section.

SAMPLE INTRODUCTION SYSTEM

The sample introduction system, comprising the peristaltic/pneumatic pump, nebulizer, spray chamber, and drain system, takes the initial abuse from the sample matrix and, as a result, is an area of the ICP mass spectrometer that requires a great deal of attention. Brennan et al. published a very informative article on how best to maintain the sample introduction system.[1] The principles of the sample introduction area have been described in great detail in Chapter 6, so let us now examine what kind of routine maintenance it requires.

PERISTALTIC PUMP TUBING

If the instrument uses a peristaltic pump, the sample is pumped at about 1 mL/min into the nebulizer. The constant motion and pressure of the pump rollers on the pump tubing, which is typically made from a polymer-based material, ensure a continuous flow of liquid to the nebulizer. However, over time, this constant pressure of the rollers on the pump tubing has the tendency to stretch it, which changes its internal diameter and, therefore, the amount of sample being delivered to the nebulizer. The impact could be a change in the analyte intensity and, therefore, a degradation in short-term stability.

Therefore, the condition of the pump tubing should be examined every few days, particularly if your laboratory has a high sample workload or if extremely corrosive solutions are being analyzed. The peristaltic pump tubing is probably one of the most neglected areas, so it is absolutely essential that it be a part of your routine maintenance schedule. Here are some suggested tips to reduce pump tubing-based problems:

- Manually stretch the new tubing before use.
- Maintain the proper tension on tubing.
- Ensure tubing is placed correctly in channel of the peristaltic pump.
- Periodically check flow of sample delivery and throw away tubing if in doubt.
- Replace tubing if there is any sign of wear; do not wait until it breaks.
- With high-sample workload, change tubing every day or every other day.
- Release pressure on pump tubing when instrument is not in use.
- Pump and capillary tubing can be a source of contamination.
- Pump tubing is a consumable item—keep a large supply of it on hand.

A very useful tool to diagnose any problems associated with the peristaltic pump tubing (or the nebulizer) is a digital thermoelectric flow meter. By inserting this device in the sample line, you always know the actual rate of sample uptake to your nebulizer. This enhances the day-to-day reproducibility of your results and reduces the need to repeat measurements due to a blocked nebulizer, worn pump tubing, or incorrect clamping of the pump tube. In addition, the borosilicate glass sample path ensures that there is no memory effect or sample contamination. A commercially available digital thermoelectric flow meter is shown in Figure 18.1.

NEBULIZERS

The frequency of nebulizer maintenance will primarily depend on the types of samples being analyzed and the design of nebulizer being used. For example, in a cross-flow nebulizer, the argon gas is directed at right angles to the sample capillary tip, in contrast to the concentric nebulizer, where the gas flow is parallel to the capillary. This can be seen in Figures 18.2 and 18.3, which show schematics of a concentric and cross-flow nebulizer, respectively.

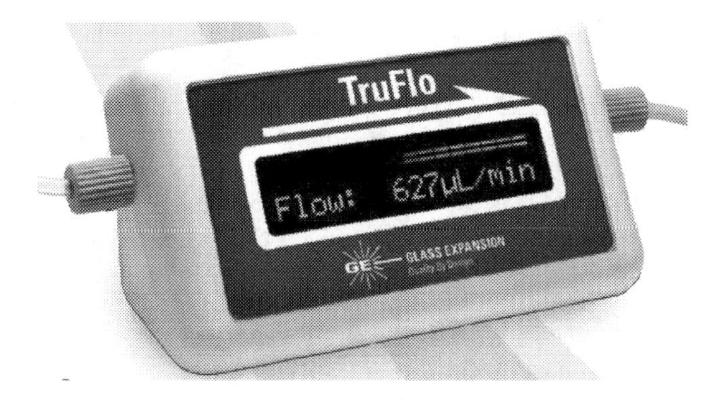

FIGURE 18.1 A commercially available digital thermoelectric flow meter to diagnose problems associated with peristaltic pump tubing and nebulizer blockages.

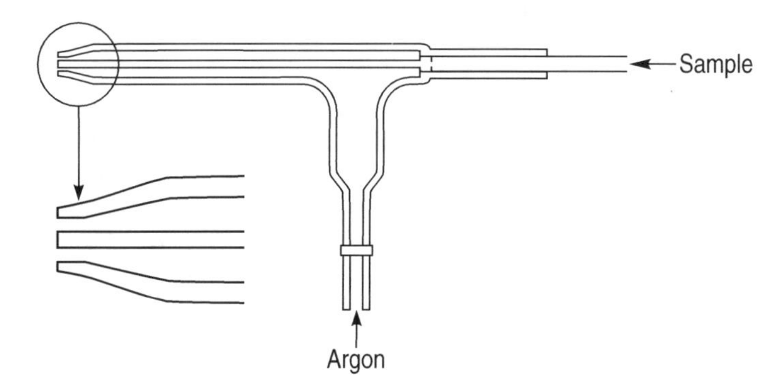

FIGURE 18.2 Schematic of a concentric nebulizer. (Courtesy of Meinhard Glass Products.)

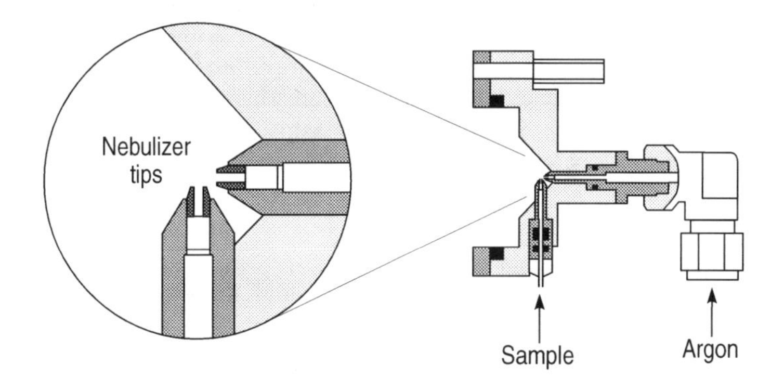

FIGURE 18.3 Schematic of a cross-flow nebulizer. (Copyright 2013, all rights reserved, PerkinElmer Inc.)

The larger diameter of the liquid capillary and longer distance between the liquid and gas tips of the cross-flow design make it far more tolerant to dissolved solids and suspended particles in the sample than the concentric design. On the other hand, aerosol generation of a cross-flow nebulizer is far less efficient than a concentric nebulizer, and therefore, it produces droplets of less optimum size than that required for the ionization process. As a result, concentric nebulizers generally produce higher sensitivity and slightly better precision than the cross-flow design but are more prone to clogging.

So, the choice of which nebulizer to use is usually based on the types of samples being aspirated and the data quality objectives of the analysis. However, whichever type is being used, attention should be paid to the tip of the nebulizer to ensure it is not getting blocked. Sometimes, microscopic particles can build up on the tip of the nebulizer without the operator noticing, which, over time, can cause a loss of sensitivity, imprecision, and poor long-term stability. In addition, O-rings and the sample capillary can be affected by the corrosive solutions being aspirated, which can also degrade performance. For these reasons, the nebulizer should always be a part of the regular maintenance schedule. Some of the most common things to check include the following:

- Visually check the nebulizer aerosol by aspirating water—a blocked nebulizer will usually result in an erratic spray pattern with lots of large droplets.
- Remove blockage by either using backpressure from argon line or dissolving the material by immersing the nebulizer in an appropriate acid or solvent—an ultrasonic bath can sometimes be used to aid dissolution, but check with manufacturer first in case it is not recommended. (Note: Never stick any wires down the end of the nebulizer, because it could do permanent damage.)
- Ensure that the nebulizer is securely seated in spray chamber end cap.
- Check all O-rings for damage or wear.
- Ensure the sample capillary is inserted correctly into the sample line of the nebulizer.
- Nebulizer should be inspected every 1–2 weeks, depending on the workload.

The digital thermoelectric flow meter described earlier is also very useful to diagnose problems with the nebulizer, even if you are using a self-aspirating nebulizer, because you are concerned about the imprecision from the pulsing of a peristaltic pump. By placing the device in-line, you always know what your sample uptake is and can take immediate corrective action if there is any change. You can also record your sample flow in order to check that you are using the same flow from day to day.

If the flow meter indicates a blocked nebulizer tip, there are also nebulizer-cleaning devices offered by most of the third-party consumables/accessories companies. Traditionally, if particulate matter from the sample lodges itself in the end of the nebulizer, cleaning wires or ultrasonic baths were the only way to remove the obstruction, which often resulted in permanent damage. These new cleaning devices are designed to efficiently deliver a pressurized cleanser through the nebulizer capillary to safely dislodge particle build-up and thoroughly clean the nebulizer, without fear of damage.

SPRAY CHAMBER

By far, the most common design of spray chamber used in commercial ICP-MS instrumentation is the double-pass design, which selects the small droplets by directing the aerosol into a central tube. The larger droplets emerge from the tube and, by gravity, exit the spray chamber via a drain tube. The liquid in the drain tube is kept at positive pressure (usually by way of a loop), which forces the small droplets back between the outer wall and the central tube; they emerge from the spray chamber into the sample injector of the plasma torch. Scott double-pass spray chambers come in a variety of shapes, sizes, and materials but are generally considered the most rugged design for routine use. Figure 18.4 shows a double-pass spray chamber (made of a polymer material) coupled to a cross-flow nebulizer.

The most important maintenance with regard to the spray chamber is to make sure that the drain is functioning properly. A malfunctioning or leaking drain can produce a change in the spray chamber backpressure, producing fluctuations in the analyte signal, resulting in erratic and imprecise data. Less frequent problems can result from degradation of O-rings between the spray chamber and

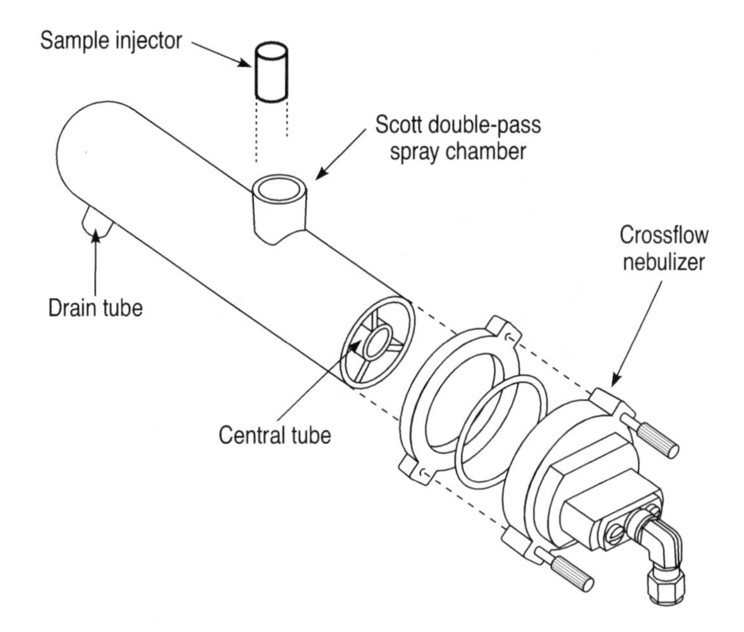

FIGURE 18.4 A double-pass spray chamber coupled to a cross-flow nebulizer. (Copyright 2013, all rights reserved, PerkinElmer Inc.)

sample injector of the plasma torch. Typical maintenance procedures regarding the spray chamber include the following:

- Make sure the drain tube fits tightly and there are no leaks.
- Ensure the waste solution is being pumped from the spray chamber into the drain properly.
- If a drain loop is being used, make sure the level of liquid in the drain tube is constant.
- Check O-ring or ball joint between spray chamber exit tube and torch sample injector—ensure the connection is snug.
- The spray chamber can be a source of contamination with some matrices/analytes, so flush thoroughly between samples.
- Empty spray chamber of liquid when instrument is not in use.
- Spray chamber and drain maintenance should be inspected every 1–2 weeks, depending on workload.

PLASMA TORCH

Not only are the plasma torch and sample injector exposed to the sample matrix and solvent but they also have to sustain the analytical plasma at approximately 10,000 K. This combination makes for a very hostile environment and, therefore, is an area of the system that requires regular inspection and maintenance. A plasma torch positioned in the radio frequency (RF) coil is shown in Figure 18.5.

As a result, one of the main problems is staining and discoloration of the outer tube of the quartz torch because of heat and the corrosiveness of the liquid sample. If the problem is serious enough, it has the potential to cause electrical arcing. Another potential problem area is blockage of the sample injector due to matrix components in the sample. As the aerosol exits the sample injector, desolvation takes place, and the sample changes from small liquid droplets to minute solid particles prior to entering the base of the plasma. Unfortunately, with some sample matrices, these particles can deposit themselves on the tip of the sample injector over time, leading to possible clogging and drift. In fact, this can be a potentially serious problem when aspirating organic solvents, because

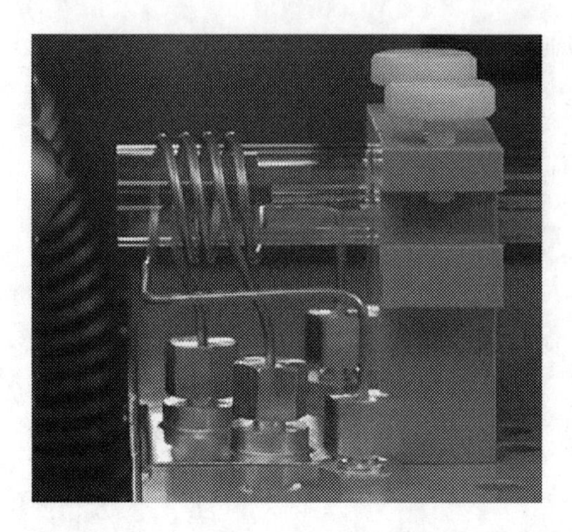

FIGURE 18.5 A plasma torch mounted in the torch box. (Courtesy of Analytic Jena.)

carbon deposits can rapidly build up on the sample injector and cones unless a small amount of oxygen is added to the nebulizer gas flow. Some torches also use metal plates or shields to reduce the secondary discharge between the plasma and the interface. These are consumable items, because of the intense heat and the effect of the RF field on the shield. A shield in poor condition can affect instrument performance, so the user should always be aware of this and replace it when necessary.

Some useful maintenance tips with regard to the torch area include the following:

- Look for discoloration or deposits on the outer tube of the quartz torch. Remove material by soaking the torch in appropriate acid or solvent if required.
- Check torch for thermal deformation. A nonconcentric torch can cause loss of signal.
- Check sample injector for blockages. If the injector is demountable, remove the material by immersing it in an appropriate acid or solvent if required (if the torch is one piece, soak the entire torch in the acid).
- Ensure that the torch is positioned in the center of the load coil and at the correct distance from the interface cone when replacing the torch assembly.
- If the coil has been removed for any reason, make sure the gap between the turns is correct as per recommendations in the operator's manual.
- Inspect any O-rings or ball joints for wear or corrosion. Replace if necessary.
- If a shield or plate is used to ground the coil, ensure it is always in good condition; otherwise, replace when necessary.
- The torch should be inspected every 1–2 weeks, depending on the workload.

INTERFACE REGION

As the name suggests, the interface is the region of the ICP mass spectrometer where the plasma discharge at atmospheric pressure is "coupled" to the mass spectrometer at 10^{-6} torr by way of two interface cones—a sampler and skimmer. This coupling of a high-temperature ionization source such as an ICP to the metallic interface of the mass spectrometer imposes demands on this region of the instrument that are unique to this AS technique. When this is combined with matrix, solvent, and analyte ions together with particulates and neutral species being directed at high velocity at the interface cones, an extremely harsh environment is the result. The most common types of problems associated with the interface are blocking or corrosion of the sampler cone and, to a lesser extent, the skimmer cone. A schematic of the interface cones showing potential areas of blockage is shown in Figure 18.6.

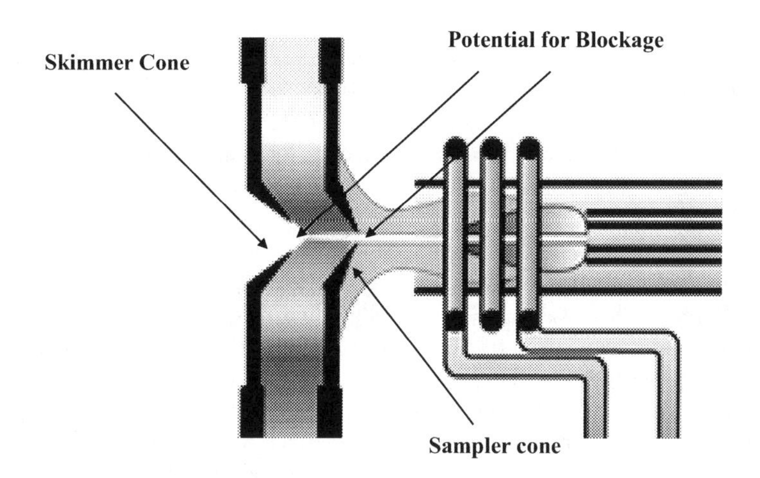

FIGURE 18.6 A schematic of the interface cones showing potential areas of blockage.

A blockage is not always obvious, because often the buildup of material on the cone or corrosion around the orifice can take a long time to reveal itself. For that reason, the sampler and skimmer interface cones have to be inspected and cleaned on a regular basis. The frequency will often depend on the types of samples being analyzed and also the design of the ICP mass spectrometer. Neufeld and coworkers wrote an excellent article on maintaining the cones and interface area to maximize productivity and uptime.[2] For example, it is well documented that a secondary discharge at the interface can prematurely discolor and degrade the sampler cone, especially when complex matrices are being analyzed or if the instrument is being used for high-sample throughput.

Besides the cones, the metal interface housing itself is also exposed to the high-temperature plasma. Therefore, it needs to be cooled by a recirculating water system, usually containing some kind of antifreeze or corrosion inhibitor or by a continuous supply of mains water. Recirculating systems are probably more widely used because the temperature of the interface can be controlled much better. There is no real routine maintenance involved with the interface housing, except maybe to check the quality of the coolant from time to time, to make sure there is no corrosion of the interface cooling system. If for any reason the interface gets too hot, there are usually built-in safety interlocks that will turn the plasma off. Some useful hints to prolong the lifetime of the interface and cones include the following:

- Check that both sampler and skimmer cone are clean and free of sample deposits. The typical frequency is weekly but will depend on sample type and workload.
- If necessary, remove and clean cones using the manufacturer's recommendations. Typical approaches include immersion in a beaker of weak acid or detergent placed in a hot water or ultrasonic bath. Abrasion with fine wire wool or a coarse polishing compound has also been used.
- Never stick any wire into the orifice; it could do permanent damage.
- Nickel cones will degrade rapidly with harsh sample matrices. Use platinum cones for highly corrosive solutions and organic solvents.
- Periodically check cone orifice diameter and shape with a magnifying glass (10–20× magnification). An irregular-shaped orifice will affect instrument performance.
- Thoroughly dry cones before installing them back in the instrument because water/solvent could be pulled back into the mass spectrometer.
- Check coolant in recirculating system for signs of interface corrosion such as copper or aluminum salts (or predominant metal of interface).

ION OPTICS

The ion optic system is usually positioned just behind or close to the skimmer cone to take advantage of the maximum number of ions entering the mass spectrometer. There are many different commercial designs and layouts, but they all have one attribute in common, and that is to transport the maximum number of analyte ions while allowing the minimum number of matrix ions through to the mass analyzer.

The ion-focusing system is not traditionally thought of as a component that needs frequent inspection, but because of its proximity to the interface region, it can accumulate minute particulates and neutral species that over time can dislodge, find their way into the mass analyzer, and affect instrument performance. Signs of a dirty or contaminated ion optic system are poor stability or a need to gradually increase lens voltages over time. For that reason, no matter what design of ion optics is used, inspection and cleaning every 3–6 months (depending on workload and sample type) should be an integral part of a preventative maintenance plan. Some useful maintenance tips for the ion optics to ensure maximum ion transmission and good stability include the following:

- Look for sensitivity loss over time, especially in complex matrices.
- If sensitivity is still low after cleaning the sample introduction system, torch, and interface cones, it could indicate that the ion lens system is becoming dirty.
- Try retuning or re-optimizing the lens voltages.
- If voltages are significantly different (usually higher than previous settings), it probably means lens components are getting dirty.
- When the lens voltages become unacceptably high, the ion lens system will probably need replacing or cleaning. Use recommended procedures outlined in the operator's manual.
- Depending on the design of the ion optics, some single-lens systems are considered consumables and are discarded after a period of time, whereas multicomponent lens systems are usually cleaned using abrasive papers or polishing compounds and rinsed with water and an organic solvent.
- If cleaning ion optics, make sure they are thoroughly dry because water or solvent could be sucked back into the mass spectrometer.
- Gloves are usually recommended when reinstalling an ion optic system because of the possibility of contamination.
- Do not forget to inspect or replace O-rings or seals when replacing ion optics.
- Depending on instrument workload, you should expect to see some deterioration in the performance of the ion lens system after 3–4 months of use. This is a good approximation of when it should be inspected and cleaned or replaced if necessary.
- With some instruments, you will need to break the vacuum to get to the ion optic region. Even though vacuum can be reestablished very quickly, this should be a consideration when carrying out your own ion lens cleaning procedures.

ROUGHING PUMPS

Typically, two roughing pumps are used in commercial instruments. One pump is used on the interface region, and the other is used as a backup to the turbomolecular pumps on the main vacuum chamber. They are usually oil-based rotary or diffusion pumps, where the oil needs to be changed on a regular basis, depending on the instrument usage. The oil in the interface pump will need changing more often than the one on the main vacuum chamber because it is pumping for a longer period. A good indication of when the oil needs to be changed is the color in the "viewing glass." If it appears dark brown, there is a good chance that heat has degraded its lubricating properties, and it needs to be changed. With the roughing pump on the interface, the oil should be changed every 1–2 months, and with the main vacuum chamber pump, it should be changed every 3–6 months.

These times are only approximations and will vary depending on the sample workload and the time the instrument is actually running. Some important tips when changing the roughing pump oil:

- Do not forget to turn the instrument and the vacuum off. If the oil is being changed from "cold," it might be useful to run the instrument for 10–15 min beforehand to get the oil to flow better.
- Drain the oil into a suitable vessel; caution, the oil might be very hot if the instrument has been running all day.
- Fill the oil to the required level in the "viewing glass."
- Check for any loose hose connections.
- Replace oil filter if necessary.
- Turn the instrument back on. Check for any oil leaks around filling cap, and tighten if necessary.

AIR FILTERS

Most of the electronic components, especially the ones in the RF generator, are air-cooled. Therefore, the air filters should be checked, cleaned, or replaced on a fairly regular basis. Although this is not carried out as routinely as the sample introduction system, a typical time frame to inspect the air filters is every 3–6 months, depending on the workload and instrument usage.

OTHER COMPONENTS TO BE PERIODICALLY CHECKED

It is also important to emphasize that other components of the ICP mass spectrometer have a finite lifetime and will need to be replaced or at least inspected from time to time. These components are not considered a part of the routine maintenance schedule and usually require a service engineer (or at least an experienced user) to clean or to change them. These areas to be cleaned are described in the following text.

THE DETECTOR

Depending on the usage and levels of ion signals measured on a routine basis, the electron multiplier should last about 12 months. A sign of a failing detector is a rapid decrease in the "gain" setting despite attempts to increase the detector voltage. The lifetime of a detector can be increased by avoiding measurements at masses that produce extremely high ion signals, such as those associated with the argon gas, solvent, or acid used to dissolve the sample (e.g., hydrogen, oxygen, and nitrogen) or any mass associated with the matrix itself. It is important to emphasize that the detector should be replaced by an experienced person wearing gloves to reduce the possibility of contamination from grease or organic/water vapor from the operator's hands. It is advisable that a spare detector be purchased with the instrument.

TURBOMOLECULAR PUMPS

Most of the instruments running today use two turbomolecular pumps to create the operating vacuum for the main mass analyzer/detector chamber and the ion optic region. However, some of the newer instruments use a single, twin-throated turbo pump. The lifetime of turbo pumps, in general, is dependent on a number of factors, including the pumping capacity of the pump (usually expressed as L/s), the size (or volume) of the vacuum chamber to be pumped, the orifice diameter of the interface cones (in mm), and the time the instrument is running. Although some instruments still use the same turbo pumps after 5–10 years of operation, the normal lifetime of a pump in an instrument that has a reasonably high sample workload is on the order of 3–4 years. This is an approximation and will obviously vary depending on the make and design of the pump (especially the type of bearings used). As the turbomolecular pump is one of the most expensive components of an ICP-MS system, this should be factored into the overall running costs of the instrument over its operating lifetime.

It is worth pointing out that although the turbo pump is not generally included in routine maintenance, most instruments use a "Penning" (or similar) gauge to monitor the vacuum in the main chamber. Unfortunately, this gauge can become dirty over time and lose its ability to measure the correct pressure. The frequency of this is almost impossible to predict but is closely related to the types and numbers of samples analyzed. A sudden drop in pressure or fluctuations in the signal are two of the most common indications of a dirty Penning gauge. When this happens, the gauge must be removed and cleaned. This should be performed by an experienced operator or service engineer because removing the gauge, cleaning it, maintaining the correct electrode geometry, and reinstalling it correctly into the instrument is a fairly complicated procedure. It is further complicated by the fact that a Penning gauge is operated at high voltage.

MASS ANALYZER AND COLLISION/REACTION CELL

Under normal circumstances, there is no need for the operator to be concerned about routine maintenance of the mass analyzer or collision/reaction cell. With modern turbomolecular pumping systems, it is highly unlikely there will be any pump contamination problems associated with the quadrupole, magnetic sector, or time-of-flight (TOF) mass analyzer. And very few sample matrix components ever make it into the mass spectrometer region, which dramatically reduces the frequency of routine maintenance tasks. This certainly was not the case with some of the early instruments that used oil-based diffusion pumps, because many researchers found that the quadrupole and pre-filters were contaminated by oil vapors from the pumps. Today, it is fairly common for turbomolecular-based mass analyzers to require no maintenance of the analyzer or the collision/reaction cell quadrupole rods over the lifetime of the instrument, other than an inspection carried out by a service engineer on an annual basis. However, in extreme cases, particularly with older instruments, removal and cleaning of the quadrupole assembly might be required to get acceptable peak resolution and abundance sensitivity performance.

FINAL THOUGHTS

The overriding message I would like to leave you with on this subject is that routine maintenance cannot be overemphasized in ICP-MS. Even though it might be considered a mundane and time-consuming chore, it can have a significant impact on the uptime of your instrument. Read the routine maintenance section of the operator's manual and understand what is required. It is essential that time be scheduled on a weekly, monthly, and quarterly basis for preventative maintenance on your instrument. In addition, you should budget for an annual preventative maintenance contract under which the service engineer checks out all the important instrumental components and systems on a regular basis to make sure they are all working correctly. This might not be as critical if you work in an academic environment, where the instrument might be down for extended periods, but in my opinion, it is absolutely critical if you work in commercial laboratory, which is using the instrument to generate revenue. There is no question that spending the time to keep your ICP mass spectrometer in good working order can mean the difference between owning an instrument whose performance could be slowly degrading without your knowledge or one that is always working in "peak" condition.

However, I must give credit to the instrument designers and accessories suppliers in making today's instrumentation extremely easy to maintain and keep clean. The technique will be 35 years old in 2018, and its commercial success has been built on acceptance by the analytical community for applying the technique to truly routine, real-world analysis. The only areas that need cleaning on a regular basis are the sample introduction system and interface cones, depending on usage and sample matrices being aspirated. And many of today's instruments have alarms which can be set to remind the operator when it's time for the few preventative maintenance tasks that are required, such as oil changes and tubing replacement. Some systems will even display how many hours various components have been used and when they might need attention. So even though routine maintenance is very important, instrument vendors have put a great deal of time and effort into making this as straightforward and seamless as possible. Refer to the websites of all the instrument vendors and sample introduction and accessories companies in Chapter 32 on useful contact information for troubleshooting hints and tips.

FURTHER READING

1. R. Brennan, J. Dulude, R. Thomas, Approaches to maximize performance and reduce the frequency of routine maintenance in ICP-MS, *Spectroscopy Magazine*, Volume 30, Issue 10, 12–25, 2015.
2. L. Neufeld, ICP-MS interface cones: Maintaining the critical interface between the mass spectrometer and the plasma discharge to optimize performance and maximize instrument productivity, *Spectroscopy Magazine*, Volume 34, Issue 7, 12–17, 2019.

19 Sampling and Sample Preparation Techniques

This chapter gives an overview of the fundamental principles of sampling and sample digestion procedures for trace element analysis, as well as focusing on sample preparation approaches for analyzing cannabis and cannabis-related products by plasma spectrochemical techniques. It has been written with a great deal of knowledge learned from the pharmaceutical and dietary/herbal supplement industries.[1] However, botanical-type samples such as cannabis and hemp are notoriously heterogeneous and can vary from one flower and one plant to the next, especially with regard to their heavy metal content. For that reason, it is critical to take enough (and at different locations) of the sample lot for testing, so it is representative of the entire batch being processed. In addition, cannabis and hemp contain high levels of organic components and, as a result, are not straightforward to achieve complete dissolution. So unless the material under test is completely digested when presented to the instrument, an accurate measurement cannot be achieved. Only then can the analyst be assured that the data generated is indicative of the elemental contaminants in the original sample matrix. This chapter discusses the many challenges being faced by the industry related to the correct sampling procedures and digestion techniques for cannabis-related materials.

With very few recognized standard methods/analytical procedures specifically written for the measurement of heavy metals in cannabis-related samples, modified United States Pharmacopeia (USP) methodology is widely used, based on USP General Chapter <233>: Elemental Impurities—Analytical Procedures.[2] This procedure was developed with pharmaceutical samples in mind, so it should be emphasized that the cannabis industry has taken this methodology and customized it for use with cannabis-related samples. The essence of this USP chapter is described below.

SAMPLE PREPARATION PROCEDURES AS DESCRIBED IN USP CHAPTER <233>

The selection of the appropriate sample preparation procedure will be dependent on the material being analyzed and is the responsibility of the analyst. The procedures described below have all shown to be appropriate for pharmaceutical and herbal supplement-based matrices, whether liquid or solid. It should also be pointed out that all liquid samples should be weighed

- Neat: This approach is applicable for liquids that can be analyzed with no sample dilution
- Direct aqueous solution: This procedure is used when the sample is soluble in an aqueous solvent
- Direct organic solution: This procedure is appropriate where the sample is soluble in an organic solvent.

- Indirect solution: This is used when a material is not directly soluble in aqueous or organic solvents. It is preferred that a total metal extraction sample preparation be carried out in order to obtain an indirect solution, such as open vessel acid dissolution or a closed vessel approach, such as microwave digestion, similar to the one described below. The sample preparation scheme should yield sufficient sample to allow quantification of each element at the elemental impurity limits specified in Chapter <232>.
- Closed vessel digestion: The benefit of closed vessel digestion is that it minimizes the loss of volatile impurities. The choice of what concentrated mineral acid to use depends on the sample matrix and its impact of any potential interferences on the analytical technique being used. An example procedure is given in the chapter, which is described below.

Weigh accurately 0.5 g of the dried sample in an appropriate flask and add 5 mL of the appropriate concentrated acids. Allow the flask to sit loosely covered for 30 min in a fume hood then add an additional 10 mL of the acid, and digest using a closed vessel technique, until digestion is complete (please follow the manufacturer's recommended procedures to ensure safe use). Make up to an appropriate volume and analyze using the technique of choice. Alternatively, a leaching extraction may be appropriate with justification following scientifically validated metal dissolution studies of the specific metal in the drug product under test.

Although not specifically mentioned in Chapter <233>, the use of a grinding mill might be beneficial for getting certain types of pharmaceutical materials into solution. Let's take a more detailed look at grinding procedures in general and how they can impact the collection of the sample.

GRINDING SOLID SAMPLES

If the sample is not in a convenient form to be dissolved, it has to be ground to a smaller particle size, mainly to improve the homogeneity of the original sample taken and make it more representative when taking a subsample. The ideal particle size will vary depending on the sample, but the sample is typically ground to pass through a fine-mesh sieve (0.1 mm^2 mesh). This uniform particle size ensures that the particles in the test portion are the same size as the particles in the rest of the ground sample. Another reason for grinding the sample into small uniform particles is that they are easier to dissolve.

The process of grinding a sample with mortar and pestle or ball mill and passing it through a metallic sieve can be time-consuming and a major cause of contamination. This can occur from the remains of a previous sample that had been prepared earlier or from materials used in the manufacture of the grinding or sieving equipment. For example, sieves, which are made from stainless steel, bronze, or nickel, can also introduce metallic contamination into the sample. In order to minimize some of these problems, plastic/polymer sieves are often used. However, still remaining is the problem of contamination from the grinding equipment. For this reason, with traditional sampling, it is usual to discard the first portion of the sample or even to use different grinding and sieving equipment for different kinds of samples.

CRYOGENIC GRINDING

Over the past few years, a new type of sample grinding techniques has become available. These grinders are based on the principle of cryogenically freezing the sample in liquid nitrogen.[3] At this kind of temperature (–196°C), many materials including pharmaceutical and cannabis materials and products become very brittle and a result can be ground to a fine powder ready for a dissolution procedure. The technology incorporates an insulated tub into which liquid nitrogen is poured. The grinding mechanism is a magnetic coil assembly suspended in the liquid nitrogen bath. Cooling materials to temperatures approaching –200°C makes samples extremely brittle, so they can be pulverized quickly by impact milling. This ability allows difficult-to-process samples (bone, rocks, polymers, metals, and food, pharmaceutical capsules) to be more efficiently processed before analysis. The sample is placed in a closed grinding vial and thoroughly cooled before grinding by the

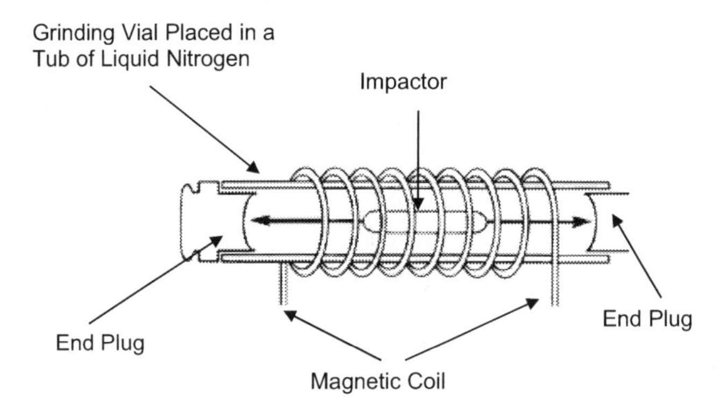

FIGURE 19.1 Schematic of a cryogenic freezer mill. (Courtesy of Spex SamplePrep.)

magnetic coil shuttling the impactor rapidly back and forth, pulverizing the sample against the end plugs of the vial, as shown in Figure 19.1. The type of vial and impactor for grinding samples is selected to reduce any potential cross-contamination of metals. For example, for pharmaceutical or cannabis samples, it might be better to use polymer vials and polymer-encased impactors instead of metal ones to reduce the possibility of metal contamination.[3]

Note: It should also be noted that if any trace element spiked additions need to be made to comply with the validation protocols, they are carried out before the digestion/sample preparation step to ensure that there is no loss of analyte or contamination from the microwave dissolution procedure. This could also indicate if any matrix suppression or enhancement effects are occurring from the dissolution acids/chemicals.

COLLECTING THE SAMPLE

Let's now take a closer look at sampling procedures. Collecting the sample and maintaining its integrity is a science of its own and is beyond the scope of this book. However, it is worth discussing briefly, in order to understand its importance in the overall scheme of collecting, preparing, and analyzing a cannabis sample for analysis. The objective of sampling is to collect a portion of the material that is small enough in size to be conveniently transported and handled and at the same time accurately represents the bulk material being sampled. Depending on the sampling requirements and the type of sample, there are basically three main types of sampling procedures that can apply to cannabis materials.

- **Random sampling** is the most basic type of sampling and only represents the composition of the bulk material at the time and place it was sampled. If the composition of the material is known to vary with time, individual samples collected at suitable intervals and analyzed separately can reflect the extent, frequency, and duration of these variations.
- **Composite sampling** is when a number of samples are collected at the same point but at different times and mixed together before being analyzed.
- **Integrated sampling** is achieved by mixing together a number of samples that have been collected simultaneously from different points.

It is not the intent of this chapter to discuss which type of sampling is the most effective, but it must be emphasized that unless the correct sampling or subsampling procedure is used, the analytical data generated by the inductively coupled plasma mass spectrometry (ICP-MS) instrumentation may be seriously flawed because it may not represent the original bulk material. If the cannabis sample is a liquid, it is also important to collect the sample in clean containers that have been thoroughly washed. In addition, if the sample is to be kept for a long period of time before analysis, it is essential that the analytes stay in solution in a preservative such as a dilute acid (this will also help stop the analytes from being absorbed into the walls of the container). It is also important to keep the samples as cool as possible (preferably refrigerated) to avoid losses through evaporation. This is especially important if mercury is being measured because mercury ions could be reduced to elemental mercury if the solutions are left for extended periods of time at room temperature before being analyzed. If this happens, the mercury can be absorbed into the walls of the containers or lost when the cap is removed. Kratochvil and Taylor give an excellent review of the importance of sampling for chemical analysis.[4]

TYPICAL SAMPLING PROCEDURES FOR CANNABIS

Let's now take a closer look at sampling procedures for cannabis-related materials. Like any plant-based, botanical sample, it's important to emphasize that the sampling of the cannabis plants/flowers is critical, particularly that the portion sampled represents the entire batch. The fundamental problem lies in the fact that cannabis is inherently heterogeneous because the hair-like trichomes in the flowers, with their sticky resin, contain the most abundant cannabinoids and terpenes. As a result, they are difficult to blend with the remainder of the plant material. Upon grinding, the rich oily consistency of the trichomes may affect homogeneity of various analytes, which makes it very challenging to replicate one sample to the next.

Sampling requirements for testing cannabis are slightly different for every state. But they all have one thing in common and that is to ensure that's what being tested is representative of what's being grown across the entire harvest. Below is a description of "Harvest Batch" sampling procedure from the California Bureau of Cannabis Control, Code of Regulations, which probably has the strictest requirements of all the states where cannabis is legal.[5] However, many of the other states have similar requirements.

A. The sampler shall obtain a representative sample from each prepacked or unpacked harvest batch. The representative sample must weigh 0.35% of the total harvest batch weight.
B. A sampler may collect a representative sample greater than 0.35% of the total harvest batch weight of a prepacked or unpacked harvest batch if necessary to perform the required testing or to ensure that the samples obtained are representative.
C. The prepacked or unpacked harvest batch from which a sample is obtained shall weigh no more than 50.0 pounds. Laboratory analyses of a sample collected from a harvest batch weighing more than 50.0 pounds shall be deemed invalid and the harvest batch from which the sample was obtained shall not be released for retail sale.
D. When the sampler obtains a representative sample from an unpacked harvest batch, the sampler shall do all the following:
    1. Collect the number of sample increments relative to the unpacked harvest batch size as listed in Table 19.1.

TABLE 19.1
Guidance on Sampling Increments for Harvested Cannabis in the State of California

Unpacked Harvest Batch Size (Pounds)	Number of Increments (per Sample)
≤10.0	8
10.1–20.0	16
20.1–30.0	23
30.1–40.0	29
40.1–50.0	34

2. Obtain sample increments from random and varying locations of the unpacked harvest batch, both vertically and horizontally. To the extent practicable, the sample increments obtained from an unpacked harvest batch shall be of equal weight.
3. To the extent practicable, collect an equal number of sample increments from each container if the unpacked harvest batch is stored in multiple containers.

SAMPLE DISSOLUTION

Let's now take a closer look at how the digestion technique described in USP Chapter <233> can be optimized for different cannabis sample types, including edibles, foods, cookies, tablets, capsules, patches, supplements, oils, and vaping liquids. If the sample under investigation is in a liquid form (such as oils, vaping liquids, or tinctures), it can be analyzed by direct aspiration or perhaps by simply diluting in an aqueous or organic solvent (see section on sample introduction techniques). However, with many cannabis-related samples, this will not be the case, particularly with the dried/milled flowers, edibles, or tablet formulations. So if the sample is a solid, it will have to be brought into solution either via a hot plate dissolution technique using concentrated mineral acids or a closed vessel, microwave digestion procedure.

REASONS FOR DISSOLVING SAMPLES

Sample dissolution using acid digestion techniques can add a significant amount of time to the overall analytical procedure. For that reason, it's important to fully understand the benefits of working with a solution, which are outlined below:

- Solid sampling analytical techniques are notoriously prone to sampling inhomogeneity—taking multiple portions of the solid material under test and dissolving them represent the best option for working with a homogeneous sample.
- Solution-based analytical techniques need a homogenous sample, which is representative of the sample matrix under test—taking a one-off solid sample such as a tablet formulation, a cookie, or gummy bear can produce erroneous results, because it may not be truly representative of the batch of samples.
- With notoriously inhomogeneous and varied botanical samples like cannabis flowers, it is important that the portion of the digested solid sample being presented to the instrument is representative of the bulk sample. For that reason, it is critical that enough of the bulk material is sampled, so that increments can be taken at random locations throughout the bulk sample.
- Measurements take a finite amount of time where the signal must stay constant—dissolving the sample and obtaining a clear solution is the best way to achieve signal stability.

DIGESTED SAMPLE WEIGHTS

It's also important to understand that the sample weight and final volume will be dictated by the expected contamination levels and total dissolved solids' (TDS) limitations of the instrumental technique being used. Typical sample weights for cannabis-related matrices are in the order of 0.2–0.5 g, based on the type of sample and mineral acid used. However, it's fair to say that when using microwave digestion procedures, the dilution factor used in the sample preparation step will ultimately have an impact on the ability of the technique to detect the expected contamination levels. So it is inevitable that there will have to be a certain level of compromise with the dissolution of the sample, based on the level of acceptable TDS for the analytical technique, particularly if ICP-MS is being used, to ensure that the analyte is measurable above the limit of quantitation (LOQ) of the instrument.

MICROWAVE DIGESTION CONSIDERATIONS

Even though USP Chapter <233> addresses different types of dissolution techniques, it actually recommends the use of closed-vessel microwave digestion for pharmaceutical raw materials, drug materials, and dietary/herbal supplements in order to completely destroy and solubilize the sample matrix. Microwave digestion systems are commonly used for trace elemental analysis studies in a multitude of application areas to get the samples into solution because they are easy to use and can rapidly process many samples at a time, which makes them ideally suited for high-sample throughput environments.

WHY USE MICROWAVE DIGESTION

So let's remind ourselves why closed or pressurized microwave digestion offers the best way to get cannabis samples into solution:

- Dissolution temperatures above boiling point of the solvent can be achieved
- The oxidation potential of reagents is higher at elevated temperatures, which means digestion is faster and more complete
- Under these conditions, concentrated nitric acid and/or hydrochloric can be used for the majority of pharmaceutical, herbal, or cannabis-related materials
- Microwave dissolution conditions and parameters can be reproduced from one sample to the next
- Safe for laboratory personnel, as there is less need to handle hot acids
- Samples can be dissolved very rapidly
- The digestion process can be fully automated
- High-sample throughput can be achieved
- Less hazardous fumes in the laboratory.

Typically, 0.2–0.5 g of sample is weighed and placed into a plastic vessel along with the appropriate acids, which are then sealed with a tight fitting cap to create a pressurized environment. Once samples are digested, which takes 10–30 min, depending on the matrix, the resulting liquid is then transferred to a volumetric flask and made up to the required volume using high-purity water. Note: In its heavy metals testing regulations, the state of California requires a minimum of 0.5 g of sample (dried to constant weight) to be digested, which is on the high side for a microwave digestion procedure, so care should be taken to ensure the microwave digestion equipment can handle the increase in pressure from this amount of sample (refer to "Commercial Microwave Technology" section later in this chapter about commercial systems).

CHOICE OF ACIDS

The choice of acids used for the preparation of digested samples is also important. Typically concentrated nitric and/or hydrochloric acids are used in various concentrations, depending on the sample type. The presence of hydrochloric acid is useful for stabilization of many elements including mercury but can sometimes produce insoluble chlorides, particularly if there is any silver in the sample. The presence of chloride can also be detrimental when ICP-MS is the chosen technique as the chloride ions combine with other ions in the sample matrix and the argon plasma to generate polyatomic spectral interferences. An example of this is the formation of the $^{40}Ar^{35}Cl$ polyatomic ion in the determination of ^{75}As and $^{35}Cl^{16}O$ in the determination of ^{51}V. These polyatomic interferences can usually be removed by the use of collision or reaction cell (CRC) technology if the ICP-MS system offers that capability. However, CRC can slow the analysis down because stabilization times have to be built into a multielement method to determine analytes that require both cell and non-cell conditions.

Nitric acid and hydrogen peroxide are often used for the dissolution of organic matrices as they are both strong oxidizing agents that effectively destroy the organic matter. In some cases, hydrofluoric acid (HF) may need to be used to dissolve certain silicate- or titanium dioxide-based materials, particularly if fillers or excipients have been used in a final tablet formulation or lotions or creams are being analyzed. In cases where HF is required, specialized plastic (PTFE) sample introduction components need to be used, including the use of buffering agents like boric acid to dissolve insoluble fluorides and neutralize excess HF. It should be emphasized that HF is a highly corrosive acid and extreme caution should be taken whenever it is used.[6]

However, it's important to understand that the more complex the sample preparation, the longer the analytical procedure will become, which will have a negative impact on the overall analysis time, particularly in a lab with a high-sample workload. In addition, the sample preparation steps could potentially have an effect on the overall TDS levels, so it is important to consider this when looking at the preparation of these samples.

COMMERCIAL MICROWAVE TECHNOLOGY

There are a number of different microwave digestion technologies on the market. Depending on the types and variety of samples being digested and the degree of automation, they have their own strengths and weaknesses. However, if a suitable digestion procedure has been optimized for a particular type of pharmaceutical or cannabis-related material, the digestion chemistry should be applicable to any type of systems. For example, many pharmaceutical, herbal supplements, or cannabis-related matrices can be digested by taking 0.2–0.5g of sample, adding 10mL of acid mixture (9mL HNO_3 and 1mL of HCl), and placing in the microwave digestion vessel. Depending on the sample and program, it can be fully dissolved to a clear/colorless liquid in approximately 15–20min.[7,8]

So let's take a closer look at the major types of microwave digestion technology on the market and specifically outline the differences for cannabis-related matrices. The optimum choice will often depend on the workload and sample diversity of the cannabis testing lab carrying out the analysis.[9,10] This section will describe the basic principles of the major types of microwave digestion technology and offer suggestions as to which might be the best approach, based on sample matrix, digestion efficiency, sample throughput, productivity, and overall cost of analysis.

DIGESTION STRATEGIES FOR CANNABIS

With increased state regulations, cannabis growers are required to conduct trace metals testing to ensure a safe and high-quality product. This testing encompasses a wide variety of samples from growers and the processing industry including soils, fertilizers, plant material, edible products, concentrates, and topicals. Obtaining analytical data required to ensure quality products starts with the crucial step of preparing the sample for analysis. Reducing handling steps, eliminating outside contamination, and minimizing reagent blank contribution are all necessary for good sample preparation. It is well recognized that closed-vessel microwave digestion offers the best approach for getting your samples into solution for analysis by ICP optical emission spectrometry (OES) or ICP-MS. However, there are basically three different commercially available designs.

- Sequential systems
- Rotor-based systems
- Single Reaction Chamber (SRC) technology.

So how do you go about selecting the optimum technology for your sample workload? What types of mineral acids will be best suited for your elements of interest, and what temperature and pressure will be required for the digestion process of your sample matrices. It's only when you have a good understanding of these issues, can you begin to look more closely at the pros and cons of the different commercially available microwave technology. So first, let's take a closer look at the fundamental principles of microwave digestion.

FUNDAMENTAL PRINCIPLES OF MICROWAVE DIGESTION TECHNOLOGY

First, it should be emphasized that sample digestion of organic, botanical matrices such as cannabis and its many products should be carried out using reagents compatible with ICP OES and ICP-MS instrumentation. For example, the chemical/physical properties and concentration of the mineral acids used and how they affect the sample introduction nebulization processes and the potential matrix suppression effects in the plasma should be taken into consideration. It is therefore well recognized that the most plasma spectrochemical-friendly reagents are typically strong oxidizing agents such as nitric acid (HNO_3) and hydrogen peroxide (H_2O_2), which are extremely efficient, but tend to generate large amounts of carbon dioxide (CO_2) and various oxides of nitrogen (NO_x) when they react with the samples. The microwave system and its components will therefore not only need to accommodate the high temperature required to digest all the different organic sample types but also be able to handle the subsequent increase in pressure produced by the generation of large volumes of these gases. For some samples, the addition of small amounts of hydrochloric acid will also help to stabilize some elements particularly mercury (Hg) and the platinum group elements. However, it should be noted that if ICP-MS is being used as the analytical technique, the $^{40}Ar^{35}Cl$ polyatomic species could potentially interfere with monoisotopic arsenic (As) at 75 atomic mass units (amu). This can be alleviated using a CRC, but it is important to be aware of this, so that the optimum instrumental conditions are used.

SEQUENTIAL SYSTEMS

The microwave cavity used in this technology produces an extremely homogeneous focused microwave field, which enables the system to digest samples faster and with greater reproducibility than batch style systems. However, it carries out the digestion in a sequential manner, one sample type at a time, which obviously limits their sample throughput capabilities. To carry out multi-sample analysis, they can be automated to move samples into the cavity automatically when the digestion of the previous sample has been completed. Another limitation of this approach is that because of their inherent design and the sample vials used, they operate at a much lower pressure, which means the achievable temperature is much lower than with other microwave approaches.

ROTOR-BASED TECHNOLOGY

With rotor-based technology, microwaves are directed onto vessels containing the sample and the digestion reagents, which are placed in a rotating carousel. The digestion process is accomplished by raising the pressure and temperature through microwave irradiation, as the carousel is rotating. This increase in temperature and pressure, together with the optimum reagent, increases both the speed of thermal decomposition of the sample and the solubility of metals in solution. Digestion conditions inside the vessels are monitored using fiber optic technology, infrared temperature sensors, and/or internal pressure controls to ensure the samples are digesting in a safe and controlled manner. A typical rotor-based microwave digestion system is shown in Figure 19.2.

FIGURE 19.2 A sample carousel being loaded into a rotor-based microwave digestion system. (Courtesy of CEM Corp, Mathews, NC.)

Rotor-based systems work extremely well for similar matrices by batching all the samples together that react in the same way. By carrying out the digestion process using the same microwave power/temperature/pressure conditions, it will ensure similar digestion quality in all positions. To increase throughput, different sized carousels can be used depending on the sample workload. However, when many different sample matrices have to be digested, productivity could be sacrificed, because each sample type has to be batched together, which unfortunately limits the ability to digest completely different samples together in the same sample run.

However, recent developments with vessel design, sensor technology, temperature measurement/control as well as improved software algorithms have meant that power levels can be optimized so that each individual sample vessel reaches a similar temperature. This means that different matrices can be mixed in the same batch to achieve an efficient digestion. This method of control is now becoming standard with rotor-based systems and has enhanced the capability of this technology from earlier designs.[11,12]

Traditional rotor-based systems can achieve relatively high pressures of around 100 bar but cannot really go much higher because of their design. For that reason, the strength of any rotor-based system is the ability to safely vent the excess pressure caused by the buildup of CO_2 and NO_x during the digestion process. There are basically three different approaches used to carry out this process. They are:

- Burst disk
- Vent and reseal
- Self-regulating.

Let's take a closer look at how they work.

Burst disk: This is the simplest method, which employs a burst disk in the cap designed to fail in an over-pressure situation, instantly releasing all pressure in the vessel. There is no danger of liquid or fumes escaping from the microwave, so it is completely safe for people working in the area around the microwave. However, when this happens, the sample contents are usually lost and the run has to be manually stopped. The result is that clean-up of the cavity is typically required, depending on the extent of the spill. There is also a strong possibility that corrosion of internal components will occur if high-concentration mineral acids are used for sample digestion.

Vent and reseal: This type of technology eliminates vessel failure in the case of an out-of-control exothermic reaction. In the "vent-and-reseal" method, the vessel cap is held in place by a dome-shaped spring. In the case of over pressure due to a highly exothermic reaction, the spring is flattened, allowing the cap to lift up slightly releasing excess pressure. Immediately, the excess pressure is released, the spring reseals the vessel, the digestion continues, and the microwave program continues to completion with no loss of sample.

Self-regulating: This third approach is typically used in high-throughput rotors, which were developed to address the needs of labs that process larger sample volumes on a routine basis. Self-regulating vessels are very easy to assemble/disassemble and rely on the Teflon sealing plug inside the cap deforming to release pressure. Their compact design results in more moderate temperature and pressure capabilities but allows for a large number of vessels to fit onto the rotor. These characteristics make self-regulating rotors ideal for high-workload labs with relatively straightforward applications such as clinical, environmental, and food. Most recently, the self-regulating approach has been applied to high-pressure rotors designed specifically for more challenging sample types. Since self-regulating vessels are designed to vent, incomplete digestions may be observed given that the pressure loss does not allow the required temperatures necessary for complete digestion to be achieved.

SINGLE REACTION CHAMBER TECHNOLOGY

So let's take a more detailed look at the principles of SRC technology and how it differs from the rotor-based system.[13] Instead of a rotor with discrete sample vessel, the samples are put into vials with loose fitting caps which are sitting in a rack that is lowered into a larger vessel containing a base load of acidified water. It's this base load that absorbs the microwave energy and transfers it to the vial. This allows every vial to react independently within the base load and ensures that all samples can reach a maximum pressure of up to 200 bar and achieve a digestion temperature of around 300°C. This means that no batching of samples is necessary, and any combination of sample type and acid chemistry can be run simultaneously in the same chamber. In addition, this typically allows SRC technology to use higher sample weights than the other approaches.

This high-pressure/high-temperature capability is particularly useful for plant-type samples like cannabis, because of their high carbon content. At lower digestion temperatures, it is well recognized that the oxidizing conditions are not aggressive enough to completely digest many organic-based samples. As a result, they often have a cloudiness to them, which indicates the carbonaceous material is not completely in solution. This is problematic for not only the buildup of carbon on the interface cones but also the negative impact of carbon-based polyatomic interferences on many analyte elements by ICP-MS.

A schematic of the SRC is shown on the left in Figure 19.3, while an actual photograph of sample vials being lowered into the chamber is shown on the right.

As previously mentioned, loose fitting caps are used to seal the vials. This is possible because they are pre-pressurized with 40 bar of nitrogen prior to the start of the microwave program, which acts as a gas cap and keeps all the vials independently closed. As the pressure builds, equilibrium is achieved both inside and outside the vial. As a result, a variety of vial types including disposable glass, quartz, and Teflon or another material in any combination can be used.

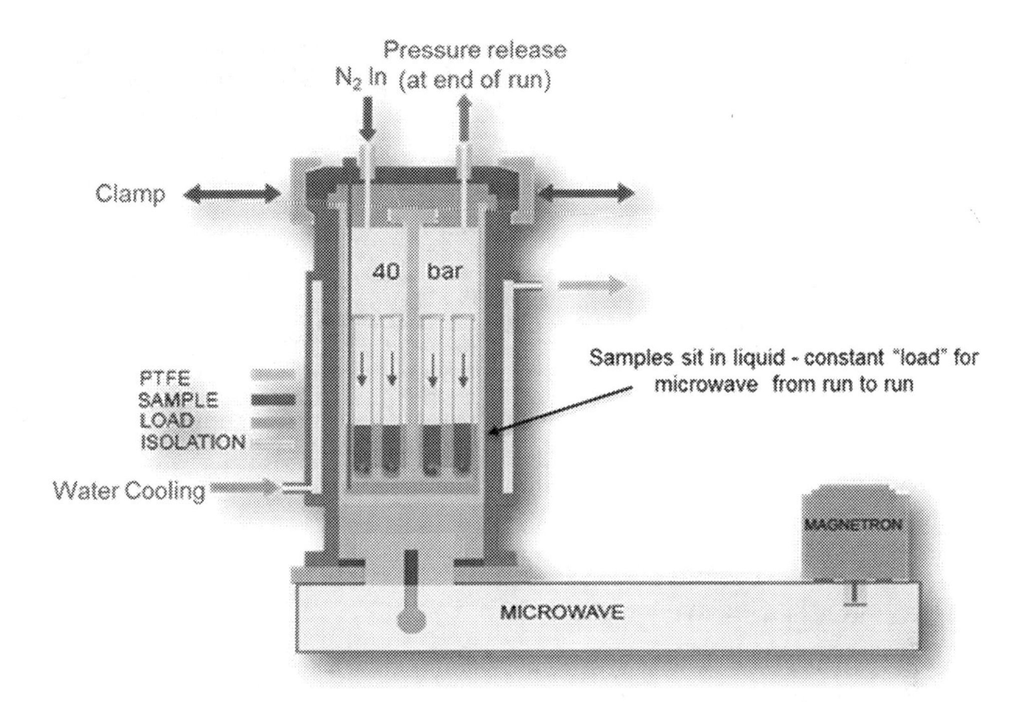

FIGURE 19.3 Single reaction chamber microwave digestion technology. (Courtesy of Milestone Inc. Shelton, CT.)

NITROGEN-PRESSURIZED CAPS

To exemplify that 40 bar of nitrogen is sufficient to seal the vials, an experiment was carried out where vials containing 110 ppm of mercury were placed right next to blank solutions in a 15-rack sample holder. In other words, every alternate sample vial in the rack was either 110 or 0 ppm mercury. It is well recognized that mercury is highly volatile, particularly when heated and would contaminate any surrounding vessels if not capped tightly. However, it can be seen in Figure 19.4 that the measurement of every alternate blank sample by ICP-MS is actually at the limit of detection for the technique, which is a clear indication that SRC technology using pressurized nitrogen gas as a sealant eliminates the potential of cross-contamination in the sample chamber.

Every lab's trace element analysis workload and sample digestion requirements are different. Rotor-based systems work extremely well, but their main limitation has traditionally been that they require batching of similar matrices and chemistries, because control of the power and therefore temperature is based on the reaction of one vessel at a time. By batching similar samples, under-digestion of some samples can be minimized due to the pressure and temperature required by others. In addition, with rotor-based systems, the vessels are typically made from PTFE-type materials, which put limits on the achievable maximum pressure and therefore the available digestion temperature.

It's important to emphasize that rotor-based technology is perfectly adequate for digesting most routine-type cannabis samples. However, if the lab knows it will be handling an extremely wide variety of different sample types, especially if they are very difficult to get into solution, SRC technology offers some tangible benefits in being able to digest many different matrices (and using larger weights) at the same time, particularly in a high-sample throughput environment. Unfortunately, the price tag of an SRC system might be prohibitive for many smaller testing labs that don't have the sample diversity and throughput to justify the higher cost.

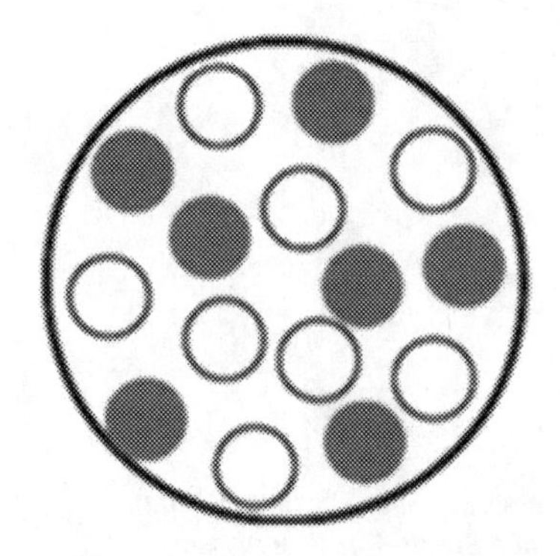

(15 Position) White vials: blanks.
Red vials: 110 ppm Hg

	Hg (ppt)
Bk position 01	0.020
Bk position 03	0.032
Bk position 05	0.001
Bk position 07	0.002
Bk position 09	-0.003
Bk position 11	-0.007
Bk position 13	-0.007
Bk position 15	-0.006

FIGURE 19.4 Study of sealing capability of nitrogen gas shows no sign of contamination from mercury.

SAMPLING PROCEDURES FOR MERCURY

It's also worth pointing out that element-specific sample preparation techniques might also be necessary. For example, when preparing samples for the determination of mercury, care must be taken not to lose the analyte because of its volatility. This is especially relevant when carrying out microwave digestion. Some of the steps taken to minimize losses of mercury include the use of hydrochloric acid in the dissolution step to produce an excess of chloride ions or the addition of gold (typically a few ppm) to stabilize the mercury in solution.[14]

Under the right chemistry conditions, mercury can also be determined by the cold vapor (CV) generation technique. This technique is used in commercially available mercury analyzers, where the mercury in solution is reduced to its atomic state and the elemental mercury vapor is detected using either atomic absorption (AA) or atomic fluorescence. These instruments are extremely sensitive and are capable of carrying out both the chemistry and detection steps online, in an automated manner. In conventional mercury analyzers, the samples must either be liquids or brought into solution using a dissolution step, which limits their applicability for the analysis of solid or powdered pharmaceutical materials. However, there is a variation of this method, which can handle solids directly. In this approach, the sample is first heated up to 900°C in a combustion furnace to volatilize the sample, then swept into a catalyst to release the mercury vapor, and concentrated onto the surfaces of a gold amalgamation trap, where it is eventually heated and swept into an AA for detection and quantitation.[15]

REAGENT BLANKS

The increasing demand for lower and lower detection limits is a major consideration for any laboratory-performing trace metals analysis, particularly in the cannabis industry, where maximum contaminant limits are extremely low (and probably going lower). The development of sophisticated instrumentation, such as ICP-OES and ICP-MS, provides the capability to achieve limits of detection previously unobtainable. However, the analytical instrument is just one component in achieving the lowest limits of detection using plasma spectrochemistry.

The other critical area that puts significant demands on a laboratory's overall detection capability is to ensure that the sample preparation procedure does not contribute any additional sources of contamination. There are several factors to consider when looking to minimize contamination and reduce reagent blank levels when preparing samples for analysis by plasma spectrochemistry, including general lab cleanliness (reducing atmospheric dust), vessel cleanliness (acid washing), reagent choice (use of high-purity acids and chemicals), quality of materials, and consumables (low in trace metals), not to mention the digestion procedure itself. All these factors make a contribution to achieving the lowest possible detection capability when carrying out ultratrace elemental determinations by ICP-OES and ICP-MS. However, this not something many lab personnel necessarily think about, particularly if they do not have experience of carrying out ultratrace analysis. Unfortunately, the industry is so new that typical instrument operators only have a couple of years' experience at most and many of them just a few months of running very sophisticated instrumentation. In addition, they may have been used to running less sensitive techniques such as ICP-OES and AA, where the cleanliness requirements are not so strict. For that reason, it's always going to be challenging for operators who are not used to working in the ultratrace environment to fully understand all the analytical issues. And if that wasn't enough of a challenge, they are also probably working in labs that in most cases were not designed for ultratrace elemental analysis.

FINAL THOUGHTS

Modern plasma spectrochemical techniques will generate trace element data of the highest quality. However, with any solution technique, the data will only be as good as the sample presented to the instrument. For that reason, it is very important to ensure that the portion of the sample being analyzed is reflective of the bulk sample, especially with cannabis flowers which are typically not homogeneous in nature.

This chapter has shown that robust closed-vessel microwave digestion techniques are critically important to the overall analytical procedure. There is no question that traditional rotor-based technology demonstrates good recovery for many elements found in cannabis and related products. Highly reactive samples such as edibles, tinctures, and oils can be completely digested even in large sample amounts, ensuring reliable analysis. In addition, by using a high-pressure, rotor-based system with a multi-position carousel, a high level of reproducibility can be achieved even for volatile elements such as As and Hg.

However, for labs that might be handling a more diverse and complex range of samples such as herbal supplements, cannabis, pharmaceuticals, or environmental-type samples, a rotor-based system might be somewhat restrictive, because similar samples have to be batched together to ensure they are being digested under the optimum conditions. Therefore, if there is a need to digest many differing sample matrices simultaneously in the same run, an SRC system could be the best solution.

Finally, it's important to emphasize that the sample preparation and sample digestion steps are littered with additional sources of contamination and errors, which are exaggerated with cannabis samples because they are such a natural absorbers of elemental contaminants from the growing, manufacturing, and sample processing steps.

It is not the intent to favor one microwave digestion approach over the other. There are many companies that offer microwave digestion equipment to offer advice to help users make the right decision based on their sample diversity and workload. For more detailed application material, refer to the contact information for microwave digestion companies in Chapter 32 "Useful Contact Information," where you will find a multitude of application notes for many different types of cannabis-related materials.

REFERENCES

1. N. Lewen, Preparation of pharmaceutical samples for elemental impurities analysis: Some potential approaches, *Spectroscopy Magazine*, 31 (4), 36–43, 2016 http://www.spectroscopyonline.com/preparation-pharmaceutical-samples-elemental-impurities-analysis-some-potential-approaches.
2. USP, United States Pharmacopeia General Chapter <233> Elemental Impurities in Pharmaceutical Materials – Procedures: Second Supplement to USP 37-NF 32, 2014.

3. Spex SamplePrep, Sample Preparation Techniques for Pharmaceutical Labs: The Benefits of Cryogenic Freezing. Spex SamplePrep: Metuchen, NJ, https://www.spexsampleprep.com/freezermill-for-cryogenic-grinding
4. B. Kratochvil and J. K. Taylor, *Analytical Chemistry*, 53 (8), 925–938A, 1981.
5. BCC, California Bureau of Cannabis Control: Code of Regulations, https://bcc.ca.gov/.
6. CDC. The National Institute for Occupational Safety and Health (NIOSH): Safe Use of Hydrogen Fluoride and Hydrofluoric Acid, http://www.cdc.gov/niosh/ershdb/emergencyresponsecard_29750030.html.
7. USP, Microwave Digestion of Pharmaceutical Samples Followed by ICP-MS Analysis for USP Chapters <232> and <233>, CEM Corp. Application Note: https://cem.sharefile.com/download.aspx?id=sc6d12fed0084ab4b#.
8. S. Hussein and T. Michel, The application of single-reaction-chamber microwave digestion to the preparation of pharmaceutical samples in accordance with USP <232> and <233>, *Spectroscopy Magazine*, Special Issue, 27 (10), 2012, http://www.spectroscopyonline.com/application-single-reaction-chamber-microwave-digestion-preparation-pharmaceutical-samples-accordanc?id=&sk=&date=&pageID=3.
9. CEM, Microwave Digestion and Trace Metals Analysis of Cannabis & Hemp Products, CEM Application Note, August, 2019, http://cem.com/media/contenttype/media/literature/ApNote_MARS6_Cannabis_Foods_ap0159_2.pdf.
10. R. Boyle and E. Farrell, Selecting microwave digestion technology for measuring heavy metals in cannabis products, *Cannabis Science and Technology*, 1 (3), 2018, http://www.cannabissciencetech.com/metals/selecting-microwave-digestion-technology-measuring-heavy-metals-cannabis-products.
11. R. Lockerman et al., Sample Preparation and Trace Elemental Analysis of Cannabis and Cannabis Products, CEM Corp, Poster #ThP11, *Plasma Winter Conference*, Tucson, AZ, 2020.
12. MILESTONE, Ethos UP Product Note/Brochure, Milestone Inc., https://milestonesci.com/ethos-up-microwave-digestion-system/.
13. MILESTONE, Single Reaction Chamber, US Patent Number 5,270,010: Milestone Inc., Shelton, CT.
14. EPA, Mercury Preservation Techniques: US Environmental Protection Agency (EPA), https://www.inorganicventures.com/pub/media/wysiwyg/files/mercury_preservation_techniques.pdf
15. R. J. Thomas, The impact of illegal Artisanal gold mining on the Peruvian amazon: Benefits of taking a direct mercury analyzer into the rain forest to monitor mercury contamination, *Spectroscopy Magazine*, 2019, http://www.spectroscopyonline.com/impact-illegal-artisanal-gold-mining-peruvian-amazon-benefits-taking-direct-mercury-analyzer-rain-fo.

20 Performance and Productivity Enhancement Techniques

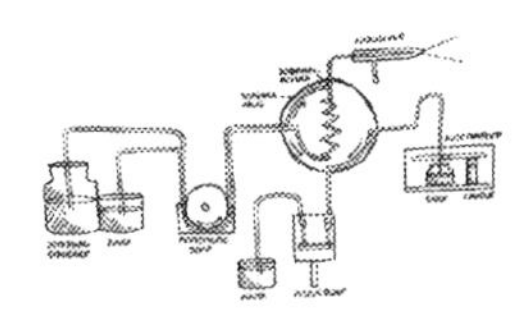

Conventional sample introduction systems using a spray chamber and nebulizer account for the majority of inductively coupled plasma mass spectrometry (ICP-MS) applications being carried out today. However, nonstandard sampling accessories such as laser ablation (LA) systems, flow injection analyzers (FIAs), electrothermal vaporizers (ETVs), cooled spray chambers, desolvation equipment, direct injection nebulizers (DINs), automated sample delivery systems, autodilutors, and online chemistry procedures are considered critical to enhancing the practical capabilities of the technique. Initially regarded as novel sampling devices, they have since proved themselves to be invaluable for solving real-world application problems by enhancing the flexibility, performance, and productivity of the technique. This chapter describes the basic principles of these accessories and gives an overview of their practical capabilities. Although not all are suitable for cannabis-type samples, desolvation devices, autodilution equipment, automated sample delivery accessories, including intelligent autosamplers and enhanced productivity sampling tools, can be very beneficial for testing labs that are looking to analyze a suite of samples in the most efficient and cost-effective manner.

It is recognized that standard ICP-MS instrumentation using a traditional sample introduction system comprising a spray chamber and nebulizer has certain limitations, particularly when it comes to the analysis of complex samples. Some of these known limitations include the following:

- Inability to analyze solids directly.
- Contamination issues with samples requiring multiple sample preparation steps.
- Liquid aerosol can impact ionization process.
- Total dissolved solids must be kept below 0.2%.
- If matrix has to be removed, it has to be done offline.
- Long washout times required for samples with a heavy matrix.
- Dilutions and addition of internal standards can be labor intensive and time-consuming.
- Matrix components can generate severe spectral overlaps on many analytes.
- The analysis of slurries is very difficult.
- Matrix suppression can be quite severe with some samples.
- Spectral interferences generated by solvent-induced species can limit detection capability.
- Organic solvents can present unique problems.
- Sample throughput is limited by the sample introduction process.
- Not suitable for the determination of elemental species or oxidation states.

Such were the demands of real-world users to overcome these kinds of problem areas that instrument manufacturers developed different strategies based on the type of samples being analyzed.

Some of these strategies involved parameter optimization or modification of instrument components, but it was clear that this approach alone was not going to solve every conceivable problem. For this reason, they turned their attention to the development of sampling accessories, which were optimized for a particular application problem or sample type. Over the past 10–15 years, this demand has led to the commercialization of specialized performance and productivity enhancement tools, manufactured not only by the instrument manufacturers themselves but also by third-party vendors specializing in these kinds of sampling techniques. The most common ones used today include the following:

- LA systems
- FIAs systems
- ETVs systems
- Chilled spray chambers and desolvation systems
- DINs systems
- Fast automated sampling procedures
- Autodilution and calibration systems
- Automated sample identification and tracking.

Although many of them are not well-suited for the analysis of pharmaceutical materials according to USP Chapter <233>, many of them are applicable to contract labs that may be analyzing different materials in addition to pharmaceutical samples. Let us take a closer look at some of these enhanced techniques to understand their basic principles and what benefits they bring to ICP-MS and ICP-OES.

LASER ABLATION

The limitation of ICP-MS to analyze solids, without dissolving the material, led to the development of LA. The principle behind this approach is the use of a high-powered laser to ablate the surface of a solid and sweep the sample aerosol into the ICP mass spectrometer for analysis in the conventional way.[1]

Before I go on to describe some typical applications suited to LA ICP-MS, let us first take a brief look at the history of analytical lasers and how they eventually became such a useful sampling tool. The use of lasers as vaporization devices was first investigated in the early 1960s. When light energy with an extremely high-power density interacts with a solid material, the photon-induced energy is converted into thermal energy, resulting in vaporization and removal of the material from the surface of the solid.[2] Some of the early researchers used ruby lasers to induce a plasma discharge on the surface of the sample and measure the emitted light with an atomic emission spectrometer.[3] Although this proved useful for certain applications, the technique suffered from low sensitivity, poor precision, and severe matrix effects caused by non-reproducible excitation characteristics. Over the years, various improvements were made to this basic design with very little success,[4] because the sampling process and the ionization/excitation process (both under vacuum) were still intimately connected and interacted strongly with each other.

This limitation led to the development of LA as a sampling device for atomic spectroscopy instrumentation, where the sampling step was completely separated from the excitation or ionization step. The major benefit is that each step can be independently controlled and optimized. These early devices used a high-energy laser to ablate the surface of a solid sample, and the resulting aerosol was swept into some kind of atomic spectrometer for analysis. Although initially used with atomic absorption[5,6] and plasma-based emission techniques,[7,8] it was not until the mid-1980s, when lasers were coupled with ICP-MS, that the analytical community sat up and took notice.[9] For the first time, researchers were coming up with evidence that virtually any type of solid could be vaporized, irrespective of electrical characteristics, surface topography, size, or shape, and be transported into the ICP for analysis by atomic emission or mass spectrometry. This was an exciting breakthrough for ICP-MS, because it meant the technique could be used for the bulk sampling of solids, or if required, for the analysis of small spots or micro-inclusions, in addition to being used for the analysis of solutions.

COMMERCIAL LASER ABLATION SYSTEMS FOR ICP-MS

The first LA systems developed for ICP instrumentation were based on solid-state ruby lasers, operating at 694nm. These were developed in the early 1980s but did not prove to be successful for a number of reasons, including poor stability, low power density, low repetition rate, and large beam diameter, which made them limited in their scope and flexibility as a sample introduction device for trace element analysis. It was at least another 5 years before any commercial instrumentation became available. These early commercial LA systems, which were specifically developed for ICP-MS, used the Nd:YAG (neodymium-doped yttrium aluminum garnet) design, operated at the primary wavelength of 1,064 nm—in the infrared (IR).[10] They initially showed a great deal of promise because analysts were finally able to determine trace levels directly in the solid without sample dissolution. However, it soon became apparent that they did not meet the expectations of the analytical community, for many reasons, including complex ablation characteristics, poor precision, non-optimization for microanalysis, and because of poor laser coupling; they were unsuitable for many types of solids. By the early 1990s, most of the LAs purchased were viewed as novel and interesting but not suited to solving real-world application problems.

These basic limitations in IR laser technology led researchers to investigate the benefits of shorter wavelengths. Systems were developed that were based on Nd:YAG technology at the 1,064nm primary wavelength but utilizing optical components to double (532nm), quadruple (266nm), and quintuple (213nm) the frequency. Innovations in lasing materials and electronic design together with better thermal characteristics produced higher energy with higher pulse-to-pulse stability. These more advanced UV lasers showed significant improvements, particularly in the area of coupling efficiency, making them more suitable for a wider array of sample types. In addition, the use of higher-quality optics allowed for a more homogeneous laser beam profile, which provided the optimum energy density to couple with the sample matrix. This resulted in the ability to make spots much smaller and with more controlled ablations irrespective of sample material, which were critical for the analysis of surface defects, spots, and micro-inclusions. Figure 20.1 shows the optical layout of a commercially available frequency-quintupled 213nm Nd:YAG LA system.

EXCIMER LASERS

The successful trend towards shorter wavelengths and the improvements in the quality of optical components also drove the development of UV gas-filled lasers, such as XeCl (308nm), KrF (248nm), and ArF (193nm) excimer lasers. These showed great promise, especially the ones that operated at shorter wavelengths and were specifically designed for ICP-MS. Unfortunately, they necessitated a more sophisticated beam delivery system, which tended to make them more expensive. In addition, the complex nature of the optics and the fact that gases had to be changed on a routine basis made them a little more difficult to use and maintain and as a result required a more skilled operator to run them. However, their complexity was far outweighed by their better absorption capabilities for UV-transparent materials such as calcites, fluorites, and silicates, smaller particle size, and higher flow of ablated material. There was also evidence to suggest that the shorter-wavelength excimer lasers exhibit better elemental fractionation characteristics (typically defined as the intensity of certain elements varying with time, relative to the dry aerosol volume) than the longer-wavelength Nd:YAG design, because they produce smaller particles that are easier to volatilize.

Even though excimer lasers are optically more complex than other designs, it's worth mentioning that today's instruments are far more rugged and robust compared to the earlier-designed systems

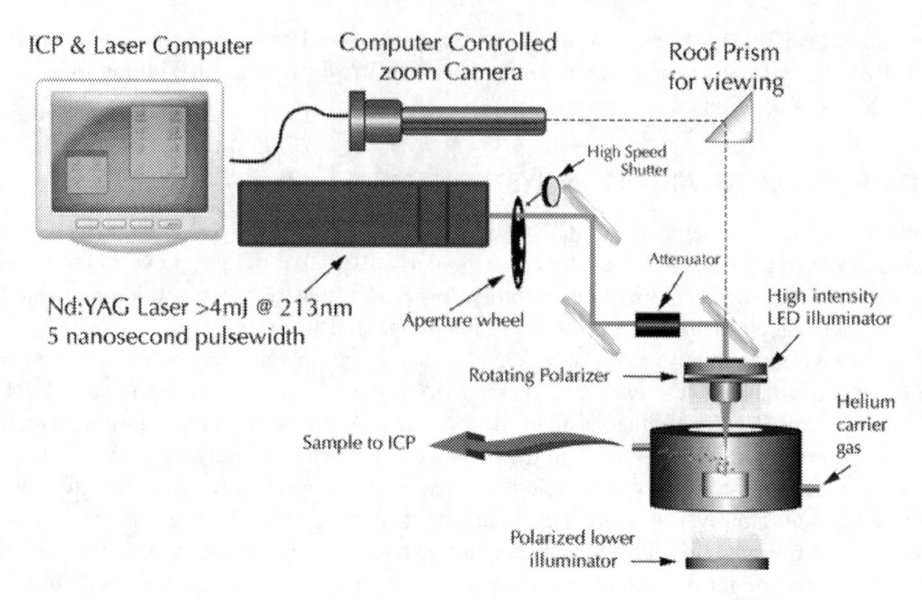

FIGURE 20.1 Schematic of a commercially available frequency-quintupled 213 nm Nd:YAG LA system. (Courtesy of Teledyne Cetac Technologies.)

and as a result are being used for more and more routine applications. And because the higher-grade optical components used in the excimer technology are available at a more realistic cost, their price has come down quite significantly in the past few years.

BENEFITS OF LASER ABLATION FOR ICP-MS

Today, there are a number of commercial LA systems on the market designed specifically for ICP-MS, including 266 and 213 nm Nd:YAG and 193 nm ArF excimer lasers. They all have varying output energy, power density, and beam profiles, and even though each one has different ablation characteristics, they all work extremely well, depending on the types of samples being analyzed and the data quality requirements. LA is now considered a very reliable sampling technique for ICP-MS, which is capable of producing data of the very highest quality directly on solid samples and powders. Some of the many benefits offered by this technique include the following:

- Direct analysis of solids without dissolution.
- Ability to analyze virtually any kind of solid material, including rocks, minerals, metals, ceramics, polymers, plastics, plant material, and biological specimens.
- Ability to analyze a wide variety of powders by pelletizing with a binding agent.
- No requirement for sample to be electrically conductive.
- Sensitivity in the ppb to ppt range, directly in the solid.
- Labor-intensive sample preparation steps are eliminated, especially for samples such as plastics and ceramics that are extremely difficult to get into solution.
- Contamination is minimized because there are no digestion/dilution steps.
- Reduced polyatomic spectral interferences compared to solution nebulization.
- Examination of small spots, inclusions, defects, or micro-features on surface of sample.
- Elemental mapping across the surface of a mineral.
- Depth profiling to characterize thin films or coatings.

Let us now take a closer look at the strengths and weaknesses of the different laser designs with respect to application requirements.

OPTIMUM LASER DESIGN BASED ON THE APPLICATION REQUIREMENTS

The commercial success of LA was initially driven by its ability to directly analyze solid materials such as rocks, minerals, ceramics, plastics, and metals, without going through a sample dissolution stage. Table 20.1 represents some typical multielement detection limits in NIST 612 glass generated with a 266 nm Nd:YAG design coupled to an ICP-MS system.[11] It can be seen that sub-ppb detection limits in the solid material are achievable for most of the elements. This kind of performance is typically obtained using larger spot sizes on the order of 100–1,000 μm in diameter, which is ideally suited to 266 nm laser technology.

However, the desire for ultratrace analysis of optically challenging materials, such as calcite, quartz, glass, and fluorite, combined with the capability to characterize small spots and micro-inclusions, proved very challenging for the 266 nm design. The major reason is that the ablation process is not very controlled and precise, and as a result, it is difficult to ablate a minute area without removing some of the surrounding material. In addition, erratic ablating of the sample initially generates larger particles (>1 μm size), which are not efficiently ionized in the plasma and therefore contribute to poor precision.[12] Even though modifications helped improve ablation behavior, it was not totally successful because of the basic limitation of the 266 nm to couple efficiently to UV-transparent materials. The drawbacks in 266 nm technology eventually led to the development of 213 nm lasers[13] because of the recognized superiority of shorter wavelengths to exhibit a higher degree of absorbance in transparent materials.[14]

Analytical chemists, particularly in the geochemical community, welcomed 213 nm UV lasers with great enthusiasm, because they now had a sampling tool that offered much better control of the ablation process, even for easily fractured minerals. This is demonstrated in Figures 20.2 and 20.3, which show the ablation differences between 266 and 213 nm, respectively, for NIST 612 glass

TABLE 20.1
Typical Detection Limits Achievable in NIST 612 SRM Glass Using a 266 nm Nd:YAG Laser Ablation System Coupled to an ICP Mass Spectrometer

Element	3σ detection limits (DLs) (ppb)	Element	3σ DLs (ppb)
B	3.0	Ce	0.05
Sc	3.4	Pr	0.05
Ti	9.1	Nd	0.5
V	0.4	Sm	0.1
Fe	13.6	Eu	0.1
Co	0.05	Gd	1.5
Ni	0.7	Dy	0.5
Ga	0.2	Ho	0.01
Rb	0.1	Er	0.2
Sr	0.07	Yb	0.4
Y	0.04	Lu	0.04
Zr	0.2	Hf	0.4
Nb	0.5	Ta	0.1
Cs	0.2	Th	0.02
Ba	0.04	U	0.02
La	0.05		

Source: Courtesy of Teledyne Cetac Technologies.

FIGURE 20.2 A 200 μm crater produced by the ablation of an NIST 612 glass SRM, using a 266 nm LA system, showing excess ablated material around the edges of the crater. (Courtesy of Teledyne Cetac Technologies.)

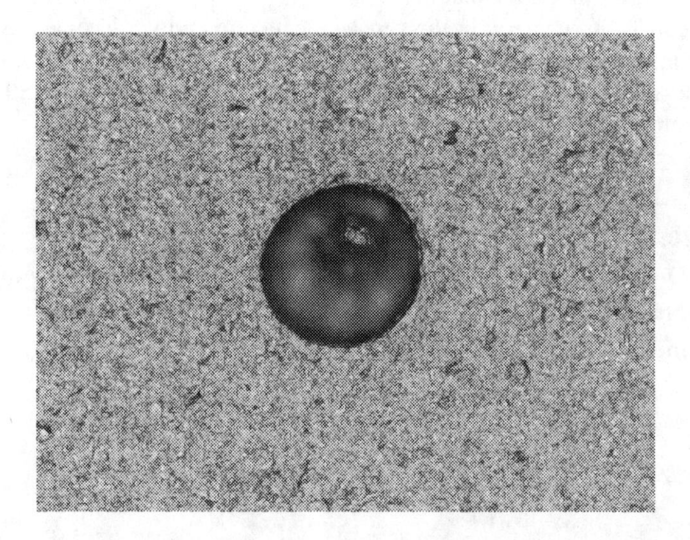

FIGURE 20.3 A 200 μm crater produced by the ablation of an NIST 612 glass SRM using a 213 nm laser system, showing a symmetrical, well-defined crater. (Courtesy of Teledyne Cetac Technologies.)

SRM. It can be seen that the 200 μm ablation crater produced with the 266 nm laser is irregular and shows redeposited ablated material around the edges of the crater, whereas the crater with the 213 nm system is very clean and symmetrical, with no ablated material around the edges. The absence of any redeposited material with the 213 nm laser means that a higher proportion of ablated material actually makes it to the plasma. Both craters are shown at 10× magnification.

This significant difference in crater geometry between the two systems is predominantly a result of the effective absorption of laser energy and the difference in the delivered power per unit area—also known as laser irradiance (or fluence per laser pulse width).[15] The result is a difference in depth penetration and the size/volume of particles reaching the plasma. With the 266 nm laser system, a high-volume burst of material is initially observed, whereas with the 213 nm laser, the signal gradually increases and levels off quickly, indicating a more consistent stream of small particles being delivered to the plasma and the mass spectrometer. Therefore, when analyzing this type of mineral with the 266 nm design, it is typical that the first 100–200 shots of the ablation process are filtered out to ensure that no data are taken during the initial burst of material. This can be somewhat problematic when analyzing small spots or inclusions, because of the limited amount of sample being ablated.

193 nm LASER TECHNOLOGY

The benefits of 213 nm lasers emphasize that matrix independence, high-spatial resolution, and the ability to couple with UV-transparent materials without fracturing (particularly for small spots or depth analysis studies) were very important for geochemical-type applications. These findings led researchers to study even shorter wavelengths, in particular, 193 nm ArF excimer technology. Besides their accepted superiority in coupling efficiency, a major advantage of the 193 nm design is that it utilizes a fundamental wavelength and therefore achieves much higher energy transfer, compared to an Nd:YAG solid-state system that utilizes crystals to quadruple or quintuple the frequency. Additionally, the less coherent nature of the excimer beam enables better optical homogenization, resulting in an even flatter beam profile. The overall benefit is that cleaner, flatter craters are produced down to approximately 3–4 μm in diameter, at energy densities up to 45 J/cm^2. This provides far better control of the ablation process, which is especially important for depth profiling and fluid inclusion analysis. This is demonstrated in Figure 20.4, which shows a scanning electron microscope (SEM) image (1,200 magnification) of a NIST 1612 glass SRM ablated with a 193 nm ArF excimer laser using a highly homogenized, flat-top optical beam profile. It can be seen that the 160 μm ablation crater produced by 50 laser pulses is extremely flat and smooth around the edges.

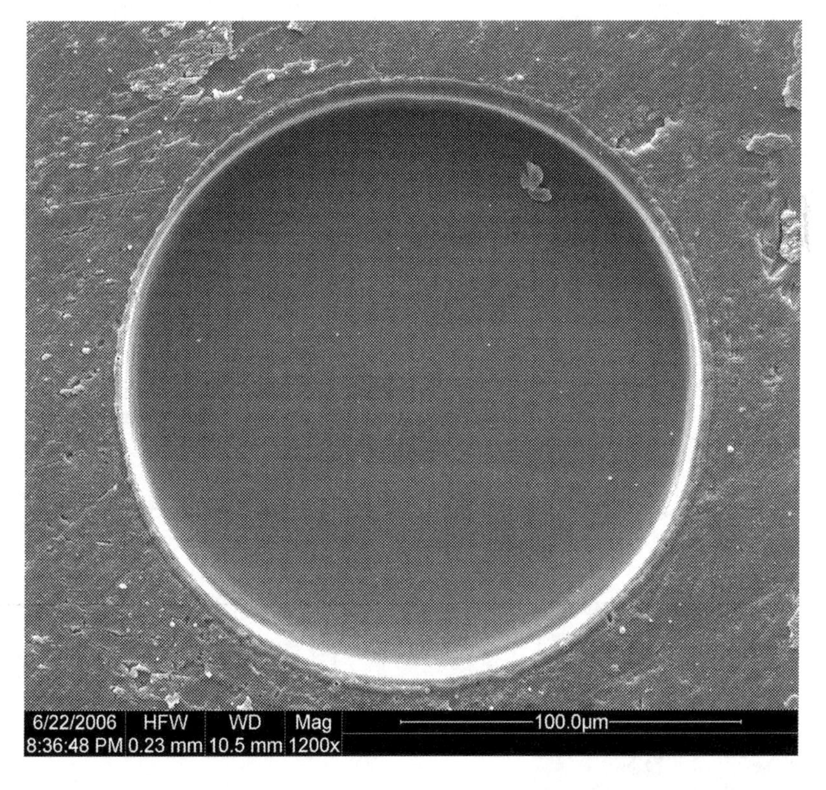

FIGURE 20.4 SEM image (1,200 magnification) of a 160 μm crater produced by the ablation (50 pulses) of NIST 1612 glass SRM using an optically homogenized flat beam, 193 nm ArF excimer laser system. (SEM image courtesy of Cetac Technologies and Dr. Honglin Yuan, Northwest University, Xi'an, China.)

The 213 nm Nd:YAG design just would not be capable of the kind of high precision required for depth analysis and small spot/fluid inclusion studies carried out by the geological community.

It is very important point to emphasize that the geochemical community initially drove the design of excimer lasers, because they were interested in characterizing extremely small spots and inclusions on the surface of minerals and geological specimens, which was very difficult using either of the other approaches. As a result, the performance and the analytical capabilities were focused on their needs, which included the requirement for finely controlled, "homogenizer-flat" ablations with high sensitivity and split-second response. Also fire-on-the-fly lasing synchronized to the stage's motion, combined with fast washout ablation cells, made high-precision depth profiling of spots, lines, and areas possible, allowing for high-spatial resolution elemental mapping. Additionally, the combination of ultra-short pulse length and the 193 nm wavelength produced very high coupling efficiency. This meant higher absorbance in a wide range of materials, which produced smaller particles on average than those produced by the 213 nm YAG design. This resulted in greater ionization within the plasma, leading to better sensitivity and less deposition at the ionization source. Today's commercial excimer lasers can ablate all materials, from opaque to highly transparent, including delicate powders, hard quartz, and resilient carbonates with depth penetration in the tens of nanometers per shot. The beam energy profile is homogenized to ensure uniform ablations across the entire range of spot sizes and on a wide range of materials.

The benefits of LA coupled with ICP-MS are now well documented by the large number of application references in the public domain, which describe the analysis of metals, ceramics, polymers, minerals, biological tissue, pharmaceutical tablets, and many other sample types.[16–21] These references should be investigated further to better understand the optimum configuration, design, and wavelength of LA equipment for different types of sample matrices. It should also be emphasized that there are many overlapping areas when selecting the optimum laser system for the sample type. Roy and Neufeld published a very useful article that offered some guidelines on the importance of matching the laser hardware to the application.[22]

Note: A dedicated LA, laser ionization time-of-flight mass spectrometer is now commercially available that does not use an ICP as an excitation source. For more information about this technique, refer to Chapter 27 "Other Traditional and Emerging Atomic Spectroscopy Techniques."

FLOW INJECTION ANALYSIS

Flow injection (FI) is a powerful front-end sampling accessory for ICP-MS that can be used for preparation, pretreatment, and delivery of the sample. Originally described by Ruzicka and Hansen,[23] FI involves the introduction of a discrete sample aliquot into a flowing carrier stream. Using a series of automated pumps and valves, procedures can be carried out online to physically or chemically change the sample or analyte before introduction into the mass spectrometer for detection. There are many benefits of coupling FI procedures to ICP-MS, including the following:

- Automation of online sampling procedures, including dilution and additions of reagents.
- Minimum sample-handling translates into less chance of sample contamination.
- Ability to introduce low sample/reagent volumes.
- Improved stability with harsh matrices.
- Extremely high-sample throughput using multiple loops.

In its simplest form, FI-ICP-MS consists of a series of pumps and an injection valve preceding the sample introduction system of the ICP mass spectrometer. A typical manifold used for microsampling is shown in Figure 20.5.

In the fill position, the valve is filled with the sample. In the inject position, the sample is swept from the valve and carried to the ICP by means of a carrier stream. The measurement is usually a transient profile of signal versus time, as shown by the signal profile in Figure 20.5.

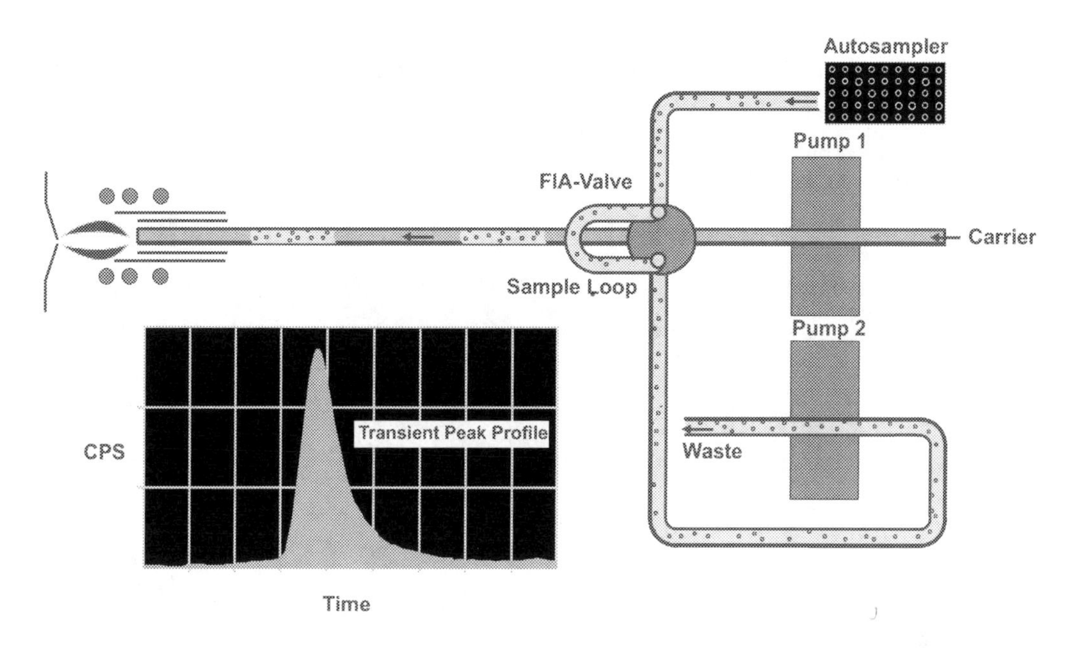

FIGURE 20.5 Schematic of an FI system used for the process of microsampling.

The area of the signal profile measured is greater for larger injection volumes, but for volumes of 500 μL or greater, the signal peak height reaches a maximum equal to that observed using continuous solution aspiration. The length of a transient peak in FI is typically 20–60 s, depending on the size of the loop. This means that if multielement determinations are a requirement, all the data quality objectives for the analysis, including detection limits, precision, dynamic range, and number of elements, must be achieved in this time frame. Similar to LA, if a sequential mass analyzer such as a quadrupole or single-collector magnetic sector system is used, the electronic scanning, dwelling, and settling times must be optimized in order to capture the maximum amount of multielement data in the duration of the transient event.[24] This can be seen in greater detail in Figure 20.6, which shows a 3D transient plot of intensity versus mass in the time domain for the determination of a group of elements.

- Some of the many online procedures that are applicable to FI-ICP-MS include the following:
- Microsampling for improved stability with heavy matrices[25]
- Automatic dilution of samples/standards[26]
- Standards addition[27]
- Cold vapor and hydride generation for enhanced detection capability for elements such as Hg, As, Sb, Bi, Te, and Se[28]
- Matrix separation and analyte preconcentration using ion exchange procedures[29]
- Elemental speciation[30]
- Maximize sample throughput.

FI coupled to ICP-MS has shown itself to be very diverse and flexible in meeting the demands presented by complex samples, as indicated in the foregoing references. However, one of the most interesting areas of research is in the direct analysis of seawater by FI-ICP-MS. Traditionally, seawater is very difficult to analyze by ICP-MS because of two major problems. First, the high NaCl content will block the sampler cone orifice over time, unless a 10- to 20-fold dilution is made of the sample. This is not such a major problem with coastal waters, because the levels are high enough. However, if the sample is open-ocean seawater, this is not an option, because the trace metals are

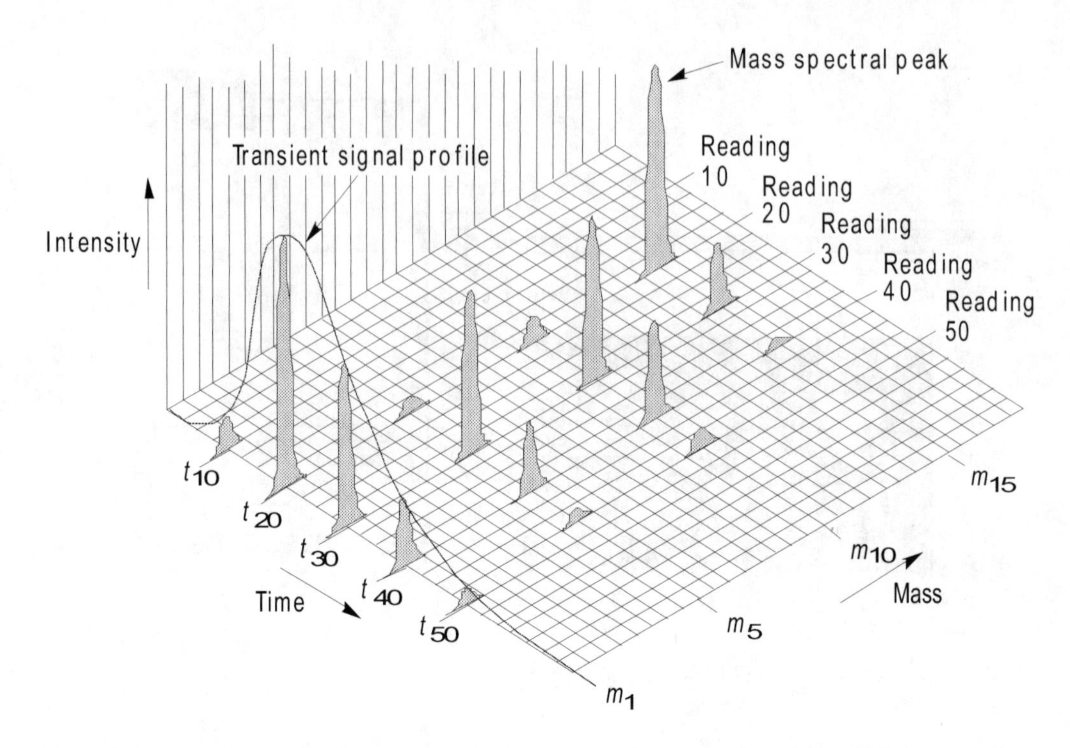

FIGURE 20.6 A 3D plot of intensity versus mass in the time domain for the determination of a group of elements in a transient peak. (Copyright 2013, all rights reserved, PerkinElmer Inc.)

at a much lower level. The other difficulty associated with the analysis of seawater is that ions from the water, chloride matrix, and the plasma gas can combine to generate polyatomic spectral interferences, which are a problem, particularly for the first-row transition metals.

Attempts have been made over the years to remove the NaCl matrix and preconcentrate the analytes using various types of chromatography and ion exchange column technology. One such early approach was to use an high-performance liquid chromatography (HPLC) system coupled to an ICP mass spectrometer utilizing a column packed with silica-immobilized 8-hydroxyquinoline.[31] This worked reasonably well but was not considered a routine method, because silica-immobilized 8-hydroxyquinoline was not commercially available and also spectral interferences produced by HCl and HNO_3 (used to elute the analytes) precluded determination of a number of the elements, such as Cu, As, and V. More recently, chelating agents based on the iminodiacetate acid functionality group have gained wider success but are still not considered truly routine for a number of reasons, including the necessity for calibration using standard additions, the requirement of large volumes of buffer to wash the column after loading the sample, and the need for conditioning between samples because some ion exchange resins swell with changes in pH.[32–34]

However, a research group at the NRC in Canada has developed a very practical online approach, using an FI sampling system coupled to an ICP mass spectrometer.[29] Using a special formulation of a commercially available, iminodiacetate ion exchange resin (with a macroporous methacrylate backbone), trace elements can be separated from the high concentrations of matrix components in the seawater, with a pH 5.2 buffered solution. The trace metals are subsequently eluted into the plasma with 1M HNO_3, after the column has been washed out with deionized water. The column material has sufficient selectivity and capacity to allow accurate determinations at ppt levels using simple aqueous standards, even for elements such as V and Cu, which are notoriously difficult in a chloride matrix. This can be seen in Figure 20.7, which shows spectral scans for a selected group

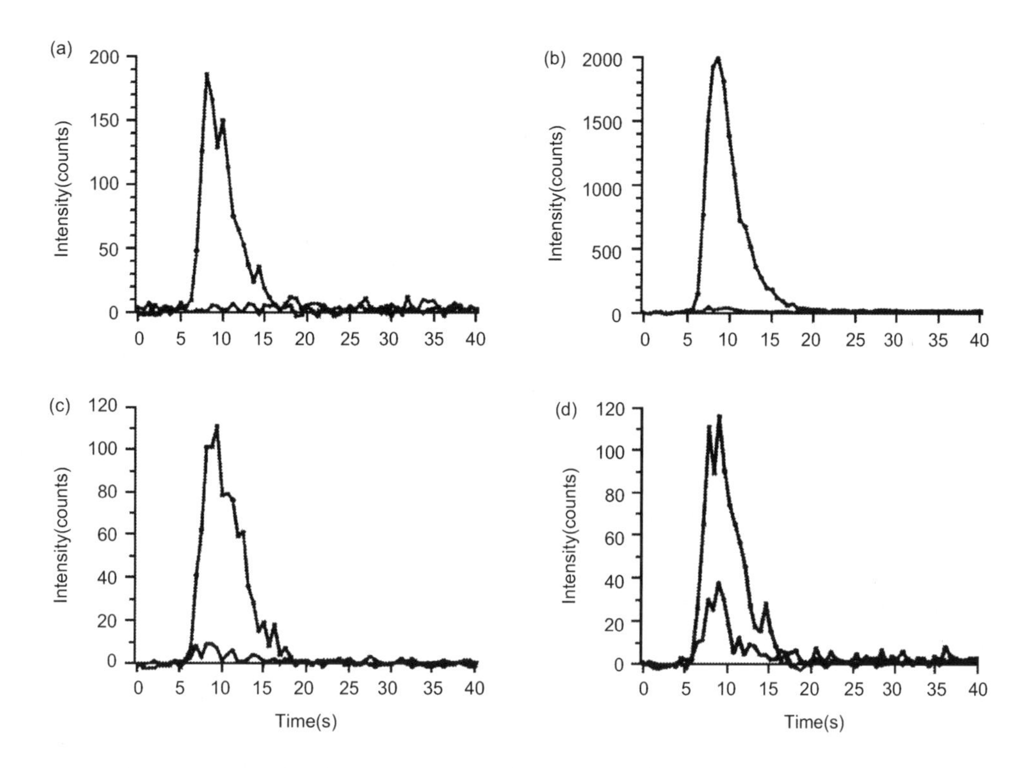

FIGURE 20.7 Analyte and blank spectral scans of (a) Co, (b) Cu, (c) Cd, and (d) Pb in NASS-4 open-ocean seawater certified reference material, using FI coupled to ICP-MS. (From S. N. Willie, Y. Iida and J. W. McLaren, *Atomic Spectroscopy*, **19**(3), 67, 1998.)

of elements in a certified reference material open-ocean seawater sample (NASS-4) and Table 20.2, which compares the results for this methodology with the certified values, together with the limits of detection (LOD). Using this online method, the turnaround time is less than 4 min per sample, which is considerably faster than other high-pressure chelation techniques reported in the literature.

TABLE 20.2
Analytical Results for NASS-4 Open-Ocean Seawater Certified Reference Material, Using Flow Injection ICP-MS Methodology

		NASS-4 (ppb)	
Isotope	**LOD (ppt)**	**Determined**	**Certified**
$^{51}V^+$	4.3	1.20 ± 0.04	Not certified
$^{63}Cu^+$	1.2	0.210 ± 0.008	0.228 ± 0.011
$^{60}Ni^+$	5	0.227 ± 0.027	0.228 ± 0.009
$^{66}Zn^+$	9	0.139 ± 0.017	0.115 ± 0.018
$^{55}Mn^+$	Not reported	0.338 ± 0.023	0.380 ± 0.023
$^{59}Co^+$	0.5	0.0086 ± 0.0011	0.009 ± 0.001
$^{208}Pb^+$	1.2	0.0090 ± 0.0014	0.013 ± 0.005
$^{114}Cd^+$	0.7	0.0149 ± 0.0014	0.016 ± 0.003

Source: From S. N. Willie, Y. Iida and J. W. McLaren, *Atomic Spectroscopy*, **19**(3) 67, 1998.

ELECTROTHERMAL VAPORIZATION (ETV)

ETA for use with atomic absorption (AA) has proved to be a very sensitive technique for trace element analysis over the last three decades. However, the possibility of using the atomization/heating device for ETV sample introduction into both ICP-OES and ICP-MS was identified in the late 1980s.[35] The ETV sampling process relies on the basic principle that a carbon furnace or metal filament can be used to thermally separate the analytes from the matrix components and then sweep them into the ICP mass spectrometer for analysis. This is achieved by injecting a small amount of the sample (usually 20–50 µL via an autosampler) into a graphite tube or onto a metal filament. After the sample is introduced, drying, charring, and vaporization are achieved by slowly heating the graphite tube/metal filament. The sample material is vaporized into a flowing stream of carrier gas, which passes through the furnace or over the filament during the heating cycle. The analyte vapor recondenses in the carrier gas and is then swept into the plasma for ionization.

One of the attractive characteristics of ETV for ICP-MS in particular is that the vaporization and ionization steps are carried out separately, which allows for the optimization of each process. This is particularly true when a heated graphite tube is used as the vaporization device, because the analyst typically has more control of the heating process and as a result can modify the sample by means of a very precise thermal program before it is introduced to the ICP for excitation and/or ionization. By boiling off and sweeping the solvent and volatile matrix components out of the graphite tube, spectral interferences arising from the sample matrix can be reduced or eliminated. The ETV sampling process consists of six discrete stages: sample introduction, drying, charring (matrix removal), vaporization, condensation, and transport. Once the sample has been introduced, the graphite tube is slowly heated to drive off the solvent. Opposed gas flows, entering from each end of the graphite tube, and then purges the sample cell by forcing the evolving vapors out the dosing hole. As the temperature increases, volatile matrix components are vented during the charring steps. Just prior to vaporization, the gas flows within the sample cell are changed. The central channel (nebulizer) gas then enters from one end of the furnace, passes through the tube, and exits out of the other end. The sample-dosing hole is then automatically closed, usually by means of a graphite tip, to ensure no analyte vapors escape. After this gas flow pattern has been established, the temperature of the graphite tube is ramped up very quickly, vaporizing the residual components of the sample. The vaporized analytes either recondense in the rapidly moving gas stream or remain in the vapor phase. These particulates and vapors are then transported to the ICP in the carrier gas, where they are ionized by the ICP for analysis in the mass spectrometer.

Another benefit of decoupling the sampling and ionization processes is the opportunity for chemical modification of the sample. The graphite furnace itself can serve as a high-temperature reaction vessel where the chemical nature of compounds within it can be altered. In a manner similar to that used in atomic absorption, chemical modifiers can change the volatility of species to enhance matrix removal and increase elemental sensitivity.[36] An alternative gas such as oxygen may also be introduced into the sample cell to aid in the charring of the carbon in organic matrices such as biological or petrochemical samples. Here, the organically bound carbon reacts with the oxygen gas to produce CO_2, which is then vented from the system. A typical ETV sampling device, showing the two major steps of sample pretreatment (drying and ashing) and vaporization into the plasma, is seen schematically in Figure 20.8.

Over the past 20 years, ETV sampling for ICP-MS has mainly been used for the analysis of complex matrices including geological materials,[37] biological fluids,[38] and seawater,[39] which have proved difficult or impossible by conventional nebulization. By removal of the matrix components, the potential for severe spectral and matrix-induced interferences is dramatically reduced. Even though ETV-ICP-MS was initially applied to the analysis of very small sample volumes, the advent of low-flow nebulizers has limited its use for this type of work.

An example of the benefit of ETV sampling is in the analysis of samples containing high concentrations of mineral acids such as HCl, HNO_3, and H_2SO_4. Besides physically suppressing analyte

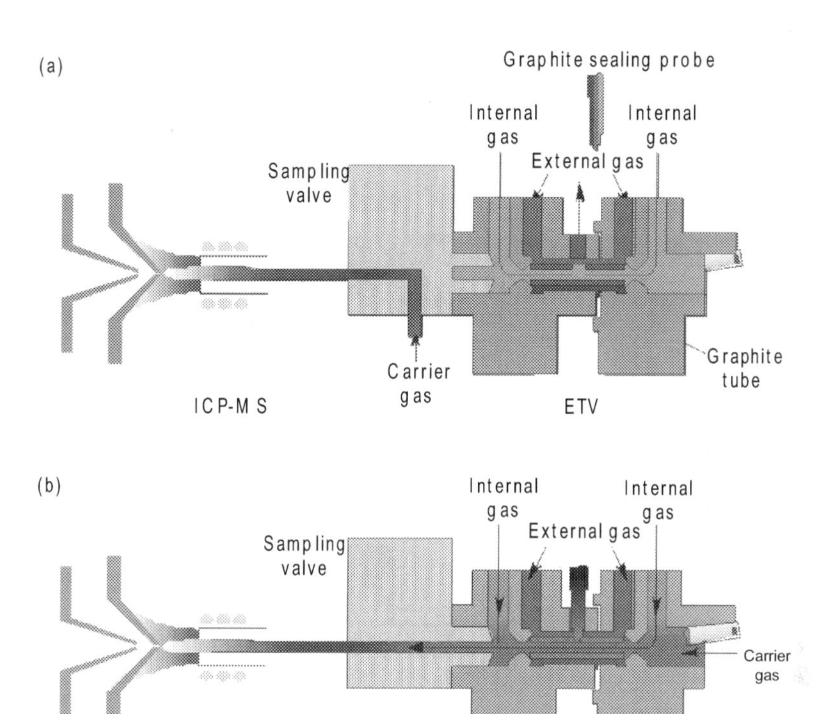

FIGURE 20.8 A graphite furnace ETV sampling device for ICP-MS, showing the two distinct steps of sample pretreatment (a) and vaporization (b) into the plasma. (Copyright 2013, all rights reserved, PerkinElmer Inc.)

signals, these acids generate massive polyatomic spectral overlaps, which interfere with many analytes, including As, V, Fe, K, Si, Zn, and Ti. By carefully removing the matrix components with the ETV device, the determination of these elements becomes relatively straightforward. This is illustrated in Figure 20.9, which shows a spectral display in the time domain for 50 pg spikes of a selected group of elements in concentrated hydrochloric acid (37% w/w), using a graphite furnace-based ETV-ICP-MS.[40] It can be seen in particular that good sensitivity is obtained for $^{51}V^+$, $^{56}Fe^+$, and $^{75}As^+$, which would have been virtually impossible by direct aspiration because of spectral overlaps from $^{39}ArH^+$, $^{35}Cl^{16}O^+$,$^{40}Ar^{16}O^+$, and $^{40}Ar^{35}Cl^+$, respectively. The removal of the chloride and water from the matrix translates into ppt detection limits directly in 37% HCl, as shown in Table 20.3.

It can also be seen in Figure 20.9 that the elements are vaporized off the graphite tube in order of their boiling points. In other words, magnesium, which is the most volatile, is driven off first, whereas V and Mo, which are the most refractory, come off last. However, even though they emerge at different times, the complete transient event lasts less than 3 s. This physical time limitation, imposed by the duration of the transient signal, makes it imperative that all isotopes of interest be measured under the highest signal-to-noise conditions throughout the entire event.

The rapid nature of the transient has also limited the usefulness of ETV sampling for routine multielement analysis, because realistically only a small number of elements can be quantified with good accuracy and precision in less than 3 s. In addition, the development of low-flow nebulizers, desolvation devices, automated online chemistry, cool plasma technology, and collision/reaction cells and interfaces has meant that multielement analysis can now be carried out on difficult

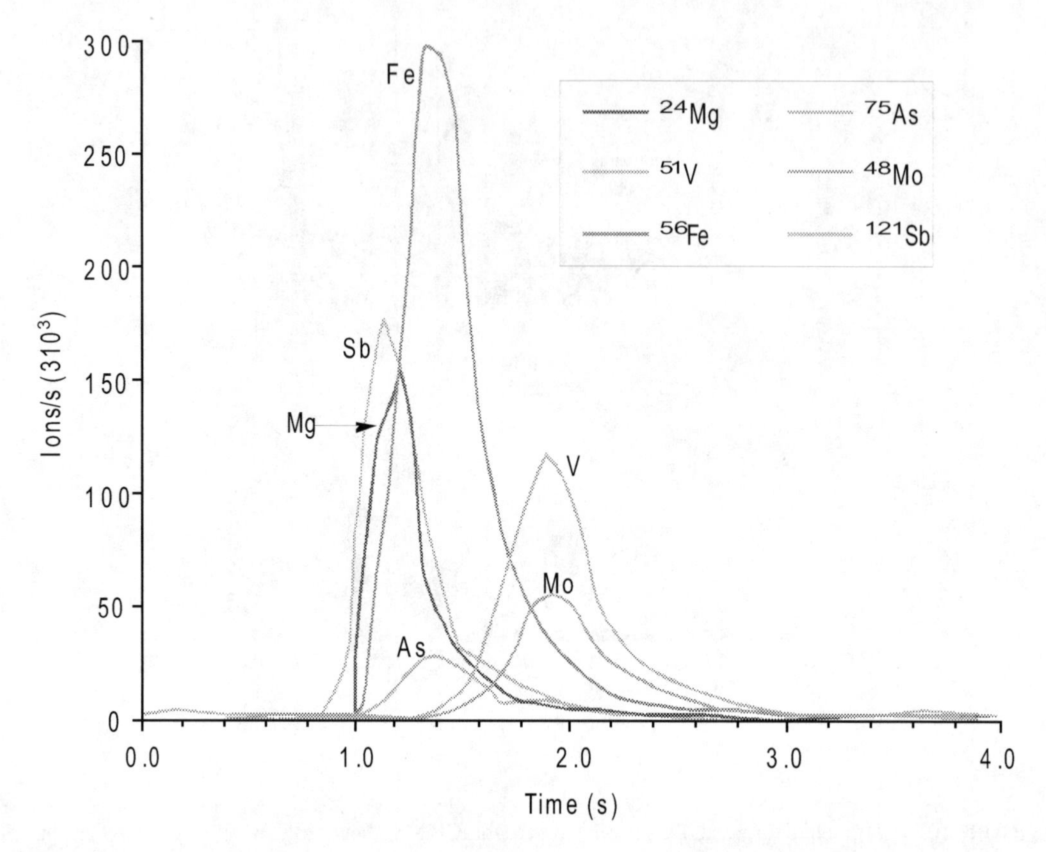

FIGURE 20.9 A temporal display of 50 pg of Mg, Sb, As, Fe, V, and Mo in 37% hydrochloric acid by ETV-ICP-MS. (From S. A. Beres, E. R. Denoyer, R. Thomas, P. Bruckner, *Spectroscopy*, **9**(1), 20–26, 1994.)

TABLE 20.3
Detection Limits for V, Fe, and As in 37% Hydrochloric Acid by ETV-ICP-MS

Element	DL (ppt)
$^{51}V^+$	50
$^{56}Fe^+$	20
$^{75}As^+$	40

Source: S. A. Beres, E. R. Denoyer, R. Thomas, P. Bruckner; *Spectroscopy*, **9**(1), 20–26, 1994.

matrices without the need for ETV sample introduction. This has limited its use with ICP-MS to more research-oriented applications.

However, more recently, ETV has been coupled with ICP-OES for the determination of heavy metals in cannabis. The major benefit is that no sample digestion is required and the cannabis sample can be simply weighed directly into a sampling boat and automatically inserted into furnace where it is thermally pretreated before the analyte particles are swept into the plasma for excitation and detection by optical emission. The benefit of ICP-OES detection over ICP-MS is that larger sample weights can be taken generating a longer temporal transient signal resulting in an optimized measurement protocol. In addition, with ICP-OES, the analyte particles are not going directly into the spectrometer (as with ions going into an ICP-MS) but instead are generating excited atoms and photons in the plasma, which are then being detected and identified by the optical system. Its main drawback is that ICP optical emission is being used for the measurement, which means the detection capability will be compromised compared to ICP-MS. For more information about this novel approach, it's worth checking out this webinar on the benefits of ETV-ICP-OES for measuring heavy metals directly in cannabis-related samples.[41]

CHILLED SPRAY CHAMBERS AND DESOLVATION DEVICES

Chilled/cooled spray chambers and desolvation devices are becoming more and more common in ICP-MS, primarily to cut down the amount of liquid entering the plasma in order to reduce the severity of the solvent-induced spectral interferences such as oxides, hydrides, hydroxides, and argon/solvent-based polyatomic interferences. They are very useful for aqueous-type samples but probably more important for volatile organic solvents, because there is a strong possibility that the sample aerosol would extinguish the plasma unless modifications are made to the sampling procedure. The most common chilled spray chambers and desolvation systems being used today include the following:

- Water-cooled spray chambers
- Peltier-cooled spray chambers
- Ultrasonic nebulizers (USNs)
- USNs coupled with membrane desolvation
- Specialized microflow nebulizers coupled with desolvation techniques.

Let us take a closer look at these devices.

WATER-COOLED AND PELTIER-COOLED SPRAY CHAMBERS

Water-cooled spray chambers have been used in ICP-MS for many years and are standard on a number of today's commercial instrumentation to reduce the amount of water or solvent entering the plasma. However, the trend today is to cool the sample using a thermoelectric device called a Peltier cooler. Thermoelectric cooling (or heating) uses the principle of generating a hot or cold environment by creating a temperature gradient between two different materials. It uses electrical energy via a solid-state heat pump to transfer heat from a material on one side of the device to a different material on the other side, thus producing a temperature gradient across the device (similar to a household air conditioning system). Peltier cooling devices, which are typically air-cooled (but water cooling is an option), can be used with any kind of spray chamber and nebulizer, but commercial products for use with ICP-MS are normally equipped with a cyclonic spray chamber and a low-flow pneumatic nebulizer.

The main purpose of cooling the sample aerosol is to reduce the amount of water or solvent entering the plasma by lowering the temperature of the spray chamber. This can be a few degrees below ambient or as low –20°C, depending on the type of samples being analyzed. This can have a threefold effect: First, it helps to minimize solvent-based spectral interferences, such as oxides and hydroxides formed in the plasma, and second, because very little plasma energy is needed to vaporize the solvent, it allows more energy to be available to excite and ionize the analyte ions. There is also evidence to suggest that cooling the spray chamber will help minimize signal drift due to external environmental temperature changes in the laboratory.

Cooling the spray chamber to as low as –20°C by either Peltier cooling or a recirculating system using ethylene glycol as the coolant is particularly useful when it comes to analyzing some volatile organic samples. It has the effect of reducing the amount of organic solvent entering the interface, and when combined with the addition of a small amount of oxygen into the nebulizer gas flow, it is beneficial in reducing the buildup of carbon deposits on the sampler cone orifice and also minimizing the problematic carbon-based spectral interferences.[42]

Some systems also have the ability to increase the temperature of the spray chamber above ambient. Studies have shown that the sensitivity for many analytes is enhanced by a factor of two- to threefold by running the spray chamber as high as 60°C, a feature that is particularly important for samples with limited volume or viscous matrices such as engine or edible oils.

ULTRASONIC NEBULIZERS

Ultrasonic nebulization was first developed in the late 1980s for use with ICP optical emission.[43] Its major benefit was that it offered an approximately ten times improvement in detection limits, because of its more efficient aerosol generation. However, this was not such an obvious benefit for ICP-MS, because more matrix entered the system compared to a conventional nebulizer, increasing the potential for signal drift, matrix suppression, and spectral interferences. This was not such a major problem for simple aqueous-type samples but was problematic for real-world matrices. The elements that showed the most improvement were the ones that benefited from lower solvent-based spectral interferences. Unfortunately, many of the other elements exhibited higher background levels and as a result showed no significant improvement in detection limit. In addition, the increased amount of matrix entering the mass spectrometer usually necessitated the need for larger dilutions of the sample, which again negated the benefit of using a USN with ICP-MS for samples with a heavier matrix. This limitation led to the development of a USN fitted with an additional membrane desolvator. This design virtually removed all the solvent from the sample, which dramatically improved detection limits for a large number of the problematic elements and also lowered metal oxide levels by at least an order of magnitude.[44]

The principle of aerosol generation using a USN is based on a sample being pumped onto a quartz plate of a piezoelectric transducer. Electrical energy of 1–2 MHz frequency is coupled to the transducer, which causes it to vibrate at high frequency. These vibrations disperse the sample into a fine-droplet aerosol, which is carried in a stream of argon. With a conventional USN, the aerosol is passed through a heating tube and a cooling chamber, where most of the sample solvent is removed as a condensate before it enters the plasma. If a membrane desolvation system is fitted to the USN, it is positioned after the cooling unit. The sample aerosol enters the membrane desolvator, where the remaining solvent vapor passes through the walls of a tubular micro-porous membrane. A flow of argon gas removes the volatile vapor from the exterior of the membrane, while the analyte aerosol remains inside the tube and is carried into the plasma for ionization. Membrane desolvation systems also have the capability to add a secondary gas such as nitrogen, which has shown to be very beneficial in changing the ionization conditions to reduce levels of oxides in the plasma. The combination of membrane desolvation with a USN can be seen more clearly in Figure 20.10, and Figure 20.11 shows the principles of membrane desolvation with water vapor as the solvent.

For ICP-MS, the system is best operated with both desolvation stages working, although for less demanding ICP-OES analysis, the membrane desolvation stage can be bypassed if required. The power of the system when coupled to an ICP mass spectrometer can be in Table 20.4, which compares the sensitivity (counts per second) and signal-to-background ratio of a membrane desolvation USN with a conventional cross-flow nebulizer for two classic solvent-based polyatomic interferences, $^{12}C^{16}O_2^+$ on $^{44}Ca^+$ and $^{40}Ar^{16}O^+$ on $^{56}Fe^+$, using a quadrupole ICP-MS system.

It can be seen that for the two analyte masses, the signal-to-background ratio is significantly better with the membrane-desolvated USN than with the cross-flow design, which is a direct result of the reduction of the solvent-related spectral background levels. This approach is even more beneficial for the analysis of organic solvents because when they are analyzed by conventional nebulization, modifications have to be made to the sampling process, such as the addition of oxygen to the nebulizer gas flow, use of a low-flow nebulizer, and probably external cooling of the spray chamber.

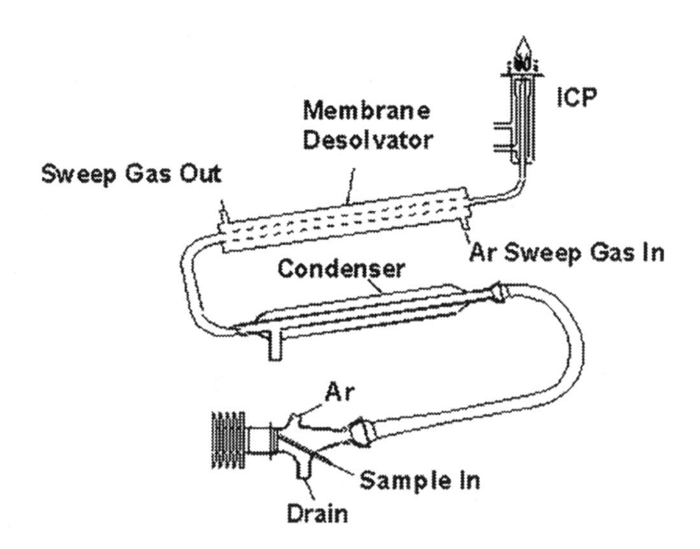

FIGURE 20.10 Schematic of a USN fitted with a membrane desolvation system. (Courtesy of Teledyne Cetac Technologies.)

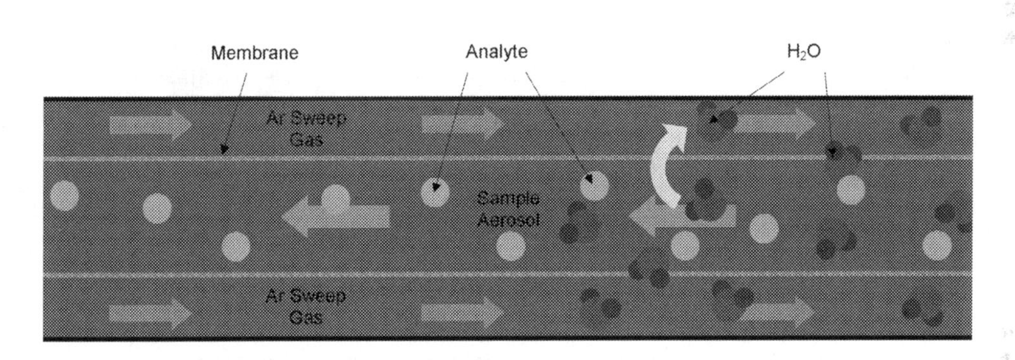

FIGURE 20.11 Principles of membrane desolvation showing the water molecules passing through a microporous membrane and being swept away by the argon gas, while the analyte is transported through the tube to the plasma. (Courtesy of Elemental Scientific Inc.)

TABLE 20.4
Comparison of Sensitivity and Signal/Background Ratios for Two Analyte Masses—44Ca+ and 56Fe+—Using a Conventional Cross-Flow Nebulizer and a USN Fitted with a Membrane Desolvation System

Analytical Mass	Cross-Flow Nebulizer (cps)	Signal/BG	USN with Membrane Desolvation (cps)	Signal/BG
25 ppb $^{44}Ca^+$ (BG subtracted)	2,300	2,300/7,640 = **0.30**	20,800	20,800/1730 = **12.0**
$^{12}C^{16}O_2^+$ (BG)	7,640		1,730	
10 ppb $^{56}Fe^+$ (BG subtracted)	95,400	95,400/868,00 = **0.11**	262,000	262,000/8,200 = **32.0**
$^{40}Ar^{16}O^+$ (BG)	868,000		8,200	

Note: Signal/BG is calculated as the background subtracted signal divided by the background.
Source: Courtesy of Cetac Technologies.

With a membrane desolvation USN system, volatile solvents such as isopropanol can be directly aspirated into the plasma with relative ease. However, it should be mentioned that depending on the sample type, this approach does not work for analytes that are bound to organic molecules. For example, the high volatility of certain mercury and boron organometallic species means that they could pass through the micro-porous PTFE membrane and never make it into the ICP-MS. In addition, samples with high-dissolved solids, especially ones that are biological in nature, could possibly result in clogging the micro-porous membrane unless substantial dilutions are made. For these reasons, caution must be used when using a membrane desolvation system for the analysis of certain types of sample matrices.

SPECIALIZED MICROFLOW NEBULIZERS WITH DESOLVATION TECHNIQUES

Microflow or low-flow nebulizers, which were described in greater detail in Chapter 6 "ICP-MS Sample Introduction," are being used more and more for routine applications. The most common ones used in ICP-MS are based on the microconcentric design, which operate at sample flows of 20–500 μL/min. Besides being ideal for small sample volumes, the major benefit of microconcentric nebulizers is that they are more efficient and produce smaller droplets than a conventional nebulizer. In addition, many microflow nebulizers use chemically inert plastic capillaries, which make them well suited for the analysis of highly corrosive chemicals. This kind of flexibility has made low-flow nebulizers very popular, particularly in the semiconductor industry, where it is essential to analyze high-purity materials using a sample introduction system that is free of contamination.[45]

Such is the added capability and widespread use of these nebulizers across all application areas that manufacturers are developing application-specific integrated systems that include the spray chamber and a choice of different desolvation techniques to reduce the amount of solvent aerosol entering the plasma. Depending on the types of samples being analyzed, some of these systems include a low-flow nebulizer, Peltier-cooled spray chambers, heated spray chambers, Peltier-cooled condensers, and membrane desolvation technology. Some of the commercially available equipment include the following:

- **Microflow nebulizer coupled with a Peltier-cooled spray chamber:** An example of this is the PC3 from Elemental Scientific Inc. (ESI). This system is offered with or without the nebulizer and utilizes a Peltier-cooled cyclonic spray chamber made from either quartz, borosilicate glass, or a fluoropolymer. Options include the ability to reduce the temperature to −20°C for analyzing organics and a dual-spray chamber for improved stability.
- **Microflow nebulizer with heated spray chamber and Peltier-cooled condenser:** An example of this design is the Apex inlet system from ESI. This unit includes a microflow nebulizer, heated cyclonic spray chamber (up to 140°C), and a Peltier multipass condenser/cooler (down to −5°C). A number of different spray chamber and nebulizer options and materials are available, depending on the application requirements. Also, the system is available with Teflon or Nafion micro-porous membrane desolvation, depending on the types of samples being analyzed. Figure 20.12 shows a schematic of the Apex sample inlet system with the cross-flow nebulizer.
- **Microflow nebulizer coupled with membrane desolvation:** An example of this is the Aridus system from Cetac Technologies. The aerosol from the nebulizer is either self-aspirated or pumped into a heated perfluoroalkoxy resin (PFA) spray chamber (up to 110°C) to maintain the sample in a vapor phase. The sample vapor then enters a heated PTFE membrane desolvation unit, where a counterflow of argon sweep gas is added to remove solvent vapors that permeate the micro-porous walls of the membrane. Nonvolatile sample components do not pass through the membrane walls but are transported to the ICP-MS for analysis. Figure 20.13 shows a schematic of the Aridus II microflow nebulizer with membrane desolvation.

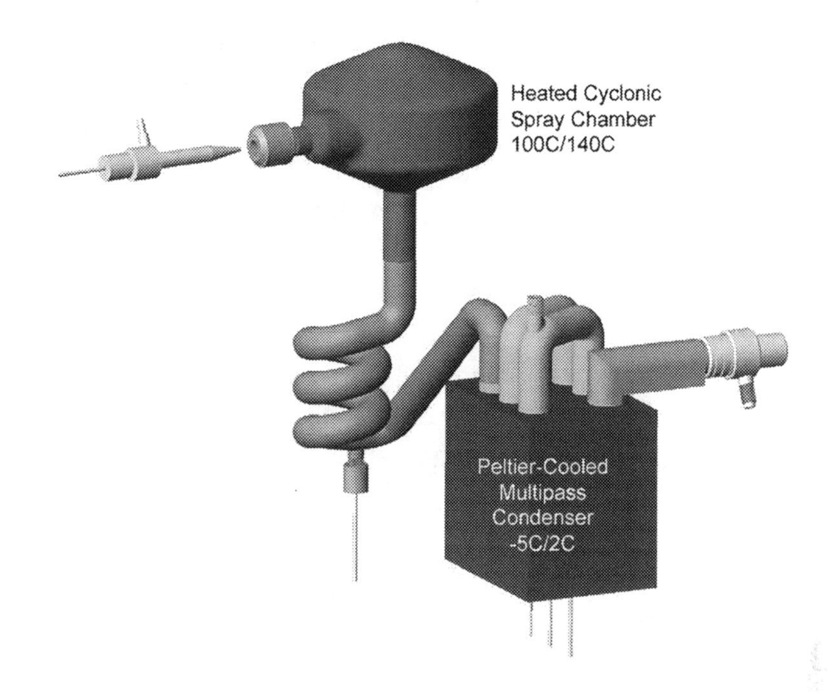

FIGURE 20.12 A schematic of the Apex sample inlet system. (Courtesy of Elemental Scientific Inc.)

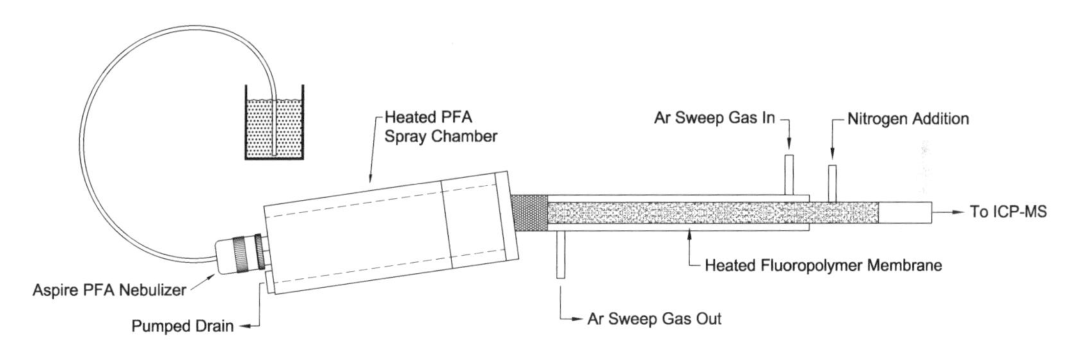

FIGURE 20.13 A schematic of the Aridus II microflow nebulizer with membrane desolvation. (Courtesy of Teledyne Cetac Technologies.)

There is an extremely large selection of these specialized sample introduction techniques, so it is critical that you talk to the vendors so they can suggest the best solution for your application problem. They may not be required for the majority of your application work, but there is no question they can be very beneficial for analyzing certain types of sample matrices and for elements that might be prone to solvent-based spectral overlaps.

DIRECT INJECTION NEBULIZERS

Direct injection nebulization is based on the principle of injecting a liquid sample under high pressure directly into the base of the plasma torch.[46] The benefit of this approach is that no spray chamber is required, which means that an extremely small volume of sample can be introduced directly

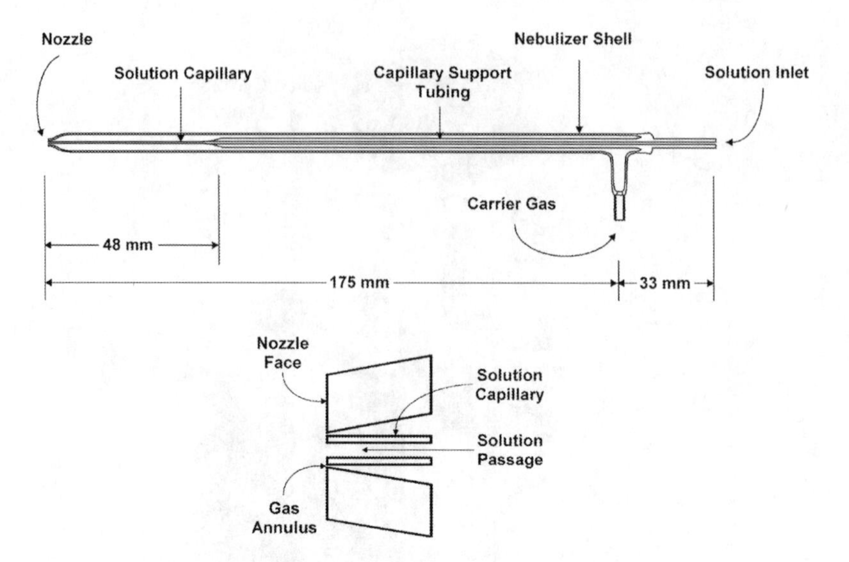

FIGURE 20.14 A schematic of a commercially available DIHEN system. (Courtesy of Meinhard Glass Products, a part of Elemental Scientific.)

into the ICP-MS with virtually no carryover or memory effects from the previous sample. Because they are capable of injecting less than 5 μL of liquid, they have found a use in applications where sample volume is limited or where the material is highly toxic or expensive.

They were initially developed over 15 years ago and found some success in certain niche applications that could not be adequately addressed by other nebulization systems, such as introducing samples from a chromatography separation device into an ICP-MS or the determination of mercury by ICP-MS, which is prone to severe memory effects. Unfortunately, they were not considered particularly user-friendly and as a result became less popular when other sample introduction devices were developed to handle microliter sample volumes. More recently, a refinement of the direct injection nebulizer has been developed, called the direct inject high-efficiency nebulizer (DIHEN), which appears to have overcome many of the limitations of the original design.[47] The advantage of the DIHEN is its ability to introduce microliter volumes into the plasma at extremely low sample flow rates (1–100 μL/min), with an aerosol droplet size similar to a concentric nebulizer fitted with a spray chamber. The added benefit is that it is almost 100% efficient and has extremely low memory characteristics. A schematic of a commercially available DIHEN system is shown in Figure 20.14.

PRODUCTIVITY ENHANCING TECHNIQUES

With the increasing demand to analyze more and more samples, carry out automated dilutions and additions of internal standards, as well as performing online chemistry procedures, manufacturers of autosamplers, and sample introduction accessories are designing automated sampling systems to maximize sample throughput, minimize sample preparation times, and increase productivity.[48–50] Depending on the application requirements, this is being achieved in a number of different ways using a variety of components including multiport/switching valves, loops, vacuum/piston/syringe pumps, mixing chambers, and ion-exchange/preconcentration columns. There are basically three approaches to enhancing productivity in ICP-MS, depending on the application requirements:

- Achieve faster analysis times by optimizing sample delivery to the instrument
- Perform online dilutions, internal standard additions, and calibrations to save manual operations

- Carry out automated chemistry online to remove sample matrices and/or preconcentrate the samples to reduce interferences and minimize labor intensive, manual sample preparation steps.

Let's take a more detailed look at each of these approaches.

FASTER ANALYSIS TIMES

This is basically a rapid sampling approach integrated into an intelligent autosampler, which significantly reduces analysis times by optimizing the sample delivery process to reduce the pre- and post-measurement time. There are a number of these systems on the market, which work slightly differently, but basically, they all use piston/syringe/vacuum pumps and switching valves and loops to control the delivery of the sample and standards to and from the ICP-MS. Besides significantly faster analysis times, other benefits include improved precision/accuracy, reduced carryover, and longer lifetime of sample introduction consumables. Depending on the design of the system, some typical areas of optimization include:

Autosampler response is the time it takes for the instrument to send a signal to the autosampler to move the sample probe to the next sample. By moving the autosampler probe over to the next sample while the previous sample is being analyzed, a significant amount of time will be saved over the entire automated run.

Sample uptake is the time taken for a sample to be drawn into the autosampler probe and pass through the capillary and pump tubing into the nebulizer. By using a small vacuum pump to rapidly fill the sample loop, which is positioned in close proximity of the sample loop to the nebulizer, sample uptake time is minimized.

Signal stabilization is the time required to allow the plasma to stabilize after air has entered the line from the autosampler probe dipping in and out of the sample tubes (this can also be exaggerated if the pump speed is increased to help in sample delivery). However, if the pump delivering the sample to the plasma remains at a constant flow rate, and the injection valve ensures no air is introduced into the sample line, very little stabilization time is required.

Rinse-out is the time required to remove the previous sample from the sample tubing and sample introduction system. So, if the probe is being rinsed during the sample analysis, minimal rinse time is needed.

Overhead time is the time spent by the ICP performing calculations and printing results, so if this time is used to ensure the previous sample has reached baseline, minimal rinse time is required for the next sample.

Another slightly different approach is to use a rapid-rinse accessory, based on an FI loop and a positive displacement piston pump coupled to an autosampler. With this system, the multiport valve switches between two positions. In the first position, a loop of capillary tubing is filled with the sample, while in the second position, the sample is delivered to the nebulizer. This is seen in greater detail in Figure 20.15a, which shows the piston pump rapidly filling the sample loop. At the same time, rinse and internal standard solutions are delivered to the nebulizer, washing out the nebulizer and spray chamber, and ensuring that plasma stability is maintained. Figure 20.15b shows the actual aspiration process where the valve switches position so that the rinse solution pushes the sample into the nebulizer. The internal standard is then mixed with the sample within the valve. At the same time, the autosampler probe and sample uptake tubing are rinsed by the piston pump.

There are a number of these enhanced productivity sampling systems on the market that all work in a slightly different way. However, they all have one thing in common and that is they offer at least a twofold reduction in analysis time compared to traditional autosamplers. Some of the other benefits that are realized with this time saving include:

- Improved precision because of no pulsing from peristaltic pump
- Better accuracy due to online dilution and addition of internal standards

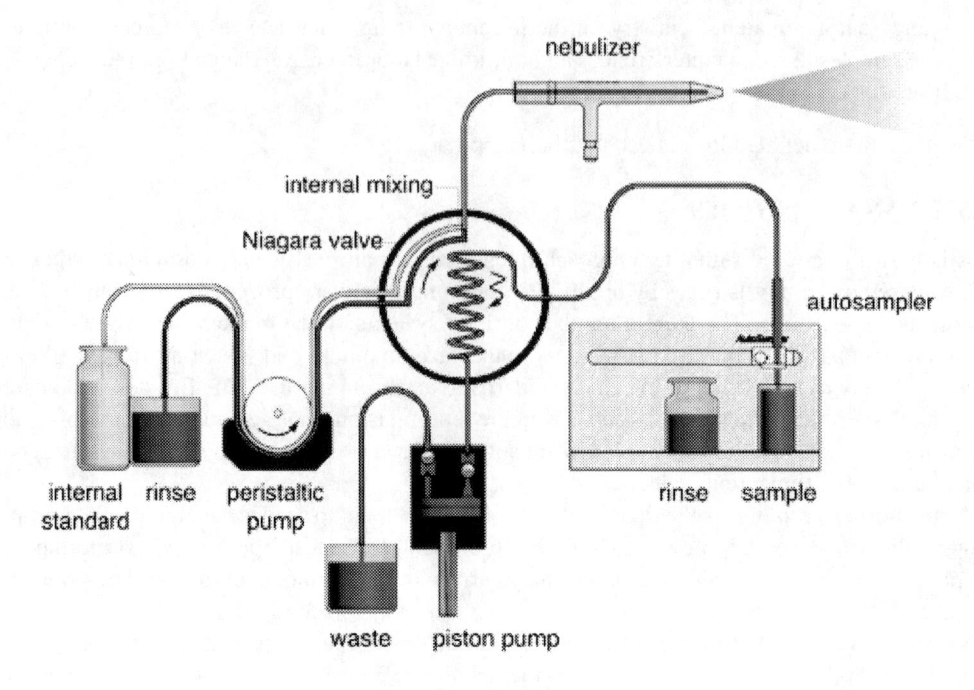

(a)

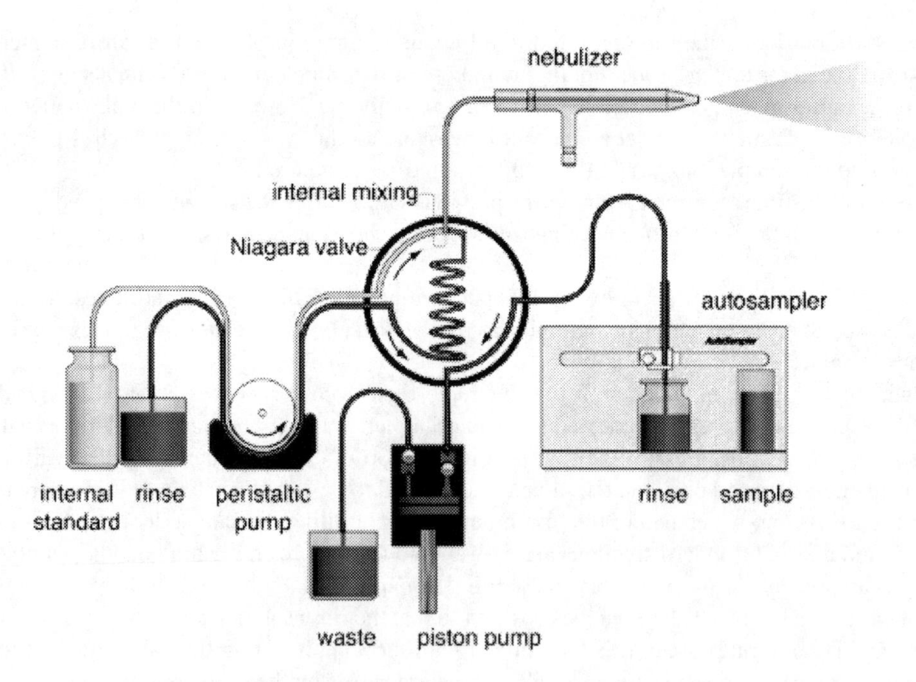

(b)

FIGURE 20.15 (a) The piston pump rapidly fills the sample loop, while at the same time, rinse and internal standard solutions are delivered to the nebulizer. (Courtesy of Glass Expansion Inc.) (b) The valve switches position so that the rinse solution pushes the sample into the nebulizer, while the internal standard is mixed with the sample and the autosampler probe and sample uptake tubing are rinsed by the piston pump. (Courtesy of Glass Expansion Inc.)

- Constant flow of solutions to plasma reduces stabilization times
- Less sample volume used
- Lower argon consumption
- Reduced cost of consumables
- Less routine maintenance
- Much lower chemical waste.

There is no question that all these benefits can make a significant improvement in the overall cost of analysis, especially in high-workload routine environmental laboratories, where high-sample throughput is an absolute requirement.[51,52]

AUTOMATED IN-LINE AUTODILUTION AND AUTOCALIBRATION SYSTEMS

A new range of automated sampling accessories has recently been developed, which performs very precise and accurate online autodilutions and autocalibration procedures using syringe/piston pumps.[53] Samples are rapidly and reproducibly loaded from each autosampler location into a sample loop. From there, the sample is injected into a diluent liquid stream and transported to a tee located between the valve and nebulizer. The internal standard is added in the tee to obtain final dilution factors defined by the operator. At the heart of the system is a syringe pump, which delivers the sample over a wide range flow rates ensuring rapid and reliable in-line dilutions. The benefits of fully-automated in-line autodilution and autocalibration include:

- Real-time dilutions
- Dilution in valve head and tee
- No additional tube or reagents required
- Eliminates manual dilutions
- Rapid uptake and wash out
- Lowers risk of contamination
- Sample analysis time constant, independent of dilution factor.

AUTOMATED SAMPLE IDENTIFICATION AND TRACKING SYSTEMS

A recent development in automation is in the area of advanced, automated sample identification and tracking systems that can accurately associate stored information with a sample throughout the sample collection/preparation/introduction process. Through a series of four distinct stages, this technology uses barcodes to enter, store, and reference data associated with a sample from initial collection to taring and final dilution.

For example, from the point of sample collection, the technology associates collection time and Global Positioning System (GPS) location with sample container barcode. This then relates the information from its location and tracks it through the dilution process, inputting weight measurements before the sample preparation process. The software then has the ability to automatically apply sample digestion procedures according to sample ID code. Finally, the system presents the samples for ICP-MS/ICP-OES analysis while confirming sample identity and providing legally defensible data for regulatory inspection.

This technology is ideally suited to any application area, such as the pharmaceutical or cannabis industry, where it is critical to keep track of the sample, as it moves from collection to preparation and dilution, etc. to finally being analyzed by the instrumentation of choice that is generating the data to make a decision about the maximum levels of elemental contaminants.

FURTHER READING

1. E. R. Denoyer, K. J. Fredeen, and J. W. Hager, *Analytical Chemistry*, **63**(8), 445–457A, 1991.
2. J. F. Ready, *Effects of High Power Laser Radiation*, Academic Press, New York, Chapters 3–4, 1972.
3. L. Moenke-Blankenburg, *Laser Microanalysis*, Wiley, New York, 1989.
4. E. R. Denoyer, R. Van Grieken, and F. Adams, D. F. S. Natusch, *Analytical Chemistry*, **54**, 26–30A, 1982.
5. J. W. Carr and G. Horlick, *Spectrochimica Acta*, **37B**, 1, 1982.
6. T. Kantor et al. *Talanta,* **23**, 585–590, 1979.
7. H. C. G. Human et al. *Analyst*, **106**, 265, 1976.
8. M. Thompson, J. E. Goulter, and F. Seiper, *Analyst*, **106**, 32–35, 1981.
9. A. L. Gray, *Analyst,* **110**, 551–555, 1985.
10. P. A. Arrowsmith and S. K. Hughes, *Applied Spectroscopy*, **42**, 1231–1239, 1988.
11. T. Howe, J. Shkolnik, and R. Thomas, *Spectroscopy*, **16** (2), 54–66 2001.
12. D. Günther and B. Hattendorf, *Mineralogical Association of Canada—Short Course Series*, **29**, 83–91, 2001.
13. T. E. Jeffries, S. E. Jackson, and H. P. Longerich, *Journal of Analytical Atomic Spectrometry*, **13**, 935–940, 1998.
14. R. E. Russo, X. L. Mao, O. V. Borisov, and L. Haichen, *Journal of Analytical Atomic Spectrometry*, **15**, 1115–1120, 2000.
15. H. Liu, O. V. Borisov, X. Mao, S. Shuttleworth, and R. Russo, *Applied Spectroscopy*, **54**(10), 1435–1440, 2000.
16. S. E. Jackson, H. P. Longerich, G. R. Dunning, and B. J. Fryer, *Canadian Mineralogist,* **30**, 1049–1064, 1992.
17. D. Günther and C. A. Heinrich, *Journal of Analytical Atomic Spectrometry*, **14**, 1361–1366, 1999.
18. D. Günther, I. Horn, and B. Hattendorf, *Fresenius Journal of Analytical Chemistry*, **368**, 4–14, 2000.
19. R. E. Wolf, C. Thomas, and A. Bohlke, *Applied Surface Science*, **127–129**, 299–303, 1998.
20. J. Gonzalez, X. L. Mao, J. Roy, S. S. Mao, and R. E. Russo, *Journal of Analytical Atomic Spectrometry* **17**, 1108–1113, 2002.
21. R. Lam and E.D. Salin, *Journal of Analytical Atomic Spectrometry,* **19**, 938–940, 2004.
22. J. Roy and L. Neufeld, *Spectroscopy,* **19**(1), 16–28, 2004.
23. J. Ruzicka and E. H. Hansen, *Analytic Chimica Acta,* **78**, 145–150, 1975.
24. R. Thomas, *Spectroscopy*, **17**(5), 54–66, 2002.
25. A. Stroh, U. Voellkopf, and E. Denoyer, *Journal of Analytical Atomic Spectrometry*, **7**, 1201–1207, 1992.
26. Y. Israel, A. Lasztity, and R. M. Barnes, *Analyst,* **114**, 1259–1264, 1989.
27. Y. Israel and R. M. Barnes, *Analyst,* **114**, 843–850, 1989.
28. M. J. Powell, D. W. Boomer, and R. J. McVicars, *Analytical Chemistry*, **58**, 2864–2871, 1986.
29. S. N. Willie, Y. Iida, and J. W. McLaren, *Atomic Spectroscopy*, **19**(3), 67–73, 1998.
30. R. Roehl and M. M. Alforque, *Atomic Spectroscopy*, **11**(6), 210, 1990.
31. J. W. McLaren, J. W. H. Lam, S. S. Berman, K. Akatsuka, and M. A. Azeredo, *Journal of Analytical Atomic Spectrometry*, **8**, 279–286, 1993.
32. L. Ebdon, A. Fisher, H. Handley, and P. Jones, *Journal of Analytical Atomic Spectrometry,* **8**, 979–981, 1993.
33. D. B. Taylor, H. M. Kingston, D. J. Nogay, D. Koller, and R. Hutton, *Journal of Analytical Atomic Spectrometry,* **11**, 187–191, 1996.
34. S. M. Nelms, G. M. Greenway, and D. Koller, *Journal of Analytical Atomic Spectrometry,* **11**, 907–912, 1996.
35. C. J. Park, J. C. Van Loon, P. Arrowsmith, and J. B. French, *Analytical Chemistry*, **59**, 2191–2196, 1987.
36. R. D. Ediger and S. A. Beres, *Spectrochimica Acta,* **47B**, 907–913, 1992.
37. C. J. Park and M. Hall, *Journal of Analytical Atomic Spectrometry,* **2**, 473–480, 1987.
38. C. J. Park and J. C. Van Loon, *Trace Elements in Medicine*, **7**, 103–107, 1990.
39. G. Chapple and J. P. Byrne, *Journal of Analytical Atomic Spectrometry,* **11**, 549–553, 1996.
40. S. A. Beres, E. R. Denoyer, R. Thomas, and P. Bruckner, *Spectroscopy*, **9**(1), 20–26, 1994.
41. Spectro Instruments, Advances in XRF and ICP-OES analysis for plant tissue and soil analysis including cannabis, Spectro Instruments, Live Webcast, Feb 26, 2020, https://www.spectro.com/landingpages-noindex/webinar-pages/icp-xrf-agronomy.
42. F. McElroy, A. Mennito, E. Debrah, and R. Thomas, *Spectroscopy,* **13**(2), 42–53, 1998.

43. K. W. Olson, W. J. Haas, Jr, and V. A. Fassel, *Analytical Chemistry*, **49**(4), 632–637, 1977.
44. J. Kunze, S. Koelling, M. Reich, and M. A. Wimmer, *Atomic Spectroscopy,* **19**, 5–13, 1998.
45. G. Settembre and E. Debrah, *Micro,* **23**(5), 67–71, 1998.
46. D. R. Wiederin and R. S. Houk, *Applied Spectroscopy*, **45**(9), 1408–1411, 1991.
47. J. A. McLean, H. Zhang, and A. Montaser, *Analytical Chemistry*, **70**, 1012–1020, 1998.
48. Glass Expansion, Characterization of a customized valve for enhanced productivity in ICP/ICP-MS, Glass Expansion Application Note, http://www.geicp.com/site/images/application_notes/NiagaraApplicationsWhitePaper_Feb2010.pdf.
49. ESI, Automated Sample Introduction with the SC Fast, ESI Application Note, http://www.icpms.com/pdf/SC-FAST.pdf.
50. Cetac Technologies, ASXpress Plus Rapid Sample Introduction System, Cetac Technologies, http://www.teledynecetac.com/products/automation/asxpress-plus
51. ESI, Improving throughput of environmental samples by ICP-MS following EPA method 200.8, ESI Application Note, http://www.elementalscientific.com/products/SC-FAST_enviro.html.
52. M. P. Field, M. LaVigne, K. R. Murphy, G. M. Ruiz, and R. M. Sherrell, *Journal of Analytical Atomic Spectrometry*, **22**, 1145, 2007.
53. ESI, The evolution of automation: Prepfast data sheet, Elemental Scientific Inc., http://www.icpms.com/products/prepfast.php.
54. ESI, PlasmaTrax: Automated sample identification and tracking system, Application Note: Elemental Scientific Inc., Omaha, NE, http://www.icpms.com/pdfv1/15054-1%20PlasmaTRAX-2D%20Barcode%20Automation.pdf.

21 Coupling ICP-MS with Chromatographic Separation Techniques for Speciation Studies

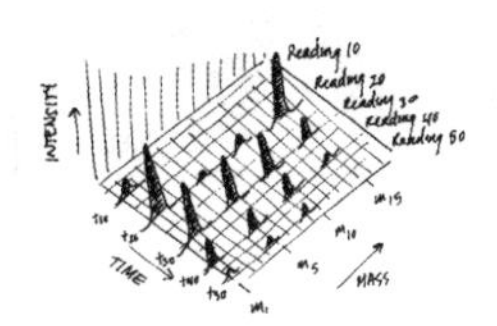

The specialized sample introduction techniques described in Chapter 20 were mainly developed as a result of a basic limitation of ICP-MS in carrying out elemental determinations on complex sample matrices or to enhance sample throughput. However, even though all of these sampling accessories significantly improved the flexibility, performance, and productivity of the technique, they were still being used to measure the total metal content of the samples being analyzed. If the requirement is to learn more about the oxidation state or speciated form of the element, the trace metal analytical community had to look elsewhere for answers. Then, in the early 1990s, researchers started investigating the use of ICP-MS as a detector for chromatography systems, which triggered an explosion of interest in this exciting new hyphenated technique, especially for environmental and biomedical applications. In this chapter, we look at what drove this research and discuss, the use of chromatographic separation techniques including high-performance liquid chromatography (HPLC) with ICP-MS to carry out trace element speciation determinations. This chapter will be of particular interest, if there is a future requirement to measure different species or oxidation state of arsenic and/or mercury (or any other elemental contaminant) in a cannabis-related sample.

ICP-MS has gained popularity over the years, mainly because of its ability to rapidly quantitate ultratrace metal contamination levels. However, in its basic design, ICP-MS cannot reveal anything about the metal's oxidation state, alkylated form, how it is bound to a biomolecule, or how it interacts at the cellular level. The desire to understand in what form or species an element exists led researchers to investigate the combination of chromatographic separation devices with ICP-MS. The ICP mass spectrometer becomes a very sensitive detector for trace element speciation studies when coupled to a chromatographic separation device based on HPLC, low-pressure liquid chromatography, ion chromatography (IC), gas chromatography (GC), size exclusion chromatography (SEC), capillary electrophoresis (CE), etc. In these hyphenated techniques, elemental species are separated based on their chromatographic retention, mobility, or molecular size, and then eluted/passed into the ICP mass spectrometer for detection.[1] The intensities of the eluted peaks are then displayed for each isotopic mass of interest, in the time domain, as shown in Figure 21.1. The figure shows a typical time-resolved chromatogram for a selected group of masses.

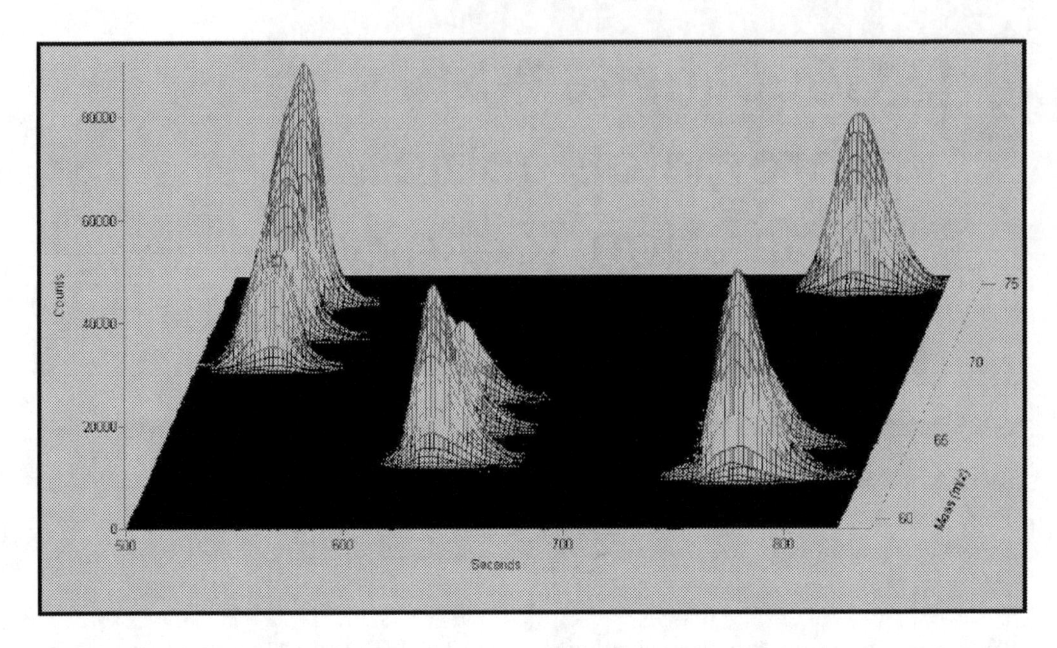

FIGURE 21.1 A typical time-resolved chromatogram generated using chromatography coupled with ICP-MS, showing a temporal display of intensity against mass. (Copyright 2013, all rights reserved, PerkinElmer Inc.)

There is no question that ICP-MS has allowed researchers in the environmental, biomedical, geochemical, and nutritional fields to gain a much better insight into the impact of different elemental species on humans and their environment. Even though elemental speciation studies were being carried out using other atomic spectrometry (AS) detection techniques, it was the commercialization of ICP-MS in the early 1980s, with its extremely low-detection capability, that saw a dramatic increase in the number of trace element speciation studies being carried out. Today, the majority of these studies are being driven by environmental regulations. In fact, the US Environmental Protection Agency (EPA) has published a number of speciation methods involving chromatographic separation with ICP-MS, including Method 321.8 for the speciation of bromine compounds in drinking water and wastewater and Method 6800 for the measurement of various metal species in potable and wastewaters by isotope dilution mass spectrometry. However, other important areas of interest include nutritional and metabolic studies, toxicity testing, bioavailability measurements, and now with the new USP/ICH elemental impurity guidelines, the measurement of different arsenic and mercury species in pharmaceutical and nutraceutical materials. Speciation studies cross over many different application areas, but the majority of determinations being carried out can be classified into three major categories:

- **Measurement of different oxidation states**: For example, hexavalent chromium, Cr(VI), is a powerful oxidant and is extremely toxic, but in soil and water systems, it reacts with organic matter to form trivalent chromium, Cr(III), which is the more common form of the element and an essential micronutrient for plants and animals.[2]
- **Measurement of alkylated forms**: Very often the natural form of an element can be toxic, although its alkylated form is relatively harmless or vice versa. A good example of this is the element arsenic. Inorganic forms of the element such as As(III) and As(V) are toxic, whereas many of its alkylated forms such as monomethylarsonic acid (MMA) and dimethylarsonic acid (DMA) are relatively innocuous.[3]
- **Measurement of metallobiomolecules**: These molecules are formed by the interaction of trace metals with complex biological molecules. For example, in animal farming studies,

TABLE 21.1
Some Typical Inorganic and Organic Species That Have Been Studied by Researchers Using Chromatographic Separation Techniques

Oxidation States	Alkylated Forms	Biomolecules
Se^{+4}	Methyl—Hg, Ge, Sn, Pb, As, Sb, Se, Te, Zn, Cd, Cr	Organometallic complexes—As, Se, Cd
Se^{+6}	Ethyl—Pb, Hg	Metalloporphyrines
As^{+3}	Butyl, phenyl, cyclohexyl—Sn	Metalloproteins
As^{+5}		Metallodrugs
Sn^{+2}		Metalloenzymes
Sn^{+4}		Metals at the cellular level
Cr^{+3}		
Cr^{+6}		
Fe^{+2}		
Fe^{+3}		

the activity and mobility of an innocuous arsenic-based growth promoter are determined by studying its metabolic impact and excretion characteristics. So, measurement of the biochemical form of arsenic is crucial in order to know its growth potential.[4]

Table 21.1 represents a small cross section of both inorganic and organic species of interest classified under these three categories.

There are 400–500 speciation papers published every year, the majority of which are based on toxicologically significant elements such as As, Cr, Hg, Se, and Sn.[5] The following is a small selection of some of the most recent research that can be found in the public domain:

- Determination of chromium(VI) in drinking water samples, using HPLC-ICP-MS[6]
- Determination of trivalent and hexavalent chromium in pharmaceutical and biological materials by IC-ICP-MS[7]
- The use of LC-ICP-MS in better understanding the role of inorganic and organic forms of selenium in biological processes[8]
- Identification of selenium compounds in contaminated estuarine waters using IC-ICP-MS[9]
- Determination of organoarsenic species in marine samples, using cation exchange HPLC-ICP-MS[10]
- Determination of biomolecular forms of arsenic in chicken manure by CE-ICP-MS[11]
- Analysis of tributyltin (TBT) in marine samples using HPLC-ICP-MS[12]
- Measurement of anticancer platinum compounds in human serum by HPLC-ICP-MS[13]
- Bioavailability of cadmium and lead in beverages and foodstuffs using SEC-ICP-MS[14]
- Investigation of sulfur speciation in petroleum products by capillary gas chromatography with ICP-MS detection[15]
- Analysis of methyl mercury in water and soils by HPLC-ICP-MS[16]
- Analysis of arsenic and mercury species from botanicals and dietary supplements using LC-ICP-MS.[17]

HPLC COUPLED WITH ICP-MS

It can be seen from this brief snapshot of speciation publications that by far the most common chromatographic separation techniques being used with ICP-MS are the many different types of HPLC, such as adsorption, ion exchange, size exclusion, gel permeation, and normal- or reverse-phase chromatography. To get a better understanding of how the technique works, particularly when

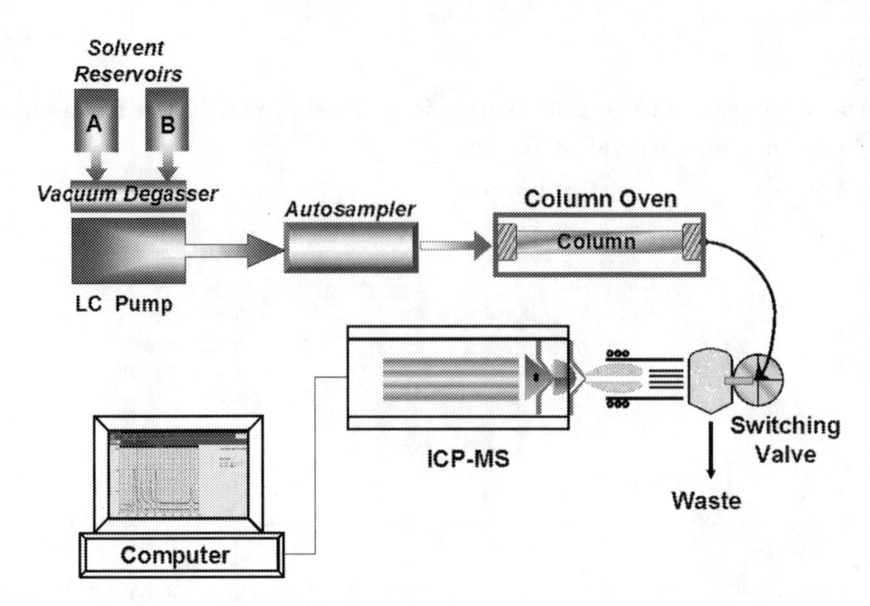

FIGURE 21.2 A typical configuration of an HPLC system interfaced with an ICP mass spectrometer. (Copyright 2013, all rights reserved, PerkinElmer Inc.)

attempting to develop a routine method to simultaneously measure multiple species in the same analytical run, let us take a more detailed look at how the HPLC system is coupled to the ICP mass spectrometer. Figure 21.2 shows a typical setup of the hardware components.

The coupling of the ICP-MS system to the liquid chromatograph hardware components is relatively straightforward, connecting a capillary tube from the end of the HPLC column, through a switching valve, to the sample introduction system of the ICP mass spectrometer. However, matching the column flow with the uptake of the ICP-MS sample introduction system is not a trivial task. Therefore, to develop a successful trace element speciation method, it is important to optimize not only the chromatographic separation but also selection of the nebulization process to match the flow of the sample being eluted off the column, together with finding the best ICP-MS operating conditions for the analytes or species of interest. Let us take a closer look at this.

CHROMATOGRAPHIC SEPARATION REQUIREMENTS

Traditionally, the measurement of trace levels of elemental species by HPLC has been accomplished by separating the species using column separation technology and detecting them, one element at a time, as they elute. This approach works well for one element or species but is extremely slow and time-consuming for the determination of multiple species or elements, because the chromatographic separation process has to be optimized for each species being measured. Therefore, the limitation to multielement speciation is rarely the detection of the elements but is usually the separation of the species. The inherent problem is that liquid chromatography works on the principle of equilibration between the species of interest, the mobile phase, and the column material. Because the chemistries of elements and their species differ, it is difficult to find common conditions capable of separating species of more than one element simultaneously. For example, toxic elements of environmental interest, such as arsenic (As), chromium (Cr), and selenium (Se), have different reaction chemistries, thus requiring different chromatographic conditions to separate them. So, to achieve the analytical goal of fast, simultaneous measurement of different species of these elements in a single sample injection, the chromatographic separation has to be optimized.

Let us first examine the chromatography, focusing on two common forms of separation (ion exchange and reversed-phase ion-pairing chromatography), together with two commonly employed HPLC elution techniques (isocratic and gradient). Each process will be described, and the advantages and disadvantages of each will be discussed in the context of choosing a separation scheme to achieve our analytical goals. One final note before we discuss the separation process: it is extremely important when preparing the sample that the speciated form of the element not be changed or altered in any way. This is not such a serious problem with a simple matrix such as drinking water, which typically involves straightforward acidification. However, when more complex samples such as soils, biological samples, or pharmaceutical matrices are being analyzed, it is quite common to use strong acids, oxidizing agents, or high temperatures to get the samples into solution. So, maintaining the integrity of the valency/oxidation state or species should always be an extremely important decision when preparing a sample for speciation analysis.

ION EXCHANGE CHROMATOGRAPHY (IEC)

In this technique, separation is based on the exchange of ions (anions or cations) between the mobile phase and the ionic sites on a stationary phase bound to a support material in the column. A charged species is covalently bound to the surface of the stationary phase. The mobile phase, typically a buffer solution, contains a large number of ions that have a charge opposite to that of the surface-bound ions. These mobile-phase ions, referred to as counter ions, establish equilibrium with the stationary phase. Sample ions passing through the column and having the same ionic charge as the counterion compete with the mobile-phase ions for sites on the column material, resulting in a disruption of the equilibrium and retention of that analyte. It is the differential competition of various analytes with the mobile-phase ions that ultimately produces species separation and chromatographic peaks, as detected by ICP-MS. Therefore, analyte retention is based on the affinity of different ions for the support material and on other solution parameters, including counterion type, ionic strength (buffer concentration), and pH. Varying these mobile-phase parameters changes the separation.

The principle of IEC is schematically represented in Figure 21.3 for an anion exchange separation. In this figure, the anions are denoted by A^- and represent the analyte species, whereas the counterions are symbolized by X^-. The main benefit of this approach is that it has a high tolerance to matrix components, and the main disadvantages are that these columns tend to be expensive and somewhat fragile.

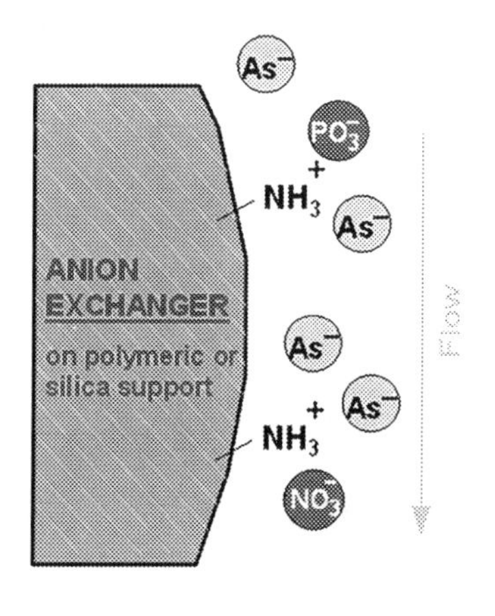

FIGURE 21.3 Principle of separation using anion exchange chromatography.

REVERSED-PHASE ION PAIR CHROMATOGRAPHY (RP-IPC)

Ion pair chromatography typically uses a reversed-phase column in conjunction with a special type of chemical in the mobile phase called an *ion-pairing reagent.* "Reversed phase" essentially means that the column's stationary phase is nonpolar (less polar, more organic) than the mobile-phase solvents. For reversed-phase ion pair chromatography (RP-IPC), the stationary phase is a carbon chain, most often consisting of 8 or 18 carbon atoms (C8, C18) bonded to a silica support. The mobile-phase solvents usually consist of water mixed with a water-miscible organic solvent, such as acetonitrile or methanol.

The ion-pairing reagent is a compound that has both an organic and an ionic end. To promote ionic interaction, the ionic end should have a charge opposite to that of the analytes of interest. For anionic IPC, a commonly used ion-pairing reagent is tetrabutylammonium hydroxide (TBAOH); for cationic IPC, hexanesulfonic acid (HSA) is often chosen. In principle, the ionized/ionizable analytes interact with the ionic end of the ion-pairing reagent, whereas the ion-pairing reagent's organic end interacts with the C18 stationary phase. Retention and separation selectivity are primarily affected by characteristics of the mobile phase, including pH, selection/concentration of ion-pairing reagent, and ionic strength, as well as the use of additional mobile-phase modifiers. With this scheme, a reversed-phase column acts like an ion exchange column.

The benefits of this type of separation are that the columns are generally less expensive and tend to be more rugged than ion exchange columns. The main disadvantages, compared to IEC, are that the separation may not be as good, and there may be less tolerance to high-sample matrix concentrations. A simplified representation of this approach is shown in Figure 21.4.

COLUMN MATERIAL

As previous explained, several variables of the mobile phase can be modified to affect separations, both by IEC and RP-IPC. A very important consideration when choosing a column is the pH of the mobile phase used to achieve the separation. The pH is critical in the selection of the column for both IEC and RP-IPC because the support materials may be pH sensitive. One of the most common column materials is silica, because it is inexpensive, rugged, and has been used for many years. The main disadvantage of silica columns is that they are useful only over a pH range of 2–8. Outside this range, the silica dissolves. For applications requiring pHs outside of this range, polymer columns are commonly used. Polymer columns are useful over the entire pH range. The downside to these columns is that they are typically more expensive and fragile. Columns consisting of other support materials are also available.

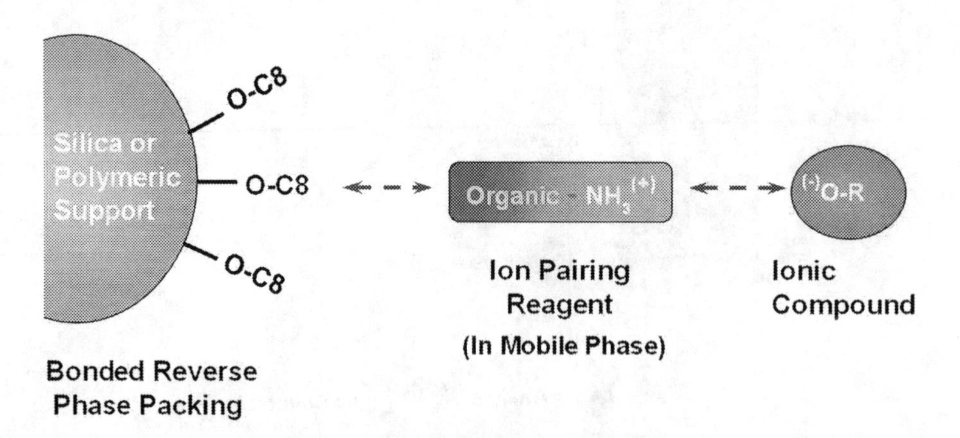

FIGURE 21.4 Principle of ion pair chromatography (anionic mode).

ISOCRATIC OR GRADIENT ELUTION

Another consideration is how best to elute the species from the column. This can be accomplished using either an isocratic or gradient elution scheme. Isocratic refers to using the same solvent throughout the analysis. Because of this, samples can be injected immediately after the preceding one has finished eluting. This type of elution can be performed on all HPLC pumping systems. The advantages of isocratic elutions are simplicity and higher sample throughput.

Gradient elution involves variation of the mobile-phase composition over time by a number of steps such as changing the organic content, altering the pH, changing concentration of the buffer, or using a completely different buffer. This is usually accomplished by having two or more bottles of mobile phases connected to the HPLC pump. The pump is then programmed to vary the amount of each mobile-phase component. The mobile-phase components are mixed online before reaching the column, and therefore, a more complex pumping system is required to handle this task.

The main reason for using gradient elution is that the mobile-phase composition can be varied to fine-tune the separation. As a result, component separations are better, and chromatograms are generally shorter than with isocratic elutions. However, overall sample throughput is much lower than with isocratic separations because of the variation in mobile-phase composition. So, after an elution is complete, the mobile phase must be changed back to its original composition as at the start of the chromatogram. This means that the equilibrium between the mobile phase and the column must be reestablished before the next sample can be injected. If the equilibrium is not established, the peaks will elute at different times compared to the previous sample. The equilibration time varies with column, but it is usually between 5 and 30 min.

Therefore, if a fast, automated, routine method for the measurement of multispecies/elements is the desired analytical goal, it is often best to attempt an isocratic separation method first, because of the complexity of method development and the low-sample throughput of gradient elution methods. In fact, a simultaneous method for the separation of As, Cr, and Se species in drinking water samples was demonstrated by Neubauer and coworkers; they developed a method to determine inorganic forms of arsenic (As^{+3}, As^{+5}), chromium (Cr^{+3}, Cr^{+6}), and selenium (Se^{+4}, Se^{+6}, and $SeCN^{-}$) by reverse-phase ion-pairing chromatography with isocratic elution.[18] Details of the HPLC separation parameters/conditions they used are shown in Table 21.2.

It is important to emphasize that if these species were being measured on a single-element basis, the optimum chromatographic conditions would be different. However, the goal of this study was to determine all the species in a single multielement run so that the method could be applied to a

TABLE 21.2
HPLC Separation Parameters/Conditions for Measuring Inorganic As, Cr, and Se Species in Potable Waters

HPLC configuration	Quaternary pump, column oven, and autosampler
Column	C8, reduced activity, 3.3 cm × 0.46 cm (3 μm packing)
Column temperature	35°C
Mobile phase	mM TBAOH + 0.15 mM NH_4CH_2COOH + 0.15 mM EDTA (K salt) + 5% MeOH
pH	7.5
pH adjustment	Dilute HNO_3, NH_4OH
Injection volume	50 mL
Sample flow rate	1.5 mL/min
Samples	Various potable waters
Sample preparation	Dilute with mobile phase (2–10); heat at 50°C–55°C for 10 min

Source: K. R. Neubauer, P. A. Perrone, W. Reuter, R. Thomas, *Current Trends in Mass Spectroscopy*, May 2006.

routine, high-throughput environment. The 3.3-cm-long column was packed with a C8 hydrocarbon material (3 μm particle size). The mobile phase was a mixture of TBAOH, ammonium acetate (NH_4CH_2COOH), the potassium salt of EDTA, and methanol. The pH was adjusted to 7.5 (prior to the addition of methanol) using 10% nitric acid and 10% ammonium hydroxide. The sample preparation consisted of dilution with the mobile phase and heating at 50°C–55°C to speed the formation of the Cr III–EDTA complex. For each analysis, 50 μL of sample was injected into the column, which resulted in a sample flow of 1.5 mL/min being eluted off the column into the ICP mass spectrometer.

SAMPLE INTRODUCTION REQUIREMENTS

When coupling an HPLC system to an ICP mass spectrometer, it is very important to match the flow of sample being eluted off the column with the ICP-MS nebulization system. With today's choice of sample introduction components, there are specialized nebulizers and spray chambers on the market that can handle sample flows from 20 μL/min up to 3,000 μL/min. The most common type of nebulizer used for chromatography applications is the concentric design, because it is self-aspirating and it generates an aerosol with extremely small droplets, which tends to produce better signal stability compared to a cross-flow design. The choice of which type of concentric nebulizer to use should therefore be based on the sample flow coming off the column. If the sample flow is on the order of 1 mL/min, a higher-flow concentric nebulizer should be used, and if the flow is much lower, such as in nano- or microflow LC work, a specialized low-flow nebulizer should be used.

It is also very important that the dead volume of the sample introduction process be kept to an absolute minimum to optimize peak integration of the separated species over the length of the transient signal. For this reason, the length of sample capillary from the end of the column to the nebulizer should be kept to a minimum; the internal volume of the nebulizer should be as small as possible; the connectors/fittings/valves should all have low dead volume; and a self-aspirating nebulizer should be used to avoid the need for peristaltic pump tubing. In addition, a spray chamber with a short aerosol path should be selected, which will not add additional dead volume to the method. However, depending on the total flow of the sample and the type of nebulizer, a spray chamber may not even be required. Cutting down on the sample introduction dead volume and minimizing peak broadening by careful selection of column technology will ultimately dictate the number of species that can be separated in a given time—an important consideration when developing a routine method for high-sample throughput. Figure 21.5 shows a high-efficiency concentric nebulizer (HEN) designed for HPLC-ICP-MS work, and Figure 21.6 shows the difference between the capillary of this nebulizer (on the right) and a standard concentric type A nebulizer (on the left).

Another important reason to match the nebulizer with the flow coming off the column is that concentric nebulizers are mainly self-aspirating. For this reason, the column flow must be high

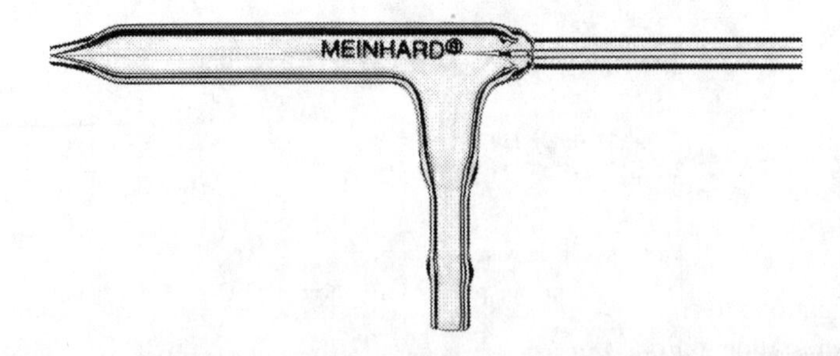

FIGURE 21.5 An HEN designed for HPLC-ICP-MS work. (Courtesy of Meinhard Glass Products.)

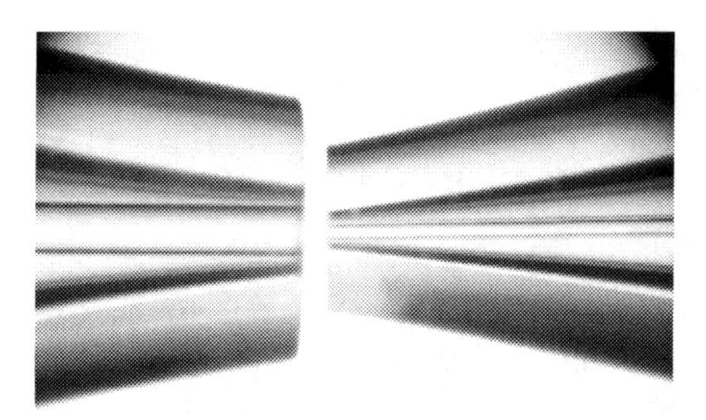

FIGURE 21.6 The difference between the capillary of an HEN for HPLC work (on the right) and a standard concentric type A nebulizer (on the left). (Courtesy of Meinhard Glass Products.)

enough to ensure that the nebulizer can sustain a consistent and reproducible aerosol. On the other hand, if the column flow is too low, a makeup flow might need to be added to the column flow to meet the flow requirements of the nebulizer being used. This has the additional benefit of being able to add an internal standard after the column with another pump to correct for instrument drift or matrix effects in gradient elution work.

OPTIMIZATION OF ICP-MS PARAMETERS

In the early days of trace element speciation studies using chromatography coupled with ICP-MS, researchers had no choice but to interface their own LC pumps, columns, and autosamplers, etc., to the ICP mass spectrometer, because off-the-shelf systems were not commercially available. However, the analytical objectives of a research project are a little different from the requirements for routine analysis. With a research project, there are fewer time constraints to optimize the chromatography and detection parameters, whereas in a commercial environment, there are often financial penalties if the laboratory cannot be up, running real samples and generating revenue as quickly as possible. This demand, especially from commercial laboratories in the environmental and biomedical communities, for routine trace element speciation methods convinced the instrument manufacturers and vendors to develop fully integrated HPLC-ICP-MS systems.

The availability of these off-the-shelf systems rapidly drove the growth of this hyphenated technique, so much so that vendors who did not offer it with full application and hardware/software support were at a disadvantage. As this technique is maturing and is being used more as a routine analytical tool, it is becoming clear that the requirements of the ICP-MS system doing speciation analysis are different from those of an instrument carrying out trace element determinations using conventional nebulization. With that in mind, let us take a closer look at the typical requirements of an ICP mass spectrometer that is being utilized as a multielement detector for trace element speciation studies.

COMPATIBILITY WITH ORGANIC SOLVENTS

The requirements of the sample introduction system, and in particular the nebulization process, have been described earlier in this chapter. However, some reverse-phase HPLC separations use gradient elution with mixtures of organic solvents such as methanol or acetonitrile. If this is the case, consideration must be given to the fact that some volatile organic solvents will extinguish the plasma.[19] Therefore, modifications to the sample introduction might need to be made, such as adding small amounts of oxygen to the nebulizer gas flow or perhaps using a cooled spray chamber or a desolvation device to stop the buildup of carbon deposits on the sampler cone. Other approaches such as direct injection nebulization have been used to introduce the sample eluent into the ICP-MS, but historically, they have not gained widespread acceptance because of usability issues.

COLLISION/REACTION CELL OR INTERFACE CAPABILITY

Another requirement of the ICP-MS system for speciation work is the collision/reaction cell/interface capability. It is becoming clear that as more and more speciation methods are being developed, the ability to minimize polyatomic spectral interferences generated by the solvent, buffer, mobile phase, pH-adjusting acids/bases, and the plasma gas, etc., is of crucial importance. Take, for example, the elements discussed earlier. As, Cr, and Se are notoriously difficult elements for ICP-MS analysis because their major isotopes suffer from argon- and sample-based polyatomic interferences. Arsenic has only one isotope (*m/z* 75), which is difficult to quantify in chloride-containing samples because of the presence of $^{40}Ar^{35}Cl^+$. Low-level chromium analysis is difficult because of the presence of the $^{40}Ar^{12}C^+$ and $^{40}Ar^{13}C^+$ interferences, which overlap the two major isotopes of chromium at masses 52 and 53. These interferences are nearly always present but are especially strong in samples with organic content. The argon dimers ($^{40}Ar^{40}Ar^+$,$^{40}Ar^{38}Ar^+$) at masses 80 and 78 interfere with the major isotopes of Se at mass 80 and 78, respectively, and bromine, which is usually present in natural waters, forms $^{79}BrH^+$ and $^{81}BrH^+$, which interferes with the Se masses at 80 and 82, respectively. The effects of these and other interferences have been reduced somewhat with conventional ICP-MS instrumentation, using alternate masses, interference correction equations, cool plasma technology, and desolvation techniques, but these approaches have not shown themselves to be particularly useful for these elements, especially at ultratrace levels.

For these reasons, it will be very beneficial if the ICP-MS instrumentation is fitted with a collision or reaction cell or interface and has the capability to minimize the formation of these undesired polyatomic interferences, using either collisional mechanisms with kinetic energy discrimination,[20,21] or ion–molecule reaction kinetics with mass bandpass tuning,[22] or by introduction of a collision/reaction gas into the interface region.[23] The best approach will depend on the type of sample being analyzed, the number of species being determined, and the detection limit requirements, but there is no doubt that for multielement work, it is beneficial if one gas can be used for all the analyte species. This is exemplified by the research of Neubauer and coworkers described earlier, who used oxygen as the reaction gas to reduce the polyatomic spectral interferences described earlier to determine seven different species of As, Cr, and Se in potable water. In fact, even though the oxygen was used to remove the argon–carbide and argon dimer interferences to quantify the chromium and selenium species, respectively, it was used in a different way to quantify the arsenic species. It was used to react with the arsenic ion to form the arsenic–oxygen molecular ion ($^{75}As^{16}O$) at mass 91 and move it away from the argon–chloride interference at mass 75. This novel approach has been reported many times in the literature[22,24] and has in fact been approved by the EPA in the recent ILM05.4 analytical procedure update for their Superfund, Contract Laboratory.

The instrumental conditions for the speciation analysis are shown in Table 21.3, and a plot of signal intensity versus time of the simultaneous separation of a 1 µg/L standard of As, Cr, and Se is shown in Figure 21.7.

It should be emphasized that even though oxygen was used as the reaction gas in this study, many other collision/reaction strategies using inert gases such as helium and low-reactivity gases such as hydrogen have been successfully used for the determination of As, Cr, and Se. The question is whether a single gas can be used to remove all the interferences to an acceptable level and allow quantitation at the trace level. If more than one gas is required, that is not such a major hardship, especially as all gas flows are under computer control and can be changed in a multielement run. However, if the method needs to be automated for a high-sample-throughput environment, it will be significantly faster if only one gas is used for interference removal, so that the collision/reaction cell conditions need not be changed for each element.

TABLE 21.3
ICP-MS Instrumental Conditions for the Speciation Analysis of As^{+3}, As^{+5}, Cr^{+3}, Cr^{+6}, Se^{+4}, Se^{+6}, and SeCN in Potable Water Samples

Nebulizer	Quartz concentric
Spray chamber	Quartz cyclonic
RF power	1,500 W
Collision/reaction cell technology	Dynamic reaction cell
Reaction gas	O_2 = 0.7 mL/min
Analytical masses	AsO^+ (*m/z* 91), Se^+ (*m/z* 78), Cr^+ (*m/z* 52)
Analyte species	As^{+3}, As^{+5}, Cr^{+3}, Cr^{+6}, Se^{+4}, Se^{+6}, $SeCN^-$
Dwell time	330 ms (per analyte)
Analysis time	5.5 min

Source: K. R. Neubauer, P. A. Perrone, W. Reuter, R. Thomas, *Current Trends in Mass Spectroscopy*, May 2006.

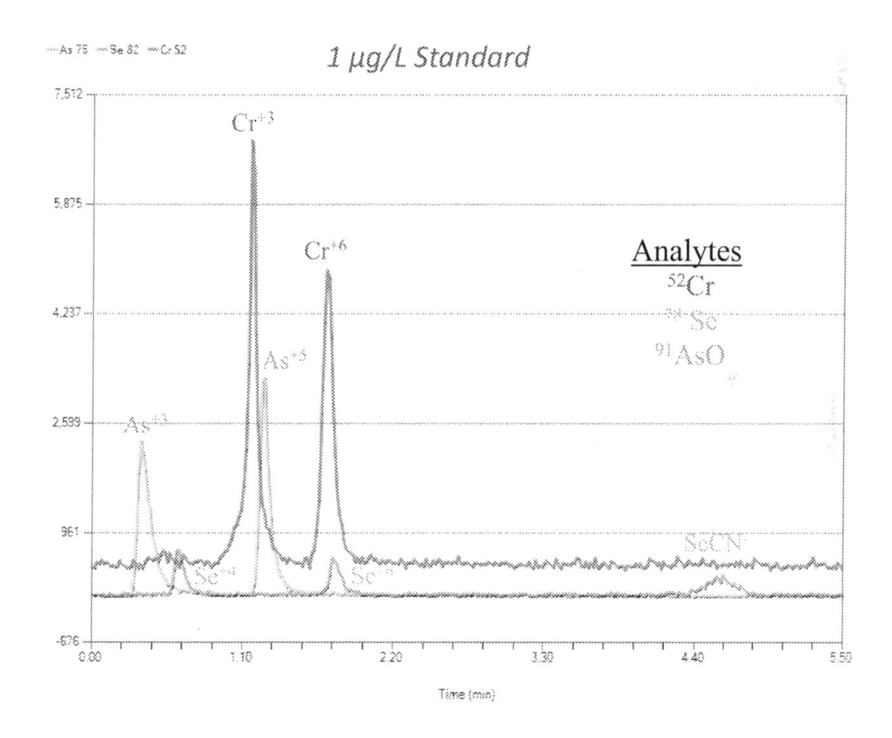

FIGURE 21.7 A plot of signal intensity versus time of the simultaneous separation of a 1 µg/L standard of As, Cr, and Se species. (From K. R. Neubauer, P. A. Perrone, W. Reuter, R. Thomas, *Current Trends in Mass Spectroscopy*, May 2006.)

OPTIMIZATION OF PEAK MEASUREMENT PROTOCOL

It can be seen from the chromatogram that the total separation time for the seven inorganic species is on the order of 2 min for the common oxidation states of the elements and about 5 min if there is any selenocyanide in the sample. This means that all the peaks have to be integrated and quantified in a transient signal lasting 2–5 min. So, even though this is not considered a short transient such as an electrothermal vaporizer (ETV) or small-spot laser ablation work that typically lasts 2–5 s,

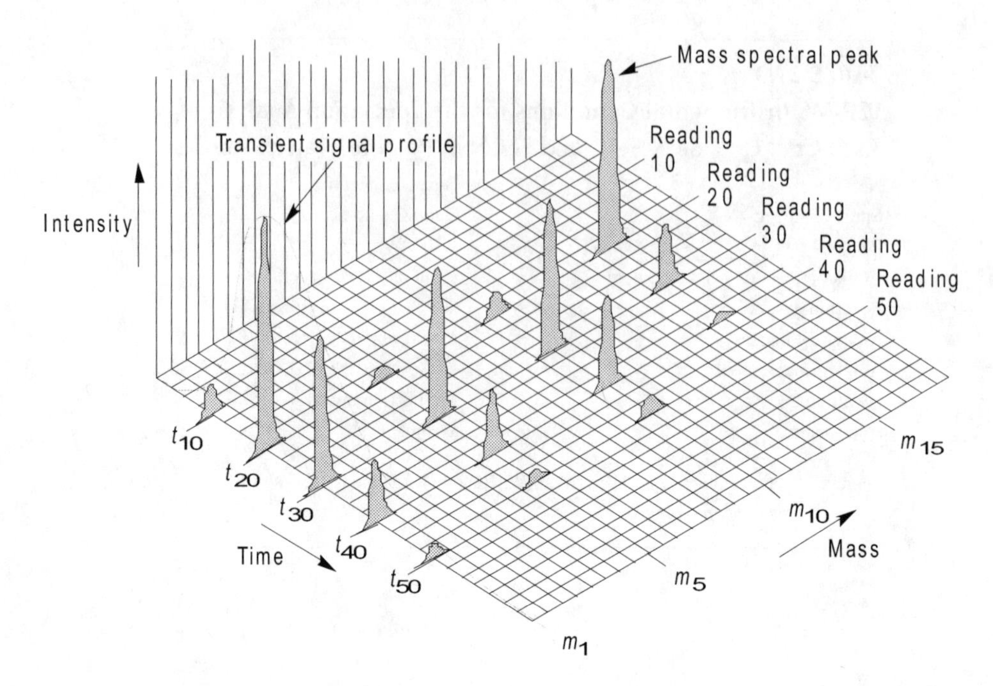

FIGURE 21.8 The temporal separation of a group of elemental species in a chromatographic transient peak.

it is not a continuous signal generated by a pneumatic nebulizer. For that reason, it is critical to optimize the measurement time in order to achieve the best multielement signal-to-noise ratio in the sampling time available. This is demonstrated in Figure 21.8, which shows the temporal separation of a group of elemental species in a chromatographic transient peak. The plot represents signal intensity against mass over the time period of the chromatogram. To get the best detection limits for this group of elements or species, it is very important to spend all the available time quantifying the peaks of interest.

For this reason, the quadrupole scanning/settling time and the time spent measuring the analyte peaks must be optimized to achieve the highest signal quality. This is described in greater detail in Chapter 15 "Peak Measurement Protocol" and basically involves optimizing the number of sweeps, selecting the best dwell time, and using short settling times to achieve the highest measurement duty cycle and maximize the peak signal-to-noise ratio over the duration of the transient event. If this is not done, there is a strong likelihood that the quality of the speciation data could be compromised. In addition, if the extended dynamic range is used to determine higher concentrations of elemental species, the scanning and settling time of the detector will also have an impact on the quality of the signal. For that reason, detectors that require two scans to characterize an unknown sample will use up valuable time in the quantitation process. This is somewhat of a disadvantage when doing multi-element speciation on a chromatographic transient signal, especially if you have limited knowledge of the analyte concentration levels in your samples.

FULL SOFTWARE CONTROL AND INTEGRATION

In the early days of coupling HPLC components with ICP-MS, there were very few sophisticated communication protocols between the two devices. The sample was injected into the chromatographic separation system, and when the analyte species was close to being eluted off the column, the read cycle of the ICP mass spectrometer was initiated manually to capture the data using the instrument's time-resolved software. Processing and manipulation of the data was then carried out after the chromatogram had been captured, sometimes by a completely different software program. The HPLC and the ICP-MS were considered almost two distinctly different hardware and software devices, with very little communication between them. This was even more surprising considering that many of the ICP-MS vendors also offered LC equipment. As you can imagine, this was not the ideal scenario for a laboratory that wanted to carry out automated speciated analysis. As a result of this demand, manufacturers realized that unless they offered fully integrated hardware and software solutions for trace element speciation work, it was never going to be accepted as a routine analytical tool. Today, just about all the ICP mass spectrograph manufacturers offer fully integrated HPLC-ICP-MS systems. Some of the features available on today's instrumentation include the following:

- Full software and hardware control of both the chromatograph and the ICP-MS system from one computer
- Ability to check the status of the ICP-MS system when setting up the chromatography method or vice versa
- Computer control of the switching valve to allow the HPLC to be used in tandem with the ICP-MS normal sample introduction system
- Real-time display of spectral data, including peak identification and quantitation, while the chromatogram is being generated
- Full data-handling capability, including manipulation of spectral peaks and calibration curves
- Comprehensive reporting options or exportable data to third-party programs.

It is also important to emphasize that most vendors also offer integrated systems with turnkey application methods that are almost ready to run samples as soon the instrument is installed. Although this is not available for all applications, it is becoming a standard offering for some of the more routine environmental applications. Manufacturers are also realizing that most analytical chemists who are experienced at trace element analysis may have very little expertise in chromatography. For that reason, they are providing full backup and customer support with HPLC application specialists as well as with the traditional ICP-MS product specialists.

FINAL THOUGHTS

Combining chromatography with ICP-MS has revolutionized trace metal speciation analysis. In particular, when HPLC or even low-pressure LC systems are coupled with the selectivity and sensitivity of ICP-MS, many elemental species at sub-ppb levels can now be determined in a single sample injection. When interference reduction methods such as collision/reaction cell or interface technology are available, it is possible to separate and detect different inorganic species of environmental significance in one automated run. By using a fully integrated, computer-controlled HPLC/ICP-MS system and optimizing the chromatographic separation, sample introduction, and ICP-MS detection parameters, simultaneous quantitation can be carried out in a few minutes. There is no question that this kind of sample throughput will arouse the interest of the environmental, biomedical, nutritional, and other application communities interested in trace element speciation studies and help them realize that it is feasible to carry out this kind of analysis in a truly routine manner.

However, it should be emphasized that historically there has limited demand for the determination of elemental species in cannabis-related products, pharmaceutical materials, or dietary supplements. As a result, there is very little application material in the public domain describing the separation and quantitation of inorganic/organic forms of arsenic and mercury in these kinds of sample matrices. For that reason, if there is a need to determine these species, it is strongly advised that a person or a laboratory with expertise in speciated techniques initially develops the method, as many interlaboratory round robin studies have indicated that an optimized sample preparation procedure and chromatographic separation technology are critical to get high-quality data.[25]

FURTHER READING

1. R. Lobinski, I. R. Pereiro, H. Chassaigne, A. Wasik, and J. Szpunar, *Journal of Analytical Atomic Spectrometry,* **13**, 860–867, 1998.
2. G. Cox and C. W. McLeod, *Mikrochimica Acta,* **109**, 161–164, 1992.
3. S. Branch, L. Ebdon, and P. O'Neill, *Journal of Analytical Atomic Spectrometry,* **9**, 33–37, 1994.
4. J. R. Dean, L. Ebdon, M. E. Foulkes, H. M. Crews, and R. C. Massey, *Journal of Analytical Atomic Spectrometry,* **9**, 615–618, 1994.
5. Z. A. Grosser and K. Neubauer, *Today's Chemist at Work*, American Chemical Society: Washington, DC, May 2004.
6. Y. L. Chang and S. J. Jiang, *Journal of Analytical Atomic Spectrometry,* **9**, 858–865, 2001.
7. K. E. Lokits, D. D. Richardson, and J. A. Caruso, *Handbook of Hyphenated ICP-MS Applications*, Agilent Technologies, Page 34, August, 2007, https://www.postnova.com/images/pdf/Applications.pdf
8. K. DeNicola, D. D. Richardson, and J. A. Caruso, *Spectroscopy,* **21**(2), 18–24, 2006.
9. D. Wallschlager and N. Bloom, *Journal of Analytical Atomic Spectrometry,* **16**, 1322–1326, 2001.
10. J. J. Sloth, E. H. Larsen, and K. Julshamn, *Journal of Analytical Atomic Spectrometry* **18**, 452–459, 2003.
11. Rosal, G. Momplaisir, and E. Heithmar, *Electrophoresis,* **26**, 1606–1614, 2005.
12. L. Yang, Z. Mester, and R. E. Sturgeon, *Analytical Chemistry*, **74**, 2968–2975, 2002.
13. V. Vacchina, L. Torti, C. Allievi, and R. Lobinski, *Journal of Analytical Atomic Spectrometry,* **18**, 884–890, 2003.
14. S. Mounicou, J. Szpunar, R. Lobinski, D. Andrey, and C. J. Blake, *Journal of Analytical Atomic Spectrometry,* **17**, 880–886, 2002.
15. B. P. Leonhard, D. Pröfrock, F. Baco, C. Lopez Garcia, S. Wilbur, and A. Prange, *Journal of Analytical Atomic Spectrometry*, **5**, 700–702, 2004.
16. C. M. Jing, and S. Wang, *Handbook of Hyphenated ICP-MS Applications*, Agilent Technologies, Page 12, August, 2007, https://www.postnova.com/images/pdf/Applications.pdf
17. B. Avula, Y.H. Wang, M Wang, A. Khan, *Planta Medica*, **78**, 136–144, 2012.
18. K. R. Neubauer, P. A. Perrone, W. Reuter, and R. Thomas, *Current Trends in Mass Spectroscopy*, MJH Associates, Iselin, NJ, May 2006.
19. M. A. Mennito, E. Debrah, and R. Thomas, *Spectroscopy,* **13**(2), 42–53, 1998.
20. M. Bueno, F. Pannier, M. Potin-Gautier, and J. Darrouzes, *Determination of Organic and Inorganic Selenium Using HPLC-ICP-MS*, Agilent Technologies 7073EN, 2007, https://www.postnova.com/images/pdf/Applications.pdf
21. S. McSheehy and M. Nash, *Determination of Selenomethionine in Nutritional Supplements Using HPLC Coupled to the XSeriesII ICP-MS with CCT*, Thermo Scientific Application Note-40745, 2005, http://www.thermo.com/eThermo/CMA/PDFs/Articles/articlesFile_26474.pdf.
22. J. Di Bussolo, W. Reuter, L. Davidowski, and K. Neubauer, *Speciation of Five Arsenic Compounds in Urine by HPLC-ICP-MS*, Perkin Elmer Inc Application Note-D-6736, 2004, https://www.perkinelmer.com/PDFs/Downloads/app_speciationfivearseniccompounds.pdf.
23. M. Leist and A. Toms, *Low Level Speciation of Chromium in Drinking Waters Using LC-ICP-MS*, The Application Notebook, Spectroscopy Online, March, 2006, http://files.alfresco.mjh.group/alfresco_images/pharma//2014/08/22/e24804f7-20db-4068-97f8-13061098e05a/article-352912.pdf
24. D. S. Bollinger and A. J. Schleisman, *Atomic Spectroscopy,* **20**(2), 60–63, 1999.
25. M. L. Briscoe, T. M. Ugrai, J. Creswell, A. T. Carter, *Spectroscopy Magazine*, **30**(5), 48–61, 2015.

22 A Practical Guide to Reducing Errors and Contamination Using Plasma Spectrochemistry

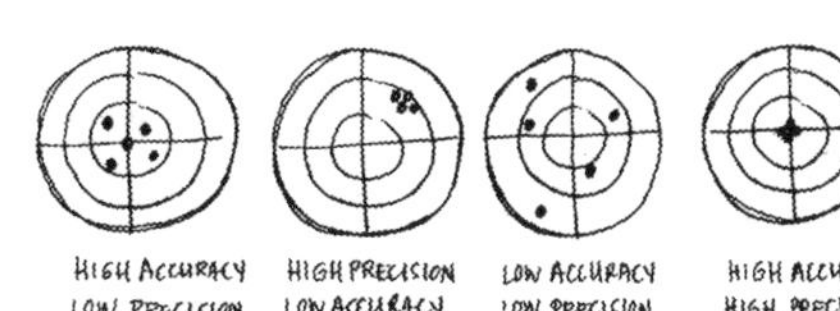

There are many factors that influence the ability to get the correct result with any trace element technique, particularly in today's high-throughput cannabis testing laboratories. Unfortunately with plasma spectrochemistry techniques, the problem is magnified even more because of their extremely high sensitivity. So, in order to ensure that the data reported is an accurate reflection of the sample in its natural state, the analyst must be aware of all the potential sources of contamination that could negatively impact the quality of the measurements. This chapter addresses this topic and suggests ways to minimize sources of errors in the generated data.

Analytical laboratories face more challenges and regulations than ever before as accreditation bodies issue an increasing number of guidelines and regulatory agencies increase the number of elements that need to be reported while the levels of detection required decrease. A lot of times, it's the effort and money invested in deciphering the data and determining its validity and accuracy. This is probably more challenging in the cannabis testing industry where there are few regulated methods and where labs are being set up almost overnight and equipped with very sophisticated instrumentation, sometimes with very little understanding of the ultratrace analytical environment they are being used in.

The methodology to reduce laboratory error and reduce contamination is composed of the three primary steps: First, understanding data validity criteria, the application of appropriate statistics, and the understanding of the use and terminology of analytical processes. The second step is correct selection and use of standards appropriate for the type and range of analysis being performed. The third step is the understanding of sample preparation and sources of contamination which can cause error to be introduced into analyses.

UNDERSTANDING DATA ACCURACY AND PRECISION

All analytical laboratories pursue "good" data and "true" values. The reality is that true values are never absolute. True values are obtained by perfect and error-free measurements that do not exist in reality. Instead, the expected, specified, or theoretical value becomes the accepted true value. Analysts then compare the observed or measured values against that accepted true value to determine accuracy or "trueness" of the data set.

Often accuracy and precision are used in the same context when discussing data quality. Accuracy and precision are very different assessments of data and the acquisition process. Accuracy

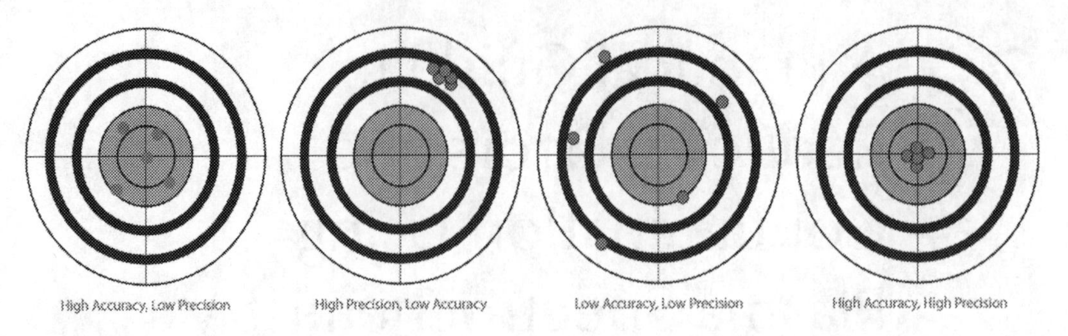

FIGURE 22.1 Representations of accuracy and precision.

TABLE 22.1
Conditions for Precision

	Repeatability	Intermediate Precision	Reproducibility
Laboratory	Same	Same	Different
Operator	Same	Different	Different
Apparatus	Same	Same in type or actual apparatus	Different
Time between replicates	Less than a day	Multiple days	Not specified

is the measurement of individual or groups of data points in relationship to the "true" value. In essence, accuracy is how close your data gets to the target and is often expressed as either a form of numerical or percent difference of the observed result and the target or "true" value.

Precision, on the other hand, is the measurement of a data set for how well the data points relate to each other. It is the measure of how clustered the data points fall within the target range and is often expressed as a calculation of standard deviation of the data set in some form. Precision is an important tool for the evaluation of instrumentation and methodologies by determining how data is produced after varied replications. The differences between accuracy and precision are exemplified in Figure 22.1.

Repeatability and reproducibility measure the quality of the data, method, or instrumentation by examining the precision under the same (minimal difference) or different (maximal difference) test conditions. Repeatability (or test–retest reliability) is the measurement of variation arising when all the measurement conditions are kept constant, such as the same location, the same procedure, the same operator, or the same instrument under the same conditions in repetition over a short period of time. Several standards organizations such as ASTM define the parameters of repeatability, intermediate precision, and reproducibility in order to create and publish test methods (Table 22.1).[1,2]

Reproducibility is the measurement of variation arising in the same measurement process occurring across different conditions such as location, operator, or instruments, and over long periods of time.

Another way of looking at accuracy and precision is in terms of measurement of different types of error.

ESTIMATING ERROR

The most common misconception regarding analytical data revolves around the concepts of error, mistakes, and uncertainty. In general, an error is a deviation or difference between the estimated or measured value and the true, specified, or theoretically correct value. If accuracy is the measurement

TABLE 22.2
Accuracy and Precision for Absolute and Relative Errors

	Absolute	Relative
Accuracy	$E_{abs} = X_o - X_t$	$E_{rel} = X_o - X_t/X_t$
Precision	σ of data set or value taken from a curve	RSD or CV of data set

of the difference between a result and a "true" value, then error is the actual difference or the cause of the difference. The estimation of error can be calculated in two ways, either as an absolute or relative error. Absolute errors are expressed in the same units as the data set and relative errors are expressed as ratios such as percent and fractions.

Absolute accuracy error is the true value subtracted from an observed value and is expressed in the same units as the data. For example, if a stated expected true value of an analysis is 5 ppm but the resulting value is 6 ppm, then the absolute error for that data point is 1 ppm. Relative accuracy error is the true value subtracted from the observed value and the result is divided by the true value. Errors in precision data are most commonly calculated as some variation of the standard deviation of the data set. An absolute precision error calculation is based on either the standard deviation of a data set or values taken from a plotted curve. A relative precision error is most commonly expressed as relative standard deviation (RSD) or the coefficient of variance (CV or %RSD) of the data set (Table 22.2).

TYPES OF ERRORS

There are many types of errors associated with scientific and statistical analyses. The most common error in regards to data is **observational or measurement errors** which are the difference between a measured valve and its true value. Most measured values contain an inherent aspect of variability as part of the measurement process which can be classified as either systematic error or random error.

Systematic (or determinate) errors are introduced inaccuracies from the measurement process or analytical system. There are some basic sources for systematic error in data. These sources are operator or analyst, apparatus and laboratory environment, and method or procedure. Systematic errors can often be reduced or eliminated. Operator or analyst errors can occur due to inattentiveness, lack of training, or misinformation. Operator or analyst errors are most often called mistakes. Apparatus or laboratory environment errors can occur with improper maintenance, substandard laboratory environment, and materials (i.e., improper volumetric, improper calibration, poor environmental temperature, and humidity controls). Method or procedure errors can occur with poor method validation or lack of periodic updates as equipment or materials change. In cases where systematic errors lead to results in a data set that trend higher or lower than the "true" value, the difference is considered to be a bias. A positive bias creates a trend where results are higher than the expected value, while a negative bias displays results lower than the expected value. Determinate errors and bias can often either be corrected or adjusted for in the instrumentation or procedure.

Random (or indeterminate) errors lead to measured values that are inconsistent with repeated measurements. Random or indeterminate errors arise from random fluctuations and variances in the measured quantities and occur even in tightly controlled analysis systems or conditions. It is not possible to eliminate all sources of random error from a method or system. Random errors can, however, be minimized by experimental or method design. For instance, while it is impossible to keep an absolute temperature in a laboratory at all times, it is possible to limit the range of temperature changes. In instrumentation, small changes to the electrical systems from fluctuations in current, voltage, and resistance cause small continuing variations which can be seen as instrumental noise. The measurement of these random errors is often determined by the examination of the precision of the generated data set. Precision is a measure of statistical variability in the description of random errors. Precision analyzes the data set for the relationship and distance between each of the data points independent of the "true" or estimated value of the data to identify and quantify the variability of the data.

Accuracy is the description of systematic errors and is a measure of statistical bias which causes a difference between a result and the "true" value (trueness). A second definition, recognized by ISO, defines accuracy as a combination of random and systematic errors which then requires high accuracy to also have high precision and high "trueness." An ideal measurement method, procedure, experiment, or instrument is both accurate and precise with measurements that are all close to and clustered around the target or "true" value. The accuracy and precision of a measurement value is a process validated by the repeated measurements of a traceable reference standard or reference material.

STANDARDS AND REFERENCE MATERIALS

A standard in a laboratory setting often referred to as an actual chemical or physical material that is a known or characterized material used to confirm identity, concentration, purity, or quality called a metrological standard. A metrological standard is the fundamental example or reference for a unit of measure. Simply stated, a standard is the "known" to which an "unknown" can be measured. Metrological standards fall into different hierarchical levels.

1. **Primary standard**: The definitive example of its measurement unit to which all other standards are compared and whose property value is accepted without reference to other standards of the same property or quantity.[3,4] Primary standards of measure such as weight are created and maintained by metrological agencies and bureaus around the world (NIST, etc.).
2. **Secondary standards**: Close representations of primary standards that are measured against primary standards. Many chemical standards companies create chemical standards against a primary weight set in order to create secondary standards traceable to that primary standard.
3. **Working standards:** Created against or with secondary standards to calibrate equipment.

There are also many standards designated as reference materials, reference standards, or certified reference materials (CRMs) which are materials manufactured or characterized for a set of properties and are traceable to a primary or secondary. If the material is a CRM, then it must be accompanied by a certificate that includes information on the material's stability, homogeneity, traceability, and uncertainty.[4,5]

Stability is when a chemical substance is nonreactive in its environment during normal use. A stable material or standard will retain its chemical properties within the designated "shelf life" or within its expiration date if it is maintained under the expected and outlined stability conditions. A material is considered to be unstable if it can decompose, volatilize (burn or explode), or oxidize (corrode) under normal stated conditions.

Homogeneity is the state of being of uniform composition or character. Reference materials can have two types of homogeneity: within-unit homogeneity or between-unit (or lot) homogeneity. Within-unit homogeneity means there is no precipitation or stratification of the material that cannot be rectified by following instructions for use. Some reference materials can settle out of solution but are still considered homogeneous if they can be re-dissolved into the solution by following the instructions for use (i.e., sonicate, heat, shake). Between-unit or lot homogeneity is the homogeneity found between separate packaging units.

Traceability is the ability to trace a product or service from the point of origin from the manufacturing or service process through to final analysis, delivery, and receipt. Reference material producers must ensure that the material can be traced back to a primary standard.

Uncertainty is the estimate attached to a certified value that characterizes the range of values where the "true value" lies within a stated confidence level. Uncertainty can encompass random effects such as changes in temperature, humidity, drift accounted for by corrections, and variability in performance of an instrument or analyst. Uncertainty also includes the contributions from within-unit and between-unit homogeneity, changes due to storage and transportation conditions, and any uncertainties arising from the manufacture or testing of the reference material.

USING STANDARDS AND REFERENCE MATERIALS

Certified standards or CRMs are materials produced by standards providers that have one or more certified values with uncertainty established using validated methods and are accompanied by a certificate. The uncertainty characterizes the range of the dispersion of values that occurs through the determinate variation of all the components which are part of the process for creating the standard.

CRMs have a number of uses including validation of methods, standardization or calibration of instrument or materials, and for use in quality control and assurance procedures. A calibration procedure establishes the relationship between a concentration of an element and the instrumental or procedural response to that element.

A calibration curve is the plotting of multiple points within a dynamic range to establish the element response within a system during the collection of data points. One element of the correct interpretation of data from instrumental systems is the effect of a sample matrix upon an instrumental analytical response. The matrix effect can be responsible for either element suppression or enhancement. In analysis where the matrix can influence the response of an element, it is common to match the matrix of analytical standards or reference materials to the matrix of the target sample to compensate for matrix effects.

Different approaches to using calibration standards may need to be employed to compensate for the possible variability within a procedure or analytical system.

- Internal standards are compounds that are either similar in character or analogs of the target elements that have a similar analytical response Internal standards are *added to the sample prior to analysis*. This type of standard allows the variation of instrument response to be compensated for by the use of a relative response ratio established between the internal standard and the target element.
  - Examples: Isotopic forms of elements, elements similar to the target which are not in the sample
  - Standard addition or spiking standard – an internal standard added to overcome matrix responses, instrument responses and the elemental responses that are indistinguishable from each other as the element concentration nears the lower limit of detection or quantitation. A target standard can then be added in known concentration to compensate for the matrix or instrument effects to bring the signal of the target element into a quantitative range.
- External standards are multiple calibration points (customarily three or more points) that contain standards or known concentrations of the target elements and matrix components and exist outside of the test samples. Depending on the type of analytical techniques, linear calibration curves can be generated between response and concentration which can be calculated for the degree of linearity or the correlation coefficient (r). An R-value approaching 1 reflects a higher degree of linearity. Most analysts accept values of>0.999 or better as acceptable correlation.

CALIBRATION CURVES

Calibration curves are often affected by the limitations of the instrumentation. Data can become biased by calibration points, the instrument's limits of detection, quantitation and linearity, and by the response of the system versus its baseline (signal-to-noise)

- **Limit of detection (LOD)**: lower limit of a method or system which the target can be detected as different from a blank with a high confidence level (usually over three standard deviations from the blank response).
- **Limit of quantitation (LOQ)**: lower limit of a method or system in which the target can be reasonably calculated where two distinct values between the target and blank can be observed (usually over ten standard deviations from the blank response) (Figure 22.2.)
- **Signal-to-noise (*s/n*)**: response of an element measured on an instrument as a ratio of that response to the baseline variation (noise) of the system. Limits of detection are often recognized as target responses which have three times the response of baseline noise or $s/n \geq 3$. Limits of quantitation are recognized as target responses which have ten times the response of baseline noise or $s/n \geq 10$.
- **Limits of linearity (LOL)**: upper limits of a system or calibration curve where the linearity of the calibration curve starts to be skewed creating a loss of linearity (Figure 22.3). This loss of linearity can be a sign that the instrumental detection source is approaching saturation.
- **Dynamic Range**: an array of data values between the LOQ and the LOL is where the greatest potential for accurate measurements will occur.

The understanding of a system's dynamic range, the accurate bracketing of calibration curves within the range and around the target element concentration increases the accuracy of the measurements. If a calibration curve is created that does not potentially bracket all the possible target data points

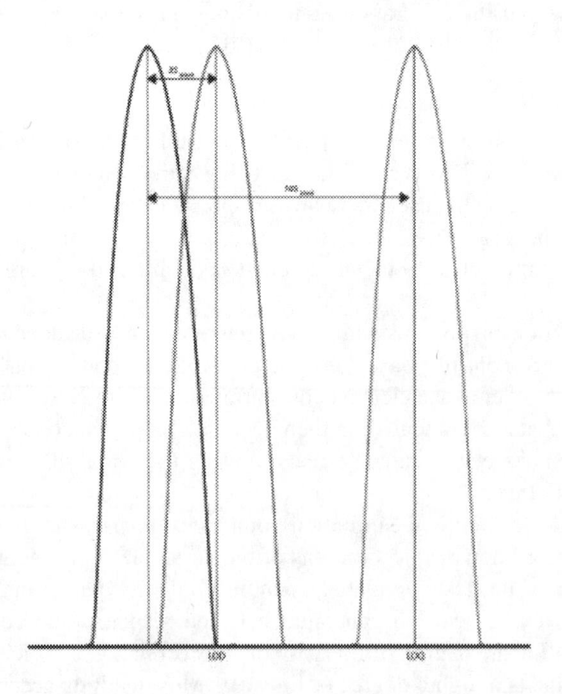

FIGURE 22.2 Limits of detection and quantitation.

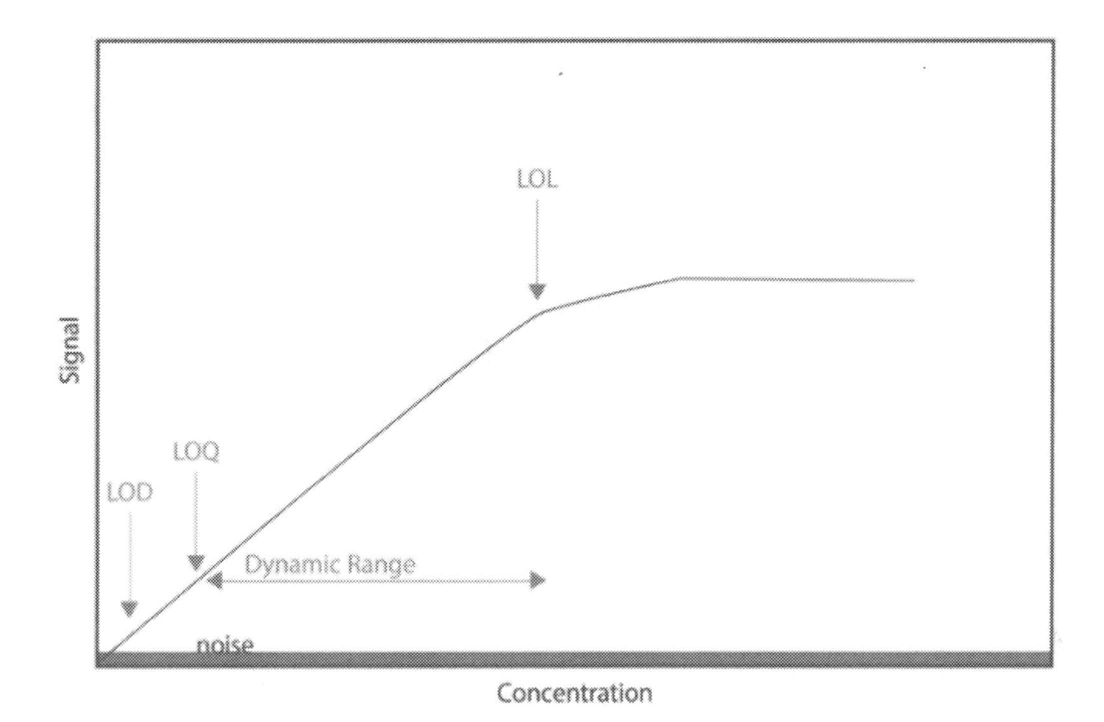

FIGURE 22.3 Calibration curve limits and range.

then the calibration curve can be biased to artificially increase or decrease the results and create error. To create calibration curves and working standards, there must be an accurate process of converting units, calculating dilution targets and preparing dilutions.

DYNAMIC RANGE, CONCENTRATION & ERROR

The first step in creating standards, working solutions, and dilutions is to understand the dynamic range your analysis is targeting – is it ppb, ppm, percent?

If you are looking for a major component of the sample, then your standards have to be in percent levels or the samples must be diluted down to the correct concentrations. If target elements are trace elements or trace contaminants, then standards and calibration curves have to be diluted down to match the target within the dynamic range of the instrument technique.

Tables 22.3 and 22.4 show the basic conversions between different concentrations based on mass or volume.

LABORATORY SOURCES OF ERROR & CONTAMINATION

Calibration curves are created by diluting standards into several target points along the dynamic range to cover the possible target results. Proper dilution of standards and samples is based on the understanding of basic dilution and volumetric procedures and dilution factors. Volumetric measurement is a commonly repeated daily activity in most analytical laboratories. Many processes in the laboratory from sample preparation to standards calculation depend on accurate and contamination-free volumetric measurements. Unfortunately, laboratory volumetric labware, syringes and pipettes are some of the most common sources of contamination, carryover and error in the laboratory.

TABLE 22.3
Weight to Weight Concentrations

Name	Symbol	Equivalence			
Parts per thousand[a]	ppt[a]	g/kg	mg/g	μg/mg	ng/μg
Parts per million	ppm	mg/kg	μg/g	ng/mg	pg/μg
Parts per billion	ppb	μg/kg	ng/g	pg/mg	fg/μg
Parts per trillion	ppt	ng/kg	pg/g	fg/mg	ag/μg

[a] Parts per thousand and parts per trillion both can use the ppt abbreviation; so use care to select appropriate unit of measure.

TABLE 22.4
Weight to Volume Concentrations

Name	Symbol	Equivalence			
Parts per thousand[a]	ppt[a]	g/L	mg/mL	μg/μL	ng/nL
Parts per million	ppm	mg/L	μg/mL	ng/μL	pg/nL
Parts per billion	ppb	μg/L	ng/mL	pg/μL	fg/nL
Parts per trillion	Ppt	ng/L	pg/mL	fg/μL	ag/nL

[a] Parts per thousand and parts per trillion both can use the ppt abbreviation; so use care to select appropriate unit of measure.

The root of these errors is based on the four "I" errors of volumetric container:

1. **Improper use:** measuring tool is not used correctly
2. **Incorrect choice:** measurement tool is not appropriate for the volume or type of measurement
3. **Inadequate cleaning:** carryover causes contamination
4. **Infrequent calibration:** measuring tool is not calibrated for use

These four "I's" can lead to error and contamination which negate all intent of careful measurement processes.

Many errors can be avoided by understanding the markings displayed on the volumetric containers and choosing the proper tool for the job. There is a lot of information displayed on volumetric labware. Most labware, especially glassware, is designated as either Class A (analytical or quantitative) or Class B (general use) labware. If a critical measurement process is needed, then only Class A glassware should be used for measurement.

Other information that can be found on labware is the name of the manufacturer, country of origin, tolerance or uncertainty of the measurement of the labware, and a series of descriptors as to how the glassware should be used. Labware can be marked with letters which designate the purpose of the container. If a volumetric is designed to contain liquid, it will be marked by either the letters TC or IN. Labware which is designated to deliver liquid will be marked by either the letters TD or EX. Sometimes there are additional designations such as wait time or delivery time inscribed on the labware. The delivery time refers to a period of time required for the meniscus to flow from the upper volume mark to the lower volume mark. The wait time refers to the time needed for the meniscus to come to rest after the residual liquid has finished flowing down from the wall of the pipette or vessel.

The second type of improper use and incorrect choice can be seen in the selection of pipettes and syringes for analytical measurements. Many syringe manufacturers recommend a minimum dispensing volume of approximately 10% of the total volume of the syringe or pipette. A study by SPEX CertiPrep showed that dispensing such a small percentage of the syringe's total volume created a large amount of error. The largest rates of error were seen in the smaller syringes of 10 and 25 μL. Dispensing 20% of the 10 μL syringe created a 23% error. Error only dropped down to below 5% as the volume dispensed approached 100%. In the larger syringes, measurements over 25% were able to see error in and around 1%. The larger syringes were able to get closer to the 10% manufacturer's dispensing minimum without a large amount of error, but the error did drop as the dispensed volume approached 100%.[6]

The third "I" of volumetric error is inadequate cleaning. Many volumetric containers can be subject to memory effects and carryover. In critical laboratory experiments, labware sometimes needs to be separated by purpose and use. Labware subject to high levels of organic compounds or persistent inorganic compounds can develop chemical interactions and memory effects. It is sometimes difficult to eliminate carryover from labware and syringes even when using a manufacturer's stated instructions. For example, many syringes are cleaned by several repeated solvent rinses prior to use. A study of syringe carryover by SPEX CertiPrep showed that some syringes are subject to high levels of chemical carryover despite repeated rinses.

The final source of error is infrequent calibration. Many laboratories have schedules of maintenance for equipment, such as balances and automatic pipettes, but often overlook calibration of reusable burettes, pipettes, syringes, and labware. Under most normal use, labware often does not need frequent calibration but there are some instances where a schedule of recalibration should be employed. Any glassware or labware in continuous use for years should be checked for calibration. Glass manufacturers suggest that any glassware used or cleaned at high temperatures, used for corrosive chemicals or autoclaved should be recalibrated more frequently.

It is also suggested that under normal conditions soda-lime glass be checked or recalibrated every five years and borosilicate glass after it has been in use for ten years. The error associated with the use of volumetric containers can be greatly reduced by choosing the correct volumetric for the task, using the tool properly and making sure the volumetric containers are properly cleaned and calibrated before use.

Inorganic analysts know that glassware is a source of contamination. Even clean glassware can contaminate samples with elements such as boron, silicon, and sodium. If glassware, such as pipettes and beakers, are reused, the potential for contamination escalates. At SPEX CertiPrep, a study was conducted of residual contamination of our pipettes after being manually and automatically cleaned using a pipette washer.[6,9]

An aliquot of 5% nitric acid was drawn through a 5 mL pipette after the pipette was manually cleaned according to standard procedures. The aliquots were analyzed by ICP-MS. The results showed significant residual contamination still persisted in the pipettes despite a thorough manual cleaning procedure.

The experiment was repeated using a pipette washer especially made for use in parts-per-trillion analysis. The pipette washer repeated forced deionized water through the pipettes for a set time period. The pipettes were cleaned in the pipette washer then the same aliquot of 5% nitric acid was drawn through the 5 mL pipettes. The aliquot was analyzed by ICP-MS. The automated washer reduced the contamination significantly as compared to the manual cleaning of the pipettes. The reduction of contamination by moving from manual cleaning to an automated cleaning process was clear. High levels of contamination of sodium and calcium (almost 20 ppb) dropped to <0.01 ppb. Other common contaminants including lead and iron dropped from 5.4 and 1.6 ppb respectively to less than 0.01 ppb.

The reduction of contamination in labware can depend on the material of the labware and its use. Different materials contain many types of elemental and organic potential contamination as seen in Table 22.5.[7] Trace inorganic analysis are best performed in polymer or high purity quartz vessels or

TABLE 22.5
Major Elemental Impurities Found in Laboratory Container Materials[7]

Material	# Elements	Total ppm	Major Impurities
Polystyrene—PS	8	4	Na, Ti, Al
Tetrafluoroethylene (TFE)	24	19	Ca, Pb, Fe, Cu
Low-density PE–LDPE	18	23	Ca, Cl, K, Ti, Zn
Polycarbonate—PC	10	85	Cl, Br, Al
Polymethyl pentene—PMP	14	178	Ca, Mg, Zn
Fluorinated ethylene propylene (FEP)	25	241	K, Ca, Mg
Borosilicate glass	14	497	Si, B, Na
Polypropylene (PP)	21	519	Cl, Mg, Ca
High-density PE (HDPE)	22	654	Ca, Zn, Si

TABLE 22.6
Changes in Element Concentration after Storage (ppb)

Element	At Manufacture	After 1 year
Al	1–10	10–40
Ca	1–10	20–100
Fe	0–5	5–50
Mg	0–10	10–40
Na	1–10	5–40
Si	5–15	15–5,000

fluorinated ethylene propylene (FEP), to minimize contact with borosilicate glass. Metals, such as Pb and Cr, are highly absorbed by glass but not by plastics. On the other hand, samples containing low levels of Hg (ppb levels) must be stored in glass or fluoropolymer because Hg vapors diffuse through polyethylene bottles.

It is always important to know the best conditions for storage for a standard. The expiration dates on most standards reflect the manufacturer's confidence level of continued accuracy at proper storage conditions for the shelf life listed on the product. If a standard is not stored properly, it can affect the quality and accuracy of the standard. Elements within the packaging can slowly be leached from the packaging over time and change the value of the samples or standards. Some of the common packaging elements were examined in samples at the time of manufacturing and bottling and again after one year to determine if the packaging contributed to the contamination of the samples over time. After one year, the amounts of common elements, such as aluminum, calcium, iron, magnesium, sodium, and silica, more than doubled in the solutions. Table 22.6 shows the changes of elemental concentrations in six common elements after a year's storage.

SOURCES OF LABORATORY CONTAMINATION & ERROR

Contamination and error can occur at almost any point of the process and then can be magnified as the method and analysis runs its course. In the past, issues of laboratory contamination were problematic but now contaminants, even in trace amounts, can severely alter results. It is hard to imagine that such small amounts of contamination can dramatically change laboratory values.

Most questions about contamination come in the form of an inquiry about a particularly high result for some common contaminant. In most cases, the root of that contamination can be traced to a common source. The most common sources of standard and sample contamination, in addition to the previously discussed volumetric labware are water, reagents and acids, laboratory environment, storage, and personnel.

WATER QUALITY

Water is one of the most basic yet most essential laboratory components. There are many types, grades, and intended uses for water. Water is most often used in two ways in the laboratory: as a cleaning solution and as a transfer solution for volumetric or gravimetric calibrations or dilutions. In both of these uses, the water must be clean in order to reduce contamination and introduce error into the process. Poor quality water can cause a host of problems from creating deposits in labware or inadvertently increasing a target element or element concentration in solution.

The confusion starts when laboratories are unsure about which type of water they get from their water filtration system. ASTM has guidelines that designate different grades of water. Table 22.7 shows the parameters for the four ASTM types of water.[8]

The actual type of water produced by a commercial laboratory water filtration system can vary in pH, solutes, and soluble silica. Critical analytical processes should always require a minimum quality of ASTM Type I water. All trace analysis standards, dilutions, dissolutions, extractions, and digestions should be conducted with the highest purity water. Analysts who use CRMs and perform quantitative analysis need to use quality water in order not to contaminate their CRMs, standards, and samples with poor quality water.

High-purity water is often achieved in several stages in multiple processes that remove physically and chemically potential contaminating substances. Municipal water supplies often test their own water sources on an annual basis but that does not mean it is applicable for use in laboratory applications. Municipal water can become contaminated from its distribution point especially when left static sitting in pipes, tubing, and hoses. Water left stationary in a laboratory water system can be exposed to the leaching of elements and compounds from the pipes and hoses.

REAGENTS

An analytical laboratory often uses a large amount of various reagents, solvents, and acids of varying quality and contamination levels. These chemical components of the laboratory can be a large economic investment for a laboratory but also a large source of potential contamination. Just as in the case of water, there are different types or grades of chemicals, reagents, acids, and solvents. Some designations are set forth by standards set by the US Pharmacopeia (USP) or the American

TABLE 22.7
ASTM Designations for Reagent Laboratory Water[8]

	ASTM Type			
Requirement	**I**	**II**	**III**	**IV**
Use	Critical laboratory applications and processes	General lab grade used for pH, buffers, feeding other water-polishing systems	Cleaning glassware, feeding water polishing systems, water baths	Not good for lab use
Specific resistance (megohm/cm) (max)	18	1	4	0.2
pH	N/A	N/A	N/A	5.0–8.0
Sodium (max)	1 µg/L	5 µg/L	10 µg/L	50 µg/L
Total silica (max)	3 µg/L	3 µg/L	500 µg/L	High
Total organic carbon (max)	50 µg/L	50 µg/L	200 µg/L	N/A

Chemical Society (ACS). Other types or grades of material are designated by individual manufacturers based on intended use. Some grades are general laboratory grades with intended use for non-critical applications, while other grades usually high in purity and low in contaminants are designated for more critical analyses.

There are many persistent solvents that can be found in the lab which can cross-contaminate samples by their presence. Some persistent solvents include dichloromethane which can cause chlorine contamination, DMSO, and carbon disulfide, which can add sulfur residues. There are solvents that react with air to form peroxides which can cause contamination and safety issues in the laboratory. Another lab reagent that can become both a potential danger and a potential contaminant in the lab are acids.

Acids are, by their nature, oxidizers and many of the strongest acids are used in the processing of samples for inorganic analysis. Common acids in digestion and dissolution include perchloric acid, hydrofluoric acid, sulfuric acid, hydrochloric acid, and nitric acid. Many of these acids are commercially available in several grades from general laboratory or reagent grade to high-purity trace metal grade. Acid grades often reflect the number of sub-boiling distillations the acid undergoes for purification before bottling. The more an acid is distilled, the higher the purity of the acid. These high-purity acids have the lowest amount of elemental contamination but can become very costly at up to ten times the cost of the reagent grade of acids.

Often the question is asked if high-purity acids are necessary for sample preparation if the laboratory is using a high-quality inductively coupled plasma mass spectrometry (ICP-MS) grade CRM. Clean acids used in sample preparation, digestion, and preservation can be very costly. But, the difference between the amounts of contamination in a low-purity acid and a high-purity acid can be dramatic. High-quality standards for use in parts-per-billion and parts-per-trillion analyses use the highest purity acids available to reduce all possible contamination from the acid source.

An example of potential contamination is an aliquot of 5 mL of acid containing 100 ppb of Ni as a contaminant, used for diluting a sample to 100 mL can introduce 5 ppb of Ni into the sample.

To reduce contamination, it is recommended that high-purity acids be used to dilute and prepare standards and samples when possible. In addition to using pure acid, it is important that the chemist check the acid's certificate of analysis to identify the elemental contamination levels present in the acid. Some laboratories prefer to use blank subtractions to negate the background contamination, but blank subtraction for acids can only work in a range well over the instrumental level of detection. If blank subtraction causes an analytical result to fall below the instrument's level of detection, it should not be used.

LABORATORY ENVIRONMENT AND PERSONNEL

All laboratories believe they observe a level of laboratory cleanliness. Most chemists recognize there are inherent levels of contamination present in all laboratories. A common belief is that the small amounts of environmental and laboratory contamination cannot truly change the analytical results. To test the background level of contamination in a typical laboratory, samples of nitric acid were distilled in both a regular laboratory and in a cleanroom laboratory with special air-handling systems (HEPA filters). The nitric acid distilled in the regular laboratory had high amounts of aluminum, calcium, iron, sodium, and magnesium contamination. Table 22.8 shows the acid distilled in the cleanroom displayed significantly lower amounts of most contaminants.[6]

Laboratory air also can contribute to the contamination of samples and standards. Common sources of air and particulate matter contamination are from surfaces and building materials, such as ceiling tiles, paints, cement, and drywall. Surface contaminants can be found in dust and rust on shelves, equipment, and furniture. Dust contains many different earth elements such as sodium, calcium, magnesium, manganese, silicon, aluminum, and titanium. Dust can also contain elements from human activities (Ni, Pb, Zn, Cu, As) and organic compounds such as pesticides, persistent organic pollutants (POPs), and phthalates. The dust and rust particles can contaminate open containers in the lab or enter containers by charge transfer from friction by the triboelectric effect. The triboelectric effect, or triboelectric charging, is when materials become charged after coming into

TABLE 22.8
Elemental Impurities Found in Nitric Acid Distilled in Regular Labs compared to a Clean Lab[6]

Element	Regular Lab	Clean Lab
Al	60	15
As	0.17	<0.02
Ca	150	100
Cd	0.3	0.003
Cr	2.5	0.4
Cu	1.7	0.23
Fe	50	9
Mg	10	4
Mn	1.1	0.1
Mo	0.8	0.03
Pb	0.5	0.4
Sb	0.04	0.013
Zn	5.5	0.7

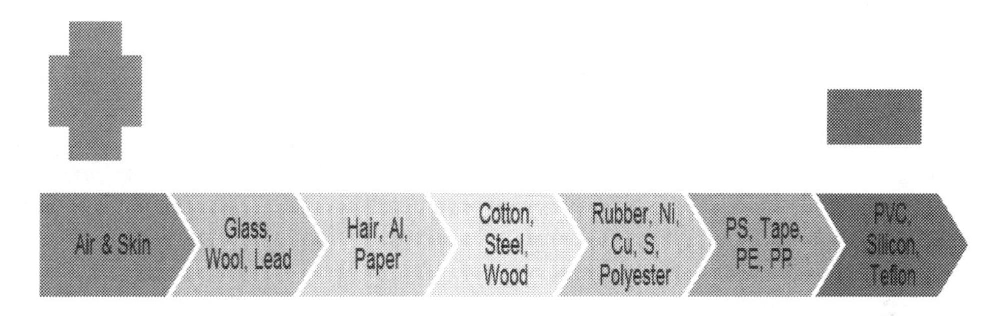

FIGURE 22.4 Triboelectric charge potential of common materials and particles in the laboratory.

contact with a second material creating friction. The most common example of this effect is seen when hair sticks to a plastic comb after a static charge is created. The polarity and the strength of the electrical charge are dependent upon the type of material and other physical characteristics. Many materials in the lab have strong positive or negative triboelectric charges as seen in Figure 22.4. In the laboratory, materials such as dust, air, skin, and lead have extreme positive charges and can be attracted to the strong negative charge of Teflon™ or other plastic bottles when the bottle is opened and creates friction inducing charge.

Laboratory personnel can add their own contamination from lab coats, makeup, perfume, and jewelry. Aluminum contamination can come from lab glassware, cosmetics, and jewelry. Many other common elements can be brought in as contamination from lotions, dyes, and cosmetics. Even sweat and hair can cause elevated levels of sodium, calcium, potassium, lead, magnesium, and many ions. If a laboratory is seeing unusually high level of cadmium in the samples, it could be from cigarettes, pigments, or batteries. If the levels of lead are out of range, contamination can be from paint, cosmetics, and hair dyes. Table 22.9 shows potential sources of common elemental contamination from outside products.

Laboratory environment and personnel contamination can be reduced by limiting the use of personal care products, jewelry, and cosmetics which could contain contamination and interfere with critical analyses. Lab coats can collect all types of contamination and should only be

TABLE 22.9
Common Elemental Contamination Sources

Element	Source
Aluminum	Lab glassware, cosmetics, dust, e-cigs
Bismuth	Cosmetics, lotions, medicines
Cadmium	Cigarettes, e-cigs, pigments, batteries
Cobalt	Surgical implants, dental prosthesis, jewelry
Copper	Algaecide used in swimming pools, e-cigs
Lead	Cosmetics, hair dye, jewelry, dust, cigarettes, e-cigs
Mercury	Mascara, dust
Selenium	Anti-dandruff shampoos
Zinc	Drugs, calamine lotion, cosmetics, powdered gloves

worn in the laboratory to avoid cross-contamination from other labs and the outside world. The laboratory surfaces should be kept clean. Deionized water can be used to wipe down work surfaces. Laboratory humidity can be kept above 50% to reduce static charge. An ethanol- or methanol-soaked laboratory wipe can be used to reduce static electricity as it evaporates.

Even with clean laboratory practices in place, erroneous results can often find their way into sample analysis. To eliminate some of these spurious results, replication of blanks and sample dilutions can be employed. The blank results should be averaged and the sample run values can either be minimally selected or averaged. The difference between the two values can then be plotted against a curve established against two more standards. A minimum of two standard points can be used if the chance of contamination is minimal, such as in the case of rare or uncommon elements. Additional standard points should be considered if the potential for contamination is high with common elements, such as aluminum, sodium, and magnesium. Multiple aliquots of blanks and dilutions can also be employed to further minimize analytical uncertainty.

GENERAL PRINCIPLES AND PRACTICES

Laboratories should follow a general regime of three runs each of wash/rinse runs, blank runs, and sample runs, as well as single runs of sample plus spike, and standard or spike runs without a sample to use as a control solution to evaluate recovery.

Analysts must realize that the cleanliness and accuracy of their procedures, equipment, and dilutions affect the quality of the standards and samples. Many laboratories will dilute CRMs to use across an array of procedures and techniques. This in-house dilution of CRMs can be a savings to the laboratory but in the final analysis can be a source of error and contamination.

CRM manufacturers design standards for particular instruments to obtain the highest level of accuracy and performance for that technique. They also use calibrated balances, glassware, and instruments to ensure the most accurate standards are delivered to customers. Certifications such as ISO 9001, ISO 17025, and ISO 17034 assure customers that procedures are being followed to ensure quality and accuracy in those standards. After those CRMs are in chemists' hands, it is then their responsibility to employ all possible practices to keep their analysis process free from contamination and error.

FURTHER READING

1. ASTM E177-19, *Standard Practice for Use of the Terms Precision and Bias in ASTM Test Methods*, ASTM International, West Conshohocken, PA, 2019, www.astm.org.
2. ASTM E456-13A(2017)e3, *Standard Terminology Relating to Quality and Statistics*, ASTM International, West Conshohocken, PA, 2017, www.astm.org.

3. BIPM, IEC, IFCC, ILAC, IUPAC, IUPAP, ISO, OIML, The international vocabulary of metrology—basic and general concepts and associated terms (VIM), 3rd edn. *JCGM*, 200, 2012. http://www.bipm.org/vim.
4. ISO Guide 30:2015. Reference materials — Selected terms and definitions, https://www.iso.org/standard/46209.html.
5. ISO 17034:2016. General requirements for the competence of reference material producers, https://www.iso.org/standard/29357.html.
6. SPEX CertiPrep Webinar, Clean laboratory techniques, https://www.spexcertiprep.com/webinar/clean-laboratory-techniques.
7. J. R. Moody and R. Lindstrom, Selection and cleaning of plastic containers for storage of trace element samples, *Analytical Chemistry*, 49, 2264, 1977.
8. ASTM D1193-06(2018), *Standard Specification for Reagent Water*, ASTM International, West Conshohocken, PA, 2018, www.astm.org.
9. SPEX CertiPrep Application Note, Understanding measurement: A guide to error, contamination and carryover in volumetric labware, syringes and pipettes, SPEX CertiPrep App Note, https://www.spexcertiprep.com/knowledge-base/files/APPNOTE-UnderstandingMeasurement.pdf

23 The Importance of Laboratory Quality Assurance

This chapter compiled by the Chemical Sciences Division of the Material Measurement Laboratory at NIST (author names in the Acknowledgement section) examines the problems faced with quality assurance (QA) when reference materials are unavailable, as is currently the case with *Cannabis* materials and what laboratories can do to ensure accurate measurements in the face of this problem. This chapter briefly presents an overview of the importance of reference materials while focusing on how to perform meaningful testing in their absence by using alternative techniques such as the production of in-house reference materials and participation in QA programs.

The passage of the Marihuana Tax Act in 1937 regulated cultivation and importation of marijuana, and industrial hemp was included in this legislation.[1,2] By the 1970s, marijuana was classified as a Schedule I drug, on the same level as narcotics, in Title 21 United States Code (USC) Controlled Substances Act.[2,3] These legislative actions effectively halted all formal research, cataloging, and control of the growth of marijuana and industrial hemp in the United States for nearly 60 years until the late 1990s when several states legalized *Cannabis* for medicinal use. Adult recreational use of *Cannabis* first became legal in Colorado and Washington in 2012. Currently, medical marijuana is legal in 33 states and the use of recreational marijuana is legal in 11 of those states as well as the District of Columbia. Section 7606 of the 2014 Farm Bill paved the way for the legal cultivation of industrial hemp followed by the passage of the 2018 Farm Bill, which removed hemp from the controlled substance list and defined it as *Cannabis* containing less than 0.3% tetrahydrocannabinol (THC) on a dry weight basis. Even with the passage of the 2018 Farm Bill, however, hemp products containing cannabidiol (CBD) *cannot* be marketed as dietary supplements and *cannot* be added to foods introduced to interstate commerce under the Federal Food, Drug, and Cosmetic Act.[4]

While purportedly aimed at protecting the ability of farmers to grow hemp and retain any federal farm program benefits, these legal changes have promoted the increased development and production of hemp products intended for human consumption. Because of this wide availability, the *Cannabis* market is booming and includes products such as hemp, medical marijuana, and marijuana for adult recreational use, as well as down-market consumables containing these plant materials or their extracts. Consumers of *Cannabis* products may not be fully aware of potential safety concerns associated with the products they are purchasing, especially medical marijuana products that are assumed to be regulated as other medicines and medical products.

Stakeholders in the *Cannabis* community are very aware of the safety needs and quality concerns surrounding the use of *Cannabis*. For example, the Food and Drug Administration (FDA) has highlighted the limited data available on the safety of CBD materials and is investigating reports of CBD containing unsafe levels of contaminants such as pesticides, THC, and toxic elements.[5] In addition to the FDA, stakeholders in the *Cannabis* industry include *Cannabis* growers, processors,

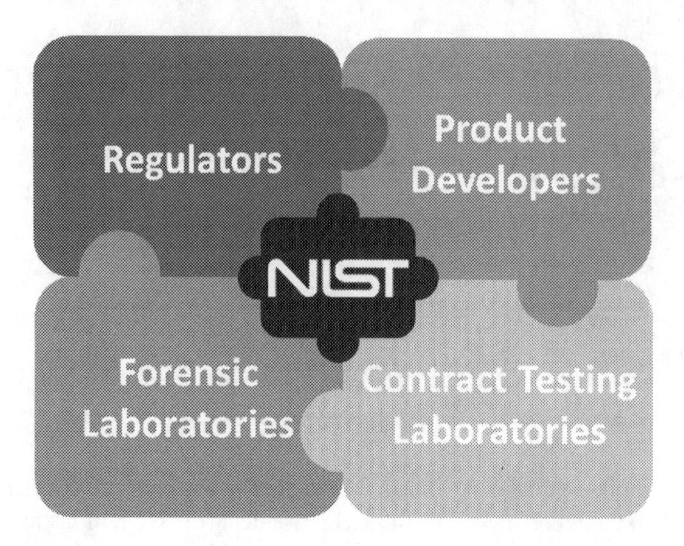

FIGURE 23.1 Cannabis industry stakeholders span a wide variety of interests.

and distributors in jurisdictions with legal *Cannabis* status; finished product manufacturers and the contract laboratories hired for analysis of their products; state and local regulators of legal *Cannabis*; forensic laboratories who must distinguish between legal hemp and illegal marijuana as well as the state and local law enforcement agencies that they support; federal law enforcement agencies supporting the federal controlled substance status of marijuana and THC; and medical researchers, including clinicians and the federal agencies that fund their work (Figure 23.1). NIST has engaged with many of these stakeholders to understand the analytical challenges facing each of their unique perspectives.

Industrial hemp (*Cannabis sativa* L.) is known as a hyper-accumulator, a class of plant historically used to remove toxic elements from soils and groundwater as a form of natural remediation.[6,7] In hyper-accumulators, the roots of the plant absorb contaminated groundwater and transport the toxic substances to various plant parts. Prior to the use of hemp for consumption, the heavy metal content of the plant biomass was of little importance when considering human exposure. Given the mechanism and known affinity of *Cannabis* for heavy metal uptake, however, potential for a significant human exposure risk results when the products are consumed. Cadmium, for example, enters the environment naturally through the application of phosphate and sewage sludge-containing fertilizers.[8] Soil contamination plays a significant role in cadmium exposure, and hemp plants grown in contaminated soil are known to accumulate cadmium.[6,9]

COMMERCIAL REFERENCE MATERIALS

Determination and subsequent regulation of the presence of toxic elements in *Cannabis*, primarily arsenic (As), cadmium (Cd), lead (Pb), and mercury (Hg), require standardized methods with accurate, quality-assured implementation. Potential contamination of *Cannabis* with other elements such as nickel (Ni), chromium (Cr), selenium (Se), uranium (U), cobalt (Co), molybdenum (Mo), vanadium (V), beryllium (Be), and manganese (Mn) is also of interest, based on a more advanced understanding of these contaminants in tobacco and tobacco smoke.[10] Stakeholders need to assess contamination and estimate consumer exposure to these metals through *Cannabis* use, which includes not only plant material but also in the large variety of other products produced using *Cannabis* including oils, chocolates, gummies, capsules, topical ointments, and pet supplements. Each of these materials presents both a unique sample preparation challenge, from the complete digestion of a complex matrix, to a unique analytical challenge, with the potential for interferences due to the low mass fraction levels of these elements. The best approach for ensuring accuracy and comparability of measurements is by using reference materials (RMs) (including certified reference materials, or CRMs, or non-certified reference materials, or RMs) to demonstrate that the measurement process is in control. Ideally, the selected reference material would match both the matrix being analyzed and the analyte mass fractions within that matrix and could be used for development and/or validation of methods used for testing these various products. Once a method has been established, the reference material can then be used for quality control to demonstrate the applicability of the method to the product being tested, such that acceptable results obtained for the tested analytes in the reference material imply that reliable results will also be obtained for unknown samples.

The large variety of matrices and low mass fraction levels of the analytes of interest present a significant analytical challenge for the *Cannabis* industry. To represent the breadth of the *Cannabis* market, various reference materials are necessary to match both the matrices being analyzed and the analyte mass fractions within those matrices for assessment of measurement accuracy and comparability. However, commercial reference materials in general are expensive and time-consuming to produce, and production of *Cannabis* reference materials involves additional challenges because of the federal legal status of marijuana. Because marijuana remains on the controlled substances list maintained and enforced by the US Drug Enforcement Administration (DEA), transport or shipping of marijuana across state lines is illegal without proper DEA licensing and would limit the distribution of an reference material beyond the state in which the reference material is prepared or licensed to laboratories. As a result, production of a hemp reference material is more feasible than production of a marijuana reference material, given the 2020 regulatory climate.

In addition to potential transport issues, gaining access to representative materials in large enough quantities for production of an reference material is difficult. Historically, *Cannabis* has been considered a small-batch, high-value crop, and many kilograms of material (25 kg or more) would be needed for commercial reference material development. Once a material is identified, homogeneity and stability must be studied during reference material production. *Cannabis* is inherently heterogeneous because the oily resin from hair-like trichomes, which produce and are most abundant in the cannabinoids, terpenes, and flavonoids, is difficult to blend with the remainder of the plant material. Upon grinding, the trichome-rich, sticky, oily powder may result in heterogeneity of various analytes, which would limit the utility of an reference material. Another challenge is the stability of a ground *Cannabis* plant material for at least 3–5 years, as would be needed for traditional commercial reference material production. While reduced temperature may extend the stability of most analytes, a rigorous evaluation of observed degradation as a function of time and temperature, as well as various packaging conditions, would be needed prior to production of an reference material.

ALTERNATE REFERENCE MATERIALS

Because of these challenges, laboratories working in the *Cannabis* industry may need interim solutions for method development and validation until a commercial *Cannabis* reference material is available. Materials with similar composition and comparable mass fractions of the toxic elements of interest should be considered as suitable QA materials, as the analytical processing would be similar. To define composition as similar, the proportions of fat, protein, and carbohydrate of the sample and reference material should be considered, as well as other factors such as how the analyte is incorporated into the matrix.

Control materials can also be developed in-house as a potential solution when a commercial reference material is not available. An in-house control material is a type of reference material produced by a laboratory for internal use and has a matrix and analyte composition similar to that of the materials routinely tested in the laboratory. Production of an in-house control material requires a large, representative supply of homogeneous and stable material. The size of the in-house control material production depends on the frequency and duration of the intended use and the amount required per analysis with the inclusion of additional material for initial evaluation and value assignment and potential losses during processing and packaging. The material should be processed, packaged, and stored to ensure homogeneity to enhance long-term stability, focusing on the least stable analyte of interest. Plant materials may require grinding and sieving to remove large stems or other pieces; food materials may require freeze-drying prior to grinding. Extremely challenging materials such as gummies may require cryo-homogenization to ensure homogeneity between large individual pieces and complex textures. The material should be stored in a convenient and stable container (e.g., glass or plastic bottle, inert pouch or packet) and may be flushed with an inert gas such as argon or nitrogen to enhance stability by removing oxygen that may cause degradation of the material or analytes of interest. The amount contained in each unit of packaged material can be designed for single use or multiple use, balancing potential instability in use of an opened package against cost savings of packaging fewer overall units. The final packaged units should be stored under consistent conditions, typically in a cool, dry, dark environment. Storage at reduced temperature such as a refrigerator or freezer, if possible, may further enhance long-term stability.

Once the material has been processed and packaged, analyte levels must be established using a selection of different sample preparation techniques, different analytical methods, and by different analysts in samples from across the production lot. The characterization may be more or less thorough depending on the intended use and the availability of appropriate CRMs or RMs to use as controls in the measurement process. This initial characterization will establish the homogeneity of the material as well as the analyte mass fraction levels, uncertainties, and minimum mass size needed for analysis.[11,12] Analyte mass fraction levels and uncertainties, and minimum sample size necessary for determination of each analyte, can be determined after homogeneity of the material is established by analyzing an established CRM or RM along with the unknown in-house control material. Homogeneity is analyte dependent, and an additional uncertainty component can be included for any analyte demonstrating possible inhomogeneity. In some cases, a larger sample size may be required to ensure homogeneity for some analytes.

With respect to in-house control materials for elements, stability is typically not a major concern. However, poorly packaged or stored materials could absorb moisture from the environment or potentially lose overall integrity, which may affect element mass fraction levels. For materials that are properly packaged and stored, cadmium and lead are expected to remain stable. For some of the chemical forms of mercury and arsenic, however, stability may be a concern. Following value assignment of the in-house control material, the stability of the material should be continually checked, and periodic testing of a CRM or RM should be used to monitor the stability of the in-house material. If the in-house material is being analyzed routinely, a control chart will reveal any trends in analyte mass fraction level over time (Figure 23.2). Observed changes in the initial assigned mass fractions for the in-house control material indicate potential issues with the sample preparation method, the analytical method, or the material itself, and further investigation is necessary.

Once a commercial CRM or RM becomes available, an in-house control material is still extremely valuable. A calibration CRM, typically a pure compound or neat solution, produced by a National Metrology Institute (NMI), can be used to establish metrological traceability of the in-house control material. Metrological traceability is a relationship to a reference through a documented unbroken chain of calibrations and may be necessary for laboratory accreditation.[13] In addition, the in-house control material may more closely represent routine samples (i.e., matrix, analytes, mass fraction levels) and may therefore better represent the analytical challenge than a commercial reference material. A commercial CRM, produced and characterized independently from the in-house control material, may also inform laboratories about potential biases in the analytical process. Finally, in-house reference materials are typically much lower in cost per analysis than a commercial reference material and can meet laboratory needs for control charting and even traceability of day-to-day analysis activities at a significantly lower cost compared to daily use of a commercial reference material.

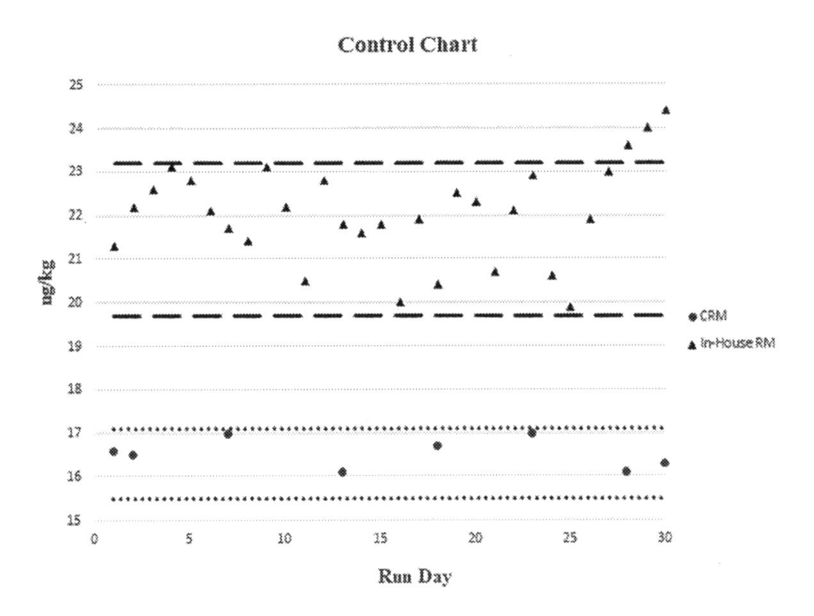

FIGURE 23.2 A value assigned in-house RM can be analyzed with every analytical run while a CRM needs to be analyzed only occasionally. Maintaining a control chart is an effective way to monitor trends in data. The dashed lines represent the uncertainty, which may be a standard deviation, associated with the in-house RM measurements (triangles). The dotted lines represent the uncertainty associated with the CRM assigned value.

QUALITY ASSURANCE PROGRAMS

Participation in QA programs allows laboratories to demonstrate accuracy-based performance and assess and improve measurement accuracy, precision, and comparability using CRMs, RMs, or other test materials. Similar to a proficiency testing (PT) program, participants evaluate test materials and report results for comparison with the results of other participants and also with a known value (if available). Unlike PT, however, a QA program provides an evaluation of performance directly to the participant without a pass/fail report, as the intention is to educate, identify, and promote the use of sound measurement practices and reference materials. QA program studies are an excellent tool for evaluating analytical methodology, identifying community wide challenges, and encouraging discussions among participants to improve the analytical methods in individual laboratories and the community as a whole. In areas where few or no standard methods are available, QA programs offer a tool for assessing and demonstrating the quality of measurements provided by a laboratory.

A recent study was conducted for toxic elements in hemp by the NIST Health Assessment Measurements QA Program (HAMQAP).[14] In the study, laboratories were asked to measure low levels of toxic elements (e.g., As, Cd, Pb, Hg) in a commercial seed-based hemp material. In a study by Angelova et al., uptake of Cd by the seeds of *Cannabis* plants was shown to be nearly equal to the uptake of Cd by the roots, stems, and flowers but higher than that by the leaves.[15] The same study demonstrated that the uptake of Pb was lower by seeds than by the other plant parts. In the HAMQAP study, the mass fractions of the toxic elements in the hemp protein powder sample ranged from 5 ng/g for Pb and Hg to 9 ng/g for As and 25 ng/g for Cd. As expected, most of the participants performed well when measuring Cd, the element with the highest mass fraction (Figure 23.3).

Discussion and evaluation of the sample preparation and analytical methods used by different laboratories is a major focus of QA programs, and extensive final reports are distributed to the participants to discuss approaches to improve performance. Final reports can identify community-wide measurement challenges and provide tools for evaluating analytical methodology and enable participants to improve accuracy and precision of measurements. Often, final reports include charts

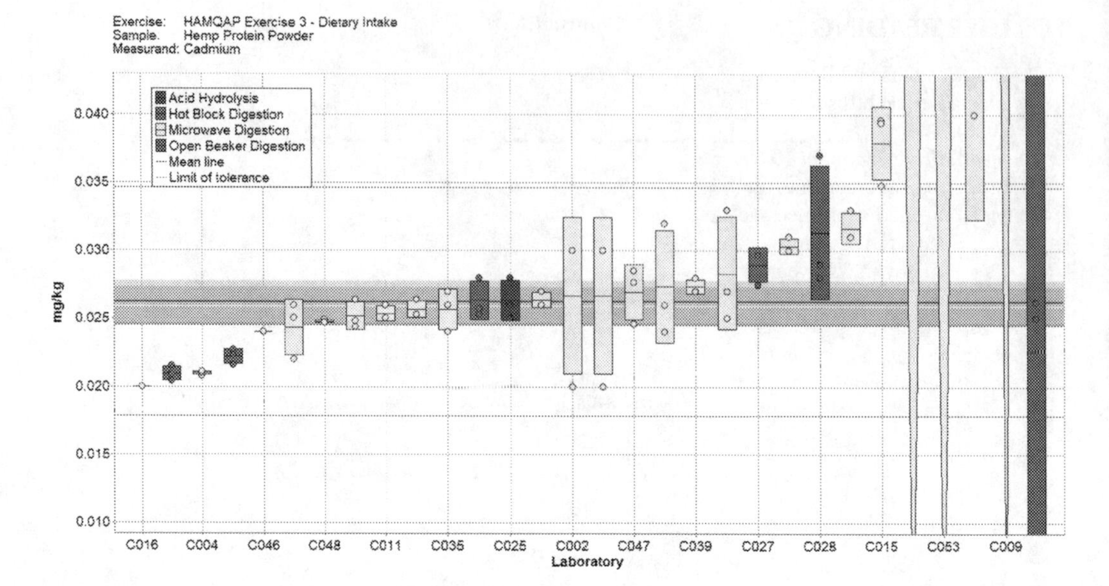

FIGURE 23.3 Results from HAMQAP Exercise 3 for cadmium in hemp protein powder. Individual laboratory data are plotted (circles) with the individual laboratory standard deviation (rectangle). The shading of the data point represents the sample preparation method employed. The solid line in the center represents the consensus mean, and the shaded regions around the consensus mean represent the 95% confidence interval for the consensus mean and the NIST range of tolerance, which encompasses the target value bounded by twice its uncertainty and represents the range that results in an acceptable Z_{NIST} score. The solid lines splitting the upper and lower portions of the graph represent the consensus range of tolerance, calculated as the values above and below the consensus mean that result in an acceptable Z_{comm} score.

and graphs showing how laboratory results compare to one another (laboratories are identified by number only). These comparability studies can be used to demonstrate performance when other methods of assessment are unavailable, and these studies can provide QA and traceability where no commercial CRMs are available.

As the demand for cannabis testing continues to increase, more QA programs will become available and the number and variety of available studies will expand. In general, participation in any type of interlaboratory comparison (e.g., internal QA program, external QA or PT program) can be beneficial to a laboratory and improve overall performance and confidence in measurements, which in turn increases the confidence in testing of routine samples for ensuring product safety.

FINAL THOUGHTS

Reference materials are an invaluable resource for evaluating measurement quality, from method development to method validation or implementation, and in routine testing to demonstrate method performance. In many cases, particularly in emerging areas such as the analysis of *Cannabis* plants and *Cannabis*-containing products, reference materials may not be available or accessible to every laboratory. In these cases, use of harmonized or standard methods from organizations such as AOAC INTERNATIONAL, ASTM, or the US Pharmacopeia may allow a regulatory requirement to be met without routine reference material use. Laboratories may also develop in-house reference materials through careful planning and characterization that will allow demonstration of method performance in the absence of National Metrology Institute (NMI)-produced or other commercial reference materials. Finally, participation in interlaboratory comparisons such as QA programs can provide an external demonstration of method performance.

FURTHER READING

1. U.S. Customs and Border Patrol, Did You Know... Marijuana Was Once a Legal Cross-Border Import? Available at https://www.cbp.gov/about/history/did-you-know/marijuana 2019, (accessed February 2020).
2. Mead, *Front Plant Sci* **10** 697, 2019.
3. U.S. Drug Enforcement Administration, The Controlled Substances Act, Available at https://www.dea.gov/controlled-substances-act (accessed February 2020).
4. U.S. Food & Drug Administration, FDA Hemp Production and the 2018 Farm Bill. Available at https://www.fda.gov/news-events/congressional-testimony/hemp-production-and-2018-farm-bill-07252019, 2019 (accessed February 2020).
5. FDA, What You Need to Know (And What We're Working to Find Out) About Products Containing Cannabis or Cannabis-Derived Compounds, Including CBD. Available at https://www.fda.gov/consumers/consumer-updates/what-you-need-know-and-what-were-working-find-out-about-products-containing-cannabis-or-cannabis, 2019 (accessed February 2020).
6. M. Girdhar, N. R. Sharma, H. Rehman, A. Kumar, and A. Mohan, *Biotech* **4**(6), 579–589, 2014.
7. P. Linger, J. Müssig, H. Fischer, and J. Kobert, *Industrial Crops and Products* **16**, 33–42, 2002.
8. U.S. Environmental Protection Agency, Cadmium Compounds Hazard Summary. Available at http://www3.epa.gov/airtoxics/hlthef/cadmium.html, 2000 (accessed February 2020).
9. Khan, S. Khan, M. A. Khan, Z. Qamar, and M. Waqas, *Environmental Science and Pollution Research* **22**, 13772–13799, 2015.
10. FDA, Harmful and Potentially Harmful Constituents in Tobacco Products and Tobacco Smoke: Established List. Available at https://www.fda.gov/tobacco-products/rules-regulations-and-guidance/harmful-and-potentially-harmful-constituents-tobacco-products-and-tobacco-smoke-established-list (accessed March 2020).
11. W.C. Cunningham, and S.G. Capar, Reference Materials in Elemental Analysis Manual for Food and Related Products. Available at http://www.fda.gov/downloads/Food/FoodScienceResearch/LaboratoryMethods/UCM421144.pdf, 2014 (accessed February 2020).
12. K.E. Sharpless, K.A. Lippa, D.L. Duewer, and A.L. Rukhin, NIST Special Publication 260-181; U.S. Government Printing Office: Washington, DC. Available at http://www.nist.gov/srm/upload/SP260-181.pdf, 2014 (accessed February 2020).
13. JCGM 200:2012, International Vocabulary of Metrology – Basic and General Concepts and Associated Terms (VIM 3rd edition); Joint Committee for Guides in Metrology (JCGM). Available at https://www.bipm.org/utils/common/documents/jcgm/JCGM_200_2012.pdf, 2012 (accessed April 2020).
14. C.A. Barber, C.Q. Burdette, M.M. Phillips, C.A. Rimmer, L.J. Wood, L. Yu, and S.P. Kotoski, Health Assessment Quality Assurance Program: Exercise 3 Final Report. Available at https://doi.org/10.6028/NIST.IR.8285, 2020 (accessed February 2020).
15. V. Angelova, R. Ivanova, V. Delibaltova, and K. Ivanov, *Industrial Crops and Products* **19** (3) 197–205, 2004.

24 Measurement of Elemental Constituents of Cannabis Vaping Liquids and Aerosols by ICP-MS

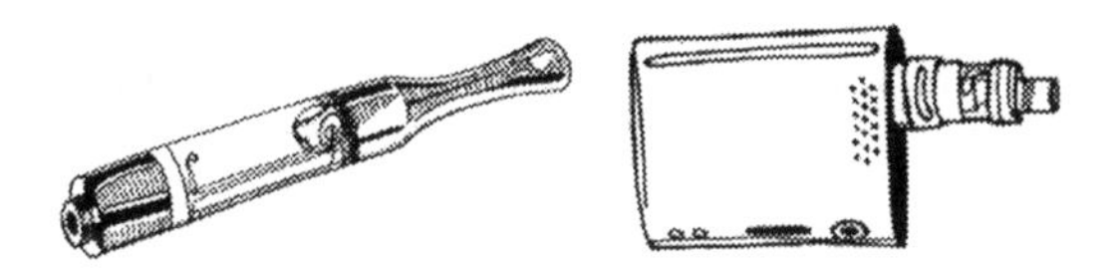

This chapter on the measurement of elemental constituents in vaping liquids and aerosol was compiled by Dr. Steve Pappas and his team at the Tobacco Inorganics Group at the Centers for Disease Control and Prevention (CDC) (more information about the authors is given in the Acknowledgment section). It was written with the intention of providing basic information that will help chemists adopt and optimize good practices, minimize poor practices in sample preparation and instrumental analysis, and improve the overall level of knowledge. This very important chapter will also address best practices and common pitfalls that will result in higher quality analyses for informational and regulatory purposes.

Since the passage of the Agriculture Improvement Act of 2018 (2018 Farm Bill), *Cannabis sativa* products derived from the hemp subspecies became legal at the federal level. Legislation has also been passed by several state governments to permit medical, or in some cases, recreational use of products derived from the subspecies that produce higher levels of Δ9-tetrahydrocannabinol (THC), commonly called marijuana.

Inhaling aerosolized extracts of hemp and marijuana, or vaping, has become a popular means of cannabinoid delivery. As cannabis products including vaping devices become more widely used, it has become important to assess the products to help address consumer safety. Several state governments have established specifications for select substances in liquids used in cannabis vaping products. In the future, setting specifications on aerosol constituents may also be established. In addition to many organic constituents, toxic inorganic substances (metals and metalloids) are an important chemical class to consider. These chemicals pose serious health risks and are often difficult to measure accurately because of their low levels and a propensity for background contamination.

If ensuring safety and compliance is the purpose for testing vaping products, it is extremely important that high-quality analyses be performed by knowledgeable analysts. Testing approaches with high false-positive or false-negative results could allow approval of unsafe products for sale to consumers or the disapproval of products that may be potentially less harmful than others. As these products are relatively new, there are key information gaps they required addressing. Unfortunately, there is also a shortage of experienced analysts who are knowledgeable about good practices related to sample preparation for inorganic analyses.

Measurements of metals in solids and liquids are increasingly performed using inductively coupled plasma mass spectrometry (ICP-MS) because of the high sensitivity and accuracy achieved with this instrumentation. There are many reports on elemental constituents of aerosol produced by Electronic Nicotine Delivery Systems (ENDS) in peer-reviewed literature with limited utility unfortunately due to some very poor analytical practices. As demand increases for information on inorganic analyses of constituents in liquids or aerosols obtained from vaping devices, it is possible that some of the poor practices previously published could be adopted, based on the erroneous assumption that the peer-review process should prevent publication of substandard applications by novice researchers.

Several critical concepts should be examined and understood before using ICP-MS to obtain reliable data. These include selection of analytes, proper sample preparation practices, use of reference materials to assure accuracy if available, determination of a meaningful procedural limit of detection (LOD), and implementing a robust quality control system. The analytical instrumentation will detect ions, count ions, and electronically calculate results of an analysis whether the ions originated from the sample or from background contamination, so it is important when performing analyses to master proper sample preparation practices in addition to instrumental control concepts.

VAPING LIQUID SOLVENT: THE NATURE OF THE SAMPLE

Propylene glycol, glycerol, or a mixture of the two and water are most commonly used as solvents for aerosol nicotine delivery in most ENDS devices.[1,2] These solvents are hydrophilic and efficiently transport fairly hydrophilic nicotine and nicotine salts. These solvents lend themselves to relatively simple sample preparation for analysis by ICP-MS. Since they are hydrophilic, aliquots of an ENDS liquid or trapped aerosol may be simply diluted in appropriate ultra-high-purity acid solutions that are suitable for maintaining metal solubility for analysis by ICP-MS.[3,4] The solvents used for generation of cannabinoid aerosols, however, have more chemically diverse characteristics, so cannabinoid vaping liquids and aerosol samples may require additional steps for preparation for analysis.

As was the case for the recent Electronic cigarette or Vaping Associated Lung Injury (EVALI) response, solvents for cannabinoid extract liquids could be selected with insufficient understanding of the health consequences of inhalation. Many substances that are "Generally Regarded as Safe" (GRAS) for ingestion are not safe for inhalation. Vitamin E (α-tocopherol) is a hydrophobic oil-soluble antioxidant vitamin that may supplement human nutrition when ingested, but it was not intended for inhalation. α-Tocopherol was a constituent detected in 17 of 46 devices submitted to the MDH Public Health Laboratory by Minnesota EVALI patients.[5] Vitamin E acetate is the acetate ester of the phenolic functional group of the α-tocopherol aromatic ring. This esterified derivative of α-tocopherol has no antioxidant properties but is an oil that was used as a cannabinoid vaping solvent. Acute or chronic inhalation exposure to oily substances, including vitamins and vitamin derivatives can induce lipoid pneumonia. The use of Electronic Cigarette Delivery Systems (ECDS) in which vitamin E acetate was the principal solvent resulted in many of more than 2,500 hospitalizations and 55 confirmed deaths in the 2019 EVALI outbreak.[6,7] Vitamin E acetate was the solvent or a constituent of the solvent used in 24 of 46 devices submitted by Minnesota EVALI patients.[5]

Medium-chain triglyceride (MCT) oils have also been used as a solvent in cannabinoid vaping devices. MCT oil has shorter fatty acid carbon chains esterified to glycerol than the higher boiling fractions obtained from oil sources such as palm oil. The shorter fatty acid carbon chain lengths reduce the viscosity and boiling point compared to oils with longer chain fatty acids. Although MCT oil was not the most commonly found etiologic agent in devices submitted to physicians or law enforcement by EVALI patients, MCT oil was present in 20 of 46 devices submitted by Minnesota EVALI patients.[1] The Minnesota report stated that some EVALI -associated products contained combinations of MCT oil, vitamin E acetate, and α-tocopherol or γ-tocopherol.

In the Minnesota study, 8 of 46 devices submitted by EVALI patients had unidentified solvents that were neither MCT oil, vitamin E acetate, nor tocopherols, so the metals analyst may be faced with the analysis of products with a wide range of solvent characteristics.

Unlike tocopherols, vitamin E acetate (identified as the main culprit in the recent EVALI outbreak) and MCT oil, propylene glycol, and glycerol are hydrophilic and miscible with water. Although cannabinoids and terpenes are hydrophobic, propylene glycol and glycerol are used as solvents in some ECDS devices as well as in ENDS devices that were discussed earlier. Hydrophilic solvents are not good solvents for cannabinoids, so when these solvents are used, the immiscible cannabinoids must be homogeneously dispersed in the solvent. One procedure for homogenous dispersion of cannabinoids and terpenes in hydrophilic solvents is nanoemulsification by sonication.[8,9] Vaping liquids made with propylene glycol or glycerol were not strongly correlated with EVALI, but inhalation of aerosols from ENDS or ECDS devices are not without health risk. Diagnoses of acute lung injury resulting from the inhalation of aerosols from ENDS devices that had propylene glycol or glycerol solvents included eosinophilic pneumonia.[10] Chronic disease may be a consequence of the inhalation of carcinogenic aldehydes formed by thermal decomposition of propylene glycol and glycerol.[11]

Since all currently used nicotine and cannabinoid aerosols have associated inhalation health risks, and the CDC has discouraged the use of vitamin E acetate as a vaping solvent,[7] manufacturers may continue to adapt and change ECDS solvents. The regulatory analyst is therefore faced with a moving target for the nature of the solvent matrix. Currently, ECDS device solvents may be either hydrophilic or hydrophobic. If the analyst knows that a solvent or aerosol consists of hydrophilic substances such as propylene glycol and glycerol, dilution of the sample in dilute acids such as have been used for the analyses of ENDS liquid and aerosol metals[3,4] is appropriate. If the analyst knows that the solvent is hydrophobic, dilution of the liquid or aerosol into aqueous acid is not an appropriate approach to sample preparation. If the analyst does not know whether the sample is hydrophilic or hydrophobic, the analytical approach for hydrophobic solvents works in both cases.

Hydrophobic samples are not miscible with dilute aqueous acids and are not ideal for analysis by ICP-MS. Therefore, the standard approach to preparation of hydrophobic samples such as these is to oxidize the organic content to carbon dioxide using microwave digestion with an oxidizing acid solution, which leaves the remainder of the sample almost entirely inorganic.

CHOICE OF LIQUID SOLUTION CONTAINERS

Oxidation of hydrophobic or unknown organic components of ECDS liquids is usually performed using nitric acid. Other acids, such as hydrochloric acid, may be necessary for recovery and stabilization of specific analytes in solution using a microwave digestion system with fluoropolymer vessels (preferably modified polytetrafluoroethylene (PTFE), known as TFM). TFM is ideal for sample preparation that requires microwave digestion, because it avoids glassware that leaches metals into acid solvents and has higher heat tolerance than vessels made from perfluoroalkoxy (PFA) resin. High-purity-fused silica quartz vessels are acceptable for some applications, but if hydrofluoric acid or phosphoric acid must be used during the digestion, quartz is not compatible with these acids. In addition, some analytes, such as chromium at trace concentrations, may be partially retained by the quartz surface.

Glass vessels or tubes should not be used as liquid containers or as digestion vessels for metals analyses.[12–14] It can be misleading that vendors sell "acid-cleaned" glass tubes for sequential microwave digestion systems to analysts. Inexpensive disposable glass sample containers could be appealing to novice analysts who are unaware of the background contamination issues. Glass is not metal free and permits ion mobility from the amorphous silicate matrix. The ion mobility is enhanced by acids that are needed to maintain metal solubility. External cleaning with acid does not eliminate extractable metals from glass. As the metals migrate to the glass silicate surface, hydrogen ions replace them and metals migrate into the acid solution. Since ions are mobile within the silicate matrix, metals from deeper in the silicate matrix migrate to the surface and into the acid solution within minutes, contaminating the sample. The well-known ion migration within glass silicate matrices is the principle behind the thin glass membrane pH electrode and selective ion electrodes.[15] For these reasons, bottletop dispensers should only be used with PFA resin, fluorinated ethylene propylene (FEP), or polypropylene (PP) bottles. The dispensers themselves should not be manufactured from glass. They should not have a ball with a metal coating such as platinum in the valve mechanism. Even 0.1% metal impurities in 99.9% pure platinum cause elevated and variable backgrounds for many metals and metalloids, depending on how long the acid has been in contact with the valve ball since the last use. Bottletop dispensers that are manufactured with only PFA in the liquid path are expensive, but available.

Analysts who had high-sample preparation throughput requirements 15 years ago had little choice in selection of autodilution devices and sometimes had to work with devices that had glass syringe barrels. Analysts who have used these devices[16] quickly learned that if the barrels were rinsed with 10–20 syringe barrel volumes of diluent acid before use, they could subsequently provide a fairly clean acid solution for analysis of metals. However, the sample preparation had to be continuous because metals from the glass barrels migrated into the acid during sample preparation. If an analyst interrupted the sample preparation for 5 min, the next diluted sample was contaminated with metals from the glass that had accumulated in the syringes unless the glass syringes were again thoroughly rinsed with acid before continuing. Months and years of use only slightly diminished acid-extractable metals from the glass syringe barrels. One manufacturer tested quartz barrels in order to address the constant equilibration of metals from glass with the acid solutions in the barrels. The quartz barrels were not high-purity-fused silica quartz, so although the metals from the quartz silicate matrices accumulated at lower levels in the acid than from the glass silicate matrices, the low-purity quartz barrels had the same problem with metals from the quartz equilibrating with the acid solution within the barrels.

"Disposable" glass tubes for microwave sample digestion should never be used for metals analysis. When using a sequential microwave system, the vendor may provide disposable PFA inserts for the sample tubes, high-purity quartz tubes, or both that could be used in place of the glass tubes.[17] Since the PFA inserts are considered consumables and are not inexpensive, this option would eliminate the appeal of "inexpensive" disposable glass tubes in favor of good analytical practice that will produce accurate and reproducible results.

PP tubes, sometimes called centrifugation tubes by manufacturers, are available in 15 mL sizes that are frequently used as containers for final dilutions of samples, blanks, and standards that are placed in instrument autosamplers. The tubes manufactured by different manufacturers have different trace metals backgrounds. Though most backgrounds are fairly low in PP tube material, some PP tubes have been observed by the authors of this chapter to have more elevated extractable aluminum or other metals than tubes from other manufacturers. It is necessary, therefore, to purchase small quantities of tubes from different manufacturer sources and test acid solutions in them for analytes of interest before making a decision on source. The aluminum, calcium, and iron backgrounds were observed to increase in "Metal Free" tubes from one vendor, when the vendor changed supplier. Even PFA tubes may be found to have higher extractable aluminum than some PP tubes. It is always a good practice to fill 15 mL tubes with 14 mL of 1% v/v nitric acid overnight and rinse them with ultrapure water the following morning, to decrease extractable metal backgrounds from the tubes.

MICROWAVE DIGESTION OF VAPING OILS

The propylene glycol and glycerol solvents used for nicotine aerosolization in ENDS devices are miscible with aqueous solutions, as previously discussed. These solvents are also used for aerosolization of cannabinoids in some ECDS devices. If liquid samples are known to have hydrophilic solvents, sample preparation with simple dilution into an appropriate acid solution for analysis without microwave digestion could be appropriate.[3] The analyst should nevertheless consider precautions for potential effects of the organic matrices on analyte response during ICP-MS analysis, since carbon in the plasma enhances the signals of high-ionization potential elements such as arsenic, selenium, and mercury relative to a solution that may not have carbon. Precautions such as the use of 1% v/v semiconductor grade 2 propanol in the final diluent will normalize the response of high-ionization potential elements between blanks, standards, and samples.

If the liquid samples are known to consist of water-immiscible solvents such as vitamin E acetate, tocopherols, or MCT oil, or if the nature of the solvent is not known, microwave digestion will oxidize the organic matrix, leaving metal analytes dissolved in an aqueous acidic solvent. Environmental Protection Agency (EPA) Method 3052 is a very good, flexible, all-purpose microwave digestion method for standard simultaneous sample digestion microwave systems. The EPA Method 3052 prescribes a 5.5-min ramp to a maximum temperature of 180°C, followed by maintaining the temperature at 180°C for at least 9.5 min. This temperature was sufficiently high to enable oxidation of most organic chemical bonds with concentrated nitric acid and was approximately a temperature limit for digestion vessels made from PFA. Presently available TFM vessels can tolerate higher temperatures, permitting more complete digestion at maximum temperatures of 190°C–200°C; therefore, a slight modification of EPA Method 3052 is advisable if TFM or quartz vessels are used. A temperature program with a 10-min ramp to 200°C and maintained at 200°C for 15 min has been used successfully for digestion of Spex Certiprep Multielement Organometallic Oil Standard and for organic substances such as smoke tar.[18] These conditions are appropriate for hydrophobic solvents such as tocopherols, vitamin E actetate, and MCT oils.

The principal solvent prescribed for different matrices in variations of EPA Method 3052 is 9 mL of ultra-high-purity concentrated nitric acid. Depending on the sample matrix and analytes, additional reagents including hydrochloric acid, hydrofluoric acid, and hydrogen peroxide are recommended.

All reagents used for sample preparation and microwave digestion should be ultrapure. After distillation, ultrapure acids for metals analyses are sold in PFA or FEP containers to avoid leaching of metals from the containers. Hydrogen peroxide (30% or greater without tin stabilizer) may be obtained in fluoropolymer or sometimes PP or well-cleaned high-density polyethylene (HDPE) containers. Acids and hydrogen peroxide sold in glass containers should not be used for metals analyses. See the discussion of glass in the previous and following sections. Due to the cost of ultra-high-purity acids, sub-boiling distillation apparati made from high-purity quartz or PFA are available for distillation of less-expensive intermediate-purity nitric acid.

LIQUID SAMPLE CONTAINERS AND AEROSOL COLLECTION MATERIALS

Acids sold in glass containers should never be used for metals analysis. See the previous section for discussion of appropriate solvent purity and materials used for acid containers. Glass tubes, watch glasses, cylinders, flasks, evaporating dishes, and other glass vessels should not be used in the analyses of metals in vaping liquids or aerosols.[12–14] Metal concentrations in ENDS aerosols have nevertheless been published after collecting the aerosols over a 5-h period in 500 mL glass flasks containing 10% nitric acid + 3% hydrochloric acid. Although sufficient method description in peer-reviewed literature should include methodological details to repeat the experiment, the volume of acid used in the flasks was not reported, nor were the method validation details.[19,20]

In order to show the impact of the use of glass vessels on analyses of metals when using these vessels, four 500 mL glass flasks were presoaked with 10% nitric acid + 3% hydrochloric acid to eliminate any surface contamination, rinsed with ultrapure water, and dried. To each flask, 100 mL of 10% nitric acid + 3% hydrochloric acid was added on each of 5 days. The acid remained in the flasks for 5 h prior to transfer of aliquots to acid-rinsed PP tubes for analysis. Analyses were performed on the same day with the same calibration in order to prevent instrumental analysis to analysis variability, so that the source of variabilities in metals concentrations in the flasks was limited to the flasks themselves.

The ranges of the concentrations of analytes in the acid mixes exposed to the glass flasks were calculated as if 60 puffs of ENDS aerosol had been collected in 400 mL acid. Results were reported in µg/10 puffs similarly to ENDS aerosols collected using glass flasks.[19] The results were compared to analyses of ENDS aerosols trapped in glass flasks[19] (Table 24.1).

While it is possible that the flasks that were used for the reported data had unusually low metals concentrations transported into solution from the glass silicate matrices, it appears that the reported aerosol concentrations of chromium, nickel, zinc, and lead were between the lowest and highest background concentrations of the five repeated analyses of acid in flasks in which no aerosol was collected. Copper and tin to a lesser extent were reported at a level that might provide analytical confidence in reported results. The differences in flask-to-flask variability and the day-to-day variability for metal concentrations in the same flask suggest that the use of different flasks for obtaining a procedural blank versus an aerosol sample or the use of the same flask on different days would impart additional variability. This would further decrease confidence in analytical results using glass flasks. If day-to-day instrument response differences were also added as a source of variability, it is apparent that the reported results are not reliable and would likely have been well below a LOD calculated based on repeated procedural blanks or according to Taylor.[21] The authors did not report an LOD in the earlier manuscript.[19] In that manuscript, the data was obtained using an inductively coupled plasma optical emission spectrometer, which does not have the sensitivity of a single quadrupole ICP-MS or a triple quadrupole ICP-MS (QQQ-ICP-MS). Data was later made available in a supplemental file that was not part of the manuscript in which limits of quantitation (LOQs) were mentioned, but there was no description of how the LOQs were calculated, and LODs for many elements were included.[20] It is possible that this list was the vendor's instrument LOD or LOQ. However, data reported in bar graph form obtained using inductively coupled plasma optical emission spectrometry (ICP-OES)[20] was still at concentrations lower than lowest reportable levels (LRLs) or LODs obtained procedurally using high-purity fluoropolymer aerosol traps and more sensitive QQQ-ICP-MS instrumentation interfaced with a sensitive desolvating introduction system.[4]

Previous reports have shown that glass or low-purity quartz fiber filters should not be used for routine collection of aerosol for metals analyses.[12–14] Metal concentrations in ENDS aerosols

TABLE 24.1
Ranges of Metal Concentrations in Acid Solution Obtained from Five Repeat Analyses of Four Acid-Cleaned Glass Flasks Compared to Results Reported Using Glass Flasks as Aerosol Collection Vessels

Concentrations as µg/10 Puffs	Cr	Ni	Cu	Zn	Sn	Pb
Flask A	0.0009–0.0247	0.0037–0.0267	0.0025–0.0299	0.0180–0.2739	0.0013–0.0097	0.0007–0.1263
Flask B	0.0002–0.0086	0.0021–0.0164	0.0007–0.0018	0.0075–0.1661	0.0010–0.0047	0.0013–0.0383
Flask C	0.0002–0.0121	0.0013–0.0065	0.0000–0.0018	0.0095–0.0676	0.0008–0.0054	0.0001–0.0064
Flask D	0.0003–0.0230	0.0083–0.0135	0.0005–0.0016	0.0111–0.0957	0.0006–0.0125	0.0012–0.0221
Reported results[19]	0.007	0.005	0.203	0.058	0.037	0.017

were nevertheless compared to mainstream smoke aerosols by blowing the aerosols into ambient room air and sucking them onto unleached low-purity quartz fiber filters.[22] The LODs were said to be calculated as "2 times the total analytical uncertainties, as the concentrations of the species approached zero." The method by which metals were obtained from the quartz fiber filters was not specifically described, but the authors cited the method of Herner et al.[23] Herner et al. did not extract metals from low-purity quartz filters. Herner et al. used microwave digestion with heat and time parameters according to EPA Method 3052 but with only 1.5 mL nitric acid, 0.5 mL hydrofluoric acid, and 0.2 mL hydrochloric acid to obtain metals from Teflon fluoropolymer (PTFE) filters. After digestion, samples were diluted to 30 mL with ultrapure water. A citation of the method of Herner et al. is insufficient information on the analytical procedure. The hydrofluoric acid used in the cited Herner et al. procedure would attack quartz and liberate high-metal concentrations from low-purity quartz fibers; therefore, it is likely that a modified Herner et al.[23] method was used by Saffari et al.[22]

Quartz fiber filters have lower metal impurities than glass fiber filters but have insufficient purity with regard to metals, limiting their utility. Quartz fiber filters have been cautiously used for mainstream cigarette smoke metals analysis when electrostatic precipitation with high-purity-fused silica quartz tubes was not available.[24,25] When quartz filters were used for cigarette smoke, random filters were screened for procedural blank testing with acid. In one study, the procedural blank variability after acid digestion was sufficiently high that only cadmium, lead, and thallium in mainstream cigarette smoke could be consistently observed at greater than background on most filters.[24] The observed average signal to background of 2:1, however, was analytically insufficient. The levels of acid-leachable impurities from low-purity quartz filters required that a rigorous three-step hot acid and ultrapure water leaching and cleaning procedure to increase the signal-to-background ratio sufficiently to be acceptable for cadmium, thallium, and lead. This labor-intensive procedure removed surface contamination so that remaining backgrounds would only come from within the matrix and not from the quartz fiber surfaces. The metals backgrounds were still too variable or too high for determination of chromium, cobalt, nickel, and arsenic even after these procedures. It is for this reason that only cadmium, thallium, and lead were reported in the cited manuscripts.[24,25] Concentrations of other metals in ENDS aerosols using quartz fiber filters that had not been preleached and rinsed have nevertheless been reported from some groups.[22,26]

To demonstrate how low-purity quartz fiber filters impact metals analyses in aerosol, twenty 44 mm circular filters were punched from filter sheets using a polyvinyl chloride punch as previously described.[24] Each filter was treated with 5 mL water and 5 mL nitric acid prior to microwave digestion (10-min ramp to 200°C, 10-min digestion at 200°C) as if aerosol had been collected on the filters.[26] After the digestion program, samples were diluted to 100 mL with ultrapure water. Same day analyses were performed using the same calibration to prevent additional analysis-to-analysis variability, so that variabilities in metals concentrations were limited to the filters selected from different levels in stacks of filters. The results from the analyses of the acid digests of the filters are summarized in Table 24.2 after converting units to ng/10 puffs, as if aerosol from 50 puffs had been collected. The results were compared to results of analyses of aerosols collected from ENDS devices using a high-purity fluoropolymer trap.[4] The mean metal background concentrations obtained from the quartz fiber filters (Table 24.2) were higher than chromium, nickel, zinc, and tin in aerosols from all 20 ENDS devices tested with the fluoropolymer system. The mean lead concentrations obtained from the quartz fiber filters were higher than in 19 of 20 aerosols. The mean copper concentrations obtained from the quartz fiber filters were higher than in 16 of 20 aerosols. The highest concentration of copper obtained from the quartz fiber filters was higher than copper in aerosols from all 20 ENDS devices tested.[4] Calculations of instrument LODs based on three times standard deviation of replicates within a single dilute high-purity acid blank are not realistic method LODs. Meaningful method LODs should take into account multiple preparations of procedural blanks analyzed on multiple days. Taylor's method[21] is commonly used to calculate LODs, but

TABLE 24.2
Metal Concentrations in Acid Solution Obtained from Analysis of 20 Quartz Fiber Filters

Blank (ng/10 Puffs)	Cr	Ni	Cu	Zn	As	Cd	Sn	Pb
Mean ± Std. Dev.	136 ± 21	40.0 ± 4.9	88.9 ± 152	794 ± 203	< 0 ± 1927	< 0 ± 0.2	60.0 ± 21.8	5.4 ± 1.5
3 × Std. Dev.	62.8	14.8	456	610	5780	0.72	65.3	4.4
Range	62–161	33–52	18–722	459–1136	< 0	< 0	17–115	3.7–9.2

Three × standard deviations of procedural blanks analyzed in a single analytical run were used as liberal approximations of method LODs for comparison with published aerosol metal concentrations.[4]

measurements of multiple procedural blanks in a minimum of seven independent analytical runs (preferably 20 runs) have also been used. Although the quartz filters discussed here were analyzed on the same day rather than in independent analytical runs, LODs were calculated based on three times standard deviations of the within-run results for each analyte. The results in Table 24.2 would suggest that all 20 aerosol chromium, nickel, zinc, and tin results, 18 of 20 aerosol copper results, and 19 of 20 aerosol lead results obtained using a high-purity fluoropolymer trap[4] would be below the calculated within-run LOD using quartz fiber filters. If the method LOD was calculated based on multiple independent runs, the calculated LODs would likely have been higher.

In a recent study using a high-purity fluoropolymer trap for aerosol capture, no vaping devices had components containing arsenic or cadmium, and no liquids or aerosols were found to contain concentrations of arsenic or cadmium greater than the method LOD.[3] Arsenic was dropped from the method as not likely to be relevant to vaping liquid or aerosol analysis, and cadmium was found to be below method detection limits calculated according to Taylor[21] in all liquids and aerosols tested.[3,4] If the few aerosol results obtained using a high-purity fluoropolymer trap[4] that were not below "within run" LODs calculated using the data from 20 quartz fiber filters, the reported results would have depended on which filter was used to trap aerosol and which was subtracted as a procedural blank. For example, if the highest mean copper concentration reported in aerosol from 20 devices[4] had resulted from subtraction of the lowest quartz filter procedural blank (18 ng/10 puffs), it would yield the 614 ng/10 puffs result. The result would have been less than 0 ng/10 puffs if the highest quartz filter copper concentration (722 ng/10 puffs) had been subtracted as the procedural blank. This strongly suggests that some reported results may have been solely driven by metals leached from inappropriate glass or low-purity quartz trapping materials. In summary, quartz fiber filters are not suitable for trapping aerosols from vaping devices and glass fiber filters even less suitable.

It is clearly evident that some investigators who have published ENDS aerosol metals data have insufficient experience with analytical method validation and quality assurance, and fewer still are unaware of the impacts of using inappropriate liquid vessels and trapping materials for inorganic analysis. Even peer reviewers have apparently been inappropriate authorities for critically reviewing some manuscripts on inorganic analyses of ENDS and ECDS aerosols. The analyst tasked with metals measurements must be aware of this and must critically review published methods prior to following procedures. It is hoped that this chapter will outline some of the common analytical pitfalls.

VAPING MACHINES AND TRAPPING MATERIALS

If vaping aerosol analyses are to provide robust data, the methodologies and instrumentation must be scientifically sound and defensible, likely within a framework of ISO 17025 accreditation or similar external accreditation source. While no vaping machine puffing regimen mimics a human vaper, it is imperative that the data collected follows a standard and reproducible procedure.

The CORESTA Method 81[27] is a widely used method with acceptable specifications, including prescribed limits for pressure drops, puff volume, puff time, and puff profile for generation and collection of reproducible ENDS aerosol results. There are no widely accepted vaping machine puff methods for generation of ECDS aerosol, so until one becomes available, CORESTA Method 81 is a defensible choice. This protocol requires a 3-s 55 mL puff every 30 s with a machine pressure drop no greater than 300 Pa and a pressure drop across the aerosol trapping assembly no greater than 900 Pa at a linear air flow of 140 mm/s. It further requires a rectangular puff profile unlike the bell-shaped puff profile specified for mainstream cigarette smoke collection in the cigarette smoking machine method, ISO 3308.[28]

Since the heating elements of some ENDS and ECDS devices are activated by sufficient air flow, the heating elements in these devices will not generate aerosol until the air flow reaches the rate that triggers the heating element. Therefore, a rectangular puff profile is required and should be verified with a 1,000 Pa restrictor in place. The sum of the times required for air flow through the device to ascend to 18.5 mL/s and to descend to baseline may not exceed 10% of the puff. This assures that air flow-activated heating elements begin heating at the beginning of the puff, since CORESTA Method 81 was written assuming the use of standard glass fiber filters for trapping organic constituents of aerosols. Thus, modification to replace the glass fiber filters with traps that are appropriate for inorganic aerosol constituents is required. Although PTFE filters and other polymer materials in PP mounts with diameters of 47 mm and smaller are readily available, all of those tested in the laboratory of the authors of this chapter have failed CORESTA Method 81 pressure drop specifications. An in-line PTFE 60 mm filter with 0.45 μm pore size in a PP mount was found to meet CORESTA Method 81 puff profile specifications and quantitatively trapped hydrophobic aerosol generated by ECDS devices. This filter probably met the specifications because of greater surface area of the 60 mm filters compared to the 47 mm and smaller diameter filters. Unfortunately, the housing shape made aerosol recovery from the internal periphery of the filter housing difficult, since liquid remained between the housing and filter at the perimeter under reverse vacuum suction. As a result, numerous rinses were required to assure quantitative aerosol recovery from within the filter housing. While impractical compared to the fluoropolymer condensation tube currently used,[4] it at least offered hope that a simple product constructed from clean materials might yield a less labor-intensive approach in the future.

The current procedure used for trapping and recovery of hydrophobic ECDS aerosols by this chapter's authors is a modification of the method used for ENDS devices.[4] The trap consists of a 518 cm length of 3.97 mm i.d. FEP tubing, thoroughly rinsed with 2% v/v nitric acid + 1% v/v hydrochloric acid before and between each use. The devices are connected to the tubing from the vaping machine syringe pump (Figure 24.1). The other end of the tubing is connected to the vaping device mouthpiece with Tygon tubing that has been thoroughly soaked in 2% v/v nitric acid + 1% v/v hydrochloric acid, since untreated Tygon has leachable metals. The FEP tubing traps aerosol and particles by condensation.[4] The trapped aerosol is rinsed from the tubing with 3 × 8 mL rinses from a PFA syringe with 2% v/v nitric acid + 1% v/v hydrochloric acid into a class A 25 mL polymethylpentene (PMP) volumetric flask and dilution to 25 mL with the same acid solution.[4] The modification of this method for oily ECDS aerosols includes an initial tubing rinse with 5 mL diethylene glycol monoethyl ether (DEGMEE). This solvent has low volatility, is an excellent solvent for oils, and is water miscible. Unfortunately, no high-purity grade with respect to trace metals is commercially available, so it must be distilled from a high-purity fused silica quartz distillation flask prior to use. DEGMEE dissolves and dilutes the oil droplets from up to 50 CORESTA Method 81 vaping puffs as it passes through the condensation tube into a 50 mL PMP volumetric flask. The remnant in the condensation tube is then followed by 4 × 8 mL rinses with 2% v/v nitric acid + 1% v/v hydrochloric acid into the same 50 mL flask and dilution to 50 mL with the same acid solution. Since DEGMEE is both oil and water miscible, it solubilizes or at least emulsifies the oil with the aqueous acid for analysis.

FIGURE 24.1 Vaping devices are connected to the vaping machine syringe pump through an acid-cleaned high-purity fluoropolymer condensation tubing trap.

PREPARING FOR ANALYSIS

Calibration standards are prepared in 2% v/v nitric acid + 1% v/v hydrochloric acid + 10% DEGMEE for the purpose of matrix matching with the same solution obtained after rinsing aerosol metals from the condensation tube. If a desolvating introduction system with a heated spray chamber is used, and if tin is one of the analytes, 0.5% hydrofluoric acid is an essential additive to the final solution that enters the spray chamber. Tin gradually oxidizes from tin(II) ion to tin(IV). If no hydrofluoric acid is used, tin(IV) forms the $SnCl_4$ complex. $SnCl_4$ is volatile below 120°C and does not efficiently enter the plasma at the 160°C temperatures of heated spray chambers in desolvating introduction systems. Because of the inefficient entry of $SnCl_4$ into the plasma at 160°C, calibration regressions are not linear. SnF_4, however, is not volatile at 160°C. Addition of 1% v/v hydrofluoric acid to the internal standard solvent (1% v/v nitric acid) converts the hard acid Sn(IV) cation to SnF_4. This results in linear calibration for tin. This is a decision point. If laboratorians are not well trained in safe handling of hydrofluoric acid, then hydrofluoric acid should not be used, and either tin should be eliminated from the list of analytes or a Peltier-cooled introduction system should be used in place of the more sensitive desolvating introduction system.

The internal standard solution is prepared in 1% v/v nitric acid with no DEGMEE. Addition of 1% v/v hydrofluoric acid to the internal standard solution is needed if hydrofluoric acid is used. See the previous discussion on hydrofluoric acid and tin. The internal standard solution (1% v/v nitric acid or 1% v/v nitric acid + 1% v/v hydrofluoric acid) is teed in with the same internal diameter peristaltic pump tubing as used for the standards and samples. Teeing in the internal standards without DEGMEE dilutes the standards and samples entering the nebulizer by one half as it cuts the organic content entering the plasma to 5% v/v. In order to further decrease the carbon load to the plasma and enhance sensitivity, a desolvating introduction system is preferred. Many labs may not have experience optimizing a desolvating introduction system, or the financial resources to purchase one, but if possible, it is a worthy investment. As an alternative, most major ICP-MS vendors now offer a standard Peltier-cooled spray chamber that is a good second choice for decreasing the solvent and carbon load to the plasma. With 5% v/v DEGMEE entering the spray chamber, the liquid is unlikely to freeze at 0°C, so Peltier-cooled spray chambers can be set to –2°C to further reduce solvent and carbon load to the plasma. Residual aqueous solvent may freeze with the Peltier-cooled spray chamber set to –2°C, if the chamber is left for extended periods. The spray chamber's power should be shut off when the instrument is not in use if the temperature is set to –2°C.

WHAT ANALYTES ARE APPROPRIATE FOR REGULATORY PURPOSES?

Because nicotine used in ENDS liquids is usually extracted from tobacco, some analysts may assume that tobacco-related analytes would also be considerations for analysis. However, toxic metal concentrations in ENDS liquids come predominantly from corrosion of internal components of the vaping device.[3] Liquid nicotine extracts that were sold for vaping in polymer containers rather than inside a cartridge or a tank had no metals at concentrations higher than the method LODs.[3] However, liquid metal concentrations can increase as corrosion occurs.[3] Metals in the liquids corresponded to metals in internal device components.[3] Scanning electron microscopy with energy-dispersive X-ray spectroscopy (SEM-EDS) has provided information for selection of metals of concern by determining the compositions of device components that contact the liquid. Not every laboratory has access to such instrumentation, so some helpful information about compositions of components that have been used, and are currently being used in vaping devices, will be provided here.

Metal components of ENDS vaping devices and exposed to the liquid inside the device, cartridge, or pod include nickel, copper, tin or tin/lead solder (less used in later generation devices), steel (mostly iron, chromium, and nickel), nichrome (nickel and chromium), Kanthal (iron, chromium, and aluminum), brass (copper and zinc), occasionally silver coating on a wire, and in some recent ENDS devices, silica heating elements and gold alloy coatings on electrical contacts.

Arsenic, selenium, barium, and mercury are notably absent from the elements used in these materials, although they have been included in some regulatory lists of analytes of concern in some states. Trace arsenic has been reported in aerosols from some ENDS devices.[26,29] However, the method LODs were not well described, and the materials used for trapping the aerosol, including low-purity quartz filters or Tygon tubing that may or may not have been sufficiently cleaned, are possible causes of the reported arsenic. The authors of this chapter had not observed concentrations of arsenic greater than method LODs among repeated analyses of ENDS liquids and aerosols during method development and even dropped arsenic from their analyte list. Selenium is present at low levels in plant materials including tobacco[30] but is unlikely to be efficiently extracted with nicotine from tobacco or cannabinoids from cannabis. If a small amount were extracted, the levels would be exceedingly low. It is unlikely that the selenium concentrations in aerosols from ENDS or ECDS devices will reach a level high enough to contribute measurably to inhalation toxicity. Barium is ubiquitous with higher concentrations than many trace metals. Wherever calcium is found environmentally, barium is also likely to be present. Although barium is toxic, the exposure to barium must be very high for clinical toxicity to be observed.[31] Mercury is a highly neurotoxic trace element, but high concentrations are rarely observed in nature. Sources of mercury in vaping devices may include trace contamination in copper or low-purity gold.

Silver wire coatings are potential exposure sources that could contribute to low silver levels in aerosols, but the toxicity of silver, like selenium and barium, is low. The authors of this chapter do not discourage the inclusion of any elements that may be of interest in its product monitoring list, but these are at least unlikely candidates for concern compared to the more prevalent elements found in device components.

Kanthal and silica ceramic materials are used in some heating elements, it would seem appropriate to include aluminum and silicon as analytes. However, both aluminum and silicon are so environmentally ubiquitous that they are analytically challenging levels even in ultrapure acids and calibration solutions from instrument manufacturers. Silicon also forms silicon oxide in the plasma that forms a refractory coating on sampler and skimmer cones. The buildup of the refractory silicon oxide on the cones causes a gradually increasing silicon background. Because of these considerations, analysis of aluminum and silicon should only be considered by laboratorians with expertise in working with such problematic background levels of these elements. Aluminum and silicon would be more appropriate choices of analytes for experienced labs than for routine analytical purposes.

Iron, though not as ubiquitous as aluminum and silicon, can also be a problematic analyte unless utmost care is taken to avoid contamination. For this reason, iron may also be an element that would be a problematic choice for routine purposes by non-experienced laboratories. Chromium and nickel are not as environmentally ubiquitous as iron, but analysts should remove any external steel ejectors from pipettes prior to using them with acid solutions or use pipettes that do not have steel pipette tip ejectors for trace metal analyses to prevent contamination issues.

Internal device component corrosion is a main contributor to metals in aerosols. A subset of the metals used in the manufacture of device components listed above (chromium, iron, nickel, copper, tin, and lead) are reasonable candidates for routine monitoring. The authors of this chapter have found nickel wires, brass (copper and zinc) and various nickel/copper and iron/copper electrical connectors, tin and tin/lead solder joints, steel substrates for electrical connectors, steel aerosol tubes, and housings as components of ENDS devices. The presence of calcium is a potential issue for cobalt. If abundant levels of calcium are present, or if hydrofluoric acid is used, a less experienced analyst may interpret polyatomic interferences from $^{43}Ca^{16}O^{+}$, $^{40}Ca^{19}F^{+}$, or $^{40}Ar^{19}F^{+}$ as cobalt and report a false positive.

If particles containing cobalt were inhaled, the health risk would be similar to that of nickel with the exception of giant cell interstitial pneumonia that is closely associated with cobalt inhalation.[32] The authors of this chapter have found nickel, which is more geologically plentiful and inexpensive than cobalt, to be prevalent in device components examined by SEM-EDS thus far, whereas no specific components manufactured from cobalt-containing alloys have been found. Therefore, it is important that analysts be cautious in methodology and interpreting results on which accurate medical diagnoses may be dependent. Cobalt should be included in the analyte panel only with attention to detail in testing for false positives during method development and in establishment of rugged procedural method LODs. Since nickel is frequently used in device components, nickel is likely to pose the greater health risk. Inhalation of nickel causes a chemical or hypersensitivity pneumonitis,[33] as does cobalt.[32]

The chromium, nickel, and iron found in Kanthal and nichrome are also found in steel components of ENDS and ECDS devices. The inclusion of chromium, nickel, copper, tin, and lead in analytical methods is supported by SEM-EDS analysis of vaping device components and by analysis of ENDS liquids and aerosols by ICP-MS.[3,4] Although additional metals have been found in device components, some metals have limited utility for routine analytical purposes, given the challenges described above and in some cases their low health impacts.

ICP-MS INSTRUMENTATION

This section focuses on single quadrupole and triple quadrupole ICP-MS instruments for ENDS and ECDS liquid and aerosol analyses for metals and metalloids because they are the most commonly used ICP-MS instrumentation in modern analytical labs.

There is a wide variety of ICP-MS nebulizers. For the analysis of ENDS and ECDS liquids and aerosols, a relatively low flow concentric nebulizer designed for approximately 400 µL per minute delivery is recommended. As described previously, a desolvating introduction system is recommended for aerosol analysis because of the low metal concentrations, but at minimum, a Peltier-cooled spray chamber is strongly recommended for analyses of aerosols and liquids. Manufacturer default peristaltic pump tubing internal diameters may be excessive for an optimum liquid flow rate to the spray chamber and for decreased organic loading of the plasma. In our experience, peristaltic pump speed and tubing diameter should be chosen that will deliver approximately 400 µL liquid per minute to the spray chamber since this is commonly the maximum liquid flow that can be efficiently desolvated by Peltier-cooled or desolvating liquid introduction systems. A combination of 0.4 rps (24 rpm) peristaltic pump speed and dual 0.38 mm internal diameter tubing teed together after the peristaltic pump works well.[4] The pump speed may vary depending on the diameter of the pump roller assembly. One pump tube delivers the internal standard solution with no organic content (no DEGMEE or 2-propanol) for analysis of hydrophobic aerosols. The second pump tube delivers blanks, standards, samples, and QCs from the autosampler which has approximately 10% v/v DEGMEE added during solution preparation or when rinsing hydrophobic aerosols from the condensation tube.

When using a desolvating introduction system, one that does not include membranes is recommended, as the membranes can be coated with organic components and with salts dissolved in samples.

Standard nickel sampler and skimmer cones may be used, but though more expensive, platinum-tipped cones require less frequent cleaning maintenance and run slightly hotter in the plasma, resulting in lower oxides and polyatomic ion formation at a given liquid and sample gas flow. Platinum-tipped cones also resist cone orifice wear longer, so they last much longer than nickel cones, providing good return on investment.

INTRODUCTION SYSTEMS AND OPTIMIZATION

If a desolvating introduction system that employs a high-temperature spray chamber is used, problems with tin calibration are likely to develop unless hydrofluoric acid is added to the acid mix used for standards and samples. Tin is gradually oxidized from tin(II) to tin(IV) ion. When tin(IV) is used with hydrochloric acid, it forms an $SnCl_4$ complex that boils below the 160°C temperature commonly used with desolvating introduction systems. As the volatile $SnCl_4$ complex vaporizes, the calibration becomes curved rather than linear. There are two solutions to this problem. The first is to add 0.5% v/v to 1.0% v/v final concentration of hydrofluoric acid to the standard and sample solvents.[4] Tin(IV) will be more strongly coordinated with fluoride ions from hydrofluoric acid than with chloride ions from hydrofluoric acid. The SnF_4 complex is not volatile at 160°C like the $SnCl_4$ complex, so the calibration will be linear. If analysts are not well trained in safety with handling hydrofluoric acid, however, it is unsafe for them to work with this acid. Serious neurological and muscular injury may occur from dermal or inhalation exposure to hydrofluoric acid. Tin should be eliminated from the analyte list when using a desolvating introduction system if analysts are not trained on hydrofluoric acid safety. The second solution is to use a Peltier-cooled spray chamber instead of a desolvating introduction system. Analytical sensitivity will suffer, but since this system does not have a heated chamber, the $SnCl_4$ complex is not volatilized.

Teeing in an organic-free internal standard solution with the same tubing size as used for standards and samples is a good way to accomplish the final dilution step and to decrease the organic content loading of the plasma, as discussed earlier.

Instrumental radio frequency (RF) power and plasma gas flow are presumed to be optimally set during instrument installation for practical purposes in this chapter. Sampling position and sample gas flow should be optimized slightly favoring low cerium or barium oxide formation over low doubly charged cerium or barium formation during tuning. Although cerium optimization is the default metal for optimization in the software for instruments manufactured by several vendors, barium is actually a better choice. When hydrochloric acid is one of the acids used in the preparation of analytical standards and samples,[35] Cl_2^+ may form sufficiently in the plasma to contribute to the total mass 70 signal intensity presumed to be caused by $^{140}Ce^{2+}$. If the apparent percentage formation of the doubly charged ion ($^{140}Ce^{2+}$) is thought to be too high, the usual solution is to increase the sample gas flow. This should decrease the formation of $^{140}Ce^{2+}$ and increase the formation of $^{140}CeO^+$ until an appropriate balance is achieved. However, in the presence of hydrochloric acid, increasing the sample gas flow increases not only the formation of the $^{140}CeO^+$ polyatomic ion but the $^{35}Cl_2^+$ polyatomic ion as well. The result is that it may appear that as sample gas flow is increased, both the $^{140}Ce^{2+}$ and the $^{140}CeO^+$ are increasing, which is both incorrect and counterintuitive. Once the sample gas flow is optimized, the auxiliary gas flow (sheath gas that flows in the second torch channel between the central sample gas channel and the outer plasma gas channel) may be reoptimized.

If a desolvating introduction system is used, the optimization signal may appear to represent an oddly shaped periodic wave. If this is the case, the likely cause is differences in the sample tubing peristaltic pump speed and the desolvating system waste tubing pump speed causing roller contact with the tubing to be fully in phase or out of phase with each other. Adjusting the waste pump speed to a speed with minimum signal fluctuation should solve this problem. After this adjustment, the analyst should monitor the waste liquid coming from the desolvating system and the plasma for flickering for about 30 min to assure that there is sufficient waste liquid pump speed to prevent accumulation of liquid inside the system.

Extraction lens voltage optimization is often possible using a software auto-optimization procedure, but sometimes these optima, especially the optimum for the initial lens, may be matrix dependent. See the manufacturer's instrument manual to verify appropriate extraction voltages for a given instrument and matrix. Mass calibration and, if necessary, detector voltage optimizations may generally be performed using software procedures. See the manufacturer's instrument manual.

Optimization of quadrupole-related voltages should be performed in the analytical mode used for a given set of analytes. If tin and lead are to be analyzed in evacuated cell (no gas) mode, quadrupole-related voltages should be optimized in this mode. Cell gases are required to eliminate or suppress polyatomic interferences for quantitation of some isotopes. If chromium, nickel, and copper are to be analyzed using kinetic energy discrimination (KED) conditions with helium gas, cell and quadrupole-related voltages should be optimized in this mode. If they are to be analyzed using a reactive gas, then cell and quadrupole-related voltages should be optimized in this mode.

SINGLE QUADRUPOLE-SPECIFIC PARAMETERS

If a single quadrupole ICP-MS is used for routine analytical analyses, one or more cell gases for suppression or removal of polyatomic ion interferences may be required. Polyatomic species that form in the plasma can interfere with some isotopes, such as ^{52}Cr, ^{60}Ni, and ^{63}Cu. The most abundant polyatomic ion interferences with $^{52}Cr^+$ analysis are $^{40}Ar^{12}C^+$, $^{36}Ar^{16}O^+$, and a very small amount of $^{46}Ca^{16}O^+$. Contribution from the latter may be negligible if the calcium concentrations are low in ENDS and ECDS liquids and aerosols. All of these polyatomic ions have the same total mass as the analyte isotope, $^{52}Cr^+$, so they may interfere with the analysis of the analyte. Some instrument vendors provide options for specific reactive gases to suppress or eliminate interferences. To save analysis time, all of the polyatomic interferences may be highly suppressed or eliminated with the use of one low mass gas or gas combination and KED.

Optimum settings vary from instrument to instrument, but under circumstances in which there are negligible polyatomic interferences for a specific isotope, the optimum voltages for cell and quadrupole biases are likely to be negative. The effect of negative biases on both the cell multipole and the analyzer quadrupole is to accelerate positive ions, resulting in fewer ion losses due to low ion kinetic energy. Analyte sensitivity is enhanced under these conditions. Expanding the allowable voltage range beyond the default limits in the manufacturer software in order to tune the biases increasingly negative may appear to provide markedly greater sensitivity, but a point is reached under which conditions, the intended purpose of the cell and analyzer quadrupoles to discriminate between isotopes of different masses is negated, and selectivity for a given mass is compromised.

When reactive gases are used to eliminate polyatomic interferences in the cell, collisions between the cell gas atoms or molecules and analyte ions cause loss of kinetic energy required for the analyte ions to proceed beyond the cell into the analyzer quadrupole. Under these conditions, it is likely that the quadrupole bias will be more negative relative to the cell bias in order to compensate for the loss of analyte ion kinetic energy in the cell while eliminating polyatomic interferences. When the cell multipole is a quadrupole, an additional "q" voltage setting optimization will narrow the mass selectivity bandpass below the selected mass for ions passing through the cell. Although the cell quadrupole is not intended to provide single mass unit resolution, the optimized voltage will be tuned for a compromise between excluding ions below a given mass range, while providing a sufficiently wide mass range selectivity to permit high-analyte isotope transmission. Hexapoles and octopoles provide greater ion transmission than quadrupoles, but they do not have the mass bandpass selectivities of quadrupoles in the cell. An additional "axial field" voltage may also be applied under reactive gas conditions to provide additional ion acceleration and further compensate for loss of kinetic energy by analyte ions due to collisions with gas in the cell.

KED is a technique where one or more cell gases eliminate polyatomic interferences, but the quadrupole bias is significantly less negative than the cell bias. This combination of settings provides less acceleration of ions passing through the cell multipole into the quadrupole. While passing through the cell multipole, polyatomic ions have a greater cross section than monatomic analyte ions. The greater cross section of polyatomic interference ions increases the probability of collisions with helium or other cell gas than that for monatomic analyte ions. The more frequent collisions result in reduced kinetic energy of most polyatomic ions relative to monatomic analyte ions. Under ideal optimum cell and quadrupole bias voltage and cell gas flow conditions, polyatomic ions are eliminated, leaving only monatomic analyte ions with sufficient kinetic energy to pass into the analyzer quadrupole. Axial field voltage is not used under KED conditions, because acceleration of ions would compensate for the loss of kinetic energy by polyatomic ions and enhance transmission of these interferences through the cell.

In practice, the ideal total elimination of polyatomic interferences using either reactive gases or KED with helium is not always achieved. KED was used to eliminate most of the polyatomic interferences for ^{52}Cr, and ^{60}Ni[35] during analysis of tobacco smoke particulate matter. For example, at 5.5 mL/min helium cell gas flow, −18 V octopole bias, −15 V quadrupole bias, the background was very low, but there were random false-positive results for both ^{52}Cr and ^{60}Ni. The presumption was that at least at mass 52, $^{36}Ar^{16}O^+$ signal was fluctuating to some degree. Since hydrogen reacts to eliminate ArO^+ interference well, 0.5 mL/min hydrogen was added to the cell gas. The addition of hydrogen to the cell gas eliminated the false positives, supporting the presumption that ArO was a random polyatomic interference, at least for ^{52}Cr.

"TRIPLE QUADRUPOLE"-SPECIFIC PARAMETERS

"Triple quadrupole" ICP-MS is a term used loosely, since the instrument may actually have two quadrupoles with quadrupole or octopole between the two main quadrupoles. At the time in which this chapter was written, two instrument vendors provide "triple quadrupole" ICP-MS instruments. One vendor's instrument utilizes a smaller initial quadrupole operated at higher alternating voltage RF than the analyzer quadrupole. Both the initial quadrupole and the cell quadrupole are generally run with "q" (low mass settings) intended to exclude low mass ions below a certain level, while being permissive to higher mass ions ("a" settings at 0 V) in order to maintain higher ion transmission through the first quadrupole and the cell, while the final analyzer quadrupole is operated at unit mass resolution. The other vendor's instrument is run with settings to provide 1 mass unit resolution on identical full-size quadrupoles. The octopole in the cell between them acts as an ion guide for high-ion transmission, since the two full quadrupoles provide the single mass unit resolution.

Triple quadrupole instruments can be operated similarly to single quadrupole instruments. However, they have several advantages. Instead of using a reactive gas to specifically eliminate interferences, the more common practice is using a reactive gas to produce a "mass shift" of the analyte ion by forming one or more complexes of the analyte ion with a reactive gas. The first quadrupole, set to one mass unit resolution, excludes all higher and lower masses except the mass of the analyte isotope, such as $^{52}Cr^+$. The reaction of NH_3 with $^{52}Cr^+$ results in the addition of two NH_3 ligands to $^{52}Cr^+$ ion in the cell to form $^{52}Cr(NH_3)_2^+$ ion at mass 86. The analyzer quadrupole is set at mass 86, the mass of the newly formed analyte ion complex. Since quadrupole 1 eliminated all substances except at mass 52, there are no species to interfere at mass 86, leaving a nearly 0 background. However, the cell bias and analyzer quadrupole bias need to be adjusted for this purpose. The triple quadrupole system utilizes mass shift to eliminate interferences. The mass shifted species, $^{52}Cr(NH_3)_2^+$ ion, is a polyatomic ion. For this species to have analytical sensitivity, reverse KED conditions need to be provided to enhance ion transmission of this polyatomic ion. Instrument settings, for illustration of how the mass shift mode functions, were −18 V octopole bias, −8 V energy discrimination. This difference implies that the quadrupole bias was 8 V more negative than the cell (octopole) bias. The specific bias optima, however, are manufacturer and model dependent.[4] The more negative quadrupole bias is the opposite of the KED arrangement. These "reverse KED" cell and quadrupole biases increase the transmission of the polyatomic analyte ions when using the mass shift approach. If one optimizes the cell (octopole) bias and energy discrimination settings (quadrupole bias) first, other electrical settings including the deflect voltage will be optimized with respect to these voltages. However, if the deflect voltage is optimized first, different optimum octopole bias and energy discrimination settings will result. Gas modes used for "triple quadrupole" ICP-MS analysis of analytes discussed in the section on appropriate analytes have been published.[4]

FINAL THOUGHTS

It is hoped that this chapter has provided appropriate information on sample preparation for ICP-MS analysis to help avoid pitfalls and improve analytical practices. The information provided is by no means exhaustive. In addition to discussing good sample preparation practices, some basic information on how different instrumental analytical modes function was included, and general settings can be used to optimize results. Multiple analytical modes are available with commercially available ICP-MS instrumentation. The choices for operating the instruments are often situation dependent. Our hope is that the information provided will provide the analyst with a better understanding of the basic principles, instrumentation modes, help troubleshoot problems, and insights for method development from a more knowledgeable based perspective, if needed.

This chapter was intended to provide sufficient information to provide a rationale for meeting minimal validation criteria such as determining a meaningful method LOD and selecting a suitable approach for measuring a target set of analyses with minimal false positives and negatives. Experienced analysts are aware of additional considerations during method validation including calibration linearity and dynamic range that are not specific to ICP-MS analysis and were not discussed here. Data should be reviewed by competent trained or experienced authority prior to reporting. Taylor has discussed these concepts thoroughly in Quality Assurance of Chemical Measurements,[21] a strongly advised resource for analytical laboratories.

FURTHER READING

1. R. Jensen, R. Strongin, D. Peyton. Solvent chemistry in the electronic cigarette reaction vessel. *Sci. Rep.* **7**, 42549, 2017. https://doi.org/10.1038/srep42549.
2. A. El-Hellani, R. Salman, R. El-Hage, S. Talih, N. Malek, R. Baalbaki, N. Karaoghlanian, R. Nakkash, A. Shihadeh, N.A. Saliba. Nicotine and carbonyl emissions from popular electronic cigarette products: Correlation to liquid composition and design characteristics. *Nicotine Tob. Res.* **20** (2), 215–223, 2018. https://doi.org/10.1093/ntr/ntw280.
3. N. Gray, M. Halstead, N. Gonzalez-Jimenez, L. Valentin-Blasini, C. Watson, R.S. Pappas. Analysis of toxic metals in liquid from electronic cigarettes. *Int. J. Environ. Res. Public Health.* **16**, 2019, https://doi.org/10.3390/ijerph16224450.
4. M. Halstead, N. Gray, N. Gonzalez-Jimenez, M. Fresquez, L. Valentin-Blasini, C. Watson, R.S. Pappas. Analysis of toxic metals in electronic cigarette aerosols using a novel trap design. *J. Anal. Toxicol.* **44**, 2020, https://doi:10.1093/jat/bkz078.

5. J. Taylor, T. Wiens, J. Peterson, S. Saravia, M. Lunda, K. Hanson, M. Wogen, P. D'Heilly, J. Margetta, M. Bye, C. Cole, E. Mumm, L. Schwerzler, R. Makhtal; R. Danila, R. Lynfield, S. Holzbauer. Characteristics of e-cigarette, or vaping, products used by patients with associated lung injury and products seized by law enforcement — Minnesota, 2018 and 2019. *MMWR Morbid. Mortal. Wkly. Rep.* **68** (47), 1096–1100, 2019.
6. Centers for Disease Control and Prevention (CDC), States Update Number of Hospitalized EVALI Cases and EVALI Deaths. https://www.cdc.gov/media/releases/2019/s1231-evali-cases-update.html, 2019, last viewed 5 February, 2020.
7. B.C. Blount, M.P. Karwowski, M. Morel-Espinosa, J. Rees, C. Sosnoff, C. Cowan, M. Gardner, L. Wang, L. Valentin-Blasini, L. Silva, V.R. De Jesús, Z. Kuklenyik, C. Watson, T. Seyler, B. Xia, D. Chambers, P. Briss, B.A. King, L. Delaney, C.M. Jones, G.T. Baldwin, J.R. Barr, L. Thomas, J.L. Pirkle. Evaluation of bronchoalveolar lavage fluid from patients in an outbreak of E-cigarette, or Vaping, Product Use–Associated Lung Injury — 10 states, August–October 2019. *MMWR Morbid. Mortal. Wkly. Rep.* **68** (45), 1040–1041, 2019.
8. S.M. Jafari, Y. He, B. Bhandari. Nano-emulsion production by sonication and microfluidization - A comparison. *Int. J. Food Prop.* **9**, 475–485, 2006.
9. T. Bresler. Nano-emulsification. What is it, and what does it have to do with cannabinoids? *Extraction Magazine* May 18, 2019, https://extractionmagazine.com/category/applied-technology/nano-emulsification/.
10. D. Christiani. Vaping-induced lung injury. *New Engl. J. Med.* 2019, https://doi.org/10.1056/NEJMe1912032.
11. L. Kosmider, A. Sobczak, M.Fik, J. Knysak, M. Zaciera, J. Kurek, M.L. Goniewicz. Carbonyl compounds in electronic cigarette vapors: Effects of nicotine solvent and battery output voltage. *Nicotine Tob. Res.* **16**(10), 1319–1326, 2014.
12. T.R. Dulski. *A Manual for the Chemical Analysis of Metals.* ASTM International: West Conshohocken, PA, 1996; p. 15.
13. G. Knapp, Decomposition methods in elemental trace analysis. *Trends Anal. Chem.* **3**, 182–185, 1984.
14. T.J. Murphy. *Accuracy in Trace Analysis: Sampling, Sample Handling, Analysis, Volume I, Publication 422*, National Bureau of Standards: Gaithersburg, MD, 1976, pp. 530–533.
15. G.D. Christian. *Analytical Chemistry.* John Wiley and Sons: Hoboken, NJ, 2004, pp. 384–395.
16. W.J. McShane, R.S. Pappas, V. Wilson-McElprang, D. Paschal. A rugged and transferable method for determining blood cadmium, mercury, and lead with inductively coupled plasma-mass spectrometry. *Spectrochim. Acta B* **63**, 638–644, 2008.
17. M.R. Fresquez, R.S. Pappas, C.H.Watson. Establishment of toxic metal reference range in tobacco from US cigarettes. *J. Anal. Toxicol.* **37**, 298–304, 2013.
18. R.S. Pappas, N. Gray, N. Gonzalez-Jimenez, M. Fresquez, C.H. Watson. Triple quad-ICP-MS measurement of toxic metals in mainstream cigarette smoke from Spectrum research cigarettes. *J. Anal. Toxicol.* **40**, 43–48, 2015.
19. M. Williams, A. Villarreal, K. Bozhilov, S. Lin, P. Talbot. Metal and silicate particles including nanoparticles are present in electronic cigarette cartomizer fluid and aerosol. *PLOS ONE.* **8**(3), e57987, 2013, https://doi.org/10.1371/journal.pone.0057987.
20. M. Williams, A. To, K. Bozhilov, P. Talbot. Strategies to reduce tin and other metals in electronic cigarette aerosol. *PLOS ONE.* **10**(9), e57987, 2015, https://doi.org/10.1371/journal.pone.0138933.
21. J.K. Taylor. *Quality Assurance of Chemical Measurements.* Lewis Publishers, Boca Raton, LA, 1987, pp. 79–81.
22. A. Saffari, N. Daher, A. Ruprecht, C. De Marco, P. Pozzi, R. Boffi, S.H. Hamad, M.M. Shafer, J.J. Schauer, D. Westerdahle, C. Sioutas. Particulate metals and organic compounds from electronic and tobacco-containing cigarettes: comparison of emission rates and secondhand exposure. *Environ. Sci. Processes Impacts.* **16**, 2259–2267, 2014.
23. J.D. Herner, P.G. Green, M.J. Kleeman. Measuring the trace elemental composition of size-resolved airborne particles. *Environ. Sci. Technol.* **40**, 1925–1933, 2006.
24. R.S. Pappas, G.M. Polzin, L. Zhang, C.H. Watson, D.C. Paschal, D.L. Ashley. Cadmium, lead, and thallium in mainstream tobacco smoke particulate. *Food Chem. Toxicol.* **44**(5), 714–723, 2006.
25. R.S. Pappas, G.M. Polzin, C.H. Watson, D.L. Ashley. Cadmium, lead, and thallium in smoke particulate from counterfeit cigarettes compared to authentic U.S. brands. *Food Chem. Toxicol.* **45**(2), 202–209, 2007.
26. V.B. Mikheev, M.C. Brinkman, C.A. Granville, S.M. Gordon, P.I. Clark. Real-time measurement of electronic cigarette aerosol size distribution and metals content analysis. *Nicotine Tob. Res.* **18**(9), 1895–1902, 2016.

27. CORESTA. Method Number 81. Routine analytical machine for e-cigarette aerosol generation and collection - definitions and standard conditions. CORESTA, Paris, France, 2015, pp. 1–6.
28. International Organization for Standardization. Routine analytical cigarette-smoking machine — Definitions and standard conditions. ISO 3308, 2000, pp. 1–23.
29. P. Olmedo, W. Goessler, S. Tanda, M. Grau-Perez, S. Jarmul, A. Aherrera, R. Chen, M. Hilpert, J.E. Cohen, A. Navas-Acien, A.M. Rule. Metal concentrations in e-cigarette liquid and aerosol samples: The contribution of metallic coils. *Environ. Health Perspect.* **126**(2), 2018, https://doi.org/10.1289/EHP2175.
30. P. Richter, R.S. Pappas, R. Bravo, et al. Characterization of SPECTRUM variable nicotine research cigarettes. *Tob. Regul. Sci.* **2**(2), 94–105, 2016.
31. Agency for Toxic Substances and Disease Registry. Toxicological profile for barium and barium compounds, 2007. pp. 10–13. https://www.atsdr.cdc.gov/toxprofiles/tp24.pdf
31. A.H. Naqvi, A. Hunt, B.R. Burnett, J.L. Abraham. Pathologic spectrum and lung dust burden in giant cell interstitial pneumonia (hard metal disease/cobalt pneumonitis): review of 100 cases. *Arch. Environ. Occup. Health.* **63**, 51–70, 2008.
32. K. Kunimasa, M. Arita, H. Tachibana, K. Tsubouchi, S. Konishi, Y. Korogi, A. Nishiyama, T. Ishida. Chemical pneumonitis and acute lung injury caused by inhalation of nickel fumes. *Intern. Med.* **50**, 2035–2038, 2011.
33. A. Khoor, A.C. Roden, T.V. Colby, V.L. Roggli, M. Elrefaei, F. Alvarez, D.B. Erasmus, J.M. Mallea, D.L. Murray, C.A. Keller. Giant cell interstitial pneumonia in patients without hard metal exposure: analysis of 3 cases and review of the literature. *Hum. Pathol.* **50**, 176–182, 2016.

25 Fundamental Principles, Method Development Optimization and Operational Requirements of ICP-Optical Emission

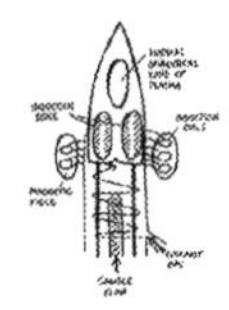

Since its introduction over 40 years ago, inductively coupled plasma optical emission spectroscopy (ICP-OES) has significantly changed the capabilities of elemental analysis. This technique combined the energy of an argon-based plasma with an optical spectrometer and detection system capable of measuring low-level emission signals, which allowed laboratories to perform rapid, automated, multielement analyses at trace concentrations.[1] This was approximately 10 years before the introduction of the first commercial inductively coupled plasma mass spectrometer, so ICP-OES became the workhorse instruments in many laboratories required to perform elemental analysis at trace level concentrations. This chapter, written by Maura Rury from the Applied Testing Reference Materials Division of LGC, gives a detailed description of the fundamental principles together with method development optimization procedures and operational requirements of this technique.

The advantage in using an atmospheric pressure ICP source for making optical emission measurements was first published in 1964,[1] and the sensitivity, speed of analysis, ease of use, and tolerance to high levels of dissolved solids are advantages that laboratories continue to rely on more than half a century later.[2,3] The success of the technique itself can be measured by the fact that thousands of ICP-OES instruments have been installed between 1983 and 2020, which have resulted in approximately 59,000 publications, with over 28,000 published since 2012 (results courtesy of Google Scholar search). That published literature features elemental determinations in a variety of sample matrices in industries including: environmental, nuclear, mining and geochemistry, materials testing, semiconductor, industrial, petrochemical, clinical and toxicological, food safety, and pharmaceutical.

BASIC DEFINITIONS

A full glossary exists at the end of this book for purposes of defining terms used throughout the text; however, several terms are defined here to ensure clarity while reading this chapter. Several optical emission techniques exist, based on atmospheric discharges, which include: inductively coupled plasmas (ICPs), direct coupled plasmas (DCPs), microwave-induced plasmas (MIPs), DC arcs, and

AC sparks. Each discharge is generated via a different mechanism and has its own inherent advantages and disadvantages; however, a comparative discussion of these techniques is outside the scope of this text. The remainder of this chapter will focus solely on ICP-OES.

It should be noted that ICP-AES and ICP-OES are terms that are sometimes used interchangeably; however, the former term can be a source of error and confusion. The term ICP-AES refers to "atomic emission spectroscopy" which nominally excludes emission contributions from other species such as ions and molecules. The latter term refers to "optical emission spectroscopy" and is more commonly used as it includes emission from multiple contributors. Only the term ICP-OES will be used in this text.

PRINCIPLES OF EMISSION

For most ICP-OES applications, a sample is delivered to the instrument's plasma in the form of an aerosol. As the aerosol travels from the base of the plasma to its tail, it travels through a variety of heated zones where it gets desolvated (unless delivered as a dry aerosol), vaporized, atomized, and ionized. Further time spent in the plasma allows the atoms and ions to absorb additional energy which excites an outer electron and produces excited state species. Relaxation back to a ground-state atom produces energy in the form of a photon. This production of photons from excited atoms and ions forms the basis for atomic emission measurements. There are many species in a sample that may absorb energy from the plasma and produce emission spectra. These species include atoms, ions, and molecules. For the purpose of this section, the contribution from molecular emission will be excluded. All references to emission will include the contribution from atoms and ions only.

ATOMIC AND IONIC EMISSION

Elemental analysis by ICP-OES relies on the emission from excited atoms and ions within a sample. Argon plasmas contain ~15.8 eV of energy, which is sufficient to remove one or two electrons from the outer orbital of most atoms. This results in the presence of both atoms and ions in the plasma, all of which are in their ground (lowest level) energy state. Excitation, and subsequent emission, occurs when a species' absorbed energy from the plasma is released in the form of wavelength-specific photons.

A simplified schematic of atomic absorption (AA) and emission is illustrated in Figure 25.1. The horizontal lines represent energy levels in an atom. The lowest horizontal line and the four remaining horizontal lines represent the ground state and excited states, respectively. If included

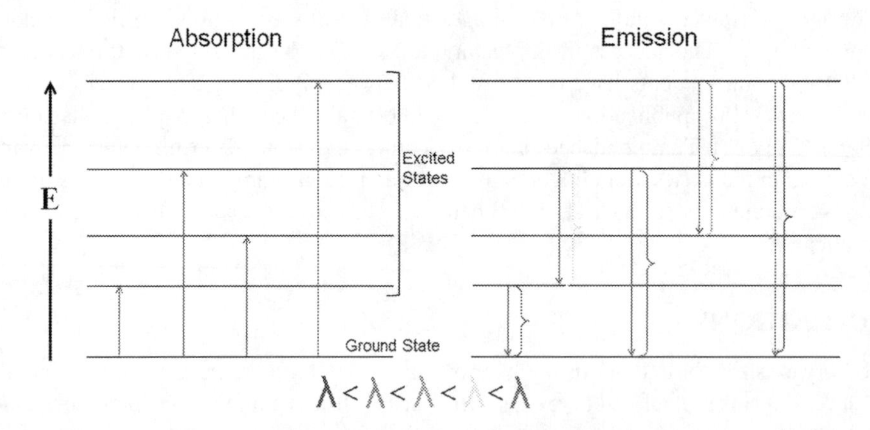

FIGURE 25.1 Diagram depicting energy transitions involved in an atom's absorption and emission of energy.

in the schematic, additional horizontal lines to represent ionic ground and excited states would be illustrated above the atomic excited states. The vertical arrows represent an energy transition for an electron, following the absorption or emission of a photon. The length of each vertical arrow correlates to the amount of energy involved in the transition.

As the schematic indicates, absorbed energy can shift electrons to different excited states, both atomic and ionic. Relaxation of these excited electrons produces energy in the form of photons. Photons vary in energy and can be correlated to their associated emission wavelength using Einstein's equation[4] which relates the energy of light and its frequency according to:

$$E = hv$$

where E represents the energy of light, h represents Planck's constant, and ν represents the frequency of light. In OES, it is more practical to speak in terms of wavelength, so the term c/λ can be substituted to yield:

$$E = hc/\lambda$$

where E represents energy in Joules, h represents Planck's constant in units of Joule seconds, c represents the speed of light in meters per second, and λ represents the wavelength in units of meters. From this equation, it becomes clear that each emitted photon is wavelength specific and represents the inverse relationship between energy and wavelength. These emission wavelengths represent the energy levels that are characteristic to each element, thus making OES a useful technique for identifying and quantifying elements in unknown samples.

INSTRUMENTATION

Commercially available spectrometers collect emission at wavelengths from 165 to 1,100 nm. Early elemental analysis did not include the collection of optical emission in the vacuum ultraviolet (VUV) region. Wavelengths below 190 nm are absorbed by oxygen, water vapor, and other components in the ambient atmosphere, and early instrument designs did not sufficiently purge the optical path between the torch and the entrance slit of the spectrometer. The use of a completely purged optical path was first proposed in the early 1970s where the use of high-purity nitrogen made it feasible to collect optical emission in the VUV wavelength region.[5,6]

SAMPLE INTRODUCTION

The "Achilles heel" of most plasma spectrochemical techniques lies with sample introduction.[7] This fundamental drawback was noted in the literature in the 1980s; however, it is an issue that holds true even decades later. Improving sample introduction systems is a research topic that has been investigated for many years, the outcome of which has produced many nebulizer and spray chamber designs, as well as introduction systems which incorporate flow injection and internal standardization techniques. Research in this area is ongoing as there are still a number of areas in which sample introduction falls short. Introduction system designs have made little progress in the way of improving sample throughput and reducing carryover from memory-prone elements. A number of systems allow for online addition of internal standards; however, many of those systems suffer from inconsistent mixing between the internal standard and carrier solutions. Flow injection techniques have been incorporated into a number of introduction systems; however, the use of a relatively small, defined sample volume at the sample flow rates typically used with ICP-MS and ICP-OES results in short-lived transient signals that do not allow for the measurement of more than a few analytes at a time.

As with all analytical atomic spectroscopic techniques, sample introduction is a critical step that strongly influences, and often dictates, analytical figures of merit such as sensitivity, precision, and stability.[8] Improving the efficiency of sample introduction is an ongoing research topic; however, it has proven to be a non-trivial task as the processes involved within both the nebulizer and spray chamber are numerous and complicated.

Alternative methods of sample introduction, including flow injection, flow injection with hydride generation, and laser ablation, have been employed with the intent to improve plasma spectrochemical measurements. Flow injection is a technique that provides a continuous flow of solution to the nebulizer which allows sample uptake and stabilization in the plasma to take place more rapidly, thereby increasing sample throughput. Furthermore, injection of small sample volumes into a relatively clean carrier stream reduces the amount of salt introduced into the instrument which can reduce matrix effects and improve limits of detection (LODs).[9] A drawback to this setup is that the introduced sample is of finite volume and the measured signal becomes transient. This is an undesirable situation when measuring a large suite of elements with a sequential instrument as many of the elements will be measured off the peak center which degrades the signal-to-noise ratio and, therefore, the sensitivity of the measurement.[10,11] In an effort to retain both steady-state signal analysis and rapid wash-in and washout, a relatively large-volume sample loop should be employed for analysis.

Hydride generation provides a way to remove interfering matrix components to allow only the analyte/s of interest to be presented to the analytical instrument. Hydride generation methods are well established[12–15] and rely on the generation of volatile gaseous hydrides, following a reaction with a reducing agent at an appropriate pH. Hydride generation can be performed online or offline prior to analysis and it is a technique that can be used for preconcentration as well matrix removal for the determination of hydride-active elements.[16,17]

Laser ablation, when combined with ICP-OES or ICP-MS, is a sampling technique that allows elemental impurities to be quantified in their native solid sample. In this technique, a laser is focused on a prepared sample to remove (ablate) surface material in the form of fine particles. The particles are swept into the instrument using an argon or helium carrier gas where they are atomized/ionized and excited, similar to the processes that would occur with an aerosol from a liquid sample.[18]

The popularity of this technique stems from its ability to provide a sample aerosol from materials that are difficult to dissolve such as metals, refractory materials, glasses, and insoluble alloys. Additional advantages include simplified sample preparation procedures with a reduced risk for contamination or loss, a reduced sample size requirement, and the ability to determine the spatial distribution of the quantified elements.[19,20] A significant challenge with laser ablation is in obtaining calibration standards which are matrix-matched to the unknown samples and contain elements at concentrations which bracket those of the analytes of interest.[21] The development of synthetic liquid standards is relatively straightforward; however, this task is non-trivial for solid-based techniques.

The driving forces behind the design of sample introduction systems include: improved sensitivity and detection limits, improved precision across the working mass range, and fewer interferences from matrix effects. Much of the design efforts have centered around improvements in nebulizers and spray chambers; however, significant progress has diminished in the recent years.[16]

AEROSOL GENERATION

Most ICP-OES applications require the delivery of a sample to the instrument's plasma in the form of an aerosol. The components in the sample introduction system, which typically include a pump, nebulizer, spray chamber, and torch, must work in a complementary fashion to convert a bulk sample solution into a body of micrometer-sized droplets. The pump (peristaltic or syringe) consistently delivers solution to the nebulizer which uses a high-velocity gas stream to break the solution into

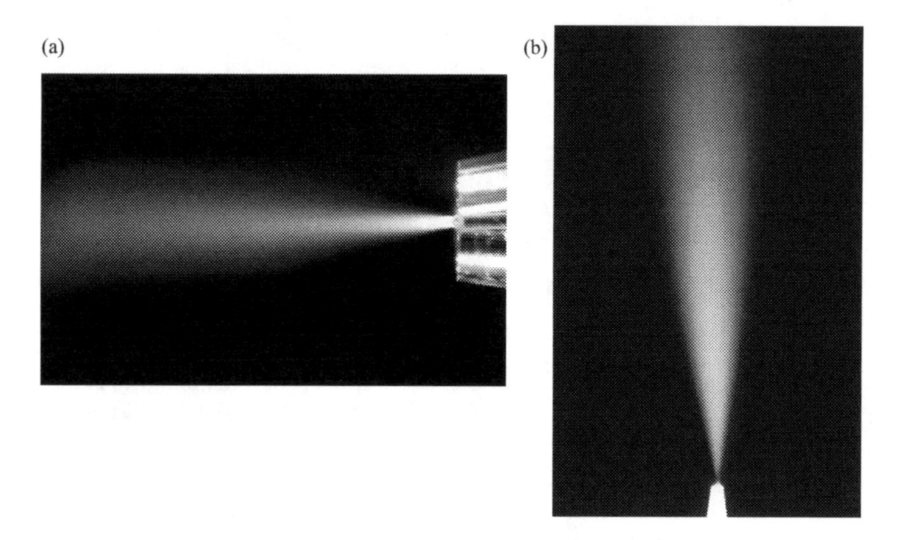

FIGURE 25.2 (a) Examples of proper sample aerosols, generated by a concentric nebulizer. (With kind permission from Meinhard Glass Products.) (b) A PFA microflow concentric nebulizer. (With kind permission from Elemental Scientific Inc.)

small droplets. The spray chamber then filters the droplets by diameter, allowing only the smallest droplets to be transferred to the torch where they are desolvated, vaporized, and atomized in the plasma. A proper aerosol should consist of as many small (<10 μm diameter) droplets as is possible, to maximize the efficiency of desolvation once the sample reaches the plasma.

The generation of a proper aerosol is a significant factor in determining the quality of the resulting emission signals and data.[22] Sample aerosols may vary in their density of droplets; however, a proper aerosol should maintain a consistent shape and density during data collection to promote analytical measurements with optimal precision. Two example aerosols are pictured in Figure 25.2 for illustrative purposes. They are generated from nebulizers with two different structural designs which produce differences in their shape and droplet density.

NEBULIZERS

The main role of the nebulizer is to generate an aerosol of small droplets from the interaction between a sample stream and a high-velocity gas stream.[23] Ideally, the nebulizer will transport an aerosol with a narrow drop size distribution into the spray chamber to allow the spray chamber to more efficiently filter out the large aerosol droplets.[24] Since the droplets must be desolvated, vaporized, and atomized during their short residence time in the plasma, the spray chamber allows passage of only small droplets (typical diameter <10 μm).[25,26]

In an effort to overcoming the limits associated with sample introduction, much effort has been put into the design of nebulizers and spray chambers. Typical introduction systems consist of a nebulizer–spray chamber arrangement in which the nebulizer is pneumatic.[27] The relative simplicity and low cost of this setup make it the preferred choice for ICP sample introduction; however, its associated drawbacks include low-analyte transport efficiency (~1%–2%), high-sample consumption (1–2 mL/min), and relatively high retention of some elements.[28] An image of a typical pneumatic nebulizer is illustrated in Figure 25.3.

Microconcentric nebulizers (MCNs)[29–34] were designed to improve the gas–liquid interaction and reduce the size distribution of droplets formed in the aerosol.[35,36] These nebulizers, which operate at lower sample flow rates compared to those typically used with pneumatic nebulizers, include high-efficiency nebulizers (HENs),[37–40] oscillating capillary nebulizers (OCNs),[41–43] and sonic spray

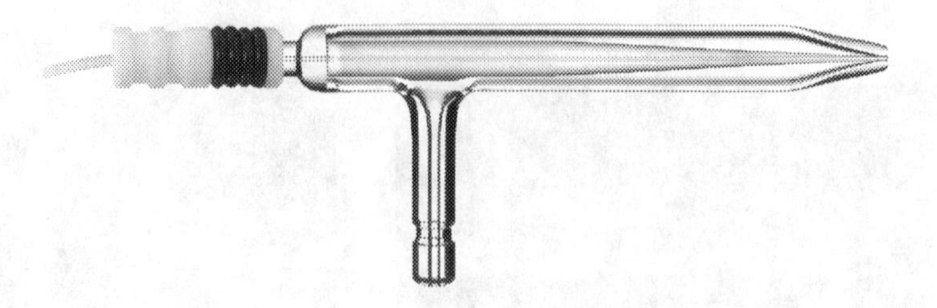

FIGURE 25.3 Image of a typical pneumatic nebulizer. (With kind permission from Glass Expansion,)

nebulizers (SSNs).[44] These nebulizers have a relatively low dead volume and operate at normal nebulizer gas pressures which improve analyte transport efficiency regardless of whether organic or aqueous solvents are used.[45,46]

Ultrasonic nebulizers (USNs)[47–51] were developed in an effort to improve efficiency of aerosol generation. These nebulizers are highly efficient at producing a large volume of small droplets with a narrow size distribution.[52] The aerosol volume is sufficiently large that most USNs are used with desolvation systems to reduce the introduction of water vapor in the case of aqueous sample analyses[53,54] and to reduce solvent loading and carbon deposition in the analysis of samples containing organic solvents or high concentrations of dissolved salts.[55,56] A schematic of a USN is shown in Figure 25.4.

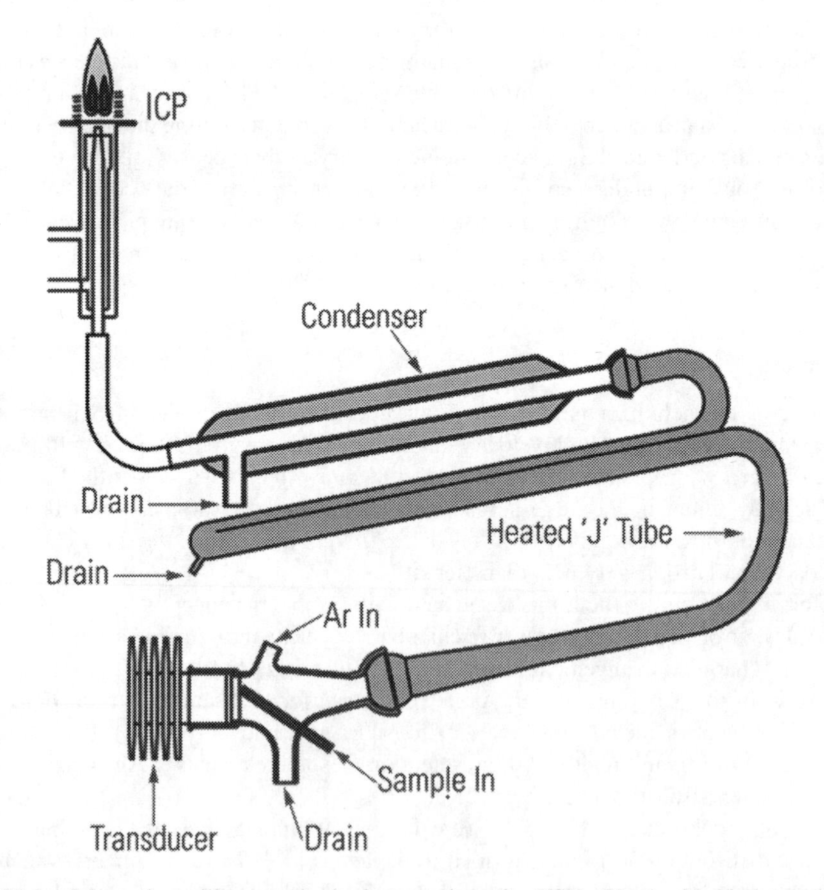

FIGURE 25.4 Schematic to illustrate the basic operation of a USN. (Used with kind permission from Teledyne Cetac Technologies.)

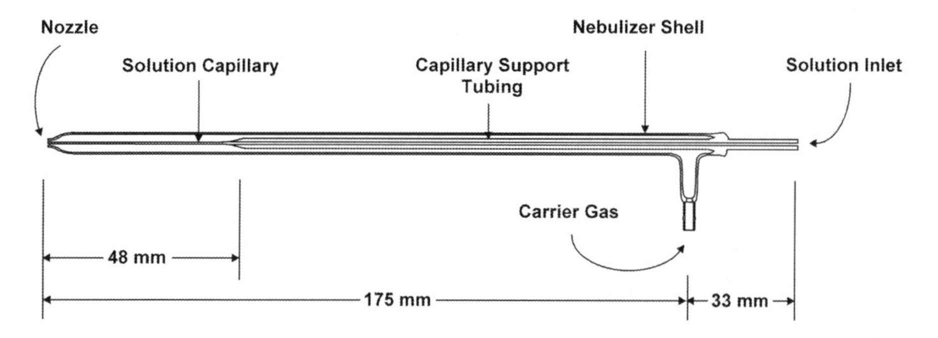

FIGURE 25.5 Schematic DIHEN nebulizer. (Used with kind permission from Meinhard.)

Nebulizers have also been designed in which no spray chamber is present and samples are injected directly into the plasma. These direct injection nebulizers (DINs)[57–61] and direct injection high-efficiency nebulizers (DIHEN)[62–65] improve the sample transport efficiency to 100%, even at relatively low flow rates. Direct injection also reduces the dead volume associated with spray chambers which increases the response time of the measurement and reduces memory effects.[63,66–68] A significant drawback of DINs is their vulnerability towards samples containing high concentrations of dissolved salts or volatile solvents which cause plasma instability and tip clogging.[69] The large bore (LB) DIHEN[70] was designed to reduce its susceptibility to blockage; however, the larger inside diameter of the nebulizer capillary produces a relatively large droplet size distribution which degrades both the precision and detection limits.[71] An example of a DIN, specifically, a quartz DIHEN, is pictured in Figure 25.5.

SPRAY CHAMBERS

Much effort has been put into spray chamber design, as spray chambers are responsible for the loss of >90% of the aerosol produced by the nebulizer.[72] Improvements have focused on directing the flow of aerosol from the nebulizer to maximize the efficiency with which larger droplets are filtered out.[73] Reverse flow, commonly called Scott-type double-pass,[74,75] spray chambers are the popular choice for relatively simple, low-cost introduction. The double-pass design is particularly useful for samples containing high concentrations of dissolved salts as the transport efficiency is relatively low compared to other spray chamber designs.[76] An example of double-pass spray chamber is illustrated in Figure 25.6.

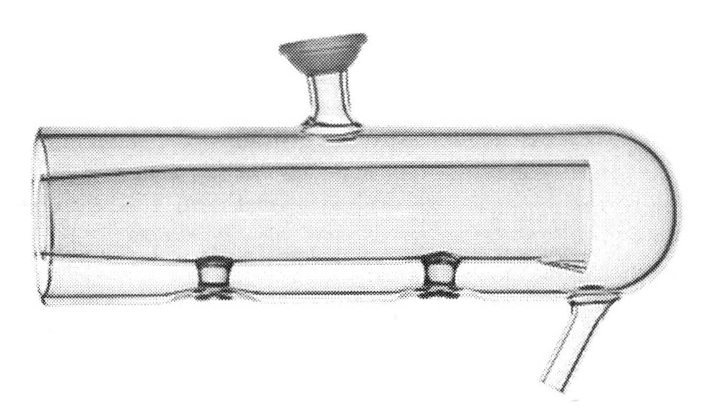

FIGURE 25.6 Image and schematic of a typical Scott-type double-pass spray chamber. (With permission from Precision Glassblowing.)

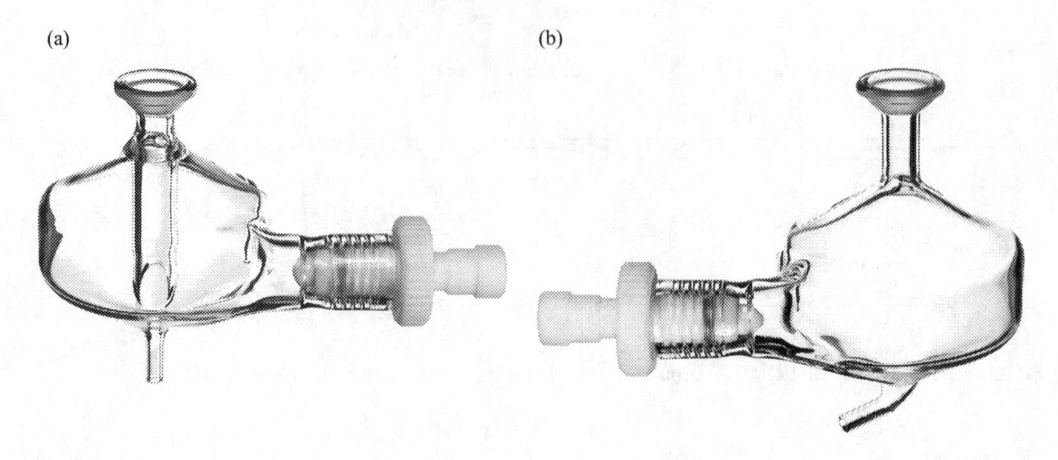

FIGURE 25.7 Cyclonic spray chamber without (a) and with (b) a knockout tube. (With kind permission from Glass Expansion.)

Cyclonic designs have been employed to improve transport efficiency, precision, and detection limits.[77,78] Popular cyclonic spray chamber designs include a flow spoiler or dimple[79] to disrupt the flow of aerosol within the spray chamber. Computer modeling of the fluid dynamics within cyclonic spray chambers suggests that the presence of three spoilers creates a "virtual cyclone"[80] which reduces interaction between the aerosol and the walls of the spray chamber, thereby improving transport efficiency and reducing memory effects. A commonly employed cyclonic spray chamber is illustrated in Figure 25.7. Spray chambers, without (Figure 25.7a) and with (Figure 25.7b) a flow spoiler, are illustrated for comparison.

Single-pass or cylindrical-type spray chambers have been designed for use in low-flow introduction.[81–83] These spray chambers provide high efficiency and reduced memory effects[83]; however, this design is unsuitable for conventional ICP analysis and is typically used when electrophoretic or chromatographic separations are employed in conjunction with ICP detection.[84,85] Spray chambers have also been blamed for the retention of elements such as B and Hg.[86–89] As discussed above, one approach to solving this problem has been in the removal of the spray chamber. Other approaches have involved the use of a thermostated spray chamber. Water-cooled spray chambers have been used to reduce aerosol desolvation and deposition along the walls of the spray chamber, thereby reducing memory effects and reducing oxide formation in the plasma.[90–92] This logic has been used in the application of heated spray chambers for desolvating and, therefore, improving the efficiency of aerosol generation in aqueous samples.[93–95]

TORCHES

The torch is the final stop in the sample's journey from sample cup to plasma and often has a major influence on the analytical performance of the instrument in terms of sensitivity, detection limits, and plasma robustness. Torch design for ICPs has been investigated for more than 50 years. The original ICP torch was based on Reed's design[96–98] and introduced the coolant and auxiliary gas flows in a tangential direction. This method of gas introduction was thought to produce fluid dynamics with maximum stability and ion density.[99]

Reed's torch design was modified by Greenfield and colleagues[100,101] who constructed a torch containing three concentric tubes to generate and sustain an annular-shaped plasma. The outer and intermediate tubes were made out of silica and the center tube was constructed out of borosilicate glass. High-purity gas was introduced tangentially, similar to the Reed torch; however, gas was not

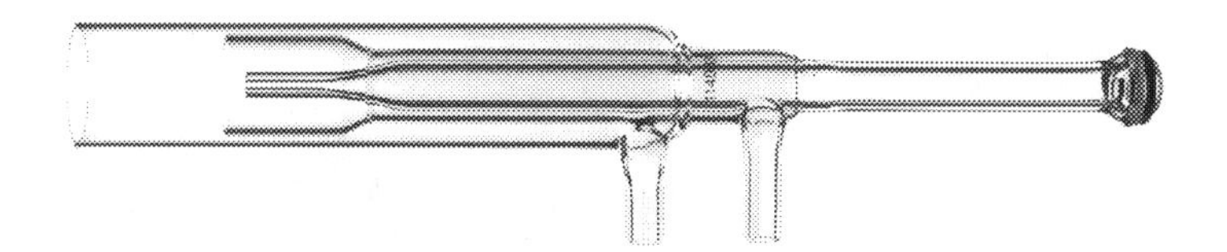

FIGURE 25.8 Schematic for a standard, commercially available ICP-OES torch. (With kind permission from Glass Expansion.)

introduced into the outer tube until after the plasma had formed. Only the intermediate tube was purged during ignition.

Fassell made minor modifications, and the resulting torch design[102] is used by most commercial instrument manufacturers, even today. The Fassell torch design evolved to a design similar to that shown in Figure 25.8. The torch consists of three concentric quartz tubes: the outer tube, the intermediate tube, and the center tube (or injector). The geometry of these tubes and their positioning with respect to each other and to the load coil, along with the gas flow settings, directly affect the formation, stability, and sustainability of the plasma.[103] This makes torch design critical to data quality in elemental analysis.

The inner tube, or injector, provides an inlet for introducing the sample aerosol into the plasma. The center tube provides a structured path for the auxiliary gas which is used to shift the base of the plasma further away from the injector. The outer tube or coolant houses the largest volume of gas which is used to generate and sustain the plasma. This gas also provides sufficient cooling to prevent the torch from melting or contributing background emission signals.

Since the introduction of the Fassell torch, minor changes have been made to torch design. Injector tubes have been constructed with both a tapered and a parallel end and they are available with a variety of inside diameters. Small diameter injectors are typically used for applications involving solvents that have a relatively high vapor pressure and are likely to overload and extinguish the plasma. Larger diameter injectors are typically used for the analysis of relatively clean samples to increase the aerosol volume delivered to the plasma which increases the resulting emission signals.

The intermediate tube has been produced with both a parallel and a tulip-shaped design. Research has been done to investigate the effect of the intermediate tube shape on the shape and stability of the plasma; however, both designs are still employed.[104] A parallel intermediate tube allows the auxiliary gas flow to remain constant along the length of the torch to reduce turbulence at the base of the plasma and to allow the auxiliary gas to reach a true laminar flow. This reduction in turbulence produces a plasma with a relatively flat base and aids in the penetration of the nebulizer gas and, therefore, the sample, into the plasma.

Torches are available as a single piece or in a demountable design. The single piece torch (refer to Figure 25.8 for example) is generally easier for operator use and ensures that the injector will be fully centered in the intermediate tube; however, the individual parts of the torch cannot be replaced and the injector diameter cannot be changed. Demountable torches must be manually assembled and aligned to ensure the injector is centered in the torch body. This style of torch is advantageous and more cost effective for applications which benefit from the use of multiple injector diameters or which are likely to clog the torch injector.

Ceramic torches have been designed for use in measuring organic solvents, extremely high levels of dissolved solids, and other samples that devitrify a standard quartz torch.[105] Commercial ceramic torches are available in fully demountable designs, which allow quartz and ceramic pieces to be combined to meet the needs of each specific application. An example of commercially available ceramic torch components is illustrated in Figure 25.9.

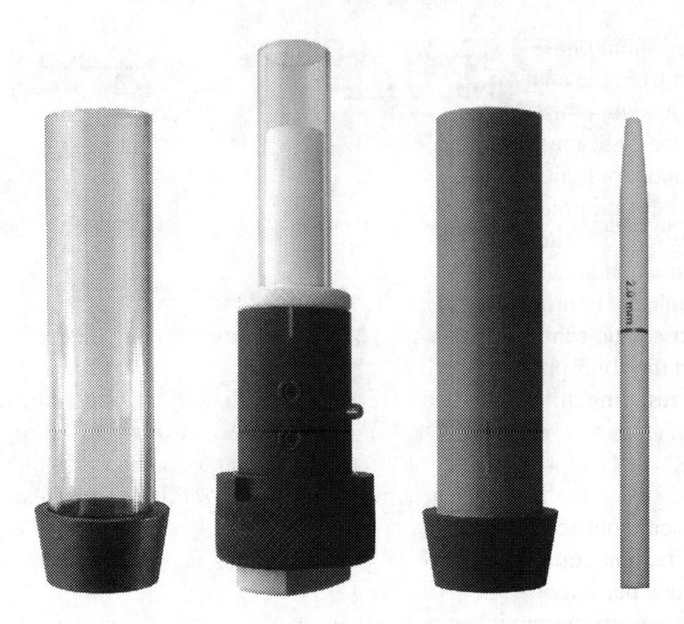

FIGURE 25.9 Demountable ceramic torch parts. (With kind permission from Glass Expansion.)

SPECTROMETERS

The optical spectrometer is the heart of the modern ICP instrument and is comprised of the fore optics and the polychromator, with the detector attached. The purpose of the optical system is to separate the ICP source emission into element-specific wavelengths and to focus the resolved light onto the detector with high efficiency and with minimal stray light contributions.[106] This should be accomplished with a minimum amount of absorption, scattering, and optical aberration to maximize the light throughput to the detector.

FORE OPTICS

The transfer of light from the plasma into the spectrometer is a critical first step in obtaining results with maximum sensitivity and signal-to-background ratios. Of equal importance is the transfer of emission with minimal contributions from molecular emission and stray light sources. The gap between the plasma and the spectrometer's entrance optics can be a challenging environment to control. In addition to protecting the entrance optics from the plasma's intense heat, the plasma–spectrometer interface must effectively replace the ambient air to prevent it from absorbing emission wavelengths below ~200 nm. For a horizontally mounted torch, the interface must also remove the relatively cool portion of the plasma ("tail") to minimize interferences from self-absorbed analytes and atoms/ions that recombine and produce molecular emission.[107]

The purpose of the fore optics is to focus emitted light from the plasma onto the entrance slit as efficiently as possible and to provide a light path suitable for transmitting in either radial or axial plasma views, or both, for systems which have that capability. In dual-view systems, changing plasma views should be a fast, automated process to avoid undesirable increases in sample analysis time and subsequent degradation of productivity and cost efficiency.

Properly designed fore optics should exhibit a number of characteristics to ensure optimal instrument performance. The light from the plasma should be transferred into the instrument with a high level of efficiency to maximize the sensitivity of the instrument. This requires a minimum number of optics as each light-reflecting surface will absorb or scatter a small amount of light. Each additional optic reduces the transfer efficiency which reduces the instrument's achievable detection limits.[108]

The fore optics should have optimum focus for better sensitivity and detection limits. Photons must be collected from the analytical zone of the plasma—the area of the plasma where the maximum amount of analyte emission and the minimum amount of emission from the plasma exists. Once collected, the light must be properly focused onto the entrance slit of the spectrometer to maximize the amount of light that travels through the spectrometer.

There should be a minimal number of moving parts for greater stability. A small amount of movement is required to optimize the viewing height of a radially configured plasma and to switch between axial and radial light collection in a dual-view instrument. The movement needs to be rapid and highly reproducible to minimize the degradation on sample throughput and precision.

The fore optics should be robust and provide thermal protection from external heat sources such as the plasma and the atmosphere in the torchbox. Excessive heat transfer to the optics will produce instrument drift, resulting in quality control (QC) failures and require frequent recalibration.

OPTICAL DESIGNS

The earliest ICP spectrometers are based on the Paschen–Rünge design.[109] In this configuration, all of the optics (diffraction grating, entrance slit, exit slit) are permanently attached and the exit slits are mounted along a portion of a Rowland circle.[110] In its original design, this type of spectrometer was referred to as a direct reader and consisted of a polychromator with photomultiplier tube (PMT) detectors positioned on the focal plane behind a series of exit slits. One exit slit and PMT was used for each analytical wavelength being measured.[111] This optical layout, which is illustrated in Figure 25.10, allowed several emission wavelengths to be measured simultaneously and in a matter of a few seconds.

Despite the ability to acquire emission signals with remarkable speed, early spectrometers with this optical configuration had relatively poor spectral resolution. Furthermore, most spectrometers utilized PMT detectors at that time, which were responsive over narrow wavelength ranges. This restriction required the end user's desired elements to be selected prior to the instrument being built and limited the instrument to a maximum of 20–30 emission lines due to the space required for having an exit slit and PMT for each wavelength.[112] This optical design poses additional challenges

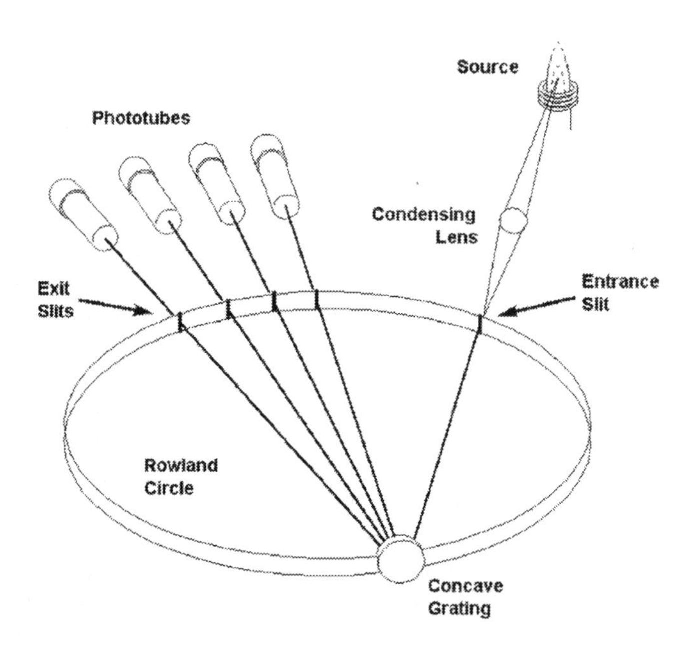

FIGURE 25.10 Schematic of a Paschen–Rünge optical design.[112]

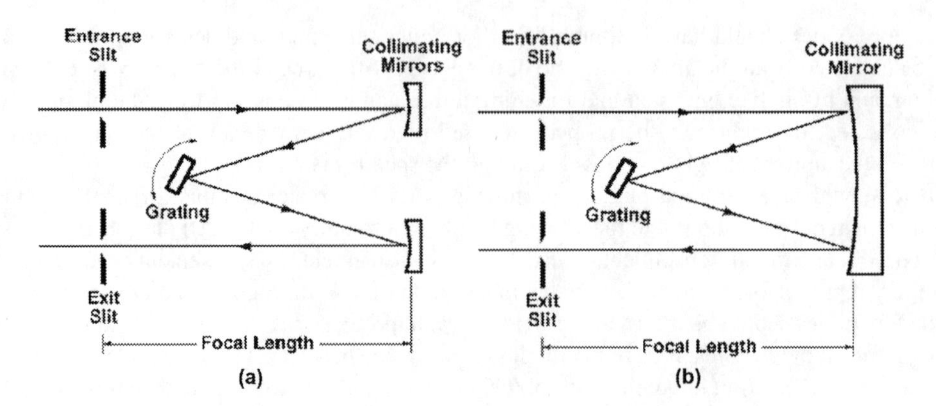

FIGURE 25.11 Schematic of Czerny–Turner (a) and Ebert (b) optical designs.[111]

with internal standardization and interference corrections. The wavelengths being used for internal standard measurements and for interference calculations must have been included in the instrument at the time of manufacture.

The interference challenges and wavelength restrictions associated with the Paschen–Rünge spectrometer led to the development of Czerny–Turner and Ebert designs. The fixed, concave grating layout of the Paschen–Rünge spectrometer is replaced with a scanning, dispersive optic in the Czerny–Turner and Ebert designs.[113–116] In this arrangement, a dispersive optic such as a plane grating diffracts light onto either one or two collimating mirrors (Ebert and Czerny–Turner, respectively). The focused light is then directed onto a single exit slit. This layout is represented in Figure 25.11.

The wavelength flexibility of these monochromator-based systems is an advantage over polychromators; however, their measurement versatility if offset by the relatively long sample analysis times. The use of a scanning optic dictates that wavelengths be accessed and measured sequentially, which means a single sample could require several minutes to analyze.

An additional challenge with this spectrometer design was in the spectral aberrations resulting from the use of spherical collimating mirrors.[117] Image distortions such as astigmatism, spherical aberration, coma aberration, or elongated slit images are inherent to the traditional design.[118] Light from the grating strikes the mirror at an angle that prevents the rays from meeting at a common focal point. To prevent these image distortions from creating errors in the spectral results, compensating optics are often used. Example schematics of typical Ebert and Czerny–Turner spectrometers are illustrated below:

The early 1980s saw the development of an echelle-based optical spectrometer.[119] This design utilizes a high-dispersion echelle grating in addition to a second dispersive optic (such as a grating or a prism) to separate incident polychromatic light into a two-dimensional spectrum, known as an echellogram. Unlike conventional diffraction gratings, echelle gratings are blazed with a relatively low groove density and relatively high angles of incidence to disperse light into its constituent wavelengths at higher spectral orders. While this dispersive behavior produces exceptional resolution, the resulting spectral orders overlap and are of no use without a second, cross-dispersive optic.

In 1983, the first commercially available, benchtop echelle optical system was released, significantly improving the optical dispersion and resolution capabilities of optical emission spectrometers.[120] Due to the high resolution of these systems over relatively short focal lengths, echellograms could be enlarged to fill the relatively large surface area of a PMT detector or reduced to fill the small active surface area of a solid-state array detector. Further advances in instrument design resulted in spectrometers that could access more than one wavelength at the same time, allowing simultaneous and sequential measurements, or a combination of both, using a single optical platform.[121] Examples of simultaneous and sequential echelle-based optical spectrometers are illustrated in Figure 25.12.

(a)

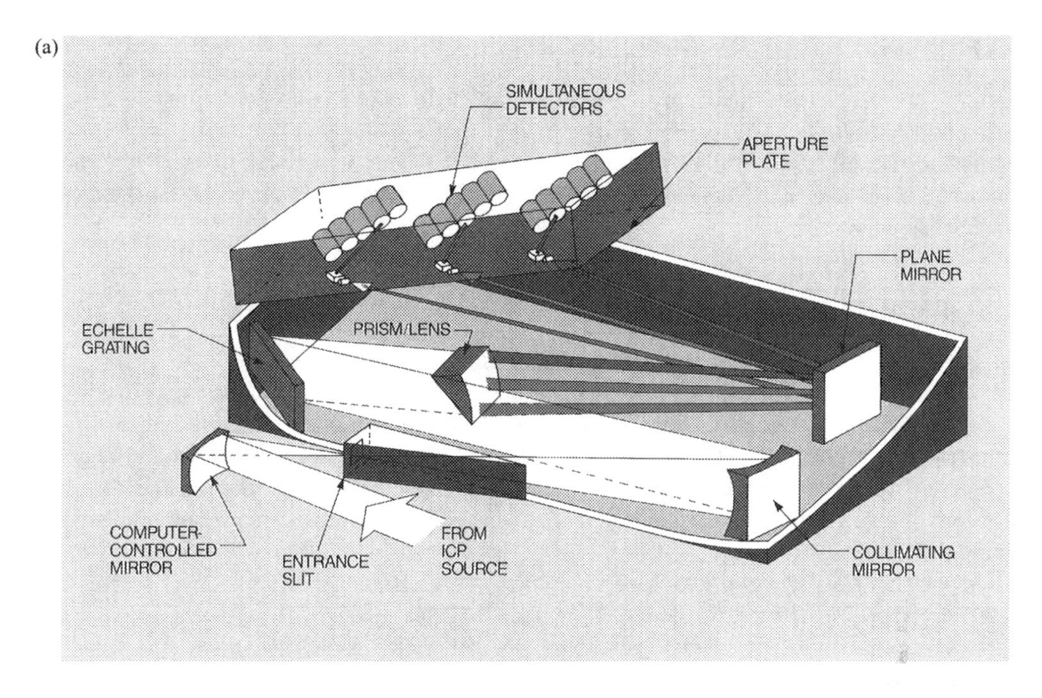

(b)

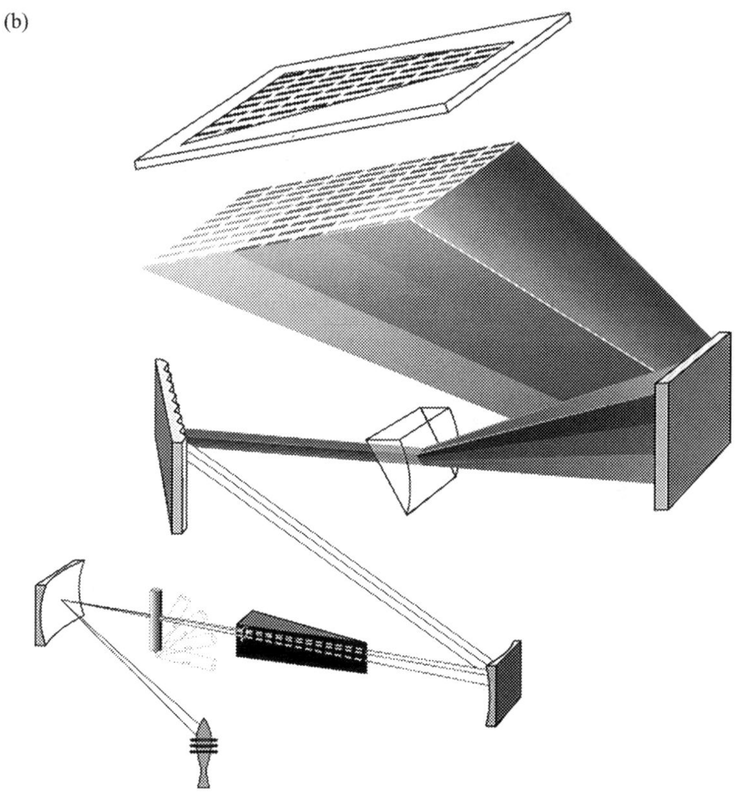

FIGURE 25.12 Schematic of a simultaneous (a) and sequential (b) echelle optical system. (With kind permission from Teledyne Leeman Labs.)

DETECTORS

Detectors are a crucial component to the success of optical emission measurements. Once the plasma transfers sufficient energy to the sample to generate photons, the emitted light is collected, collimated, and diffracted by the spectrometer where it is then transferred to the detector. The efficiency and performance of the detector then determine the quality and sensitivity of the resulting measured signals.

HISTORICAL PERSPECTIVE

The earliest published research involving optical emission measurements utilized photographic plate detectors. Photographic plate detectors operate via the photoelectric effect, so-called because the light emitted from the sample is of sufficient energy to eject electrons from the surface of a photo-reactive solid surface. The resulting image contains a series of dark bands and represents the emission spectrum for the sample being analyzed. The position and width of the dark bands can be used to determine which elements, and at what concentrations, were present in the original sample.[122]

Photographic plate spectrometers have been utilized since the early 1900s and were successful in analyzing complex spectra such as steel. Spectrographs were designed with a variety of optical systems to achieve various levels of resolution and to accommodate photographic plates of various sizes.[123]

Photographic plates offered the first permanent record of emission spectra from a variety of excitation sources including arc, spark, and ICP. The photographic emulsion coated on its surface reacts over a wide energy range, allowing a single plate to record emission from most (or all) analytes of interest in a single sample run. A drawback to the detector technology, however, was its non-linear response and its relatively low quantum efficiency (1%–3%).[124]

The early 1930s saw the development of PMT detectors, and by the late 1970s, PMTs were being used fairly routinely in atomic emission spectrometers.[125–127] By the early 1980s, technological advances were made to produce photodiode array (PDA) detectors which possessed multichannel measurement capabilities. These photodiodes were responsive in both the ultraviolet and visible wavelength ranges and had a wider dynamic range compared to their photomultiplier predecessors. This technology made a fundamental deviation in the way in which it measured spectral emission. Unlike its PMT predecessor which produces a measurable current from incoming radiation, the PDA is able to generate and store that photogenerated charge.[128,129]

During the late 1980s and early 1990s, charge transfer devices (CTDs) such as charge-coupled devices (CCDs) and charge-injection devices (CIDs) began to replace PMTs and PDAs in analytical spectroscopy. These devices respond more uniformly across relatively wide wavelength ranges, and they have the ability to perform simultaneous background measurements. Today, CTD detectors are used almost exclusively in optical spectrometers to quantify spectral emission from a variety of analytes over a wide range of wavelengths.[130–133]

PHOTOMULTIPLIER TUBES

Over the years, PMTs have been the most commonly used detectors for ICP-OES instruments. These detectors typically consist of a sealed vessel, with a series of emission dynodes placed between an anode and a large surface-area cathode. The dynodes are mounted in series and kept at voltage potentials that get progressively more positive than that of the cathode. A photon which strikes the cathode with a sufficient amount of energy dislodges an electron. The ejected electrons accelerate towards the first dynode in the series, which causes a cascade of electrons to be ejected and accelerated towards the next dynode. The process repeats until the multitude of electrons reaches the anode, which generates a measurable electrical pulse.[121] An example of a PMT detector is illustrated in Figure 25.13.

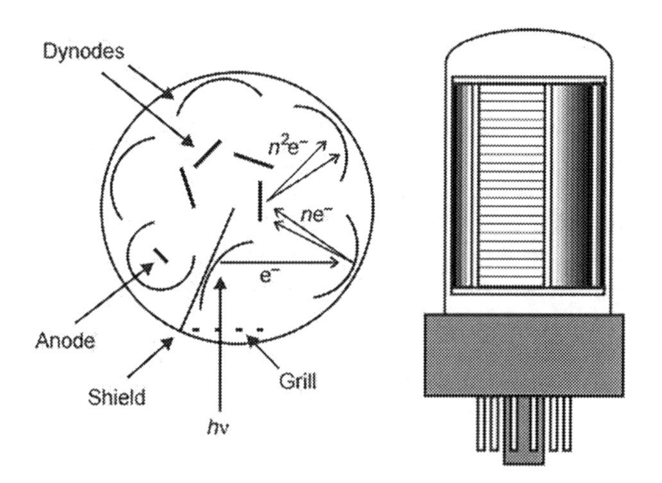

FIGURE 25.13 Schematic and basic function of a PMT.[121]

This type of detector has long been known for its long linear working range and its near noiseless gain. The detector's design gives it the ability to generate more than a million electrons from a single photon, which results in a signal gain of 10^6–10^8. The gain is achieved with almost no measurable dark current or background noise, making it well suited for small signals, such as those from trace level concentrations of analytes.

These detectors lack simultaneous multielement collection capability; however, their single-channel design allows the signal-to-noise ratio to be optimized for each analyte that is being measured. If simultaneous emission detection is desired, multiple PMT detectors can be positioned behind multiple exit slits and used in parallel. The metal oxide construction of the cathode makes the PMT responsive over a relatively short wavelength range; however, a combination of materials can be used to widen the response range.

PHOTODIODE ARRAYS

PDAs were available for use in making spectrochemical measurements by 1980.[134] Similar to PMTs, these devices generate an electrical current but do not have the same gain and signal response as PMTs; however, they offer the advantage of being able to store photogenerated charge which allows them to integrate emission signals.[128]

PDAs do not have the sensitivity and signal gain achievable with PMTs; however, their responses are linear over a relatively wide range and they can be operated in a high speed mode that allows for parallel readout (similar to the random access readout feature of charge injection device detectors).[135] These devices also require cooling to reduce contributions from dark current and fixed pattern noises; however, signal-to-noise ratios could be improved if measurements could be made with longer integration times and fewer signal averages.[136]

CHARGE TRANSFER DEVICES

CTDs detectors are solid-state devices, consisting of an array of photoactive elements, which convert incident radiation to an electrical charge which is stored and measured. These photoactive elements, also known as pixels, can be configured as a linear or two-dimensional array, and can be manufactured in a variety of sizes, to meet the requirements of the instrument.[137] Among other benefits, the structure of these array-based devices provides them with the ability to measure more than one wavelength simultaneously, which allows for true simultaneous background correction and real-time internal standardization.[138] In some applications, the efficiency of the background subtraction is sufficient to remove the contribution from background emission that is significantly larger than that for the analyte/s being measured.

As the name implies, these devices employ a charge transfer process to read out the accumulated charge. Charge transfer is executed via one of two methods: (1) intercell charge transfer, which shifts charge from where the charge accumulated to a single output amplifier, or (2) intracell charge transfer, which measures the voltage change induced by shifting charge within the detector element where it was first accumulated.[139] CTDs generate dark current when in use; however, operating them under cooled conditions reduces the dark current contribution to a level that is almost immeasurable, even when emission signals are integrated for long periods of time.[140]

F. L. J. Sangster and K. Teer invented the first CTD at Philips Research Laboratories almost 50 years ago. Unbeknownst to them, it would become a device that had a significant impact on future developments and advancements in atomic spectroscopy.[141] These devices were originally referred to as bucket brigade devices (BBDs) to analogize the method of storing and transferring signals to a line of people passing buckets of water to fill a storage container or to put out a small fire. The potential for using these devices in imaging sensors was quickly realized and they became a memory storage device with significant use.[142]

In less than a year, Willard Boyle and George Smith, working at Bell Laboratories at the time, improved the BBD's method for processing signals and introduced the CCD.[143] The BBD, which used transistors to transfer charge between a series of capacitors, was improved with the CCD, which transferred charge between capacitive bins (also known as "wells") across the surface of a metal-oxide semiconductor (MOS). This invention would earn them a Nobel Prize for Physics and would find mainstream use in imaging devices such as telescopes, digital cameras, and video cameras.[144,145]

By 1973, Hubert Burke and Gerald Michon, working at General Electric, had introduced the CID detector. For the next two decades, CIDs were used as imagers in machine vision applications where digital image processing capabilities were required. By 1990, CID devices were adapted for use in astronomy, astrophysics, and microscopy, where detector requirements included high sensitivity and spectral accuracy, large dynamic range, and low dark current levels, even during integration periods lasting several hours.[146] The most significant difference between CCDs and CIDs is their readout mechanism, which leads to unique benefits and challenges in atomic spectroscopy applications. Each type of device will be discussed separately in the following sections.

CHARGE-COUPLED DEVICES

CCDs are compact CTDs that have found use as solid-state imagers in a variety of applications. In its original design, the CCD was constructed for the storage of digital information for use in devices such as digital cameras and video recorders. Once in use for routine consumer applications, it became clear that the CCD possessed the ability to convert digital charges to analog signals, broadening its utility to include industrial and scientific applications.[147,148] Figure 25.14 illustrates the structure of a typical CCD device.

The mechanism for transferring charge through a typical CCD detector is illustrated in Figure 25.15. When the device is read out, charge that has accumulated across the entire device must be transferred and measured, even if only a small portion of the array needs to be read out. To accomplish this, a small bias voltage is applied to one of the pixels, which creates an area of depleted charge in the substrate directly below. A greater bias voltage is applied to the adjacent pixel, which creates a similar depletion area and promotes the transfer of charge. The process repeats itself until the accumulated charge is transferred through the entire device until the charge is shifted to the output amplifier for analog-to-digital conversion and measurement with a digital processor. This readout process destroys the stored charge and prevents it from being measured again in the future.[150]

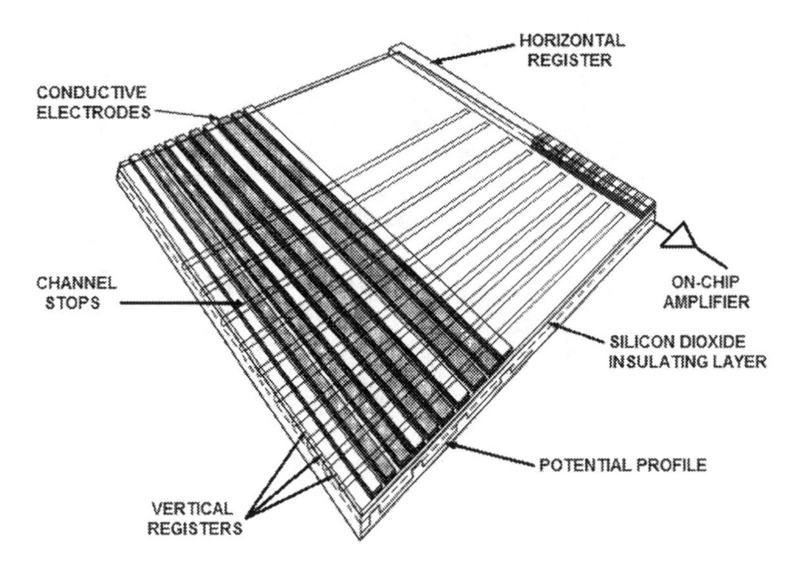

FIGURE 25.14 Basic schematic of a CCD.[149]

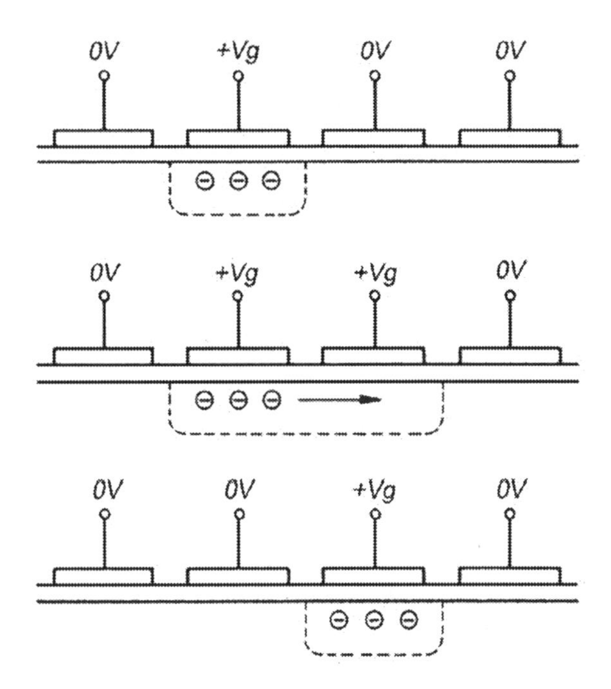

FIGURE 25.15 Charge transfer mechanism in a typical CCD.[150]

The structure of a CCD, combined with its readout mechanism, allows it to read its stored charge with a high level of uniformity, which translates to high-quality spectral images. However, it is this structure and readout that make the CCD susceptible to a phenomenon known as blooming. Blooming occurs when one of the devices detector elements (pixels) accumulates more charge than it can store and spills the excess charge into nearby pixels. This occurrence can significantly impact measurements for emission signals that are being measured simultaneously with intense emission from nearby wavelengths.[138]

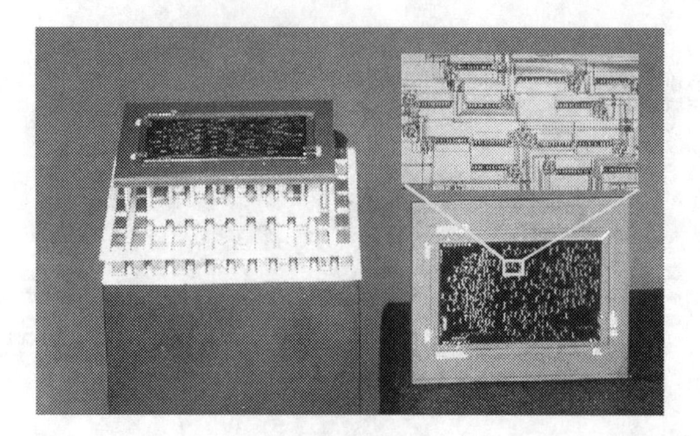

FIGURE 25.16 An SCD detector showing a photo-micrograph close-up of the separate photo-sensitive linear arrays.[137]

To reduce the risk of blooming, a variety of modifications have been implemented. Some CCD devices have been assembled to include structures that are designed to remove and drain excess charge from pixels before they saturate and spill that charge into nearby pixels. These structures, often referred to as charge drains or anti-blooming gates, surround and isolate active pixels to isolate to allow low-level and high-level emission signals to be accurately measured, even if it requires storage in nearby pixels on the same detector.[151] These charge drains have been shown to be effective enough to prevent blooming, even when reading charge that's adjacent to pixels that accumulated charge 1,000 times its saturation level.[131]

Another method employed to reduce the likelihood of charge spillage was to use a series of separate linear CCD arrays, working as a cohesive unit. This device, known as a segmented charge-coupled device (SCD) detector, is constructed such that a separate linear array is used for the measurement of one wavelength (up to three wavelengths if they occur in close proximity) for each of the elements on the periodic table. Using arrays that are physically separate from one another mimics the structure of a single CCD detector with built-in charge drains.[152,153] Size constraints limit the number of arrays that can be used; however, over 200 arrays can be combined for use in covering important analytical wavelengths across both the ultraviolet and visible wavelength regions.[154]

In addition to overcoming potential effects from blooming, the structure of an SCD allows it to measure a small number of wavelengths (or even a single wavelength) without reading the charge across the entire detector, in a method known as random access integration (RAI). This feature increases the readout speed and dynamic range, compared to a single (non-segmented) CCD detector.[137] Figure 25.16 is an image of an SCD detector showing a photo-micrograph close-up of the separate photo-sensitive linear arrays.[137]

CHARGE-INJECTION DEVICES

CID detectors fall into the category of complementary metal-oxide semiconductor (CMOS) technology. CIDs contain an array of detector elements which can store photogenerated charge. Each detector contains a light sensitive area, as well as row and column electrodes which are addressable for purposes of storing and reading out charge. The electrodes are constructed from a conductive silicon material laid over an insulating layer which provides a region for charge to be stored.[139] An example of a CID detector element is illustrated in Figure 25.17.

The mechanism for transferring charge through a typical CID detector is illustrated in Figure 25.18a–d. The four images represent a single pixel during the four-step charge readout process. The pixel contains two photogates (labeled "Row" and "Column") that are used for storing

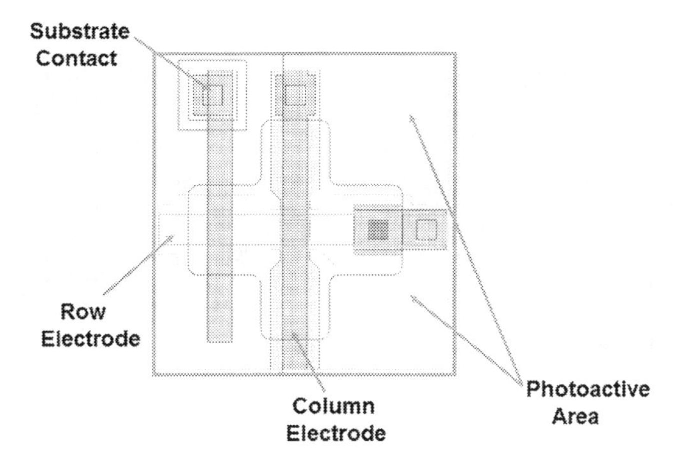

FIGURE 25.17 Schematic of an example detector element (pixel) within a CID detector. (With kind permission from Teledyne Leeman Labs.)

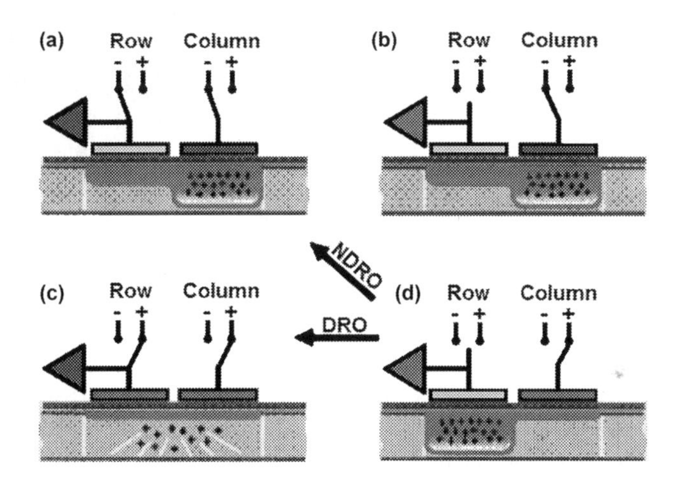

FIGURE 25.18 Schematic and operation of a CID pixel.[137] (a) Integrate, (b) first sample, (c) second sample, and (d) inject.

and measuring photogenerated charge. When an analytical measurement commences, the detector starts collecting charge, based on the integration time set by the user (Step A). During this step, the incident radiation is converted to charge and collected in the column photogate. Once integration has begun, the detector measures the accumulated charge by shifting it between the row and column photogates (Steps B and C).

If the pixel has accumulated enough charge to approach saturation, the charge can be cleared by injecting it into the substrate (Step D). This is a destructive readout (DRO) and resets the pixel to allow it to resume charge collection. If the pixel's charge is measured and is far from reaching its saturation point, the charge can be transferred back to the column photogate, where the pixel can continue collecting charge. This process is known as a non-destructive readout (NDRO), as the detector shifts the charge without destroying it.[137]

The structure of each pixel allows CID detectors to perform NDROs, which is the main functional difference between CID and CCD detectors. Measuring charge without destroying it allows each pixel to perform an NDRO repeatedly during an analytical measurement, as needed. Since each pixel can sense and measure its own charge, the pixels within a CID can be randomly accessed

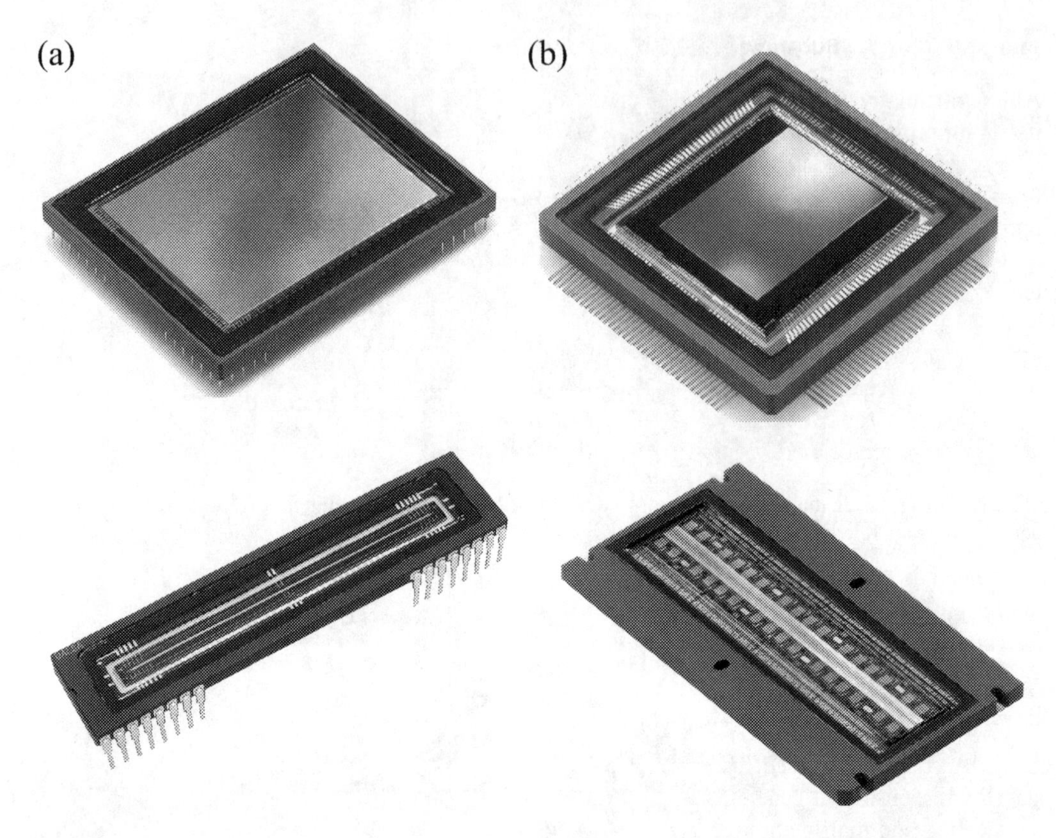

FIGURE 25.19 CCD (a) and CMOS (b) imagers. (With kind permission from Teledyne DALSA, Billerica, MA.)

for charge integration. These NDRO and RAI features make these detectors well suited for a number of applications in spectroscopy.[155,156]

The presence of photogates on each detector element reduces the size of the photoactive area of each pixel which decreases the total amount of charge that each pixel can accumulate before reaching saturation. This construction also decreases the uniformity of charge storage and readout across the array. Benefits to the structure of this detector are that it's inherently resistant to charge blooming, it allows for simultaneous wavelength measurements and background correction, plus simultaneous internal standardization for real-time drift correction.[157]

These devices can rapidly access all the addressable pixels within the array, which allows the detector to optimize the signal-to-noise ratio for each wavelength being measured. This optimization also increases the device's dynamic range to make it functional over several orders of magnitude. This feature is beneficial for applications in which both low- and high-intensity wavelengths must be measured simultaneously.[158] Figure 25.19 illustrates examples of typical CCD (Figure 25.19a) and CID (Figure 25.19b) devices.

ANALYTICAL PERFORMANCE

The quality of any developed and optimized analytical method can be assessed by a number of different figures of merit including accuracy, precision, sensitivity, speed, detection limits, working range, ease-of-use, and reproducibility. Both the instrument operating conditions and method parameters can affect these performance characteristics, so it is important to be familiar with these factors, to understand their effects, on the analytical data being collected.

DEPENDENCE ON ENVIRONMENTAL OPERATING CONDITIONS

All commercially available ICP-OES instruments have cooling, venting, electrical, and gas requirements for proper and safe operation. These instruments also prefer a certain set of room conditions with specified temperature, pressure, humidity, and room vibration requirements. The environmental conditions will vary slightly between instruments; however, instruments should generally be operated under the following range of conditions.

The laboratory should be maintained at a temperature between 15°C and 35°C. Ideally, the temperature within the room will not change, particularly when the instrument is being used for analysis. If the room temperature cannot be maintained over the course of a day, the rate of temperature change should be minimal. Most instruments can tolerate temperature changes of 1–3 degrees per hour; however, significant temperature changes will produce wavelength drift.

The relative room humidity should be kept between 20% and 80% in a non-condensing and non-corrosive atmosphere. Most spectrometers remain constantly purged with a low flow of dry, high-purity gas; however, if the purge is disrupted or stopped and the gas inside the spectrometer exchanges with the air in the room, moisture in the air could deposit on the components inside the spectrometer and fog the optics.

Though ICP-OES instruments are designed with a certain level of robustness built in, the manufacturer's recommendations for environmental conditions should be followed for optimal instrument performance.

EXHAUST REQUIREMENTS

An exhaust system is used for removing heat and gases from the instrument for both safety and stability reasons. Heat generated by the system electronics must be removed to prevent them from overheating and malfunctioning while excess heat from the plasma must be withdrawn to prevent the temperature inside the torchbox from rising and producing background emission from the torch.

The torchbox area must be properly vented to remove gases generated in the plasma which may contain ozone and other noxious substances. The positioning of the exhaust vent/s, the minimum and maximum draw required, and the diameter of the vent tubing vary between instruments so it is important to follow manufacturer recommendations.

ELECTRICAL REQUIREMENTS

Optical emission instruments require high voltage power to operate properly. If the instrument operates with a water-cooled load coil, a chiller or water recirculator may be used which has a voltage requirement that is different from that for the instrument. A computer, which likely controls the instrument, may have its own set of electrical requirements. The power requirements, including voltage, frequency, and phase, will vary slightly between instruments and the manufacturer requirements should be closely followed.

The performance and longevity of an ICP-OES can be affected by the quality of the power being delivered to the instrument. In severe cases, power fluctuations can cause severe damage to electronic components in the instrument. A power conditioner may be employed to protect the instrument from voltage sags, transient voltage issues, and general regulation problems. An uninterruptible power supply (UPS) will allow the instrument to be properly shut down in the event of a power outage.

TEMPERATURE AND PRESSURE REQUIREMENTS

The spectrometer must be maintained at a constant temperature and pressure to prevent excessive wavelength drift during analysis. Spectrometers are typically heated to a temperature between 35°C and 38°C and the set point is maintained to a precision of approximately ±0.1°C. Maintaining the spectrometer at a temperature well above that in the laboratory reduces the effect of small temperature changes in the room on wavelength drift.

The spectrometer is typically purged with a relatively dry, high-purity gas to maintain a constant environment inside the spectrometer. A wide range of flow rates can be used for the purge gas, from 0.1 to 15 L/min; however, the pressure inside the spectrometer should be at equilibrium prior to starting a sample run to avoid wavelength drift due to pressure changes.

MAINTENANCE

Routine preventive maintenance is one of the best ways to ensure proper instrument performance as poor quality data can often be traced back to the sample introduction system. General good laboratory practices should be followed in terms of maintaining a clean laboratory, and any spills should be cleaned up immediately. Solutions should be properly stored and labeled, and standards and samples should be kept in covered vials or containers at the completion of a sample run to minimize spills, contamination, or loss from volatilization. A clean rinse solution should be run through the sample introduction system for a suitable length of time at the end of each sample run to help reduce the frequency with which the sample introduction system must be disassembled and thoroughly cleaned.

If a peristaltic pump is being used, all pump tubing should be regularly checked. Solution must move smoothly through the tubing, both into and out of the instrument, to minimize pulsations in the plasma. Tubing should be checked for distortions, flat spots, and tears and should be replaced if any are discovered. The platens on the pump should be properly tightened down on the pump tubing to allow solution to move through the tubing without any jerking movements.

The nebulizer should periodically be checked for cracks, blockages, or leakages. The performance should be checked by pumping a clean solution through the nebulizer and visually inspecting the resulting aerosol to ensure it has the proper shape and lacks irregular pulsations. The nebulizer should be cleaned according to the manufacturer instructions and should be carefully stored when not in use to prevent breakages.

The spray chamber requires little to no maintenance; however, it should be visually checked for cracks on occasion. The spray chamber should be viewed during aerosol generation to ensure the inside surface is being properly wetted and large droplets are not beading up and disrupting the aerosol transfer. The drain tubing should fit tightly onto the spray chamber to allow waste solution to be properly removed.

The torch should be checked for leaks, cracks, or other physical damage. The injector should be checked for deposits or blockages and thoroughly cleaned if any are present. If the torch has gas connections or O-rings, those should be inspected for damage or leakages. Torches will accumulate deposits during normal sample analysis and should be cleaned as necessary to remove them. Over time, small deposits will accumulate which do not get removed during cleaning. These deposits should not be of concern unless the instrument's performance is affected.

Most of the remaining parts of the instrument, including the radio frequency (RF) generator, the spectrometer, and the detector, require little or no maintenance other than an optical entrance window that requires occasional cleaning or air filters that need to be cleaned or replaced.

DEPENDENCE ON PLASMA OPERATING CONDITIONS

Optimum plasma operating conditions are critical in obtaining the highest quality data. Plasma parameters will impact the stability and energy of the plasma and, therefore, the behavior of the elements, and are typically optimized to maximize the emission intensity for the analyte wavelengths in the method. Plasma conditions that are optimal for one type of sample are often different from those for another sample type so the operating conditions should be carefully evaluated for each application.

Plasma parameters to consider for optimization are:

- RF power
- Nebulizer gas flow
- Auxiliary gas flow
- Coolant gas flow
- Pump settings
- Radial viewing height.

It should be noted that, while important, plasma optimization is not as critical as the other method development steps. ICP-OES methods typically are developed for the analysis of multiple elements. Each element prefers a slightly different set of plasma parameters to achieve its maximum intensity for the wavelength selected. Since ICP-OES instruments collect emission intensities from groups of elements simultaneously, it is not practical to select different plasma settings for each individual element. Therefore, the plasma settings for multielement analysis will be somewhat of a compromise for most of the elements in the method. Most modern ICP-OES instrumentation provides an automated optimization feature to rapidly and automatically select the best plasma parameters to maximize the emission intensity for as many elements as is possible in a multielement method.[159]

RF POWER

The quality of the RF generator and the RF power selected for use during analysis will determine the quality and precision of the resulting analytical measurements. While there may be more than one RF power setting that will produce high-quality data, this is an important plasma parameter because the RF field, in combination with the argon gas, is responsible for generating and maintaining the plasma source.

RF generators operate at a frequency of either 27.12 or 40.68 MHz. These are the only two frequencies approved by the Federal Communications Commission (FCC) for RF generator operation, as other frequencies will interfere with established communication activities. There has been much debate over the relationship between an RF generator's operating frequency and its resulting performance[160,161]; however, there is no evidence to indicate that one frequency is beneficial over the other.

The stability of the plasma largely relies on the generator's ability to adjust to changing plasma conditions. Each time a new sample is introduced, the composition of the plasma rapidly changes. As the composition changes, the power required to adjust to the new plasma conditions changes and the RF generator must react accordingly. Matching the power being output by the generator to that required by the new plasma conditions is known as impedance-matching and can be accomplished with two basic oscillator designs: (1) crystal-controlled (fixed-frequency) and (2) free-running (variable frequency).[162] There are benefits and detriments to both designs; however, the variable frequency design of free-running oscillators makes them better able to generate a plasma from a cold start and better able to adapt to changes incurred by challenging sample matrices. A stable, successfully formed plasma is illustrated in Figure 25.20.

In addition to having a well-designed generator, an appropriate RF power must be selected for the application. Generally speaking, RF power is proportional to plasma temperature. Given the same set of gas flows and sample uptake parameters, a higher RF power produces a hotter, more energetic plasma. This is particularly advantageous for sample matrices that contain organic solvents or high levels of dissolved solids which are notorious for causing plasma instabilities or blowouts. While a higher RF power produces a more robust plasma, this setting should still be carefully selected to avoid using a power that is higher than necessary. Increasing the RF power increases the amount of emission that gets produced from all components in the sample solution, which can increase

FIGURE 25.20 A close-up of a Thermo Jarrell Ash Atomscan 16 inductively coupled argon plasma's torch. (With permission from Wblanchard.)

the level of background emission, degrade the signal-to-noise ratio for the analytes of interest, and degrade the background equivalent concentration (BEC).[163,164]

PLASMA GASES

While the plasma gases play an important role in generating and sustaining a stable plasma, the individual gas flow settings can have an effect on the emission of some elements and on the overall data quality. The coolant gas serves two major purposes when the plasma is being operated: plasma generation and torch cooling. The coolant gas flows through the outer tube in a swirled pattern to help the plasma maintain its annular shape. A relatively high flow rate is used to sustain a stable plasma and to prevent the torch from overheating. Since the main function of the coolant gas is related to the operation of the plasma, the flow rate does not have a significant effect on the emission of most elements. Having said that, the flow rate should be chosen with logic and care. A setting that is too low will provide insufficient cooling of the torch. A slight overheating can cause an increase in background emission as blackbody radiation from the torch itself is emitted. A severely overheated torch will melt and will require replacement. A flow setting that is too high is wasteful of the coolant gas and may provide more turbulence in the plasma than is necessary. Flow rates for the coolant gas can range from 8 to 20 L/min; however, flow rates between 10 and 14 L/min are appropriate for most ICP-OES applications.

The auxiliary gas flows through the intermediate tube of the torch and supplements the coolant gas to shift the base of the plasma further away from the end of the inner tube (injector). Not all applications or torch configurations require the use of an auxiliary gas; however, it is useful during the analysis of sample matrices with high levels of dissolved solids, organic solvents, or other matrix components that are known to cause a buildup of material inside the injector. The auxiliary gas adds turbulence to the plasma which degrades the penetration of the nebulizer gas into the plasma. Therefore, ideal circumstances would dictate that minimal or no auxiliary gas is used. However, if the application requires an auxiliary gas, the flow rate should be set such that the plasma is far enough from the injector to avoid a blockage from forming, yet not so far that the plasma becomes unstable and at risk for being extinguished. Flow rates for the auxiliary gas, if being used, are typically between 0.1 and 1.5 L/min.

Of the three plasma gases, the nebulizer gas typically has the most significant effect on analyte emission and should be optimized carefully. The nebulizer gas flows through the center tube (injector) of the torch and carries the sample into the plasma. The gas must be able to penetrate the plasma and travel along its central channel to maximize the efficiency with which the sample is desolvated, vaporized, atomized, ionized, and excited. Since the nebulizer gas carries the sample into the plasma, it stands to reason that increasing the nebulizer gas will transport more sample into the plasma which will produce larger emission signals and more sensitive measurements. This is not the case, however. The nebulizer gas must be carefully selected based on the challenge of the application and analytes being measured. In addition to generating some turbulence which affects plasma stability, the nebulizer gas transports the nebulized sample aerosol which affects the load on the plasma. If the change in plasma load is too severe, the RF generator will not be able to adjust its power output to compensate and severe plasma instability will result. In severe cases, the plasma load will be great enough to extinguish the plasma completely. Conversely, if the nebulizer gas flow is too low, the sample will not adequately penetrate the plasma and emission signals will significantly degrade. Most instruments will allow the nebulizer gas flow to be set between 0.1 and 2 L/min; however, a setting of approximately 1 L/min (or the equivalent pressure if the nebulizer is not mass flow-controlled) is typically used for most applications. Elements from different groups on the periodic table will produce optimal emission with slightly different nebulizer gas settings. Since most applications involve the analysis of multiple elements, a compromise must be made and a nebulizer setting should be selected that will produce optimal sensitivity and detection limits for the overall method.

PUMP SETTINGS

Bulk solution is delivered to the instrument via either a peristaltic or a syringe pump. The speed and precision with which the solution is introduced to the nebulizer significantly affects the quality of the resulting aerosol. The pump speed, combined with the inside diameter of the pump tubing, dictates the rate of solution that is delivered to the nebulizer. While each nebulizer will have a recommended flow rate for sample introduction, an optimal pump speed should be selected such that a consistent, stable aerosol is produced. If the pump speed is set such that the nebulizer aspirates solution faster than it is supplied via the pump, the nebulizer is being underfed ("starved"). Conversely, if the pump delivers solution faster than the nebulizer can aspirate it and produce an aerosol, the nebulizer is being flooded. Most nebulizers operate optimally when they are being starved for solution. Flooded nebulizers will produce an aerosol with a pulsing stream of large droplets that spill out the end. This "spitting" is from the excess solution that was pumped into the nebulizer but not converted to an aerosol. Some nebulizers, known as self-aspirating nebulizers, are designed to operate by drawing the solution through the pump tubing without assistance from a pump. Since these nebulizers do not require a pump for sample uptake, self-aspirating nebulizers are excluded from the text in this section.

The precision of the delivered solution will affect the precision of the aerosol. This is dictated by the style of the pump. A peristaltic pump contains a number of rollers which push solution through a flexible piece of tubing by alternatively compressing and relaxing the walls of the tubing. Each time the tubing is decompressed, a vacuum is created and a miniscule amount of solution is drawn backward into the pump tubing. This causes a fluctuation in the volume of solution being delivered to the nebulizer which produces an aerosol and resulting emission signals with a measurable pulsation.

A syringe pump can also be used for solution delivery to the nebulizer. A syringe pump is a type of infusion pump which delivers solution via small, rapid pulses. The pulsations are smaller and more frequent than those incurred with a peristaltic pump, which results in an aerosol and emission signals with smaller pump-related fluctuations. This concept is illustrated in Figure 25.21. A syringe pump will produce an aerosol with fewer fluctuations; however, pulsations from a peristaltic pump can be minimized with the use of a small-diameter pump and small pump rollers.

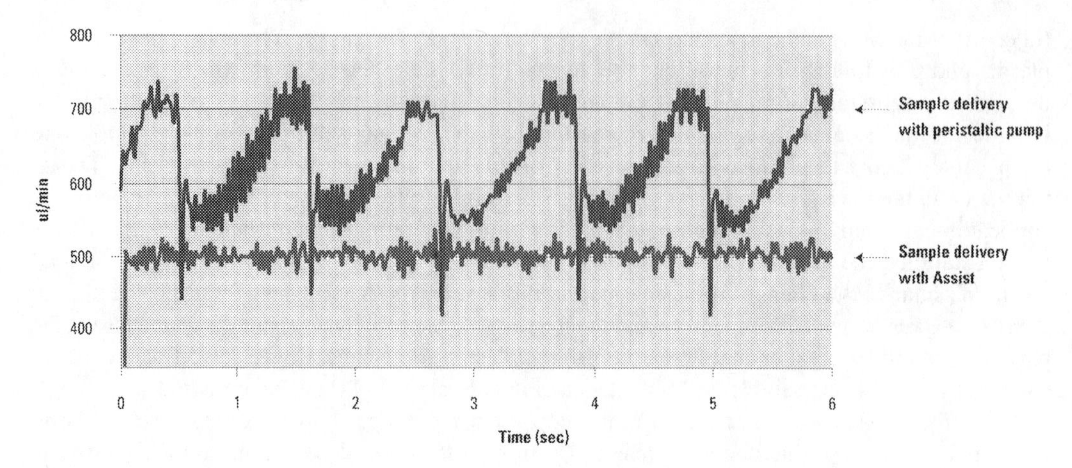

FIGURE 25.21 Signal fluctuation produced with a peristaltic pump versus a syringe pump. (With kind permission from Glass Expansion.)

PLASMA VIEWING HEIGHT

Emission signals should be collected from a specific part of the plasma to maximize the signal-to-background ratio and sensitivity of each measurement. Often referred to as the "normal analytical zone," this section of the plasma consists of a temperature zone that is optimal for atomic and ionic emission. Axial measurements are made along the central channel of the plasma which prevents emission from being collected from only one of the temperature zone. Radial measurements are made perpendicularly to the direction of the plasma so the observation height, also known as the viewing height, should be carefully optimized.[165]

Viewing height affects the emission strength of different elements in different ways. Some would prefer a hotter plasma which would render an optimal viewing height that is closer to the load coil. Other elements would prefer slightly cooler conditions which would put the optimal viewing height further from the load coil and closer to the tail of the plasma. Unless the instrument is being used for single-element analysis, the plasma viewing height will be a compromise. This process is computer-controlled through the instrument's software; however, manual adjustments can sometimes be made to fully optimize the observation height of the measurement.

PRECISION AND ACCURACY

When discussing analytical performance, precision and accuracy are figures of merit that indicate the robustness and reliability of the instrument and of the method developed for analysis. The terms are often used together; however, it is important not to use them incorrectly or interchangeably. Precision is a term that describes the agreement between replicate results which provides an indication of the reproducibility of the method. It is a figure of merit that can be obtained by repeating the measurement of a particular standard or sample and calculating the variation between the replicate results. To properly calculate precision, data for the same elements must be collected the exact same way each time.

Both short- and long-term precisions can be calculated for a method. Short-term precision typically represents the variation between back-to-back measurements of a small number of standards and/or samples. This represents the reliability of the method for a particular application, and depending upon which solutions are included in the measurement, this can include the reliability of the sample preparation procedure. Long-term precision, which typically represents the robustness and/or transferability of a method or procedure, can be calculated using a number of different approaches. Results can be collected for a chosen standard or sample periodically over the course of a sample run to determine the reliability of the method during a typical 8-h day. Results can also be collected on non-consecutive days, with two different laboratory technicians or in two different laboratories (as long as both laboratories are using the same instrumentation). Results collected under these conditions will determine the robustness and transferability of the method.

Accuracy is a term that describes how close the experimental results are to the known ("true") results. In other words, the accuracy of the method determines the correctness of the results. This is expressed as either an absolute or relative error and the calculation should be based on the measurement of a certified or other well-characterized standard. The acceptable level of inaccuracy should be determined during the method development and optimization stage. All calculated results are going to possess a small amount of inaccuracy due to basic errors and uncertainties in preparing and analyzing samples. If the calculated accuracy does not meet the requirements of the method, the source of the error should be investigated. Errors can consist of two different types: random or systematic. Each type of error could be due to a number of sources and a combination of errors could be contributing to the overall issue, so troubleshooting inaccuracies should be approached with care.

Precision and accuracy should not be used interchangeably as they describe significantly different figures of merit. While it's desirable for a method to have excellent precision and accuracy, it is possible to have one without the other. Results that are precise but inaccurate could indicate that the instrument settings are properly chosen, but an issue exists with the calibration curve or there are interferences that have not been properly accounted for.

If a method has good accuracy, but poor precision, it is sometimes referred to as being "accurate in the mean." In this situation, the accuracy is a result of systematic imprecision and typically indicates an issue with the sample introduction system (steering mirror is moving too slowly (dual-view methods only), sample uptake time is too short, sample flow rate is too high, platens on the peristaltic pump are not tightened properly).

DETECTION LIMITS

One of the most important, yet one of the more highly debated figures of merit for an analytical method is the LOD. For a given analyte, International Union of Pure and Applied Chemists (IUPAC) defines the detection limit as the smallest change in the emission signal (x_L) that can be detected with statistical certainty.[166] In other words, this is the lowest signal that can be detected above the blank and is defined according to the following equation:

$$x_L = k\sigma_B$$

where x_L is the net intensity, k represents a numerical multiplier that is selected according to the desired statistical confidence level, and σ_B represents the standard deviation of replicate measurements of the blank solution. The net intensity can be converted to a detection limit concentration using the following equation[167]:

$$\text{LOD} = (k\sigma_B)/S$$

where k represents the same multiplier as that in the previous equation, σ_B represents the standard deviation of the blank, and S is the slope of the calibration curve for that analyte (i.e., the sensitivity). Kaiser[168] asserts that an appropriate value for k is 3 as that will represent a 95% confidence interval for the calculated detection limit for most applications, so 3 is the most commonly used value in this calculation.

The debate over the formula for calculating detection limits stems from its definition which indicates that the concentrations must be calculated with statistical certainty.[169] Most accepted formulas for calculating detection limits are based on the IUPAC equation listed above. The resulting concentrations are typically referred to as "instrumental detection limits" and allow detection limits to be compared between different instruments or between different conditions on the same instrument. While useful for comparison purposes, detection limits based on this formula are unrealistic and do not reflect those that would be obtained when measuring analyte signals in the presence of the sample matrix under investigation.

Suggested modifications have been made to reflect more conservative estimates of detection limits. The Environmental Protection Agency (EPA) suggests the calculation of a "method detection limit" based on seven replicate analyses of a blank fortified with the analytes of interest at two or three times the concentration of the calculated instrument detection limit.[170] This is a more conservative estimate of the detection limits and accounts for any signal degradation caused by the sample matrix. Currie[171] suggests calculating a "determination limit" by using the IUPAC definition of LOD with a multiplier of 10 instead of 3. Currie defines this as "a determination limit at which a given procedure will be sufficiently precise to yield a satisfactory quantitative estimate." Long and Winefordner[172] recommend multiplying the standard deviation of the blank by 6 to calculate a "limit of guarantee."

LIMIT OF QUANTITATION

The limits of quantitation (LOQs) are often calculated along with the LODs. The LOQ is a conservative estimate of the LOD and is sometimes used as the reporting limit for a method.[173] For each analyte, the LOQ represents the lowest concentration that can be measured and reported with a sufficient amount of precision and accuracy. The desirable level of precision and accuracy is somewhat arbitrarily defined so there is no concrete formula for calculating the LOQ. However, the desired precision level is often around 10% (reported as a relative standard deviation), which means the LOQ is calculated using ten times the standard deviation of the blank.[174] By definition, the LOQ cannot be lower than the LOD and it should not be equal to the LOD.

BACKGROUND EQUIVALENT CONCENTRATION

A figure of merit that relates to the instrument detection limit is the BEC. For a given analyte, the BEC is the concentration of that analyte which produces a net signal (emission signal minus the contribution from the background) equal to the background signal at that wavelength. In other words, this is the analyte concentration which yields a signal-to-background ratio of 1.[175] There are a number of different formulas that exist for calculating BEC values; however, they can easily be determined using the calculated calibration curves. If the curves are plotted as measured emission intensity (x-axis) versus concentration (y-axis), the BEC for each analyte is represented by the y-intercept value. Since the BEC is based on the noise associated with the background signal (which is typically ~1% of the signal intensity of the background), a rough BEC calculation can be performed by multiplying the instrument detection limit by a factor of 30.

SENSITIVITY

Sensitivity is generally referred to as the ability to confidently measure small differences in analyte concentrations. The acceptable definition of sensitivity is outlined by the IUPAC and refers to calibration sensitivity. Sensitivity can vary by element and is represented by the calculated calibration curve. Most calibrations that are used with ICP-OES applications are linear and can be represented using the following equation:

$$y = mx + b$$

where y represents the instrument response for a given analyte at a given concentration, m represents the calibration sensitivity (i.e., the slop of the calibration curve), x represents the analyte concentration, and b is the measured signal for a blank solution. If the concentration of the signal is zero, the equation becomes $y = mx$ and the calibration sensitivity becomes independent of concentration.

Mandel and Stiehler[176] published a slightly different definition of sensitivity. Referred to as analytical sensitivity, this definition includes the precision of the measurement according to the following equation:

$$\gamma = m / s_S$$

where γ represents the analytical sensitivity, m represents the slope of the calibration curve, and s_S represents the standard deviation of the measured signal. Since this term is based on the slope and standard deviation of the measurement, the sensitivity value is independent of the units that were used to measure the emission signals. A drawback to this definition is the inclusion of the signal standard deviation. Since s_S can vary with analyte concentration, analytical sensitivity is typically concentration dependent.

METHOD DEVELOPMENT CONSIDERATIONS

As with many techniques, its success in the laboratory depends upon the quality of the methods that are developed for the application work being conducted. While ICP-OES instruments provide automated, intuitive operation with ppb-level detection limits and working ranges that extend over several orders of magnitude, developing a useful, well-optimized analytical method can be a manual, labor-intensive, and time-consuming process.

Whether a method is being developed from scratch or an existing method is being further optimized, there are a number of parameters that should be taken into consideration to ensure the method is best suited for the intended application. These include:

- Analytical wavelengths
- Interferences
- Plasma parameters
- Data acquisition parameters
- Validation of method.

Let's take a look at each of these considerations, in turn, to understand how each factor plays a role in the quality of the final optimized method. It should be noted that the theory behind this approach is applicable to the development of an analytical method for any elemental analysis technique; however, some of the specific considerations will change. For example, if developing an analytical method for an ICP-MS application, one would be choosing isotopes instead of emission wavelengths for measuring the analytes and internal standards.

ANALYTICAL WAVELENGTH CONSIDERATIONS

When choosing appropriate wavelengths for the elements of interest, it's important to keep in mind the method's desired performance attributes. For example, if method development is taking place for quantifying trace elements in drinking water samples, the main performance attributes will likely be detection capability and accuracy. Therefore, the most sensitive wavelength should be chosen for each element and all potential interferences should be identified and carefully corrected. Alternatively, if a method is being developed to quantify elements in oil additives, a wider working range may be desired. In this case, wavelengths with lower sensitivity may need to be chosen to maximize the concentration range over which the elements can be calibrated.

A critical factor in determining the suitability of a wavelength is whether it suffers from interferences. Several types of interferences exist in ICP-OES and their severity is dependent on the analyte wavelength, other elements present in the sample, and the sample matrix itself.[177] Evaluating interferences is not a trivial task and must be carried out with a great deal of care and thought. Interferences that are erroneously identified or not compensated for correctly will result in poor quality data. With the number of interference types and the variety of correction approaches possible, interference correction can quickly become an overwhelming task, particularly to an inexperienced operator. For this reason, when addressing interferences, the following wavelength selection procedure is recommended.

1. Choose several wavelengths for each element in the method
2. Collect analyte wavelength scans
3. Visually inspect the peaks for the data collected
4. Review data
5. Eliminate unsuitable wavelengths.

Let's review this process in more detail. In the first step, it is recommended to choose two or three wavelengths for each element of interest. This includes both the analyte elements and any applicable internal standard elements. Typically, the most sensitive wavelength for each element would be chosen, along with two wavelengths of slightly lower sensitivity. If a wide range of concentrations is expected to be encountered in the samples, a high-sensitivity wavelength might be chosen along with a low-sensitivity wavelength. In some cases, the sample matrix will produce an emission spectrum that is so complex, there may be only one or two suitable wavelengths (interference free and providing the desired working concentration range) to choose for the analysis.

Once a set of wavelengths has been chosen for evaluation, data should be collected to determine which wavelengths are best suited, based on the requirements of the application. It is recommended that three wavelength measurements be collected for each of the following solutions:

- A blank (calibration and method blanks, where applicable)
- A low-concentration calibration standard
- A high-concentration calibration standard
- A sample that represents each type of sample matrix to be analyzed.

Data from the blank provides an emission profile of the matrix in the absence of analytes. Data from the calibration standards provides profiles for the elements of interest at low and high concentrations. Both sets of measurements are required for the data inspection outlined in steps 3 and 4 described earlier. The highest concentration calibration standard should be used for this data collection to ensure that none of the selected wavelengths suffer from peak broadening or self-absorption.[178] Sample measurements produce emission profiles for the elements of interest in the presence of the sample matrix. One sample for each sample matrix type should be collected to ensure emission profiles are examined in every applicable sample matrix. For laboratories analyzing a wide range of sample matrices, collecting wavelength measurement scans for each matrix type could require data collection for a large number of samples.

After wavelength data has been collected, they should be visually inspected to determine that the emission peaks have the proper shape and size. An ideal wavelength would produce the series of emission signals illustrated in Figure 25.22 where the emission from the blank produces a flat emission signal with a relatively low intensity. The emission signals for the calibration standards and representative samples should produce a symmetric, Gaussian-like shape with a single peak. Each peak should exhibit a proportional change in magnitude to match the difference in concentrations. For example, data from a standard containing elements at 1.0 ppm would be expected to have roughly twice the signal of that for a 0.5 ppm standard, while maintaining approximately the same shape.

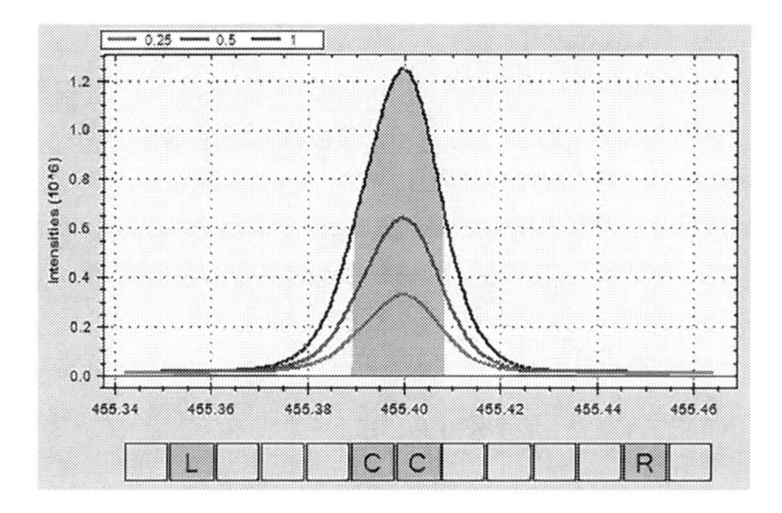

FIGURE 25.22 Examples of emission profiles for three calibration standards and a blank. (With kind permission from Thermo Fisher Scientific.)

Numerical data should be inspected to determine the intensity and precision at each wavelength. The intensity for each standard should increase at a rate that is proportional to the increase in its concentration. If a set of three wavelength measurements was collected for each solution, the measured intensities can be used to calculate the approximate precision for each measured standard. Calculated precision values should be within the acceptable limits for the application.

INTERFERENCES

Interferences are common with any plasma-based technique, particularly when trying to measure trace level concentrations in a sample matrix which contains high concentrations of elements that produce line rich spectra, such as Fe, Ca, and Si.[179] Sample matrices that are known to generate these types of interferences include geological, metallurgical, and high-matrix environmental samples, such as soils or waste waters. The three common types of interferences in ICP-OES are: physical, chemical, and spectral.

PHYSICAL INTERFERENCES

Physical interferences occur when the nebulization and/or transport efficiency of the standards differs from that of the samples. These differences in the physical characteristics of the matrices (density, viscosity, level of dissolved solids) can produce errors in the measured sample concentrations. Physical interferences are not wavelength specific and can be overcome by utilizing internal standards and preparing the calibration standards in a matrix that matches that of the samples.[159]

CHEMICAL INTERFERENCES

Chemical interferences occur when the standards behave differently from the samples as they enter the plasma. These types of interferences typically result from changes in temperature within the plasma and include easily ionized element (EIE) effects, molecular emission, and plasma loading. Referring to Figure 25.23,[180] the plasma consists of several temperature zones, which translate to varying amounts of energy available for excitation of the ground state atoms. The plasma is hottest and contains the greatest amount of energy at its base, which is where the sample is introduced. In this region, the plasma contains sufficient energy to atomize and ionize elements before they are

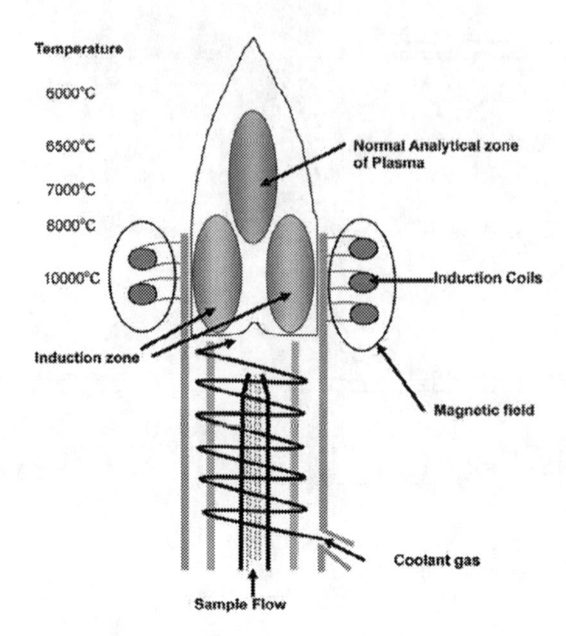

FIGURE 25.23 Different temperature zones within a plasma discharge.[180]

excited. Ionization is particularly prevalent for elements in the first two groups of the periodic table which all have relatively low first ionization potentials. These are often referred to as "easily ionized elements" (EIEs) and include elements such as Na, K, and Li.

When easily ionized elements are present at low concentrations, very few of the atoms become ionized which means emission will take place from excited atoms. At high concentrations, a significant number of the atoms will become ionized, which shifts the emission wavelength such that it will predominantly occur from the excited ions. If left uncorrected, this interference will reduce the linear dynamic range of the calibration curve and can potentially produce highly inaccurate results.[181]

This interference can be corrected by adding an ionization suppressant (also known as an ionization buffer) to all the solutions prior to analysis. An ionization suppressant is a solution containing a high concentration of an EIE (1,000 ppm Cs, for example). The solution will produce an extremely high concentration of cesium ions in the plasma which, according to Le Châtelier's principle,[182] will reduce the concentration of ions and increase the concentration of atoms to provide a chemical equilibrium balance in the plasma. Therefore, the element of interest will remain in its atomic state in the plasma.

If a dual-view instrument is being used, an additional step to take in correcting for this type of interference is to also measure emission from a radially configured plasma. EIE effects are much more significant for axial plasmas as emission is being measured from species that are present in all temperature zones within the central channel of the plasma. If emission is measured perpendicularly to the direction of the plasma (radial view), emission will be measured from a single temperature zone.[183]

Converse to the formation of EIE interferences are the formation of molecular interferences. If we refer again to Figure 25.23, the tip (also known as the "tail") of the plasma is the coolest and least energetic part of the plasma.[154] In this region, atoms and ions that were formed in the hotter part of the plasma can combine to form molecules. These molecules can become excited and emit light, which will produce broadband, molecular emission spectra.

This type of interference is relatively straightforward to correct and requires no intervention from the analyst. Since the coolest part of the plasma is in its tail, removing that portion of the

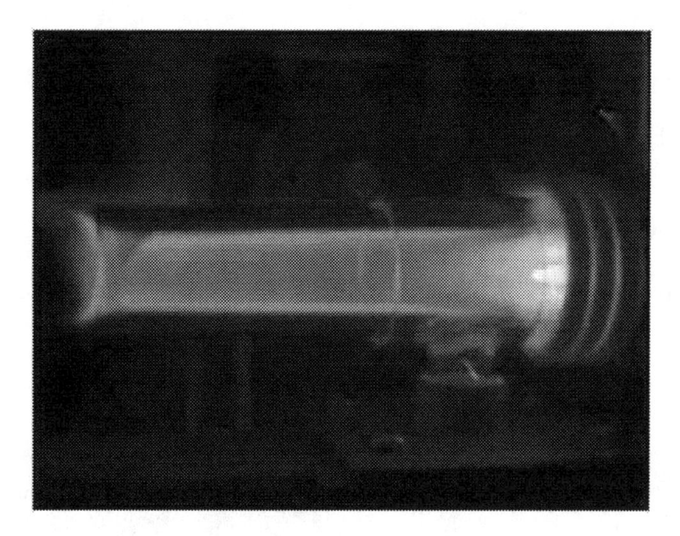

FIGURE 25.24 Image of a dual-view plasma to demonstrate removal of plasma tail. (With kind permission from Thermo Fisher Scientific.)

plasma prevents atoms and ions from traveling through a zone that's cool enough to allow them to combine and form molecular species. All modern ICP-OES instruments are set up to automatically and effectively remove the tail of the plasma. There are a variety of methods in which this is accomplished—a flow of inert gas that is counter to the direction of the plasma, a high-velocity shear gas, or a cooled cone interface; however, all are effective in cutting off the tail of the plasma. An example of the way in which the tail is removed is illustrated in Figure 25.24.

As with chemical interferences, another option for eliminating molecular interferences is to measure emission from a radially configured plasma. If emission is collected at 90° to the plasma and the observation height is optimized, emission should be collected from the "normal analytical zone" which is well separated from the cooler tail of the plasma.

Plasma loading is an interference that is sometimes experienced when samples contain organic solvents or high levels of dissolved solids.[184] When a significant amount of material is introduced into the plasma, it cools the plasma and, in severe cases, can extinguish it completely. Cooling the plasma reduces the amount of available energy which affects the emission intensity of the elements present. As a result, if samples contain concentrations of dissolved solids that are different from those in the standards, the emission intensity for the same analyte concentration will be different in the samples and standards which could produce data inaccuracies.

This interference can usually be addressed by preparing standards in a matrix that matches that of the samples, utilizing proper internal standards, choosing sample introduction components that were designed for the application, and selecting sample introduction conditions that minimize plasma loading.

SPECTRAL INTERFERENCES

The third, and sometimes most challenging, interferences are spectral. These interferences occur when emission from one or more species in the sample matrix overlaps with the emission for the analyte of interest. These interferences can result in a background shift, a partial peak overlap, or a direct overlap. If the analyte is suffering from a direct spectral overlap, the interfering element must either be removed from all solutions prior to analysis or an alternative wavelength must be selected. If the interfering element produces a simple background shift (either a flat, raised baseline, or a sloping background), the analyte can be accurately quantified by the use of optimized background correction points.

If the analyte suffers from partial spectral overlap, a combination of background correction and mathematical correction may need to be employed. An example of partial spectral overlap is illustrated in Figure 25.25. The figure depicts data for a solution containing 1 ppm B in a matrix which contains 100 ppm Fe.

A shoulder is clearly visible on the right side of the main peak, indicating the presence of a spectral overlap on B at 249.773 nm. If the emission profile for the sample solution is overlaid with that for single-element solutions containing 1 ppm B and 100 ppm Fe, the image in Figure 25.26 is produced.

In this figure, the red peak represents the emission profile for 1 ppm B at 249.773 nm, whereas the blue trace represents the emission peak for 100 ppm Fe at 249.782 nm. The black outline represents the combined emission from a sample containing 1 ppm B and 100 ppm Fe. The center shaded region represents the area under the peak which will be included for data collection. This area

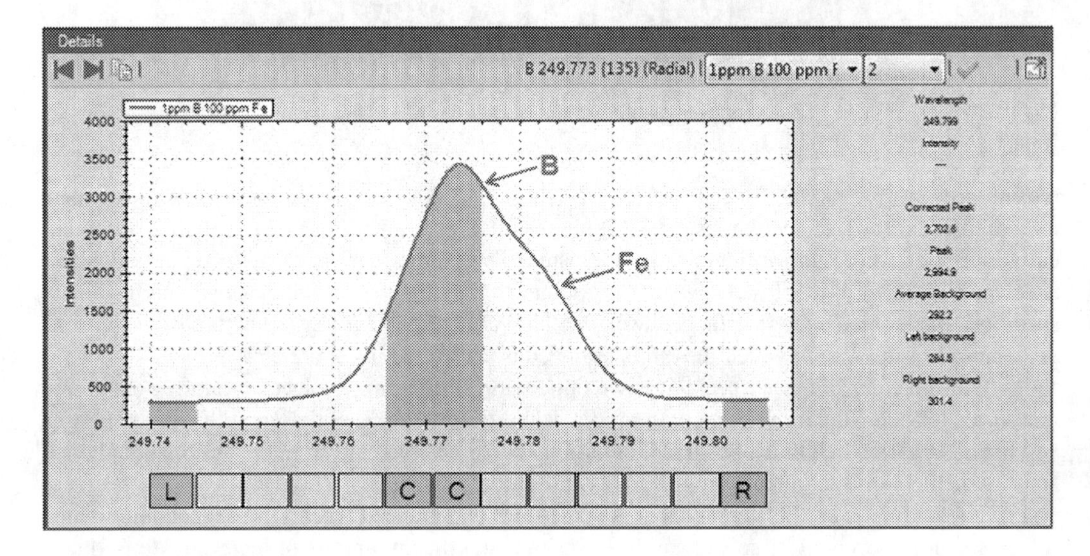

FIGURE 25.25 Emission profile of 1 ppm B in a matrix containing 100 ppm Fe.[159]

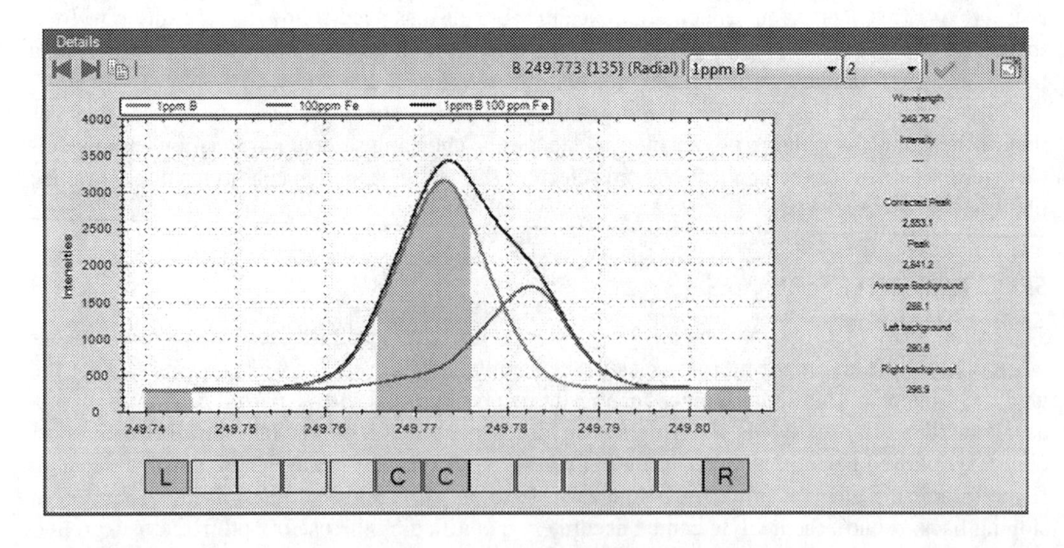

FIGURE 25.26 Emission profiles to demonstrate spectral overlap.[159]

includes a portion of the emission peak from Fe, and, if left uncorrected, this analytical measurement might produce erroneous data.

The best way to correct for this interference is to select an alternative emission wavelength for B that is free from interferences. If an alternative wavelength cannot be used, an interelement correction (IEC) factor should be calculated to correct for the overlap of Fe on B. If IEC factors are to be used, they must be carefully calculated and applied to ensure spectral overlaps are being accurately corrected. Improperly calculated IEC factors can produce data that is more inaccurate than if no IEC factors were used.[185]

DATA ACQUISITION

Data acquisition parameters play a crucial role in determining the quality of the resulting data. Even if the instrument is performing optimally, sensitive wavelengths have been selected for all analytes of interest and all interferences have been identified and removed, the analytical data will suffer significantly if the proper data acquisition parameters are not chosen. For the purposes of this text, data acquisition parameters will be restricted to: the choice of axial or radial plasma viewing (if using a dual-view instrument), the integration time, and the number of integrations. In general, an axial plasma is used for determining low, ppb-level concentrations, while a radial plasma view is used for higher (ppm-level) concentrations or to help compensate for chemical interferences.[107] The integration time for each plasma viewing configuration must be long enough to collect sufficient emission from all elements being measured. This would usually be dictated by the element with the weakest emission line or the element that is present at the lowest concentration. An axial integration time of 15 s and a radial integration time of 5 s are fairly typical to use. Finally, the number of integrations must be set. Keep in mind that if every sample is collected using a single integration, statistical analysis and precision data and LOQs cannot be calculated and reported. Therefore, two or more integrations must be collected for each sample in order to report this kind of information.

METHOD VALIDATION

The final step in the method development process is to validate that the method will produce results that meet the figures of merit required for the application. Methodology that achieves good detection limits, yet cannot obtain the required accuracy, is unlikely to meet the data quality objectives of the analysis.

A detection limit study is always encouraged for any analytical method. Calculated detection limits and LOQs allow reporting limits to be calculated, in order to determine the lowest concentrations that can be measured and reported for a suite of elements in a given sample matrix.[186]

Regardless of whether a detection limit study is performed, the developed method needs to be validated prior to use with real samples. The best approach to validating an analytical method is to obtain a certified reference material (CRM) that is appropriate for the application. If an analytical method is valid for the application, data acquired for a suitable CRM should match the certified concentrations that accompany the CRM (within defined error limits).

If a suitable CRM is not available for a given application, a number of other check standards can be utilized. A calibration standard can be prepared and analyzed as a sample to verify that the measured concentration matches the concentration at which it was prepared. A calibration or QC standard can be purchased from an external source and analyzed as a sample. Samples can be prepared and measured in duplicate to determine the precision of the method. Samples can be spiked with a known concentration of the elements of interest to determine whether the spiked elements can be measured and recovered at their expected concentrations. If interference corrections are being used, interference check standards can be prepared or purchased and analyzed to ensure that the correct results are obtained.

FINAL THOUGHTS

It should be emphasized that even though ICP-OES is a very useful analytical technique for measuring sub-ppm and ppb levels in different sample matrices, it might have limited use to determine ultra-low-level elemental impurities in many cannabis-related materials. This is mainly due to the fact that the sample digestion procedure and dilution factors involved can take the analyte levels in solution below the detection capability of the technique. If a cannabis product is a liquid or already in solution, it could be a viable approach to use. In addition, for state regulations that are based on orally delivered products like edibles, an ICP-OES coupled with a USN might be a good option. However, with many states using the ultra-low permitted daily exposure (PDE) limits for pharmaceutical inhalation products as guidelines, ICP-MS is probably the safest choice to use. Chapter 28 will give this a little more clarity, by comparing the detection capability of ICP-MS, ICP-OES, AA, and atomic fluorescence (AF) for the four commonly regulated heavy metals in cannabis, Pb, Cd, As, and Hg.

FURTHER READING

1. K. Ohls and B. Bogdain, *Journal of Analytical Atomic Spectrometry*, **31**, 22–31, 2016.
2. S. Greenfield, I. L. I. Jones, and C. T. Berry, *Analyst,* **89**, 713–720, 1964.
3. R.H. Wendt and V.A. Fassel, *Analytical Chemistry,* **37**(7), 920–922, 1965.
4. D. A. Skoog, F. J. Holler and T.A. Nieman, *Principles of Instrumental Analysis*, Harcourt Brace & Company, Orlando, FL, 1998.
5. G. F. Kirkbright, A. F. Ward, and T. S. West, *Analytica Chimica Acta* **62**(2), 241–251, 1972.
6. G. F. Kirkbright, A. F. Ward, and T. S. West, *Analytica Chimica Acta* **64**(3), 353–362, 1973.
7. R. F. Browner and A. W. Boorn, *Analytical Chemistry*, **56**, A786–A798, 1984.
8. A. Montaser, M. G. Minnich, J. A. McLean, H. Liu, J. A. Caruso, C. W. McLeod, *Sample Introduction in ICP-MS*, Wiley-VCH, New York, 1998.
9. J. F. Tyson, *Analytica Chimica Acta*, **234**, 3–12, 1990.
10. E. R. Denoyer, *Atomic Spectroscopy*, **13**, 93–98, 1992.
11. E. R. Denoyer, *Atomic Spectroscopy*, **15**, 7–16, 1994.
12. G. G. Bortoleto, S. Cadore, *Talanta*, **67**, 169–174, 2005.
13. J. Muller, *Fresenius' Journal of Analytical Chemistry*, **363**, 572–576, 1999.
14. A. U. Shaikh, D. E. Tallman, *Analytical Chemistry*, **49**, 1093–1096, 1977.
15. M. A. Wahed, D. Chowdhury, B. Nermell, S. I. Khan, M. Ilias, M. Rahman, L. A. Persson, M. Vahter, *Journal of Health, Population and Nutrition*, **24**, 36–41, 2006.
16. E.H. Evans, J.A. Day, C. Palmer, W.J. Price, C.M.M. Smith and J.F. Tyson, *Journal of Analytical Atomic Spectrometry*, **22**, 663–696, 2007.
17. Z. Long, Y. Luo, C. Zheng, P. Deng and X. Hou, *Applied Spectroscopy Reviews*, **47**, 382–413, 2012.
18. R. E. Russo, X. Mao, H. Liu, J. Gonzalez and S. S. Mao, *Talanta*, **57**, 425–451, 2002.
19. J. R. Bacon, K. L. Linge, R. R. Parrish and L. Van Vaeck, *Journal of Analytical Atomic Spectrometry*, **21**(8), 785–818, 2006,
20. O. T. Butler, J. M. Cook, C. F. Harrington, S. J. Hill, J. Rieuwerts and D. L. Miles, *Journal of Analytical Atomic Spectrometry*, **21**(2), 217–243, 2006.
21. S. A. Wilson, W. I. Ridley and A. E. Koenig, *Journal of Analytical Atomic Spectrometry*, **17**, 406–409, 2002.
22. J. Mora, S. Maestre, V. Hernandis and J. L. Todoli, *Trends in Analytical Chemistry*, **22**(3), 123–132, 2003.
23. B. L. Sharp, *Journal of Analytical Atomic Spectrometry*, **3**, 613–652, 1988.
24. R. F. Browner and A. W. Boorn, *Analytical Chemistry*, **56**, A875–A888, 1984.
25. A. Montaser, M. G. Minnich, H. Liu, A. G. T. Gustavsson and R. F. Browner, *Fundamental Aspects of Sample Introduction in ICP Spectrometry*, Wiley-VCH, New York, 1998.
26. G. Schaldach, L. Berger, I. Razilov and H. Berndt, *Journal of Analytical Atomic Spectrometry*, **17**, 334–344, 2002.
27. J. W. Olesik and L. C. Bates, *Spectrochimica Acta Part B*, **50**, 285–303, 1995.
28. L. Ebdon and M. R. Cave, *Analyst*, **107**, 172–178, 1982.

29. S. Augagneur, B. Medina, J. Szpunar and R. Lobinski, *Journal of Analytical Atomic Spectrometry*, **11**, 713–721, 1996.
30. E. Debrah, S. A. Beres, T. J. Gluodenis, R. J. Thomas and E. R. Denoyer, *Atomic Spectroscopy*, **16**, 197–202, 1995.
31. F. Vanhaecke, M. VanHolderbeke, L. Moens and R. Dams, *Journal of Analytical Atomic Spectrometry*, **11**, 543–548, 1996.
32. J. W. Olesik and S. E. Hobbs, *Analytical Chemistry*, **66**, 3371–3378, 1994.
33. K. E. Lawrence, G. W. Rice and V. A. Fassel, *Analytical Chemistry*, **56**, 289–292, 1984.
34. J. L. Todoli and J. M. Mermet, *Journal of Analytical Atomic Spectrometry*, **13**, 727–734, 1998.
35. A. Gustavsson, *Spectrochimica Acta Part B*, **39**, 743–746, 1984.
36. A. Gustavsson, *Spectrochimica Acta Part B*, **39**, 85–94, 1984.
37. H. Y. Liu and A. Montaser, *Analytical Chemistry*, **66**, 3233–3242, 1994.
38. H. Y. Liu, R. H. Clifford, S. P. Dolan and A. Montaser, *Spectrochimica Acta Part B*, **51**, 27–40, 1996.
39. S. H. Nam, J. S. Lim and A. Montaser, *Journal of Analytical Atomic Spectrometry*, **9**, 1357–1362, 1994.
40. H. Y. Liu, A. Montaser, S. P. Dolan and R. S. Schwartz, *Journal of Analytical Atomic Spectrometry*, **11**, 307–311, 1996.
41. T. T. Hoang, S. W. May and R. F. Browner, *Journal of Analytical Atomic Spectrometry*, **17**, 1575–1581, 2002.
42. P. W. Kirlew and J. A. Caruso, *Applied Spectroscopy*, **52**, 770–772, 1998.
43. L. Q. Wang, S. W. May, R. F. Browner and S. H. Pollock, *Journal of Analytical Atomic Spectrometry*, **11**, 1137–1146, 1996.
44. M. Huang, H. Kojima, A. Hirabayashi and H. Koizumi, *Analytical Sciences*, **15**, 265–268, 1999.
45. E. Debrah, S. A. Beres, T. J. Gluodenis, R. J. Thomas and E. R. Denoyer, *Atomic Spectroscopy*, **16**, 197–202, 1995.
46. J. L. Todoli and V. Hernandis, *Journal of Analytical Atomic Spectrometry*, **14**, 1289–1295, 1999.
47. R. I. Botto and J. J. Zhu, *Journal of Analytical Atomic Spectrometry*, **9**, 905–912, 1994.
48. B. Budic, *Journal of Analytical Atomic Spectrometry*, **16**, 129–134, 2001.
49. P. Masson, A. Vives, D. Orignac and T. Prunet, *Journal of Analytical Atomic Spectrometry*, **15**, 543–547, 2000.
50. M. A. Tarr, G. X. Zhu and R. F. Browner, *Applied Spectroscopy*, **45**, 1424–1432, 1991.
51. R. J. Thomas and C. Anderau, *Atomic Spectroscopy*, **10**, 71–73, 1989.
52. Q. H. Jin, F. Liang, Y. F. Huan, Y. B. Cao, J. G. Zhou, H. Q. Zhang and W. Yang, *Laboratory Robotics and Automation*, **12**, 76–80, 2000.
53. S. Yamasaki and A. Tsumura, *Water Science & Technology*, **25**, 205–212, 1992.
54. T. T. Nham, *American Laboratory*, **27**, 48L–48V, 1995.
55. I. B. Brenner, J. Zhu and A. Zander, *Fresenius' Journal of Analytical Chemistry*, **355**, 774–777, 1996.
56. J. Kunze, S. Koelling, M. Reich and M. A. Wimmer, *Atomic Spectroscopy*, **19**, 164–167, 1998.
57. S. C. K. Shum and R. S. Houk, *Analytical Chemistry*, **65**, 2972–2976, 1993.
58. S. C. K. Shum, R. Neddersen and R. S. Houk, *Analyst*, **117**, 577–582, 1992.
59. S. C. K. Shum, H. M. Pang and R. S. Houk, *Analytical Chemistry*, **64**, 2444–2450, 1992.
60. T. W. Avery, C. Chakrabarty and J. J. Thompson, *Applied Spectroscopy*, **44**, 1690–1698, 1990.
61. D. R. Wiederin, F. G. Smith and R. S. Houk, *Analytical Chemistry*, **63**, 219–225, 1991.
62. J. A. McLean, H. Zhang and A. Montaser, *Analytical Chemistry*, **70**, 1012–1020, 1998.
63. M. G. Minnich and A. Montaser, *Applied Spectroscopy*, **54**, 1261–1269, 2000.
64. J. L. Todoli and J. M. Mermet, *Journal of Analytical Atomic Spectrometry*, **16**, 514–520, 2001.
65. E. Bjorn and W. Frech, *Journal of Analytical Atomic Spectrometry*, **16**, 4–11, 2001.
66. A. C. S. Bellato, M. F. Gine, A. A. Menegario, *Microchemical Journal*, **77**, 119–122, 2004.
67. M. J. Powell, E. S. K. Quan, D. W. Boomer and D. R. Wiederin, *Analytical Chemistry*, **64**, 2253–2257, 1992.
68. S. E. O'Brien, J. A. McLean, B. W. Acon, B. J. Eshelman, W. F. Bauer and A. Montaser, *Applied Spectroscopy*, **56**, 1006–1012, 2002.
69. J. A. McLean, M. G. Minnich, L. A. Iacone, H. Y. Liu and A. Montaser, *Journal of Analytical Atomic Spectrometry*, **13**, 829–842, 1998.
70. B. W. Acon, J. A. McLean and A. Montaser, *Analytical Chemistry*, **72**, 1885–1893, 2000.
71. C. S. Westphal, K. Kahen, W. E. Rutkowski, B. W. Acon and A. Montaser, *Spectrochimica Acta Part B*, **59**, 353–368, 2004.
72. G. Schaldach, L. Berger, I. Razilov and H. Berndt, *Journal of Analytical Atomic Spectrometry*, **17**, 334–344, 2002.

73. B. L. Sharp, *Journal of Analytical Atomic Spectrometry*, **3**, 939–963, 1988.
74. C. Rivas, L. Ebdon, and S. J. Hill, *Journal of Analytical Atomic Spectrometry*, **11**, 1147–1150, 1996.
75. R. H. Scott, V. A. Fassel, R. N. Kniseley and D. E. Nixon, *Analytical Chemistry*, **46**, 75–81, 1974.
76. D. R. Luffer and E. D. Salin, *Analytical Chemistry*, **58**, 654–656, 1986.
77. J. L. Todoli, S. Maestre, J. Mora, A. Canals and V. Hernandis, *Fresenius' Journal of Analytical Chemistry*, **368**, 773–779, 2000.
78. X. H. Zhang, H. F. Li and Y. F. Yang, *Talanta*, **42**, 1959–1963, 1995.
79. M. Wu and G. M. Hieftje, *Applied Spectroscopy*, **46**, 1912–1918, 1992.
80. G. Schaldach, H. Berndt and B. L. Sharp, *Journal of Analytical Atomic Spectrometry*, **18**, 742–750, 2003.
81. H. Isoyama, T. Uchida, C. Iida and G. Nakagawa, *Journal of Analytical Atomic Spectrometry*, **5**, 307–310, 1990.
82. H. Isoyama, T. Uchida, T. Niwa, C. Iida and G. Nakagawa, *Journal of Analytical Atomic Spectrometry*, **4**, 351–355, 1989.
83. B. Bouyssiere, Y. N. Ordonez, C. P. Lienemann, D. Schaumloffel and R. Lobinski, *Spectrochimica Acta Part B*, **61**, 1063–1068, 2006.
84. A. Prange and D. Schaumloffel, *Journal of Analytical Atomic Spectrometry*, **14**, 1329–1332, 1999.
85. D. Schaumloffel, J. R. Encinar and R. Lobinski, *Analytical Chemistry*, **75**, 6837–6842, 2003.
86. A. Woller, H. Garraud, F. Martin, O. F. X. Donard and P. Fodor, *Journal of Analytical Atomic Spectrometry*, **12**, 53–56, 1997.
87. Y. F. Li, C. Y. Chen, B. Li, J. Sun, J. X. Wang, Y. X. Gao, Y. L. Zhao and Z. F. Chai, *Journal of Analytical Atomic Spectrometry*, **21**, 94–96, 2006.
88. A. Al-Ammar, R. K. Gupta and R. M. Barnes, *Spectrochimica Acta Part B*, **54**, 1077–1084, 1999.
89. A. Al-Ammar, R. K. Gupta and R. M. Barnes, *Spectrochimica Acta Part B*, **55**, 629–635, 2000.
90. R.L. Sutton, *Journal of Analytical Atomic Spectrometry*, **9**, 1079–1083, 1994.
91. H. Naka and H. Kurayasu, *Bunseki Kagaku*, **45**, 1139–1144, 1996.
92. P. Schramel, *Fresenius' Journal of Analytical Chemistry*, **320**, 233–236, 1985.
93. J. H. D. Hartley, S. J. Hill and L. Ebdon, *Spectrochimica Acta Part B*, **48**, 1421–1433, 1993.
94. W. Schron and U. Muller, *Fresenius' Journal of Analytical Chemistry*, **357**, 22–26, 1997.
95. A. R. Eastgate, R. C. Fry and G. H. Gower, *Journal of Analytical Atomic Spectrometry*, **8**, 305–308, 1993.
96. T.B. Reed, *Journal of Applied Physics,* **32**, 821–824, 1961.
97. T.B. Reed *Journal of Applied Physics,* **32**, 2534–2535, 1961.
98. T.B. Reed, 57th Annual Meeting American Institute of Chemical Engineers (December, 1964).
99. C.D. Allemand and R.M. Barnes, *Applied Spectroscopy*, **31**, 434–443, 1977.
100. S. Greenfield, I. Ll. Jones, and C. T. Berry, *Analyst*, **89**, 713, 1964.
101. S. Greenfield, U.S. Patent No. 3,467,471 (September 16, 1969).
102. R. H. Scott, V. A. Fassel, R. N. Kniseley, and D. E. Nixon, *Analytical Chemistry*, **46**, 75, 1975.
103. R. Rezaaiyaan, G. M. Hieftje, H. Anderson, H. Kaiser and B. Meddings, *Applied Spectroscopy*, **36**, 627–631, 1982.
104. P.W.J.M. Boumans, *Fresenius' Journal of Analytical Chemistry*, **299**, 337–361, 1979.
105. Thermo Scientific, Radial Demountable Ceramic Torch for the Thermo Scientific iCAP 6000 Series ICP spectrometer, Thermo Fisher Product Technical Note: 43053, 2010, http://tools.thermofisher.com/content/sfs/brochures/D01563~.pdf.
106. G.F. Larson, V.A. Fassel, R.K. Winge and R.N. Kniseley, *Applied Spectroscopy*, **30**, 384–391, 1976.
107. F.V. Silva, L.C. Trevizan, C.S. Silva, A.R.A. Nogueira and J.A. Nóbrega, *Spectrochimica Acta Part B*, **57**, 1905–1913, 2002.
108. Thermo Scientific, Thermo Scientific iCAP 7000 Plus Series ICP-OES: Innovative ICP-OES Optical Design, Thermo Fisher Scientific Product Technical Note: 43333, 2016, https://tools.thermofisher.com/content/sfs/brochures/TN-43333-ICP-OES-Optical-Design-iCAP-7000-Plus-Series-TN43333-EN.pdf.
109. C.R. Runge and F. Paschen, *Abhandlungen der Königlich Preussischen Akademie der Wissenschaften*, **1** (1902).
110. G. L. Clark, *The Encyclopedia of Spectroscopy*, Reinhold Publishing Corporation, New York, 1960.
111. J.F. James and R.S. Sternberg, *The Design of Optical Spectrometers*, Chapman and Hall Ltd, London, 1969.
112. C.B. Boss and K.J. Fredeen, *Concepts, Instrumentation, and Techniques in Inductively Coupled Plasma Optical Emission Spectrometry*, 2nd edition, Perkin-Elmer Corporation, Waltham, MA, 1997.
113. K. Jankowski, A. Jackowska, A.P. Ramsza and E. Reszke, *Journal of Analytical Atomic Spectrometry*, **23**, 1234–1238, 2008.

114. Y. Okamoto, *Analytical Sciences*, **7**, 283–288, 1991.
115. T. Maeda and K. Wagatsuma, *Microchemical Journal*, **76**, 53–60, 2004.
116. U. Engel, C. Prokisch, E. Voges, G.M. Hieftje and J.A.C. Broekaert, *Journal of Analytical Atomic Spectrometry*, **13**, 955–961, 1998.
117. Q. Xue, *Applied Optics*, **50**, 1338–1344, 2011.
118. Q. Xue, S. Wang and F. Lu, *Applied Optics*, **48**, 11–16, 2009.
119. J. A. C. Broekaert, Instrument Column, *Spectrochimica Acta Part B*, **37B**, 359, 1982.
120. Plasma-Spec., The next generation in plasma spectrometry. Technical Bulletin of Leeman Labs, Inc., Lowell, MA.
121. X. Hou and B.T. Jones, Inductively coupled plasma/optical emission spectrometry, in *Encyclopedia of Analytical Chemistry*, R.A. Meyers (ed.), John Wiley & Sons Ltd, Chichester, pp. 9468–9485, 2000.
122. K. Paech and M.V. Tracey, *Modern Methods of Plant Analysis, 1*, Springer, Berlin, p. 177, 1956.
123. R.F. Jarrell, F. Brech and M.J. Gustafson, *Journal of Chemical Education*, **77**, 592–598, 2000.
124. A. Scheeline, C.A. Bye, D.L. Miller, S.W. Rynders and R.C. Owen, Jr., *Applied Spectroscopy*, **45**, 334–346, 1991.
125. A.T. Zander and P.N. Keliher, *Applied Spectroscopy*, **33**, 499–502, 1979.
126. C. Allemand, *ICP Information Newsletter*, **2**, 1, 1976.
127. C. Allemand, *ICP Information Newsletter*, **4**, 44, 1978.
128. R.B. Bilhorn, P.M. Epperson, J.V. Sweedler and M.B. Denton, *Applied Spectroscopy*, **41**, 1125–1136, 1987.
129. Y. Talmi and R.W. Simpson, *Applied Optics*, **19**, 1401–1414, 1980.
130. F.M. Pennebaker, D.A. Jones, C.A. Gresham, R.H. Williams, R.E. Simon, M.F. Schappert and M.B. Denton, *Journal of Analytical Atomic Spectrometry*, **13**, 821–827, 1998.
131. J. Marshall, A. Fisher, S. Chenery and S.T. Sparkes, *Journal of Analytical Atomic Spectrometry*, **11**, 213R–238R, 1996.
132. Q.S. Hanley, C.W. Earle, F.M. Pennebaker, S.P. Madden and M.B. Denton, *Analytical Chemistry*, **68**, 661A–667A, 1996.
133. A. Sweedler, J. V., Ratzlaff, K. L., Denton, M. B., Eds. *Charge Transfer Devices in Spectroscopy*, VCH Publishers, New York, 1994.
134. Y. Talmi and R. W. Simpson, *Applied Optics*, **19**, 1401–1414, 1980.
135. Y. Talmi, *Applied Spectroscopy*, **36**, 1–18, 1982.
136. E.D. Salin and G. Horlick, *Analytical Chemistry*, **52**, 1578–1582, 1980.
137. J.M. Harnly and R.E. Fields, *Applied Spectroscopy*, **51**, 334A–351A, 1997.
138. J.V. Sweedler, R.D. Jalkian, R.S. Pomeroy and M.B. Denton, *Spectrochimica Acta Part B*, **44B**, 683–692, 1989.
139. R.B. Bilhorn, J.V. Sweedler, P.M. Epperson and M.B. Denton, *Applied Spectroscopy*, **41**, 1114–1125, 1987.
140. Q. Xue, S. Wang and F. Lu, *Applied Optics*, **48**, 11–16, 2009.
141. F.L.J. Sangster and K. Teer, *IEEE Journal of Solid-state Circuits*, **SC-4**, 131–136, 1969.
142. A.J.P. Theuwissen, *Solid-State Imaging with Charge-Coupled Devices*, Kluwer Academic Publishers, New York, 2002.
143. W. Boyle and G. Smith, *Bell System Technical Journal*, **49**, 587–593, 1970.
144. Boyle, W. and G. Smith, U.S. Patent 3792322 "Buried channel charge coupled devices," Feb 12, 1974.
145. Boyle, W. and G. Smith, U.S. Patent 3796927 "Three dimensional charge coupled devices," March 12, 1974.
146. P.M. Epperson, J.V. Sweedler, R.B. Bilhorn, G.R. Sims and M.B. Denton, *Analytical Chemistry*, **60**, 327A–335A, 1988.
147. A.G. Milnes, *Charge-Transfer Devices in Semiconductor Devices and Integrated Electronics*, Van Nostrand Reinhold Company Regional Offices, New York, pp. 590–642, 1980.
148. G.C. Holst and T.S. Lomheim, *CMOS/CCD Sensors and Camera Systems*, JCD Publishing, Oviedo, FL, 2007.
149. J.R. Janesick, *Scientific Charge-Coupled Devices*, SPIE – The International Society for Optical Engineering, Bellingham, WA, p. 24, 2001.
150. N. Waltham, CCD and CMOS sensors in *Observing Photons in Space: A Guide to Experimental Space Astronomy*, M.C.E. Huber, A. Pauluhn, J.L. Culhane, J.G. Timothy, K. Wilhelm and A. Zehnder (eds.), Springer Science & Business Media, New York, pp. 423–442, 2013.
151. J.M. Mermet, A. Cosnier, Y. Danthez, C. Dubuisson, E. Fretel, O. Rogerieux and S. Vélasquez, *Spectroscopy*, **20**, 60–66, 2005.

152. T.W. Barnhard, M.I. Crockett, J.C. Ivaldi and P.L. Lundberg, *Analytical Chemistry*, **65**, 1225–1230, 1993.
153. T.W. Barnhard, M.I. Crockett, J.C. Ivaldi, P.L. Lundberg, D.A. Yates, P.A. Levine and D.J. Sauer, *Analytical Chemistry*, **65**, 1231–1239, 1993.
154. I.B. Brenner and A.T. Zander, *Spectrochimica Acta Part B*, **55**, 1195–1240, 2000.
155. P. M. Epperson, J. V. Sweedler, R. B. Bilhorn, G. R. Sims and M. B. Denton, *Analytical Chemistry*, **60**, 327A–335A, 1988.
156. G. R. Sims and M. B. Denton, *Multichannel Image Detectors*, Y. Talmi, (ed.), ACS Symposium Series No. 236, American Chemical Society, Washington, D.C., Vol. 2, Chap. 5, 1983.
157. S. Bhaskaran, T. Chapman, M. Pilon and S. VanGorden, *SPIE Proceedings, Infrared Systems and Photoelectronic Technology III*, **7055**, 70550R, 2008.
158. R.B. Bilhorn and M.B. Denton, *Applied Spectroscopy*, **44**, 1538–1546, 1990.
159. Thermo Scientific, Overcoming Interferences with the Thermo Scientific iCAP 7000 Plus Series ICP-OES, Thermo Fisher Scientific Product Technical Note: 43332, 2016, https://tools.thermofisher.com/content/sfs/brochures/TN-43332-ICP-OES-Overcoming-Interferences-iCAP-7000-Plus-Series-TN43332-EN.pdf.
160. K. E. Jarvis, P. Mason, T. Platzner, and J. G. Williams, *Journal of Analytical Atomic Spectrometry*, **13**, 689–696, 1998.
161. G. H. Vickers, D. A. Wilson, and G. M. Hieftje, *Journal of Analytical Atomic Spectrometry*, **4**, 749–754, 1989.
162. H.E. Taylor, *Inductively Coupled Plasma-Mass Spectrometry: Practices and Techniques*, Academic Press, San Diego, CA, 2001.
163. I.B. Brenner, A. Zander, M. Cole and A. Wiseman, *Journal of Analytical Atomic Spectrometry*, **12**, 897–906, 1997.
164. I.B. Brenner, M. Zischka, B. Maichin and G. Knapp, *Journal of Analytical Atomic Spectrometry*, **13**, 1257–1264, 1998.
165. L.C. Trevizan and J.A. Nobrega, *Journal of the Brazilian Chemical Society*, **18**, 1678–4790, 2007.
166. G.L. Long and J.D. Winefordner, *Analytical Chemistry*, **55**, 712A–724A, 1983.
167. J. Mermet and E. Poussel, *Applied Spectroscopy*, **49**, 12A–18A, 1995.
168. H. Kaiser, *Analytical Chemistry*, **42**, 53A, 1987.
169. V. Thomsen, D. Schatzlein, and David Mercuro, *Spectroscopy*, **18**, 112–114, 2003.
170. EPA Method 200.7, Revision 4.4, 1994, https://www.epa.gov/sites/production/files/2015-08/documents/method_200-7_rev_4-4_1994.pdf.
171. L.A. Currie, *Analytical Chemistry*, **40**, 586–593, 1968.
172. G.L. Long and J.D. Winefordner, *Analytical Chemistry*, **55**, 713A–724A, 1983.
173. J.M. Mermet, G. Granier and P. Fichet, *Spectrochimica Acta Part B*, **76**, 221–225, 2012.
174. F.C. Garner and G.L. Robertson, *Chemometrics and Intelligent Laboratory Systems*, **3**, 53–59, 1988.
175. V. Thompsen, *Spectroscopy*, **27**, 3, 2012.
176. J. Mandel and R.D. Stiehler, *Journal of Research of the National Bureau of Standards*, **155**, A53, 1964.
177. G. F. Larson, V. A. Fassel, R.H. Scott and R.N. Kniseley, *Analytical Chemistry,* **47**, 238–243, 1975.
178. Thermo Scientific, High Performance Radio Frequency Generator Technology for the Thermo Scientific iCAP 7000 Plus Series ICP-OES, Thermo Fisher Scientific Product Technical Note: 43334, 2016, https://tools.thermofisher.com/content/sfs/brochures/TN-43334-ICP-OES-RF-Generator-iCAP-7000-Plus-Series-TN43334-EN.pdf.
179. H.G.C. Human and R.H. Scott, *Spectrochimica Acta Part B,* **31**, 459–473, 1976.
180. R.F. Browner, *Fundamental Aspects of Aerosol Generation and Transport, in "Inductively Coupled Plasma Emission Spectrometry," Part II, "Applications and Fundamentals,"* Wiley-Interscience, New York, 1987.
181. M.W. Blades and G. Horlick, *Spectrochimica Acta Part B,* **36**, 881–900, 1981.
182. A. J. Miller, *Journal of Chemical Education*, **31**, 455, 1954.
183. M.H. Abdallah, R. Diemiaszonek, J. Jarosz, J.M. Mermet, J. Robin and C. Trassy, *Analytica Chimica Acta,* **84**, 271–282, 1976.
184. A.W. Boorn and R.F. Browner, *Analytical Chemistry,* **54**, 1402–1410, 1982.
185. V. Thomsen, D. Mercuro and D. Schatzlein, *Spectroscopy,* **21**(7), 2006, available at: http://www.spectroscopyonline.com/interelement-corrections-spectrochemistry.
186. T.R. Dulski, Statistics and Specifications in *A Manual for the Chemical Analysis of Metals*, ASTM International, West Conshohocken, PA, pp. 200–201, 1996.

26 Atomic Absorption and Atomic Fluorescence

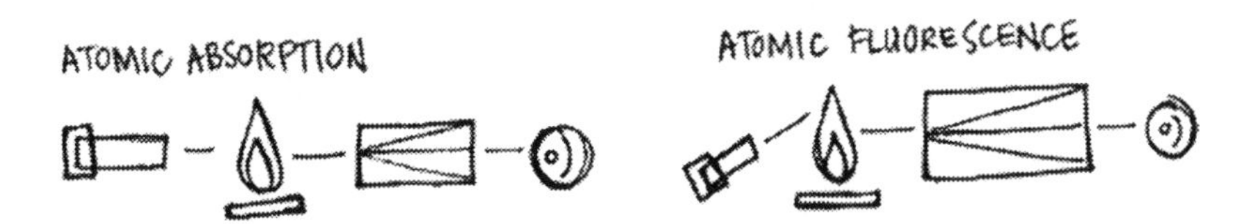

Atomic absorption spectrometry (AAS) is the most mature of all the atomic spectroscopic techniques having been developed and commercialized in the late 1950s. There are literally hundreds of thousands of instruments being used for many diverse and varied applications across the globe. Atomic fluorescence spectrometry (AFS) is a more recent development with commercially available systems on the market for the determination of Pb, Cd, As, and Hg in cannabis and hemp. AAS is predominantly single element techniques, which limits their use for multielement analysis. However, AFS can offer multielement capability by surrounding the atomization source with a number of element-specific photon sources.

When a small suite of elements is required, or the sample workload is not so high, AAS can be a very cost-effective technique to use. Conventional atomic absorption (AA) typically uses an air (or nitrous oxide) acetylene flame as the atomization source, but by using a graphite furnace in place of a flame, lower detection limits can be achieved, in some cases as good as ICP-MS. Additionally, dedicated AFS system using vapor generation atomization can be used for measuring elements that readily form volatile hydrides/cold vapor (Pb, As, Cd, Hg Sb, Se, Sn, Te). This chapter will present the fundamental principles of both AA and atomic fluorescence and summarize their strengths and weaknesses with a particular emphasis on their use for the determination of heavy metals in cannabis and hemp.

In the world of atomic spectroscopy instrumentation, there are four techniques used quantitatively: AA, atomic emission, mass spectrometry, and atomic fluorescence. Unlike the plasma techniques, AA and atomic fluorescence require element-specific light sources for the energy required to cause atoms of the selected elements in the light path to go through excitation.

The spectrometers for AA can support a variety of atomizers: flames, electrically heated furnace, and vapor generators. In terms of the instruments, flame AA is the oldest of the techniques. With a team led by Sir Alan Walsh, the current form of the flame AA instrument was developed in the 1950s. This work revolutionized the quantitative determination of more than 65 elements and led to the variations of the instruments in use today.

Unlike plasmas, in most cases, the temperature of the atomizer does not supply sufficient energy to cause the resulting atoms to go through excitation. Instead, element-specific lamps are required to provide this extra energy. The lamp emits light energy at the wavelengths specific to the element and that energy is absorbed if there are atoms of that element in the light path. This absorption causes the atoms to go through excitation. The excited state is not a stable state for the atoms so they also go through decay and light energy is emitted. The process that occurs in the flame is shown in Figure 26.1.

The amount of light absorbed is defined by comparing the light intensity before the atomizer, I_o, to the amount of light intensity after the atomizer, I_f. The relationship is shown in Figure 26.2. As the number of atoms increases due to increased concentration in the sample, the amount of light absorbed increases. It is this difference in the light intensity absorbed from the lamp that is actually

$$MX \xrightarrow[\text{bonds broken}]{\text{heat}} M + X \xrightarrow[\text{absorption}]{\text{light energy}} M^* \xrightarrow[\text{emission}]{\text{decay}} M + \text{light}$$

FIGURE 26.1 Excitation processes that occur in a flame.

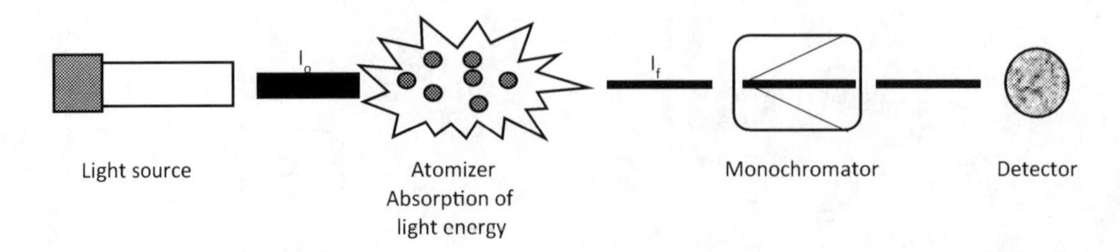

FIGURE 26.2 Basic components of AA.

being measured with each reading. The amount of light absorbed by solutions of known concentrations can be compared to the amount of light absorbed by unknown solutions, and hence, the concentration of the analyte in real samples can be determined.

Transmittance is defined as a ratio of the lamp intensity after the atomizer, I_f, to the lamp intensity before the atomizer, I_o, or I_f/I_o, and

$$\%T = 100 \times I_f / I_o$$

Therefore, the percent absorption is

$$\%A = 100 - \%T$$

Absorbance is defined as the log (I_o/I_f) and is a convenient form of measurement. This term follows a linear relationship with concentration according to Beer's law:

$$\text{Absorbance} = (\text{concentration, } c) \times (\text{an absorption coefficient, } a) \times (\text{path length, } b).$$

The absorption coefficient is dependent on the element, the wavelength chosen, the instrument configuration, and operational parameters. The path length is the length of the atomizer.

While Beer's law defines a linear relationship between concentration and absorbance, when the absorbance becomes too high, the relationship becomes non-linear. The point at which this non-linearity occurs is specific to the element and the wavelength at which the measurement is made. As long as the absorbance increases with the increase in concentration, a non-linear curve fit can be used for calibration.

FLAME AAS

As with plasma techniques, flame AAS (FLAAS) requires the sample in liquid form and utilizes nebulizers and spray chambers. The nebulizer mixes the liquid sample with air to form an aerosol mist that is sprayed into the spray chamber. A small part of the aerosol is dried as it transported into the flame via a burner head aligned in the light path. The heat of the flame is used primarily to break molecular bonds. The most commonly used flames are air–acetylene and nitrous oxide–acetylene. A few elements (such as Cs, K, and Na) will excite merely by the heat of the flame and not require a lamp. In these cases, the flame atomic emission signal can be read as the excited atoms return to ground state. But most elements are determined in the AA mode.

ADVANTAGES OF FLAAS

There are many advantages to using flame AA. The cost of the instrumentation and consumables required are much less than the other atomic spectroscopy techniques. Consumables consist primarily of the lamps, which have to be replaced periodically, and the gas supplies. Nebulizers, spray chambers, and burner heads are also considered consumables, but with care, they can be used for many years. FLAAS is also a technique that is easy to use and does not require a great deal of operator skill to produce accurate results. There is minimal maintenance of the hardware required.

While there are some multielement AA instruments on the market, the number of elements that can be determined in one method is generally limited. Most AA spectrometers can determine only one element at a time. While this is considered by many to be a disadvantage, analysis by FLAAS is very quick. Literally, per element, results for each sample can be obtained in a matter of a few seconds. On most modern instruments, progressing from one element to the next can be automated with the use of autosamplers. Therefore, a great deal of data for multiple elements can be obtained in a relatively short time, with minimal analyst interaction. It is reasonable to expect 125–150 results per hour.

For the laboratory that needs to determine high concentrations, FLAAS is very versatile. By choosing a less sensitive wavelength or by merely turning the angle of the burner head (thus, reducing the path length), the analyst can easily calibrate to and determine high concentrations.

On the lower concentration end, there are many elements that can produce acceptable instrument detection limit (IDL) values (10 µg/L or less): Ag, Be, Ca, Cr, Cu, K, Mg, and Na to name a few. Actual reporting limits will vary with sample matrix, as well as the instrumental conditions.

As with all the atomic spectroscopy techniques, there are interferences associated with FLAAS. These are well understood and documented so that the technique has become "cookbook" in nature. It is up to the analyst to pay attention to the information provided and follow the appropriate guidelines.

FLAAS INTERFERENCES AND THEIR CONTROL

In AA, interferences are divided into two main categories: non-spectral and spectral. Non-spectral interferences have to do with how the atom population is achieved and include transport (or physical), chemical, and ionization. Spectral interferences are those which contribute to the element signal producing an incorrect result and are primarily background absorption and emission interferences.

Transport (physical) interferences occur when the sample matrix is sufficiently different from that of the standards used for calibration causing different amount of sample per time being transported to the flame. This can result in different absorbance readings for the same concentration of an analyte in sample versus the calibration solutions as shown in Figure 26.3. This is the same affect we see with the plasma techniques but are typically less severe in FLAAS.

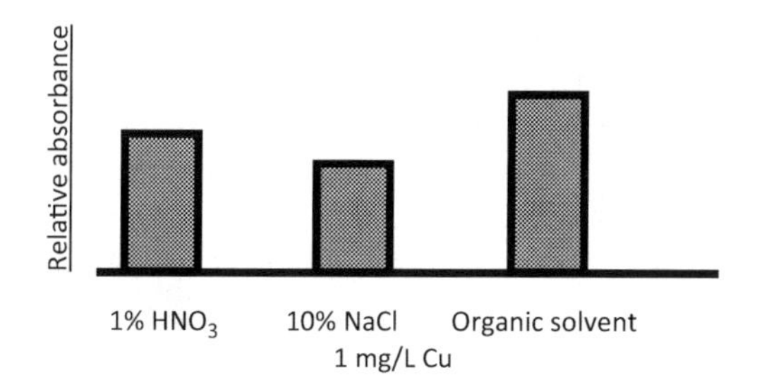

FIGURE 26.3 Inaccuracy produced by transport interferences in FLAAS.

There are two ways of alleviating transport issues when using FLAAS. If possible, match the matrix of the calibration solutions to those of the samples. For example, if the sample is an organic solvent (such as xylenes), then prepare the standards from an organometallic stock in the same solvent.

Matrix matching is not always possible or reasonable. The other option is to do a method of standard additions, where the calibration standards are prepared in the presence of the sample matrix. The preparation for this is more time-consuming and does require that all the absorbances are in the linear range. Transport interferences are sample specific and will affect all the elements determined for that matrix.

Chemical interferences occur when the heat of the flame is not sufficient to break the analyte molecular bonds to create the free atom population. The most commonly used flame, air–acetylene reaches temperatures in the range of 2,100°C–2,400°C. Unfortunately, not all chemical bonds are easily broken at these temperatures and low results are obtained.

One example is the determination of Ca, which forms strong molecular bonds with a number of sample constituents. One of these strong bonds is $Ca_3(PO_4)_2$. These bonds are not easily broken by the air–acetylene flame so the atom population is not formed. However, a bond that is easily broken in the air–acetylene flame is $CaCl_2$. So for this interference, the $Ca_3(PO_4)_2$ can be converted to $CaCl_2$ by adding an excess of a *releasing reagent* ($LaCl_3$). This is added to all solutions (calibration blanks and standards, samples, quality control (QC) solutions, etc.) to be analyzed to convert the form of the Ca to the form where the molecular bonds are broken to create the free atoms:

$$Ca_3(PO_4)_2 + 2LaCl_3 \rightarrow 3CaCl_2 + 2LaPO_4$$

There are many commonly determined elements which form these strong molecular bonds: Al, Ba, Ca, Ge, Mg, Mo, Si, Sn, V, and Zr. In most cases, addition of the releasing reagent does not change the chemistry of the element sufficiently. So another option is to use a hotter flame, nitrous oxide–acetylene. This flame reaches higher temperatures in the range of 2,600°C–2,800°C and is sufficient to break the stronger bonds.

The use of the hotter flame can lead to another interference once the molecular bonds are broken. With the heat of the nitrous oxide–acetylene flame, there can be sufficient energy to cause some of the free atoms to go through ionization, reducing the atom population, and leading to incorrect results. The solution for this type of interference is to add an *ionization buffer*, an excess of an element that is easily ionized, to all solutions to be analyzed. The elements used for the buffer generally will also ionize in the air–acetylene flame and include Cs, K, La, Li, and Na.

Chemical and ionization interferences are element specific and well documented for FLAAS. Even the new user should have no trouble compensating for them by using the information provided by the instrument manufacturer.

Spectral interferences: Due to the specificity of the bandwidth setting for each element at its wavelengths, spectral interferences in terms of spectral overlaps are generally not a problem in AA. The most common type of spectral interference in AA is *background absorption*, where something other than the analyte prevents the light from the lamp getting to the detector. Most often in FLAAS this is undissociated molecules and tiny particulates that can occur in the flame when analyzing samples with high-dissolved solids. This is not a major problem in FLAAS unless (1) the sample matrix has more than 1% total dissolved solids (TDS), (2) the element wavelength is below 250 nm, and (3) the concentration of the element determined is less than 1 mg/L. An example of the necessity of using background correction is the determination of Zn, at 214.9 nm, in seawater samples. Naturally occurring Zn in seawater is typically <50 µg/L. The absorbance due to background may be as much as 30 µg/L. Without background correction, the matrix would contribute to a large portion of the reading.

When the background readings do interfere with obtaining accurate analyte readings, most instruments have background correction systems. The most common of these for FLAAS is continuum source background correction. This correction technique utilizes a secondary lamp, usually deuterium (D2), installed in the instrument. The D2 lamp emits continuously over a broad wavelength range (180–350 nm) rather than at discreet wavelengths as do the element lamps. The instrument then monitors intensities from both the D2 lamp and the element lamp. Absorption of the D2 lamp is essentially background only. Absorption of the element lamp is background and analyte. The instrument electronics then performs a correction as follows:

$$\underset{(\text{background} + \text{analyte})}{\text{Element lamp absorbance}} - \underset{(\text{background only})}{\text{D2 lamp absorbance}} = \text{Corrected Analyte Absorbance}$$

Background readings, if present, are generally small in FLAAS and continuum source typically works well. Two other types of correction, Zeeman and Smith–Hieftje, can also be used but are more commonly used for graphite furnace analysis.

Another potential spectral interference comes from the brightness of the flame when determining elements at visible wavelengths, particularly when the sample matrix causes an increase in the brightness as well. An example where this can be a problem is the determination of barium in a high sodium sample. In order to break molecular bonds, barium requires a nitrous oxide–acetylene flame, which is already bright. The most sensitive wavelength for barium for FLAAS is at 553 nm, the visible region. And high Na adds additional brightness to the flame. A combination of these three factors can "blind" the detector, making the analysis difficult. Diluting the sample, narrowing the bandwidth, or choosing a different wavelength if possible can help with this particular problem.

DISADVANTAGES OF FLAAS

One potential disadvantage of FLAAS is the volume of sample required for the determination of the elements of interest. While much depends on a number of different factors (shown below), uptake volumes are usually 2–7 mL of sample per minute.

1. Number of replicate readings desired for a mean value
2. Read time settings
3. Type of nebulizer and spray chamber
4. Whether an autosampler is used.

So for the lab that is sample limited, this could pose a problem when determining several elements.

Another disadvantage is the flame itself. While instruments today have many safety interlocks, it is still an open flame using flammable gases and should never be left completely unattended when the flame is on. There should always be an analyst in the area that can "keep an eye" on the instrument during analysis, even when using an autosampler.

The greatest limitation of FLAAS for many of the elements of interest is poor sensitivity. As reporting levels are being forced to go lower, FLAAS cannot achieve the levels required for many elements. A number of them determined that require reporting levels lower than it is possible to achieve with FLAAS include but are not limited to: As, Hg, Pb, Sb, and Se. To achieve lower levels of quantitation, the flame portion of the instrument can be replaced with alternate ways of creating and measuring the atomic signal. Two of these ways are to use graphite furnace or vapor generation.

GRAPHITE FURNACE AAS

When lower levels of quantitation are required, one option is to replace the flame with an electrically heated furnace, an alternate technique developed in the 1960s. Because of the nature of the furnace materials, this is commonly known as graphite furnace AAS (GFAAS) sometimes known as electrothermal atomization (ETA). Some instruments sold today are manufactured for furnace analysis only and some for flame analysis only. Other instruments have capabilities of both, though only one at a time can be used.

When using GFAAS, sample is pipetted into a graphite tube through an opening, the dosing hole of the tube, and then heated through a series of programmable steps to: dry the sample solvent, eliminate some of the sample matrix, without losing any of the analyte (the pyrolysis step), break the molecular bonds so that the AA signal can be read (the atomization step), and clean out any remaining sample residue. A typical furnace cycle is shown in Figure 26.4. There are many designs of furnace systems and the associated graphite components. The steps can vary not only with design but with the element determined and the sample matrix analyzed.

It is important that the sample be dried completely with no boiling or spattering. While in some cases there may be only one dry step, a two-step drying program can insure complete drying of the sample. The times and temperatures used are dependent on the sample matrix and the furnace design. The furnace is ramped up to these temperatures and then held to insure dryness. Argon flows into the ends of the graphite tube to expedite the elimination of the moisture out the dosing hole of the tube.

The pyrolysis step has also been called ash, char, and pretreatment. During pyrolysis, the goal is to eliminate as much of the sample matrix as possible before the atomization step, without losing any of the element. Again there is a ramping up to the desired temperature. Times and temperatures are dependent on the sample matrix and the element determined. Argon continues to flow into the tube ends to eliminate the excess matrix by pushing it out the dosing hole.

The atomization temperature must be sufficient to break the molecular bonds to form the free atom population. To minimize interference effects, it is important that the element volatilizes into a thermally stable environment. So in this step, to heat the tube as quickly as possible, there is no ramping time to the set temperature. Typically, the gas flow is stopped in the furnace during atomization to give the longest residence time, and the highest signal, possible.

The last step, cleanout, is to eliminate any remaining matrix (and in some cases analyte) from the graphite tube before injection of the next sample. Again, argon flows in the tube during this step.

In addition to lamps, consumables now include the graphite components, particularly the graphite tube (cuvette) which degrades with repeated cycles and must be changed on a routine basis to insure consistent results. High temperatures, particularly held for extended times, high acid concentrations, and high dissolved solids in the samples are contributing factors for how fast the graphite

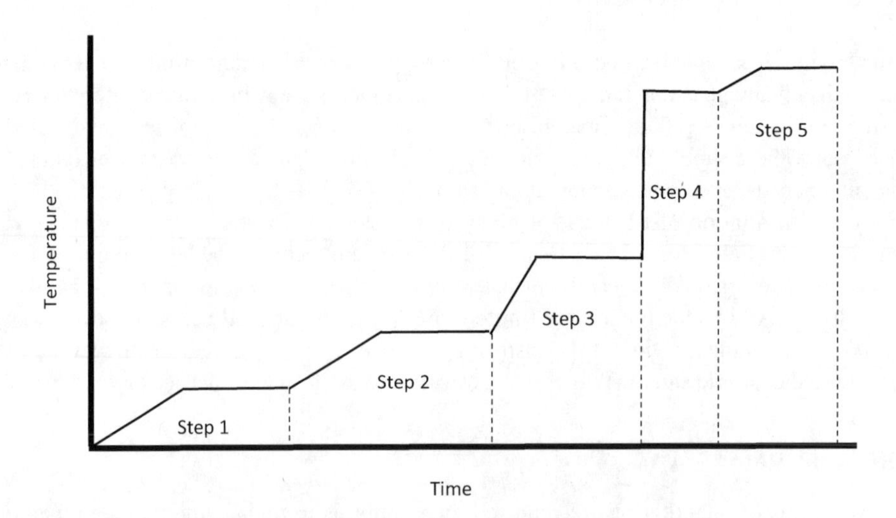

FIGURE 26.4 Typical furnace program cycle. Steps 1 and 2 dry the sample, step 3 is pyrolysis, step 4 is atomization and the AA signal is read, and step 5 is the cleanout step.

tube degrades. Depending on these parameters, the tube may last for only a few hundred to over 1,000 cycles. The contact cylinders housing the graphite tube (and also made of graphite) will also have to be changed periodically but much less frequently than the tube itself.

It is imperative when running a furnace analysis that each step of the analysis from injection to injection be as reproducible as possible. This includes sample deposition. Autosamplers in graphite furnace are not just a convenience but a necessity for good reproducibility of results.

GFAAS INTERFERENCES AND THEIR CONTROL

For very simple matrices, method development for graphite furnace analysis can be relatively simple. But as the matrix becomes more complicated with higher dissolved solids in the samples, method development can be time-consuming. This is because, unlike FLAAS, interference effects in graphite furnace can vary with each sample matrix and the element determined.

Interferences again are separated into non-spectral, having to do with how the atom population is formed, and spectral, where something other than the analyte prevents the light from the lamp getting to the detector. The latter, unlike in FLAAS, can be very severe in furnace analysis.

The non-spectral interferences, which affect how the atom population is achieved, are typically controlled using a set of conditions that allow the accurate analysis of even very complicated sample matrices. The first of these conditions is to use a good quality graphite tube. The tubes are generally coated with pyrolytic graphite to prevent interaction of the element with the graphite surface. Poor quality of graphite leads to contamination issues, poor reproducibility, and short tube life.

It is necessary to delay the volatilization (breaking molecular bonds) of the analyte until the environment in the tube has become thermally stable. One way to accomplish this is to pipette the sample onto a secondary surface, a platform, inside the tube. The platform surface heats more slowly than the tube wall allowing the temperature inside the tube to stabilize before volatilization of the analyte. Another style tube is thicker in the middle than at the ends. The thicker portion will again heat more slowly.

Two modes of signal collection can be used: peak height and peak area. While peak height certainly is indicative of the concentration of atoms in the vapor phase during the atomization step, it is also very dependent on the rate at which the molecular bonds are broken. The element may be bound in different molecular states as compared to the calibration solution, causing the molecular bonds to break more quickly or more slowly than the standards. Therefore, it is better to use peak area, which is rate independent, for furnace determinations.

Many elements are volatile, their molecular bonds broken at relatively low temperatures. Arsenic, selenium, cadmium, and lead are just a few of these. In order to use a pyrolysis temperature high enough to eliminate troublesome matrix components during this step, it becomes necessary to add a "modifier" to stabilize these elements so that no losses of the element occur before the atomization step.[1] Modifiers have been used since the early days of graphite furnace, and most elements will require them, except when analyzing the simplest of solutions. Most modifiers and the amount of modifier used are dependent on the element determined, but some may also depend on the sample matrix analyzed. The most common analyte stabilizing modifiers currently used are Pd and Pd mixed with $Mg(NO_3)_2$ as these are effective for a variety of elements typically determined by GFAAS.[2] Some typical pyrolysis temperatures with and without a modifier are shown in Table 26.1.[3]

There are also modifiers that can change the nature of the sample matrix, generally to make it more volatile so that it can be eliminated at a lower temperature. One example is the use of NH_4NO_3 as a modifier to react with samples with a high salt concentration. NaCl has a melting point of 1,400°C, so it is more thermally stable than many elements and difficult to eliminate without analyte loss. When NH_4NO_3 is added in excess to all solutions analyzed, it will react with the NaCl to produce NH_4Cl and $NaNO_3$, both of which have melting points below 400°C. The matrix now is less stable than the elements determined and can be eliminated without analyte loss. Note that it

TABLE 26.1
Recommended Pyrolysis Temperatures with and without a Modifier for Some Elements

Element	Pyrolysis °C No Modifier Used	Pyrolysis °C $Pd/Mg(NO_3)_2$ Modifier
Arsenic	200	1,200
Cadmium	400	500
Chromium	1,200	1,500
Copper	900	1,200
Lead	450	1,200
Selenium	200	1,100

Note: These temperatures are for standards and the addition of a real sample matrix can change the recommendation. These temperatures also vary with instrument design and are shown for comparisons only.[4,5]

may still be necessary to stabilize the analyte, so two modifiers may be required: one to stabilize the analyte and the other to make the matrix more volatile.

Spectral interferences fall into two categories: emission intensity from the tube wall and background absorption. Emission intensity is minimized by making sure the furnace is aligned properly in the light path. As the analyte signal is increased using graphite furnace, so the background signal also increases. Therefore, background correction is always used with furnace analysis. While continuum source correction can be used as it is with FLAAS, there are limitations to its effectiveness for GFAAS. With continuum source, there is not full wavelength coverage using D2 lamps. Continuum source also does not correct accurately for cases of structured background, and it cannot correct for very high background signals (greater than absorbances of about 1.7).

As samples analyzed by graphite furnace have increased in complexity, instrument manufacturers offer more efficient background correction systems to be able to handle samples that produce complicated background signals. The Smith–Hieftje type of correction is based on the principle that as the lamp is pulsed to very high currents, there is a loss of analyte sensitivity due to self-absorption. Readings are taken during atomization as the lamp is pulsed between normal and high currents and the following correction is made:

$$\underset{\left(\text{Analyte} + \text{Background}\right)}{\text{Normal Current}} - \underset{\left(\text{Background only}\right)}{\text{High Current}} = \text{Corrected Analyte Signal}$$

This type of correction does not require any additional specialized equipment as only the element lamp is involved. Of course, pulsing the lamp to high currents does affect lamp life. Another limitation of this type of correction is that not all elements go through self-absorption as efficiently as others and there can be a significant loss of sensitivity. As measurements are made slightly off the element's exact wavelength, it also does not correct well for cases of structured background within the bandpass.

The third type of background correction, Zeeman, is based on the principle that when atomic spectra are placed in a magnetic field, it becomes split into three or more π and σ components, while molecular structures remain unchanged. To date, this is the most effective form of background correction for AA instruments as it can correct for even structured background within the bandpass.

The atomizer is placed in a strong magnetic field and quickly pulsed on and off. When the magnet is off, light is absorbed by both the element and background. When the magnet is on, there is only background absorption. Correction is made as follows:

$$\underset{(\text{Analyte} + \text{Background})}{\text{Magnet Off}} - \underset{(\text{Background only})}{\text{Magnet On}} = \text{Corrected Analyte Signal}$$

This type of correction requires the addition of a strong magnet and the electronics to control it, adding additional cost to the instrumentation. There is some loss of sensitivity due to elemental splitting patterns. In most cases this is minor, but in a few cases it is significant, as in the case of Cu where the loss is more than 40%.

ADVANTAGES OF GFAAS

Because all of the analyte in the volume pipetted into the tube is atomized and the atoms are retained in the tube for an extended time, sensitivity and reportable levels are very good with graphite furnace. Detection limits are typically 100–1,000 times lower than with FLAAS.

For applications where the sample size is limited, another advantage is the small sample volumes required for analysis. Typical sample volumes range from 1 to 100 μL per injection. When less sensitivity is required, less sample volume can be used. When more sensitivity is required, more volume can be used.

With GFAAS, no flammable gases are used but rather argon. So this is an analysis that can be left completely unattended. To use their instruments to full advantage, many labs with instruments that have both flame and furnace capabilities choose to run their flame analysis during the day and their furnace analysis overnight.

Another advantage of GFAAS is the ability to determine solid samples directly, without dissolution.[6] While a weighed portion of the solid can be inserted into the furnace in a variety of ways, the easiest way is to mix the solid with a liquid portion and pipette directly in the form of a "slurry." The advantage of this method of introducing the solid into the furnace is that calibration is achieved with simple aqueous standards.

DISADVANTAGES OF GFAAS

A disadvantage of GFAAS can be the time required for method development. While manufacturers give a set of conditions to be used for each element, unless the samples analyzed have very low dissolved solids, these are just starting conditions. The times and temperatures used for drying the samples and the pyrolysis step generally have to be optimized for the matrix. The modifier strength, or even which modifier to use, may have to be changed. Less frequent changes for the atomization step are required, but again, this is always dependent on the sample matrix.

Probably the greatest disadvantage of the graphite furnace technique is the time it takes to do an analysis once the method has been developed. In contrast to flame, this is a very slow technique. The cycle time from reading to reading can range from 1 to 5 min, dependent primarily on the sample matrix. Most of this time is spent in the dry and pyrolysis steps. The greater the volume pipetted and the higher the dissolved solids, the longer these steps become. If the analyst has only one or two elements to determine with few samples to be analyzed, perhaps this is not an issue. But when eight to ten elements (or more) have to be determined or there are many samples to be analyzed, this is very time-consuming. Many laboratories which have this situation may have several furnace systems running at the same time.

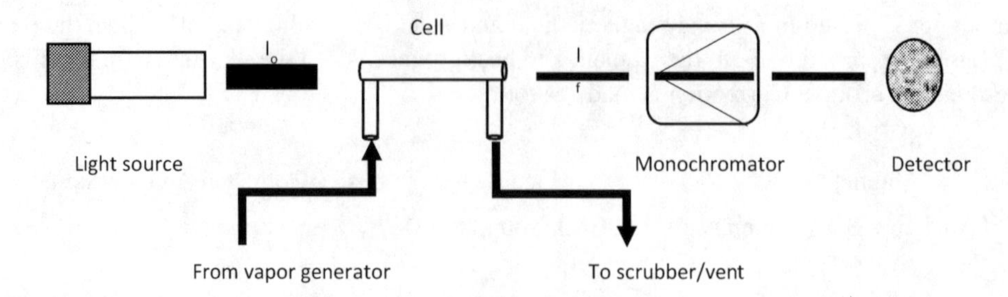

FIGURE 26.5 Basic configuration to use vapor generation AA.

VAPOR GENERATION AAS

In addition to using graphite furnace to improve detection limits, for some elements there is also vapor generation. This falls into two categories, cold vapor and hydride generation. In both cases, the element of interest is reacted with a strong reducing reagent and then stripped into a vapor phase (usually air or argon). The vapor is swept through a cell (glass or quartz) in the light path and the AA measurement is made using peak height or peak area mode as shown in Figure 26.5.

Cold vapor is specific for one element: mercury. It is the only element that can exist as free ground-state atoms at room temperature. The reducing reagent, either stannous chloride or sodium borohydride, is reacted with the sample. The gas is then bubbled through this solution to release the free mercury atoms into the gas phase, which is then swept into the cell and the AA signal is read.

ADVANTAGES OF COLD VAPOR AAS

As FLAAS and even graphite furnace are not very sensitive for mercury, this technique allows the analyst to achieve reportable levels of approximately 0.02 µg/L. This can be improved further by sweeping the mercury-rich vapor over gold or gold–platinum wire or gauze. The Hg bonds to the gold, thereby trapping and concentrating the mercury. The gold is then heated to release a preconcentrated amount of mercury. While early systems were very manual, this type of analysis can now be automated.

DISADVANTAGES OF COLD VAPOR AAS

The obvious limitation of this is that it is only applicable to the determination of mercury since it is the only element to release free atoms at room temperature. It also is necessary to have the vapor generation system, an additional expense.

HYDRIDE GENERATION AAS

This technique is similar to cold vapor in that the element of interest is reacted with a reductant, typically sodium borohydride, then stripped into the gas phase. The difference is that the element is not released as free atoms but rather bound in a hydride bond. Therefore, the quartz cell through which the vapor will travel must be heated to break these bonds. Heating of the cell can be achieved with the flame portion of the instrument or the cell can be inserted into an electrically heated device.

ADVANTAGES OF HYDRIDE GENERATION AAS

The elements that can be determined using this technique are: As, Bi, Ge, Pb, Sb, Se, Sn, and Te. These elements can also be determined using graphite furnace with similar detection limits. However, samples with high-dissolved solids are often challenging to eliminate the interferences they produce without losing some of the analyte during the furnace pyrolysis step. Vapor generation

TABLE 26.2
Instrument Detection Limit Comparisons (µg/L)

Element	FLAAS IDL µg/L	GFAAS IDL µg/L	Vapor Generation IDL µg/L
Arsenic	150	0.05	0.03
Bismuth	50	0.05	0.03
Mercury	300	0.6	0.009
Antimony	45	0.05	0.15
Selenium	100	0.05	0.03
Tellurium	30	0.1	0.03
Lead[a]	15	0.05	–
Cadmium[a]	0.8	0.002	–

Data from PerkinElmer "Atomic Spectroscopy: A Guide to Selecting the Appropriate Technique and System,"[7]

[a] Pb and Cd are not typically hydride forming elements by AA, but were included, because together with As and Hg, they are the panel of heavy metals required by state regulators for cannabis and hemp.

separates the matrix from the analyte to produce a cleaner analysis. Table 26.2 lists a comparison of potential flame, furnace, and vapor generation IDL values. These values are shown for comparison only and are dependent on instrument and instrumental parameters.

DISADVANTAGES OF HYDRIDE GENERATION AAS

Again, the most obvious limitation is the limited number of elements that can be determined using hydride generation. In addition, results for each element are very dependent on the valence state of the element. Therefore, proper preparation of the samples is critical. Unfortunately, the same preparation does not work for all the hydride elements so different preps may be needed for the different elements determined.

HYPHENATED TECHNIQUES

AA can also be used by coupling more than one technique. To achieve lower limits of quantitation using FLAAS, samples can be first run through a metal chelating column to preconcentrate elements of interest as the bulk of the matrix passes through. The preconcentrated elements can then be stripped from the column with a mobile phase and introduced for flame analysis. The degree of improved limits of quantitation is dependent on the total volume passed through the column and the capacity of the column resin.

Graphite furnace can be coupled with vapor generation to improve reportable concentrations. This can be particularly advantageous when the sample matrix is very high in dissolved solids and difficult to eliminate in the pyrolysis step, resulting in overwhelming background signals during the atomization step. In this combined technique, the vapor that has stripped the analyte from the matrix is collected, "trapped," in the graphite tube, and can be preconcentrated in this fashion. Once the desired volume of vapor has been collected, the furnace is raised to an appropriate atomization temperature to drive it off the tube surface and breaking molecular bonds and the signal is read.

ATOMIC FLUORESCENCE

Another technique in atomic spectroscopy is atomic fluorescence. It shares properties with AA in that it has a light source, atomizer, monochromator, and detector. The element-specific energy from the light source is absorbed by the ground-state atoms, causing them to be excited. Rather than

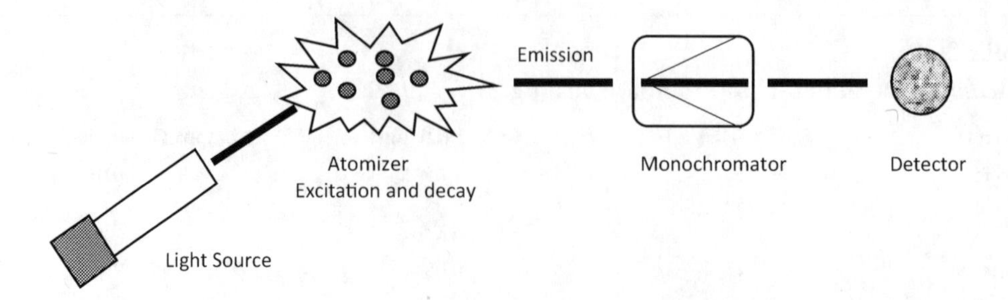

FIGURE 26.6 Basic components of atomic fluorescence.

determining the amount of light being absorbed by the element in the atomizer, as the excited atoms return to ground state, they emit photons of light and this emission, or fluorescence, resulting from the decay is measured. As shown in Figure 26.6, the element-specific light source is at an angle so as not to interfere with the fluorescence process. To carry out multielement analysis using AFS, many lamps are positioned around the atomization/excitation source.

TABLE 26.3
IDL Values Using Vapor Generation with Atomic Fluorescence, µg/L[8,9]

Elements	IDL (µg/L)
As, Bi, Pb, Sb, Se, Sn, Te	0.01
Cd, Hg	0.001
Ge	0.05
Zn	1

ADVANTAGES AND DISADVANTAGES OF AFS

Background signals in AFS are very low, resulting in improved sensitivity as compared to absorption. However, the fluorescence measurement can be quenched by many molecular forms, so separation of the matrix from the analyte may be desirable. Commercially available vapor generation AFS systems have been used successfully for the analysis of a variety of sample types, including water, soil, plants, pharmaceuticals, food, biologicals, petrochemicals, and cannabis.[8,9] However, it's limited to the determination of those elements which work well using vapor generation such as As, Bi, Cd, Ge, Hg, Pb, Sb, Se, Sn, Te, and Zn. Table 26.3 reflects typical average IDL values that have been obtained using commercially available AFS systems. Actual performance will be dependent on lab conditions, sample type, reagents, and instrument setup.

FINAL THOUGHTS

AA and atomic fluorescence are both viable options for measuring heavy metals in cannabis and hemp, particularly for labs that don't have a high-sample workload. It's unlikely that flame AA has the sensitivity for most state-based limits, but graphite furnace and hydride generation AA would offer the required detection capability and could be very attractive options. Furthermore, a dedicated AFS combined with a vapor generation system could also be a viable option for labs with a limited budget. However, it should be emphasized that if the suite of heavy metals in cannabis and hemp increases due to future stricter federal regulations, these techniques would be limited in their elemental range.

FURTHER READING

1. "Atomic absorption analysis with the graphite furnace using matrix modification", R. Ediger, *At. Absorp. Newsl.*, Vol. 14, 127–145 (1975).
2. "Investigations of a reduced palladium chemical modifier for graphite furnace atomic absorption spectrometry", L.M. Voth-Beach, D.E. Shrader, *JAAS*, Vol. 2, 45–50 (1987).
3. "Palladium and magnesium nitrates, a more universal modifier for graphite furnace atomic absorption spectrometry", G. Schlemmer, B. Welz, *Spectrochim. Acta B*, Vol. 41, Issue 11, 1157–1165 (1986).
4. "Palladium nitrate-magnesium nitrate modifier for graphite furnace atomic absorption spectrometry. Part 1. Determination of arsenic, antimony, selenium and thallium in airborne particulate matter", B. Welz, G. Schlemmer, J.R. Mudakavi, *JAAS*, Vol. 3, 93–97 (1988).
5. "Palladium nitrate-magnesium nitrate modifier for graphite furnace atomic absorption spectrometry. Part 2. Determination of arsenic, cadmium, copper, manganese, lead, antimony, selenium and thallium in water", B. Welz, G. Schlemmer, J.R. Mudakavi, *JAAS*, Vol. 3, 695–701 (1988).
6. "Solids Analysis By GFAAS", N.J. Miller-Ihli, *Anal. Chem.*, Vol. 64, Issue 20, 964A–968A (1992).
7. Guide to Techniques and Applications, PerkinElmer Inc., https://www.perkinelmer.com/PDFs/Downloads/BRO_WorldLeaderAAICPMSICPMS.pdf
8. Lumina 3500 Atomic Fluorescence Spectrometer, Aurora Illuminating Systems Elemental Analysis Instruments, 2020, https://www.aurorabiomed.com/lumina-3500/
9. PF7 Heavy Metals Analyzer, Persee Analytics Inc., http://perseena.com/index/prolist/pid/33/c_id/43.html

27 Other Traditional and Emerging Atomic Spectroscopy Techniques

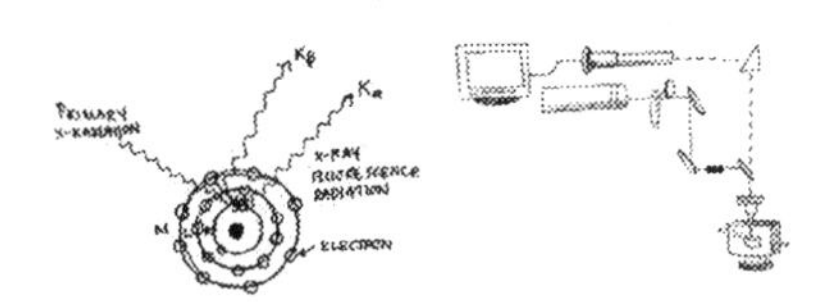

Even though atomic absorption (AA), inductively couple plasma optical emission (ICP-OES), and inductively couple plasma mass spectrometry (ICP-MS) are considered the most common atomic spectroscopic (AS) techniques, there are others which are worth discussing. In this chapter, we will introduce the reader to four solid sampling techniques, X-ray fluorescence (XRF), X-ray diffraction (XRD), laser-induced breakdown spectrometry (LIBS), laser ablation, laser ionization time-of-flight mass spectrometry (LALI-TOFMS), together with a novel emission technique called microwave-induced atomic emission spectrometry (MIP-AES) for solution work. They all have their own strengths and weaknesses but could be used for testing cannabis products, depending on the state requirements and the type of the cannabinoid being tested. Additionally, three of the solid sampling techniques offer the possibility of testing the products without having to digest the sample or alternatively could be used for the on-site characterization of heavy metals in the plant material or the soil where the cannabis or hemp is being grown.

X-RAY FLUORESCENCE

Over the past 30 years, XRF has become the elemental technique of choice for the determination of low parts-per-million to percentage concentration levels in solid samples including metals, ores, rocks, soils, glasses, powders, plastics, ceramics, foodstuffs, pharmaceuticals, and plant materials.[1]

The principles of XRF spectrometry are well documented in the literature.[2] A sample is irradiated with a beam of high-energy X-rays. As the excited electrons in the atom fall back to a ground state, they emit X-rays that are characteristic of those elements present in the sample, as shown in Figure 27.1.

The individual X-ray wavelengths are then separated and measured via a system of crystals, optics, and detectors. Elemental concentrations in unknown samples are quantified by comparing the X-ray intensities against known calibration standards. The major benefit of XRF over other solid sampling techniques such as arc or spark emission is that it can analyze both conducting and non-conducting materials as well as inorganic and organic matrices with minimal sample preparation.[4,5]

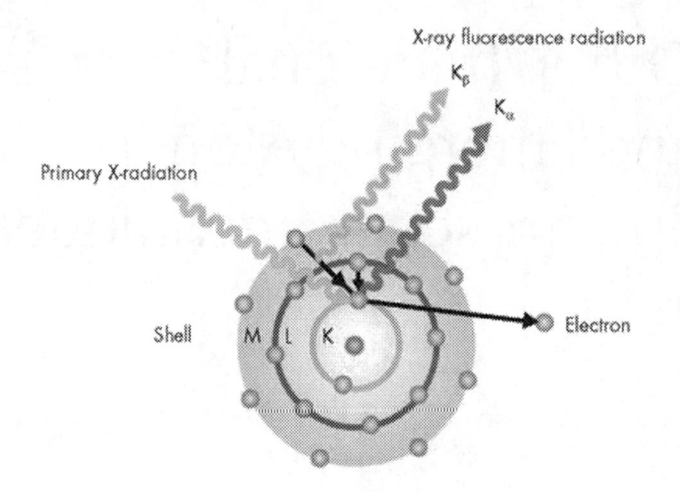

FIGURE 27.1 Principles of XRF.[3]

XRF INSTRUMENTAL CONFIGURATION

The XRF technique is available in two separate configurations—energy-dispersive (EDXRF) and wavelength-dispersive (WDXRF). In the EDXRF approach, the intensity of the photon energy of the individual X-rays is detected and measured simultaneously using multichannel data electronics (Figure 27.2).

Whereas in WDXRF spectrometry, the polychromatic beam emerging from a sample surface is dispersed into its monochromatic components or wavelengths with an analyzing crystal. A specific wavelength is then calculated from knowledge of the dispersion characteristics of the X-ray crystal. This design is shown in Figure 27.3.

It is generally recognized that WDXRF offers several advantages over EDXRF, including better spectral resolution, superior detection limits (particularly for the low mass elements), and the ability to determine major concentrations of elements with very high precision and accuracy. As a result, the higher performance and better capabilities of WDXRF means they are typically used for high-end research-type applications.[7]

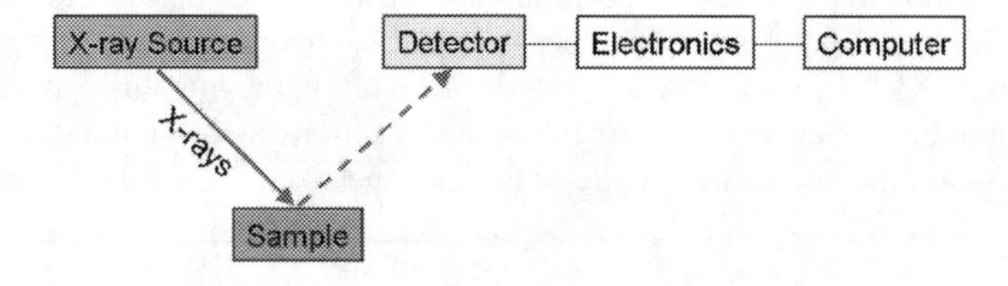

FIGURE 27.2 Simple schematic of EDXRF system.[3]

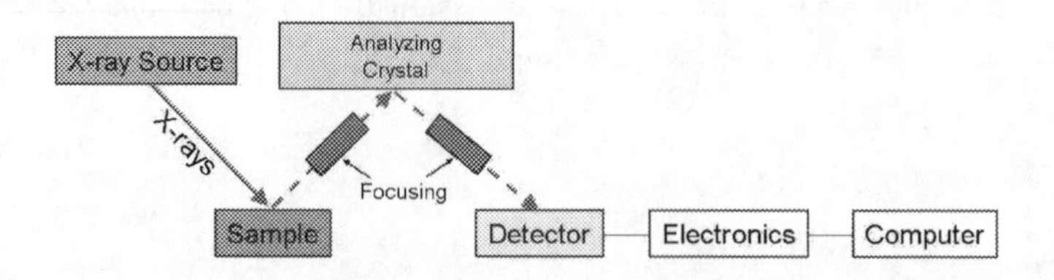

FIGURE 27.3 Simple schematic of WDXRF.[6]

On the other hand, EDXRF systems are smaller than WDX, they do not require any external utilities, such as chillers or gases, they typically use a smaller power supply, and they have no moving parts. As a result of this compact footprint, EDXRF systems can be placed on a small bench, or using portable hand-held devices, can even be taken into the field to carry out remote site evaluations.[8]

QUANTITATION BY XRF

XRF techniques do not require chemical pretreatment such as dissolution and being a non-destructive analysis can identify and determine a wide elemental range from low-ppm up to percentage levels directly in solid, powdered, or liquid samples. However, because the energy of the fluorescent X-rays corresponds directly to the atomic number of the element, the capability drops off at the low mass end, particularly for hand-held EDXRF systems.

Quantitation is conducted using external calibration standards containing varying concentrations of elements. XRF calibrations have the advantage of being suitable stable for long periods of time before requiring recalibration, in contrast to solution-based AS techniques, which require recalibration procedures to be carried out at regular intervals. However, it should be emphasized that to ensure the highest accuracy in XRF, calibration standards should be of a similar matrix to the samples being measured.

EDXRF has been applied to the bulk analysis of a wide variety of organic-based samples such as pharmaceuticals botanicals, foods, plant materials, and soil samples without the need for sample dissolution or chemical pretreatment required by ICP-OES or ICP-MS, which can be very labor intensive, prone to human error and contamination of the sample. Because of these sample digestion issues, EDXRF technique is often used as a preliminary screening approach to determine if additional chemical analyses are required by one of the more accurate solution techniques ... this is quite common in the pharmaceutical industry.[9] In fact this could be a great opportunity for portable units to be used as an initial semiquantitative tool by the cannabis/hemp growing community before the samples are sent to the independent testing lab for a more accurate assessment of the heavy metal content by ICP-MS.

XRF DETECTION LIMITS

Detection capability of ED XRF is typically in the low-ppm (µg/g) range in the solid material. Furthermore, it can achieve lower detection limits if the measurement time is extended. Typical measurement times are in the range of 10–30 s, but a fivefold to tenfold increase in integration time can show an improvement. However, it should be emphasized that this represents detection capability directly in the solid material. For plasma spectrochemistry to achieve similar performance, the solution detection limit must be in the order of 10 ppb (µg/L), if the EDXRF detection limit is 1 ppm, assuming a sample weight of 1 g is digested and made up to 100 mL (100-fold dilution factor). Typical detection limits of some heavy metals in selected matrices are shown in Table 27.1.

TABLE 27.1
Typical Heavy Metal Detection Limits in ppm (µg/g) in Selected Matrices

	Pb	Cd	As	Hg	Cr	Ni
Plant materials[10]	1	N/M	N/M	N/M	2	1
Candy[11]	6	N/M	N/M	N/M	N/M	N/M
Soil[12]	15	8	7	13	30	40

Note: Measurement time 60 s, N/M, not measured.

SAMPLE PREPARATION FOR XRF

Solid, liquid, and powder samples can be analyzed by XRF with the minimum of sample preparation. The only preparation required for XRF is reducing the sample to a size that fits in the sample cell or sample chamber. So basically, the larger the sample volume, the smaller the sampling error. Almost all solid samples can be analyzed directly by simply placing them in the sample chamber as is. Viscous liquid samples like oils are poured in a sample cell, with a supporting film at the bottom. Typical films are made from polypropylene of a few micrometers thickness.[13] The majority of pharmaceutical or botanical-type materials are organic matrices and have relatively low X-ray absorption, allowing for the relatively straightforward measurement of the elemental impurities. Powder samples are placed directly into the sample cell using the tapping method to remove voids in the sample. Coarse powders must be ground to a fine particle size, while non-homogenous samples should be ground by means of mortar and pestle or a grinding device, such as a ball mill.[9] Pelletizing using a laboratory press and a binding agent is often a good way to keep powdered samples together.

X-RAY DIFFRACTION

A comparative technique to XRF is XRD, which is an analytical technique that looks at the X-ray scattering from crystalline or polymorphic materials.[14] Each material produces a unique X-ray "fingerprint" of X-ray intensity versus scattering angle that is characteristic of its crystalline atomic structure. Qualitative analysis is possible by comparing the XRD pattern of an unknown material with a library of known patterns. Although its principles are different, XRD can be considered complementary to XRF. For example, XRF can tell you that a material is composed of iron and sulfur, while XRD can tell you that both iron sulfide (FeS) and elemental iron (Fe) are present. Furthermore, because XRD works with any crystalline solid, there is almost no limit to the types of materials that can be studied. Some of the materials that are typically characterized by XRD include chemicals, dusts, rocks, minerals, metals, cements, pigments, and forensic samples.

One of the major applications of XRD is to identify and quantify specific compounds or phases in materials that are organic in nature, such as foodstuffs or pharmaceutical compounds.[15] These kinds of applications have evolved over the last few years mainly because of a new generation of solid-state detectors, higher quality optics, and the availability of sampling attachments. These technological improvements have led to both higher sensitivity, which is needed for the detection of minor and trace levels of crystalline compounds in pharmaceutical compounds, and higher resolution for improved spectral specificity in the presence of interfering species.

LASER-INDUCED BREAKDOWN SPECTROSCOPY

LIBS has been commercially available since the early 2000s, but it's only been in the past 5 years where its application potential has been fully explored. Initially developed by a group of government scientists in the early 1980s,[16] it was first applied to the analysis of soils and hazardous waste sites, because of its remote sampling capabilities.[17] However, since it has been in the hands of the routine analytical community, it is now being used to solve real-world application problems.[18] Although the analytical capability, performance, and elemental range of LIBS compares quite favorably with other AS techniques, it should be considered complimentary (and a competitor) to both ICP-OES and XRF. The unique benefits of LIBS are in its ability to sample a very diverse range of matrices, both in a laboratory environment and at remote locations out in the field. Let's take a closer look at the fundamental principles and capabilities of LIBS in order to get a better understanding of where it fits into the AS toolbox.

LIBS FUNDAMENTAL PRINCIPLES

LIBS is an atomic emission spectroscopic technique that utilizes a small plasma generated by a focused pulsed laser beam (typically from an Nd:YAG laser) as the emission source. The plasma formed is about 10× hotter than inductively coupled plasma and as a result will vaporize just about any material it comes in contact with. The energy from the plasma excites the vaporized sample which results in a characteristic emission spectrum of the elements present in the sample, which is then optically dispersed and detected by optical components traditionally used in an emission spectrometer.[19] A typical optical layout of an LIBS system is shown in Figure 27.4, while Figure 27.5 exemplifies a generated emission spectrum.

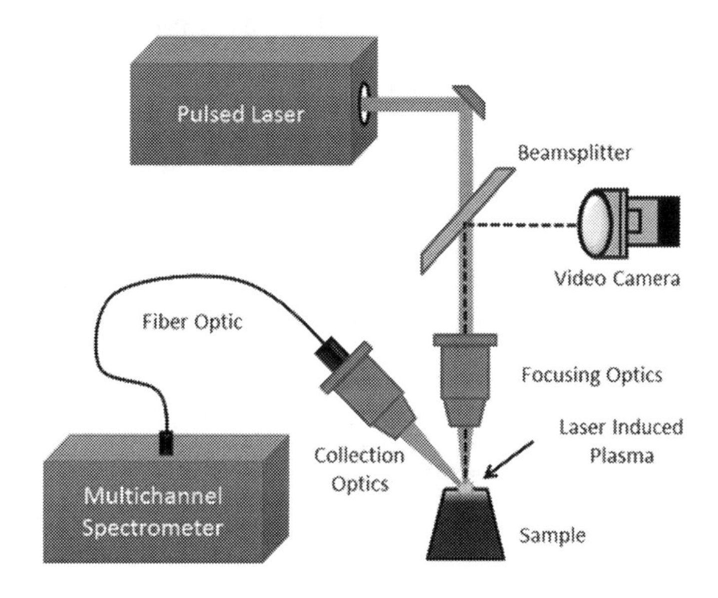

FIGURE 27.4 Optical configuration of an LIBS system. (Courtesy of TSI Inc.)

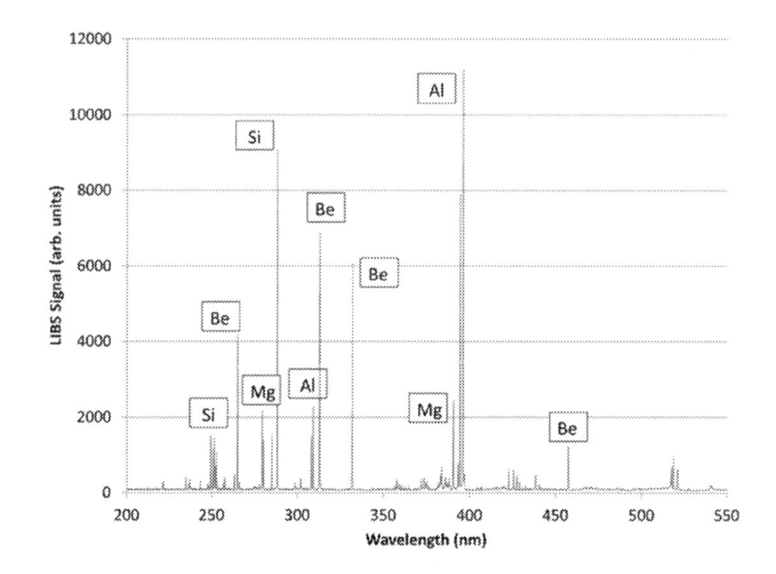

FIGURE 27.5 A typical LIBS emission spectrum. (Courtesy of TSI Inc.)

LIBS CAPABILITIES

This capability has given LIBS some unique characteristics over other AS techniques including:

- Any type of solid, liquid, or gaseous sample can be vaporized and excited and is not limited to liquids like ICP-OES or conducting solids like arc/spark emission
- Unlike XRF, the majority of the periodic table is available, even the low mass elements like B and Li
- The laser-induced plasma is readily generated in the open atmosphere negating the need for large argon flows or expensive consumables used in ICP instrumentation
- The laser beam can be directed almost anywhere by the appropriate optical components or through a fiber optic making it suitable for remote sampling applications such as hazardous waste sites or studies on other planets
- The optical design makes it ideally suited to be put into a briefcase that can be carried around for field applications
- Precise control of spot size and laser penetration allows surface mapping and depth profiling studies.

LIBS APPLICATION AREAS

Some of the many applications that are being carried out today which have contributed to the growing popularity of LIBS include:

- The determination of major and minor elements in metallurgical samples
- Identification and classification of different materials using multivariate processing techniques
- Microanalysis of gemstones and precious minerals
- Trace element analysis of glass and quartz materials
- Minimally destructive analysis of forensic samples
- Uniformity of pharmaceutical ingredients
- Analysis of tablet coatings
- Analysis of botanicals and plant materials
- Direct analysis of coals
- Measurement of the metallic content of liquid baths or flowing process streams
- Remote field sampling of environmental, geological materials, and soils.

LIBS DETECTION CAPABILITY

Detection limits for LIBS are very similar to EDXF, with LIBS having an advantage of being able to measure lower levels of the lighter elements (below Mg in the periodic table). Additionally, similar to XRF, LIBS is predominantly a solid sampling technique. So for a solution technique like ICP-MS to achieve similar performance (assuming a sample weight of 1 g is digested and made up to 100 mL), the detection limit must be in the order of 10 ppb (μg/L), if the LIBS detection limit is 1 ppm. Table 27.2 exemplifies some typical LIBS heavy metal limits of detection in ppm (μg/g) in selected matrices.

LIBS ON MARS

There is no question that the most high-profile LIBS system is the ChemCam instrument onboard NASA's Curiosity Rover, which landed on the surface of Mars in the fall of 2012. Once the rock or mineral has been prepared for analysis by Curiosity's robotic arm, the ChemCam system can

TABLE 27.2
LIBS Heavy Metal Detection Limits in Selected Matrices

	Pb	Cd	As	Cr	Cu
Fe-mineral[20]	N/M	N/M	N/M	4	14
Slag[20]	N/M	10	N/M	7	N/M
Soil[21,22]	43	N/M	85	N/M	N/M
Fertilizer[23,24]	15	1	N/M	2	N/M
Rice[25]	44	3	N/M	N/M	N/M

N/M, not measured.

analyze the sample remotely from a distance of up to 20 feet away, by firing the laser pulses and directing the resulting emission from the generated plasma back to the onboard spectrometer via fiber optics for detection and quantitation.[26] This work will be extremely relevant to better understand the geological composition of Mars and ultimately how our solar system came into existence.[27]

MICROWAVE-INDUCED PLASMA OPTICAL EMISSION SPECTROSCOPY

ICP-OES has become the dominant solution technique for carrying out the multielement analysis of samples that require high-ppb and low-ppm analyte levels. It was first developed for commercial use in the late 1970s as a response to a demand from the application community that flame atomic absorption (FAA) was limited in its ability to determine a large suite of elements in a timely manner. Initially available as a simultaneous spectrometer and then in the more flexible sequential configuration, ICP-OES rapidly became the technique of choice for laboratories where high-sample throughput, multielement analysis was the overriding requirement. There was still a place for FAA in laboratories where the sample workload was significantly lower, but there still appeared to be applications where either the analyte or sample throughput requirements were too high for FAA or they were not high enough to justify the cost of an ICP-OES. Instrument vendors attempted to fill those gaps with either high-cost, fully-automated AAs or low-cost ICP-OES systems, but each approach tended to be a compromise.

However, this gap in the application toolbox has been filled with the development and commercialization of a microwave-induced plasma optical emission spectrometer (MIP-OES). Microwave-generated plasmas have been used as gas chromatography detectors for decades and were explored as potential excitation sources for emission studies back in the 1970s, when the ICP was being investigated. However, it was found they had limited applicability, because they were not considered robust enough for introducing liquids. As a result of this limitation in MIP technology, the ICP became the dominant excitation source. To better understand this early limitation, let's take a closer look at how an MIP is generated.

A microwave-induced plasma basically consists of a quartz tube surrounded by a microwave waveguide or cavity. Microwaves produced from a magnetron fill the cavity and cause the electrons in the plasma support gas to oscillate. The oscillating electrons collide with other atoms in the flowing gas to create and maintain a high-temperature plasma. As in the inductively coupled plasmas, a high-voltage spark is needed to create the initial electrons to create the plasma, which achieves temperature of approximately 5,000 K.

The limiting factor to their use was that with the low power and high frequency of the MIP, it was very difficult to maintain the stability of the plasma when aspirating liquid samples containing high levels of dissolved solids. Various attempts had been made over the years to couple desolvation techniques to the MIP but only managed to achieve limited success.[28] However, the latest MIP technology appears to have overcome many of the limitations of the earlier designs. Let's take a closer look its fundamental principles.

BASIC PRINCIPLES OF THE MP-AES TECHNOLOGY

There is only one vendor of this technology (Agilent Technologies: 4210 MP). At the heart of this technology is a 2.5 GHz magnetron coupled into nitrogen plasma, which is created using compressed air and a nitrogen generator. By using the magnetic field rather than the electric field for excitation, extremely robust plasma is formed, which is capable of handling much higher dissolved solids than previous MIP designs. The microwave waveguide concentrates both the axial magnetic and radial electrical fields around the torch, which creates a conventional-looking plasma, allowing for a traditional inert concentric nebulizer and double-pass spray chamber to be connected to the torch.

The optical system uses an axial viewing configuration, where the emission from the plasma is directed into a fast-scanning, 600 mm focal length, Czerny–Turner monochromator with a wavelength range 178–780 nm. The 2,400 lines/mm holographic grating, blazed at 250 nm for optimum UV performance, offers a resolution of 0.050 nm. The detection system is a back-lit charge-coupled device (CCD) array detector, which is cooled to 0°C using a thermoelectric Peltier device, and collects the analyte wavelengths and surrounding background spectra, allowing for simultaneous background correction.

BENEFITS OF MP-AES

One of the biggest advantages of this particular design over ICP-OES is that it runs on compressed air, which means that no gaseous or liquid argon is required. This makes it very attractive for laboratories that are on a tight consumables' budget. However, the benefits over FAA are probably where it will receive the most attention[29]: These benefits include:

- Lower detection limits
- Wider linear dynamic range means larger concentration ranges can be determined
- Increased sample throughput
- Higher productivity, as it can be run unattended overnight
- No requirement to purchase or replace multiple hollow cathode lamps
- The ability to add other analytes to the suite of elements whenever the demand arises
- No safety concerns as neither acetylene nor nitrous oxide is required.

The detection limit improvement over FAA is exemplified in Table 27.3.

TABLE 27.3
Detection Limit Comparison (µg/L) of MP-AES with FAA

Element	FAA	MP-AES	Element	FAA	MP-AES
K	0.8	0.65	As	60	57
Ca	0.4	0.04	Cd	1.5	1.4
Mg	0.3	0.09	Cr	5	0.3
Na	0.3	0.12	Mn	1.0	1.05
Au	5	2.1	Pb	14	2.5
Pt	76	6.1	Sb	37	12
Pd	15	1.6	Se	500	77
Ag	1.7	1.2	Zn	1.6	3.1
Rh	4	0.5			

Note: 10 s integration time, except for As and Se, which were 30 s.[30]

TYPICAL APPLICATIONS OF MP-AES

There is no question, based on the application material in the public domain, that the MP-AES system has been designed to fill the gap between FAA and ICP-AES and particularly to address the many limitations of AA. Some of the most common applications being addressed by this technique include the analysis of geological materials,[31] particularly if there is a requirement for remote sampling or when field studies are being carried out; the analysis of petrochemical samples using the addition of air to reduce carbon buildup on the torch; analysis of major, minor, and trace elements in crops and plant materials[32]; and non-regulated applications, such as the environmental monitoring of industrial waste streams; and the analysis of fruit juices[33] and agricultural samples appears to be some of the most common applications that can be readily addressed by using the MP-AES. For a full suite of application literature on this technique, check out the following reference.[34]

LASER ABLATION LASER IONIZATION TIME-OF-FLIGHT MASS SPECTROMETRY

LALI-TOFMS is a brand new development that offers virtually the entire periodic table of the elements at mass spec detection capability directly on solid samples. It has applicability far beyond the cannabis industry but would be well suited for a testing lab that was looking to avoid sample digestion procedures to measure heavy metals in cannabis and hemp. In addition, when the regulatory elemental suite is expanded, as is likely when the FDA gets involved, the technology will be ideally positioned to measure other elemental contaminants at sub-ppm levels directly in cannabis and cannabis-related samples. Currently, Exum Instruments is the only manufacturer of this technique with the Massbox™ LALI-TOFMS. So let's take a closer look at this novel technology.

The Massbox LALI-TOFMS is the first commercial instrument equipped with a LALI source. The patented, dual-laser technology extracts and subsequently ionizes material in two discrete steps—first ablation and then ionization. Historically, LALI has been referred to by numerous acronyms in the literature, that is, SALI,[35,36] LDI,[37,38] L^2MS,[39] LD-LPI-TOFMS.[40] Most of the research regarding LALI was purely academic and performed between the late 1980s and early 1990s. At the time, cost and electronic limitations prohibited commercialization of LALI technology, and other techniques took center stage. However, recent advances in computing technology and miniaturized, high-powered, solid-state lasers make LALI much more commercially viable today.

BASIC PRINCIPLES LALI-TOFMS

The first step of LALI uses a focused laser beam to ablate/desorb material from a solid sample surface (for more detailed information about laser ablation, refer to Chapter 20). The Massbox utilizes an Nd:YAG desorption laser with an adjustable wavelength. Depending on the application, the desorption/ablation laser can be set to the fourth harmonic (266 nm) or fifth harmonic (213 nm) for ablation or the fundamental 1,064 nm wavelength for desorption. In a process exactly analogous to laser ionization mass spectrometry (LIMS) and LIBS, the ablation/desorption laser generates an initial set of ion/electron pairs from a temporal plasma along with a neutral particle cloud that migrates normal to the sample surface.

A short delay (<1 µs) allows plasma extinction and the dispersion of plasma-generated ions before a second laser is triggered for ionization (also Nd:YAG set to 266 nm). As shown in Figure 27.6, the ionization laser is aligned parallel to the sample surface and its beam is focused inside the neutral particle cloud. The focused beam from the ionization laser has an energy density $>10^9$ W/cm^2, which allows for ionization of neutral particles via multiphoton ionization (MPI). MPI differs

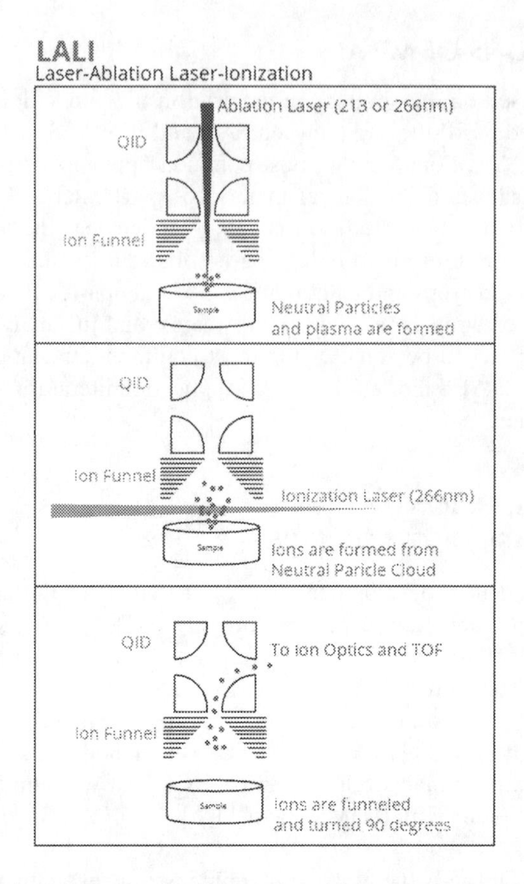

FIGURE 27.6 Laser ablation laser ionization process flow. (Courtesy of Exum Instruments.)

from resonant-enhanced multiphoton ionization (REMPI) in that with MPI the laser is not tuned to a specific elemental or molecular frequency for ionization.

By ionizing elements across a wide range of ionization energies, MPI serves as a highly efficient ion source and replaces the Ar plasma of ICP instruments. Once ionized, an ion funnel collects and focuses the ions in a low-pressure (0.2–0.3 mbar) environment. After exiting the ion funnel, a quadrupole ion deflector (QID) turns the ions 90° and directs the ion beam through an Einzel Lens Stack and a quadrupole to further improve the beam shape. After the transfer quadrupole, the Massbox is equipped with a notch quadrupole filter. For applications that require high sensitivity, the notch filter increases dynamic range by selectively reducing the signal of up to four different ion masses (typically the most abundant matrix elements). Ions are then transferred to the reflectron TOF mass spectrometer which completes the mass analysis as exemplified in Figure 27.7. (For more information about TOF technology refer to Chapter 12.)

MATRIX EFFECTS

Compared to LIBS and LIMS, which both use only the plasma generated by the laser, LALI has more efficient ionization and reduced matrix effects. LALI's ability to reduce/eliminate matrix effects makes quantification more straightforward. In LIBS, for example, matrix-matched standards are a critical requirement for reliable quantification. For applications that require quantification across a variety of matrices, this is a severe limitation of LIBS.

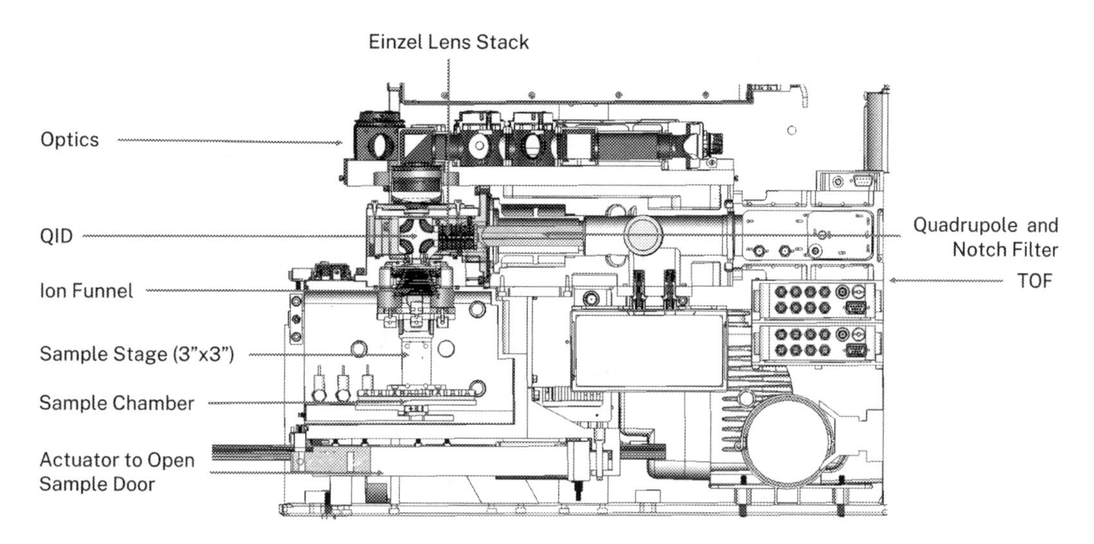

FIGURE 27.7 Simplified schematic diagram of Massbox instrument. (Courtesy of Exum Instruments.)

The Massbox's LALI source (patent-pending) ionizes gas-phase particles within the neutral particle cloud instead of relying on the plasma generated during ablation. Compared to material ionized by the initial ablation plume, material in the neutral cloud is significantly less variable across different matrices. Ionization of neutral cloud material also results in stoichiometric accuracy, which enables quantification of a variety of sample matrices without the need for matrix-matched standards.

DIFFUSION AND TRANSPORT

Another major advantage of this design is that ionization occurs under vacuum in the sample chamber. The Massbox is a completely static system held at high vacuum ($\sim 10^{-7}$ mbar) in the TOF and a pressure gradient to $\sim 10^{-4}$ mbar in quadrupoles and the ion optics. The pressure in the sample chamber is maintained at ~0.2 mbar with an inert helium cooling gas system. The low pressure of the ion source results in a significant improvement in sensitivity because it greatly reduces losses associated with gas transport from atmospheric pressure to a vacuum system. Furthermore, without an inductively coupled plasma source, the Massbox does not require a carrier gas, which eliminates most polyatomic molecular isobaric interferences. Interference reduction improves the detection for a whole host of elements, including Si, K, Ca, and Fe, which are traditionally problematic by ICP-MS. The removal of the plasma source also has the advantage that thermal emission of contaminant ions from the cones and/or injector is eliminated, which greatly improves the ability to determine many of the volatile elements including Na and Pb. Additionally, without a plasma source or any carrier gasses, LALI does not rely on components with dynamic fluctuations, which leads to long-term signal stability without regular instrument tuning.

INTERFERENCES

Figure 27.8 shows an LALI mass spectrum from a GSE-2G standard glass which is free of polyatomic or multiple-charged interferences. As can be seen in the marked zoomed insets (27–30, 80–100, 130–150, 170–180 amu), peak area ratios of the LALI mass spectra are isotopically correct across the entire periodic table, which makes identification easier and quantification more accurate.

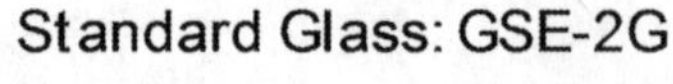

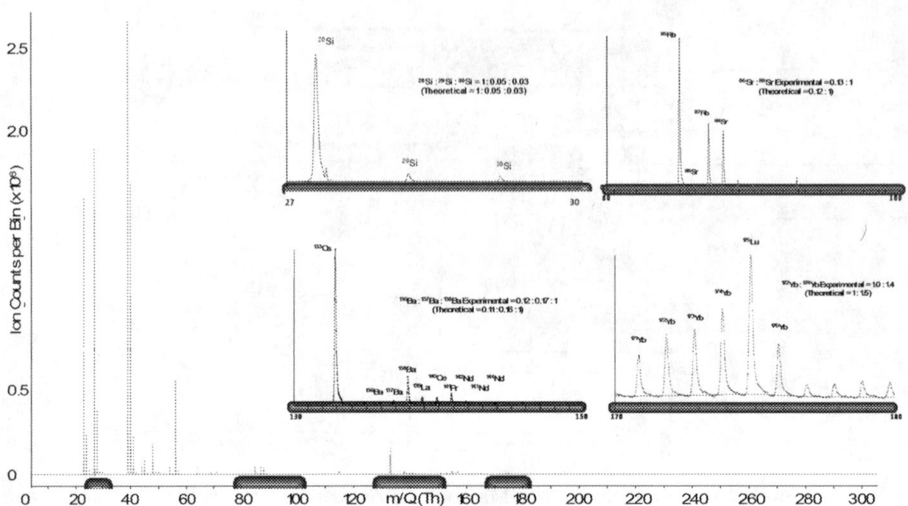

FIGURE 27.8 LALI mass spectrum and selected zoomed insets from GSE-2G standard glass. (Courtesy of Exum Instruments.)

TRANSMISSION EFFICIENCY

LALI provides considerable improvements in transmission efficiency of an ion beam compared to techniques that generate ions at atmospheric pressure. Ions generated by a plasma at atmospheric pressure are generally transferred in several stages before reaching the high-vacuum mass analyzer. Each stage transition of cones and/or lenses removes a significant portion of ions. For instance, LA-ICP-MS has a very high-ionization efficiency for elements with a first ionization potential (FIP) less than 8 eV, but only ~1 in every 10^5–10^6 ions reach the detector (~0.01%–0.001% transmission efficiency). The Massbox's LALI source is already under vacuum, so it does not suffer from transmission loss going from atmospheric pressure to vacuum. Removing the atmospheric/vacuum interface greatly improves transmission efficiency, allows for higher sensitivity, and further reduces matrix effects by removing plasma–ion spatial interaction effects.

INORGANIC AND ORGANIC ANALYSIS

LALI is capable of analyzing both organic and inorganic samples, which is difficult to carry out with other techniques that suffer from the problems previously discussed.[41] The ability to analyze organic compounds by LALI has been studied by many groups for applications including planetary missions and crude oil analysis.[39,42] The analysis of organics utilizes the infrared component of the Nd:YAG laser (1,064 nm). The intense IR pulse flash heats the sample (10^8 K/s) to desorb intact material from the sample surface.[43] Following desorption, organic compounds are ionized via MPI using the secondary ionization laser as previously described. The ability to analyze both organics and inorganics in the same analytical run enables mapping (or bulk characterization) of both in the same sample almost simultaneously after a quick mode switch.

OPERATIONAL USE

Apart from its technical performance, this technology has been developed to be easily deployed across multiple industries and applications, which is a step-change in analytical capability over other lab-based approaches for trace element analysis. Because of its compact footprint (28″×24″×24″), it is field and lab portable, which can be deployed on a table top with no additional equipment required. In addition, its simplicity of design allows for a lower purchase price and operational costs compared to typical laser ablation ICP-MS systems.

USER INTERFACE

The instrument has a very straightforward user interface (UI). Figure 27.9 shows the sample chamber opening to allow the sample tray to be loaded and accommodating a 3″×3″ stage with nanometer (nm) precision.

A macro-camera then opens to take a high-precision, spatially located picture. After the sample door closes, the chamber begins to pump down to vacuum. Meanwhile, the high-resolution image from the macro-camera is loaded onto the touch screen interface. The macro-image is used to enable navigation around the sample(s) as shown in Figure 27.10. Moving back and forth on the

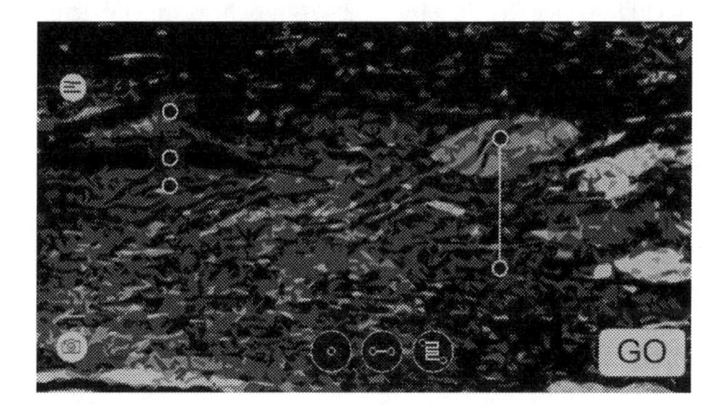

FIGURE 27.9 Schematic of sample loading, macro-imaging, and LALI source. (Courtesy of Exum Instruments.)

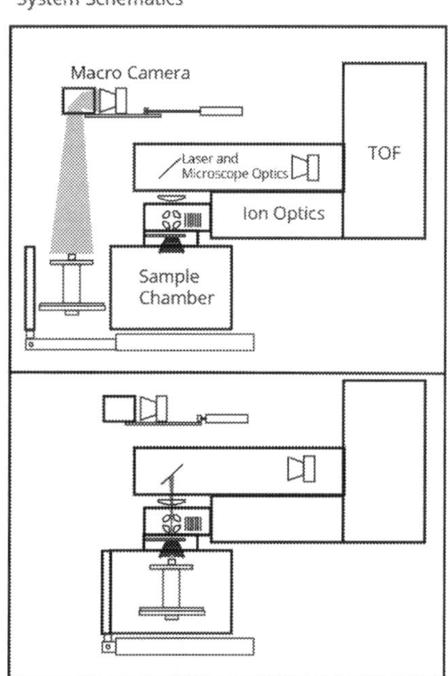

FIGURE 27.10 UI used to identify where characterization occurs. (Courtesy of Exum Instruments.)

screen physically moves the sample stage within the chamber and aligns the desired area properly with the lasers. "Pinching" on the image zooms in and switches to a live microscopy image that allows precision when choosing an area to analyze. From the live view, spots, lines, and/or rasters for maps are chosen. After selecting the type and number sampling areas, analysis and data processing are automated, and all that remains is to interpret the results.

PERFORMANCE CAPABILITIES

The Massbox is the only LALI-TOFMS instrument on the market, which can be used for both organic and inorganic analyses, which provides quantitative chemical mapping and/or bulk analysis without the need for complex quantification schemes or matrix-matched standard materials. The instrument's simple design enables field portability, and its easy-to-use, intuitive interface makes it seamless to train novice users. These factors open up a wide range of potential applications for this exciting technology including meeting the present and future heavy metal demands of the cannabis industry.

The performance specifications of the instrument are shown in Table 27.4.

FINAL THOUGHTS

The AS tool kit seems to expand every few years with new, innovative technology and developments. It is a very active area of research for many vendors, which makes it more difficult for the user community to know which is the best one to use to solve their specific application problems. Today's cannabis testing labs have many choices depending on their budget and sampling and detection requirements, including ICP-MS, ICP-OES, sampling accessories, graphite furnace AAS, or vapor

TABLE 27.4
Performance Specifications of the Massbox

Massbox Specifications	
Analysis type	Inorganic and organic
Sample type	Solid or pseudo-solid
Elemental coverage	Li–U (3–238 amu)
Detection limits	10 ppb–1 ppm (application dependent)
Mass range	1–20,000 mass units
Mass resolution	700–1,100 or 6,000–14,000
Mass accuracy	4 ppm
Notch filter channels	4
Mapping	80 mm × 80 mm × 30 mm Three-axis with nm repeatability and resolution
Laser spot size	Adjustable from 1 to 250 μm
Repetition rate	1–50 Hz
Ablation laser power (266 nm)	10 mJ
Ablation laser power (213 nm)	3 mJ
Ionization laser power (266 nm)	10 mJ
Beam characteristics	Flat-top profile
Total analysis time (including pump down)	<5 min
Sample holder	Customizable depending on application. Standard holder has nine sample slots
Quantification	Variable matrix materials resolved without complex standardization

generation AFS. However, this chapter has presented some alternative and innovative approaches to consider, even though it might not be immediately obvious which technique would be best suited to characterizing a suite of heavy metals in cannabis- and hemp-related samples.

This was the exact dilemma the pharmaceutical industry found themselves in when they were evaluating AS techniques to measure elemental impurities in drug compounds. Even though the United States Pharmacopeia (USP) initially recommended plasma spectrochemistry, other techniques could be used as long as they met the required validation protocols. As a result, there is much application material in the public domain using other techniques, and in fact Davies and coworkers showed that EDXRF could meet the permitted daily exposure (PDE) limits for many of the impurities in pharmaceutical materials.[9] Additionally, there are now commercially available LIBS instruments that are being used for QA/QC checking of incoming raw materials. Finally, LALI-TOFMS could quickly begin to get the attention of the cannabis industry, because it can achieve extremely low detection limits without having to digest the samples.

So I see an opportunity for solid sampling techniques, particularly as a rapid qualitative tool to assess the purity of the cannabis and hemp at the growing site. In addition, they definitely are well suited to characterize the soil and growing medium even before the cannabis or hemp is planted to minimize their uptake of heavy metals.

FURTHER READING

1. R. Jenkins, R.W. Gould, and D. Gedcke, *Quantitative X-Ray Spectrometry* (Marcel Dekker, Inc., New York, 1995).
2. E.P. Bertin, *Introduction to X-Ray Spectrometric Analysis* (Plenum Press, New York, 1978).
3. Wikipedia Commons File, https://commons.wikimedia.org/wiki/File:Dmedxrfschematic.jpg
4. H.W. Major and B.J. Price, *Plastics Engineering* 46(8), 37–39 (1990).
5. V. Thomsen and D. Schatzlein, *Spectroscopy* 17(7), 22–27 (2002).
6. Wikipedia Commons File, https://commons.wikimedia.org/wiki/File:Dmwdxrfschematic.jpg
7. R. Yellipedi and R. Thomas, New developments in wavelength-dispersive X-ray fluorescence and X-ray diffraction for the analysis of foodstuffs and pharmaceutical materials. *Spectroscopy* 21(9), 2–6 (September 2006).
8. Thermo Fisher, The Applications of Hand-Held XRF Technology, https://assets.thermofisher.com/TFS-Assets/CAD/Specification-Sheets/Niton%20XL5-Brochure.pdf
9. D. Davis and H. Furakawa, Using XRF as an alternative technique to plasma spectrochemistry for the new USP and ICH directives on elemental impurities in pharmaceutical materials. *Spectroscopy* 32(7), 12–17 (July 2017).
10. H.L. Byers, L.J. McHenry, and T.J. Grund, XRF techniques to quantify heavy metals in vegetables at low detection limits. *Food Chemistry* 1 (30 March 2019). doi: 10.1016/j.fochx.2018.100001
11. A.M. Phipps, Evaluation of X-ray fluorescence for a lead detection method for candy, UNLV, Dissertation, Environment of Public Health Commons, 2009, https://digitalscholarship.unlv.edu/thesesdissertations/963/
12. R.L. Johnson, Real time demonstration of XRF performance for the Paducha Gas Diffusion Plant, Kentucky Research Consortium for Energy and Environment, April 3, 2008, http://www.ukrcee.org/Challenges/Documents/Soil/RTD_XRF_Perf_Eval_Results_040708fin.pdf
13. L. Oelofse, Measuring the metal content of crude and residual oils by EDXRF: a rapid alternative approach to ICP-OES. *Spectroscopy* 33(7), 12–17 (July 2018).
14. Basics of X-ray Diffraction, http://www.thermo.com/eThermo/CMA/PDFs/Product/product-PDF_11602.pdf
15. L. Perring, D. Andrey, M. Basic-Dvorzak, and J. Blanc, *Journal of Agricultural and Food Chemistry* 53, 4696–4700 (2005).
16. L.J. Radziemski, D.A. Cremers, and T.R. Loree, Detection of beryllium by laser-induced breakdown spectroscopy. *Spectrochimica Acta B* 38(12), 349–355 (1983).
17. K.Y. Yamamoto, D.A. Cremers, M.J. Ferris, and L.E. Foster, Detection of metals in the environment using a portable laser-induced breakdown spectroscopy instrument. *Applied Spectroscopy* 50(2), 222–233 (1996).
18. R.S. Harmon, R.E. Russo, and R.R. Hark, Applications of laser-induced breakdown spectroscopy for geochemical and environmental analysis: a comprehensive review. *Spectrochimica Acta, Part B* 87, 11–26 (2013).

19. D. Cremers and L. Radziemski, *Handbook of Laser Induced Breakdown Spectroscopy* (John Wiley & Sons, Ltd, Hoboken, NJ, 2006).
20. T. Hussain and M.A. Gondal, Laser induced breakdown spectroscopy (LIBS) as a rapid tool for materials analysis. *Journal of Physics: Conference Series* 439, 012050 (2013). https://iopscience.iop.org/article/10.1088/1742-6596/439/1/012050/pdf
21. C. Wu et al., Quantitative analysis of Pb in soil samples by laser-induced breakdown spectroscopy with a simplified standard addition method. *Journal of Analytical Atomic Spectrometry*, 34(14), 1478–1484 (2019). https://pubs.rsc.org/en/content/articlepdf/2019/ja/c9ja00059c.
22. J.H. Kwaka et al., Quantitative analysis of arsenic in mine tailing soils using double pulse-laser induced breakdown spectroscopy. *Spectrochimica Acta Part B* 64, 1105–1110 (2009).
23. J.J. Mortvedt, Heavy metal contaminants in inorganic and organic fertilizers. *Fertilizer Research* 43(1–3), 55–61 (January 1995).
24. L.C. Nunez et al., Determination of Cd, Cr and Pb in phosphate fertilizers by laser-induced breakdown spectroscopy. *Spectrochimica Acta Part B: Atomic Spectroscopy* 97, 42–48 (1 July 2014).
25. P. Yang et al., High-sensitivity determination of cadmium and lead in rice using laser-induced breakdown spectroscopy. *Food Chemistry* 272, 323–328 (January 30, 2019). doi: 10.1016/j.foodchem.2018.07.214. Epub 2018 Aug 11, https://www.ncbi.nlm.nih.gov/pubmed/30309550
26. NASA's Mars Science Laboratory website, http://mars.nasa.gov/msl/mission/instruments/spectrometers/chemcam/
27. R.C. Weins and S. Maurice, The ChemCam Instrument Suite on the mars science laboratory rover curiosity: remote sensing by laser-induced plasmas. *Geochemical News* (June 2011). https://www.geochemsoc.org/publications/geochemicalnews/gn145jun11/chemcaminstrumentsuite
28. K. Jankowski, Evaluation of analytical performance of low-power MIP-AES with direct solution nebulization for environmental analysis. *Journal of Analytical Atomic Spectrometry* 14, 1419–1423 (1999).
29. Agilent Technologies Application Note, Benefits of Transitioning from FAA to the 4200 MP-AES, http://www.chem.agilent.com/Library/technicaloverviews/Entitled%20Partner/5991-3807EN.pdf
30. S. Elliott, Overview of 4200 MPAES, Agilent Webinar, 2014, https://www.agilent.com/cs/library/eseminars/public/4200%20MP-AES%20for%20Mining%20Playlist.pdf
31. Agilent Technologies Application Note, Determination of Major and Minor Elements in Geological Samples Using the 4200 MP-AES, http://www.chem.agilent.com/Library/applications/5991-3772EN.pdf
32. Agilent Technologies Application Note, Analysis of Major, Minor and Trace Elements in Rice Flour Using the 4200 MP-AES, http://www.chem.agilent.com/Library/applications/5991-3777EN.pdf
33. Agilent Technologies Application Note, Analysis of Major Elements in Fruit Juices Using the Agilent 4200 MP-AES with the Agilent 4107Nitrogen Generator. https://www.agilent.com/cs/library/applications/5991-3613EN.pdf
34. Agilent Technologies, 4210 MP-AES Application Library, https://www.agilent.com/en/products/mp-aes/mp-aes-systems/4210-mp-aes#literature
35. C.H. Becker and K.T. Gillen, Can nonresonant multiphoton ionization be ultrasensitive? *Journal of the Optical Society of America* B2, 1438 (1985).
36. C.H. Becker and K.T. Gillen, Surface analysis by nonresonant multiphoton ionization of desorbed or sputtered species. *Analytical Chemistry* 56, 1671–1674 (1984).
37. G.R. Kinsel and D.H. Russell, Design and calibration of an electrostatic energy analyzer-time-of-flight mass spectrometer for measurement of laser-desorbed ion kinetic energies. *Journal of the American Society for Mass Spectrometry* 6, 619–626 (1995).
38. S.A. Getty et al., Compact two-step laser time-of-flight mass spectrometer for in situ analyses of aromatic organics on planetary missions. *Rapid Communications in Mass Spectrometry* 26, 2786–2790 (2012).
39. P. Hurtado, F. Gamez, and B. Martinez-Haya. One- and two-step ultraviolet and infrared laser desorption ionization mass spectrometry of asphaltenes. *Energy & Fuels* 24, 6067–6073 (2010).
40. H. Sabbah et al., Laser Desorption single-photon ionization of asphaltenes: mass range, compound sensitivity, and matrix effects. *Energy & Fuels* 26, 3521–3526 (2012).
41. B. Schueler and R.W. Odom, Nonresonant multiphoton ionization of the neutrals ablated in laser microprobe mass spectrometry analysis of GaAs and Hg0.78Cd0.22Te. *Journal of Applied Physics* 61, 4652–4661 (1987).
42. J.H. Hahn, R. Zenobi, and R.N. Zare, Subfemtomole quantitation of molecular adsorbates by two-step laser mass spectrometry. *Journal of the American Chemical Society* 109, 2842–2843 (1987).
43. L.V. Vaeck and R. Gijbels, Laser microprobe mass spectrometry: potential and limitations for inorganic and organic micro-analysis. *Fresenius Journal of Analytical Chemistry* 337, 743–754 (1990).

28 What Atomic Spectroscopic Technique Is Right for Your Lab?

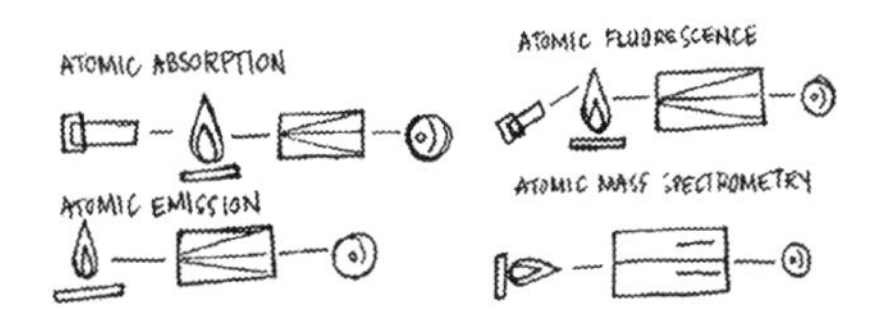

Since the introduction of the first commercially available atomic absorption spectrophotometer (AAS) in the early 1960s, there has been an increasing demand for better, faster, easier-to-use, and more flexible trace element instrumentation. A conservative estimate shows that in 2019, the market for atomic spectroscopy (AS)-based instruments will be well in excess of $1 billion in annual revenue and that doesn't include after-market sales and service costs, which could increase this number by an additional $500 million. As a result of this growth, we have seen a rapid emergence of more sophisticated equipment and easier-to-use software. Moreover, with an increase in the number of manufacturers of AS instrumentation and its sampling accessories, together with the availability traditional solid sampling techniques such as X-ray fluorescence (XRF) and newer techniques like microwave-induced optical emission spectroscopy (MIP-OES) and laser-induced breakdown spectroscopy (LIBS), the choice of which technique to use is often unclear. It also becomes even more complicated when budgetary restrictions allow only one analytical technique to be purchased to solve a particular application problem. This chapter takes a look at the major AS techniques and compares their performance characteristics and in particular examines the limit of quantitation (LOQ) for the four heavy metals (Pb, As, Cd, and Hg) currently being regulated in cannabis and hemp by most states in the United States.

In order to select the best technique for a particular analytical problem, it is important to understand exactly what the problem is and how it is going to be solved. For example if the requirement is to monitor copper at percentage levels in a copper plating bath and it is only going to be done once per shift, flame atomic absorption (FAA) would adequately fill this role. Alternatively, when selecting an instrument to measure 24 elemental impurities in pharmaceutical products, this application is clearly better suited for a multielement techniques such as inductively coupled plasma optical emission spectrometry (ICP-OES). However, if the demand is for four heavy metals in cannabis or hemp, the decision is not so clear. ICP-MS could be the most suitable technique, but depending on the state maximum limits, the cannabis product being tested, the sample workload, or budgetary restrictions, other techniques such as graphite furnace AA or vapor generation atomic fluorescence might be the best fit.

So when choosing a technique, it is important to understand not only the application problem, but also the strengths and weaknesses of the technology being applied to solve the problem. However, there are many overlapping areas between the major AS techniques, so it is highly likely that for some applications, more than one technique would be suitable. For that reason, it is important to go through a carefully thought-out evaluation process when selecting a piece of equipment.[1]

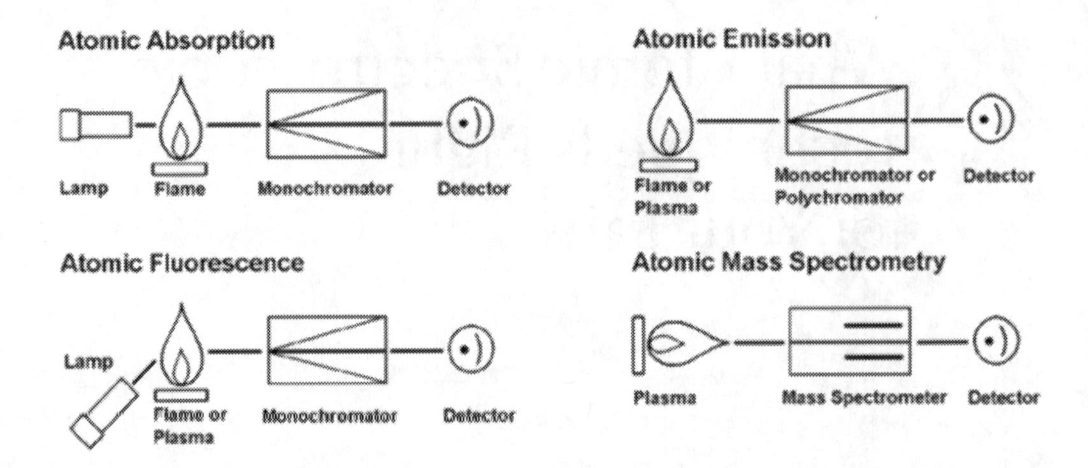

FIGURE 28.1 Simplified schematic of AA, AF, ICP-OES, and ICP-MS.

The main intent of this chapter is therefore to look at the application requirements for the determination of the big four heavy metals in cannabis and cannabis products to offer some insight as to which might be the most suitable atomic spectroscopic approach. We will not be discussing XRF, MIP-OES or LIBS techniques in this evaluation. They are all very useful techniques for many sample types and have been described at length in the open literature.[2] Chapter <233> does allow the use of an alternative technique, as long as it meets the validation and verification protocols described in the analytical procedure section. In fact, it has been shown that XRF is a valid technique to use as a screening tool to reduce the number of samples that need to be analyzed by one of the spectrochemical techniques.[3] However for the purpose of this comparison, we will be focusing on the most commonly used AS techniques—FAA, electrothermal atomization (ETA), hydride generation atomic absorption (HGAA), vapor generation atomic fluorescence (VGAF), ICP-OES, and ICP-MS.

In order to set the stage, let's first take a brief look at their fundamental principles. We have covered the techniques in greater detail in previous chapters, but it's useful to briefly compare the performance of these two plasma spectrochemical techniques (ICP-OES, ICP-MS) with FAA, ETA, and VGAF. Simplified schematics of each of these techniques are shown in Figure 28.1.

FLAME ATOMIC ABSORPTION

This is predominantly a single-element technique for the analysis of liquid samples that uses a flame to generate ground-state atoms. The sample is aspirated into the flame via a nebulizer and a spray chamber. The ground-state atoms of the sample absorb light of a particular wavelength, from either an element-specific, hollow cathode lamp or a continuum source lamp. The amount of light absorbed is measured by a monochromator (optical system) and detected by a photomultiplier tube or solid-state detector, which converts the photons into an electrical signal. As in all AS techniques, this signal is used to determine the concentration of that element in the sample, by comparing it to calibration or reference standards.

Flame AA typically uses about 2–5 mL/min of liquid sample and is capable handling in excess of 10% total dissolved solids, although for optimum performance, it is best to keep the solids down below 2%. For the majority of elements, it's detection capability is in the order of 1–100 parts per billion (ppb) levels with an analytical range up to 10–1,000 parts per million (ppm), depending on the absorption wavelength used. However, it is not really suitable for the determination of the halogens and non-metals such as carbon, sulfur, and phosphorus and has very poor detection limits for the refractory, rare earth, and transuranic elements. Sample throughput for 15 elements per sample is in the order of ten samples an hour.

ELECTROTHERMAL ATOMIZATION

This is also mainly a single-element technique, although multielement instrumentation is now available. It works on the same principle as flame AA, except that the flame is replaced by a small heated tungsten filament or graphite tube. The other major difference is that in ETA, a very small sample (typically, 50 μL) is injected onto the filament or into the tube, and not aspirated via a nebulizer and a spray chamber. Because the ground-state atoms are concentrated in a smaller area, more absorption takes place. The result is that ETA offers detection capability at the 0.01–1 parts per billion (ppb) level, with an analytical range up to 10–100 parts per billion (ppb).

The elemental coverage limitations of the technique are similar to flame AA technique. However, because a heated graphite tube is used for atomization in most commercial instruments, it cannot determine the refractory, rare earth, and transuranic elements, because they tend to form stable carbides that cannot be readily atomized. One of the added benefits is that ETA can also analyze slurries and some solids due to the fact that no nebulization process is involved in introducing the sample. This technique is not ideally suited for multielement analysis, because it takes 3–4 min to determine one element per sample. As a result, sample throughput for four elements is in the order of three samples per hour.

HYDRIDE/VAPOR GENERATION AA

Hydride generation (HG) is a very useful analytical technique to determine the hydride forming metals, such as As, Bi, Sb, Se, and Te, usually by AA (although detection by either ICP-OES or ICP-MS can also be used). In this technique, the analytes in the sample matrix are first reacted with a very strong reducing agent, such as sodium borohydride to release their volatile hydrides, which are then swept into a heated quartz tube for atomization. The tube is heated by either a flame or a small oven which creates the ground-state atoms of the element of interest and then measured by atomic absorption. When used with ICP-OES or ICP-MS, the volatile hydrides are passed directly in the plasma for excitation or ionization.

By choosing the optimum chemistry, mercury can also be reduced in solution in this way to generate elemental mercury. This is known as the cold vapor (CV) technique. HG and CV AA can improve the detection for these elements over flame AA by up to 3 orders of magnitude, achieving detection capability in the order of 0.005–0.1 parts per billion (ppb) levels with an analytical range of 5–100 parts per billion (ppb), depending on the element of interest. It should also be pointed out that dedicated mercury analyzers (some using gold amalgamation techniques), coupled with atomic absorption or atomic fluorescence, are capable of better detection limits. Because of the online chemistry involved, these techniques are very time-consuming and are normally used in conjunction with FAA, so they will most likely impact the overall sample throughput.

ATOMIC FLUORESCENCE

Hydride/vapor generation can also be used with atomic fluorescence, to improve the detection capability of the volatile elements/heavy metals. Atomic fluorescence shares properties with atomic absorption in that it has a light source, atomizer, monochromator, and detector. The element-specific light energy is absorbed, taking the element through excitation. Rather than determining the amount of light being absorbed by the element in the atomizer, as the excited atoms return to ground state, they emit photons of light and this emission, or fluorescence, resulting from the decay is measured. The major difference is the element-specific light source is at an angle so as not to interfere with the fluorescence process. To carry out multielement analysis using AFS, many lamps are positioned around the atomization/excitation source.

Background signals in AFS are very low, resulting in improved sensitivity and detection limit as compared to atomic absorption. However, the fluorescence measurement can be quenched by many molecular forms, so separation of the matrix from the analyte may be desirable. There are commercially available vapor generation AFS system on the market, which have been used successfully used for the measurement of Pb, Cd, As, and Hg in cannabis samples.[4,5]

RADIAL ICP-OES

Radially viewed ICP-OES is a multielement technique that uses a traditional radial (side-view) inductively coupled plasma to excite ground-state atoms to the point where they emit wavelength-specific photons of light that are characteristic of a particular element. The number of photons produced at an element-specific wavelength is measured by high-resolution optics and a photon-sensitive device such as a photomultiplier tube or a solid-state detector. This emission signal is directly related to the concentration of that element in the sample. The analytical temperature of an ICP is about 6,000°K–7,000°K, compared to that of a flame or a graphite furnace, which is typically 2,500°K–3,500°K.

For the majority of elements, a radial ICP instrument can achieve detection capability in the order of 0.1–100 parts per billion (ppb) levels with an analytical range up to 10–1,000 parts per million (ppm), depending on the emission wavelength used. The technique can determine a similar number of elements to FAA but has the advantage of offering the capability of non-metals such as sulfur and phosphorus, together much better performance for the refractory, rare earth, and transuranic elements. The sample requirement for ICP-OES is approximately 1 mL/min and is capable of aspirating samples containing up to 10% total dissolved solids but, for optimum performance, is usually kept below 2%. ICP-OES is a rapid multielement technique, so sample throughput for 15 elements per sample is in the order of 20 samples an hour.

AXIAL ICP-OES

The principle is exactly the same as radial ICP-OES, except that in the axial view, the plasma is viewed horizontally (end-on). The benefit is that more photons are seen by the detector and for this reason; detection limits can be as much as an order of magnitude lower, depending on the design of the instrument. The disadvantage is that the working range is also reduced by an order of magnitude. As a result, for the majority of elements, an axial ICP instrument can achieve detection capability in the order of 0.01–10 parts per billion (ppb) levels with an analytical range up to 1–100 parts per million (ppm), depending on the emission wavelength used. The other disadvantage of viewing axially is that more severe matrix interferences are observed, which means that the total dissolved solids content of the sample needs to be kept much lower. Sample flow requirements are the same as for radial ICP-OES.

It's also important to point out that most commercial ICP-OES instruments have both radial and axial capability built in. However, because of the different hardware configurations available, some instruments work by carrying out the analysis using either the radial view or the axial view (sequentially). Others can use both the radial view and axial view at exactly the same time (simultaneously), which for some applications may be advantageous. Strategic use of both radial and axial views, together with the optimum wavelength selection, can extend the analytical range by 2–3 orders of magnitude. Sample throughput will be approximately 20 samples per hour, the same as radial ICP-OES.

ICP-MS

The fundamental difference between ICP-OES and ICP-MS is that in ICP-MS, the plasma is not used to generate photons, but to generate positively charged ions. The ions produced are transported and separated by their atomic mass-to-charge ratio using a mass-filtering device such as a quadrupole or a magnetic sector. The generation of such large numbers of positively charged ions allows ICP-MS to achieve detection limits approximately 3 orders of magnitude lower than ICP-OES, even for the refractory, rare earth, and transuranic elements. As a result, for the majority of elements, an ICP-MS instrument can achieve detection capability in the order of 0.0001–1 parts per billion (ppb) with an analytical range up to 0.1–100 parts per million (ppm), using pulse counting measurement, but can be extended even further up to 100–100,000 ppm by using analog counting techniques. However, it should be emphasized that if such large analyte concentrations are being measured, expectations should be realistic about also carrying out ultratrace determinations of the same element in the same sample run.

The sample requirement for ICP-MS is approximately 1 mL/min and is capable of aspirating samples containing up to 5% total dissolved solids for short periods with the use of specialized sampling accessories. However, because the sample is being aspirated into the mass spectrometer, for optimum performance, matrix components should ideally be kept below 0.2%. This is particularly relevant for laboratories that experience a high sample workload. Sample throughput will be approximately 15 samples per hour, for the determination of 24 elements in a sample.

COMPARISON HIGHLIGHTS

There is no question that the multielement techniques like ICP-OES and ICP-MS are better suited for the determination of the elements defined in USP Chapters <232> and <2232>, especially in labs that are expected to be carrying out the analysis in a high-throughput, routine environment. If the best detection limits are required, ICP-MS offers the best choice followed by graphite furnace AA (ETA). Axial ICP-OES offers very good detection limits for most elements, but generally not as good as ETA. Radial ICP-OES and flame AA show approximately the same detection limit performance, except for the refractory, rare earth, and the transuranic elements, for which performance is much better by ICP-OES. For mercury and those elements that form volatile hydrides, such as As, Bi, Sb, Se, and Te, the CV or HG techniques offer exceptional detection limits. Figures 28.2 and 28.3 and Table 28.1 show an overview of the detection capability, working analytical range, and sample throughput of the major AS approaches, which are three of the metrics most commonly

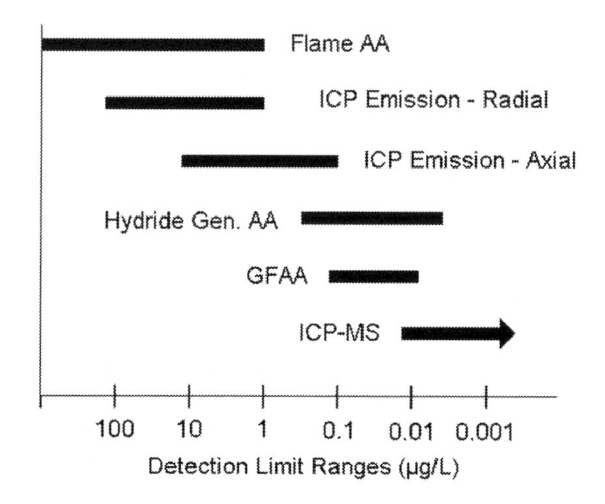

FIGURE 28.2 Typical AS detection limit ranges.

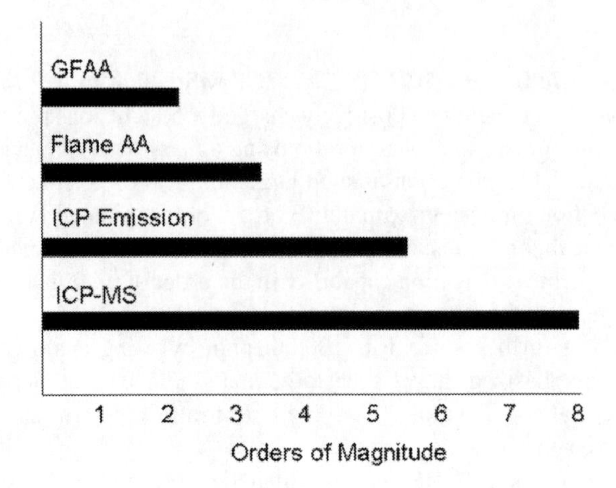

FIGURE 28.3 Typical AS analytical working ranges.

TABLE 28.1
Approximate Sample Throughput Capability of AS Techniques

Technique	Elements at a Time	Duplicate Analysis (min)	Samples per Hour (1 Element)	Samples per Hour (4 Elements)	Samples per Hour (15 Elements)
FAA	1	0.3	150	30	10
ETA	1	5	12	3	1
AF	Up to 4	3	20	20	5
ICP-OES	Up to 70	3	20	20	20
ICP-MS	Up to 70	3	20	20	20

used to select a suitable technique. They should be considered an approximation and only used for guidance purposes.

It is also worth mentioning that the detection limits of quadrupole-based ICP-MS, when used in conjunction with collision/reaction cell/interface and/or multiple mass separation/selection devices, are now capable of subparts per trillion (ppt) detection limits, even for the elements such as Fe, K, Ca, Se, As, Cr, Mg, V, and Mn, which traditionally suffer from plasma- and solvent-based polyatomic interference.

This is not meant to be a detailed description of each technique, but a basic understanding as to how they differ from each other. For a more detailed comparison of the fundamental principles and application strengths and weaknesses of ICP-MS and ICP-OES, please refer to the previous chapters. Let's now turn our attention to selecting the best technique based on the demands of pharmaceutical application.

DEMANDS OF THE CANNABIS INDUSTRY

Comparing an instrumental technique based on its performance specifications with simple standards is important, but it bears little relevance to how that instrument is going to be used a real-world situation. For example, instrument detection limits (IDL) are important to know, but how are they impacted by the matrix and sample preparation procedure? So what is the real-world method detection limits (MDLs) and is the technique capable of quantifying the maximum concentration values expected for this analysis? And what kind of precision and accuracy can be expected if working close to the LOQ for the overall methodology being used. Additionally on the sample throughput side, how many samples are expected and at what frequency will they be coming into the lab. How much time can be spent on sample preparation and how quickly must they be analyzed. Sometimes, the level of interferences from the matrix will have a major impact on the selection process or even the amount/volume of sample available for analysis.

Understanding the demands of an application is therefore of critical importance when a technique is being purchased and particularly if there is a minimum amount expertise/experience in house on how best to use it to solve a particular application problem. This could be the likely scenario in a cannabis testing laboratory that is being asked to check the four elemental contaminants in cannabis, or one of its many cannabinoid products. As there are currently very few regulated methods available, the industry is turning to the highly regulated pharmaceutical industry for guidance. As a result, they use the new USP Chapters <232> and/or ICH Q3D guidelines to carry out the measurements, which recommend the use of a plasma-based spectroscopic technique or any other AS technique as long as it meets the validation protocols described in USP Chapter <233>.

The expertise of the operator should never be underestimated, because if ICP-MS is being seriously considered, it generally requires an analyst with a higher skill level to develop rugged methodology free of interferences that can eventually be put in the hands of an inexperienced user to operate on a routine basis. This again is a real concern if the technique is being used by novice users who have limited expertise in running analytical instrumentation, which may be the case in the pharmaceutical or cannabis industries. So let's take a closer look at the permitted daily exposure (PDE) limits defined in USP chapters to understand what AS techniques might best meet the demands of measuring the four heavy metals in cannabis.

SUITABILITY OF TECHNIQUE

USP Chapter <233> is being adopted by most cannabis testing labs as the standardized methodology to test all cannabinoid-based products for heavy metals. To get a better understanding of the suitability of the technique being used and whether its detection capability is appropriate for the samples being tested, it's important to know the PDE limits for each of the four target elements and, in particular, what the USP calls the J-value, as described in Chapter <233>. As mentioned in previous chapter, the J-value is defined as the PDE concentration of the element of interest, appropriately diluted to the working range of the instrument, after the sample preparation procedure to get the sample into solution is completed. So let's take Pb as an example. The PDE limit for Pb in an oral medication (or an orally delivered cannabinoid product) defined in Chapter <232> is 5 μg/day.

Based on a suggested dosage of 10 g of the product/day, that's equivalent to 0.5 μg/g Pb. If 1.0 g of sample is digested/dissolved and made up to 500 mL (that's 0.2 g in 100 mL, which is used for many cannabis samples), that's a 500-fold dilution, which is equivalent to 1.0 μg/L. So the J value for Pb in this example is equal to1.0 μg/L.

The method then suggests using a calibration made up of two standards: Standard 1 = 1.5 J, Standard 2 = 0.5 J. So for Pb, that's equivalent to 1.5 μg/L for Standard 1and 0.5 μg/L for Standard 2.

The suitability of a technique is then determined by measuring the calibration drift and comparing results for Standard 1 before and after the analysis of all the sample solutions under test. This calibration drift should be <20% for each target element. However, once the suitability of the technique has been determined, further validation protocols described previously must be carried out to show compliance to the regulatory agency if required.

It should also be pointed out that no specific instrumental parameters are suggested in Chapter <233>, but only to analyze according to the manufacturer's suggested conditions and to calculate and report results based on the original sample size. However, it does say that appropriate measures must be taken to correct for interferences, such as matrix-induced wavelength overlaps in ICP-OES and argon-based polyatomic interference with ICP-MS.

Let's exemplify this by taking an analytical scenarios of measuring four elemental contaminants (Pb, Cd, As, and Hg) in a cannabis product that is being regulated according to the USP Chapter <232>/ICH Q3D inhalation PDEs (many states use the inhalation maximum limits, even though a cannabis product might be consumed orally, because the inhalation values are much lower) and calculate the J-values for each heavy metal and compare them with the LOQ for each technique to give us an assessment of their suitability. For this analytical scenario, we'll take the LOQ for the technique as 10x the IDL. These LOQs were calculated by taking the average of published IDLs from three instrument vendors' application material and multiplying them by 10 to get an approximation of LOQ. In practice, a method LOQ is typically determined by processing the matrix blank through the entire sample preparation procedure and taking ten replicate measurements. The method LOQ, sometimes referred to as the MDL, is then calculated as 3–7 × standard deviations of these ten measurements, depending on the percentage of confidence level required.

To make this comparison valid, the sample weight was adjusted for each technique, based on the detection limit and analytical working range. So for AA, AF, and ICP-OES, we based the calculation on a dilution of 2 g/100 mL (or 0.5 g/25 mL), whereas for ICP-MS, we used 0.2 g/100 mL. AA/ICP-OES could definitely use larger sample weights, but for high-throughput routine analysis, we are probably at the optimum dilution for ICP-MS. Tables 28.2–28.4 show the comparison of AA (FAA and ETA), AF, ICP-OES, and ICP-MS, respectively, for four elemental contaminants in an inhaled cannabis product according to Chapter <232>. The relevant comparative data to consider is in the final column, labeled "Factor Improvement," which is the J-value, divided by the LOQ. Generally speaking, the higher this number is, the more suitable the technique is.

RELATIONSHIP BETWEEN LOQ AND J-VALUE

It should be emphasized again that LOQ in these examples is just a guideline as to the real-world detection capability of the technique for this method. However, it does offer a very good approximation as to whether the technique is suitable based on the factor improvement number compared to the J-values for each elemental impurity. Clearly, if this improvement number is less than 1, as it is with Pb and Cd by FAA, the technique is just not going to be suitable. Although, As and Hg could be improved using hydride or CV techniques. On the other hand, ETA would be suitable for all the heavy metals. However, ETA technique is very time-consuming and labor-intensive, so it probably wouldn't be a practical solution in a high-throughput cannabis testing laboratory. On the other hand, VGAF looks to be very promising as the factor improvement for all four heavy metals is between 40 and 600. The comparison between FAA, ETA, and VGAF is exemplified in Table 28.2.

TABLE 28.2
USP Chapter <232> J-Values Compared to Limits of Quantitation for FAA and ETA

Element	Conc. Limits for an Inhaled Product with a Max Daily Dose of ≤10 g/day (µg/g)	J-Value with a Sample Dilution of 2g/100 mL (µg/L)	AA LOQ (IDL × 10) (µg/L)		VGAF LOQ (IDL × 10) (µg/L)	Factor Improvement (J-Value/LOQ)		
			FAA	ETA	HGAF	FAA	ETA	HGAF
Cadmium	0.3	6	7	0.02	0.01	0.8	300	600
Lead	0.5	10	150	0.5	0.1	0.07	20	100
Arsenic	0.2	4	0.2[a]	0.5	0.1	20[a]	8	40
Mercury	0.1	2	0.1[b]	5	0.01	20[b]	12	200

[a] HGAA used for As.
[b] CVAA used for Hg.

TABLE 28.3
USP Chapter <232> J-Values Compared to Limits of Quantitation for Axially Viewed ICP-OES

Element	Conc. Limits for an Inhaled Product with a Max Daily Dose of ≤10 g/day (μg/g)	J-Value with a Sample Dilution of 2g/100 mL (μg/L)	~Axial ICP-OES LOQ (IDL × 10) (μg/L)	Factor Improvement (J-Value/LOQ)
Cadmium	0.3	6	0.02	300
Lead	0.5	10	10	1
Arsenic	0.2	4	10	0.4
Mercury	0.1	2	10	0.2

TABLE 28.4
USP Chapter <232> J-Values Compared to Limits of Quantitation for ICP-MS

Element	Conc. Limits for an Inhaled Product with a Max Daily Dose of ≤10 g/day (μg/g)	J-Value with a Sample Dilution of 0.2g/100 mL (μg/L)	~ICP-MS LOQ (IDL × 10) (μg/L)	Factor Improvement (J-Value/LOQ)
Cadmium	0.3	0.6	0.0007	857
Lead	0.5	1	0.0004	2,500
Arsenic	0.2	0.4	0.004	100
Mercury	0.1	0.2	0.01	20

Table 28.3 shows that axial-ICP-OES doesn't look to be a good candidate because aside from Cd, the factor improvements are all around 1 or less. These numbers could be further improved, by using a higher sample weight in the sample preparation procedure without compromising the method, or possibly using a sampling accessory such as an ultrasonic nebulizer, which could improve the detection capability by a factor of fivefold to tenfold. (Note: as most commercial ICP-OES instrumentation offers both axial and radial capability, it was felt that the axial performance was most appropriate for this comparison.)

However, it can be seen in Table 28.4 that ICP-MS shows significant improvement factors for all contaminants which are not offered by any other technique. The added benefit of using ICP-MS is that if/when the elemental suite expands to include other heavy metal contaminants, the additional workload should not pose too much of a problem. Additionally, if arsenic or mercury levels were found to be higher than the PDE levels, it would be relatively straight-forward to couple ICP-MS with HPLC to monitor the speciated forms of these elements if required.

FINAL THOUGHTS

It is important to understand that there are many factors to consider when selecting a trace element technique most suited to the demands of your application. Sometimes, one technique stands out as being the clear choice, whereas other times, it is not quite so obvious. And as is true with many applications, more than one technique is often suitable. However, the current state regulations for cannabis products are based on a mix of USP Chapter <232> inhalation and oral PDEs which presents some unique challenges, not only from a perspective of performance capability, but also because testing labs have to meet the validation protocols defined in USP Chapter <233> to show suitability of the technique to the analytical procedure being used. From a practical perspective, there is no question that to meet the lowest limits for inhaled products (used by many state regulators), ICP-MS is probably the most appropriate technique, followed closely by vapor generation atomic fluorescence. However for oral cannabis products or transdermal applications, where the maximum limits are higher, axial-ICP-OES combined with performance enhancing sampling accessories could offer a more cost-effective approach. And if the sample workload requirements are not so demanding, ETA could provide a solution, particularly as solid samples can be analyzed. However ICP-MS has shown it has the detection limits and throughput capability to be the optimum technique of choice for these types of samples, particularly if the suite of required heavy metals expands due to stricter federal regulations.

FURTHER READING

1. R.J. Thomas, Choosing the Right Trace Element Technique: Do you know what to look for, *Today Chemist at Work*, October, 1999.
2. R.J. Thomas, Emerging Technology Trends in Atomic Spectroscopy Are Solving Real-World Application Problems, *Spectroscopy Magazine*, 29 (3), 42–51, 2014.
3. D. Davis and H. Furukawa, Using XRF as an Alternative Technique to Plasma Spectrochemistry for the New USP and ICH Directives on Elemental Impurities in Pharmaceutical Materials, *Spectroscopy Magazine*, 32 (7), 12–17, 2017.
4. Lumina 3500 Atomic Fluorescence Spectrometer, Aurora Illuminating Systems Elemental Analysis Instruments, 2020, https://www.aurorabiomed.com/lumina-3500/
5. PF7 Heavy Metals Analyzer, Persee Analytics Inc., http://perseena.com/index/prolist/pid/33/c_id/43.html

29 Do You Know What It Costs to Run Your AS System?

Chapter 28 gave an overview of the performance differences of the major atomic spectroscopic techniques, but selection of the optimum approach might also involve the cost of the instrumentation, particularly if funds are limited. This chapter takes a looks at the running costs of the four AS techniques, particularly from the standpoint of the cost of consumables, gases, and electricity.

For the purpose of this evaluation, let us make the assumption that the major operating costs associated with running AS instrumentation are the gases, electricity, and consumable supplies. For comparison purposes, the exercise will be based on a typical laboratory running their instrument for 2½ days (20 h) per week and 50 weeks a year (1,000 h per year). These data are based on the cost of gases, electricity, and instrument consumables in the United States in 2017. They have been obtained from a number of publically available commercial sources, including suppliers of industrial and high-purity gases, independent utilities companies, a number ICP-MS instrument vendors, and sample introduction/consumable suppliers. It's also important to emphasize that these costs might vary slightly based on the different vendor instrument design/technology/ being used.

GASES

FAA: Most flame AA systems use acetylene (C_2H_2) as the combustion gas, and air or nitrous oxide (N_2O) as the oxidant. Air is usually generated by an air compressor, but the C_2H_2 and N_2O come in high-pressure cylinders. Normal atomic absorption grade C_2H_2 cylinders contain 380 ft^3 (10,760 L) of gas. N_2O is purchased by weight and comes in cylinders containing 56 lb of gas, which is equivalent to 490 ft^3 (13,830 L). A cylinder of C_2H_2 costs approximately $200, whereas a cylinder of N_2O costs about $70. These prices have remained fairly stable over the past few years. Normal C_2H_2 gas flows in FAA are typically 2 L/min when air is the oxidant and 5 L/min when N_2O is the oxidant. N_2O gas flows are on the order of 10 L/min.

Air–C_2H_2 mixtures are used for the majority of elements, whereas an N_2O–C_2H_2 mixture has traditionally been used for the more refractory elements. So, for this costing exercise, we will assume that half the work is done using air–C_2H_2, and for the other half N_2O–C_2H_2 is being used. Therefore, a typical laboratory running the instrument for 1,000 h per year will consume 16 cylinders of C_2H_2, which is equivalent to $3,200 per year, and 22 cylinders of N_2O costing $1,500, making a total of $4,700.

ETA: The only gas that the electrothermal atomization process uses on a routine basis is high-purity argon, which costs about $100 for a 340 ft^3 (9,630 L) cylinder. Typically, argon gas flows of up to 300 mL/min are required to keep an inert atmosphere in the graphite tube.

At these flow rates, 540h of use can be expected from one cylinder. Therefore, a typical laboratory running their instrument for 1,000h per year would consume almost two cylinders costing $200.

ICP-OES and ICP-MS: The consumption of gases in ICP-OES and ICP-MS is very similar. They both use a total of approximately 15–20 L/min (~1,000 L/h) of gaseous argon (including plasma, nebulizer, auxiliary, and purge flows), which means a cylinder of argon (9,630L) would last only about 10h. For this reason, most users install a Dewar vessel containing a liquid supply of argon. Liquid argon tanks come in a variety of different sizes, but a typical Dewar system used for ICP-OES/ICP-MS holds about 240L of liquid gas, which is equivalent to 6,300 ft^3 (178,000L) of gaseous argon. (Note: The Dewar vessel can be bought outright, but are normally rented.) It costs about $350 to fill a 240-liter Dewar vessel with liquid argon. At a typical argon flow rate 17 L/min total gas flow, a full vessel would last for almost 175h. Again, assuming a typical laboratory runs their instrument for 1,000h per year, this translates to 6 fills at approximately $350 each, which is equivalent to about $2,100 per year. If cylinders were used, about 100 would be required, which would elevate the cost to almost $10,000 per year.

Note: When liquid argon is stored in a Dewar vessel, there is a natural bleed-off to the atmosphere when the gas reaches a certain pressure. For this reason, a bank of argon cylinders is probably the best option for laboratories that do not use their instruments on a regular basis. Some of the newer ICP-OES instruments operate at approximately 60%–70% argon consumption compared to older instruments. So this should be taken into consideration if this technology is being used.

Another added expense with ICP-MS is that if it is fitted with collision/reaction cell technology, the cost of the collision or reaction gas will have to be added to the running costs of the instrument. Fortunately, for most applications, the gas flow is usually less than 5 mL/min, but for the collision/reaction interface approach, typical gas flows are 100–150 mL/min. The most common collision/reaction gases used are hydrogen, helium, and ammonia. The cost of high-purity helium is on the order of $400 for a 300 ft^3 (8,500L) cylinder, whereas that of a cylinder of hydrogen or ammonia is approximately $250. One cylinder of either gas should be enough to last 1,000h at these kinds of flow rates. So, for this costing exercise, we will assume that the laboratory is running a collision/reaction cell/interface instrument, with an additional expense of $650. If other collision/reaction gases are being used, these should be factored into the calculation.

It should also be pointed out that some collision/reaction cells require high-purity gases with extremely low impurity levels, because of the potential of the contaminants in the gas to create additional by-product ions. This can be achieved either by purchasing laboratory-grade gases and cleaning them up with a gas purification system (getter), or by purchasing ultra-high-purity gases directly from the gas supplier. If the latter option is chosen, you should be aware that ultra-high-purity helium (99.9999%) is approximately twice the price of laboratory-grade helium (99.99%), whereas ultra-high-purity hydrogen is approximately four times the cost of laboratory-grade hydrogen.

ELECTRICITY

Calculations for power consumption are based on the average cost of electricity, which is currently about $0.10 kW/h in the United States. The cost will vary depending on the location and demand, but it represents a good approximation for this exercise. So the following formula has been used for calculating the cost of electricity usage for each technique:

$$\text{Cost per kW hour}\,(\$) \times \text{Power consumption}\,(\text{kW}) \times \text{Annual usage}\,(\text{h})$$

FAA: The power in a flame AA system is basically used for the hollow cathode lamps and the onboard microprocessor that controls functions such as burner head position, lamp selection, photo multiplier tube voltage, and grating position. A typical instrument requires less than 1 kW of power. If it is used for 1,000 h per year, it will be drawing less than 1,000 kW total power, which is ~$100 per year.

ETA: A graphite furnace system uses considerably more power than a flame AA system because a separate power unit is used to heat the graphite tube. In routine operation, there is a slow ramp heating of the tube for ~3 min until it reaches an atomization temperature of about 2,700°C, requiring a maximum power of ~3 kW. This heating cycle combined with the power requirements for the rest of the instrument costs ~$300, for a system that is run 1,000 h per year.

ICP-OES and ICP-MS: Both these techniques can be considered the same with regard to power requirements as the RF generators are of very similar design. Based on the voltage, magnitude of the electric current, and the number of lines used, the majority of modern instruments draw about 5 kW total power. This works out to be ~$500 for an instrument that is run 1,000 h per year.

CONSUMABLES

Because of the fundamental differences between the four AS techniques, it is important to understand that there are considerable differences in the cost of consumables. In addition, the cost of the same component used in different techniques can vary significantly between different vendors and suppliers. So, the data has been taken from a number of different sources and averaged.

FAA: The major consumable supplies used in flame AA are the hollow cathode lamps. Depending on usage, you should plan to replace three of them every year, at a cost of $300–500 for a good quality, single-element lamp. However, if a continuum source AA system is being used, there will not be a requirement to replace lamps on a regular basis. Other minor costs are nebulizer tubing and autosampler tubes. These are relatively inexpensive, but should be planned for. The total cost of lamps, nebulizer tubing, and a sufficient supply of autosampler tubes should not exceed $1,500–2,000 per year, based on 1,000 h of instrument usage.

ETA: As long as the sample type is not too corrosive, a graphite furnace AA tube should last about 300 heating cycles (firings). Based on a normal heating program of 3 min per replicate, this represents 20 firings per hour. If the laboratory is running the instrument 1,000 h per year, it will carry out a total of 20,000 firings and use 70 graphite tubes in the process. There are many designs of graphite tubes, but for this exercise, we will base the calculation on platform-based tubes that cost about $50 each when bought in bulk. If we add the cost of graphite contact cylinders, hollow cathode lamps, and a sufficient supply of autosampler cups, the total cost of consumables for a graphite furnace will be approximately $5,000–6,000 per year.

ICP-OES: The main consumable supplies in ICP-OES are in the plasma torch and in the sample introduction area. The major consumable is the torch itself, which consists of two concentric quartz tubes and a sample injector made of either quartz or some ceramic material. In addition, a quartz bonnet normally protects the torch from the RF coil. There are many different demountable torch designs available, but they all cost about $600–700 for a complete system. Depending on sample workload and matrices being analyzed, it is normal to go through a torch every 4–6 months. In addition to the torch, other parts that need to be replaced or at least need to have spares include the nebulizer, spray chamber, and sample capillary and pump tubing. When all these items are added together, the annual cost of consumables for ICP-OES is on the order of $3,000–3,200.

TABLE 29.1
Annual Operating Costs ($US) for the Four AS Techniques for a Laboratory Running an Instrument 1,000 h/year (20 h/week)

Technique	Gases ($)	Power ($)	Consumable Supplies ($)	Total ($)
FAA	4,700	100	1,750	6,550
ETA	200	300	5,500	6,000
ICP-OES	2,100[a]	500	3,100	5,700
ICP-MS	2,750[a,b]	500	10,000	13,250

[a] Using a liquid argon supply.
[b] Using a collision/reaction cell.

ICP-MS: In addition to the plasma torch and sample introduction supplies, ICP-MS requires consumables that are situated inside the mass spectrometer. The first area is the interface region between the plasma and the mass spectrometer, which contains the sampler and skimmer cones. These are traditionally made of nickel, which is recommended for most matrices, or platinum for highly corrosive samples and organic matrices. A set of nickel cones costs $700–1,000, whereas a set of platinum cones costs about $3,000–4,000. Two sets of nickel cones and perhaps one set of platinum cones would be required per year. The other major consumable in ICP-MS is the detector, which has a lifetime of approximately 1 year, and costs about $1,200–1,800. When all these are added together with the torch, the sample introduction components, and the vacuum pump consumables, investing in ICP-MS supplies represents an annual cost of $9,000–11,000.

The approximate annual cost of gases, power, and consumable supplies of the four AS techniques being operated for 1,000 h/year, is shown in Table 29.1.

COST PER SAMPLE

We can take the data given in Table 29.1 a step further and use these numbers to calculate the operating costs per individual sample, assuming that a laboratory is determining ten analytes per sample. Let us now take a look at each technique to see how many samples can be analyzed, assuming the instrument runs 1,000 h per year.

FAA: A duplicate analysis for a single analyte in flame AA takes about 20 s. This is equivalent to 180 analytes per hour or 180,000 analytes per year. For 10 analytes, this represents 18,000 samples per year. Based on an annual operating cost of $6,550, this equates to $0.36 per sample.

ETA: A single analyte by ETA takes about 5–6 min for a duplicate analysis, which is equivalent to approximately 10 analytes per hour or 10,000 analytes per year. For 10 analytes per sample, this represents 1,000 samples per year. Based on an annual operating cost of $6,000, this equates to $6.00 per sample.

ICP-OES: A duplicate ICP-OES analysis for as many analytes as you require takes about 3 min. So for 10 analytes, this is equivalent to 20 samples per hour or 20,000 samples per year. Based on an annual operating cost of $5,700, this equates to $0.30 per sample.

ICP-MS: ICP-MS also takes about 3 min to carry out a duplicate analysis for 10 analytes, which is equivalent to 20,000 samples per year. Based on an annual operating cost of $13,250, this equates to $0.66 per sample.

TABLE 29.2
Operating Costs for a Sample Requiring 10 Analytes, Based on the Instrument Being Used for 1,000 h/year

Technique	Operating Cost for 10 Analytes per Sample ($US)
FAA	0.36
ETA	6.50
ICP-OES[a]	0.29
ICP-MS[a,b]	0.66

[a] Using a liquid argon supply.
[b] Using a collision/reaction cell.

Operating costs for all four AS techniques for the determination of 10 analytes per sample are summarized in Table 29.2.

It must also be emphasized that this comparison does not take into account the detection limit requirements, but is based on instrument-operating costs alone. These figures have been generated for a typical workload using what would be considered the average cost of gases, power, and consumables in the United States. Even though there will be geographical differences in the cost of these items in other parts of the world, the comparative costs should be very similar. Every laboratory's workload and analytical needs are unique, so this costing exercise should be treated with caution and only be used as a guideline for comparison purposes.

For a more accurate assessment of your lab's workload, the costing exercise should be carried using your sample throughput and analyte requirements. However, whatever the workload, it is a good exercise to show that there are running cost differences between the major AS techniques. If required, it can be taken a step further by also including the purchase price of the instrument, the cost of installing a clean room, the cost of sample preparation, and the salary of the operator. This would be a very useful exercise as it would give a good approximation of the overall cost of analysis, and therefore, it could be used as a guideline for calculating what a laboratory might charge for running samples on a commercial basis.

RUNNING COSTS OF ATOMIC FLUORESCENCE

Atomic fluorescence has intentionally not been used in this comparison, because there are so many variables to consider. Even though it uses hollow cathode-type lamps similar to FAA and ETA, it will also depend on the type of heating used in the atomization source, particularly if the system is dedicated to hydride/cold vapor generation. Some instruments use a flame to heat the quartz cell, whereas others use an electrically heated furnace. If combustion gases are used to generate the flame, running costs will vary based on whether the gas is acetylene or hydrogen (one system uses a hydrogen diffusion flame). If resistive electrical heating is used, it will also depend on the current used to generate the ground-state atoms of the analyte elements. Additionally, if mercury is one of the analytes, atomization occurs at room temperature so there is no requirement for heating the cell. Finally, argon gas is typically used to sweep the gaseous hydride from the reaction vessel into the atomization cell, so this also has to be considered as a cost factor.

As a result, all these variables make it very complicated to calculate the cost of running an atomic fluorescence spectrometry (AFS) system, especially when reducing chemicals such as stannous chloride or sodium borohydride have to be built into the calculation, which will clearly depend on the number of samples being run. This makes it very difficult to make a comparison with the other AS techniques, particularly as AFS can only measure a small group of elements at the same time. However, as a guideline, I think it's fair to say that the annual running costs of AFS will be more in line with FAA and ETA than either of the plasma spectrochemical techniques exemplified in Table 29.1.

FINAL THOUGHTS

It can be seen from this evaluation that based on the annual operating costs, FAA, ETA, and ICP-OES are all very similar, with ICP-MS being approximately twice as expensive to operate. However, when the number of samples/analytes is taken into consideration, the picture changes quite dramatically. It is also important to remember that there are many criteria to consider when selecting a trace element technique. Operating costs are just one of them, and they should not prevent you from choosing an instrument if your analytical requirements change, such as the need for lower detection limits. It's also important to emphasize that detection capability could be the most important selection criterion, particularly if you are trying to achieve the very low regulated limits of inhaled cannabinoid products. However, if more than one of these techniques fulfills your analytical demands, then knowledge of the operating costs should help you make the right decision.

FURTHER READING

1. R. Thomas, *Practical Guide to ICP-MS: A Tutorial for Beginners*, Third Edition, ISBN: 978-1-4665-5543-3 (CRC Press, Boca Raton, FL, 2013). https://www.crcpress.com/Practical-Guide-to-ICP-MS-A-Tutorial-for-Beginners-Third Edition/Thomas/9781466555433

30 How to Select an ICP Mass Spectrometer

Some Important Analytical Considerations

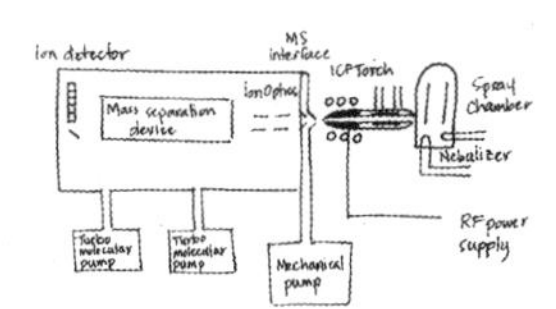

When sample homogeneity, sample preparation, and dilution factors are taken into consideration, inductively coupled plasma mass spectrometry (ICP-MS) is probably going to be the most suitable technique for carrying out the determination of elemental contaminants in cannabis and cannabis products because of its extremely low detection capability. Understanding the basic principles of ICP-MS is important but not absolutely essential to operate and use an instrument on a routine basis. However, understanding how these basic principles affect the performance of an instrument is a real benefit when evaluating the analytical capabilities of the technique. There are excellent commercial instruments on the market and are all capable of generating high-quality data. However, they all have their own strengths and weaknesses. For that reason, the better informed you are going into an instrument evaluation, the better chance you have of selecting the right one for your organization. Having been involved in demonstrating ICP-MS equipment and running customer samples for over almost 25 years, I know the mistakes that people make when they get into the evaluation process. So, in Chapter 30, I present a set of evaluation guidelines that hopefully will help you make the right decision. They are not specifically aimed at users in the cannabis industry, but are a systematic approach to comparing different commercial instrumentation.

OK, you have convinced your boss that ICP-MS is the ideal technique to meet the analytical demands of cannabis testing. Hopefully, the chapters on the fundamental principles have given you the basic knowledge and a good platform on which to go out and evaluate the marketplace. However, they do not really give you an insight into how to compare instrument designs, hardware components, and software features, which are of critical importance when you have to make a decision regarding which instrument to purchase. There are a number of high-quality commercial systems available in the marketplace, which look very similar and have very similar specifications, but how do you know which is the best one that fits your needs? This chapter, supported by the other chapters in this book, presents a set of evaluation guidelines to help you decide the most important analytical figures of merit for your application. You might not need to run all the tests, but experience has told me over the years that each one will give you valuable information about the performance of the instrument, depending on your evaluation objectives.

EVALUATION OBJECTIVES

It is very important before you begin the selection process to decide what your analytical objectives are. This is particularly important if you are part of an evaluation committee. It is fine to have more than one objective, but it is essential that all the members of the group begin the evaluation process with the objectives clearly defined. For example, is detection limit (DL) performance an important objective for your application, or is it more important to have an instrument that is easy to use? If the instrument is being used on a routine basis, maybe good reliability is also very critical. On the other hand, if the instrument is being used to generate revenue as is likely in the cannabis testing industry, perhaps sample throughput and cost of analysis are of greater importance. Every laboratory's scenario is unique, so it is important to prioritize before you begin the evaluation process. So, as well as looking at instrument features and components, the comparison should also be made with your analytical objectives in mind. Let us take a look at the most common ones that are used in the selection process. They typically include the following:

- Analytical performance
- Usability aspects
- Reliability issues
- Financial considerations.

Let us examine these in greater detail.

ANALYTICAL PERFORMANCE

Analytical performance can mean different things to different people. The major reason that the trace element community was attracted to ICP-MS over 30 years ago was its extremely low multielement DLs. Other multielement techniques, such as inductively coupled plasma optical emission spectroscopy (ICP-OES), offered very high throughput but just could not get down to ultratrace levels of ICP-MS. Even though electrothermal atomization (ETA) offered much better detection capability than ICP-OES, it did not offer the sample throughput capability that many applications demanded. In addition, ETA was predominantly a single-element technique and so was impractical for carrying out rapid multielement analysis. These limitations quickly led to the commercialization and acceptance of ICP-MS as a tool for rapid ultratrace element analysis. However, there are certain areas where ICP-MS is known to have weaknesses. For example, dissolved solids for most sample matrices must be kept below 0.2%; otherwise, it can lead to serious drift problems and poor precision.

Polyatomic and isobaric interferences, even in simple acid matrices, can produce unexpected spectral overlaps, which will have a negative impact on your data. High-resolution instrumentation and collision/reaction cell/interface technology are helping to alleviate these spectral problems, but they also have their limitations. Depending on the types of samples being analyzed, matrix components can dramatically suppress analyte sensitivity and affect accuracy. These potential problems can all be reduced to a certain extent, but different instruments compensate for these problematic areas in different ways. With a novice, it is often a basic lack of understanding of how a particular instrument works that makes the selection process more complicated than it really should be. So, any information that can help you prepare for the evaluation will put you in a much stronger position.

It should be emphasized that these evaluation guidelines are based on my personal experience and should be used in conjunction with other material in the open literature that has presented broad guidelines to compare figures of merit for commercial instrumentation.[1–3] In addition, you should talk with colleagues in the cannabis, pharmaceutical, or dietary supplement industries, who might have carried out an evaluation or are using a particular instrument for the analysis. For example, if they have gone through a lengthy evaluation process, they can give you valuable pointers or even suggest an instrument that is better suited to your needs. Finally, before we begin, it is strongly suggested that you narrow the actual evaluation to two, or maybe three, commercial products. By carrying out some preevaluation research, you will have a better understanding of what ICP-MS technology or instrument to focus on. For example, if funds are limited and you are purchasing ICP-MS for the very first time to carry out high-throughput environmental testing, it is probably more cost-effective to focus on single quadrupole technology. On the other hand, if you are investing in a second system to enhance the capabilities of your quadrupole instrument, it might be worth taking a look at magnetic sector or "triple quadrupole" collision/reaction cell technology to enhance your lab capabilities. Or, if fast multielement transient signal analysis is one of your major reasons for investing in ICP-MS, time-of-flight (TOF) technology should be given serious consideration (speciation studies would be included in this category). One final note I would like to add, although it is not strictly a technical issue. If you are prepared to forego an instrument demonstration or do not need any samples run, you will be in a much stronger position to negotiate price with the instrument vendor. You should keep that in mind before you decide to get involved in a lengthy selection process.

So, let us begin by looking at the most important aspects of instrument performance. Depending on the application, the major performance issues that need to be addressed include the following:

- Detection capability
- Precision/signal stability
- Accuracy
- Dynamic range
- Interference reduction
- Sample throughput
- Transient signal capability.

Detection Capability

Detection capability is a term used to assess the overall detection performance of an ICP mass spectrometer. There are a number of different ways of looking at detection capability, including instrument detection limit (IDL), elemental sensitivity, background signal, and background equivalent concentration (BEC). Of these four criteria, the IDL is generally thought to be the most accurate way of assessing instrument detection capability. It is often referred to as signal-to-background noise, and for a 99% confidence level is typically defined as 3× standard deviation (SD) of n replicates ($n = \sim 10$) of the sample blank and is calculated in the following manner:

$$\text{IDL} = \frac{3 \times \text{Standard deviation of background signal}}{\text{Analyte intensity} - \text{Background signal}} \times \text{Analyte concentration}$$

However, there are slight variations of both the definition and the calculation of IDLs, so it is important to understand how different manufacturers quote their DLs if a comparison is to be made. They are usually run in single-element mode, using extremely long integration times (5–10 s) to achieve the highest-quality data. So, when comparing DLs of different instruments, it is important to know the measurement protocol used.

A more realistic way of calculating analyte DL performance in your sample matrices is to use method detection limit (MDL). The MDL is broadly defined as the minimum concentration of analyte that can be determined from zero with 99% confidence. MDLs are calculated in a similar manner to IDLs, except that the test solution is taken through the entire sample preparation procedure before the analyte concentration is measured multiple times. This difference between MDL and IDL is exemplified in EPA Method 200.8, where a sample solution at two to five times the estimated IDL is taken through all the preparation steps and analyzed. The MDL is then calculated in the following manner:

$$\text{MDL} = (t) \times (S)$$

where t = Student's "t" value for a 95% confidence level and specifies a SD estimate with $n - 1$ degrees of freedom ($t = 3.14$ for seven replicates) and S = the SD of the replicate analyses.

Both IDL and MDL are very useful to understand the capability of ICP-MS. However, whatever method is used to compare DLs of different manufacturers' instrumentation, it is essential to carry out the test using realistic measurement times that reflect your analytical situation. For example, if you are determining a group of elements across the mass range in digested soil samples, it is important to know how much the sample matrix suppresses the analyte sensitivity, because the DL of each analyte will be impacted by the amount of suppression across the mass range. On the other hand, if you are carrying out high-throughput multielement analysis of drinking or wastewater samples, you probably need to be using relatively short integration times (1–2 s per analyte) to achieve the desired sample throughput. Or if you are dealing with transient signal from a chromatographic separation system, that only lasts 5–10 s, it is absolutely critical that you understand the impact that the time has on DLs compared to a continuous signal generated with a conventional nebulizer. (In fact, analysis time and DLs are very closely related to each other and will be discussed later on in this chapter.) In other words, when comparing IDLs, it is absolutely critical that the tests represent your real-world analytical situation.

Elemental sensitivity is also a useful assessment of instrument performance, but it should be viewed with caution. It is usually a measurement of background-corrected intensity at a defined mass and is typically specified as counts per second (cps) per concentration (ppb or ppm) of a mid-mass element such as $^{103}Rh^+$ or $^{115}In^+$. However, unlike DL, raw intensity usually does not tell you anything about the intensity of the background or the level of the background noise. It should be emphasized that instrument sensitivity can be enhanced by optimization of operating parameters such as RF power, nebulizer gas flows, torch-sampling position, interface pressure, and sampler/skimmer cone geometry, but usually comes at the expense of other performance criteria, including oxide levels, matrix tolerance, or background intensity. So, be very cautious when you see an extremely high-sensitivity specification, because there is a strong probability that the oxide or background specs might also be high. For this reason, it is unlikely there will be an improvement in DL unless the increase in sensitivity comes with no compromise in the background level. It is also important to understand the difference between background and background noise when comparing specifications (the background noise is a measure of the stability of the background and is defined as the square root of the background signal). Most modern quadrupole instruments today specify 150–200 million cps per ppm of rhodium ($^{103}Rh^+$) or indium ($^{115}In^+$) and <1–2 cps of background (usually at 220 amu), whereas magnetic sector instrument sensitivity specifications are typically 10–20 times the higher with 10 times the lower background.

Another figure of merit that is being used more routinely nowadays is BEC. BEC is defined as the intensity of the background at the analyte mass, expressed as an apparent concentration, and is typically calculated in the following manner:

$$\text{BEC} = \frac{\text{Intensity of background signal}}{\text{Analyte intensity} - \text{Background intensity}} \times \text{Analyte concentration}$$

It is considered more of a realistic assessment of instrument performance in real-world sample matrices (especially if the analyte mass sits on a high background), because it gives an indication of the level of the background—defined as a concentration value. DLs alone can sometimes be misleading because they are influenced by the number of readings taken, integration time, cleanliness of the blank, and at what mass the background is measured, and are rarely achievable in a real-world situation. Figure 30.1 emphasizes the difference between DL and BEC.

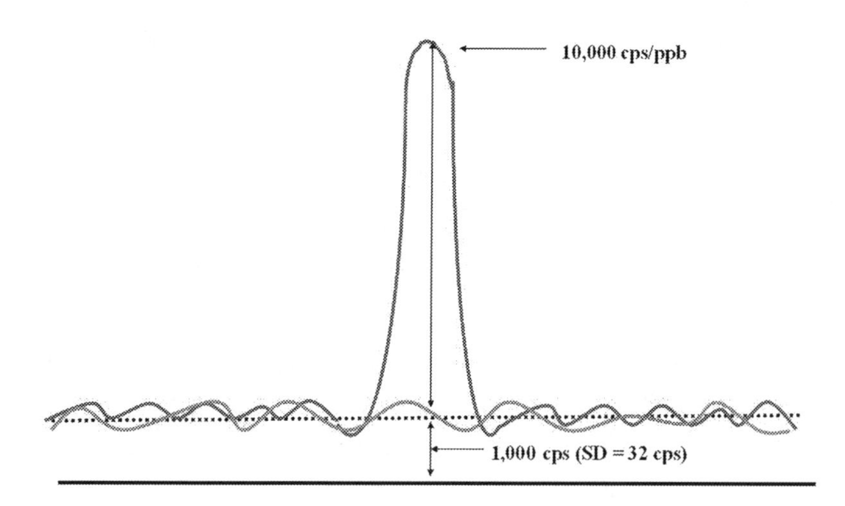

FIGURE 30.1 DL is calculated using the noise of the background, whereas BEC is calculated using the intensity of the background.

In this example, 1 ppb of an analyte produces a signal of 10,000 cps and a background of 1,000 cps. Based on the calculations defined earlier, the BEC is equal to 0.11 ppb because it is expressing the background intensity as a concentration value. On the other hand, the DL is ten times lower because it is using the SD of the background (i.e., the noise) in the calculation. For this reason, BECs are particularly useful when it comes to comparing the detection capabilities of techniques such as cool plasma and collision/reaction cell technology, because it gives you a very good indication of how efficient the background reduction process is.

It is also important to remember that peak measurement protocol will also have an impact on detection capability. As mentioned in Chapter 15, there are basically two approaches to measuring an isotopic signal in ICP-MS. There is the multichannel scanning approach, which uses a continuous smooth ramp of 1–20 channels per mass across the peak profile, and peak-hopping approach, where the mass analyzer power supply is driven to a discrete position on the peak, allowed to settle, and a measurement taken for a fixed period of time. This is usually at the peak maximum but can be as many points as the operator selects. This process is simplistically shown in Figure 30.2.

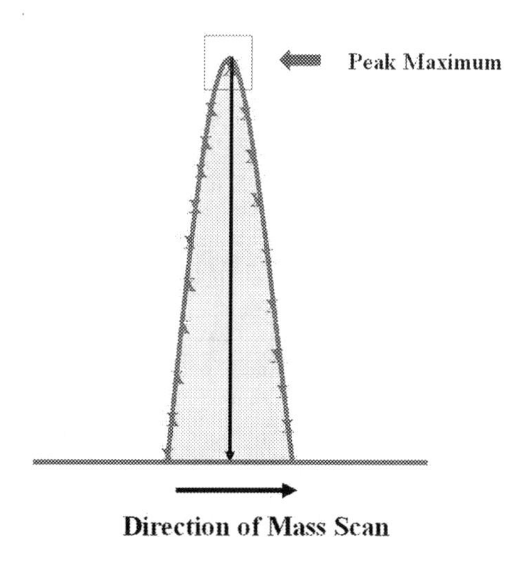

FIGURE 30.2 There are typically two approaches to peak quantitation—peak hopping (usually at peak maximum) and multichannel scanning (across the full width of the peak).

The scanning approach is best for accumulating spectral and peak shape information when doing mass calibration and resolution scans. It is traditionally used as a classical method development tool to find out what elements are present in the sample and to assess spectral interferences on the masses of interest. However, when the best possible DLs are required, it is clear that the peak-hopping approach is best. It is important to understand that to get the full benefit of peak hopping; the best DLs are achieved when single-point peak hopping at the peak maximum is chosen. It is well accepted that measuring the signal at the peak maximum will always give the best signal-to-background noise for a given integration time, and there is no benefit in spreading your available integration time over more than one measurement point per mass.[4] Instruments that use more than one point per peak for quantitation are sacrificing measurement time on the sides of the peak, where the signal-to-noise ratio is worse. However, the ability of the mass analyzer to repeatedly scan to the same mass position every time during a multielement run is of paramount importance for peak hopping. If multiple points per peak are recommended, it is a strong indication that the spectrometer has poor mass calibration stability, because it cannot guarantee that it will always find the peak maximum with just one point. Mass calibration specification, which is normally defined as a shift in peak position (in amu) over an 8-hour period, is a good indication of mass stability. However, it is not always the best way to compare systems, because peak algorithms using multiple points are often used to calculate the peak position. A more accurate way is to assess the short-term and long-term mass stability by looking at relative peak positions over time. The short-term stability can be determined by aspirating a multielement solution containing four elements (across the mass range) and recording the spectral profiles using multichannel ramp scanning of 20 points per peak. Now repeat the multielement scan ten times and record the peak position of every individual scan. Calculate the average and relative standard deviation (RSD) of the scan positions. The long-term mass stability can then be determined by repeating the test 8 h later to see how far the peaks have moved. It is important, of course, that the mass calibration procedure not be carried out during this time. Figure 30.3 shows what might happen to the peak position over time, if the analyzer's mass stability is poor.

Precision

Short- and long-term precision specifications are usually a good indication of how stable an instrument is (refer to Chapter 15). Short-term precision is typically specified as %RSD of ten replicates of 1–10 ppb of three elements across the mass range using 2–3-second integration times, whereas

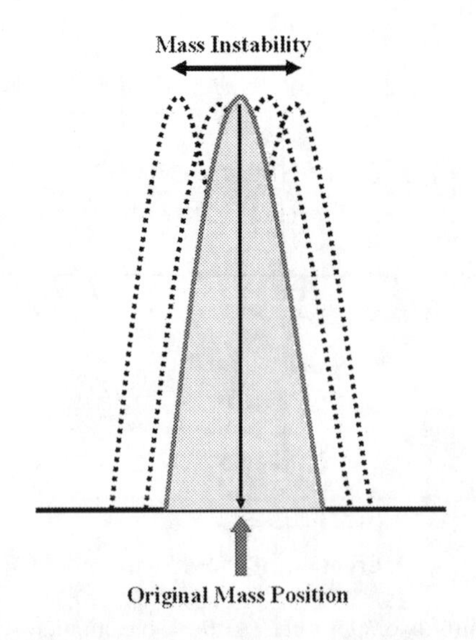

FIGURE 30.3 Good mass stability is critical for single-point, peak-hopping quantitation.

long-term precision is a similar test, but normally carried out every 5–10 min over a 4–8-hour period. Typical short-term precision, assuming an instrument warm-up time of 30–40 min, should be approximately 1%–3%, whereas long-term precision should be on the order of 3%–5%—both determined without using internal standards. However, it should be emphasized that under these measurement protocols, it is unlikely you will see a big difference in the performance between different instruments in simple aqueous standards. A more accurate reflection of the stability of an instrument is to carry out the tests using a typical matrix that would be run in your laboratory at the concentrations you expect. It is also important that stability should be measured without the use of an internal standard. This will enable you to evaluate the instrument drift characteristics, without any type of signal correction method being applied.

It is recognized that the major source of drift and imprecision in ICP-MS, particularly with real-world samples, is associated with either the sample introduction area, design of the interface, or the ion optics system. Some of the most common problems encountered with plant materials, organic matrices, agricultural samples, and food-based products are as follows:

- Pulsations and fluctuations in the peristaltic pump (if one is used), leading to increased signal noise
- Blockage of the nebulizer over time, resulting in signal drift—especially if the nebulizer does not have a tolerance for high dissolved solids
- Poor drainage, producing pressure changes in the spray chamber and resulting in spikes in the signal
- Buildup of solids in the sample injector, producing signal drift
- Changes in the electrical characteristics of the plasma, generating a secondary discharge and increasing ion energies
- Blockage of the sampler and skimmer cone orifice with sample material, causing instability, particularly challenging when aspirating carbon-based samples
- Erosion of the sampler and skimmer cone orifice with high-concentration acids
- Coating of the ion optics with matrix components, resulting in slight changes in the electrical characteristics the of ion lens system.

These can all be somewhat problematic depending on the types of samples being analyzed. However, the most common and potentially serious problem with real-world matrices is the deposition of sample material on the interface cones and the ion optics over time. It does not impact short-term precision that much, because careful selection of internal standards matched to the analyte masses can compensate for slight instability problems. However, sample material, particularly matrix components found in environmental, botanical, biological, and soil samples, can have a dramatic effect on long-term stability. The problem is exaggerated even more in a high-throughput laboratory, because poor stability will necessitate more regular recalibration and might even require some samples to be rerun if QC standards fall outside certain limits. There is no question that if an instrument has poor drift characteristics, it will take much longer to run an autosampler tray full of samples, and in the long term, this will result in much higher argon consumption.

For these reasons, it is critical that when short- and long-term precisions are evaluated, you know all the potential sources of imprecision and drift. It is therefore important that you choose a matrix that is representative of your cannabis-related samples, and will genuinely test the instrument out.

Whatever matrices are chosen, it must be emphasized that for the stability test to be meaningful, no internal standards should be used, the sample should contain less than 0.2% total dissolved solids, and the representative elements should be at a reasonably high concentration (1–10 ppb) and be spread across the mass range. In addition, no recalibration should be carried out for the length of the test, which should reflect your real-world situation.[5] For example, if you plan to run your instrument in a high-throughput environment, you might want to carry out an 8-hour or even an overnight (12–16 h) stability test. If you are not interested in such long runs, a 2–4-hour stability test will probably suffice. However, just remember, plan the test beforehand and make sure you know how to evaluate the vast amount of data that will be generated. It will be hard work, but I guarantee it will be worth it in order to fully understand the short- and long-term drift characteristics of the instruments you are evaluating.

One word of caution. If a syringe or piston pump is being used to deliver the sample, it will almost certainly give you better precision than a peristaltic pump. So make sure you are comparing like with like. It is almost pointless comparing the stability of an instrument that uses a peristaltic pump with that of an instrument that uses a syringe pump. If you are unsure, make sure that the same pumping system is being used on both instruments.

Isotope Ratio Precision

An important aspect of ICP-MS is its ability to carry out rapid isotope precision analysis, particularly if multiple isotopes are being used for quantitation. There is much information in the public domain about using carbon isotopes to determine whether the cannabis has been grown indoors or outside. By monitoring the C^{12}/C^{13} isotopic ratios by ICP-MS, it can be determined if the plant was grown outdoors, indoors using natural respiration, or indoors using artificial CO_2 fertilization. With this example, two different isotopes of carbon are continuously measured over a fixed period of time. The ratio of the signal of one isotope to the other isotope is taken, and the precision of the ratios is then calculated. This ratio will be a very good indicator as to how the plant was grown (https://atlasofscience.org/carbon-isotopes-the-chemists-tool-to-trace-marijuana-cultivation-environment/).

Analysts interested in isotope ratios are usually looking for the ultimate in precision. The optimum way to achieve this to get the best counting statistics would be to carry out the measurement simultaneously with a multicollector magnetic sector instrument, or a TOF ICP-MS system. However, a quadrupole mass spectrometer is a rapid sequential system, so the two isotopes are never measured at exactly the same moment. This means that the measurement protocol must be optimized to get the best precision. As discussed earlier, the best and most efficient use of measurement time is to carry out single-point peak hopping between the two isotopes. In addition, it is also beneficial to be able to vary the total measurement time of each isotope, depending on their relative abundance. The ability to optimize the dwell time and the number of sweeps of the mass analyzer ensures that the maximum amount of time is being spent on the top of each individual peak where the signal-to-noise ratio is at its highest.[6]

It is also critical to optimize the efficiency cycle of the measurement. With every sequential mass analyzer, there is an overhead time called a *settling time* to allow the power supply to settle before taking a measurement. This time is often called non-analytical time, because it does not contribute to the quality of the analytical signal. The only time that contributes to the analytical signal is the *dwell time* or the time that is actually spent measuring the peak. The measurement efficiency cycle (MEC) is a ratio of the dwell time to the total analytical time (which includes settling time) and is expressed as follows:

$$\text{MEC}\ (\%) = \frac{\text{No. of sweeps} \times \text{Dwell time} \times 100}{\left\{\text{No. of sweeps} \times \left(\text{Dwell time} + \text{Settling time}\right)\right\}}$$

It is therefore obvious that to get the best precision over a fixed period of time, the settling time must be kept to an absolute minimum. The dwell time and the number of sweeps are operator selectable, but the settling time is usually fixed because it is a function of the mass analyzer electronics. For this reason, it is important to know what the settling time of the mass spectrometer is when carrying out peak hopping. Remember, a shorter settling time is more desirable because it will increase the MEC and improve the quality of the analytical signal.[7]

In addition, if isotope ratios are being determined on vastly different concentrations of major and minor isotopes using the extended dynamic range of the system, it is important to know the settling time of the detector electronics. This settling time will affect the detector's ability to detect the analog and pulse signals (or in dynamic attenuation mode with a pulse-only EDR system) when switching between measurement of the major and minor isotopes, which could have a serious impact on the accuracy and precision of the isotope ratio. So, for that reason, no matter how the higher concentrations are handled, shorter settling times are more desirable, so that the switching or attenuation can be carried out as quickly as possible.

This is shown in Figure 30.4, which shows a spectral scan of $^{63}Cu^+$ and $^{65}Cu^+$ using an automated pulse/analog EDR detection system. The natural abundance of these two isotopes is 69.17% and 30.83%, respectively. However, the ratio of these isotopes has been artificially altered to be 0.39% for $^{63}Cu^+$ and 99.61% for $^{65}Cu^+$. The intensity of ^{63}Cu is about 70,000 cps, which requires pulse counting, whereas the intensity of the $^{65}Cu^+$ is about 10 million cps, which necessitates analog counting. There is no question that the counting circuitry would miss many of the ions and generate erroneous concentration data if the switching between pulse and analog modes was not fast enough.

So, when evaluating isotopic ratio precision with a scanning device like a quadrupole, it is important that the measurement protocol and peak quantitation procedure are optimized. Isotope precision specifications are a good indication regarding what the instrument is capable of, but once again, these will be defined in aqueous-type standards, using relatively short total measurement times (typically 5 min). For that reason, if the test is to be meaningful, it should be optimized to reflect your real-world analytical situation. That is why if very high precision (low RSDs) is a requirement of the analysis, then quadrupole ICP-MS is probably not the optimum technique to use.

Accuracy

Accuracy is a very difficult aspect of instrument performance to evaluate because it often reflects the skill of the person developing the method and analyzing the samples, instead of the capabilities of the instrument itself. If handled correctly, it is a very useful exercise to go through, particularly

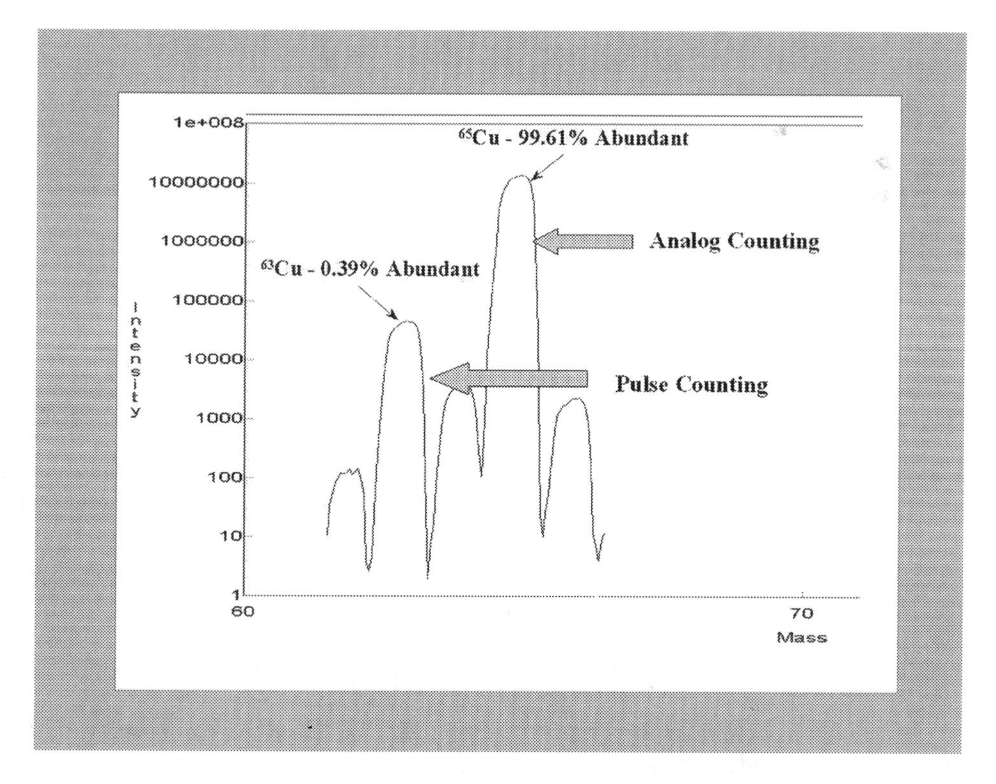

FIGURE 30.4 The detector electronics must be able to switch fast enough to detect isotope ratios that require both pulse and analog counting modes. (Copyright 2013, all rights reserved, PerkinElmer Inc.)

if you can get hold of reference material (ideally of similar matrices to your own) whose values are well defined. However, when attempting to compare the accuracy of different instruments, it is essential that you prepare every sample yourself, including the calibration standards, blanks, unknown samples, QC standards, or certified reference material (CRM) if available. I suggest that you make up enough of each solution to give to each vendor for analysis. By doing this, you eliminate the uncertainty and errors associated with different people making up different solutions. It then becomes more of an assessment of the capability of the instrument, including its sample introduction system, interface region, ion optics, mass analyzer, detector, and measurement circuitry, to handle the unknown samples, minimize interferences, and get the correct results.

A word of caution should be expressed at this point. Having worked in the field of ICP-MS for over 35 years, I know that the experience of the person developing the method, running the samples, and performing the demonstration has a direct impact on the quality of the data generated in ICP-MS. There is no question in my mind that the analyst with the most application expertise has a much better chance of getting the right answer than someone who is either inexperienced or not familiar with a particular type of sample. I think it is quite valid to compare the ability of the application specialist because this might be the person who is supporting you. However, if you want to assess the capabilities of the instrument alone, it is essential to take the skill of the operator out of the equation. This is not as straightforward as it sounds, but I have found that the best way to "level the playing field" is to send some of your sample matrices to each vendor before the actual demonstration. This allows the application person to spend time developing the method and become familiar with the samples. You can certainly hold back on your CRM or QC standards until you get to the demonstration, but at least it gives each vendor some uninterrupted time with your samples. This also allows you to spend most of the time at the demonstration evaluating the instrument, assessing hardware components, comparing features, and getting a good look at the software. It is my opinion that most instruments on the market should get the right answer—at least for the majority of routine applications. So, even though the accuracy of different instruments should be compared, it is more important to understand how the result was generated, especially when it comes to the analysis of very difficult samples. This is especially true with a triple quad ICP-MS system fitted with a collision/reaction cell. Method development with these kinds of instruments, whether it's carried out by a skilled operator, or using sophisticated decision-making software, is critical to achieving high-quality data.

Dynamic Range

When ICP-MS was first commercialized, it was primarily used to determine very low analyte concentrations. As a result, detection systems were only asked to measure concentration levels up to approximately five orders of magnitude. However, as the demand for greater flexibility grew, such systems were being called upon to extend their dynamic range to determine higher and higher concentrations. Today, the majority of commercial systems come standard with detectors that can measure signals up to ten orders of magnitude.

As mentioned in Chapter 14, there are subtle differences between how various detectors and detection systems achieve this, so it is important to understand how different instruments extend the dynamic range. The majority of quadrupole-based systems on the market extend the dynamic range by using a discrete dynode detector operated either in pulse-only mode or a combination of pulse and analog mode. When evaluating this feature, it is important to know whether this is done in one or two scans because it will have an impact on the types of samples you can analyze. The different approaches have been described earlier, but it is worth briefly going through them again:

- **Two-scan approach:** Basically, two types of two-scan or prescan approaches have been used to extend the dynamic range. In the first one, a survey or prescan is used to determine what masses are at high concentrations and what masses are at trace levels. Then, the second scan actually measures the signals by switching rapidly between analog and pulse counting. In the second two-scan approach, the detector is first run in the analog mode to measure the high signals and then rescanned in pulse-counting mode to measure the trace levels.

- **One-scan approach:** This approach is used to measure both the high levels and trace concentrations simultaneously in one scan. This is typically achieved by measuring the ion flux as an analog signal at some midpoint on the detector. When more than a threshold number of ions are detected, the ions are processed through the analog circuitry. When less than a threshold number of ions are detected, the ions cascade through the rest of the detector and are measured as a pulse signal in the conventional way.
- **Using pulse-only mode:** The most recent development in extending the dynamic range is to use the pulse-only signal. This is achieved by monitoring the ion flux at one of the first few dynodes of the detector (before extensive electron multiplication has taken place) and then attenuating the signal by applying a control voltage. Electron pulses passed by the attenuation section are then amplified to yield pulse heights that are typical in normal pulse-counting applications. Under normal circumstances, this approach requires only one scan, but if the samples are complete unknowns, dynamic attenuation might need to be performed, where an additional pre-measurement time is built into the settling time to determine the optimum detector attenuation for the selected dwell times used.

The methods that use a prescan or pre-measurement time work very well, but they do have certain limitations for some applications, compared to the one-scan approach. Some of these include the following:

- The additional scan/measurement time means it will use more of the sample. Ordinarily this will not pose a problem, but if sample volume is limited to a few hundred microliters, it might be an issue.
- If concentrations of analytes are vastly different, the measurement circuitry reaction time of a prescan system might struggle to switch quickly enough between high- and low-concentration elements. This is not such a major problem, unless the measurement circuitry has to switch rapidly between consecutive masses in a multielement run, or there are large differences in the concentrations of two isotopes of the same element when carrying out ratio studies. In both these situations, there is a possibility that the detection system will miss counting some of the ions and produce erroneous data.
- The other advantage of the one-scan approach is that more time can be spent measuring the peaks of interest in a transient peak generated by a flow injection or laser ablation system. With a detector that uses two scans or a prescan approach, much of the time will be used just to characterize the sample. It is exaggerated even more with a transient peak, especially if the analyst has no prior knowledge of elemental concentrations in the sample.

This final point is exemplified in Figure 30.5, which shows the measurement of a flow injection peak of NIST 1643C potable water CRM, using an automated simultaneous pulse/analog EDR system. It can be seen that the K and Ca are at ppm levels, which requires the use of the analog counting circuitry, whereas the Pb and Cd are at ppb levels, which requires pulse counting.[8] This would not be such a difficult analysis for a detector, except that the transient peak has only lasted 10 s. This means that to get the highest-quality data, you want to be spending all the available time quantifying the peak. In other words, you cannot afford the luxury of doing a pre-measurement, especially if you have no prior knowledge of the analyte concentrations.

For these reasons, it is important to understand how the detector handles high concentrations to evaluate them correctly. If you are truly interested in using ICP-MS to determine higher concentrations, you should check out the linearity of different masses across the mass range by measuring low ppt (~10 ppt), low ppb (~10 ppb), and high ppm (~100 ppm) levels. Do not be hesitant to analyze

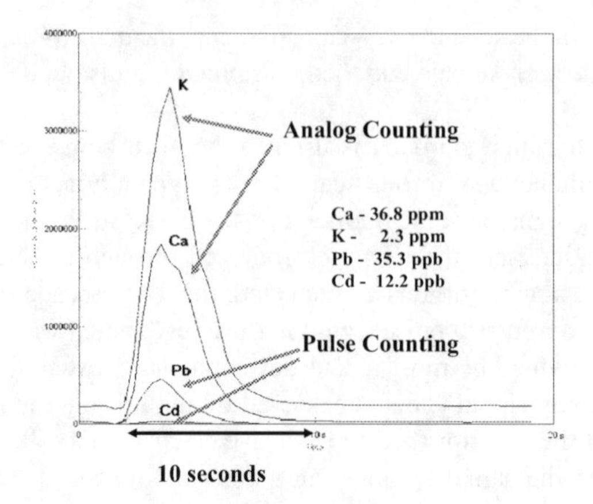

FIGURE 30.5 A one-scan approach to extending the dynamic range is more advantageous for handling a fast transient flow injection peak, in this example, generated by NIST 1643C. (From E. R. Denoyer and Q. H. Lu, *Atomic Spectroscopy,* **14**(6), 162–169, 1993.)

a standard reference material (SRM) sample such as one of the NIST 1643 series of drinking water reference standards, which has both high (ppm) and low (ppb) levels. Finally, if you know there are large concentration differences between the same analytes, make sure the detector is able to determine them with good accuracy and precision. On the other hand, if your instrument is only going to be used to carry out ultratrace analysis, it probably is not worth spending the time to evaluate the capability of the extended dynamic range feature.

However, it should be strongly emphasized that irrespective of which extended range technology is used, if low and high concentrations of the same analyte are expected in a suite of samples, it is unrealistic to think you can accurately quantitate down at the low end and at the top end of the linear range with the same calibration graph. If you want to achieve accurate and precise data at or near the limit of quanitation, you must run a set of appropriate calibration standards to cover your low-level samples. In addition, if you are expecting high and low concentrations in the same suite of samples, you have to be absolutely sure that a high-concentration sample has been thoroughly washed out from the spray chamber/nebulizer system, before a low-level sample is introduced. For this reason, caution must be taken when setting up the method with an autosampler, because if the read delay/integration times are not optimized for a suite of samples, erroneous results can be generated, which might necessitate a rerun under the manual supervision of the instrument operator.

Interference Reduction

As mentioned in Chapter 17, there are two major types of interferences that have to be compensated for spectral and matrix (space charge and physical). Although most instruments approach the principles of interference reduction in a similar way, the practical aspect of compensating for them will be different, depending on the differences in hardware components and instrument design. Let us look at interference reduction in greater detail and compare the different approaches used.

Reducing Spectral Interferences

The majority of spectral interferences seen in ICP-MS are produced by the sample matrix, solvent, plasma gas, or various combinations of them. If the interference is caused by the sample, the best approach might be to remove the matrix by some kind of ion exchange column. However, this can be cumbersome and time-consuming to do on a routine basis. If the interference is caused by solvent ions, simply desolvating the sample will have a positive effect on reducing the interference. For that reason, systems that come standard with chilled spray chambers to remove much of the solvent usually generate less sample-based oxide-, hydroxide-, and hydride-induced spectral interferences.

There are alternative ways to reduce these types of interferences, but cooling the spray chamber or using a membrane desolvation system can be a very effective way of reducing the intensity of the solvent-based ionic species.

Spectral interferences are an unfortunate reality in ICP-MS, and it is now generally accepted that instead of trying to reduce or minimize them, the best way is to resolve the problem away using high-resolution technology such as a double-focusing magnetic sector mass analyzer.[9] Even though they are not considered ideal for a routine, high-throughput laboratory, they offer the ultimate in resolving power and have found a niche in applications that require ultratrace detection and a high degree of flexibility for the analysis of complex sample matrices. If you use a quadrupole-based instrument and are looking to purchase a second system to enhance the flexibility of your laboratory, it might be worth taking a serious look at magnetic sector technology. The full benefits of this type of mass analyzer for ICP-MS have been described in Chapter 11.

Let us now turn our attention to the different approaches used to reduce spectral interferences using quadrupole-based technology. Each approach should be evaluated on the basis of its suitability for the demands of your particular application.

Resolution Improvement

As described in Chapter 10, there are two very important performance specifications of a quadrupole—resolution and abundance sensitivity.[10] Although they both define the ability of a quadrupole to separate an analyte peak from a spectral interference, they are measured differently. Resolution reflects the shape of the top of the peak and is normally defined as the width of a peak at 10% of its height. Most instruments on the market have similar resolution specifications of 0.3–3.0 amu and typically use a nominal setting of 0.7–1.0 amu for all masses in a multielement run. For this reason, it is unlikely you will see any measurable difference when you make your comparison.

However, some systems allow you to change resolution settings on the fly on individual masses during a multielement analysis. Under normal analytical scenarios, this is rarely required, but at times, it can be advantageous to improve the resolution for an analyte mass, particularly if it is close to a large interference and there is no other mass or isotope available for quantitation. This can be seen in Figure 30.6, which shows a spectral scan of 10 ppb $^{55}Mn^{+}$, which is monoisotopic

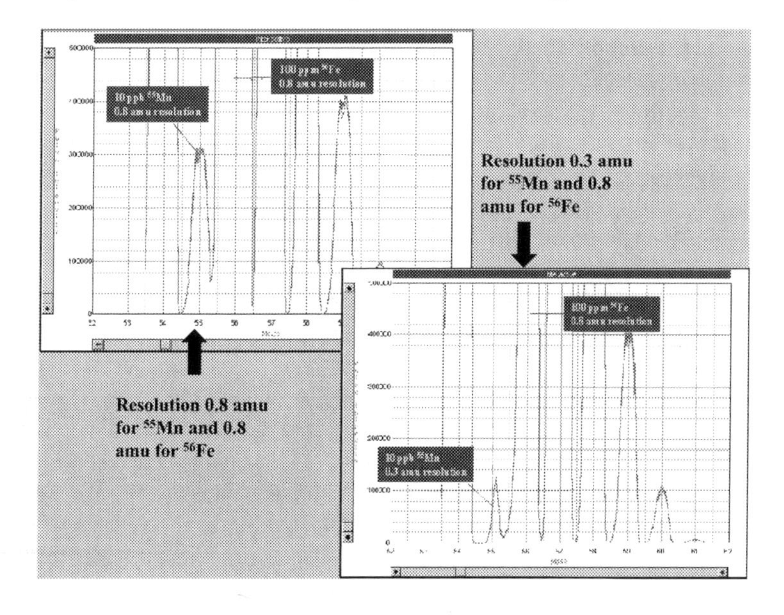

FIGURE 30.6 A resolution setting of 0.3 amu will improve the DL for $^{55}Mn^{+}$ in the presence of high concentrations of $^{56}Fe^{+}$. (Copyright 2013, all rights reserved, PerkinElmer Inc.)

and 100 ppm of $^{56}Fe^+$. The left-hand spectra shows the scan using a resolution setting of 0.8 amu for both $^{55}Mn^+$ and $^{56}Fe^+$, whereas the right-hand spectra shows the same scan, but using a resolution setting of 0.3 amu for $^{55}Mn^+$ and 0.8 amu for $^{56}Fe^+$. Even though the $^{55}Mn^+$ peak intensity is about three times lower at 0.3 amu resolution, the background from the tail of the large $^{56}Fe^+$ is about 7 times less, which translates into a fivefold improvement in the $^{55}Mn^+$ DL at a resolution of 0.3 amu compared to 0.8 amu.

Higher Abundance Sensitivity Specifications

The second important specification of a mass analyzer is abundance sensitivity, which is a measure of the width of a peak at its base. It is defined as the signal contribution of the tail of a peak at one mass lower and one mass higher than the analyte peak, and generally speaking, the lower the specification, the better the performance of the mass analyzer. The abundance sensitivity of a quadrupole is determined by a combination of factors, including shape, diameter, and length of the rods; frequency of quadrupole power supply; and slope of the applied RF/DC voltages. Even though there are differences between designs of quadrupoles in commercial ICP-MS systems, there appears to be very little difference in their practical performance.

When comparing abundance sensitivity, it is important to understand what the numbers mean. The trajectory of an ion through the analyzer means that the shape of the peak at one mass lower than the mass M, i.e., $(M - 1)$, is slightly different from the other side of the peak at one mass higher, i.e., $(M + 1)$. For this reason, the abundance sensitivity specification for all quadrupoles is always worse on the low-mass side (–) than the high-mass side (+), and is typically 1×10^{-6} at $M - 1$ and 1×10^{-7} at $M + 1$. In other words, an interfering peak of 1 million cps at $M - 1$ would produce a background of 1 cps at M, whereas it would take an interference of 10 million cps at $M + 1$ to produce a background of 1 cps at M. In theory, hyperbolic rods will demonstrate better abundance sensitivity than round ones, as will a quadrupole with longer rods and a power supply with higher frequency. However, you have to evaluate whether this produces any tangible benefits when it comes to the analysis of your real-world samples.

Use of Cool Plasma Technology

All of the instruments on the market can be set up to operate under cool or cold plasma to achieve very low DLs for elements such as K, Ca, and Fe. Cool plasma conditions are achieved when the temperature of the plasma is cooled sufficiently low enough to reduce the formation of argon-induced polyatomic species.[11] This is typically achieved with a decrease in the RF power, an increase in the nebulizer gas flow, and sometimes a change in the sampling position of the plasma torch. Under these conditions, the formation of species such as $^{40}Ar^+$, $^{38}ArH^+$, and $^{40}Ar^{16}O^+$ is dramatically reduced, which allows the determination of low levels of $^{40}Ca^+$, $^{39}K^+$, and $^{56}Fe^+$, respectively.[12]

Under normal hot plasma conditions (typically, RF power of 1200–1600 W and a nebulizer gas flow of 0.8–1.0 L/min), these isotopes would not be available for quantitation because of the argon-based interferences. Under cool plasma conditions (typically, RF power of 600–800 W and a nebulizer gas flow of 1.2–1.6 L/min), the most sensitive isotopes can be used, offering low ppt detection in aqueous matrices. However, not all instruments offer the same level of cool plasma performance, so if these elements are important to you, it is critical to understand what kind of detection capability is achievable. A simple way to test cool plasma performance is to look at the BEC for iron at mass 56 with respect to cobalt at mass 59. This enables the background at mass 56 to be compared to a surrogate element such as Co, which has a similar ionization potential to Fe, without actually introducing Fe into the system and contributing to the ArO^+ background signal. When carrying out this test, it is important to use the cleanest deionized water to guarantee that there is no Fe in the blank. First, measure the background in counts/second at mass 56 aspirating deionized water. Then, record the analyte intensity of a 1 ppb Co solution at mass 59. The ArO^+ BEC can be calculated as follows:

$$\text{BEC (ArO}^+) = \frac{\text{Intensity of deionized water background at mass } 56 \times 1 \text{ ppb}}{\text{Intensity of 1 ppb of Co at mass } 59 - \text{background at mass } 56}$$

The ArO^+ BEC at mass 56 will be a good indication of the DL for $^{56}Fe^+$ under cool plasma conditions. The BEC value will typically be about an order of magnitude greater than the DL.

Although most instruments offer cool plasma capability, there are subtle differences in the way it is implemented. It is therefore very important to evaluate the ease of setup and how easy it is to switch from cool to normal plasma conditions and back in an automated multielement run. Also, remember that there will be an equilibrium time in switching from normal to cool plasma conditions. Make sure you know what this is, because an equivalent read delay will have to be built into the method, which could be an issue if speed of analysis is important to you. If in doubt, set up a test to determine the equilibrium time by carrying out a short stability run while switching back and forth between normal and cool plasma conditions.

It is also critical to be aware that the electrical characteristics of a cool plasma are different from normal plasma. This means that unless there is a good grounding mechanism between the plasma and the RF coil, a secondary discharge can easily occur between the plasma and sampler cone. The result is an increased spread in kinetic energy of the ions entering the mass spectrometer, making them more difficult to control and steer through the ion optics into the mass analyzer. So, understand how this grounding mechanism is implemented and whether any hardware changes need to be made when going from cool to normal plasma conditions and vice versa (testing for a secondary discharge will be discussed later).

It should be noted that one of the disadvantages of the cool plasma approach is that cool plasma contains much less energy than a normal, high-temperature plasma. As a result, elemental sensitivity for the majority of elements is severely affected by the matrix, which basically precludes its use for the analysis of samples with a real matrix, unless the necessary steps are taken. This is shown in Figure 30.7, which shows cool plasma sensitivity for a selected group of elements in varying concentrations of nitric acid, and Figure 30.8, which shows the same group of elements under normal plasma conditions. It can be seen clearly that analyte sensitivity is dramatically reduced in a cool plasma as the acid concentration is increased, whereas, under normal plasma conditions, the sensitivity for most of the elements varies only slightly with increasing acid concentration.[13]

In addition, because a cool plasma contains much less energy than a normal plasma, chemical matrices and acids with a high boiling point are often difficult to decompose in the plasma, which

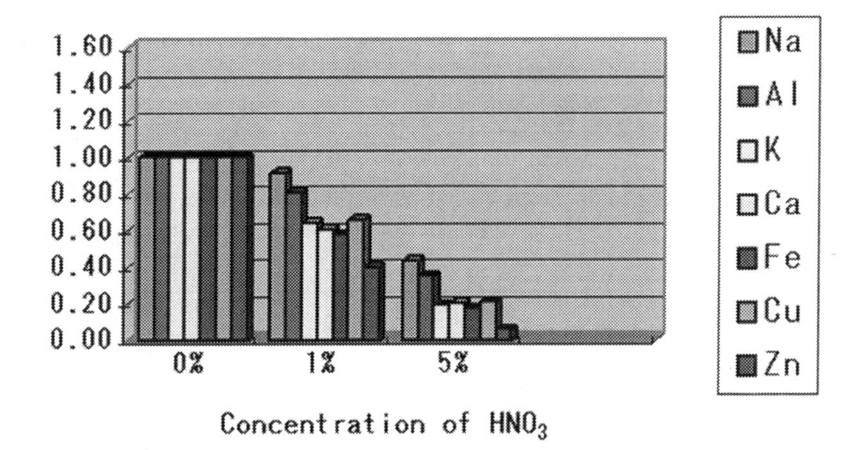

FIGURE 30.7 Sensitivity for a selected group of elements in varying concentrations of nitric acid, using cool plasma conditions (RF power—800 W, nebulizer gas—-1.5 L/min). (From J. M. Collard, K. Kawabata, Y. Kishi, and R. Thomas, *Micro*, **20**(1), 39–46, January 2002.)

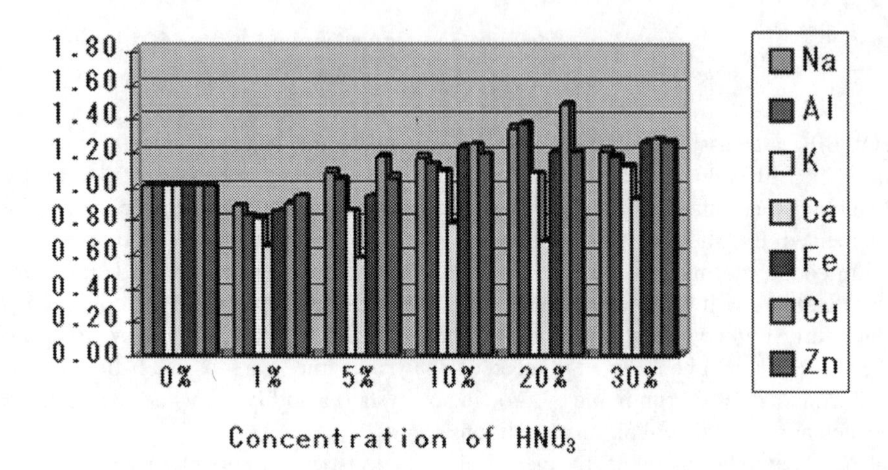

FIGURE 30.8 Sensitivity for a selected group of elements in varying concentrations of nitric acid, using normal plasma conditions (RF power—1,600 W, nebulizer gas—1.0 L/min). (From J. M. Collard, K. Kawabata, Y. Kishi, and R. Thomas, *Micro*, **20**(1), 39–46, January 2002.)

has the potential to cause corrosion problems on the interface of the mass spectrometer. This is the inherent weakness of the cool plasma approach—instrument performance is highly dependent on the sample being analyzed. As a result, unless simple aqueous-type samples are being analyzed, cool plasma operation often requires the use of standard additions or matrix matching to achieve satisfactory results. Additionally, to obtain the best performance for a full suite of elements, a multielement analysis often necessitates the use of two sets of operating conditions—one run for the cool plasma elements and another for normal plasma elements—which can be both time- and sample-consuming.

In fact, these application limitations have led some vendors to reject the cool plasma approach in favor of collision/reaction cell technology. So, it could be that the cool plasma capability of an instrument may not be that important if the equivalent elements are superior using the collision/reaction cell option. However, you should proceed with caution in this area, because on the current evidence, not all collision/reaction cell instruments offer the same kind of performance. For some instruments, cool plasma DLs are superior to the same group of elements determined in the collision cell mode. For that reason, an assessment of the suitability of using cool plasma conditions or collision/reaction cell technology for a particular application problem has to be made based on the vendor's recommendations.

For example, the recent development of a novel 34 MHz free-running designed RF generator using solid-state electronics has enhanced the capability of ICP-MS to analyze some real-world samples, particularly when using cool plasma conditions (refer to Chapter 17 on Review of Interferences). This new design, which is based on an air-cooled plasma load coil, allows the matching network electronics to rapidly respond to changes in the plasma impedance produced by different sampling conditions and sample matrices, while still maintaining low plasma potential at the interface region. For these reasons, this technology appears to offer some benefits over traditional RF technology for some applications.

Using Collision/Reaction Cell and Interface Technology

Collision and reaction cells and interfaces are an available option on all quadrupole-based instruments today and are used to reduce the formation of harmful polyatomic spectral interferences, such as $^{38}ArH^+$, $^{40}Ar^+$, $^{40}Ar^{12}C^+$, $^{40}Ar^{16}O^+$, and $^{40}Ar_2^+$, to improve detection capability for elements such as K, Ca, Cr, Fe, and Se. However, when comparing systems, it is important to understand how the interference reduction is carried out, what types of collision/reaction gases are used, and how the collision/reaction cell or interface deals with the many complex side reactions that take place—reactions that can potentially generate brand new interfering species and cause significant problems at other mass regions. The difference between collision cells and reaction cells and interfaces has been described in detail in Chapter 13. Two different approaches are used to reject these undesirable species. It can be done by either kinetic energy discrimination (KED) or mass discrimination, depending on the type of multipole and the reaction gas used in the cell.

Unfortunately, the higher-order multipoles such as hexapoles or octopoles have less defined mass stability boundaries than lower-order multipoles, making them less than ideal to intercept these side reactions by mass discrimination. This means that some other mechanism has to be used to reject these unwanted species. The approach that has been traditionally used is to discriminate between them by kinetic energy. This is a well-accepted technique that is typically achieved by setting the collision cell potential slightly more negative than the mass filter potential. This means that the collision-product ions generated in the cell, which have a lower kinetic energy as a result of the collision process, are rejected, whereas the analyte ions, which have a higher kinetic energy, are transmitted. This method works very well but restricts their use to inert gases such as helium and less reactive gases such as hydrogen because of the limitations of higher-order multipoles in efficiently controlling the multitude of side reactions.

However, the use of highly reactive gases such as ammonia and methane can lead to more side reactions and potentially more interferences unless the by-products from these side reactions are rejected. The way around this problem is to utilize a lower-order multipole, such as a quadrupole, inside the reaction/collision cell and use it as a mass discrimination device. The advantages of using a quadrupole are that the stability boundaries are much better defined than a hexapole or an octapole, so it is relatively straightforward to operate the quadrupole inside the reaction cell as a mass or bandpass filter. Therefore, by careful optimization of the quadrupole electrical fields, unwanted reactions between the gas and the sample matrix or solvent, which could potentially lead to new interferences, are prevented. This means that every time an analyte and interfering ions enter the reaction cell, the bandpass of the quadrupole can be optimized for that specific problem and then changed on the fly for the next one.[14]

When assessing the capabilities of collision and reaction cells and interfaces, it is important to understand the level of interference rejection that is achievable, which will be reflected in the instrument's DL and BEC values for the particular analytes being determined. This has been described in greater detail in Chapter 13, but depending on the nature of interference being reduced, there will be differences between the collision/reaction cell methods as well as with the collision/reaction interface approach. It is therefore critical to evaluate the capabilities of commercial instrumentation on the basis of your sample matrices and particular analytes of interest.

On the evidence published to date, it seems that the use of highly reactive gases appears to offer a more efficient way of reducing some interfering ion background levels because the optimum reaction gas can be selected to create the most favorable ion–molecule reaction conditions for each analyte. In other words, the choice and flow of the reaction gas can be optimized for each and every application problem.

The benefit of using highly reactive gases to reduce interferences has been confirmed by the recent development of the "triple quadrupole" collision/reaction cell instruments, where an additional quadrupole is placed prior to the collision/reaction cell multipole and the analyzer quadrupole. This first quadrupole acts as a simple mass filter to allow only the analyte masses to enter the cell, while rejecting all other masses. With all non-analyte, plasma and sample matrix ions excluded from the cell, interference removal is then carried out using highly reactive gases in the collision/reaction cell. The analyte mass, free of the interference, is then passed into the analyzer quadrupole for separation and detection. The use of any kind of ion–molecule reaction chemistry is not as straightforward when it comes to developing methods, especially when new samples are encountered. In addition, if more than one reaction gas needs to be used, they might not be ideally suited for a high-sample-throughput environment because of the lengthy analysis times involved.

However, I think it's fair to say that with the advent of intelligent, decision-making software routines, the choice of gases, cell conditions, and the overall method development process using a reaction cell has become relatively straight forward and user-friendly. This is true not only with a single quadrupole but also using reaction chemistry with a triple quad system.

On the other hand, the use of inert or low-reactivity gases and KED appears to offer a much simpler approach to reducing polyatomic spectral interferences. Normally, only one gas is used for a particular application problem, which is much better suited to routine analysis. It is possible to use other low gases such as hydrogen when helium does not work, but for the majority of elements, one gas is sufficient. For some applications, the collision gas is kept flowing all the time, even for elements that do not need a collision cell. However, its major analytical disadvantage is that its interference reduction capabilities are generally not as good as a system that uses highly reactive gases. Because more collisions are required with an inert gas to suppress the interfering ions, the analyte will also undergo more collisions, and as a result, fewer of the analyte ions will make it through the kinetic energy barrier at the exit of the cell. For that reason, DLs for the majority of the elements that benefit from a collision/reaction cell are generally poorer using inert gases and KED than with a system that uses highly reactive gases and selective bandpass (mass) tuning.

However, it should be emphasized that when you are comparing systems, it should be done with your particular analytical problem in mind. In other words, evaluate the interference suppression capabilities of the different collision and reaction cell interface approaches by measuring BEC and DL performance for the suite of elements and sample matrices you are interested in. In other words, make sure it works for your application problem. This is even more important with the newer "triple quadrupole" collision/reaction cell approach because it is complicated and at this present moment in time, there are very few applications in the public domain.

Every laboratory's analytical scenario is different, so it is almost impossible to determine which approach is better for a particular application problem. If you are not pushing DLs but are looking for a simplified approach to running samples on a routine basis, then maybe a collision cell using KED best suits your needs. However, if your samples are spectrally more complex and you are looking for more performance and flexibility because your DL requirements are more challenging, then either the dynamic reaction cell using bandpass tuning or the "triple quadrupole" is probably the best way to go. Also, be aware that systems that use collisional mechanisms and KED will have to use either higher-purity gases or a gas purifier (getter) because of the potential for impurities in the gas generating unexpected reaction by-product ions, which could potentially interfere with other analyte ions (refer to Chapter 12).[15] This not only has the potential to affect the detection capability, but ultra-high-purity gases are typically two to three times more expensive than industrial- or laboratory-grade gases. But at the end of the day, if you are investing in brand-new technology, it also depends on what kinds of funds are available to solve a particular application problem.

Reduction of Matrix-Induced Interferences

As discussed in Chapter 16, there are three major sources of matrix-induced problems in ICP-MS. The first and simplest to overcome is often called a *sample transport* or *viscosity effect* and is a physical suppression of the analyte signal brought about by the matrix components. It is caused by the sample matrices' impact on droplet formation in the nebulizer or droplet size selection in the spray chamber. In some samples, it can also be caused by the variation in sample flow through the peristaltic pump. The second type of signal suppression is caused by the impact of the sample matrix on the ionization temperature of the plasma discharge. This typically occurs when different levels of matrix components or acids are aspirated into a cool/cold plasma. The ionization conditions in the low-temperature plasma are so fragile that higher concentrations of matrix components result in severe suppression of the analyte signal. The third major cause of matrix suppression is the result of poor transmission of ions through the ion optics owing to matrix-induced space charge effects.[16] This has the effect of defocusing the ion beam, which leads to poor sensitivity and DLs, especially when trace levels of low-mass elements are being determined in the presence of large concentrations of high-mass matrix elements. Unless an electrostatic compensation is made in the ion optic region, the high-mass element will dominate the ion beam, resulting in severe matrix suppression on the lighter ones. All these types of matrix interferences are compensated to varying degrees by the use of internal standardization, where the intensity of a spiked element that is not present in the sample is monitored in samples, standards, and blank.

The single biggest difference in commercial instrumentation to focus the analyte ions into the mass analyzer is in the design of the ion lens system. Although they all basically do the same job of transporting the maximum number of analyte ions through the system, there have been many different ways of implementing this fundamental process, including the use of an extraction lens, multicomponent lens systems, single ion cylinder lens, right-angled reflectors, or multipole ion guide systems. First, it is important to know how many lens voltages have to be optimized. If a system has many lens components, it is probably going to be more complex to carry out optimization on a routine basis. In addition, the cleaning and maintenance of a multicomponent lens system might be a little more time-consuming. All of these are possible concerns, especially in a routine environment where maybe the skill level of the operator is not so high.

However, the design of the ion-focusing system or the number of lens components used is not as important as its ability to handle real-world matrices.[17] Most lens systems can operate in a simple aqueous sample because there are relatively few matrix ions to suppress the analyte ions. The test of the ion optics comes when samples with a real matrix are encountered. When a large number of matrix ions are present in the system, they can physically "knock" the analyte ions out of the ion beam. This shows itself as a suppression of the analyte ions, which means that less analyte ions are transmitted to the detector in the presence of a matrix. For this reason, it is important to measure the degree of matrix suppression of the instrument being evaluated across the full mass range. The best way to do this is to choose three or four of your typical analyte elements spread across the mass range (e.g., $^{7}Li^{+}$, $^{63}Cu^{+}$, $^{103}Rh^{+}$, and $^{138}Ba^{+}$). Run a calibration of a 20 ppb multielement standard in 1% HNO_3. Then, make up a synthetic sample of 20 ppb of the same elements in one of your typical matrices. Measure this sample against the original calibration.

The percentage matrix suppression at each mass can then be calculated as follows:

$$\frac{20\text{ ppb} - \text{Apparent concentration of 20 ppb analytes in your matrix}}{20\text{ ppb}} \times 100$$

There is a strong possibility that your own samples will not really test the matrix suppression performance of the instrument, particularly if they are simple aqueous-type samples. If this is the case and you really would like to understand the matrix capabilities of your instrument, then make up a synthetic sample of your analytes in 500 ppm of a high-mass element such as thallium, lead, or uranium. For this test to be meaningful, you should tell the manufacturers to set up the ion optic voltages that are best suited for multielement analysis across the full mass range. If the ion optics are designed correctly for minimum matrix interferences, it should not matter if it incorporates an extraction lens, uses a photon stop, has an off-axis mass analyzer, or even whether it utilizes a single, multicomponent, or right-angled ion lens system.

It is also important to understand that an additional role of the ion optic system is to stop particulates and neutral species from making it through to the detector, which would increase the noise of the background signal. This will certainly impact the instrument's detection capability in the presence of complex matrices. Therefore, it is definitely worth carrying out a DL test in a difficult matrix such as rock digests, soil samples, biological specimens, or metallurgical alloys, which tests the ability of the ion optics to transport the maximum number of analyte ions while rejecting the maximum number of matrix ions, neutral species, and particulates.

Another aspect of an instrument's matrix capability is its ability to aspirate lots of different types of samples, using both conventional nebulization and sampling accessories that generate a dryer aerosol, such as chromatographic separation system or a desolvation device. When changing sample types similar to this on a regular basis, parameters such as RF power, nebulizer gas flow, and sampling depth usually have to be changed. When this is done, there is an increased chance of altering the electrical characteristics of the plasma and producing a secondary discharge at the interface. All instruments should be able to handle this to some extent, but depending on how they compensate for the increase in plasma potential, parameters might need to be re-optimized because of the change in the spread of kinetic energy of the ions entering the mass spectrometer.[18] This may not be such a serious problem, but once again, it is important for you to be aware of this, especially if the instrument is running many different sample matrices on a routine basis.

Some of the repercussions of a secondary discharge, including increased doubly charged species, erosion of material from the skimmer cone, shorter lifetime of the sampler cone, a significantly different full-mass range response curve with laser ablation, and the occurrence of two signal maximums when optimizing nebulizer gas flow, have been well reported in the literature.[19–21] On the other hand, systems that do not show signs of this phenomena have reported an absence of these deleterious effects.[22]

A simple way of testing for the possibility of a secondary discharge is to aspirate one of your typical matrices containing approximately 1 ppb of a small group of elements across the mass range (such as $^{7}Li^{+}$, $^{115}In^{+}$, and $^{208}Pb^{+}$) and continuously monitor the signals while changing the nebulizer gas flow. In the absence of a secondary discharge, all three elements, which have widely different masses and ion energies, should track each other and have the same optimum nebulizer gas flow. This can be seen in Figure 30.9, which shows the signals for $^{7}Li^{+}$, $^{115}In^{+}$, and $^{208}Pb^{+}$ changing as the nebulizer gas flow is first increased and then decreased.

If the signals do not track each other or there is an erratic behavior in the signals, it could indicate that the normal kinetic energy of the ions has been altered by the change in the nebulizer gas flow. There are many reasons for this kind of behavior, but it could point to a possible secondary discharge at the interface or that the RF coil-grounding mechanism is not working correctly.[23] It should be emphasized that Figure 30.9 is just a graphical representation of what the relative signals might look like and might not exactly be the same for all instruments.

Sample Throughput

In laboratories where high sample throughput is a requirement, the overall cost of analysis is a significant driving force determining what type of instrument is purchased. However, in a high-workload laboratory there sometimes has to be a compromise between the number of samples analyzed and the DL performance required. For example, if the laboratory wants to analyze as many samples as possible, relatively short integration times have to be used for the suite of elements being

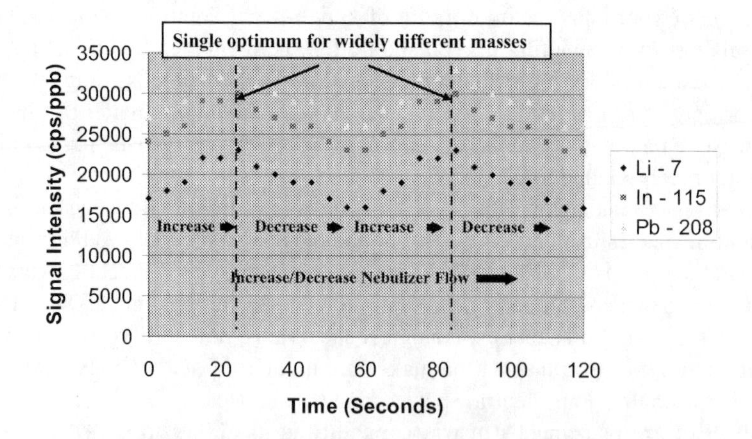

FIGURE 30.9 If the interface is grounded correctly, signals for 1 ppb $^{7}Li^{+}$, $^{115}In^{+}$, and $^{208}Pb^{+}$ should all track each other and have similar optimum values as the nebulizer gas flow is changed.

determined. On the other hand, if DL performance is the driving force, longer integration times need to be used, which will significantly impact the total number of samples that can be analyzed in a given time. This was described in detail in Chapter 15, but it is worth revisiting to understand the full implications of achieving high sample throughput.

It is generally accepted that for a fixed integration time, peak hopping will always give the best DLs. As discussed earlier, measurement time is a combination of time spent on the peak taking measurements (dwell time) and the time taken to settle (settling time) before the measurement is taken. The ratio of the dwell time to the overall measurement time is often called the *measurement efficiency*. The settling time, as we now know, does not contribute to the analytical signal but definitely contributes to the analysis time. This means that every time the quadrupole sweeps to a mass and sits on the mass for the selected dwell time, there is an associated settling time. The greater the number of points that have been selected to quantitate the mass, the longer the total settling time and the worse the overall measurement efficiency.

For example, let us take a scenario where 20 elements need to be determined in duplicate. For argument's sake, let us use an integration time of 1 s per mass, comprising 20 sweeps at 50 ms per sweep. The total integration time that contributes to the analytical signal and the DL is therefore 20 s per replicate. However, every time the analyzer is swept to a mass, the associated scanning and settling time must be added to the dwell time. The greater the numbers of points that are taken to quantify the peak, the greater the magnitude of the settling time that must be added. For this scenario, let us assume that three points/peak are being used to quantify the peaks. Let us also assume for this case that the quadrupole and detector has a settling time of 5 ms. This means that a 15-ms settling time will be associated with every sweep of each individual mass. So, for 20 sweeps of 20 masses, this is equivalent to 6 s of non-analytical time every replicate, which translates into 12 s (plus 40 s of actual measurement time) for every duplicate analysis. This is equivalent to a $40/(12 + 40) \times 100\%$ or a 77% MEC. It does not take long to realize that the fewer the number of points taken per peak and the shorter the settling time, the better the measurement cycle. Just by reducing the number of points to one per peak and cutting the detector settling time by half, the non-analytical time is reduced to 4 s, which is a $40/(2 + 40) \times 100\%$ or a 95% measurement efficiency per duplicate analysis. It is therefore very clear that the measurement protocol has a big impact on the speed of analysis and the number of the samples that can be analyzed in a given time. So, if sample throughput is important, you should understand how peak quantitation is carried out on each instrument.

The other aspect of sample throughput is the time taken for the sample to be aspirated through the sample introduction system into the mass spectrometer, reach a steady-state signal, and then be washed out when the analysis is complete. The wash-in and wash-out characteristics of the instrument will most definitely impact its sample throughput capabilities. Therefore, it is important for you to know what these times are for the system you are evaluating. You should also be aware that if the instrument uses a computer-controlled peristaltic pump to deliver the sample to the nebulizer and spray chamber, it can be speeded up to reduce the wash-in and wash-out times. So, this should also be taken into account when evaluating the memory characteristics of the sample introduction system.

Therefore, if speed of analysis is important to your evaluation criteria, it is worth carrying out a sample throughput test. Choose a suite of elements that represents your analytical challenge. Assuming you are also interested in achieving good detection capability, let the manufacturer set the measurement protocol (integration time, dwell time, settling time, number of sweeps, points/peak, sample introduction wash-in/wash-out times, etc.) to get their best DLs. If you are interested in measuring high and low concentrations, also make sure that the extended dynamic range feature is implemented. Then, time how long it takes to achieve DL levels in duplicate from the time the sample probe goes into the sample to the time a result comes out on the screen or printer. If you have time, it might also be worth carrying out this test in an autosampler with a small number of your typical samples. It is important that the DL measurement protocol be used because factors such as integration times and wash-out times can be compromised to reduce the analysis time.

All the measurement time issues discussed in this section and the memory characteristics of the sample introduction system will be fully evaluated with this kind of test. (Note: If high sample throughput is important, there are automated productivity enhancement systems on the market that by efficient delivery and wash-out of the sample are realizing a twofold to threefold improvement in multielement analysis times—refer to Chapter 20 for details.)

Transient Signal Capability

The demands on an instrument to handle transient signals generated by sampling accessories, such as laser ablation, chromatography separation devices, flow injection, or electrothermal vaporization systems, are very different from conventional multielement analysis using solution nebulization. Because the duration of a sampling accessory signal is short compared to a continuous signal generated by a pneumatic nebulizer, it is critical to optimize the measurement time to achieve the best multielement signal-to-noise ratio in the sampling time available. This was addressed in greater detail in Chapter 20, but basically the optimum design to capture the maximum amount of multielement data in a transient peak is to carry out the measurement in a simultaneous manner with a multicollector magnetic sector instrument, a TOF mass spectrometer or the Mattauch-Herzog simultaneous detection sector instrument.

However, a scanning system such as a quadrupole instrument can achieve good performance on a transient peak if the measurement time is maximized to get the best multielement signal-to-noise ratio. Therefore, instruments that utilize short settling times are more advantageous, because they achieve a higher MEC. In addition, if the extended dynamic range is used to determine higher concentrations, the scanning and settling time of the detector will also have an impact on the quality of the signal. So, detectors that require two scans to characterize an unknown sample will use up valuable time in the quantitation process. For example, if a transient peak generated by a laser ablation device only lasts 10 s, a survey or prescan of 2 s will use up 20% of the available measurement time. This, of course, is a disadvantage when doing multielement analysis on a transient signal, especially if you have limited knowledge of the analyte concentration levels in your samples.

Single-Particle ICP-MS Transient Signals

Of all the applications involving transient peak measurements, single-particle ICP-MS (SP-ICP-MS) is probably the most demanding, because the transient event only lasts a few milliseconds. Let's take a closer look at this rapidly emerging application area to better understand the measurement requirements. SP-ICP-MS (SP-ICP-MS) is a new technique that has recently been developed for detecting and sizing metallic nanoparticles (NPs) at extremely low levels, in order to predict their environmental behavior. While this method is fairly new, it has shown a great deal of promise in several applications, including determining concentrations of NPs in complex matrices, such as wastewater streams, potable waters and effluents (Note: this type of analysis could be a future requirement for the cannabis industry as more studies are being published on how plants absorb NPs through their roots system). The method involves introducing a liquid sample containing the NPs at very dilute concentration into the ICP-MS. After the particles have been nebulized, ionized in the plasma, and separated by the mass analyzer, the resulting ions are detected and collected as time-resolved pulses. The number of pulses generated is directly related to the population of NPs in the sample, while the intensity of the pulses are related to the size (and mass) of the NPs.

In ICP-MS, dilute solutions of dissolved metals will produce relatively constant signals. If there are NPs of various sizes suspended in that solution, they will appear as pulses which deviates from a steady-state continuous signal generated by the background of the dissolved metals. By using relatively short dwell times of a few milliseconds, the packets of pulses can be quantified as long as the pulses can be adequately resolved in the time domain. The pulse height is then compared to the signal intensity of dissolved metal in a set of calibration standards. This is represented in Figure 30.10, which shows sample A containing dissolved metal in solution being analyzed and measured in the conventional way and sample B containing the same metal in the form of NPs

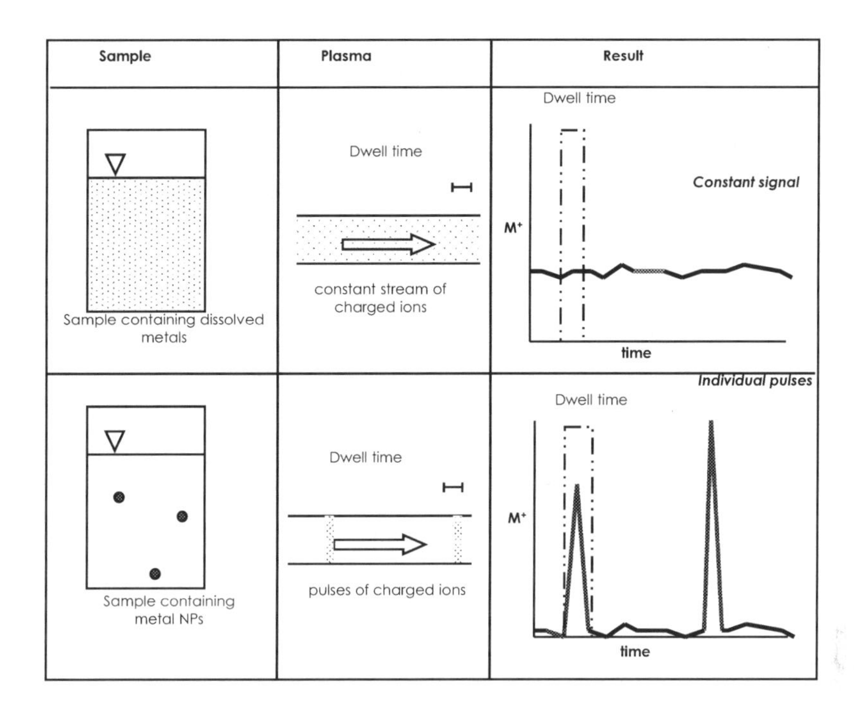

FIGURE 30.10 A comparison of a sample A containing dissolved metal in solution being analyzed and measured in the conventional way and another sample B containing the same metal in the form of NPs being analyzed and measured as a transient pulses of ions. (Courtesy of the Colorado School of Mines.)

being analyzed and measured as a transient pulses of ions. The integration (dwell time), which is the same in both examples, is shown as the dotted box. By subtracting the background produced by the dissolved analyte in the sample from the NP pulse height, and comparing the net signal intensity against the calibration standard, the concentration and therefore the size (and mass) distribution of the particle can be calculated using well-understood SP-ICP-MS theory.

However, for this approach to work effectively at low concentrations, the speed of data acquisition and the response time of the ICP-MS quadrupole and detector electronics must be fast enough to capture the time-resolved NP pulses, which typically last only a few milliseconds or less. This is emphasized in Figure 30.11, which shows a real-world example of the time-resolved analysis of 30-nm silver NPs by SP-ICP-MS. It can be seen that the silver NP pulse has been resolved with approximately 15 data points in <1 ms. For this application, the ICP-MS should be capable of using dwell times shorter than the particle transient time, thus avoiding false signals generated from clusters of particles. In practice, this means using a dwell time of a 10–100 μs so the pulse can be fully characterized.

For this kind of resolution, it is advantageous that the instrument measurement electronics are capable of very fast data acquisition rates, including dwell times that are as short as possible to capture the maximum number of data points within the transient event. It is also desirable that the quadrupole settling time is extremely short, so there is no wasted time waiting for the quadrupole power supply to stabilize. Ideally, it would beneficial if there was no settling time, so the quadrupole could just park itself on the mass of interest and just take measurements. Unfortunately, this is not typically a standard feature of the measurement protocol on most ICP-MS systems, because for multielement analysis, there has to be a built-in settling time as the quadrupole scans from one mass to the next. However, for optimum measurement conditions when one element is under investigation, there is no question that the ability to run a method with no settling time, very short dwell times, and fast data acquisition is extremely beneficial in order to increase the upper limit of the dynamic range when characterizing NPs using SP-ICP-MS measurements.[24]

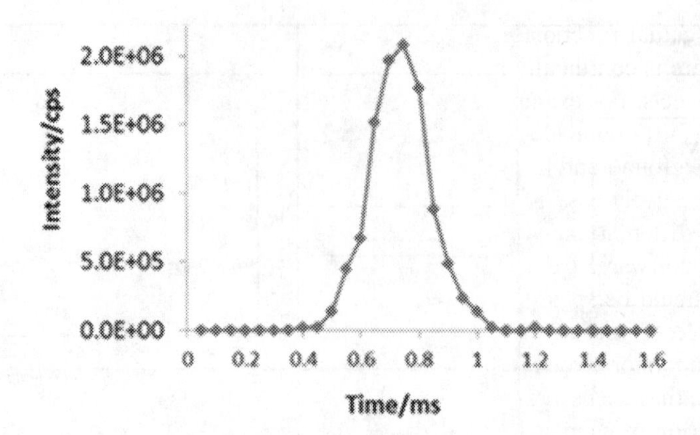

FIGURE 30.11 A time-resolved analysis of 30-nm gold particles by sp-ICP-MS showing the pulse is fully characterized in less than 1 ms. (Copyright 2013, all rights reserved, PerkinElmer Inc.)

Usability Aspects

In most applications, analytical performance is a very important consideration when deciding what instrument to purchase. However, the vast majority of instruments being used today are being operated by technician-level chemists. They usually have had some experience in the use of trace element techniques such as atomic absorption (AA) or ICP-OES, but in no way could be considered experts in ICP-MS. Therefore, the usability aspects might be competing with analytical performance as the most important selection criteria, particularly if the application does not demand the ultimate in detection capability. Even though usability is in the eye of the user, there are some general issues that need to be addressed. They include, but are not limited to, the following:

- Ease of use
- Routine maintenance
- Compatibility with sampling accessories
- Installation requirements
- Technical support
- Training.

Ease of Use

First of all, you need to determine the skill level of the operator who is going to run the instrument. If the operator is a PhD-type chemist, then maybe it is not critical that the instrument be easy to use. However, if the instrument is going to be used in a high-workload environment and possibly operated around the clock, such as the cannabis-related industry, there is a strong possibility that the operators will not be highly skilled. Therefore, you should be looking at how easy the software is to use and how similar it is to other trace element techniques that are used in your laboratory. This will definitely have an impact on the time it takes to get a person fully trained on the instrument. Another issue to consider is whether the person who runs the instrument on a routine basis is the same person who will be developing the methods. Correct method development is critical because it impacts the quality of your data, and therefore usually requires more expertise than just running routine methods. This is most definitely the case with collision/reaction cells and interfaces, especially if the method has never been done before. It can take a significant amount of time and effort to select the best gases, gas flows, and optimization of the cell parameters to maximize the reduction of interferences for certain analytes in a new sample matrix.

I am not going to get into different software features or operating systems, because it is a complicated criterion to evaluate and decisions tend to be made more on a personal preference or comfort level than on the actual functionality of ICP-MS software features. It is also a moving target, as instrument software is continually being modified and updated. However, there are differences in the way software feels. For example, if you have come from a MS background, you are probably comfortable with fairly complex research-type software. Alternatively, if you have come from a trace element background and have used AA or ICP-OES, you are probably used to more routine software that is relatively easy to use. You will find that different vendors have come to ICP-MS from a variety of different analytical chemistry backgrounds, which is often reflected in the way they design their software. Depending on the way the instrument will be used, an appropriate amount of time should be spent looking at software features that are specific to your application needs. For example, if you are working in a high-throughput environmental, pharmaceutical, or cannabis testing laboratory, you might be interested in turnkey methods such as standard operating procedures (SOPs) that are used to run a particular application methodology, such as Method 200.8 for the determination of elements in water and wastes. Alternatively with more and instruments going into to high-throughput production environments, look for customized methods for different applications. In addition, maybe you should also be looking very closely at all the features of the automated "Quality Control" (QC) software, or if you do not have the time to export your data to an external spreadsheet to create reports, you might be more interested in software with comprehensive reporting capabilities.

In highly regulated industries, the operating and reporting software needs to be compliant with a set of regulated standards and guidelines. For example, the pharmaceutical and food-related manufacturing industry is dictated by federal regulations set down by the Food and Drug Authority (FDA) in Title 21 CFR Part 11, which gives detailed requirements that computerized systems need to fulfill in order to allow electronic signatures and records in lieu of handwritten signatures on paper records. In summary, the regulations apply to validations for closed and open computerized systems, controlled access to the computerized system, content integrity, use of electronic signatures for authentication of electronic documents, audit trails for all records/signatures, and access to electronic records. Although not currently a requirement in the cannabis industry, it is only a matter of time before federal regulations become a way of life. When that time eventually arrives, it is important that any ICP-MS which is used in a regulated environment has all the necessary software to be compliant.

Routine Maintenance

ICP mass spectrometers are complex pieces of equipment that, if not maintained correctly, have the potential to fail when you least expect them to. For that reason, a major aspect of instrument usability is how often routine maintenance has to carried out, especially if complex sample matrices are being analyzed. You must not lose sight of the fact that your samples are being aspirated into the sample introduction system and the resulting ions generated in the plasma are steered into the mass analyzer via the interface and ion optics. In other words, the sample, in one form or another, is in contact with many components inside the instrument. Even though modern instruments require much less routine maintenance than older generation equipment, it is still essential to find out what components need to be changed and at what frequency to keep the instrument in good working order. Routine maintenance has been covered in great depth in Chapter 18, but you should be asking the vendor what needs to be changed or inspected on a regular basis and what type of maintenance should be done on daily, weekly, monthly, or yearly intervals. Some typical questions might include the following:

- If a peristaltic pump is being used to deliver the sample, how often should the tubing be changed?
- How often should the spray chamber drain system be checked?
- Can components be changed if a nebulizer gets damaged or blocked?
- How long does the plasma torch last?
- Are ceramic torches available, which tend to have a much longer lifetime?

- Can the torch sample injector be changed without discarding the torch?
- How is a neutral plasma maintained, and if an external shield or sleeves are used for grounding purposes, how often do they last?
- Is the RF generator solid state or does it use a power amplifier (PA) tube? This is important because PA tubes are expensive consumable items that typically need replacing every 1–2 years.
- How often do you need to clean the interface cones, and what is involved in cleaning them and keeping the cone orifices free of deposits?
- How long do the cones last?
- Do you have a platinum cone trade-in service, and what is their trade-in value?
- Which type of pump is used on the interface, and if it is a rotary-type pump, how often should the oil be changed?
- What mechanism is used to keep the ion optics free of sample particulates or deposits?
- How often should the ion optics be cleaned?
- What is the cleaning procedure for the ion optics?
- Do the turbomolecular pumps require any maintenance?
- How long do the turbomolecular pumps typically last?
- Does the mass analyzer require any cleaning or maintenance?
- How long does the detector last and how easy is it to change?
- What spare parts do you recommend to keep on hand? This can often indicate the components that are likely to fail most frequently.
- What maintenance needs to be carried out by a qualified service engineer, and how long does it take?

This is not an exhaustive list, but it should give you a good idea about what is involved in keeping an instrument in good working order. I also encourage you to talk to real-world users of the equipment to make sure you get their perspective of these maintenance issues. This is particularly important if you are investing in a brand new instrument that hasn't been on the market long. There won't be a large number of users in the field, so it's important you talk to them to get their real-world perspective of the routine maintenance issues. For more information about how to maximize performance and reduce routine maintenance, check out this reference by Brennan and coworkers.[25]

Compatibility with Productivity and Performance Enhancing Tools

Alternative sample introduction techniques to enhance performance and productivity are becoming more necessary as ICP-MS is being utilized to analyze more complex sample types. Therefore, it is important to know if the sampling accessory is made by the ICP-MS instrument company or by a third-party vendor. Obviously, if it has been made by the same company, compatibility should not be an issue. However, if it is made by a third party, you will find that some sampling accessories work much better with some instruments than with others. It might be that the physical connection of coupling the accessory to the ICP-MS torch has been better thought out, or that the software "talks" to one system better than another. You should also understand how much routine maintenance a sampling accessory needs. The benefits of a rugged ICP-MS system, which requires very little maintenance is negated if the sampling accessory needs to be cleaned every five to ten samples. You should refer to Chapter 20 on Performance and Productivity Enhancement Tools for more details on their suitability for cannabis-related analysis, but if they are required, software/hardware compatibility should be one of your evaluation objectives. You should also read Chapter 21 if you are thinking of interfacing an HPLC system to your ICP-MS for arsenic or mercury speciation studies.

Installation of Instrument

Installation of the instrument and where it is going to be located does not seem to be an obvious evaluation objective at first, but it could be important, particularly if space is limited. For example, is the instrument free-standing or bench-mounted? Maybe you have a bench available but no floor space, or vice versa? It could be that the instrument requires a temperature-controlled room to ensure good stability and mass calibration. If this is the case, have you budgeted for this kind of expense? If the instrument is being used for ultratrace detection levels, does it need to go into a class 1, 10, or 100 clean room? If it does, what is the size of the room and do the roughing pumps need to be placed in another room? In other words, it is important to fully understand the installation requirements for each instrument being evaluated and where it will be located. It is unlikely that cannabis testing labs are going to require a Class 1 or Class 10 environment. However, you cannot install an ICP-MS instrument in a warehouse and hope to get meaningful data at sub-ppb DLs.

Technical Support

Technical and application support is a very important consideration, especially if you have had no previous experience with ICP-MS. You want to know that you are not going to be left on your own after you have made the purchase. Therefore, it is important to know not only the level of expertise of the specialist who is supporting you, but also whether they are local to you or located in the manufacturer's corporate headquarters. In other words, can the vendor give technical help whenever you need it? Another important aspect related to application support is the availability of application literature. Is there a wide selection of material available for you to read, in the form of either web-based application reports or references in the open literature, to help you develop your methods? Also, find out if there are active user- or Internet-based discussion groups, because they will be an invaluable source of technical and application help. One such source of help in this area can be found on the PlasmaChem Listserver, a plasma spectrochemistry discussion group out of Syracuse University.[26]

Training

Find out what kind of training course comes with the purchase of the instrument and how often it is run. Most instruments come with 2–3-day training course for one person, but most vendors should be flexible regarding the number of people who can attend. Some manufacturers also offer application training, where they teach you how to optimize methods for major application areas. Historically, these customized application courses have included environmental, clinical, semi-conductor, and pharmaceutical analysis. However with the rapid growth of the cannabis industry today, vendors are now beginning to offer ICP-MS training courses specifically for cannabis testing labs. So talk to other users in your field about the quality of the training they received when they purchased their instruments, and also ask them what they thought of the vendor's operator manuals. You will often find that this is a good indication of how important customer training is to the vendor.

Reliability Issues

To a certain degree, instrument reliability is impacted by routine maintenance issues and the types of samples being analyzed, but it is generally considered more of a reflection of the design of an instrument. Most manufacturers will guarantee a minimum percentage uptime for their instrument, but this number (which is typically ~95%) is almost meaningless unless you really understand how it is calculated. Even when you know how it is calculated, it is still difficult to make the comparison, but at least you should understand the implications if the vendor fails to deliver. Good instrument reliability is taken for granted nowadays, but it has not always been the case. When ICP-MS was first commercialized, the early instruments were a little unpredictable, to say the least, and quite prone to frequent breakdowns. However, as the technique became more mature, the quality of instrument components improved, and therefore, the reliability improved. However, you should be aware that some components of the instrument are more problematic than others. This is particularly true when the design of an instrument is new or a model has had a major redesign. You will therefore find that in the life cycle of a newly designed instrument, the early years will be more susceptible to reliability problems than when the instrument is of an older design.

When we talk about instrument reliability, it is important to understand whether it is related to the samples being analyzed, the inexperience of the person operating the instrument, an unreliable component, or an inherent weakness in the design of the instrument. For example, how does the instrument handle highly corrosive chemicals such as concentrated mineral acids? Some sample introduction systems and interfaces will be more rugged than others and require less maintenance in this area. On the other hand, if the operator is not aware of the dissolved solid limitation of the instrument, they might attempt to aspirate a sample, which will slowly block the interface cones, causing signal drift and, in the long term, possible instrument failure. Or, it could be something as unfortunate as a major component, such as the RF generator power amplifier tube, discrete dynode detector, or turbomolecular pump (which all have a finite lifetime) failing in the first year of use.

Service Support

Instrument reliability is very difficult to assess at the evaluation stage, so you have to look very carefully at the kind of service support offered by the manufacturer. For example, how close is a qualified support engineer to you or what is the maximum amount of time you will have to wait to get a support engineer at your laboratory or at least to call you back to discuss the problem. Ask the vendor if they have the capability for remote diagnostics, where a service engineer can remotely run the instrument or check the status of a component by "talking to" your system computer via a modem. Even if this approach does not fix the problem, at least the service engineer can come to your laboratory with a very good indication of what the problem could be.

You should know up front what a service visit is going to cost you, irrespective of what component has failed. Also find out what routine maintenance jobs you can do and what requires an experienced service engineer. If it does require a service engineer, how long will it take them, because their time is not inexpensive. Most companies charge an hourly rate for a service engineer (which typically includes travel time as well), but if an overnight stay is required, fully understand what you are paying for (accommodations, meals, gas, etc.). Some companies might even charge for mileage between the service engineer's base and your laboratory. If you work in a commercial laboratory and cannot afford the instrument to be down for any length of time, find out what it is going to cost for 24/7 service coverage.

You can take a chance and just pay for each service visit, or you might want to budget for an annual preventative maintenance contract, where the service engineer checks out all the important instrumental components and systems frequently to make sure they are all working correctly. This might not be as critical if you work in an academic environment, where the instrument might be down for extended periods, but in my opinion, it is absolutely critical if you work in a commercial laboratory, which is using the instrument to generate revenue. Also find out what is included in the contract, because some also cover the cost of consumables or replacement parts, whereas others just cover the service visits. These annual preventative maintenance contracts typically make up about 10%–15% of the cost of the instrument, but are well worth it if you do not have the expertise in-house or if you just feel more comfortable having an insurance policy to cover instrument breakdowns.

Once again, talking to existing users will give you a very good perspective of the quality of the instrument and the service support offered by the manufacturer. There is no absolute guarantee that the instrument of choice is going to perform to your satisfaction 100% of the time, but if you work in a high-throughput, routine laboratory, make sure it will be down for the minimum amount of time. In other words, fully understand what it is going to cost you to maximize the uptime of all the instruments being evaluated.

FINANCIAL CONSIDERATIONS

The financial side of choosing an ICP mass spectrometer can often dominate the selection process. You may or may not have budgeted quite enough money to buy a top-of-the-line instrument or perhaps you had originally planned to buy another lower-cost trace element technique, or you could be using funds left over at the end of your financial year. All these scenarios dictate how much money you have available and what kind of instrument you can purchase. In my experience, you should proceed with caution in this kind of situation, because if only one manufacturer is willing to do a deal with you, there is no real need to carry out the evaluation process. Therefore, you should budget at least 12 months before you are going to make a purchase and add another 10%–15% for inflation and any unforeseen price increases. In other words, if you want to get the right instrument for your application, never let price be the overriding factor in your decision. Always be wary of the vendor who will undercut everyone else to get your business. There could be a very good reason why they are doing this; for example, the instrument is being discontinued for a new model, or it could be having some reliability problems that are affecting its sales.

This is not to say that price is unimportant, but what might appear to be the most expensive instrument to purchase might be the least expensive to run. Therefore, you must never forget the cost of ownership in the overall financial analysis of your purchase. So, by all means compare the price of the instrument, computer, and any accessories you buy, but also factor in the cost of consumables, gases, and electricity based on your usage. Maybe instrument consumables from one vendor are much less expensive than those from another vendor. This is particularly the case with interface cones and plasma torches. Or maybe the purity of collision/reaction gases is more critical with one cell-based instrument than another. For example, there is a factor of 4 difference between the cost of high-purity (99.999%) hydrogen gas and ultra-high-purity (99.9999%) grade. Supplies of high-purity helium gas are diminishing, so the price is increasing quite significantly, as well as becoming much more difficult to obtain from your gas supplier.

So, be diligent when you compare prices. Look at the overall picture, and not just the cost of the instrument. Also, be aware of differences in sample throughput. Perhaps you can analyze more samples with one instrument because its measurement protocol is faster or it does not need recalibrating as often (less drift). It therefore follows that if you can get through your daily allocation of samples much faster with one instrument than another, then your argon consumption will be reduced.

Another aspect that should be taken into consideration is the salary of the operator. Even though you might think that this is a constant, irrespective of the instrument, you must assess the expertise required to run it. For example, if you are thinking of purchasing more complex technology such as a magnetic sector instrument or a "triple quad" collision/reaction cell instrument for a research-type application, the operator needs to be of a much higher skill level than, say, someone who is being asked to run a routine application with a quadrupole-based instrument. As a result, the salary of that person will probably be higher.

Finally, if one instrument has to be installed in a temperature-controlled, air-conditioned, environment for stability purposes, the cost of preparing or building this kind of specialized room must be taken into consideration when doing your financial analysis. In other words, when comparing systems, never automatically reject the most expensive instrument. You will find that over the 10 years that you own the instrument, the cost of doing analysis and the overall cost of ownership are more important evaluation criteria.

THE EVALUATION PROCESS: A SUMMARY

As mentioned earlier in this chapter, it is not my intention to compare instrument designs and features, but to give you some general guidelines as to what are the most important evaluation criteria. Besides being a framework for your evaluation process, these guidelines should also be used in conjunction with the other chapters in this book and the cited referenced information.

However, if you want to find the best instrument for your application needs, be prepared to spend a few months evaluating the marketplace. Do not forget to prioritize your objectives and give each of them a weighting factor based on their degree of importance for the types of samples you analyze. Be careful to take the evaluation in the direction you want to go and not where the vendor wants to take it. In other words, it is important to compare apples with apples and not to be talked into comparing an apple with an orange that looks like an apple! However, be prepared that there might not be a clear-cut winner at the end of the evaluation. If this is the case, then decide what aspects of the evaluation are most important and ask the manufacturer to put them in writing. Some vendors might be hesitant to do this, especially if it is guaranteeing instrument performance with your samples.

Talk to as many users in your field as you possible can—not only ones given to you as references by the vendor, but ones chosen by yourself also. This will give you a very good indication of the real-world capabilities of the instrument, which can often be overlooked at a demonstration. You might find from talking to "typical" users that it becomes obvious which instrument to purchase. If that is the case and your organization allows it, ask the vendor what your options are if you do not have samples to run and you do not want a demonstration. I guarantee you will be in a much better position to negotiate a lower price.

Never forget that it is a very competitive marketplace, and your business is extremely important to each of the ICP-MS manufacturers. AS mentioned at the beginning of this chapter, it is not meant to focus exclusively on the evaluation needs of cannabis testing labs, but is meant to offer guidance about what are the most important analytical considerations and to suggest a template by which you can compare different features, approaches, and instrument designs.

Hopefully, it has not only helped you understand the fundamentals of the technique a little better but also given you some thoughts and ideas on how to find the best instrument for your needs. Refer to Chapter 32 for details on how to contact all the instrument vendors and consumables and accessories companies. And if you are still confused, I teach a half-day Short Course at the Pittsburgh Conference every year on "How to Evaluate ICP-MS: The Most Important Analytical Considerations." We talk about all these issues in more detail. But if I don't see you there … GOOD LUCK with your evaluation!

FURTHER READING

1. K. Nottingham, *Analytical Chemistry*, 76, 35A–38A, 2004, https://pubs.rsc.org/en/content/articlelanding/1997/an/a700441i/unauth#!divAbstract.
2. Royal Society of Chemistry, *The Analyst*, **122**, 393–408, 1997.
3. A. Montasser, Analytical figures of merit for ICP-MS, *Inductively Coupled Plasma Mass Spectrometry: An Introduction to ICP Spectrometries for Elemental Analysis*, A. Montasser, Ed., Wiley-VCH, Hoboken, NJ, USA, 1998, pp. 16–28, Chap. 1.4, ISBN: 978-0-471-18620-5.
4. E. R. Denoyer, *Atomic Spectroscopy*, **13**(3), 93–98, 1992.
5. M. A. Thomsen, *Atomic Spectroscopy*, **13**(3), 93–98, 2000.
6. L. Halicz, Y. Erel, and A. Veron, *Atomic Spectroscopy*, **17**(5), 186–189, 1996.
7. R. Thomas, *Spectroscopy*, **17**(7), 44–48, 2002.
8. E. R. Denoyer and Q. H. Lu, *Atomic Spectroscopy*, **14**(6), 162–169, 1993.
9. R. Hutton, A. Walsh, D. Milton, and J. Cantle, *ChemSA*, **17**, 213–215, 1991.
10. *Quadrupole Mass Spectrometry and Its Applications*, P. H. Dawson, Ed., Elsevier, Amsterdam, 1976; reissued by AIP Press, Woodbury, NY, 1995.
11. S. J. Jiang, R. S. Houk, and M. A. Stevens, *Analytical Chemistry*, **60**, 217, 1988.
12. K. Sakata and K. Kawabata, *Spectrochimica Acta*, **49B**, 1027, 1994.
13. J. M. Collard, K. Kawabata, Y. Kishi, and R. Thomas, *Micro*, **20**(1), 39–46, January 2002.
14. S. D. Tanner and V. I. Baranov, *Atomic Spectroscopy*, **20**(2), 45–52, 1999.
15. B. Hattendorf and D. Günther, *Journal of Analytical Atomic Spectrometry*, **19**, 600–606, 2004.
16. S. D. Tanner, D. J. Douglas, and J. B. French, *Applied Spectroscopy*, **48**, 1373, 1994.
17. E. R. Denoyer, D. Jacques, E. Debrah, and S. D. Tanner, *Atomic Spectroscopy*, **16**(1), 1, 1995.
18. R. C. Hutton and A. N. Eaton, *Journal of Analytical Atomic Spectrometry*, **5**, 595, 1987.
19. A. L. Gray and A. Date, *Analyst*, **106**, 1255, 1981.
20. E. J. Wyse, D. W. Koppenal, M. R. Smith, and D. R. Fisher, 18th FACSS Meeting, Anaheim, CA, October 1991, Paper No. 409.
21. W. G. Diegor and H. P. Longerich, *Atomic Spectroscopy*, **21**(3), 111, 2000.
22. D. J. Douglas and J. B. French, *Spectrochimica Acta*, **41B**(3), 197, 1986.

23. E. R. Denoyer, *Atomic Spectroscopy*, **12**, 215–224, 1991.
24. E. Heithmar and S. Pergantis, Characterizing Concentrations and Size Distributions of Metal-Containing Nanoparticles in Waste Water, EPA Report APM 272, 2010.
25. R. Brennan, G. Dulude, and R. Thomas, *Spectroscopy Magazine*, **30**(10), 12–25, 2015.
26. PlasmaChem Listserver: A discussion group for plasma spectrochemists worldwide, Syracuse University, NY, http://www.lsoft.com/scripts/wl.exe?SL1=PLASMACHEM-L&H=LISTSERV.SYR.EDU.

31 Glossary of Terms Used in Atomic Spectroscopy

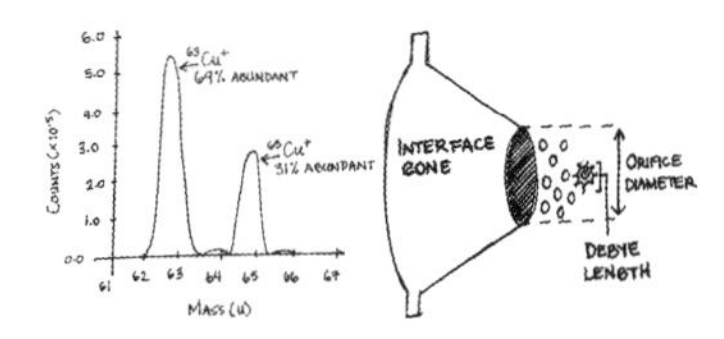

In all my years of working in the field of plasma spectrochemistry, I had never come across any written material that included a basic dictionary of terms, primarily aimed at someone new to the technique. When I first became involved in the ICP-MS technique, most of the literature I read tended to give complicated descriptions of instrument components and explanations of fundamental principles that more often than not sailed over my head. It was not until I became more familiar with the technique that I began to get a better understanding of the complex jargon used in technical journals and presentations at scientific conferences. So, when I wrote my second ICP-MS textbook, I knew that a glossary of terms was an absolute necessity and it has been in every subsequent book I've written since. In this book, the glossary has been expanded to also include all the atomic spectroscopy techniques described including ICP-OES, AA, AF, XRF, XRD, LIBS, LALI-TOFMS, and MIP-AES terms. Even though the glossary is not exhaustive, it contains explanations and definitions of the most common AS words, expressions, and terms used in this book. It should mainly be used as a quick reference guide. If you want more detailed information about the subject matter, you should use the index to find a more detailed explanation of the topic in the appropriate book chapter. Many of the same terms are used in different chapters, so to save duplication and where appropriate, those definitions have only been used once. (Note: the glossary does not include commercial names used by any of the instrument, accessories, or consumable vendors.)

INDUCTIVELY COUPLED PLASMA MASS SPECTROMETRY (ICP-MS) GLOSSARY

A

AA: An abbreviation for atomic absorption.

abundance sensitivity: A way of assessing the ability of a mass separation device, such as a quadrupole, to identify and measure a small analyte peak adjacent to a much larger interfering peak. An abundance sensitivity specification is a combination of two measurements. The first is expressed as the ratio of the intensity of the peak at 1 amu (atomic mass unit) below the analyte peak to the intensity of the analyte peak, and the second is the ratio of the peak intensity 1 amu above the analyte mass to the intensity of the analyte peak. Because of the motion of the ion through the mass filter, the abundance sensitivity specification of a mass-filtering device is always worse on the low-mass side compared to the high-mass side.

active film multipliers: Another name for discrete dynode multipliers, which are used to detect, measure, and convert ions into electrical pulses in ICP-MS. *Also refers to* channel electron multiplier (CEM) *and* discrete dynode detector (DDD).

addition calibration: A method of calibration in ICP-MS using standard additions. All samples are assumed to have a similar matrix, so spiking is only carried out on one representative sample and not the entire batch of samples, as per conventional standard additions used in graphite furnace AA analysis.

AE: An abbreviation for atomic emission.

aerosol: The result of breaking up a liquid sample into small droplets by the nebulization process in the sample introduction system. *Also refers to* nebulizer *and* sample introduction system.

aerosol dilution: A way of introducing a flow of argon gas between the nebulizer and the torch, which has the effect of reducing the sample's solvent loading on the plasma, so it can tolerate much higher total dissolved solids.

AF4: *Refers to* asymmetrical flow field flow fractionation.

alkylated metals: A metal complex containing an alkyl group. Typically detected by coupling liquid chromatography with ICP-MS. *Also refers to* speciation analysis.

alpha-counting spectrometry: A particle-counting technique that uses the measurement of the radioactive decay of alpha particles. *Also refers to* particle-counting techniques.

alternative sample introduction accessories: Alternative ways of introducing samples into an ICP mass spectrometer other than conventional nebulization. Also known as alternative sample introduction devices. Often used to describe desolvation techniques or laser ablation.

analog counting: A way of measuring high signals by changing the gain or voltage of the detector. *Also refers to* pulse counting.

argon: The gas used to generate the plasma in an ICP.

argon-based interferences: A polyatomic spectral interference generated by argon ions combining with ions from the matrix, solvent, or any elements present in the sample.

array detectors: An ion detector based on solid-state, direct charge arrays, similar to CID/CCD technology used in ICP optical emission. By projecting all the separated ions from a mass separation device onto a two-dimensional array, these detectors can view the entire mass spectrum simultaneously. Used with the Mattauch-Herzog sector technology.

ashing: A sample preparation technique that involves heating the sample (typically in a muffle furnace) until the volatile material is driven off and an ash-like substance is left.

Asymmetrical flow field flow fractionation (AF4): Field flow fractionation (FFF) is a single-phase chromatographic separation technique, where separation is achieved within a very thin channel, against which a perpendicular force field is applied. One of the most common forms of FFF is asymmetrical flow FFF (AF4), where the field is generated by a cross-flow applied perpendicular to the channel. Coupled with ICP-MS for the characterization of nanoparticles.

atom: A unit of matter. The smallest part of an element having all the characteristics of that element and consisting of a dense, central, positively charged nucleus surrounded by orbiting electrons. The entire structure has an approximate diameter of 10^{-8} cm and characteristically remains undivided in chemical reactions except for limited removal, transfer, or exchange of certain electrons.

atom-counting techniques: A generic name given to techniques that use atom or ion counting to carry out elemental quantitation. Some common ones, besides ICP-MS, include secondary ionization mass spectrometry (SIMS), thermal ionization mass spectrometry (TIMS), accelerator mass spectrometry (AMS), and fission track analysis (FTA). *Also refers to* ionizing radiation counting techniques.

atomic absorption (AA): An analytical technique for the measurement of trace elements that uses the principle of generating free atoms (of the element of interest) in a flame or electrothermal atomizer (ETA) and measuring the amount of light absorbed from a wavelength-specific light source, such as a hollow cathode lamp (HCL) or electrode discharge lamp (EDL).

atomic emission (AE): A trace element analytical technique that uses the principle of exciting atoms in a high-temperature source such as a plasma discharge and measuring the amount of light the atoms emit when electrons fall back down to a ground (stable) state.

atomic mass or weight: The average mass or weight of an atom of an element, usually expressed relative to the mass of carbon 12, which is assigned 12 atomic mass units.

atomic number: The number of protons in an atomic nucleus.

atomic structure: Describes the structural makeup of an atom. *Also refers to* neutron, proton, *and* electron.

attenuation (of the detector): Reduces the amplitude of the electrical signal generated by the detector, with little or no distortion. Usually carried out by applying a control voltage to extend the dynamic range of the detector. *Also refers to* extended dynamic range.

autocalibration: A way of carrying out calibration with an automated in-line sample delivery system.

autodilution: A way of carrying out automatic in-line dilution of large numbers of samples with no manual intervention by the operator.

autosampler: A device to automatically introduce large numbers of samples into the ICP-MS system with no manual intervention by the operator.

axial view: An ICP-OES system in which the plasma torch is positioned horizontally (end-on) to the optical system as opposed to the conventional vertical (radial) configuration. It is generally accepted that viewing the end of the plasma improves emission intensity by a factor of approximately 5- to 10-fold.

B

background equivalent concentration (BEC): Defined as the apparent concentration of the background signal based on the sensitivity of the element at a specified mass. The lower the BEC value, the more easily a signal generated by an element can be discerned from the background. Many analysts believe BEC is a more accurate indicator of the performance of an ICP-MS system than detection limit, especially when making comparisons of background reduction techniques, such as cool plasma or collision/reaction cell and interface technology.

background noise: The square root of the intensity of the blank in counts per second (cps) anywhere of analytical interest on the mass range. Detection limit (DL) is a ratio of the analyte signal to the background noise at the analyte mass. Background noise as an instrumental specification is usually measured at mass 220 amu (where there are no spectral features), while aspirating deionized water. *Also refers to* background signal, instrument background noise, *and* detection limit.

background signal: The signal intensity of the blank in counts per second (cps) anywhere of analytical interest on the mass range. Detection limit (DL) is a ratio of the analyte signal to the noise of the background at the analyte mass. Background as an instrumental specification is usually measured at mass 220 amu (where there are no spectral features), while aspirating deionized water. *Also refers to* background noise, instrument background signal, *and* detection limit.

bandpass tuning/filtering: A mechanism used in a dynamic reaction cell (DRC) to reject the by-products generated through secondary reactions utilizing the principle of mass discrimination. Achieved by optimizing the electrical fields of the reaction cell multipole (typically a quadrupole) to allow transmission of the analyte ion, while rejecting the polyatomic interfering ion.

BEC: An abbreviation for background equivalent concentration.

by-product ions: Ionic species formed as a result of secondary reactions that take place in a reaction/collision cell. *Also refers to* secondary (side) reactions.

C

calibration: A plot, function, or equation generated using calibration standards and a blank, which describes the relationship between the concentration of an element and the signal intensity produced at the analyte mass of interest. Once determined, this relationship can be used to determine the analyte concentration in an unknown sample.

calibration standard: A reference solution containing accurate and known concentrations of analytes for the purpose of generating a calibration curve or plot.

capacitive coupling: An undesired electrostatic (or capacitive) coupling between the voltage on the load coil and the plasma discharge, which produces a potential difference of a few hundred volts. This creates an electrical discharge or arcing between the plasma and sampler cone of the interface, commonly known as a "secondary discharge" or "pinch effect."

capillary electrophoresis (CE): *Refers to* capillary-zone electrophoresis (CZE).

capillary zone electrophoresis (CZE or CE): A chromatographic separation technique used to separate ionic species according to their charge and frictional forces. In traditional electrophoresis, electrically charged analytes move in a conductive liquid medium under the influence of an electric field. In capillary (zone) electrophoresis, species are separated based on their size-to-charge ratio inside a small capillary filled with an electrolyte. Its applicability to ICP-MS is mainly in the field of separation and detection of large biomolecules.

CCD: Charge-coupled device detector.

CE: *Refers to* capillary-zone electrophoresis.

cell: In ICP-MS terminology, a cell usually refers to a collision or reaction cell.

ceramic torch: A plasma torch where either (or all) the sample injector, inner tube, or outer tube is made of a ceramic material. Typically has longer lifetime than a traditional quartz torch.

certified reference material (CRM): Well-established reference matrix that comes with certified values and associated statistical data that have been analyzed by other complementary techniques. Its purpose is to check the validity of an analytical method, including sample preparation, instrument methodology, and calibration routines to achieve sample results that are as accurate and precise as possible and can be defended when subjected to intense scrutiny.

channel electron multiplier (CEM): A detector used in ICP-MS to convert ions into electrical pulses using the principle of multiplication of electrons via a potential gradient inside a sealed tube.

Channeltron®: Another name for a channel electron multiplier detector.

charge transfer reaction: Sometimes referred to as "charge exchange." This is one of the ion–molecule reaction mechanisms that take place in a collision/reaction cell. Involves the transfer of a positive charge from the interfering ion to the reaction gas molecule, forming a neutral atom that is not seen by the mass analyzer. An example of this kind of reaction:

$$H_2 + {}^{40}Ar^+ = Ar + H_2{}^+.$$

Charge-coupled device detector (CCD): A type of solid state detector technology for converting photons into an electrical signal. Typically applied to ICP-OES.

charge injection device detector (CID): A type of solid-state detector technology for converting photons into an electrical signal. Typically applied to ICP-OES.

chemical modification: The process of chemically modifying the sample in electrothermal vaporization (ETV) ICP-MS work to separate the analyte from the matrix. *Also refers to* chemical modifier *and* electrothermal vaporization.

chemical modifier: A chemical or substance that is added to the sample in an electrothermal vaporizer to change the volatility of the analyte or matrix. Typically added at the ashing stage of the heating program to separate the vaporization of the analyte away from the potential interferences of the matrix components. *Also refers to* electrothermal vaporization.

chromatographic separation device: Any device that separates analyte species according to their retention times or mobility through a stationary phase. When coupled with an ICP-MS system, it is used for the separation, detection, and quantitation of speciated forms of trace elements. Examples include liquid, ion, gas, size exclusion, and capillary electrophoresis chromatography. *Also refers to* speciation analysis.

chromatography terminology (as applied to trace element speciation): The following are some of the most important terms used in the chapter on trace element speciation. For easy access, they are contained in one section and not distributed throughout the glossary.

buffer: A mobile-phase solution that is resistant to extreme pH changes, even with additions of small amounts of acids or bases.

chromatogram: The graphical output of the chromatographic separation. It is usually a plot of peak intensity of the separated species over time.

column: The main component of the chromatographic separation. It is typically a tube containing the stationary-phase material that separates the species and an eluent that elutes the species off the column.

counter-ions: The mobile phase contains a large number of ions that have a charge opposite to that of the surface-bound ions. These are known as counterions, which establish equilibrium with the stationary phase.

dead volume: Usually refers to the volume of the mobile phase between the point of injection and the detector that is accessible to the sample species, minus the volume of mobile phase that is contained in any union or connecting tubing.

gradient elution: Involves variation of the mobile-phase composition over time through a number of steps such as changing the organic content, altering the pH, changing the concentration of the buffer, or using a completely different buffer.

ion exchange: A technique in which separation is based on the exchange of ions (anions or cations) between the mobile phase and the ionic sites on a stationary phase bound to a support material in the column.

ion pairing: A type of separation that typically uses a reversed-phase column in conjunction with a special type of chemical in the mobile phase called an "ion-pairing reagent." *Also refers to* reverse phase.

isochratic elution: An elution of the analytes or species using the same solvent throughout the analysis.

mobile phase: A combination of the sample or species being separated or analyzed and the solvent that moves the sample through the column.

retention time: The time taken for a particular analyte or species to be separated and pass through the column to the detector.

reverse phase: A type of separation that is typically combined with ion pairing, and essentially means that the column's stationary phase is less polar and more organic than the mobile-phase solvents.

stationary phase: A solid material, such as silica or a polymer, that is set in place and packed into the column for the chromatographic separation to take place.

clean room: The general description given to a dedicated room for the sample preparation and analysis of ultrapure materials. Usually associated with a number that describes the number of particulates per cubic foot of air (e.g., a class 100 clean room will contain 100 particles/ft^3 of air). It is commonly accepted that the semiconductor industry has the most stringent demands, which necessitates the use of class 10 and sometimes class 1 clean rooms.

cluster ions: Ions that are formed by two or more molecular ions combining together in a collision/reaction cell to form molecular clusters.

CMOS: Complementary metal oxide semiconductor technology used in the direct charge array detector, which is utilized in the Mattauch-Herzog simultaneous sector instrument.

cold plasma technology: Cool or cold plasma technology uses low-temperature plasma to minimize the formation of certain argon-based polyatomic species. Under normal plasma conditions (approximately 1,000 W RF power and 1.0 L/min nebulizer gas flow), argon ions combine with matrix and solvent components to generate problematic spectral interferences, such as $^{38}ArH^+$, $^{40}Ar^+$, and $^{40}Ar^{16}O^+$, which impact the detection limits of a small number of elements including K, Ca, and Fe. By using cool plasma conditions (approximately 600 W RF power and 1.6 L/min nebulizer gas flow), the ionization conditions in the plasma are changed so that many of these interferences are dramatically reduced and detection limits are improved.

cold vapor atomic absorption (CVAA): An analytical approach to determine low levels of mercury by generating mercuric vapor in a quartz cell and measuring the number of mercury atoms produced, using the principle of atomic absorption. *Also refers to* hydride generation atomic absorption.

collision cell: Specifically, a cell that predominantly uses the principle of collisional fragmentation to break apart polyatomic interfering ions generated in the plasma discharge. Collision cells typically utilize higher-order multipoles (such as hexapoles or octopoles) with inert or low-reactive gases (such as helium and hydrogen) to first stimulate ion–molecule collisions, and then kinetic energy discrimination to reject any undesirable by-product ionic species formed.

collision-induced dissociation (CID): A basic principle, first used for the study of organic molecules using tandem mass spectrometry, that relies on using a nonreactive gas in a collision cell to stimulate ion–molecule collisions. The more collision-induced daughter species that are generated, the better the chance of identifying the structure of the parent molecule.

collision/reaction cell (CRC) technology: A generic term applied to collision and reaction cells that use the principle of ion–molecule collisions and reactions to cleanse the ion beam of problematic polyatomic spectral interferences before they enter the mass analyzer. Both collision and reaction cells are positioned in the mass spectrometer vacuum chamber after the ion optics but prior to the mass analyzer. *Also refers to* collision cell *and* reaction cell.

collision/reaction interface (CRI) technology: A collision/reaction mechanism approach, which instead of using a pressurized cell injects a gas directly into the interface between the sampler and skimmer cones. The injection of the collision/reaction gas into this region of the ion beam produces high collision frequency between the argon gas and the injected gas molecules. This has the effect of removing argon-based polyatomic interferences before they are extracted into the ion optics.

collisional damping: A mechanism that describes the temporal broadening of ion packets in a quadrupole-based dynamic reaction cell to dampen out fluctuations in ion energy. By optimizing cell conditions such as gas pressure, RF stability boundary (q parameter), entrance/exit lens potentials, and cell rod offsets, it has been shown that fluctuation in ion energies can be dampened sufficiently to carry out isotope ratio precision measurements near their statistical limit.

collisional focusing: The mechanism of focusing ions towards the center of the ion beam in a collision/reaction cell. By using a neutral collision gas of lower molecular weight than the analyte, the analyte ions will lose kinetic energy and migrate towards the axis as a result of the collisions with the gaseous molecules. Therefore, the number of ions exiting the cell and reaching the detector will increase. *Also refers to* collision cell *and* reaction cell.

collisional fragmentation: The mechanism of breaking apart (fragmenting) a polyatomic interfering ion in a collision/reaction cell using collisions with a gaseous molecule. The predominant mechanism used in a collision cell, as opposed to a reaction cell. *Also refers to* collision cell *and* reaction cell.

collisional mechanisms: The mechanisms by which the interfering ion is reduced or minimized to allow the determination of the analyte ion. The most common collisional mechanisms seen in collision/reaction cells include collisional focusing, dissociation, and fragmentation, whereas the major reaction mechanisms include exothermic/endothermic associations, charge transfer, molecular associations, and proton transfer.

collisional retardation: A mechanism in a collision/reaction cell where the gas atoms/molecules undergo multiple collisions with the polyatomic interfering ion, retarding or lowering its kinetic energy. Because the interfering ion has a larger cross-sectional area than the analyte ion, it undergoes more collisions and, as a result, can be separated or discriminated from the analyte ion based on their kinetic energy differences.

concentric nebulizer: A nebulizer that uses two narrow concentric capillary tubes (one inside the other) to aspirate a liquid into the ICP-MS spray chamber. Argon gas is usually passed through the outer tube, which creates a Venturi effect, and as a result, the liquid is sucked up through the inner capillary tube.

cones: *Refers to* interface cones.

cool plasma technology: *Refers to* cold plasma technology.

cooled spray chamber: A spray chamber that is cooled in order to reduce the amount of solvent entering the plasma discharge. Used for a variety of reasons, including reducing oxide species, minimizing solvent-based spectral interferences, and allowing the trouble-free aspiration of organic solvents.

correction equation: A mathematical approach used to compensate for isobaric and polyatomic spectral overlaps. It works on the principle of measuring the intensity of the interfering species at another mass, which is ideally free of any interference. A correction is then applied by knowing the ratio of the intensity of the interfering species at the analyte mass to its intensity at the alternate mass.

counts per second (cps): Units of signal intensity used in ICP-MS. Number of detector electronic pulses counted per second.

cps: An abbreviation for counts per second.

CRC: An abbreviation for collision/reaction cell technology.

CRI: An abbreviation for collision/reaction interface technology.

CRM: An abbreviation for certified reference materials.

cross-calibration: A calibration method that is used to correlate both pulse (low levels) and analog (high levels) signals in a dual-mode detector. This is possible because the analog and pulse outputs can be defined in identical terms (of incoming pulse counts per second) based on knowing the voltage at the first analog stage, the output current, and a conversion factor defined by the detection circuitry electronics. By carrying out a cross-calibration across the mass range, a dual-mode detector is capable of achieving approximately eight to nine orders of dynamic range in one simultaneous scan.

cross-flow nebulizer: A nebulizer that is designed for samples that contain a heavier matrix or small amounts of undissolved solids. In this design, the argon gas flow is directed at right angles to the tip of a capillary tube through which the sample is drawn up with a peristaltic pump.

CVAA: An abbreviation for cold vapor atomic absorption.

cyclonic spray chamber: A spray chamber that operates using the principle of centrifugal force. Droplets are discriminated according to their size by means of a vortex produced by the tangential flow of the sample aerosol and argon gas inside the spray chamber. Smaller droplets are carried with the gas stream into the ICP-MS, while the larger droplets impinge on the walls and fall out through the drain.

cylinder lens: A type of lens component used in the ion optics.

CZE: An abbreviation for capillary-zone electrophoresis.

D

data-quality objectives: A term used to describe the quality goals of the analytical result. Typically achieved by optimizing the measurement protocol to achieve the desired accuracy/precision/sample throughput required for the analysis.

DCD: *Refers to* direct charge detector.

dead time correction: Sometimes ions hit the detector too fast for the measurement circuitry to handle in an efficient manner. This is caused by ions arriving at the detector during the output pulse of the preceding ion and not being detected by the counting system. This "dead time," as it is known, is a fundamental limitation of the multiplier detector and is typically 30–50 ns, depending on the detection system. A compensation or "dead time correction" has to be made in the measurement circuitry in order to count the maximum number of ions hitting the detector.

Debye length: The distance over which ions exert an electrostatic influence over one another as they move from the interface region into the ion optics. In the ion-sampling process, this distance is small compared to the orifice diameter of the sampler or skimmer cone. As a result, there is little electrical interaction between the ion beam and the cones, and relatively little interaction between the individual ions within the ion beam. In this way, the compositional integrity of the ion beam is maintained throughout the interface region.

desolvating microconcentric nebulizer: A microconcentric nebulizer that uses some type of desolvation system to remove the sample solvent. *Also refers to* desolvation device *and* membrane desolvation.

desolvating spray chamber: A general name given to a spray chamber that removes or reduces the amount of solvent from a sample using the principle of desolvation. Some of the approaches that are typically used include conventional water cooling, heating with cooling condensers, Peltier (thermoelectric) cooling, or membrane-based desolvation techniques.

desolvation device: A general name given to a device that removes or reduces the amount of solvent from a sample using the principle of desolvation. Some of the approaches that are typically used include conventional water cooling, heating/condensing units, Peltier (thermoelectric) cooling, or membrane-based desolvation techniques.

detection capability: A generic term used to assess the overall detection performance of an ICP mass spectrometer. There are a number of different ways of evaluating detection capability, including instrument detection limit (IDL), method detection limit (MDL), element sensitivity, and background equivalent concentration (BEC).

detection limit: Most often refers to the instrument detection limit (IDL) and is typically defined as a ratio of the analyte signal to the noise of the background at a particular mass. For a 99% confidence level, it is usually calculated as 3× standard deviation (SD) of ten replicates (measurements) of the sample blank expressed as concentration units.

detector: A generic name used for a device that converts ions into electrical pulses in ICP-MS.

detector dead time: *Refers to* dead time correction.

devitrification: Crystalline breakdown of glass or quartz by a combination of chemical attack and elevated temperatures, typically associated with the plasma torch.

digital counting: Refers to the process of counting the number of pulses generated by the conversion of ions into an electrical signal by the detector measurement circuitry.

DIHEN: An abbreviation for direct injection high-efficiency nebulizer.

DIN: An abbreviation for direct injection nebulizer.

Direct charge detector (DCD): A detector technology used to convert photons into an electric current. A type of CMOS array ion detector used in the Mattauch–Herzog simultaneous sector instrument.

direct injection high-efficiency nebulizer (DIHEN): A more recent refinement of the direct injection nebulizer (DIN), which appears to have overcome many of the limitations of the original design.

direct injection nebulizer (DIN): A nebulizer that injects a liquid sample under high pressure directly into the base of the plasma torch. The benefit of this approach is that no spray chamber is required, which means that an extremely small volume of sample can be introduced directly into the ICP-MS with virtually no carryover or memory effects from the previous sample.

discrete dynode detector (DDD): The most common type of detector used in ICP-MS. As ions emerge from the quadrupole rods onto the detector, they strike the first dynode, liberating secondary electrons. The electron-optic design of the dynode produces acceleration of these secondary electrons to the next dynode, where they generate more electrons. This process is repeated at each dynode, generating a pulse of electrons that are finally captured by the multiplier anode. *Also refers to* active film multipliers.

double-focusing magnetic sector mass spectrometer (analyzer): A mass spectrometer that uses a very powerful magnet combined with an electrostatic analyzer (ESA) to produce a system with very high resolving power. This approach, known as "double focusing," samples the ions from the plasma. The ions are accelerated in the plasma to a few kilovolts into the ion-optic region before they enter the mass analyzer. The magnetic field, which is dispersive with respect to ion energy and mass, then focuses all the ions with diverging angles of motion from the entrance slit. The ESA, which is only dispersive with respect to ion energy, then focuses all the ions onto the exit slit, where the detector is positioned. If the energy dispersions of the magnet and ESA are equal in magnitude but opposite in direction, they will focus both ion angles (first focusing) and ion energies (second focusing) when combined together. *Also refers to* electrostatic analyzer.

double-pass spray chamber: A spray chamber that comprises an inner (central) tube inside the main body of the spray chamber. The smaller droplets are selected by directing the aerosol from the nebulizer into the central tube. The aerosol emerges from the tube, where the larger droplets fall out (because of gravity) through a drain tube at the rear of the spray chamber. The smaller droplets then travel back between the outer wall and the central tube into the sample injector of the plasma torch. The most common type of double-pass spray chamber is the Scott design.

doubly charged ion: A species that is formed when an ion is generated with a double-positive charge as opposed to a normal single charge and produces an isotopic peak at half its mass. For example, the major isotope of barium at mass 138 amu also exhibits a doubly charged ion at mass 69 amu, which can potentially interfere with gallium at mass 69. Some elements such as the rare earths readily form doubly charged species, whereas others do not. Formation of doubly charged ions is also impacted by the ionization conditions (RF power, nebulizer gas flow, etc.) in the plasma discharge.

DRC: An abbreviation for dynamic reaction cell.

droplet: Refers to individual particles (either small or large) that make up an aerosol generated by the nebulizer.

dry plasma: When a sample is introduced into the plasma that does not contain any liquid or solvent, such as laser ablation, ETV, or desolvation sample introduction systems.

duty cycle (%): Also known as the "measurement duty cycle." It refers to the actual peak measurement time and is expressed as a percentage of the overall integration time. It is calculated by dividing the total peak quantitation time (dwell time × number of sweeps × replicates × elements) by the total integration time ([dwell time + settling/scanning time] × number of sweeps × replicates × elements).

dwell time: The time spent sitting (dwelling) on top of the analytical peak (mass) and taking measurements.

dynamic reaction cell (DRC): A type of collision/reaction cell. Unlike a simple collision cell, a quadrupole is used instead of a hexapole or octapole. A highly reactive gas such as ammonia or methane is bled into the cell, which is a catalyst for ion–molecule chemistry to take place. By a number of different reaction mechanisms, the gaseous molecules react with the interfering ions to convert them either into an innocuous species different from the analyte mass or a harmless neutral species. The analyte mass then emerges from the dynamic reaction cell, free of its interference, and is steered into the analyzer quadrupole for conventional mass separation. Through careful optimization of the quadrupole electrical fields, unwanted reactions between the gas and the sample matrix or solvent, which could potentially lead to new interferences, are prevented. Therefore, every time an analyte and interfering ions enter the dynamic reaction cell, the bandpass of the quadrupole can be optimized for that specific problem and then changed on the fly for the next one.

dynamically scanned ion lens: A commercial ion optic approach to focus the maximum number of ions into the mass analyzer. In this design, the voltage is dynamically ramped on the fly in concert with the mass scan of the analyzer. The benefit is that the optimum lens voltage is placed on every mass in a multielement run to allow the maximum number of analyte ions through, while keeping the matrix ions down to an absolute minimum. This is typically used in conjunction with a grounded stop acting as a physical barrier to reduce particulates, neutral species, and photons from reaching the mass analyzer and detector.

E

EDR: An abbreviation for the term "extended dynamic range," used in detector technology.

electrodynamic forces: Flow of the ion beam through the interface region, where the positively charged ions of varying mass-to-charge exert no electrical influence on each other.

electron: A negatively charged fundamental particle orbiting the nucleus of an atom. It has a mass equal to 1/1,836 of a proton's mass. Removal of an electron by excitation in the plasma discharge generates a positively charged ion.

electrostatic analyzer (ESA): An ion-focusing device (utilizing a series of electrostatic lens components) that varies the electric field to allow the passage of ions of certain energy. In ICP-MS, it is typically used in combination with a conventional electromagnet to focus ions based on their angular motion and their kinetic energy to produce very high resolving power. *Also refers to* double-focusing magnetic sector mass spectrometer (analyzer).

electrothermal atomization (ETA): An atomic absorption (AA) analytical technique that uses a heated metal filament or graphite tube (in place of the normal flame) to generate ground-state analyte atoms. The sample is first injected into the filament or tube, which is heated up slowly to remove the matrix components. Further heating then generates ground-state atoms of the analyte, which absorb light of a particular wavelength from an element-specific, hollow cathode lamp source. The amount of light absorbed is measured by a monochromator (optical system) and detected by a photomultiplier or solid-state detector, which converts the photons into an electrical pulse. This absorbance signal is used to determine the concentration of that element in the sample. Typically used for ppb-level determinations.

electrothermal vaporization (ETV): A sample pretreatment technique used in ICP-MS. Based on the principle of electrothermal atomization (ETA) used in atomic absorption (AA), ETV is not used to generate ground-state atoms but instead uses a carbon furnace (tube) or metal filament to thermally separate the analytes from the matrix components and then sweep them into the ICP mass spectrometer for analysis. This is achieved by injecting a small amount of the sample into a graphite tube or onto a metal filament. After the sample is introduced, drying, charring, and vaporization are achieved by slowly heating the graphite tube or metal filament. The sample material is vaporized into a flowing stream of carrier gas, which passes through the furnace or over the filament during the heating cycle. The analyte vapor recondenses in the carrier gas and is then swept into the plasma for ionization.

elemental fractionation: A term used in laser ablation. It is typically defined as the variation in intensity of a particular element over time compared to the total amount of dry aerosol generated by the sample. It is generally sample and element specific, but there is evidence to suggest that the shorter-wavelength excimer lasers exhibit better elemental fractionation characteristics than the longer-wavelength Nd:YAG design because they produce smaller particles that are easier to volatilize.

endothermic reaction: In thermodynamics, this describes a chemical reaction that absorbs energy in the form of heat. In ICP-MS, it generally refers to an ion–molecule reaction in a collision/reaction cell that is not allowed to proceed because the ionization potential of the analyte ion is significantly less than that of the reaction gas molecule. *Also refers to* exothermic reaction.

engineered nanomaterials (ENMs): These are man-made materials made of particles with <100 nm diameter that can be made to exhibit: greater physical strength, enhanced magnetic properties, conduction of heat or electricity, greater chemical reactivity, or size-dependent optical properties. An example of an engineered nanomaterial is silver nanoparticles, which are added to detergents as a bactericide.

ENM: *Refers to* engineered nanomaterials.

ESA: An abbreviation for electrostatic analyzer.

ETA: An abbreviation for electrothermal atomization.

ETV: An abbreviation for electrothermal vaporization.

excimer laser: A gas-filled laser in which a very short electrical pulse excites a mixture containing a halogen such as fluorine and a rare gas such as argon or krypton. It produces a brief, intense pulse of UV light. The output of an excimer laser is used for writing patterns on semiconductor chips because the short wavelength can write very fine lines. In ICP-MS, the most common excimer laser used is ArF at 193 nm and is typically used to ablate material with a very small size, such as inclusions on the surface of a geological sample.

exothermic reaction: In thermodynamics, this describes a chemical reaction that releases energy in the form of heat. In ICP-MS, it generally refers to an ion–molecule reaction in a collision/reaction cell that is spontaneous because the ionization potential of the interfering ion is much greater than the reaction gas molecule. *Also refers to* endothermic reaction.

extended dynamic range (EDR): An approach used in ICP-MS to extend the linear dynamic range of the detector from 5 orders of magnitude up to eight or 9 orders of magnitude. *Also refers to* discrete dynode detector *and* Faraday cup detector.

external standardization: The normal mode of calibration used in ICP-MS by comparing the analyte intensity of unknown samples to the intensity of known calibration or reference standards.

extraction lens: An ion lens used to electrostatically extract the ions out of the interface region.

F

FAA: An abbreviation for flame atomic absorption.

Faraday collector: Another name for a Faraday cup detector.

Faraday cup detector: A simple metal electrode detector used to measure high ion counts. When the ion beam hits the metal electrode, it will be charged, whereas the ions are neutralized. The electrode is then discharged to measure a small current equivalent to the number of discharged ions. By measuring the ion current on the metal part of the circuit, the number of ions in the circuit can be determined. Unfortunately, with this approach, there is no control over the applied voltage (gain). So, it can only be used for high ion counts and therefore is not suitable for ultratrace determinations.

field flow fractionation (FFF): Field flow fractionation (FFF) is a single-phase chromatographic separation technique, where separation is achieved within a very thin channel, against which a perpendicular force field is applied. One of the most common forms of FFF is asymmetrical flow FFF (AF4), where the field is generated by a cross-flow applied perpendicular to the channel. Coupled with ICP-MS for the characterization of nanoparticles.

FFF: *Refers to* field flow fractionation.

FGDW: *Refers to* flue gas desulfurization wastewaters.

FIA: An abbreviation for flow injection analysis.

flame atomic absorption (FAA): A n atomic absorption analytical technique that uses a flame (usually air–acetylene or nitrous oxide–acetylene) to generate ground-state atoms. The sample solution is aspirated into the flame via a nebulizer and a spray chamber. The ground-state atoms of the sample absorb light of a particular wavelength from an element-specific, hollow cathode lamp source. The amount of light absorbed is measured by a monochromator (optical system) and detected by a photomultiplier or solid-state detector, which converts the photons into an electrical pulse. This absorbance signal is used to determine the concentration of the element in the sample. Typically used for ppm-level determinations.

flatapole: A quadrupole with rods that have flat corners. Used in a particular commercial design of collision/reaction cell.

flight tube: A generic name given to the housing that contains a series of optical components which focus ions onto the detector of a time-of-flight (TOF) mass analyzer. There are basically two different kinds of flight tubes that are used in commercial TOF mass analyzers. One is the orthogonal design, where the flight tube is positioned at right angles to the sampled ion beam, and the other the axial design, where the flight tube is in the same axis as the ion beam. In both designs, all ions are sampled through the interface region, but instead of being focused into the mass filter in the conventional sequential way, packets (groups) of ions are electrostatically injected into the flight tube at exactly the same time.

flow injection analysis (FIA): A powerful front-end sampling accessory for ICP-MS that can be used for preparation, pretreatment, and delivery of the sample. It involves the introduction of a discrete sample aliquot into a flowing carrier stream. Using a series of automated pumps and valves, procedures can be carried out online to physically or chemically change the sample or analyte before introduction into the mass spectrometer for detection.

flue gas desulfurization wastewaters (FGDW): This is one of the most widely used technologies for removing pollutants such as sulfur dioxide, from flue gas emissions produced by coal-fired power plants. Sometimes called the limestone forced oxidation scrubbing system, but more commonly known as flue gas desulfurization (FGD), this process employs gas scrubbers to spray limestone slurry over the flue gas to convert gaseous sulfur dioxide to calcium sulfate.

fractogram: The separated particles that exit the outlet port of a field flow fractionation device into the detection system (e.g., UV/Vis or ICP-MS) are displayed as a temporal signal called a fractogram (similar to a chromatogram in chromatographic separation).

fringe rods: A set of four short rods operated in the RF-only mode, positioned at the entrance of a quadrupole mass analyzer. Their function is to minimize the effect of the fringing fields at the entrance of a quadrupole mass analyzer and thus improve the efficiency of transmission of ions into the mass analyzer. They are usually straight, but it has been suggested that curved fringe rods might reduce background levels.

fusion mixture: A compound or mixture added to solid samples as an aid to get them into solution. Fusion mixtures are usually alkaline salts (e.g., lithium metaborate, sodium carbonate) that are mixed with the sample (in powdered form) and heated in a muffle furnace to create a chemical/thermal reaction between the sample and the salt. The fused mixture is then dissolved in a weak mineral acid to get the analytes into solution.

G

gamma-counting spectrometry: A particle-counting technique that uses the measurement of the radioactive decay of gamma particles. *Also refers to* particle-counting techniques.

gas dynamics: In ICP-MS, it refers to the flow and velocity of the plasma gas through the interface region. It dictates that the composition of the ion beam immediately behind the sampler cone be the same as the composition in front of the cone because the expansion of the gas at this stage is not controlled by electrodynamics. This happens because the distance over which ions exert influence on one another (the Debye length) is small compared to the orifice diameter of the sampler or skimmer cone. Consequently, there is little electrical interaction between the ion beam and the cone and relatively little interaction between the individual ions in the beam. In this way, gas dynamics ensures that the compositional integrity of the ion beam is maintained throughout the interface region.

getter (gas purifier): A device that "cleans up" inorganic and organic contaminants in pure gases. The getter usually refers to a metal that oxidizes quickly, and when heated to a high temperature (usually by means of RF induction), evaporates, and absorbs/reacts with any residual impurities in the gas.

GFAA: An abbreviation for graphite furnace atomic absorption.

graphite furnace atomic absorption (GFAA): An electrothermal atomization (ETA) analytical technique that specifically uses a graphite tube (in place of the normal flame) to generate ground-state analyte atoms. The sample is first injected into the tube, which is heated up slowly to remove the matrix components. Further heating then generates ground-state atoms of the analyte, which absorb light of a particular wavelength from an element-specific, hollow cathode lamp source. The amount of light absorbed is measured by a monochromator (optical system) and detected by a photomultiplier or solid-state detector, which converts the photons into an electrical pulse. This absorbance signal is used to determine the concentration of that element in the sample. Typically used for ppb-level determinations. *Also refers to* electrothermal atomization (ETA).

grounding mechanism: A way of eliminating the secondary discharge (pinch effect) produced by capacitive (RF) coupling of the load coil to the plasma. This undesired coupling between the RF voltage on the load coil and the plasma discharge produces a potential difference of a few hundred volts, which creates an electrical discharge (arcing) between the plasma and sampler cone of the interface. This mechanism varies with different instrument designs, but basically involves grounding the load coil to make sure the interface region is maintained at zero potential.

H

half-life: The time required for half the atoms of a given amount of a radioactive substance to disintegrate. This principle is used in particle-counting measuring techniques.

heating zones: The zones that describe the different temperature regions within a plasma discharge, where the sample passes through. The most common zones include the preheating zone (PHZ), where the sample is desolvated; the initial radiation zone (IRZ), where the sample is broken down into its molecular form; and the normal analytical zone (NAZ), where the sample is first atomized and then ionized.

HEN: An abbreviation for high-efficiency nebulizer.

hexapole: A multipole containing six rods, used in collision/reaction cell technology.

HGAA: An abbreviation for hydride generation atomic absorption.

high-efficiency nebulizer (HEN): A generic name given to a nebulizer that is very efficient, with very little wastage. Usually used to describe direct injection or microconcentric-designed systems, which deliver all or a very high percentage of the sample aerosol into the plasma discharge.

high-resolution mass analyzer: A generic name given to a mass spectrometer with very high resolving power. Commercial designs are usually based on the double-focusing magnetic sector design.

high-sensitivity interface (HSI): High-sensitivity interfaces (HSIs) are offered as an option with most commercial ICP-MS systems. They all work slightly differently but share similar components. By using a slightly different cone geometry, higher vacuum at the interface, one or more extraction lenses, or modified ion optic design, they offer up to ten times the sensitivity of a traditional interface. However, their limitations are that background levels are often elevated, particularly when analyzing samples with a heavy matrix. Therefore, they are more suited for the analysis of clean solutions.

high-solids nebulizers: Nebulizers that are used to aspirate higher concentrations of dissolved solids into the ICP-MS. The most common types used are the Babbington, V-groove, and cone-spray designs. Not widely used for ICP-MS because of the dissolved-solids limitations of the technique, but are sometimes used with flow injection sample introduction techniques.

hollow ion mirror: A more recent development in ion-focusing optics. The ion mirror, which has a hollow center, creates a parabolic electrostatic field to reflect and refocus the ion beam at right angles to the ion source. This allows photons, neutrals, and solid particles to pass through it, while allowing ions to be reflected at right angles into the mass analyzer. The major benefit of this design is the highly efficient way the ions are refocused, offering extremely high sensitivity and low background across the mass range.

homogenized sample beam: The laser beam in an excimer laser, which produces a much flatter beam profile and more precise control of the ablation process.

hydride generation atomic absorption (HGAA): A very sensitive analytical technique for determining trace levels of volatile elements such as As, Bi, Sb, Se, and Te. Generation of the elemental hydride is carried out in a closed vessel by the addition of a reducing agent, such as sodium borohydride, to the acidic sample. The resulting gaseous hydride is swept into a special heated quartz cell (in place of the traditional flame burner head), where atomization occurs. Atomic absorption quantitation is then carried out in the conventional way, by comparing the absorbance of unknown samples against known calibration or reference standards.

hydrogen atom transfer: An ion–molecule reaction mechanism in a collision/reaction cell where a hydrogen atom is transferred to the interfering ion, which is converted to an ion at one mass higher.

hyperbolic fields: The four rods that make up a quadrupole are usually cylindrical or elliptical in shape. The electrical fields produced by these rods are typically hyperbolic in shape.

hyper skimmer cone: The name for an additional cone used in one commercial ICP-MS interface design, used in order to tighten the ion beam entering the ion optics.

I

ICP: An abbreviation for inductively coupled plasma.

ICP-OES: An abbreviation for inductively coupled plasma optical emission spectrometry.

IDL: *Refers to* instrument detection limit.

impact bead (nebulizer): A type of spray chamber more commonly used in atomic absorption spectrometers. The aerosol from the nebulizer is directed onto a spherical bead, where the impact breaks the sample into large and small droplets. The large droplets fall out due to gravitational force, and the smaller droplets are directed by the nebulizer gas flow into the atomization/excitation/ionization source.

inductively coupled plasma (ICP): The high-temperature source used to generate ions in ICP-MS. It is formed when a tangential (spiral) flow of argon gas is directed between the outer and middle tube of a quartz torch. A load coil (usually copper) surrounds the top end of the torch and is connected to an RF generator. When RF power (typically, 750–1,500 W) is applied to the load coil, an alternating current oscillates within the coil at a rate corresponding to the frequency of the generator. The RF oscillation of the current in the coil creates an intense electromagnetic field in the area at the top of the torch. With argon gas flowing through the torch, a high-voltage spark is applied to the gas, causing some electrons to be stripped from their argon atoms. These electrons, which are caught up and accelerated in the magnetic field, then collide with other argon atoms, stripping off still more electrons. This collision-induced ionization of the argon continues in a chain reaction, breaking down the gas into argon atoms, argon ions, and electrons, forming what is known as an "inductively coupled plasma (ICP) discharge" at the open end of the plasma torch.

inductively coupled plasma optical emission spectrometry (ICP-OES): A multielement technique that uses an inductively coupled plasma to excite ground-state atoms to the point where they emit wavelength-specific photons of light, characteristic of a particular element. The number of photons produced at an element-specific wavelength is measured using high-resolving optical components to separate the analyte wavelengths and a photon-sensitive detection system to measure the intensity of the emission signal produced. This emission signal is directly related to the concentration of that element in the sample. Commercial instrumentation comes in two configurations: a traditional radial view, where the plasma is vertical and is viewed from the side (side-on viewing), and an axial view, where the plasma is positioned horizontally and is viewed from the end (end-on viewing).

infrared (IR) lasers: Laser ablation systems that operate in the IR region of the electromagnetic spectrum, such as the Nd:YAG laser, which has its primary wavelength at 1,064 nm.

instrument background noise: Square root of the spectral background of the instrument (in cps), usually measured at mass 220 amu, where there are no spectral features. *Also refers to* background signal, background noise, *and* detection limit.

instrument background signal: Spectral background of the instrument (in cps), usually measured at mass 220 amu, where there are no spectral features. *Also refers to* background signal, background noise, *and* detection limit.

instrument detection limit (IDL): It's a way of assessing an instrument's detection capability. It is often referred to as signal-to-background noise and for a 99% confidence level is typically defined as 3× standard deviation (SD) of ten replicates of the sample blank.

integration time: The total time spent measuring an analyte mass (peak). Comprising the time spent dwelling (sitting) on the peak multiplied by the number of points used for peak quantitation multiplied by the number of scans used in the measurement protocol. *Also refers to* duty cycle, measurement duty cycle, peak measurement protocol, settling time, *and* dwell time.

interface: The plasma discharge is coupled to the mass spectrometer via the interface. The interface region comprises a water-cooled metal housing containing the sampler cone and the skimmer cone, which directs the ion beam from the central channel of the plasma into the ion optic region.

interface cones: *Refers to* the sampler and skimmer cones housed in the interface region. *Also refers to* interface *and* interface region.

interface pressure: The pressure between the sampler cone and skimmer cone. This region is maintained at a pressure of approximately 1–2 torr by a mechanical roughing pump.

interface region: A region comprising a water-cooled metal housing containing the sampler cone and the skimmer cone, which directs and focuses the ion beam from the central channel of the plasma into the ion optic region.

interferences: A generic term given to a non-analyte component that enhances or suppresses the signal intensity of the analyte mass. The most common interferences in ICP-MS are spectral, matrix, or sample transport in nature.

internal standardization (IS): A quantitation technique used to correct for changes in analyte sensitivity caused by variations in the concentration and type of matrix components found in the sample. An internal standard is a non-analyte isotope that is added to the blank solution, standards, and samples before analysis. It is typical to add three or four internal standard elements to the samples to cover all the analyte elements of interest across the mass range. The software adjusts the analyte concentration in the unknown samples by comparing the intensity values of the internal standard elements in the unknown sample to those in the calibration standards. Because ICP-MS is prone to many matrix- and sample-transport-based interferences, internal standardization is considered necessary to analyze most sample types.

ion: An electrically charged atom or group of atoms formed by the loss or gain of one or more electrons. A cation (positively charged ion) is created by the loss of an electron, and an anion (negatively charged ion) is created by the gain of an electron. The valency of an ion is equal to the number of electrons lost or gained and is indicated by a plus sign for cations and a minus sign for anions. ICP-MS typically involves the detection and measurement of positively charged ions generated in a plasma discharge.

ion chromatography (IC): A chromatographic separation technique used for determination of anionic species such as nitrates, chlorides, and sulfates. When coupled with ICP-MS, it becomes a very sensitive hyphenated technique for the determination of a wide variety of elemental ionic species.

ion energy: In ICP-MS, it refers to the kinetic energy of the ion, in electron volts (eV). It is a function of both the mass and velocity of the ion ($KE = \frac{1}{2}MV^2$). It is generally accepted that the spread of kinetic energies of all the ions in the ion beam entering the mass spectrometer must be on the order of a few electron volts to be efficiently focused by the ion optics and resolved by the mass analyzer.

ion energy spread: The variation in kinetic energy of all the ions in the ion beam emerging from the ionization source (plasma discharge). It is generally accepted that this variation (spread) of kinetic energies must be on the order of a few electron volts to be efficiently focused by the ion optics and resolved by the mass analyzer.

ion flow: The flow of ions from the interface region through the ion optics into the mass analyzer.

ion-focusing guide: An alternative name for the ion optics.

ion-focusing system: An alternative name for the ion optics.

ion formation: The transfer of energy from the plasma discharge to the sample aerosol to form an ion. By traveling through the different heating zones in the plasma, where the sample is first dried, vaporized and atomized, and then finally converted to an ion.

ion kinetic energy: *Refers to* ion energy.

ion lens: Often referred to as a single-lens component in the ion optic system. *Also refers to* ion optics.

ion lens voltages: The voltages put on one or more lens components in the ion optic system to electrostatically steer the ion beam into the mass analyzer. *Also refers to* ion optics.

ion mirror: A more recent development in ion-focusing optics. With this design, a parabolic electrostatic field is created with a hollow ion mirror to reflect and refocus the ion beam at right angles to the ion source. The ion mirror is an electrostatically charged ring, which is hollow in the center. This allows photons, neutrals, and solid particles to pass through it, while allowing ions to be reflected at right angles into the mass analyzer. The major benefit of this design is the highly efficient way the ions are refocused, offering extremely high sensitivity across the mass range with very little compromise in oxide performance. In addition, there is very little contamination of the ion optics because a vacuum pump sits behind the ion mirror to immediately remove these particles before they have a chance to penetrate further into the mass spectrometer.

ion optics: Comprises one or more electrostatically charged lens components that are positioned immediately after the skimmer cone. They are made up of a series of metallic plates, barrels, or cylinders, which have a voltage placed on them. The function of the ion optic system is to take ions after they emerge from the interface region and steer them into the mass analyzer. Another function of the ion optics is to reject the nonionic species such as particulates, neutral species, and photons and prevent them from reaching the detector. Depending on the design, this is achieved by using some kind of physical barrier, positioning the mass analyzer off axis relative to the ion beam, or electrostatically bending the ions by 90° into the mass analyzer.

ion packet: A "slice of ions" that is sampled from the ion beam in a time-of-flight (TOF) mass analyzer. In the TOF design, all ions are sampled through the interface cones, but instead of being focused into the mass filter in the conventional way, packets (groups) of ions are electrostatically injected into the flight tube at exactly the same time. Whether the orthogonal (right-angle) or axial (straight-on) approach is used, an accelerating potential is applied to the continuous ion beam. The ion beam is then "chopped" by using a pulsed voltage supply to provide repetitive voltage "slices" at a frequency of a few kilohertz. The "sliced" packets of ions are then allowed to "drift" into the flight tube, where the individual ions are temporally resolved according to their differing velocities.

ion repulsion: The degree to which positively charged ions repel each other as they enter the ion optics. The generation of a positively charged ion beam is the first stage in the charge separation process. Unfortunately, the net positive charge of the ion beam means that there is now a natural tendency for the ions to repel one another. If nothing is done to compensate for this repulsion, ions of higher mass-to-charge ratio will dominate the center of the ion beam and force the lighter ions to the outside. The degree of loss will depend on the kinetic energy of the ions—ions with high kinetic energy (high-mass elements) will be transmitted in preference to ions with medium (mid-mass elements) or low kinetic energy (low-mass elements).

ionization source: In ICP-MS, the ionization source is the plasma discharge, which reaches temperatures of up to 10,000 K to ionize the liquid sample.

ionizing radiation counting techniques: Particle-counting techniques such as alpha, gamma, and scintillation counters that are used to measure the isotopic composition of radioactive materials. However, the limitation of particle-counting techniques is that the half-life of the analyte isotope has a significant impact on the method detection limit. This implies that they are better suited for the determination of short-lived radioisotopes, because meaningful data can be obtained in a realistic amount of time. They have also been successfully applied to the quantitation of long-lived radionuclides, but unfortunately require a combination of extremely long counting times and large amounts of sample to achieve low levels of quantitation.

ion–molecule chemistry: A chemical reaction between the analyte or interfering ion and molecules of the reaction gas in a collision/reaction cell. A reactive gas, such as hydrogen, ammonia, oxygen, methane, or gas mixtures, is bled into the cell, which is a catalyst for ion–molecule chemistry to take place. By a number of different reaction mechanisms, the gaseous molecules react with the interfering ions to convert them into either an innocuous species different from the analyte mass or a harmless neutral species. The analyte mass then emerges from the cell free of its interference and is steered into the analyzer quadrupole for conventional mass separation. In some cases, the chemistry can take place between the gaseous molecule and the analyte to form a new analyte ion free of the interfering species.

isobar (or isobaric): Used in the context of atomic principles, it refers to two or more atoms with the same atomic mass (same number of neutrons) but different atomic number (different number of protons). *Also refers to* isobaric interferences.

isobaric interferences: The word "isobaric" is used in the context of atomic principles and refers to two or more atoms with the same atomic mass but different atomic number. In ICP-MS, they are a classification of spectrally induced interferences produced mainly by different isotopes of other elements in the sample, creating spectral interferences at the same mass as the analyte.

isotope: A different form of an element having the same number of protons in the nucleus (i.e., same atomic number) but a different number of neutrons (i.e., different atomic mass). There are 275 isotopes of the 81 stable elements in the periodic table, in addition to over 800 radioactive isotopes. Isotopes of a single element possess very similar properties.

isotope dilution: An absolute means of quantitation in ICP-MS based on altering the natural abundance of two isotopes of an element by adding a known amount of one of the isotopes. The principle works by spiking the sample solution with a known weight of an enriched stable isotope. By knowing the natural abundance of the two isotopes being measured, the abundance of the spiked enriched isotope, the weight of the spike, and the weight of the sample, it is possible to determine the original trace element concentration. It is considered one of the most accurate and precise quantitation techniques for elemental analysis by ICP-MS.

isotope ratio: The ability of ICP-MS to determine individual isotopes makes it suitable for an isotopic measurement technique called "isotope ratio analysis." The ratio of two or more isotopes in a sample can be used to generate very useful information, such as an indication of the age of a geological formation, a better understanding of animal metabolism, and the identification of sources of environmental contamination. Similar to isotope dilution, isotope ratio analysis uses the principle of measuring the exact ratio of two isotopes of an element in the sample. With this approach, the isotope of interest is typically compared to a reference isotope of the same element, but can also be referenced to an isotope of another element.

isotope ratio precision: The reproducibility or precision of measurement of isotope ratios is very critical for some applications. For the highest-quality isotopic ratio precision measurements, it is generally acknowledged that either magnetic sector or time-of-flight (TOF) instrumentation offers the best approach over quadrupole ICP-MS.

isotopic abundance: The percentage abundance of an isotope compared to the element's total abundance in nature. *Also refers to* natural abundance *and* relative abundance of natural isotopes.

K

KE: An abbreviation for kinetic energy.

KED: An abbreviation for kinetic energy discrimination.

kinetic energy (KE): The energy possessed by a moving body due to its motion. It is equal to one-half the mass of the body times the square of its speed (velocity): $KE = ½MV^2$. For kinetic energy as applied to moving ions, *refers to* ion energy.

kinetic energy discrimination (KED): In collision/reaction cell technology, it is one way to separate the newly formed by-product ions from the analyte ions. It is typically achieved by setting the collision cell potential (voltage) slightly more negative than the mass filter potential. This means that the collision by-product ions generated in the cell, which have a lower kinetic energy as a result of the collision process, are rejected, whereas the analyte ions, which have a higher kinetic energy, are transmitted to the mass analyzer.

L

laser ablation: A sample preparation technique that uses a high-powered laser beam to vaporize the surface of a solid sample and sweep it directly into the ICP-MS system for analysis. It is mainly used for samples that are extremely difficult to get into solution or for samples that require the analysis of small spots or inclusions on the surface.

laser absorption: The "coupling" efficiency of the sample with the laser beam in laser ablation work. The more light the sample absorbs, the more efficient the ablation process becomes. It is generally accepted that the shorter-wavelength excimer lasers have better absorption characteristics than the longer-wavelength IR laser systems for UV-transparent/opaque materials such as calcites, fluorites, and silicates and, as a result, generate smaller particle size and higher flow of ablated material.

laser fluence: A term used to describe the power density of a laser beam in laser ablation studies. It is defined as the laser pulse energy per focal spot area, measured in J/cm^2. It is related to laser irradiance, which is the ratio of the fluence to the width of the laser pulse.

laser irradiance: A term used to describe the power density of a laser beam in laser ablation studies. Laser irradiance is the ratio of the laser pulse energy per focal spot area (i.e., fluence) to the width of the laser pulse. *Also refers to* laser fluence.

laser sampling: *Refers to* laser ablation.

laser vaporization: *Refers to* laser ablation.

laser wavelength: The primary wavelength of the optical components used in the design of a laser ablation system.

linear plane array detectors: Solid-state detector technology used to measure the mass spectrum in a simultaneous manner (recently commercialized in the Mattauch–Herzog magnetic sector instrument).

load coil: Another name for the RF coil used to generate a plasma discharge. *Also refers to* RF generator.

low-mass cutoff: This is a variation on bandpass filtering in a collision/reaction cell, which uses slightly different control of the filtering process. By operating the cell in the RF-only mode, the quadrupole's stability boundaries can be tuned to cut off low masses where the majority of the interferences occur.

low-temperature plasma: An alternative name for cool or cold plasma.

M

magnetic field: A region around a magnet, an electric current, or a moving charged particle that is characterized by the existence of a detectable magnetic force at every point in the region and by the existence of magnetic poles. In ICP-MS, it usually refers to the magnetic field around the RF coil of the plasma discharge or the magnetic field produced by a quadrupole or an electromagnet.

magnetic sector mass analyzer: A design of mass spectrometer used in ICP-MS to generate very high resolving power as a way of reducing spectral interferences. Commercial designs typically utilize a very powerful magnet combined with an electrostatic analyzer (ESA). In this approach, known as the double-focusing design, the ions from the plasma are sampled. In the plasma, the ions are accelerated to a few kilovolts into the ion optic region before they enter the mass analyzer. The magnetic field, which is dispersive with respect to ion energy and mass, then focuses all the ions with diverging angles of motion from the entrance slit. The ESA, which is only dispersive with respect to ion energy, then focuses all the ions onto the exit slit, where the detector is positioned. If the energy dispersion of the magnet and ESA are equal in magnitude but opposite in direction, they will focus both ion angles (first focusing) and ion energies (second focusing) when combined together. *Also refers to* electrostatic analyzer.

mass analyzer: The part of the mass spectrometer where the separation of ions (based on their mass-to-charge ratio) takes place. In ICP-MS, the most common type of mass analyzers are quadrupole, magnetic sector, and time-of-flight (TOF) systems.

mass calibration: The ability of the mass spectrometer to repeatedly scan to the same mass position every time during a multielement analysis. Instrument manufacturers typically quote a mass calibration stability specification for their design of mass analyzer based on the drift or movement of the peak (in atomic mass units) position over a fixed period of time (usually 8 h).

mass calibration stability: *Refers to* mass calibration.

mass discrimination: Sometimes called "mass bias." In ICP-MS, it occurs when a higher-concentration isotope is suppressing the signal of the lower-concentration isotope, producing a biased result. The effect is not so obvious if the concentrations of the isotopes in the sample are similar, but can be quite significant if the concentrations of the two isotopes are vastly different. If that is the case, it is recommended to run a standard of known isotopic composition to compensate for the effects of the suppression.

mass filter: Another name for a mass analyzer.

mass-filtering discrimination: A way of discriminating between analyte ions and the unwanted by-product interference ions generated in a collision/reaction cell.

mass resolution: A measure of a mass analyzer's ability to separate an analyte peak from a spectral interference. The resolution of a quadrupole is nominally 1 amu and is traditionally defined as the width of a peak at 10% of its height.

mass scanning: The process of electronically scanning the mass separation device to the peak of interest and taking analytical measurements. Basically, two approaches are used: single-point peak hopping, in which a measurement is typically taken at the peak maximum, and the multipoint-scanning approach, in which a number of measurements are taken across the full width of the peak. *Also refers to* ramp scanning, integration time, dwell time, settling time, peak measurement protocol, *and* peak hopping.

mass separation: The process of separating the analyte ions from the non-analyte, matrix, solvent, and interfering ions with the mass analyzer.

mass separation device: Another name for a mass analyzer.

mass shift mode: A mode used with a "triple quadruple collision/reaction cell instrument." In this configuration, Q1 and Q2 are set to different masses. Similar to the "on-mass mode," Q1 is set to the precursor ion mass (analyte and on-mass polyatomic interfering ions), controlling the ions that enter octapole collision/reaction cell. However in the mass-shift mode, Q2 is then set to mass of a target reaction product ion containing the original analyte. Mass-shift mode is typically used when the analyte ion is reactive, while the interfering ions are unreactive with a particular collision/reaction cell gas.

mass spectrometer: The mass spectrometer section of an ICP-MS system is generally considered to be everything in the vacuum chamber from the interface region to the detector, including the interface cones, ion optics, mass analyzer, detector, and vacuum pumps.

matching network (RF): The matching network of the RF generator compensates for changes in impedance (a material's resistance to the flow of an electric current) produced by the sample's matrix components or differences in solvent volatility. In crystal-controlled generators, this is usually done with mechanically driven servo-type capacitors. With free-running generators, the matching network is based on electronic tuning of small changes in the RF brought about by the sample, solvent, or matrix components.

mathematical correction equations: Used to compensate or correct for spectral interference in ICP-MS. Similar to interelement corrections (IECs) used in ICP-OES, they work on the principle of measuring the intensity of the interfering isotope or interfering species at another mass, which is ideally free of any interferences. A correction is then applied, depending on the ratio of the intensity of the interfering species at the analyte mass to its intensity at the alternate mass.

Mathieu stability plot: A graphical representation of the stability of an ion as it passes through the rods of a multipole mass separation device. It is a function of the ratio of the RF to the DC current placed on each pair of rods. A plot of these ratios of multiple ions traveling through the multipole shows which ions are stable and make it through the rods to the detector and which ions are unstable and get ejected from the multipole. The most well-defined stability boundaries are obtained with a quadrupole and become more diffuse with higher-order multipoles such as hexapoles and octopoles.

matrix interferences: There are basically three types of matrix-induced interferences. The first, and simplest to overcome, is often called a "sample transport or viscosity effect" and is a physical suppression of the analyte signal brought on by the level of dissolved solids or acid concentration in the sample. The second type of matrix suppression is caused when the sample matrix affects the ionization conditions of the plasma discharge, which results in varying amounts of signal suppression depending on the concentration of the matrix components. The third type of matrix interference is often called "space-charge matrix suppression." This occurs mainly when low-mass analytes are being determined in the presence of larger concentrations of high-mass matrix components. It has the effect of defocusing the ion beam, and unless any compensation is made, the high-mass matrix element will dominate the ion beam, pushing the lighter elements out of the way, leading to low sensitivity and poor detection limits. The classical way to compensate for matrix interferences is to use internal standardization.

matrix separation: Usually refers to some kind of chromatographic column technology to remove the matrix components from the sample before it is introduced into the ICP-MS system.

Mattauch–Herzog magnetic sector design: One of the earliest designs of double-focusing magnetic sector mass spectrometers. In this design, which was named after the German scientists who invented it, two or more ions of different mass-to-charge ratios are deflected in opposite directions in the electrostatic and magnetic fields. The divergent monoenergetic ion beams are then brought together along the same focal plane. Recently commercialized using a simultaneous-based direct charge array detector.

MDL: *Refers to* method detection limits.

measurement duty cycle: Also known as the duty cycle, it refers to a percentage of actual quantitation time compared to total integration time. It is calculated by dividing the total quantitation time (dwell time × number of sweeps × replicates × elements) by total integration time ([dwell time + settling/scanning time] × number of sweeps × replicates × elements).

membrane desolvation: Can be used with any sample introduction technique to remove solvent vapors. However, it is typically used with an ultrasonic or microconcentric nebulizer to remove the solvent from a liquid sample. In this design, the sample aerosol enters the membrane desolvator, where the solvent vapor passes through the walls of a tubular microporous PTFE or Nafion membrane. A flow of argon gas removes the volatile vapor from the exterior of the membrane, while the analyte aerosol remains inside the tube and is carried into the plasma for ionization.

method detection limit (MDL): The MDL is broadly defined as the minimum concentration of analyte that can be determined from zero with 99% confidence. MDLs are calculated in a similar manner to IDLs, except that the test solution is taken through the entire sample preparation procedure before the analyte concentration is measured multiple times.

microconcentric nebulizer: Is based on the concentric nebulizer design, but operates at much lower flow rates. Conventional nebulizers have a sample uptake rate of about 1 mL/min with an argon gas pressure of 1 L/min, whereas microconcentric nebulizers typically run at less than 0.1 mL/min and typically operate at much higher gas pressure to accommodate the lower sample flow rates.

microflow nebulizer: A generic name for nebulizers that operate at much lower flow rates than conventional concentric or cross-flow designs. *Also refers to* microconcentric nebulizer.

micro-porous membrane: A tubular membrane made of an organic micro-porous material such as Teflon or Nafion, used in membrane desolvation. The sample aerosol enters the desolvation system, where the solvent vapor passes through the walls of the tubular membrane. A flow of argon gas then removes the volatile vapor from the exterior of the membrane, while the analyte aerosol remains inside the tube and is carried into the plasma for ionization.

microsampling: A generic name given to any front-end sampling device in atomic spectrometry that can be used for the preparation, pretreatment, and delivery of the sample to the spectrometric analyzer. The most common type of microsampling device used in ICP-MS is the flow injection technique, which involves the introduction of a discrete sample aliquot into a flowing carrier stream. Using a series of automated pumps and valves, procedures can be carried out online to physically or chemically change the sample or analyte, before introduction into the mass spectrometer for detection.

microwave digestion: A method of digesting difficult-to-dissolve solid samples using microwave technology. Typically, a dissolution reagent such as a concentrated mineral acid is added to the sample in a closed acid-resistant vessel contained in a specially designed microwave oven. By optimizing the current, temperature, and pressure settings, difficult samples can be dissolved in a relatively short time compared to traditional hot plate sample digestion techniques.

microwave dissolution: An alternative name for microwave digestion.

microwave-induced plasma (MIP): The most basic form of electrodeless plasma discharge. In this device, microwave energy (typically, 100–200 W) is supplied to the plasma gas from an excitation cavity around a glass/quartz tube. The plasma discharge in the form of a ring is generated inside the tube. Unfortunately, even though the discharge achieves a very high power density, the high excitation temperatures only exist along a central filament. The bulk of the MIP never goes above 2,000–3,000 K, which means it is prone to very severe matrix effects. In addition, it is easily extinguished when aspirating liquid samples, so it has a found a niche as a detection system for gas chromatography.

MIP: An abbreviation for microwave-induced plasma.

molecular association reaction: An ion–molecule reaction mechanism in a collision/reaction cell, where an interfering ion associates with a neutral species (atom or molecule) to form a molecular ion.

molecular cluster ions: Species that are formed by two or more molecular ions combining together in a reaction cell to form molecular clusters.

molecular spectral interferences: Another name for polyatomic spectral interferences, which are typically generated in the plasma by the combination of two or more atomic ions. They are caused by a variety of factors, but are usually associated with the argon plasma/nebulizer gas used, matrix components in the solvent/sample, other elements in the sample, or entrained oxygen/nitrogen from the surrounding air.

monodisperse particulates: Nanoparticles, which are predominantly one size.

M/S mode: A mode used in a "triple quadrupole" collision reaction cell where the first quadrupole acts as a simple ion guide allowing all ions through to the collision/reaction cell, similar to a traditional single-quad ICP-MS system that uses a collision cell.

M/S M/S mode: A mode used in a "triple quadrupole" collision reaction cell, where the first quadrupole is operated with a 1 amu fixed band pass window, allowing only the target ions to enter the collision/reaction cell. This process can be implemented in two different ways: either the "on-mass" or "mass shift" modes.

multicomponent ion lens: An ion lens system consisting of several lens components, all of which have a specific role to play in the transmission of the analyte ions into the mass filter. To achieve the desired analyte specificity, the voltage can be optimized on every ion lens and is usually combined with an off-axis mass analyzer to reject unwanted photons and neutral species.

multichannel analyzer: The data acquisition system that stores and counts the ions as they strike the detector. As the ions emerge from the end of the quadrupole rods, they are converted into electrical pulses by the detector and stored by the multichannel analyzer. This multichannel data acquisition system typically has 20 channels per mass, and as the electrical pulses are counted in each channel, a profile of the mass is built up over the 20 channels, corresponding to the spectral peaks of the analyte masses being determined.

multichannel data acquisition: The process of storing and counting ions in ICP-MS. *Also refers to* multichannel analyzer.

multipole: The generic name given to a mass filter that isolates an ion of interest by applying DC or RF currents to pairs of rods. The most common type of mass analyzer multipole used in ICP-MS is the quadrupole (four rods). However, other higher orders of multipoles used in collision/reaction cell technology include hexapoles (six rods) and octopoles (eight rods).

***m/z*:** Another way of expressing mass-to-charge ratio.

N

nanomaterials: Nanomaterials can occur in nature, such as clay minerals and humic acids; they can be incidentally produced by human activity such as diesel emissions or welding fumes; or they can be specifically engineered to exhibit unique optical, electrical, physical, or chemical characteristics. *Also refers to* engineered nanomaterials (ENMs).

nanometrology: The measurement and characterization of nanoparticles.

nanoparticles (NP): Particles that are released from engineered nanomaterials when they enter the environment.

natural abundance: The natural amount of an isotope occurring in nature. *Also refers to* isotopic abundance.

natural isotopes: Different isotopic forms of an element that occur naturally on or beneath the earth's crust.

Nd:YAG laser: Nd:YAG is an acronym for neodymium-doped yttrium aluminum garnet, a compound that is used as the lasing medium for certain solid-state lasers. In this design, the YAG host is typically doped with around 1% neodymium by weight. Nd:YAG lasers are optically pumped using a flashlamp or laser diodes and emit light with a wavelength of 1,064 nm in the infrared region. However, for many applications, the infrared light is frequency-doubled, frequency-tripled, frequency-quadrupled, or frequency-quintupled by using additional optical components to generate output wavelengths in the visible and UV regions. Typical wavelengths used for laser ablation/ICP-MS work include 532 nm (doubled), 266 nm (quadrupled), and 213 nm (quintupled). Pulsed Nd:YAG lasers are usually operated in the so-called "Q-switching" mode, where an optical switch is inserted in the laser cavity, waiting for a maximum population inversion in the neodymium ions before it opens. Then, the light wave can run through the cavity, depopulating the excited laser medium at maximum population inversion. In this Q-switched mode, output powers of 20 MW and pulse durations of less than 10 ns are achieved.

nebulizer: The component of the sample introduction system that takes the liquid sample and pneumatically breaks it down into an aerosol using the pressure created by a flow of argon gas. The concentric and cross-flow designs are the most common in ICP-MS.

neutral species: Species generated in the plasma torch that have no positive or negative charge associated with them. If they are not eliminated, they can find their way into the detector and produce elevated background levels.

neutron: A fundamental particle that is neutral in charge, found in the nucleus of an atom. It has a mass equal to that of a proton. The number of neutrons in the atomic nucleus defines the isotopic composition of that element.

Nier–Johnson magnetic sector design: Nier–Johnson double-focusing magnetic sector instrumentation is the technology that all modern magnetic sector instrumentation is based on. Named after the scientists who developed it, Nier–Johnson geometry comes in two different designs, the "standard" Nier–Johnson geometry and "reverse" Nier–Johnson geometry. Both these designs, which use the same basic principles, consist of two analyzers: a traditional electromagnet analyzer and an electrostatic analyzer (ESA). In the standard (sometimes called "forward") design, the ESA is positioned before the magnet, and in the reverse design, it is positioned after the magnet.

ninety (90)-degree ion lens: A design of ion optics that bends the ion beam 90°.

NP: *Refers to* nanoparticles.

O

octapole: A multipole mass-filtering device containing eight rods. In ICP-MS, octopoles are typically used in collision/reaction cell technology.

off-axis ion lens: An ion lens system that is not on the same axis as the mass analyzer. Designed to stop particulates, neutral species, and photons from hitting the detector.

on-mass mode: A mode used with the "triple quadrupole" collision/reaction cell. In this configuration, Q1 and Q2 are both set to the target mass. Q1 allows only the precursor ion mass to enter the cell (analyte and on-mass polyatomic interfering ions). The octapole collision/reaction cell then separates the analyte ion from the interferences using the reaction chemistry of a reactive gas, while Q2 measures the analyte ion at the target mass after the on-mass interferences have been removed by reactions in the cell.

oxide ions: Polyatomic ions that are formed between oxygen and other elemental components in the plasma gas, sample matrix, or solvent. They are generally not desirable because they can cause spectral overlaps on the analyte ions. Oxide formation is typically worse in the cooler zones of the plasma, and as a result, can be reduced by optimizing the RF power, nebulizer gas flow, and sampling position.

P

parabolic field: Shape of the magnetic fields produced by a quadrupole.

particle-counting techniques: Include alpha, gamma, and scintillation counters that are used to measure the isotopic composition of radioactive materials. However, the limitation of particle-counting techniques is that the half-life of the analyte isotope has a significant impact on the method's detection limit. This means that to get meaningful data in a realistic amount of time, they are better suited for the determination of short-lived radioisotopes. They have been successfully applied to the quantitation of long-lived radionuclides, but unfortunately require a combination of extremely long counting times and large amounts of sample to achieve low levels of quantitation.

peak hopping: A quantitation approach in which the quadrupole power supply is driven to a discrete position on the analyte mass (normally the maximum point), allowed to settle (settling time), and a measurement taken for a fixed amount of time (dwell time). The integration time for that peak is the dwell time multiplied by the number of scans (scan time). Multielement peak quantitation involves peak hopping to every mass in the multielement run. *Also refers to* measurement duty cycle *and* peak measurement protocol.

peak integration: The process of integrating an analytical peak (mass). *Also refers to* integration time, peak measurement protocol, *and* measurement duty cycle.

peak measurement protocol: The protocol of scanning the quadrupole and measuring a peak in ICP-MS. In multielement analysis, the quadrupole is scanned to the first mass. The electronics are allowed to settle (settling time), left to dwell for a fixed period of time at one or multiple points on the peak (dwell time), and signal intensity measurements are taken (based on the dwell time). The quadrupole is then scanned to the next mass and the measurement protocol repeated. The complete multielement measurement cycle (sweep) is repeated as many times as is needed to make up the total integration per peak and the number of required replicate measurements per sample analysis.

peak quantitation: The process of quantifying the peak in ICP-MS using calibration standards. *Also refers to* peak hopping, peak integration, *and* peak measurement protocol.

Peltier cooler: A thermoelectric cooler using the principle of generating a cold environment by creating a temperature gradient between two different materials. It uses electrical energy via a solid-state heat pump to transfer heat from a material on one side of the device to a different material on the other side, thus producing a temperature gradient across the device (similar to a household air-conditioning system).

peristaltic pump: A small pump in the sample introduction system that contains a set of mini rollers (typically, 12) all rotating at the same speed. The constant motion and pressure of the rollers on the pump tubing feeds the sample through to the nebulizer. Peristaltic pumps are usually used with cross-flow nebulizers.

photon stop: A grounded metal disk in the ion lens system that is used as a physical barrier to stop particulate matter, neutral species, and photons from getting to the detector.

physical interferences: An alternative term used to describe sample transport- or viscosity-based suppression interferences.

pinch effect: An effect caused by an undesired electrostatic (capacitive) coupling between the voltage on the load coil and the plasma discharge, which produces a potential difference of a few hundred volts. This capacitive coupling is commonly referred to as the pinch effect and shows itself as a secondary discharge (arcing) in the region where the plasma is in contact with the sampler cone.

piston pump: A pump using a small piston or syringe to introduce the sample to the nebulizer (used instead of a peristaltic pump). They typically produce a more stable signal.

plasma discharge: Another name for an inductively coupled plasma (ICP).

plasma source: Refers to the RF hardware components that create the plasma discharge, including the RF generator, matching network, plasma torch, and argon gas pneumatics.

plasma torch: Another name for the quartz torch that is used to generate the plasma discharge. The plasma torch consists of three concentric tubes: an outer tube, middle tube, and sample injector. The torch can either be one piece, where all three tubes are connected, or have a demountable design, in which the tubes and the sample injector are separate. The gas (usually argon) that is used to form the plasma (plasma gas) is passed between the outer and middle tubes at a flow rate of 12–17 L/min. A second gas flow (auxiliary gas) passes between the middle tube and the sample injector at 1 L/min, and is used to change the position of the base of the plasma relative to the tube and the injector. A third gas flow (nebulizer gas) also at 1 L/min brings the sample, in the form of a fine-droplet aerosol, from the sample introduction system and physically punches a channel through the center of the plasma. The sample injector is often made from other materials besides quartz, such as alumina, platinum, and sapphire, if highly corrosive materials need to be analyzed.

polyatomic spectral interferences: Another name for molecular-based spectral interferences, which are typically generated in the plasma by the combination of two or more atomic ions. They are caused by a variety of factors, but are usually associated with the argon plasma/nebulizer gas used, matrix components in the solvent/sample, other elements in the sample, or entrained oxygen/nitrogen from the surrounding air.

polydisperse particulates: Nanoparticles that are many different sizes and dimensions.

precursor ion: Usually refers to a polyatomic or isobaric interfering ion that is formed in the plasma as opposed to a product (or by-product) ion that is formed in the collision/reaction cell.

product ion: Usually refers to a product (or by-product) ion that is formed in the collision/reaction cell as opposed to a precursor interfering ion (polyatomic or isobaric) that is formed in the plasma.

proton: A stable, positively charged fundamental particle that shares the atomic nucleus with a neutron. It has a mass 1,836 times that of the electron.

proton transfer: A reaction mechanism in a collision/reaction cell in which the interfering polyatomic species gives up a proton, which is then transferred to the reaction gas molecule to form a neutral atom.

pulse counting: Refers to the conventional mode of counting ions with the detector measurement circuitry. Depending on the type of detection system that is used, an ion emerges from the quadrupole and strikes the ion-sensitive surface (discrete dynode, Channeltron, etc.) of the detector to generate electrons. These electrons move down the detector and generate more secondary electrons. This process is repeated at each stage of the detector, producing a pulse of electrons that is finally captured by the detector's collecting and counting circuitry.

Q

QID: *Refers to* quadrupole ion deflector.

quadrupole: The most common type of mass separation device used in commercial ICP-MS systems. It consists of four cylindrical or hyperbolic metallic rods of the same length (15–20cm) and diameter (approximately 1cm). The rods are typically made of stainless steel or molybdenum and sometimes coated with a ceramic coating for corrosion resistance. A quadrupole operates by placing both a DC field and a time-dependent AC of RF 2–3MHz on opposite pairs of the four rods. By selecting the optimum AC/DC ratio on each pair of rods, ions of a selected mass are then allowed to pass through the rods to the detector, whereas the others are unstable and ejected from the quadrupole.

quadrupole ion deflector (QID): A commercial ion optics design that bends the ion beam at right angles.

quadrupole power supply: Another name for the electronic components that control the RF and DC voltages to change the mass-filtering characteristics.

quadrupole scan rate: Scan rates of commercial quadrupole mass analyzers are on the order of 2,500 amu/s. The quadrupole scan rate and the slope at which the RF and DC voltages of the quadrupole power supply are scanned will determine the desired resolution setting. A steeper slope translates to higher resolution, whereas a shallower slope means poorer resolution.

quadrupole stability regions: The region of the Mathieu stability plot where the trajectory of an ion is stable and makes it through to the end of the quadrupole rods. All commercial ICP-MS systems that utilize quadrupole technology as the mass separation device operate in the first stability region, where resolving power is typically on the order of 500–600. If the quadrupole is operated in the second or third stability regions, resolving powers of 4,000 and 9,000, respectively, have been achieved. However, improving resolution using this approach has resulted in a significant loss of signal and higher background levels.

quantitative methods: The different kinds of quantitative analyses available in ICP-MS, which include traditional quantitative analysis (using external calibration, standard additions, or addition calibration), semiquantitative routines (semiquant), isotope dilution (ID) methods, isotope ratio (IR) measurements, and classical internal standardization (IS).

quartz torch: The standard plasma torch used in ICP-MS. *Also refers to* plasma torch.

R

radio frequency (RF) generator: The power supply used to create the plasma discharge. Hardware includes the RF generator, matching network, plasma torch, and argon gas pneumatics.

radioactive isotope: Sometimes known as "radioisotope," a radioactive isotope is a natural or artificially created isotope of an element having an unstable nucleus that decays, emitting alpha, beta, or gamma rays until stability is reached. The stable end product is typically a nonradioactive isotope of another element.

ramp scanning: One of the two approaches for quantifying a peak in ICP-MS (peak hopping being the other). In the multichannel ramp scanning approach, a continuous smooth ramp of $1 - n$ channels (where n is typically 20) per mass is made across the peak profile. Mainly used for accumulating spectral and peak shape information when doing mass scans. It is normally used for doing mass calibration and resolution checks and as a classical qualitative method development tool to find out what elements are present in the sample and to assess their spectral implications on the masses of interest. Full-peak ramp scanning is not normally used for doing rapid quantitative analysis, because valuable analytical time is wasted taking data on the wings and valleys of the peak where the signal-to-noise ratio is poorest. For this kind of work, peak hopping is normally chosen.

reaction cell: A collision/reaction cell that specifically uses ion–molecule reactions to eliminate the spectral interference. Often used to describe a dynamic reaction cell (DRC).

reaction mechanism: The mechanism by which the interfering ion is reduced or minimized to allow the determination of the analyte ion. The most common collisional mechanisms seen in collision/reaction cells include collisional focusing, dissociation, and fragmentation, whereas the major reaction mechanisms include exothermic/endothermic associations, charge transfer, molecular associations, and proton transfer.

reactive gases: In ICP-MS, the term refers to gases such as hydrogen, ammonia, oxygen, methane, or those used to stimulate ion–molecule reactions in a collision/reaction cell (CRC) or collision/reaction interface (CRI).

relative abundance of natural isotopes: The isotopic composition expressed as a percentage of the total abundance of that element found in nature.

resolution: A measure of the ability of a mass analyzer to separate an analyte peak from a spectral interference. The resolution of a quadrupole is nominally 1 amu and is traditionally defined as the width of a peak at 10% of its height.

resolving power: Although resolving power and resolution are both a measure of a mass analyzer's ability to separate an analyte peak from a spectral interference, the term "resolving power" is normally associated with magnetic sector technology and is represented by the equation $R = m/\Delta m$, where m is the nominal mass at which the peak occurs and Δm is the mass difference between two resolved peaks. The resolving power of commercial double-focusing magnetic sector mass analyzers is on the order of 1,000–10,000, depending on the resolution setting chosen.

response tables: The intensity values for known concentrations of every elemental isotope stored in the instrument's calibration software. When semiquantitative analysis is carried out, the signal intensity of an unknown sample is compared against the stored response tables. By correcting for common spectral interferences and applying heuristic, knowledge-driven routines in combination with numerical calculations, a positive or negative confirmation can be made for each element present in the sample.

reverse Nier–Johnson double-focusing magnetic sector instrumentation: The technology that all modern magnetic sector instrumentation is based on. Named after the scientists who developed it, Nier–Johnson geometry comes in two different designs: the "standard" Nier–Johnson geometry and "reverse" Nier–Johnson geometry. Both these designs, which use the same basic principles, consist of two analyzers: a traditional electromagnet analyzer and an electrostatic analyzer (ESA). In the standard (sometimes called "forward") design, the ESA is positioned before the magnet, and in the reverse design, it is positioned after the magnet.

RF generator: An alternative name for radio frequency generator.

right-angled ion lens design: A recent development in ion-focusing optics, which utilizes a parabolic ion mirror to bend and refocus the ion beam at right angles to the ion source.

The ion mirror incorporates a hollow structure that allows photons, neutrals, and solid particles to pass through it, while allowing ions to be deflected at right angles into the mass analyzer.

roughing pump: Traditional mechanical roughing or oil-based pumps are used in ICP-MS to pump the interface region down to approximately 1–2 torr and also to back up the turbomolecular pump used in the ion optics region of the mass spectrometer.

ruby laser: Ruby laser systems operate at 694 nm in the visible region of the electromagnetic spectrum.

S

S/B: An abbreviation for signal-to-background ratio.

sample aerosol: *Refers to* aerosol.

sample digestion: The process of digesting a sample by traditional hot plate, fusion techniques, or microwave technology to get the matrix and analytes into solution.

sample dissolution: The process of dissolving a sample by traditional hot plate, fusion techniques, or microwave technology to get the matrix and analytes into solution.

sample injector: The central tube of the plasma torch that carries the sample aerosol mixed with the nebulizer gas. It can be a fixed part of the quartz torch or it can be separate (demountable) and be made from other materials, such as alumina, platinum, and sapphire, for the analysis of highly corrosive materials.

sample introduction system: The part of the instrument that takes the liquid sample and puts it into the plasma torch as a fine-droplet aerosol. It comprises a nebulizer to generate the aerosol and a spray chamber to reject the larger droplets and allow only the smaller droplets into the plasma discharge.

sample preparation: The entire process of preparing the sample for aspiration into the ICP mass spectrometer.

sample throughput: The rate at which samples can be analyzed.

sample transport interferences: A term used to describe a physical suppression of the analyte signal caused by matrix components in the sample. It is more exaggerated with samples having high levels of dissolved solids, because they are transported less efficiently through the sample introduction system than aqueous-type samples. *Also refers to* physical interferences.

sampler cone: A part of the mass spectrometer interface region, where the ion beam from the plasma discharge first enters. The sampler cone, which is the first cone of the interface, is typically made of nickel or platinum and contains a small orifice of approximately 0.8–1.2 mm diameter, depending on the design. The sampler cone is much more pointed than the skimmer cone.

sampling accessories: Customized sample introduction techniques optimized for a particular application problem or sample type. The most common types used today include the following: laser ablation/sampling (LA/S), flow injection analysis (FIA), electrothermal vaporization (ETV), desolvation systems, direct injection nebulizers (DIN), and chromatography separation techniques.

scan time: The mass analyzer scan time is the time it takes to scan from one isotope to the next.

Scott spray chamber: A sealed spray chamber with an inner tube inside a larger tube. The sample aerosol from the nebulizer is first directed into the inner tube. The aerosol then travels the length of the inner tube, where the larger droplets fall out by gravity into a drain tube and the smaller droplets return between the inner and outer tube, where they eventually exit into the sample injector of the plasma torch.

secondary discharge: Another term used for the pinch effect.

secondary (side) reactions: Reactions that occur in a collision/reaction cell that are not a part of the main interference reduction mechanism. If not anticipated and compensated for, secondary reactions can lead to erroneous results.

semiquant: An abbreviated name used to describe semiquantitative analysis.

semiquantitative analysis: A method for assessing the approximate concentration of up to 70 elements in an unknown sample. It is based on comparing the intensity of a small group of elements against known response tables stored in the instrument's calibration software. By correcting for common spectral interferences and applying heuristic, knowledge-driven routines in combination with numerical calculations, a positive or negative confirmation can be made for each element present in the sample.

settling time: The time taken for the mass analyzer electronics to settle before a peak intensity measurement is taken for the operator-selected dwell time. The dwell time can usually be selected on an individual mass basis, but the settling time is normally fixed because it is a function of the mass analyzer and detector electronics.

shadow stop: A grounded metal disk that stops particulate matter, neutral species, and photons from getting to the detector. It is considered a part of the ion optics and is sometimes called a photon stop.

side reactions: Reactions that occur in a collision/reaction cell that are not a part of the main interference reduction mechanism. If not anticipated and compensated for, secondary reactions can lead to erroneous results.

signal-to-background ratio (S/B): The ratio of the signal intensity of an analyte to its background level at a particular mass. When considering the noise of the background signal (standard deviation of the signal), it is typically used as an assessment of the detection limit for that element. *Also refers to* detection limit, background signal, *and* background noise.

single-particle ICP-MS: A technique used to characterize nanoparticles. The method involves introducing nanoparticle (NP)-containing samples, at very dilute concentration, into the ICP-MS, and collecting time-resolved data.

single-point peak hopping: A quantitation in which the quadrupole power supply is driven to a discrete position on the analyte mass (normally the maximum point), allowed to settle (settling time), and a measurement taken for a fixed amount of time (dwell time). The integration time for that peak is the dwell time multiplied by the number of scans (scan time). Multielement peak quantitation involves peak hopping to every mass in the multielement run. *Also refers to* measurement duty cycle *and* peak measurement protocol.

skimmer cone: A part of the mass spectrometer interface region where the ion beam from the plasma discharge first enters. The skimmer cone, which is the second cone of the interface, is typically made of nickel or platinum and contains a small orifice of approximately 0.5–0.8 mm diameter, depending on the design. The skimmer cone is much less pointed than the sampler cone.

solvent-based interferences: Spectral interferences derived from an elemental ion in the solvent (e.g., water and acid) combining with another ion from either the sample matrix or plasma gas (argon) to produce a polyatomic ion that interferes with the analyte mass.

SOP: Standard operating procedure.

space charge effect: A type of matrix-induced interference that produces a suppression of the analyte signal. This occurs mainly when low-mass analytes are being determined in the presence of larger concentrations of high-mass matrix components. It has the effect of defocusing the ion beam, and without compensation, the high-mass matrix element will dominate the ion beam, pushing the lighter elements out of the way, leading to low sensitivity and poor detection limits. The classical way to compensate for a space-charge matrix interference is to use an internal standard of similar mass to the analyte.

speciation analysis: In ICP-MS, it is the study and quantification of different species or forms of an element using a chromatographic separation device coupled to an ICP mass spectrometer. In this configuration, the instrument becomes a very sensitive detector for trace element speciation studies when coupled with high-performance liquid chromatography (HPLC), ion chromatography (IC), gas chromatography (GC), or capillary electrophoresis (CE). In these hybrid techniques, element species are separated on the basis of their chromatograph retention/mobility times and then eluted/passed into the ICP mass spectrometer for detection. The intensity of the eluted peaks is then displayed for each isotopic mass of interest in the time domain.

spectral interferences: A generic name given to interferences that produce a spectral overlap at or near the analyte mass of interest. In ICP-MS, there are two main types of spectral interference that have to be taken into account. Polyatomic spectral interferences (or molecular-based spectral interferences) are typically generated in the plasma by the combination of two or more atomic ions. They are caused by a variety of factors, but are usually associated with the argon plasma/nebulizer gas used, matrix components in the solvent/sample, other elements in the sample, or entrained oxygen/nitrogen from the surrounding air. The other type is an isobaric spectral interference, which is caused by different isotopes of other elements in the sample creating spectral interferences at the same mass as the analyte.

SP-ICP-MS: *Refers to* single-particle ICP-MS.

spray chamber: The component of the sample introduction system that takes the aerosol generated by the nebulizer and rejects the larger droplets for the more desirable smaller droplets.

SRM: An abbreviation for standard reference materials.

stability: The ability of a measuring device to consistently replicate a measurement. In ICP-MS, it usually refers to the capability of the instrument to reproduce the signal intensity of the calibration standards over a fixed period of time without the use of internal standardization. Short-term stability is generally defined as the precision (as % RSD [relative standard deviation]) of ten replicates of a single or multielement solution, whereas long-term stability is defined as the precision (as % RSD) of a fixed number of measurements over a 4–8-hour time period of a single or multielement solution. However, stability in mass spectrometry can *also refers to* mass calibration stability, which is the ability of the mass spectrometer to repeatedly scan to the same mass position every time during a multielement analysis.

stability boundaries/regions: The RF/DC boundaries of the Mathieu stability plot where an ion is stable as it passes through a quadrupole mass-filtering device. *Also refers to* Mathieu stability plot.

standard additions: A method of calibration that provides an effective way to minimize sample-specific matrix effects by spiking samples with known concentrations of analytes. In standard addition calibration, the intensity of a blank solution is first measured. Next, the sample solution is "spiked" with known concentrations of each element to be determined. The instrument measures the response for the spiked samples and creates a calibration curve for each element for which a spike has been added. The calibration curve is a plot of the blank subtracted intensity of each spiked element against its concentration value. After creating the calibration curve, the unspiked sample solutions are then analyzed and compared to the calibration curve. Depending on the slope of the calibration curve and where it intercepts the X-axis, the instrument software determines the unspiked concentration of the analytes in the unknown samples.

standard reference materials (SRM): Well-established reference matrices that come with certified values and associated statistical data which have been analyzed by other complementary techniques. Their purpose is to check the validity of an analytical method, including sample preparation, instrument methodology, and calibration routines, to achieve sample results that are as accurate and precise as possible and can be defended under intense scrutiny.

standardization methods: Refers to the different types of calibration routines available in ICP-MS, including quantitative analysis (external calibration and standard additions), semiquantitative analysis, isotope dilution, isotope ratio, and internal standardization methods.

syringe pump: *Refers to* piston pump.

T

thermoelectric cooling device: Better known as a Peltier cooler, it generates a cold environment by creating a temperature gradient between two different materials. It uses electrical energy via a solid-state heat pump to transfer heat from a material on one side of the device to a different material on the other side, thus producing a temperature gradient across the device (similar to a household air conditioner).

thermoelectric flow meter: A device to measure the liquid flow through a nebulizer to check for any blockages or breakages.

time-of-flight mass spectrometry (TOFMS): A mass spectrometry technique based on the principle that the kinetic energy (KE) of an ion is directly proportional to its mass (m) and velocity (V), which can be represented by the equation $KE = \frac{1}{2}MV^2$. Therefore, if a population of ions with different masses is given the same KE by an accelerating voltage (U), the velocities of the ions will all be different, depending on their masses. This principle is then used to separate ions of different mass-to-charge (m/z) in the time (t) domain, over a fixed flight path distance (D), represented by the equation $m/z = 2Ut^2/D^2$. The simultaneous nature of sampling ions in TOF offers distinct advantages over traditional scanning (sequential) quadrupole technology for ICP-MS applications, where large amounts of data need to be captured in a short amount of time, such as the multielement analysis of transient peaks (laser ablation, flow injection, etc.).

time-of-flight (TOF) mass spectrometry (axial design): There are basically two different sampling approaches that are used in commercial TOF mass analyzers: the axial and orthogonal designs. In the axial design, the flight tube is in the same axis as the ion beam, whereas in the orthogonal design, the flight tube is positioned at right angles to the sampled ion beam. The axial approach applies an accelerating potential in the same axis as the incoming ion beam as it enters the extraction region. Because the ions are in the same plane as the detector, the beam has to be modulated using an electrode grid to repel the "gated" packet of ions into the flight tube. This kind of modulation generates an ion packet that is long and thin in cross section (in the horizontal plane), which is then resolved in the time domain according to the different ionic masses. *Also refers to* time-of-flight mass spectrometry.

time-of-flight (TOF) mass spectrometry (orthogonal design): There are basically two different sampling approaches that are used in commercial TOF mass analyzers, the axial and orthogonal designs. In the axial design, the flight tube is in the same axis as the ion beam, whereas in the orthogonal design, the flight tube is positioned at right angles to the sampled ion beam. With the orthogonal approach, an accelerating potential is applied at right angles to the continuous ion beam from the plasma source. The ion beam is then "chopped" by using a pulsed voltage supply coupled to the orthogonal accelerator to provide repetitive voltage "slices" at a frequency of a few kilohertz. The "sliced" packets of ions, which are typically tall and thin in cross section (in the vertical plane), are then allowed to "drift" into the flight tube, where the ions are temporally resolved according to their differing velocities. *Also refers to* time-of-flight mass spectrometry.

TOFMS: An abbreviation for time-of-flight mass spectrometry.

torch design: Refers to the different kinds of commercially available torch designs.

trace metal speciation studies: *Refers to* speciation analysis.

transient signal (peak): A signal that lasts for a finite amount of time, compared to a continuous signal that lasts for as long as the sample is being aspirated. Transient peaks are typically generated by alternative sampling devices such as laser ablation, flow injection, or chromatographic separation systems where discrete amounts of sample are introduced into the ICP mass spectrometer.

triple cone interface: A commercial design of an ICP-MS interface that uses three cones. *Also refers to* hyper skimmer cone.

"triple quadrupole" collision/reaction cell: A commercial collision/reaction cell design that has an additional quadrupole prior to the collision/reaction cell multipole and the analyzer quadrupole. This first quadrupole acts as a simple mass filter to allow only the analyte masses to enter the cell, while rejecting all other masses. With all non-analyte, plasma and sample matrix ions excluded from the cell, sensitivity and interference removal efficiency is significantly improved compared to traditional collision/reaction cell technology coupled with a single quadrupole mass analyzer.

turbomolecular pump (turbo pump): A type of vacuum pump used to maintain a high vacuum in the ion optics and mass analyzer regions of the ICP mass spectrometer. These pumps work on the principle that gas molecules can be given momentum in a desired direction by repeated collision with a moving solid surface. In a turbo pump, a rapidly spinning turbine rotor strikes gas (argon) molecules from the inlet of the pump towards the exhaust, creating and maintaining a vacuum. In the case of ICP-MS, two pumps are normally used, a large pump for the ion optic region, which creates a vacuum of approximately 10^{-3} torr, and another small pump for the mass analyzer region, which generates a vacuum of 10^{-6} torr. However, some designs use a twin-throated turbo pump, in which one powerful pump is used with two outlets, one for the ion optics and one for the mass analyzer region.

twin-throated turbomolecular pump: In some designs of ICP mass spectrometer, a single twin-throated turbo pump is used instead of two separate pumps. In this design, one powerful pump is used with two outlets, one for the ion optics and one for the mass analyzer region.

U

ultrasonic nebulizer (USN): A type of desolvating nebulizer that generates an extremely fine-droplet aerosol for introduction into the ICP mass spectrometer. The principle of aerosol generation using this approach is based on a sample being pumped onto a quartz plate of a piezoelectric transducer. Electrical energy of 1–2 MHz is coupled to the transducer, which causes it to vibrate at high frequency. These vibrations disperse the sample into a fine-droplet aerosol, which is carried in a stream of argon. With a conventional ultrasonic nebulizer, the aerosol is passed through a heating tube and a cooling chamber, where most of the sample solvent is removed as a condensate before it enters the plasma. However, commercial ultrasonic nebulizers are also available with membrane desolvation systems.

universal cell: The name applied to a commercial collision/reaction cell that offers the capability of either a collision cell using inert gases and KED or a DRC using highly reactive gases.

USN: An abbreviation for ultrasonic nebulizer.

UV laser: A generic name given to a laser ablation system that works in the ultraviolet region of the electromagnetic spectrum. The three most common wavelengths used in commercial equipment are all UV lasers. They include the 266 nm (frequency-quadrupled) Nd:YAG laser, the 213 nm (frequency-quintupled) Nd:YAG laser, and the 193 nm ArF excimer laser system. *Also refers to* excimer laser *and* Nd:YAG laser.

V

vacuum chamber: The region of the mass spectrometer that is under negative pressure created by a combination of roughing and turbomolecular pumps. As the ion beam moves from the plasma, which is at atmospheric pressure (760 torr), it enters the interface region between the sampler and skimmer cone (1–2 torr) before it is focused through the ion optic vacuum chamber region (10^{-3} torr) and eventually goes through the mass analyzer vacuum chamber (at 10^{-6} torr). *Also refers to* turbomolecular pump.

vacuum gauge: Used to measure the pressure in the different vacuum chambers of the mass spectrometer.

vacuum pump: A number of vacuum pumps are used to create the vacuum in an ICP mass spectrometer. Two roughing pumps are used, one for the interface region and another to back up the first turbomolecular pump of the ion optic region. Also, two turbomolecular pumps (or in some designs, one twin-throated pump) are used, one for the ion optics and another for the main mass analyzer region. *Also refers to* roughing pump, turbomolecular pump, *and* twin-throated turbomolecular pump.

Vapor-phase decomposition (VPD): A technique used to dissolve and collect trace metal impurities on the surface of a silicon wafer, by rolling a few hundred microliters of hydrofluoric acid (HF) over the surface.

visible laser: A laser that operates in the visible region of the electromagnetic spectrum. An example is the ruby laser, which operates at 694 nm.

W

wavelength: A name commonly used to identify an emission line in nanometers (nm) used in ICP optical emission spectroscopy. Also used to describe a type of laser ablation system (e.g., a ruby laser operating at a 694 nm wavelength) used to couple to an ICP-MS to analyze solid samples directly.

wet plasma: When a liquid sample is introduced or aspirated into the plasma, the plasma is called a "wet plasma."

X

X-rays: A part of the electromagnetic spectrum that has a wavelength in the range of 0.01–10nm.

X-ray fluorescence: An analytical technique that uses the emission of characteristic "secondary" (or fluorescent) X-rays from a material that has been excited by bombarding with high-energy X-rays or gamma rays. The phenomenon is widely used for elemental analysis and chemical analysis, particularly in the study of solid materials such as metals, glass, ceramics, soils, and rocks.

"*x*" position: Refers to the alignment of the plasma torch in the lateral (sideways) position. Typically carried out to maximize sensitivity or to optimize sampling conditions for cool plasma use.

Y

"*y*" position: Refers to the alignment of the plasma torch in the longitudinal (vertical) position. Typically carried out to maximize sensitivity or to optimize sampling conditions for cool plasma use.

Z

"*z*" position: Refers to the alignment of the plasma torch in relation to the distance from the interface cone (in and out position). Typically carried out to maximize sensitivity or to optimize sampling conditions for cool plasma use.

INDUCTIVELY COUPLED PLASMA OPTICAL EMISSION SPECTROMETRY (ICP-OES) GLOSSARY

A

aerosol: A group of small-diameter liquid droplets that are formed when the nebulizer breaks up a continuously pumped solution.

argon: One of the noble gases in the periodic table. This gas is used to generate a plasma in ICP-based instruments.

array detector: A group of photoactive elements, arranged in a linear or two-dimensional configuration, which is used to measure photons from a plasma. Arrays can consist of a variety of detectors, including photodiodes, CCD image sensors, and CID image sensors.

atomic emission (AE): This is sometimes used to *refer to* the technique ICP-AES and the emission from atoms in a sample. This term has been replaced with the term "optical emission" which is more accurate (*refers to* optical emission definition).

axial view: This describes the measurement of light down the central channel of the plasma which includes emission from the entire length of the plasma. Currently, technology allows axial measurements to be made, regardless of whether the torch is mounted vertically or horizontally in the instrument.

B

background equivalent concentration (BEC): For a given analyte, this is the concentration of that analyte which produces a net emission signal equal to the background signal.

background noise: The uncertainty affiliated with measuring a blank solution. This value is often calculated by taking the square root of the background signal.

background signal: The intensity of the signal emitted when a blank solution is measured.

C

ceramic torch: A torch for an ICP-OES that is made from a ceramic material that makes the torch more robust and typically increases its lifetime. The outer, intermediate and injector tubes can all be made out of ceramic.

charge coupled device (CCD): A type of solid-state detector that can convert photons into photoelectric signals. This type of detector is used in many commercially available ICP-OES instruments.

charge injection device (CID): A type of solid-state detector that can convert photons into photoelectric signals. This type of detector is used in many commercially available ICP-OES instruments.

chemical interferences: These occur when the standards behave differently from the samples as they enter the plasma. These types of interferences typically result from changes in temperature within the plasma. They can be overcome by utilizing ionization suppressants or using a radially configured plasma.

complementary metal oxide semiconductor (CMOS): This is a type of technology that is used to construct circuits that are used in many applications, including imaging sensors for detectors in ICP-OES instruments.

concentric nebulizer: A nebulizer design that uses a stream of high velocity argon gas to create an area of low pressure at its tip, causing a constant flow of solution to break apart into small droplets as it leaves the nebulizer. This nebulizer can be fed via a pumped solution, or it can operate via free aspiration.

Cooled Spray Chamber: A spray chamber that is housed inside a cooled chamber which lowers the temperature inside the spray chamber and reduces the amount of aerosol that gets transported from the spray chamber to the plasma.

cyclonic spray chamber: A spray chamber design that is commonly used in ICP-OES instruments. As the name implies, this spray chamber accepts aerosol from the nebulizer and forms it into a virtual cyclone before transferring the smallest droplets into the plasma. This spray chamber is manufactured with or without a center tube, which acts as an additional impact surface for the aerosol.

D

desolvating nebulizer: A nebulizer which desolvates (removes a significant amount of the solvent from) the aerosol prior to its passage into the plasma. This system utilizes a heated or cooled spray chamber and a solvent membrane to remove either aqueous or organic solvent from the nebulized aerosol.

detection limit (DL): Sometimes referred to as the LOD, the detection limit is the lowest concentration of an analyte that can be distinguished from a solution which does not contain that analyte (i.e., a blank solution).

detector: A device used to convert photons from the instrument's plasma to a digital signal that can be read out by a microprocessor.

devitrification: A chemical and temperature-related process that breaks down the structure of a glass or quartz component (typically a torch).

direct injection high-efficiency nebulizer (DIHEN): This is a modification of the original direct injection nebulizer, which incorporates some design improvements.

direct injection nebulizer (DIN): A nebulizer design that injects sample solution directly into the base of the instrument's plasma, eliminating the need for a spray chamber. This nebulizer maximizes the transport efficiency of the aerosol and minimizes solution waste and memory effects between samples.

double pass spray chamber: A spray chamber design.

droplet: A small drop of solution. During the process of introducing samples into the instrument's plasma, samples must be converted from continuous solutions to a group of small-diameter droplets.

H

high-efficiency nebulizer (HEN): A term used to describe nebulizers that have been designed to convert a significant percentage of the pumped solution into a useable aerosol. Very little solution is sent to waste.

high-solids nebulizers: These nebulizers are designed with the intent to create a useable aerosol from solutions with high levels of dissolved solids. These include V-groove and Babington nebulizer designs.

I

inductively coupled plasma (ICP): A body of gas ions that are formed and maintained via the interaction between radio frequency and magnetic fields. This type of plasma is used as an atomization/ionization source in inductively coupled plasma optical emission spectrometry.

inductively coupled plasma optical emission spectrometry (ICP-OES): An elemental analysis technique that uses an inductively coupled plasma to desolvate, vaporize, atomize, and excite liquid samples. The resulting emitted light is collected, measured, and used to quantify the elements present in the original sample.

injector: Also known as the sample injector, this is the center tube in the plasma torch and is used to transport the sample aerosol and inject it into the center of the plasma.

instrument detection limit (IDL): This is often used synonymously with the limit of detection (LOD) and refers to the signal-to-background noise capability of the instrument, for a given analyte. This is often calculated in accordance with the IUPAC definition as three times the standard deviation of ten replicate measurements of a blank solution.

integration time: The amount of time the instrument allows light to pass to the detector to collect emission from the plasma.

interferences: In plasma-based techniques, this is a broad term which refers to something that hinders the instrument's ability to precisely and accurately quantify elemental impurities in a sample. Interferences can be chemical, physical or spectral in nature and sometimes involve more than one correction technique to ensure quality results.

internal standardization: A technique used to overcome some physical and chemical interferences in ICP-based techniques. This technique involves the addition of an analyte to all blanks, standards, and samples which will behave in a similar manner to the analytes being measured. Sample results are reported as a ratio between the analyte and the internal standard, which corrects for small changes in intensity that are due to physical/chemical changes in the solutions. These small changes, if left uncorrected, would result in errors in the reported results.

L

load coil: A term which refers to the radio frequency (RF) coil in a plasma-based instrument. This coil maintains the radio frequency signal being generated by the instrument's RF generator which helps sustain the plasma.

M

method detection limit (MDL): A conservative estimate of the instrument's detection limits. The MDL can be calculated in a number of different ways; however, it is the minimum concentration of an analyte that can be distinguished from the absence of the analyte and quantified with a high degree of confidence (95%–99%).

microconcentric nebulizer: Based on the concentric design, this nebulizer operates at significantly lower solution flow rates than the standard concentric nebulizer.

N

nebulizer: One of the components in the sample introduction system, the nebulizer takes a continuous solution and breaks it up into a body of small-diameter droplets (aerosol).

P

Peltier cooler: A thermoelectric cooler which is used to cool the spray chamber and maintain a specific cooled temperature during a sample run.

peristaltic pump: A type of pump which uses rollers to positively displace sample solutions, pushing them through flexible tubing. These pumps are used to move solution through the sample uptake and drain tubing that is used on many commercially available ICP-OES instruments.

physical interferences: Occur when the nebulization and/or transport efficiency of the standards differs from that of the samples. Physical interferences are usually overcome by utilizing internal standardization and preparing the calibration standards in a matrix that matches that of the samples.

Q

quartz torch: A torch for an ICP-OES that is made from quartz. The outer, intermediate and injector tubes can all be made out of quartz.

R

radial view: This describes the measurement of light perpendicular to the direction of the plasma.

radio frequency (RF) generator: The RF generator typically consists of a DC power supply, an RF generator board, and a number of other electrical components that work together to generate a radio frequency field which is necessary for maintaining the plasma in an ICP-based instrument.

S

sample introduction system: This refers to the components that transport a sample solution into the instrument's plasma, while converting it from a continuous solution to a fine aerosol. The components that make up the sample introduction system include the nebulizer, spray chamber, and the torch, as well as the pump and affiliated pump tubing.

Scott spray chamber: This is a type of double-pass spray chamber.

signal-to-background ratio (S/B or SBR): For a given analyte, this is the intensity ratio of the signal and the background at a particular concentration. When the S/B is calculated in a blank solution, this is often used to represent the detection limit for that analyte.

spray chamber: One of the components in the sample introduction system, the spray chamber takes the aerosol from the nebulizer and filters out the larger diameter droplets such that only small diameter droplets are transferred to the plasma. The larger droplets are sent directly to waste.

T

torch: One of the components in the sample introduction system, the torch houses the plasma and receives the sample aerosol that was generated and transported by the nebulizer and spray chamber.

torch injector: Also known as the injector or the sample injector, this is the center tube in the plasma torch and is used to transport the sample aerosol and inject it into the center of the plasma.

U

ultrasonic nebulizer (USN): A nebulizer design which uses high-frequency vibrations to convert a solution into an aerosol with small-diameter droplets. Unlike conventional nebulizers, an ultrasonic nebulizer generates a higher percentage of smaller droplets which improves the transport of solution to the plasma and increases the analyte signal. Most commercially available ultrasonic nebulizers use a heated/cooled chamber and/or a membrane desolvation system to remove excess solvent from the aerosol prior to being transported to the plasma.

V

viewing height: This refers to where the emitted light is being collected from the plasma when making measurements in radial view mode. This viewing position is usually in reference to the height above the top turn of the load (RF) coil.

W

wavelength: In ICP-OES, wavelength-specific photons are emitted from excited atoms and ions and measured by the instrument's detector.

ATOMIC ABSORPTION AND ATOMIC FLUORESCENCE

A

AAS: An abbreviation for atomic absorption spectrometry, which can be either flame AAS or graphite furnace AAS.

absorbance: Defined as the log of the ratio of the lamp intensity after the atomizer, to the lamp intensity before the atomizer. This term follows a linear relationship with concentration according to Beer's law.

acetylene: Flammable gas used to create the flame used as the atomization source.

aerosol: Fine particles generated by the nebulizer transported via the spray chamber into the flame burner head.

AFS: An abbreviation for atomic fluorescence spectrometry.

B

background absorption: An artifact other than the analyte prevents the light from the hollow cathode lamp reaching the detector.

background correction: This is a way of compensating for background absorption and is carried out by either continuum source or Zeeman background correction.

Beer's law: Also called the Beer–Lambert law, defines the relationship between the absorption of radiant energy by an absorbing medium. It's expressed as an equation: Absorbance = (concentration) × (an absorption coefficient) × (path length).

C

cold vapor: A technique used for the determination of mercury using a reducing agent to generate elemental mercury vapor and passing into a heated tube for atomization.

continuum background correction: This correction technique utilizes a secondary deuterium lamp installed in the instrument, which emits continuously over a broad wavelength range, rather than at a discreet wavelength. The instrument then monitors intensities from both the deuterium lamp and the element lamp and then subtracts the one from the other.

E

electrothermal atomization: Another term used for graphite furnace AAS.

ETA: Abbreviation for electrothermal atomization.

F

flame: At atomization source used in AAS usually produced by air/acetylene or nitrous oxide/acetylene.
FAAS: An abbreviation of flame atomic absorption.
furnace program: Heating cycle used in a GFAAS.

G

GFAAS: Abbreviation for graphite furnace AAS.
graphite furnace atomic absorption: Atomization source in AAS that uses a small graphite tube to atomize the sample by resistive heating.
graphite tube: The device/tube used for atomization in GFAAS.

H

hollow cathode lamp: Source of the photons used in AAS, generated by passing an electric current between two electrodes and heating/exciting the cathode material which is made from the analyte element being determined.
hydride generation: A technique used to generate volatile hydrides to improve detection limits for a small group of elements, including Pb, As, Cd, Hg Sb, Se, Sn, and Te.

I

interferences: In atomic absorption, interferences are divided into two main categories: spectral and non-spectral.

M

matrix modifiers: A chemical that changes the volatility of either the analyte or the matrix in order to stabilize the analyte so that no losses of the element occur before the atomization or to increase volatility of the matrix to drive it off before atomization.

N

nebulizer: The nebulizer mixes the liquid sample with air to form an aerosol mist that is sprayed into the spray chamber.
nitrous oxide: Oxidant used with acetylene to create a hotter atomization flame than air/acetylene.
non-spectral interferences: Relates to how the atom population is achieved and includes transport (or physical), chemical, and ionization interferences.

P

peak area: One of the measurement protocols used in quantitation in GAAS that measures the area under the peak.
peak height: One of the measurement protocols used in quantitation in GAAS...the other being peak area.
pyrolysis: Slow decomposition of the sample in a graphite furnace, typically referred to by the pyrolysis temperature.

S

spectral interferences: Are those which contribute to the element signal producing an incorrect result and are primarily background absorption and emission interferences from other elements in the sample matrix.

spray chamber: Device used to eliminate larger particles from the aerosol generated by the nebulizer.

T

transmittance: Defined as the ratio of the lamp intensity after the atomizer, to the lamp intensity before the atomizer.

V

volatilization: This is the breaking of molecular bonds as the sample is being preheated in a graphite furnace prior to atomization.

W

wavelength: Energy of photons generated by a hollow cathode lamp, which are specific and unique to the analyte element being measured.

Z

Zeeman background correction: This mode of correction carries out the background measurement at the exact analyte wavelength when a magnetic field is applied. Since the background is measured at the analyte wavelength and not averaged as in deuterium background system, structural molecular background and spectral interferences are easily corrected.

OTHER ATOMIC SPECTROSCOPY TECHNIQUES

E

EDXRF: An abbreviation for energy-dispersive XRF.

energy-dispersive XRF (EDXRF): In EDXRF, the intensity of the photon energy of the individual X-rays is detected and measured simultaneously using multichannel data electronics.

F

fiber optics: A way of transferring photons generated by an excitation source into a spectrometer where the individual analyte wavelengths are separated and identified.

H

hand-held XRF: Small portable XRF systems for on-site or field studies.

L

LALI-TOFMS: An abbreviation for laser ablation, laser ionization TOF mass spectrometry.

laser ablation, laser ionization TOF mass spectrometry (LALI-TOFMS): This technique utilizes one laser to remove material from a solid sample and a second laser to subsequently ionize neutrals. The wavelength and power of the first laser can be selected to accommodate both ablations for inorganic elemental analysis or desorption, for organic analysis. This approach removes much of the matrix effects that plague other techniques and allows quantitative analysis across heterogenous matrices without matrix-matched reference standards.

laser-induced breakdown spectroscopy (LIBS): LIBS is an atomic emission spectroscopic technique that utilizes a small plasma generated by a focused pulsed laser beam. The energy from the plasma excites the vaporized sample which results in a characteristic emission spectrum of the elements present in the sample, which is then optically dispersed and detected by an emission spectrometer.

LIBS: An abbreviation for laser-induced breakdown spectroscopy.

M

microwave-induced plasma atomic emission spectrometry (MIP-AES): In MIP-AES, a magnetron is coupled into a nitrogen plasma. By using the magnetic field rather than the electric field (used in ICP-OES), it concentrates both the axial magnetic and radial electrical fields around the torch, and as a result an extremely robust plasma is formed, which can be used to excite the sample and to study elemental emission profiles using a conventional optical spectrometer.

MIP-AES or MP-AES: An abbreviation for microwave-induced plasma atomic emission spectrometry.

P

primary X-rays: The primary X-rays generated by the X-ray source (typically a tube) that irradiates the sample.

S

secondary X-rays: X-rays that are emitted and measured as a result of the excitation of the sample atoms.

W

wavelength-dispersive (WDXRF): In WDXRF, the polychromatic beam emerging from a sample surface is dispersed into its monochromatic components or wavelengths with an analyzing crystal. A specific wavelength is then calculated from knowledge of the dispersion characteristics of the crystal.

WDXRF: An abbreviation for wavelength-dispersive XRF.

X

X-ray diffraction (XRD): XRD is a comparative technique to XRF that looks at the X-ray scattering from crystalline or polymorphic materials. Each material produces a unique X-ray "fingerprint" of X-ray intensity versus scattering angle that is characteristic of its crystalline atomic structure.

X-ray fluorescence spectrometry (XRFS): X-ray fluorescence (XRF) is a non-destructive analytical technique used to determine the elemental composition of materials by measuring the fluorescent (or secondary) X-ray emitted from a sample when it is excited by a primary X-ray source.

XRF: An abbreviation for X-ray fluorescence.

32 Useful Contact Information

The final chapter of the book is dedicated to providing you with useful contact information related to atomic spectroscopy. It includes contact details for manufacturers of inductively coupled plasma optical emission spectrometry (ICP-OES) and inductively coupled plasma mass spectrometry (ICP-MS) instrumentation, instrument consumables, sample introduction components, and alternative sources of sampling accessories, together with suppliers of laboratory chemicals, calibration standards, certified reference materials, high-purity gases, deionized water systems, and clean-room equipment. I have also included information about the major scientific conferences, professional societies, publishing houses, Internet discussion groups, and the most popular ICP-related journals. It is sorted alphabetically by category. However, some vendors sell many different products, so they are listed under the category represented by their major product line. Please note that they represent contact information for users in North America, even for vendors whose corporate headquarters are outside the United States. Please visit the websites given, for support issues or ordering information in other parts of the world.

Certified Reference Materials/Calibration Standards

National Research Council (NRC) of Canada
1500 Montreal Road
Ottawa, Ontario, K1A 0R9, Canada
Phone: 613-993-9391
Fax: 613-952-8239
www.nrc-cnrc.gc.ca

VHG Labs
276 Abby Road
Manchester, NH 03103
Phone: 603-622-7660
Fax: 603-622-5180
www.vhglabs.com

Inorganic Ventures, Inc.
Inorganic Ventures
300 Technology Drive
Christiansburg, VA 24073
Phone: 540-585-3030
Fax: 540-585-3012
www.inorganicventures.com

SCP Science
21800 Clark Graham
Baie D'urfe, H9X 4B6, Canada
Phone: 800-361-6820
Fax: 514-457-4499
www.scpscience.com

LGC Standards USA
276 Abby Road
Manchester
NH 03103
Tel: 603 206-0799
Toll-free: 1-850-LGC-USA1
www.lgcgroup.com

NIST
100 Bureau Drive, Stop 200
Gaithersburg, MD 20899
Phone: 301-975-6776
Fax: 301-975-2183
www.nist.gov

High Purity Standards
P.O. Box 41727
Charleston, SC 29423
Phone: 843-767-7900
Fax: 843-767-7906
www.highpuritystandards.com

SPEX CertiPrep
203 Norcross Avenue
Metuchen, NJ 08840
Phone: 800-522-7739
Fax: 732-603-9647
www.spexcertiprep.com

Conostan Standards
A Division of SCP Science
348 Route 11
Champlain, NY 12919
Phone: 800-361-6820
Fax: 800-253-5549
www.conostan.com

United States Pharmacopeial Convention (USP)
12601 Twinbrook Parkway
Rockville, MD 20852-1790
Phone: 1-800-227-8772
www.usp.org/reference-standards

High Purity/Reagent Chemicals

Sigma-Aldrich Chemicals
Customer Support
P.O. Box 14508
St. Louis, MO 63178
Phone: 800-325-3010
Fax: 800-325-5052
www.sigmaaldrich.com

Fisher Scientific, Inc.
300 Industry Drive
Pittsburgh, PA 15275
Phone: 800-766-7000
Fax: 800-926-1166
www.fishersci.com

Eichrom Technologies, LLC
1955 University Lane
Lisle, IL 60532
Phone: 630-963-0320
Fax: 630-963-1928
www.eichrom.com

Mallinckrodt Baker
222 Red School Lane
Phillipsburg, NJ 08865
Phone: 908-859-9315
Fax: 908-859-9385
www.mbigloballabcatalog.com

Chromatographic Separation Equipment

Agilent Technologies
2850 Centerville Road
Wilmington, DE 19808
Phone: 302-633-8264
Fax: 302-633-8916
www.chem.agilent.com

Dionex Corporation
A Thermo Company
1228 Titan Way
P.O. Box 3603
Sunnyvale, CA 94088
Phone: 408-737-0700
Fax: 408-730-9403
www.dionex.com

PerkinElmer Life and Analytical Sciences
710 Bridgeport Avenue
Shelton, CT 06484
Phone: 800-762-4000
Fax: 203-944-4914
www.perkinelmer.com

Shimadzu Scientific Instruments (North America)
7102, Riverwood Drive
Columbia, MD 21046
Phone: 410-381-1227
Fax: 410-381-122
http://www.ssi.shimadzu.com/

Thermo Fisher Scientific
81 Wyman Street
Waltham, MA 02454
Phone: 781-622-1000
Fax: 781-622-1207
www.thermoscientific.com

Waters Corporation
34 Maple Street
Milford, MA 01757
Phone: 508-478-2000
Fax: 508-872-1990
www.waters.com

Clean Rooms and Equipment

Microzone Corp.
86 Harry Douglas Drive
Ottawa, Ontario
Canada, K2S 2C7
Phone: 613-831-8318
Fax: 613-831-8321
www.microzone.com

Clestra Hauserman
259 Veterans Lane, Suite 201
Doylestown, PA 18901
Phone: 267-880-3700
Fax: 267-880-3705
www.clestra-cleanroom.com

Consumables—ICP-MS Detectors

SGE, Inc.
2007 Kramer Lane
Austin, TX 78758
Phone: 800-945-6154
Fax: 512-836-9159
www.sge.com

Spectron, Inc.
1601 Eastman Avenue, Suite 205
Ventura, CA 93003
Phone: 805-642-0400
Fax: 805-642-0300
www.spectronus.com

Consumables (Sample Introduction/Interface Components)

Burgener Research, Inc.
1680-2 Lakeshore Rd. W.
Mississauga, Ontario, L5J 1J5, Canada
Phone: 905-823-3535
Fax: 905-823-2717
www.burgenerresearch.com

Elemental Scientific, Inc. (ESI)
2440 Cumming Street
Omaha, NE 68131
Phone: 402-991-7800
Fax: 402-997-7799
www.elementalscientific.com

Meinhard Glass Products
An Elemental Scientific Company
700 Corporate Circle, Suite A
Golden, CO 80401
Phone: 303-277-9776
Fax: 303-216-2649
www.meinhard.com

SCP Science
21800 Clark Graham
Baie D'urfe, H9X 4B6, Canada
Phone: 800-361-6820
Fax: 514-457-4499
www.scpscience.com

Savillex LLC
c/o Spectron, Inc.
1601 Eastman Avenue, Suite 205
Ventura, CA 93003
Phone: 805-642-0400
Fax: 805-642-0300
www.spectronus.com

CPI International
5580 Skylane Blvd.
Santa Rosa, CA 95403
Phone: 800-878-7654
Fax: 707-545-7901
www.cpiinternational.com

Glass Expansion Pty.
4 Barlows Landing Road, Unit #2
Pocasset, MA 02559
Phone: 505-563-1800
Fax: 505-563-1802
www.geicp.com

Precision Glassblowing
14775 E. Hinsdale Avenue
Centennial, CO 80112
Phone: 303-693-7329
Fax: 303-699-6815
www.precisionglassblowing.com

Spectron, Inc.
1601 Eastman Avenue, Suite 205
Ventura, CA 93003
Phone: 805-642-0400
Fax: 805-642-0300
www.spectronus.com

Expositions and Conferences

Eastern Analytical
P.O. Box 633
Montchanin, DE 19710
Phone: 610-485-4633
Fax: 610-485-9467
www.eas.org

Pittsburgh Conference and Exposition
300 Penn Center Blvd., Suite 332
Pittsburgh, PA 15235
Phone: 412-825-3220
Fax: 412-825-3224
www.pittcon.org

FACSS (SCIX)
1201 Don Diego Avenue
Santa Fe, NM 87505
Phone: 505-820-1648
Fax: 505-989-1073
www.facss.org

Plasma Winter Conference
c/o Dr. Ramon Barnes
ICP Information Newsletter
P.O. Box 666
Hadley, MA 01035
Phone: 239-674-9430
Fax: 239-674-9431
http://icpinformation.org/

High-Purity Gases

Air Liquide Specialty Gases LLC
Corporate Headquarters
6141 Easton Road
Plumsteadville, PA 18949
Phone: 215-766-8860
Fax: 215-766-2476
www.airliquide.com

Air Products and Chemicals
7201 Hamilton Blvd.
Allentown, PA 18195
Phone: 610-481-4911
Fax: 610-481-5900
www.airproducts.com

Praxair Inc.
Worldwide Headquarters
39 Old Ridgebury Road
Danbury, CT 06810
Phone: 800-772-9247
Fax: 716-879-2040
www.praxair.com

Scott Specialty Gases
A Division of Air Liquide
6141 Easton Road, P.O. Box 310
Plumsteadville, PA 18949
Phone: 215-766-8861
Fax: 215-766-2476
www.scottgas.com

ICP-MS Instrumentation (Magnetic Sector Technology)

Thermo Fisher Scientific
81 Wyman Street
Waltham, MA 02454
Phone: 781-622-1000
Fax: 781-622-1207
www.thermoscientific.com

Nu Instruments
Sales and Support
8 Magnolia Street
Newburyport, MA 01950
Phone: 978-465-2484
Fax: 978-465-2484
www.nu-ins.com

SPECTRO Analytical Instruments Inc.
91 McKee Drive
Mahwah, NJ 07430
Phone: 201-642-3000
Fax: 201-642-3091
www.spectro.com

ICP-MS Instrumentation (Quadrupole Technology)

Agilent Technologies
ICP-MS Systems
2850 Centerville Road
Wilmington, DE 19808
Phone: 302-633-8264
Fax: 302-633-8916
www.chem.agilent.com

PerkinElmer Inc.
710 Bridgeport Avenue
Shelton, CT 06484
Phone: 800-762-4000
Fax: 203-944-4914
www.perkinelmer.com

Thermo Fisher Scientific
81 Wyman Street
Waltham, MA 02454
Phone: 781-622-1000
Fax: 781-622-1207
www.thermoscientific.com

Analytic Jena
100 Cummings Center
Suite 234-N
Beverly, MA 01915
Phone: 781-376-9899
Fax: 781-376-9897
http://us.analytik-jena.com

Shimadzu Scientific Instruments (North America)
7102, Riverwood Drive
Columbia, MD 21046
Phone: 410-381-1227
Fax: 410-381-122
http://www.ssi.shimadzu.com/

ICP-MS Instrumentation (Time-of-Flight Technology)

GBC Scientific
151A North State Street
P.O. Box 339
Hampshire, IL 60140
Phone: 847-683-9870
Fax: 847-683-9871
www.gbcsci.com

TOFWERK, USA
2760 29th St, Suite 1F
Boulder, CO 80301
Phone: 303-524-2205
http://www.tofwerk.com/contact-us/

Laser Ablation, Laser Ionization (LALI) TOF-MS Technology

Exum Istruments
973 W Ellsworth Ave
Denver, CO, 80223
Phone: 303 993 8340
info@exuminstruments.com
https://www.exuminstruments.com/

ICP-OES/AS Instrumentation

Agilent Technologies
2850 Centerville Road
Wilmington, DE 19808
Phone: 302-633-8264
Fax: 302-633-8916
www.chem.agilent.com

Analytic Jena
100 Cummings Center
Suite 234-N
Beverly, MA 01915
Phone: 781-376-9899
Fax: 781-376-9897
http://us.analytik-jena.com

Jobin Yvon/HORIBA Scientific
3880 Park Avenue
Edison, NJ 08820
Phone: 732-494-8660
Fax: 732-549-5125
http://www.horiba.com/

Perkin Elmer Inc.
710 Bridgeport Avenue
Shelton, CT 06484
Phone: 800-762-4000
Fax: 203-944-4914
www.perkinelmer.com

Shimadzu Scientific Instruments
7102, Riverwood Drive
Columbia, MD 21046
Phone: 410-381-1227
Fax: 410-381-122
http://www.ssi.shimadzu.com/

SPECTRO Analytical Instruments Inc.
91 McKee Drive
Mahwah, NJ 07430
Phone: 201-642-3000
Fax: 201-642-3091
www.spectro.com

Teledyne Leeman Labs
110 Lowell Road
Hudson, NH 03051
Phone: 603-886-8400
Fax: 603-886-4322
www.teledyneleemanlabs.com/

Thermo Fisher Scientific
81 Wyman Street
Waltham, MA 02454
Phone: 781-622-1000
Fax: 781-622-1207
www.thermoscientific.com

Internet Discussion Group

PLASMACHEM List Server
312 Heroy Geology Laboratory
University of Syracuse
Syracuse, NY 13244
Phone: 315-443-1261 (Michael Cheatham)
Fax: 315-443-3363
To subscribe:
Email:
mmcheath@mailbox.syr.edu

European Virtual Institute for Speciation Analysis (EVISA)
A forum dedicated to trace element speciation analysis
http://www.speciation.net/

Journals/Magazines

Analytical Chemistry
1155 16th Street, NW
Washington, DC 20036
Phone: 202-872-4570
Fax: 202-872-4574
www.pubs.acs.org/ac

International Labmate Ltd.
Oak Court Business Center
Sandridge Park
Porters Wood, St. Albans
Hertfordshire, AL3 6PH, UK
44-1727-840-310
www.labmate-online.com

Pharmaceutical Technology
485 Route One South
Building F, Suite 210
Iselin, NJ 08830
Phone: 732-596-0276
Fax: 732-647-1235
http://www.pharmtech.com/

Applied Spectroscopy
201b Broadway Street
Frederick, MD 21701
Phone: 301-694-8122
Fax: 301-694-6860
www.s-a-s.org

Journal of Analytical Atomic Spectrometry (JAAS)
Thomas Graham House
Science Park
Milton Rd.
Cambridge, CB4 4WF, UK
Phone: 44-1223-420-066
Fax: 44-1223-420-247
www.rsc.org

Spectroscopy Magazine
485 Route One South
Building F, First Floor
Iselin, NJ 08830
Phone: 732-596-0276
Fax: 732-225-0211
www.spectroscopyonline.com

Professional Societies/Regulatory Agencies/Pharmacopeias

American Chemical Society (ACS)
1155 16th Street NW
Washington, DC 20036
Phone: 800-227-5558
Fax: 202-872-4615
www.pubs.acs.org

American Society for Testing Materials (ASTM)
100 Barr Harbor Drive
West Conshohocken, PA 19428
Phone: 610-832-9605
Fax: 610-834-3642
www.astm.org

Society for Applied Spectroscopy (SAS)
5320 Spectrum Drive, Suite C
Frederick, MD 21703
Phone: 301-694-8122
Fax: 301-694-6860
www.s-a-s.org

U.S. Food and Drug Administration (FDA)
10903 New Hampshire Avenue
Silver Spring, MD 20993
Phone: 1-888-463-6332
https://www.fda.gov/

American Society for Mass Spectrometry (ASMS)
2019 Galisteo St, Bldg I-1
Santa Fe, NM 87505
Phone: 505-989-4517
Fax: 505-989-1073
www.asms.org

International Society for Pharmaceutical Engineering (ISPE)
7200 Wisconsin Ave., Suite 305
Bethesda, MD 20814
Phone: 301-364-9201
Fax: 240-204-6024
http://www.ispe.org/

United States Pharmacopeial Convention (USP)
12601 Twinbrook Parkway
Rockville, MD 20852
Phone: 1-800-227-8772
www.usp.org/

European Pharmacopeia (EP)
https://www.edqm.eu/en/european-pharmacopoeia.html

Japanese Pharmacopeia (JP)
http://jpdb.nihs.go.jp/jp14e/

International Convention for Harmonization (ICH)
http://www.ich.org/home.html

European Medicines Agency (EMA)
http://www.ema.europa.eu/ema

Laser Ablation/LIBS Equipment

Applied Spectra Inc.
46665 Fremont Blvd.
Fremont, CA 94538
Phone: 510-657-7679
http://appliedspectra.com/

Electro Scientific Industries (ESI)
13900 NW Science Park Drive
Portland, OR 97229
Phone: 800-331-4708
Fax: 503-671-5551
https://www.esi.com

Teledyne CETAC Technologies
14306 Industrial Road
Omaha, NE 68144
Phone: 402-733-2829
Fax: 402-733-5292
http://www.teledynecetac.com/

TSI Inc.
500 Cardigan Road
Shoreview, MN 55126
Phone: 651-483-0900
Fax: 651-490-3824
http://www.tsi.com

Microwave Digestion Equipment

CEM Corporation
3100 Smith Farm Road
Matthews, NC 62810
Phone: 800-726-3331
Fax: 704-821-5185
www.cem.com

SCP Science
21800 Clark Graham
Baie D'urfe, H9X 4B6, Canada
Phone: 800-361-6820
Fax: 514-457-4499
www.scpscience.com

PerkinElmer Inc.
710 Bridgeport Avenue
Shelton, CT 06484
Phone: 800-762-4000
Fax: 203-944-4914
www.perkinelmer.com

Milestone, Inc.
160B Shelton Road
Monroe, CT 06468
Phone: 203-925-4240
Fax: 203-925-4241
www.milestonesci.com

Anton Paar USA Inc.
10215 Timber Ridge Drive
Ashland, VA 23005
Phone: 804-5501051
Fax: 804-5501051
www.info.us@anton-paar.com

Publishers

CRC Press/Taylor and Francis
6000 Broken Sound Parkway NW, Suite 300
Boca Raton, FL 33487
Phone: 561-994-0555
Fax: 561-989-9732
www.crcpress.com

John Wiley and Sons
Corporate Headquarters
111 River Street
Hoboken, NJ 07030
Phone: 201-748-6000
Fax: 201-748-6088
www.wiley.com

Elsevier Science Publishing
Journals Customer Service
3251 Riverport Lane
Maryland Heights, MO 63043
Phone: 800-222-9570
www.elsevier.com

American Laboratory
395 Oyster Point Blvd., #321
South San Francisco, CA 94080
Phone: 650-234-5600
http://www.americanlaboratory.com

Performance/Productivity Enhancement Technology

Teledyne CETAC Technologies
14306 Industrial Road
Omaha, NE 68144
Phone: 402-733-2829
Fax: 402-733-5292
www.cetac.com

Glass Expansion Pty.
4 Barlows Landing Road, Unit #2
Pocasset, MA 02559
Phone: 505-563-1800
Fax: 505-563-1802
www.geicp.com

Elemental Scientific, Inc. (ESI)
1500N. 24th St.
Omaha, NE 68110
Phone: 402-991-7800
Fax: 402-997-7799
www.elementalscientific.com

Sample Preparation Equipment (Cryogenic Grinding, Fusion Systems, Pelletizing Equipment)

SPEX SamplePrep
203 Norcross Avenue
Metuchen, NJ 08840
Phone: 800-522-7739
Fax: 732-603-9647
www.spexsampleprep.com

SCP Science
21800 Clark Graham
Baie D'urfe, H9X 4B6, Canada
Phone: 800-361-6820
Fax: 514-457-4499
www.scpscience.com

Vacuum Pumps and Components

Oerlikon Leybold Vacuum (USA), Inc.
5700 Mellon Road
Export, PA 15632
Phone: 800-764-5369
Fax: 724-733-1217
www.leyboldvacuum.com

Agilent Technologies
Vacuum Products Division
121 Hartwell Avenue
Lexington, MA 02421
Phone: 781-861-7200
Fax: 781-860-5405
www.chem.agilent.com/en-US/ProductsServices/Instruments-Systems/Vacuum-Technologies/Pages/default.aspx

XRF Equipment

Shimadzu Scientific Instruments (North America)
7102, Riverwood Drive
Columbia, MD 21046
Phone: 410-381-1227
Fax: 410-381-122
http://www.ssi.shimadzu.com/

Bruker Daltonics
Chemical Analysis Group
3500 West Warren Avenue
Fremont, CA 94538
Phone: 510-683-4300
Fax: 510-687-1217
www.bdal.com/chemicalanalysis

SPECTRO Analytical Instruments Inc.
91 McKee Drive
Mahwah, NJ 07430
Phone: 201-642-3000
Fax: 201-642-3091
www.spectro.com

Applied Rigaku Technologies
9825 Spectrum Drive, Bldg. 4, Suite 475
Austin, TX 78717
Phone: 512-225-1796
Fax: 512-225-1797
https://www.rigakuedxrf.com

Index